METHODS IN MOLECULAR BIOLOGY™

Series Editor
John M. Walker
School of Life Sciences
University of Hertfordshire
Hatfield, Hertfordshire, AL10 9AB, UK

For further volumes:
http://www.springer.com/series/7651

DNA Repair Protocols

Third Edition

Edited by

Lotte Bjergbæk

Department of Molecular Biology and Genetics, University of Aarhus, Aarhus, Denmark

Humana Press

Editor
Lotte Bjergbæk
Department of Molecular Biology and Genetics
University of Aarhus
Aarhus, Denmark

ISSN 1064-3745 ISSN 1940-6029 (electronic)
ISBN 978-1-4939-5954-9 ISBN 978-1-61779-998-3 (eBook)
DOI 10.1007/978-1-61779-998-3
Springer New York Heidelberg Dordrecht London

Printed on acid-free paper

Humana Press is a brand of Springer
Springer is part of Springer Science+Business Media (www.springer.com)

Preface

Our current knowledge of the mechanisms that regulate DNA repair has grown significantly over the past years with the great help of technological advances such as RNA interference, advanced proteomics, and microscopy as well as high-throughput screens. These technological advances have revealed a tight interplay between DNA repair, recombination, and DNA replication through complex protein network that allows the timely, efficient, and coordinated activation of cellular responses to genomic insult. However, we are far from understanding exactly how cellular pathways protect the genome, and current research frontiers include enzymatic details, lesion accessibility in the context of chromatin, regulatory modes such as posttranslational protein modifications of DNA repair and recombination pathways and not at least, how cells assign the respective pathways depending on the nature of damage or genotoxic replication stress.

This edition of *DNA repair protocols* should supply researchers in the field of DNA repair with a protocol handbook suitable for daily use. It provides detailed step-by-step instructions for studying the various aspects of the eukaryotic response to genomic insult including recent advanced protocols as well as standard techniques used in the field of DNA repair. Both mammalian and nonmammalian model organisms are covered in the book, and many of the techniques can be applied only with minor modifications to other systems than the one described.

DNA repair protocols is organized into five parts and begins with an introduction kindly written by Vilhelm Bohr, which discusses the history and more recent developments in the field of DNA repair. Then follows Part I, Identification of repair genes and generation of repair mutants, which covers three chapters; the first two describe how to screen for repair and/or checkpoint mutants using RNAi-based screen in *Drosophila* and *Caenorhabditis elegans*, respectively. Chapter 4 describes targeted gene disruption of repair genes in DT40 cells, which still remains a powerful tool to investigate the regulation and function of gene and protein in vertebrate.

Part II, Assessing DNA damage and repair (Chapters 5–9), describes assays for measuring the level of DNA damage and repair. Chapter 5 presents an electrophoretic mobility shift assay, which is used to study proteins that bind to DNA structures generated by DNA damaging agents. Chapter 6 describes the widely used comet assay, a microgel eletrophoresis technique for detecting overall damage in individual cells. Combination of the Comet assay with FISH (Chapter 7) allows the detection of DNA damage or repair at higher level of resolution and provides a mean for visualizing region-specific damage and repair. The use of lower eukaryotes and prokaryotes in the comet assay has been hampered by the low cellular DNA content; however, a recent improved comet assay protocol described in Chapter 8 now permits sensitive and reproducible measurement of DNA damage and repair using *Saccharomyces cerevisiae* as model system. Chapter 9 presents a general QPCR assay for measuring any kind of polymerase-blocking DNA lesion such as single-strand breaks, abasic sites, and bulky adducts.

Part III, Repair of incorrect or damaged DNA bases brings together a panel of approaches to characterize repair proteins and repair pathways. Chapters 10 and 11 detail the preparation of mismatched substrates used with nuclear extracts from human or yeast cells. Chapters 12–14 present three different methods for detection and/or repair of DNA photolesions such as cyclobutane pyrimidine dimers and (6-4) photoproducts, which are processed by the nucleotide excision repair pathway. One of the most harmful types of DNA damage is the roadblock forming interstrand crosslinks (ICL), where the repair mechanism is still poorly understood. Chapter 15 details the preparation of plasmid DNA substrates containing a single ICL at a specific site, and Chapter 16 describes several assays that examine specific steps in ICL repair making use of the plasmid DNA substate together with *Xenopus laevis* egg extract. The remaining chapters in this part are allocated to base excision repair assays either using mammalian cell extract (Chapter 17), Arabidopsis cell extracts (Chapter 18), or yeast extracts (Chapter 19), whereas Chapter 20 outlines protocols for isolating mitochondria from mammalian cells and rodent tissues and describes in vitro assays for measuring the enzymatic steps of BER in the lysates of isolated mitochondria.

Part IV, DNA strand breakage and repair, describes methods for detecting strand breakage (single- or double-strand break), for inducing specific strand breaks in the DNA and measuring repair. Chapter 21 presents a modified alkaline gel electrophoresis assay coupled with a neutral *O*-hydroxylamine to measure true single-strand break formation. This is followed by a chapter on pulse-field gel electrophoresis, which is used to analyze DNA double-strand break formation in human chromosomal DNA (Chapter 22). Chapters 23 and 24 both describe methods to detect and quantify the formation of single-stranded DNA, which is a DNA repair, recombination, and replication intermediate. Chapter 23 presents the QAOS method based on qPCR, whereas Chapter 24 describes a rapid yeast genomic DNA extraction procedure that can be used in a simple fluorescent in-gel assay to measure single-stranded DNA. Chapters 25–28 all describe in vivo systems to study repair of strand breakage. One of the most powerful in vivo systems to study double-strand break repair is the HO-endonuclease system from budding yeast described in Chapter 25. The rare-cutting I-SceI endonuclease is also exploited to analyze DSB repair and is described for DT40 cells (Chapter 26) and mammalian cells (Chapter 27). Chapter 28 presents a method to induce single-strand breakage by the Flp recombinase and at the same time to study repair of a protein–DNA adduct. Chapter 29 describes how the immunoglobulin loci of DT40 cells can be used to study homologous recombination in a chromosomal context. Chapter 30 details the use of gene targeting designed to investigate mechanisms of homologous recombination in somatic mammalian cells. Live cell microscopy has added significantly to our understanding of how a DNA damage response is orchestrated in the cells. Chapter 31 reviews useful fluorescent markers and genotoxic agents for studying the DNA damage response in living yeast cells and provides protocols for live cell imaging, time-lapse microscopy, and for induction of site-specific DNA lesions. This chapter has specific focus on proteins involved in recombinational repair. The last two chapters in this part both describe in vitro assays for analyzing double-strand break repair. Chapter 32 describes an in vitro assay that allows the study of DSB rejoining in genomic DNA using either "naked" DNA or DNA organized in chromatin, whereas Chapter 33 presents an in vitro end-joining assay employing linear oligonucleotides.

In addition to pathways involved directly in repairing DNA lesions cells have also developed strategies of damage tolerance (postreplication repair), which allows for lesion bypass and thereby completion of DNA replication in the presence of damage. Part V gathers together a few methods for studying damage tolerance and checkpoint responses. Chapters 34 and 35 both describe assays to measure translesion DNA synthesis with plasmid-based systems. Modifications of the eukaryotic sliding clamp PCNA play a significant role both for translesion synthesis and template switching, as well as for hindring unscheduled recombination. Chapters 36 and 37 deal with methods for detecting PCNA modifications in vivo as well as in vitro PCNA modification assays. It has for long time been known that DNA damage has a detrimental effect on the replication process possibly controlled by checkpoints. Chapter 38 describes a method to measure in vitro DNA replication using cytoplasmatic extracts from cells that have sustained DNA damage. Chapter 39 provides methods for detecting checkpoint activation due to DNA damage with focus on the central checkpoint kinase Chk1, whereas Chapter 40 describes a few standard protocols for detecting H2AX phosphorylation, which in recent years have attracted much attention as a specific marker for DNA damage.

Finally, I wish to take this opportunity to thank all authors for their commitment, cooperation, and contributions that made my first editing job a very enjoyable experience. A special thanks to Dr. John Walker for his excellent advice during the process.

Aarhus, Denmark ***Lotte Bjergbæk***

Contents

Part III Repair of Incorrect or Damaged DNA Bases

Part IV DNA Strand Breakage and Repair

Contributors

GERALD M. ADAIR • *Department of Molecular Carcinogenesis, University of Texas M.D. Anderson Cancer Center, Smithville, TX, USA*

FRED AHNE • *Institute for Radiation Biology, Helmholtz Zentrum für Gesundheit und Umwelt, Oberschleissheim, Germany*

ALEXANDRU ALMASAN • *Department of Cancer Biology, Lerner Research Institute Cleveland Clinic, Cleveland, OH, USA*

ANNI H. ANDERSEN • *Department of Molecular Biology and Genetics, University of Aarhus, Aarhus, Denmark*

RAFAEL R. ARIZA • *Department of Genetics, University of Córdoba and Maimónides Institute of Biomedical Research, Córdoba, Spain*

THOMAS R. BERTON • *Department of Carcinogenesis, University of Texas M.D. Anderson Cancer Center, Smithville, TX, USA*

AHMAD BESARATINIA • *Division of Biology, Beckman Research Institute of the City of Hope, Duarte, CA, USA*

AMANDA SMITH BESS • *Nicholas School of the Environment, Duke University, Durham, NC, USA*

ALEXANDER JAMES ROY BISHOP • *Creehey Children's Cancer Research Institute University of Texas Health Science Center at San Antonio, San Antonio, TX, USA; Department of cellular and Structural Biology, University of Texas Health Science Center at San Antonio, San Antonio, TX, USA*

LOTTE BJERGBÆK • *Department of Molecular Biology and Genetics, University of Aarhus, Aarhus, Denmark*

VILHELM A. BOHR • *Laboratory of Molecular Gerontology, National Institute on Aging NIH, Baltimore, MD, USA*

NIGEL J. BRADY • *Department of Cancer Studies and Molecular Medicine, University of Leicester, Leicester, UK*

HELEN E. BRYANT • *Department of Oncology, The Institute for Cancer Studies, University of Sheffield, Sheffield, UK*

NGE CHEONG • *Department of Paediatrics, National University of Singapore, Clinical Research Center, Singapore, Singapore*

GILBERT CHU • *Departments of Medicine and Biochemistry, Stanford University Medical Center, Stanford, CA, USA*

MARCUS S. COOKE • *Department of Cancer Studies and Molecular Medicine, University of Leicester, Leicester, UK*

MARILA CORDEIRO-STONE • *Department of Pathology and Laboratory Medicine, Lineberger Comprehensive Cancer Center University of North Carolina at Chapel Hill, Chapel Hill, NC, USA; Center for Environmental Health and Susceptibility, University of North Carolina at Chapel Hill, Chapel Hill, NC, USA*

DOLORES CÓRDOBA-CAÑERO • *Department of Genetics, University of Córdoba and Maimónides Institute of Biomedical Research, Córdoba, Spain*
KAMAL DATTA • *Department of Biochemistry and Molecular and Cellular Biology, Georgetown University Medical Center, Washington, USA*
ADELINA A. DAVIES • *Cancer Research UK London Research Institute, Clare Hall Laboratories, South Mimms, UK*
JAMES M. DEWAR • *Institute for Cell and Molecular Biosciences, Newcastle University, Medical School, Newcastle upon Tyne, UK*
NOAM DIAMANT • *Department of Biological Chemistry, Weizmann Institute of Science, Rehovot, Israel*
GRIGORY L. DIANOV • *Gray Institute for Radiation, Oncology and Biology, University of Oxford, Oxford, UK*
NADINE ECKERT-BOULET • *Department of Biology, University of Copenhagen, Copenhagen, Denmark*
MILICA ENOIU • *Institute of Molecular Cancer Research, University of Zürich, Zürich, Switzerland*
CHARLES M. ENSOR • *Graduate Center for Toxicology, University of Kentucky College of Medicine, Lexington, KY, USA*
MARK D. EVANS • *Department of Cancer Studies and Molecular Medicine, University of Leicester, Leicester, UK*
AMY M. FURDA • *Department of Pharmacology and Chemical Biology, University of Pittsburgh School of Medicine, Pittsburgh, PA, USA*
IRENE GALLINA • *Department of Biology, University of Copenhagen, Copenhagen, Denmark*
MICHAEL GLEI • *Department of Nutritional Toxicology, Institute for Nutrition, Friedrich-Schiller-University Jena, Jena, Germany*
RICARDO GREDILLA • *Department of Physiology, Faculty of Medicine, Complutense University, Madrid, Spain*
LIYA GU • *Graduate Center for Toxicology, University of Kentucky College of Medicine, Lexington, KY, USA*
AMANDA GUNN • *Department of Cancer Biology, Irell and Nanella Graduate School of Biological Sciences, Beckman Research Institute of the City of Hope, Duarte, CA, USA*
JAMES E. HABER • *Department of Biology, Brandeis University, Waltham, MA, USA; Rosenstiel Basic Medical Sciences Research Center, Brandeis University, Waltham, MA, USA*
AYAL HENDEL • *Department of Biological Chemistry, Weizmann Institute of Science, Rehovot, Israel*
THE VINH HO • *Department of Pharmacological Sciences, Stony Brook University, Stony Brook, USA*
EVA-MARIA HOLSTEIN • *Institute for Cell and Molecular Biosciences, Newcastle University, Medical School, Newcastle upon Tyne, UK*
BYUNG JOON HWANG • *Department of Molecular Bioscience, College of Biomedical Science, Kangwon National University, Chuncheon, Republic of Korea*
GEORGE ILIAKIS • *Institute of Medical Radiation Biology, Medical School, University of Duisburg-Essen, Essen, Germany*

MASAMICHI ISHIAI • *Laboratory of DNA Damage Signaling, Department of Late Effect Studies, Radiation Biology Center, Kyoto University, Kyoto, Japan*
BJÖRN JOHANSSON • *Department of Biology, University of Minho, Braga, Portugal*
MAHSA KARBASCHI • *Department of Cancer Studies and Molecular Medicine, University of Leicester, Leicester, UK*
PUCK KNIPSCHEER • *Hubrecht Institute, KNAW and University Medical Center Utrecht, Utrecht, The Netherlands*
GUO-MIN LI • *Graduate Center for Toxicology, University of Kentucky College of Medicine, Lexington, KY, USA*
MICHAEL LISBY • *Department of Biology, University of Copenhagen, Copenhagen, Denmark*
FENGSONG LIU • *College of Life Sciences, Hebei University, Baoding, Hebei Province, China*
ZVI LIVNEH • *Department of Biological Chemistry, Weizmann Institute of Science, Rehovot, Israel*
DAVID T. LONG • *Department of Biological Chemistry and Molecular Pharmacology, Harvard Medical School, Boston, MA, USA*
APRIL M. LUKE • *Department of Environmental Sciences and Engineering, University of North Carolina, NC, USA*
DAVID LYDALL • *Institute for Cell and Molecular Biosciences, Newcastle University, Medical School, Newcastle upon Tyne, UK*
JOEL N. MEYER • *Nicholas School of the Environment, Duke University, Durham, NC, USA*
DAVID L. MITCHELL • *Department of Carcinogenesis, The University of Texas M.D. Anderson Cancer Center, Smithville, TX, USA*
EMIL MLADENOV • *University of Duisburg-Essen, Medical School, Institute of Medical Radiation Biology, Essen, Germany*
SIMONE MOERTL • *Institute for Radiation Biology, Helmholtz Zentrum für Gesundheit und Umwelt, Oberschleissheim, Germany*
TINE HØRNING MÖLLER • *Department of Molecular Biology and Genetics, University of Aarhus, Aarhus, Denmark*
RODNEY S. NAIRN • *Department of Molecular Carcinogenesis, M.D. University of Texas Anderson Cancer Center, Smithville, TX, USA*
JUN NAKAMURA • *Department of Environmental Sciences and Engineering, University of North Carolina, Chapel Hill, NC, USA*
RONALD D. NEUMANN • *Nuclear Medicine Department, Warren Grant Magnuson Clinical Center, National Institutes of Health, Bethesda, USA*
IDA NIELSEN • *Department of Molecular Biology and Genetics, University of Aarhus, Aarhus, Denmark*
NANA NIKOLAISHVILI-FEINBERG • *Department of Pathology and Laboratory Medicine, Lineberger Comprehensive Cancer Center and Center for Environmental Health and Susceptibility, University of North Carolina at Chapel Hill, Chapel Hill, USA*
STEFFEN NÖRGAARD • *Department of Molecular Biology and Genetics, University of Aarhus, Aarhus, Denmark*
RUI OLIVEIRA • *Department of Biology, University of Minho, Braga, Portugal*

ANDERS OLSEN • *Department of Molecular Biology and Genetics, University of Aarhus, Aarhus, Denmark*
JOANNE L. PARKER • *Clare Hall Laboratories, Cancer Research UK London Research Institute, South Mimms, UK*
JASON L. PARSONS • *Gray Institute for Radiation, Oncology and Biology, University of Oxford, Oxford, UK*
GERD P. PFEIFER • *Division of Biology, Beckman Research Institute of the City of Hope, Duarte, CA, USA*
DAVID DU PLESSIS • *Medical Research Council Laboratory of Molecular Biology, Cambridge, UK*
SHUBHADEEP PURKAYASTHA • *Nuclear Medicine Department, Warren Grant Magnuson Clinical Center, National Institutes of Health, Bethesda, USA*
JENNIFER J. RAHN • *Department of Pharmaceutical and Biomedical Sciences, Medical University of South Carolina, Charleston, SC, USA*
MARKUS RÄSCHLE • *Department of Proteomics and Signal Transduction, Max Planck Institute of Biochemistry, Martinsried, Germany*
DASHNAMOORTHY RAVI • *Creehey Children's Cancer Research Institute and Department of cellular and Structural Biology, University of Texas Health Science Center at San Antonio, San Antonio, TX, USA*
TERESA ROLDÁN-ARJONA • *Department of Genetics, University of Córdoba and Maimónides Institute of Biomedical Research, Córdoba, Spain*
ANDREAS ROTHFUSS • *Pharmaceuticals Division, LaRoche AG, Basel, Switzerland*
JULIAN E. SALE • *Medical Research Council Laboratory of Molecular Biology, Cambridge, UK*
ORLANDO D. SCHÄRER • *Stony Brook University, Stony Brook, NY, USA*
WIEBKE SCHLÖRMANN • *Department of Nutritional Physiology, Institute for Nutrition, Friedrich-Schiller-University Jena, Jena, Germany*
LONE VEDEL SCHÖLER • *Department of Molecular Biology and Genetics, University of Aarhus, Aarhus, Denmark*
SIGAL SHACHAR • *Department of Biological Chemistry, Weizmann Institute of Science, Rehovot, Israel*
ARISHYA SHARMA • *Department of Cancer Biology, Lerner Research Institute Cleveland Clinic, Cleveland, OH, USA*
SONIA SILVA • *Department of Biology, University of Copenhagen, Copenhagen, Denmark*
KAMINI SINGH • *Department of Cancer Biology, Lerner Research Institute Cleveland Clinic, Cleveland, OH, USA*
VAUGHN SMIDER • *Department of Molecular Biology, The Scripps Research Institute, La Jolla, CA, USA*
LIMIN SONG • *Department of Biochemistry and Molecular Biology, University of Miami Miller School of Medicine, Miami, FL, USA*
GÜNTER SPEIT • *Institut für Humangenetik, Universität Ulm, Ulm, Germany*
JEREMY M. STARK • *Department of Cancer Biology, Irell and Nanella Graduate School of Biological Sciences, Beckman Research Institute of the City of Hope, Duarte, CA, USA*

TINNA STEVNSNER • *Department of Molecular Biology and Genetics, University of Aarhus, Aarhus, Denmark*
NEAL SUGAWARA • *Department of Biology, Brandeis University, Waltham, MA, USA; Rosenstiel Basic Medical Sciences Research Center, Brandeis University, Waltham, MA, USA*
MINORU TAKATA • *Laboratory of DNA Damage Signaling, Department of Late Effect Studies, Radiation Biology Center, Kyoto University, Kyoto, Japan*
CHUN TSAI • *Departments of Medicine and Biochemistry, Stanford University Medical Center, Stanford, CA, USA*
EMI UCHIDA • *Laboratory of DNA Damage Signaling, Department of Late Effect Studies, Radiation Biology Center Kyoto University, Kyoto, Japan*
HELLE D. ULRICH • *Cancer Research UK London Research Institute, Clare Hall Laboratories, South Mimms, UK*
BENNETT VAN HOUTEN • *Department of Pharmacology and Chemical Biology, Hillman Cancer Center, University of Pittsburgh School of Medicine, Pittsburgh, USA*
LOTTE VESTERGAARD • *Department of Molecular Biology and Genetics, University of Aarhus, Aarhus, Denmark*
JOHANNES C. WALTER • *Department of Biological Chemistry and Molecular Pharmacology, Harvard Medical School, Boston, USA*
HONG YAN WANG • *Department of Radiation Oncology, School of Medicine and Winship Cancer Institute, Emory University, Atlanta, GA, USA*
YA WANG • *Department of Radiation Oncology, School of Medicine and Winship Cancer Institute, Emory University, Atlanta, GA, USA*
THOMAS A. WINTERS • *Nuclear Medicine Department, Warren Grant Magnuson Clinical Center, National Institutes of Health, Bethesda, MD, USA*
FENGHUA YUAN • *Department of Biochemistry and Molecular Biology, University of Miami Miller School of Medicine, Miami, FL, USA*
YANBIN ZHANG • *Department of Biochemistry and Molecular Biology, University of Miami Miller School of Medicine, Miami, FL, USA*
OMER ZIV • *Department of Biological Chemistry, Weizmann Institute of Science, Rehovot, Israel*

Chapter 1

DNA Repair: Front and Center and Not Going Away!

Vilhelm A. Bohr

Abstract

This introduction to the book: DNA repair protocols: third edition, edited by Bjergbaek, discusses the history and more recent developments in the field of DNA repair. This research field started in the 1950 and developed from a small group of researchers interested in the damage caused to DNA by ultraviolet irradiation from the sun to become a large field of research today. DNA damage and its repair are now thought to play an important role in the etiologies of cancer, aging, and neurodegeneration and there is a great deal of interest in this venture. Thus, understanding of DNA processing is now a central field in molecular and cellular biology, and the field is still growing.

Key words: DNA damage, DNA repair, Aging, Cancer, Neurodegeneration

1. Historical Perspective

DNA repair first emerged as a specialized area of genetic research in the mid 1950s, when scientists began investigating the ability of cells to recover from exposure to ultraviolet (UV) light. It took approximately 10 years of intense study to analyze and characterize this process, now known to involve at least two subpathways: photoreactivation and nucleotide excision repair. Detailed understanding of several additional prototypical DNA repair pathways followed relatively quickly. Since the mid-1970s, research on DNA repair has steadily gained more and more visibility, rapidly becoming recognized as a major research area. In the "early days," only a handful of people worldwide were even interested in DNA repair: in contrast, the public health and clinical relevance of DNA repair are now widely recognized, many researchers focus on understanding one or

Lotte Bjergbæk (ed.), *DNA Repair Protocols*, Methods in Molecular Biology, vol. 920,
DOI 10.1007/978-1-61779-998-3_1, © Springer Science+Business Media New York 2012

two DNA repair pathways or subpathways, multiple international DNA repair conferences are held annually, and DNA repair research is funded by many major funding agencies, including those focused uniquely on cancer biology.

2. DNA Repair Pathways

DNA is continuously exposed to endogenous and exogenous agents that generate lesions or structural perturbations into DNA. Endogenous DNA damage is primarily caused by reactive oxygen species, formed as a by-product of mitochondrial oxidative phosphorylation. Exogenous DNA damage is caused by a large number of compounds and agents, including naturally occurring and man-made chemicals, as well as ambient and induced forms of radiation. It has been estimated that endogenous and exogenous sources combined generate approximately 100,000 lesions in the genome of a single human cell in one 24 h day (1).

All prokaryotic and eukaryotic cells express multiple DNA repair pathways, which collectively provide slightly redundant capacity to repair all types of DNA damage. However, the efficiency of repair of each DNA lesion is a function of many variables, including the nature of the lesion, the context/site of the lesion in the genome, as well as the type of cell and the specific cellular genotype, to mention a few.

Oxidative DNA damage and DNA strand breaks are repaired by *base excision repair* (*BER*) and *single-strand break repair* (*SSBR*). Bulky DNA lesions are repaired by *nucleotide excision repair* (*NER*), which includes both *global genome* (*GG-NER*) and *transcription-coupled* (*TC-NER*) subpathways. DNA double-strand breaks and interstrand cross-links are repaired by *DNA double-strand break repair* (*DSBR*), a pathway under intense investigation, with many novel pathway components and subpathways being explored and more likely to be discovered. *DNA mismatch repair* (*MMR*) recognizes and repairs insertion/deletion loops, hairpins, and base–base mismatches introduced throughout the genome during semiconservative DNA replication and in heteroduplex DNA segments generated during homologous recombination and other DNA repair reactions. UV photoproducts can also be repaired by *direct reversal*, for example, by a *DNA photolyase.*

While all of these pathways are being studied as separate entities, it is also becoming more and more evident that there are significant overlaps and that some proteins participate in more than one pathway. It is therefore necessary to keep a rather open mind about the relative pathway entities and their integrities.

As mentioned above, the cellular capacity to repair a specific DNA lesion at a specific site under specific conditions varies significantly. Initial studies of the site or region specificity of DNA repair revealed that certain genome regions are repaired more efficiently than others. Another observation was that active genes are preferentially repaired, while relatively quiescent transcriptionally inactive regions of the genome are repaired more slowly. This phenomenon, most commonly referring to repair of UV-induced photolesions, is now known as *transcription-coupled repair*, or *TC-NER*. At this time, there is no convincing evidence for transcription-coupled repair of oxidative DNA lesions, base–base mismatches, or double-strand breaks in actively transcribed regions of the genome. *TC-NER* was first characterized using assays that detect repair of specific DNA lesions in specific genes and similar assays have more recently been used to detect preferential repair in specific gene promoter regions. In this book there are chapters on gene-specific repair and regional repair; some are based on quantitative southern analysis and others on PCR approaches. As these approaches are quite technically advanced, they require specific equipment and special skills.

3. Repair of the Mitochondrial Genome

While studies of the repair and replication of the nuclear genome advanced rapidly during the years of the "molecular biology" explosion (approximately 1960–1990), the repair and replication of the mitochondrial genome is now of increasing interest to many geneticists as well to as molecular and cell biologists. In fact, a strong consensus is emerging that mitochondria and the mitochondrial genome play essential roles in many aspects of cellular biology, including nuclear genomic stability and integrity. Early studies suggested that aging-related DNA damage accumulates more rapidly in the mitochondrial than in the nuclear genome and that the mitochondrial genome is repaired very slowly, if at all. While these ideas have been strongly disputed in recent years, the technical challenges associated with studying mitochondria were at least in part responsible for slow progress in understanding mitochondrial genomics. That said, it is now clear that oxidative DNA lesions in mtDNA are repaired efficiently, and that both short and long patch BER subpathways are robustly expressed in mitochondria (Table 1). Mitochondria are also proficient in MMR, but appear to lack capacity to repair bulky DNA lesions by NER (or an NER-like process). Interestingly, many nuclear DNA repair proteins have been detected in mitochondria, despite early reports to the contrary, and despite accurate reports that these enzymes lack canonical mitochondrial targeting sequences.

Table 1
Nuclear and mitochondrial DNA repair pathways

Pathways	Subpathways	Nuclear	Mito
Base excision	Long patch	Y	Y
	Short patch	Y	Y
Nucleotide excision	General genome	Y	N
	Gene	Y	N
	Transcription associated	Y	N
Mismatch		Y	Y
Recombination	HR	Y	?
	NHEJ	Y	?

4. DNA Repair Defects: The Clinical Consequences

In 1968, James E. Cleaver published the first direct evidence that the skin cancer susceptibility disorder *xeroderma pigmentosum* is an inherited disease syndrome, in which skin pathology might be causally linked to mutations that inactivate or reduce the efficiency of NER (2, 3). Consistent with this, reduced exposure to UV light dramatically reduces skin cancer incidence in patients with *xeroderma pigmentosum*. This, as it turns out, was the "tip of the iceberg:" and at least partly because of Paul Modrich's discovery that defects in human MMR genes are linked to inherited susceptibility to human colon cancer (HNPCC), the DNA repair enzymes got "front page news" status as *Science* magazine's *Molecule of the Year* in 1994 (4–8). Since 1994, many additional links between DNA repair defects and human disease have been discovered. Moreover, DNA repair defects have not only been linked to cancer susceptibility, but also appear to play a significant role in human neurodegenerative diseases as well as normal and premature human aging.

5. DNA Repair Research: The Devil Is in the Details (i.e., The Assay)

The idea and purpose behind this book is to gather and present in one volume technical information on DNA repair assays, with the hope that the book will facilitate research progress. In their day-to-day activities, researchers who study DNA repair must choose between a very large number of well-established DNA repair assays, novel less well-characterized assays, and the option of embarking themselves on new assay development. These are important and difficult choices, and there is usually no "correct" answer, or single good approach … in fact, the best approach is to ask the same question using two or more different methods or approaches, each of which is carefully selected based on the research goal (i.e., what is the biologically important question being asked) and the strengths and limitations of each method. Because these questions are so critical and difficult, the goal of this volume is to make it a bit easier to answer them and develop a well-designed experimental approach to address any given question about DNA repair.

6. The Future of DNA Repair Research: Critical Goals

Although DNA repair defects cause several human diseases including skin and colon cancer, with minor exceptions, DNA repair assays cannot yet be used for routine clinical diagnosis or to evaluate disease progression in patients receiving chemotherapy. This is because most available DNA repair assays are either not sufficiently sensitive, or are too cumbersome (i.e., low throughput) or too expensive for clinical use. Thus, an important focus for the future is to develop DNA repair assays that can be used effectively in the clinic.

A second important future research goal is to develop highly specific inhibitors of individual DNA repair enzymes, pathways, or subpathways. Once available, it is widely believed that it will be possible to develop combination therapies using one or more DNA repair inhibitors and one or more known cancer drugs, dramatically enhancing therapeutic efficacy and potentially preventing cancer progression or even inducing cancer regression.

Clearly, well-characterized assays for specific DNA repair enzymes and pathways will be critical tools for achieving these research goals. It is hoped that this volume will be useful to researchers working on these critical research areas or asking other interesting questions about DNA repair.

References

1. Lindahl T (1993) Instability and decay of the primary structure of DNA. Nature 362: 709–715
2. Cleaver JE (1968) Defective repair replication of DNA in xeroderma pigmentosum. Nature 218:652–656
3. Setlow RB, Regan JD, German J, Carrier WL (1969) Evidence that xeroderma pigmentosum cells do not perform the first step in the repair of ultraviolet damage to their DNA. Proc Natl Acad Sci U S A 64:1035–1041
4. Koshland DE (1994) Molecule of the year: the DNA repair enzyme. Science 266:1925
5. Culotta E, Koshland DE Jr (1994) DNA repair makes its way to the top. Science 266: 1926–1929
6. Modrich P (1994) Mismatch repair, genetic stability, and cancer. Science 266:1959–1960

7. Parsons R, Myeroff LL, Liu B, Willson JK, Markowitz SD, Kinzler KW, Vogelstein B et al (1995) Microsatellite instability and mutations of the transforming growth factor beta type II receptor gene in colorectal cancer. Cancer Res 55:5548–5550
8. Marx J (1994) DNA repair comes into its own. Science 266:728–730

Part I

Identification of Repair Genes and Generation of Repair Mutants

Chapter 2

Identification of Genes Required for Damage Survival Using a Cell-Based RNAi Screen Against the *Drosophila* Genome

Dashnamoorthy Ravi and Alexander James Roy Bishop

Abstract

Exposure to DNA-damaging agents invokes biological responses necessary for damage recovery and cell survival. Despite the presence of intact DNA repair pathways, lack of certain other biological pathways has been shown to sensitize cells to DNA-damaging agents' exposure. It is likely that following DNA damage a complex interplay between DNA repair pathways and other biological pathways might be required to ensure cell survival. In this chapter, we describe a high-throughput method for the identification of genes essential for cell survival following DNA damage by using a cell-based assay to measure viability in combination with an RNA interference-based genome-wide screening experiment.

Key words: DNA damage, RNAi, Genome-wide screening, *Drosophila*, Survival

1. Introduction

Although DNA is arguably the most biologically relevant target of exogenous DNA-damaging agents, other intracellular components such as RNA, protein, and lipids can also be chemically modified. As such, it is not surprising that a DNA-damaging agent invokes a pleiotropic biological response that is not just limited to DNA repair and damage response pathways. Similarly, defects in DNA repair pathways result in sensitivity to DNA damage, though decreased viability in response to damage exposure can also be observed with loss of certain biological pathways that have no apparent role in DNA repair. It is possible that such genes and pathways could play an ancillary role in the process of response to DNA damage and DNA repair. Understanding the overall coordinated response to damage exposure not only extends our existing

Lotte Bjergbæk (ed.), *DNA Repair Protocols*, Methods in Molecular Biology, vol. 920,
DOI 10.1007/978-1-61779-998-3_2, © Springer Science+Business Media New York 2012

knowledge about DNA repair and damage survival, but also provides novel mechanisms that can be exploited in disease management, such as cancer treatment. Although microarray and proteomics are powerful exploratory tools for the analysis of genome-wide responses to DNA damage, understanding the role of such responses in terms of cell survival requires a cell-based assessment combined with loss of gene function, a reverse genetics approach. Since the discovery of RNA interference (RNAi) as a mechanism to specifically silence any gene's expression, this powerful technology has been utilized to determine loss of function phenotypes on a genomic scale. In conjunction with DNA damage, RNAi screening allows the rapid identification of genes that are essential for DNA damage survival. Although RNAi-based screens are performed with mammalian cells, such screening requires purchasing a genomic library of small interfering RNA (siRNA), which is relatively expensive. Frequently these libraries consist of combinations of four siRNAs for each target mRNA, any one of which may not be effective for the target sequence. In addition, further costs for mammalian screening result from the mode of siRNA introduction, usually lipid based, and this is often toxic to the cells itself. In contrast, RNAi with *Drosophila* cells is relatively inexpensive as these cells readily uptake long double-stranded RNA (dsRNA) from the medium by phagocytosis, a nontoxic procedure, followed by processing the dsRNA into siRNA internally within the cell effectively producing a large population pool of siRNA. This large population pool of siRNA induces a robust knockdown of the target mRNA. Since the implementation of this technology with *Drosophila* cells, there have been numerous investigations that resulted in finding novel disease mechanisms and it has facilitated the discovery of novel genes and biological pathways in various processes (1). We have utilized this technology to identify genes that are essential for cell survival following alkylation DNA damage induced by methyl methane sulfonate (MMS) and uncovered genes and pathways which were previously unknown to play a role in DNA damage response (2). *Drosophila*-based RNAi screening is a powerful tool for cell biologists aspiring to understand the complex biological response to DNA damage or any exogenous stimuli. However, it is important to consider that different types of DNA damage at different concentrations can alter cellular responses. Therefore, to obtain comprehensive knowledge about complex biology of cellular response to DNA damage, it might be necessary to conduct investigations with a variety of different DNA-damaging agents and at different concentrations using the methods described in this chapter.

2. Materials

2.1. High-Throughput Synthesis of dsRNA Library

1. Total RNA from whole flies or *Drosophila* cell line such as Kc_{167} or S2.
2. Trizol.
3. *Reverse transcription*: Impromp-II Reverse Transcriptase system (Promega, WI).
4. *cDNA synthesis*:
   (a) *Primers*: For synthesis and amplification of gene-specific cDNA in a high-throughput (HTP) scale, PCR primers and PCR reactions should be assembled in 96-well plates. To make use of the amplified cDNA for the synthesis of dsRNA, each gene-specific PCR primer (both forward and reverse) should include a T7 overhang at the 5′ end (taatacgactcactatagggaga-5′-primer sequence-3′) (see Note 1). Working stock of primers are prepared by combining forward and reverse primers in the same well and diluting the primers 1:5 with PCR grade water.
   (b) PCR master mix (cDNA, dNTPs, $MgCl_2$, PCR buffer, Taq polymerase (ROCHE, IN), and nuclease-free water).
5. Thermal cycler.
6. Agarose gel (1 % in Tris–borate–EDTA (TBE) buffer).
7. TBE buffer.
8. Biorad horizontal gel electrophoresis tank and power supply.
9. Ethidium bromide (see Note 2).
10. UV transilluminator.
11. Sterile blades.
12. Deep well, 96-well plates.
13. Gel isolation kit (Qiagen or Invitrogen, CA).
14. Multiscreen HTS PCR purification plate (Millipore, MA).
15. Millipore vacuum manifold.
16. In vitro transcription kit (T7 Ribomax Express, Promega, WI).
17. Nuclease-free water.
18. 10 mM Tris–HCl, pH 7.5.
19. UV spectrophotometer (Spectramax M2, Molecular Devices, CA, multi-detection microplate reader).
20. 96-Well, UV transparent microplate (Corning).

2.2. HTP Transfection of dsRNA in Drosophila Cells

1. *Drosophila* cell line: Kc_{167}.
2. Schneider's insect cell medium, serum free.
3. Schneider's insect cell medium, 10 % serum.
4. Schneider's insect cell medium, 15 % serum.
5. Fetal bovine serum (FBS) heat inactivated (55 °C, 10 min).
6. Antibiotics/antimycotics.
7. 384-Well plate, solid white (Costar).
8. Liquid handling system for aliquoting dsRNA (Precision XS Microplate sample processor (BioTek, Winooski, VT)).
9. Liquid dispensing system (Wellmate, Matrix Technologies/ Thermo Fisher Scientific, NH or Micro 10×, Hudson Robotics).
10. 70 % Ethanol.
11. PBS.
12. Counting chamber.
13. Centrifuge (Eppendorf 5810R).
14. Humid chamber.
15. *Drosophila* incubator, set at 25 °C.
16. Sterile 50 mL conical centrifuge tubes.
17. T75 tissue culture flask.
18. Tissue culture hood.

2.3. Induction of DNA Damage

1. Damaging agent, such as MMS, Bleomycin, UV, Cisplatin, etc.
2. Sterile 2- or 16-well aspirating wand (V & P Scientific, CA).
3. Experimental plate.
4. Liquid dispensing system.
5. Centrifuge.
6. Humid chamber.
7. *Drosophila* incubator, set at 25 °C.
8. Tissue culture hood.

2.4. Assessment of Cell Viability

1. Sterile 24- or 16-well aspirating wand.
2. Celltiter glo (Promega, WI).
3. 1× Phosphate-buffered saline (PBS).
4. Liquid dispensing system.
5. Experimental plate.
6. Centrifuge.
7. Plate reader (Spectramax M2, Molecular Devices, CA, multi-detection microplate reader).

3. Methods

3.1. HTP Synthesis of dsRNA Library

Drosophila-based genome-wide RNAi screen for DNA damage is performed using a library of long dsRNAs that are each approximately of 500 bp in length. The dsRNA for genomic screens is usually synthesized in an HTP scale by manipulating reactions in a 96-well PCR plate. First, total RNA from the *Drosophila* cells (such as Kc_{167} or S2 cell lines, or from whole flies) is isolated and used as template for generation of a cDNA pool through reverse transcription-based reaction. In the next step, the cDNA pool is used as template in the amplification of an individual cDNA, specific to a particular gene, by using gene-specific PCR primers. The amplified PCR products are then isolated individually after separation by agarose gel electrophoresis. Purified gene-specific cDNA products from this 1st round of PCR amplification are then re-amplified, purified, and used as templates for the synthesis of dsRNA by in vitro transcription. To facilitate in vitro transcription through T7 RNA polymerase-mediated catalysis, T7 overhangs are added to the 5′ end of cDNA at the PCR amplification step by adding the T7 promoter sequence included as a part in the design of gene-specific primer. dsRNA library synthesized in this manner is used for genome-wide RNAi screening experiments. Steps involved in HTP synthesis of dsRNA are summarized in Fig. 1 (see Note 3).

3.1.1. First PCR

1. Array 10 μL of primers from the working stock using multichannel pipette and array 40 μL of PCR master mix with repeater pipette into 96-well plates. The PCR master mix consists of 2 μL of cDNA, 8 μL of 1.25 mM dNTP mix, 2 μL of 50 mM $MgCl_2$, 10 μL of 10× PCR buffer, 17.58 μL of nuclease-free water, and 0.5 μL of Taq polymerase (5 U/μL).
2. Perform PCR reaction as follows: denaturation at 94 °C for 3 min, followed by two cycles of annealing at 58 °C for 30 s, extension at 72 °C for 1 min, and denaturation at 94 °C for 30 s, followed by 35 cycles of annealing at 60 °C for 30 s, extension at 72 °C for 45 s, and denaturation at 94 °C for 30 s. In the final step anneal at 56 °C for 30 s followed by an extension at 72 °C for 7 min and then the entire reaction is chilled to 4 °C.
3. Array 5 μL of 10× gel loading dye to all wells using multichannel pipette, load the entire reaction mixture, and perform electrophoresis using 1 % agarose gel with ethidium bromide. Use 100 bp ladder marker to separate either rows or columns of a 96-well plate, while performing the gel electrophoresis.
4. Take a picture of the gel using a suitable gel documentation apparatus after completion of the electrophoretic separation, verify the band size, and note the band intensity.

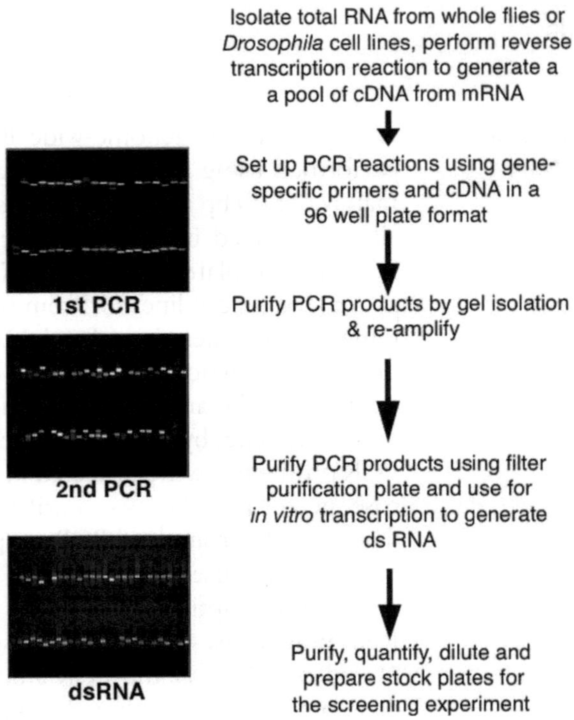

Fig. 1. Work flow for high-throughput synthesis of dsRNA. Following isolation of total RNA, reverse transcription is used to produce cDNA from which dsRNA template is amplified. Using primers specific to a target gene of interest, usually 500 bp apart, templates are PCR amplified twice, being gel purified each time, 96 targets at a time. As the primers also encode T7 primer sequences the amplified PCR products are then used for an in vitro transcription reaction to produce dsRNA, which is then purified, diluted, and aliquoted into stock plates ready for the screening experiment.

5. Equilibrate a water bath to 50 °C.
6. Slice out the area of the gel containing the amplified PCR product, by observing under UV transilluminator, and transfer into a pre-labeled microfuge tube or 96-deep-well plate. In order to minimize the risk of UV-induced DNA damage to PCR products, the agarose gel could be cut into small sections and processed separately (see Note 4).
7. Proceed to gel isolation of the PCR product.
8. For purifying the PCR products from the agarose gel slices in an HTP format, combination of PureLink HTP PCR purification kit and gel isolation buffer from Invitrogen Gel isolation kit will be useful (as described in the next step) (see Note 5).
9. Dispense 600 μL (or three times the gel volume) of gel solubilization buffer (from Gel isolation kit) for a gel that approximately weighs 200 mg (see Note 6). Dispense the buffer using a repeater pipette to all wells, if all the gel slices are of uniform size (see Note 7).

10. Place the 96-well plate containing the gel slice and gel solubilization buffer into a water bath (or incubator) set at 50 °C, incubate for 10 min, and shake the plate occasionally to ensure dissolution of the gel. Gel slices appearing pink with gel solubilization buffer will become clear once the gel is completely dissolved. After the gel slice appears dissolved, continue incubation for an additional 5 min.
11. Add 200 μL of isopropanol to the dissolved gel slice that is approximately 200 mg by weight. Mix well thoroughly and proceed to next step.
12. Set up a vacuum manifold (Multiscreen vacuum manifold, 96 wells or equivalent) and place a 96-well PCR filter plate (PureLink HTP PCR purification kit) on top of the manifold.
13. Load isopropanol mixed solubilized gel solution to the PCR filter plate. Apply vacuum for 2–3 min or until the solution is drained off completely, and then release the vacuum.
14. Dispense 600 μL wash buffer (provided with PureLink HTP PCR purification kit) consisting of ethanol to the PCR filter plate. Apply vacuum for 1–2 min or until the solution is drained off completely, and then release the vacuum (see Note 8).
15. Repeat step 14 one more time, tap the filter plate on a stack of paper towels, and allow the plate dry for about 10 min.
16. Place a 96-well Elution Plate/Receiver (PureLink™ HTP PCR purification kit, Invitrogen) under the PCR filter plate, add 100 μL elution buffer to PCR filter plate, and allow the filter to soak at room temperature for 1 min. Centrifuge this setup for 2–3 min at 2,054 × *g* to elute DNA into the elution plate, label the plate, and seal and store the PCR products at −20 °C.

3.1.2. Second PCR

1. Thaw the primer plates and PCR products and centrifuge the plates at 2,683 × *g* for 5 min to get rid of air bubbles.
2. Array 10 μL of primers from the working stock using a multichannel pipette, 38 μL of PCR master mix (minus 2 μL cDNA) as in 1st PCR step, instead of cDNA, array out 2 μL of purified 1st PCR products using a multichannel pipette.
3. Repeat PCR as in first PCR (see steps 2–4 in Subheading 3.1.1).
4. At the end of PCR, add 100 μL nuclease-free water to all wells, using repeater.
5. Transfer the entire volume using multichannel pipette to 96-well Multiscreen HTS PCR purification plate.
6. Turn on the vacuum (Multiscreen vacuum manifold, 96 wells) and set the pressure to 10 psi. To avoid damage to the filter plate, do not allow the vacuum to exceed 15 psi. Allow the wells to dry for 10–15 min until they appear shiny (see Note 9).

7. Once all the wells are dried, wash the wells with 100 μL nuclease-free water/well and continue to drain with the vacuum.
8. When the wells are close to drying and appear shining, add 30 μL nuclease-free water or 10 mM Tris–HCl pH 7.0 to the wells, seal the top and bottom of the plate with an aluminum foil, and shake it using a rocker or an Eppendorf mixer mate (800 rpm) for 10 min. Collect the cDNA elute using a multichannel pipette and transfer into a new PCR plate labeled as 2nd PCR product. *Optional*: An additional elution may increase the yield of the PCR products, by repeating this with 10–15 μL of nuclease-free water or buffer.
9. Run 1 % agarose gel, and load 3–4 μL of the PCR product to confirm the amplification.

3.1.3. *In Vitro* Transcription

1. Thaw out the Ribomax express 2× buffer at 37–42 °C, mix using vortex, and ensure that the solution is clear; if not, continue to warm the solution and resuspend the buffer completely before proceeding to next step.
2. Aliquot 8 μL of 2nd PCR product in the 96-well plate using multichannel pipette; keep the plate warm at room temperature (do not place the plate on ice). Of note, this reaction could be scaled up; however it is not recommended to scale the volume to more than two reactions/well because it could affect the downstream purification step if using Millipore multiscreen HTS plates.
3. Prepare a mix 10 μL 2× buffer plus 2 μL T7 Ribomax express mix, in bulk. Array using repeater pipette.
4. Spin the plate at 2,683 × *g* for 30 s, and proceed to the following PCR program: in vitro transcription at 37 °C for 16 h, denaturation at 80 °C for 1 min, and slow annealing (70 °C for 10 min, 65 °C for 1 min, 60 °C for 45 s, 55 °C for 30 s, 50 °C for 10 s).
5. Remove the plate from PCR machine and allow the plate to cool for 1 h at room temperature.
6. Warm nuclease-free water to 37 °C prior to the use in following steps (using warm water will enhance solubility of the salt precipitates formed in the reaction mixture, and might prevent clogging the pores present in the purification plate).
7. Add DNAse 1 μL/150 μL nuclease-free water/well. Prepare the reagents in bulk and dispense using a repeater pipette.
8. Transfer the plate to 37 °C and incubate for 30 min.
9. Transfer the reaction mixture to 96-well Millipore Multiscreen HTS PCR purification plate and repeat the steps as described in 2nd PCR; for washing the wells, use 150 μL of warm nuclease-free water.

10. For elution use 50 μL of 10 mM Tris–HCl, pH 7.0, seal the top and bottom sides of the plate with aluminum fold, and shake it on a rocker or microplate shaker (800 rpm for 10 min). A second elution with 25 μL Tris would improve the yield of dsRNA.
11. Store the dsRNA elutes in properly labeled and sealed 96-well plates at −70 °C.

3.1.4. Quantification of dsRNA

1. Use single channel if possible or if using multichannel 10 μL pipette ensure that the volumes are accurate. Spin the plate to remove air bubbles.
2. Transfer 2 μL into 96-well UV plate (can increase to 2–3 μL but remember to change the calculation in the plate reader file later on, and keep the dilution 1:100).
3. Add 198 μL of 10 mM Tris–HCl pH 7.0 (dilution factor 1:100), mix well, and take reading at 260/280. Use 1 OD = 45 for dsRNA to calculate concentration (see Note 10).
4. Adjust the concentration (by using either nuclease-free water or preferably with 1 mM Tris–HCl, pH 7.0) of dsRNA to 0.08 μg/μL or higher, if needed, and store it as a barcoded or labeled 96-well plate. It is important to perform a gel electrophoresis after diluting the dsRNA (using 1 % agarose and TBE buffer), with about 0.4 μg of dsRNA to check the quality and approximate uniformity in diluting dsRNA, before using in a screening experiment. By doing so, it is possible to identify those dsRNAs which failed or are insufficient in quantity for the experiment, and could be flagged to exclude from data analysis, which would otherwise become a source for false negative results.

3.2. HTP Transfection of dsRNA in *Drosophila* Cells

The genomic screening with dsRNA library involves transfection of dsRNA on Day 0, exposure with damaging agents on Day 3 allowing sufficient time for the RNAi to silence gene expression, and assessment of cell viability on Day 6. Steps involved in genomic screening with dsRNA are summarized in Fig. 2 and described in the following sections.

Transfection of dsRNA into *Drosophila* cells is performed by bathing *Drosophila* cells with dsRNA under serum deprived condition to facilitate the uptake of dsRNA from cell culture medium through phagocytosis. Using RNAi with *Drosophila* cells in this manner a genome-wide screen was conducted to identify essential genes required for cell viability (3). Subsequently this screening method was utilized to develop different types of HTP assays to study various cellular functions (1). We have used this screening method and the cell viability assay combined with DNA-damaging agents to identify genes essential for cell survival after MMS-induced DNA damage (2). RNAi libraries for the *Drosophila*

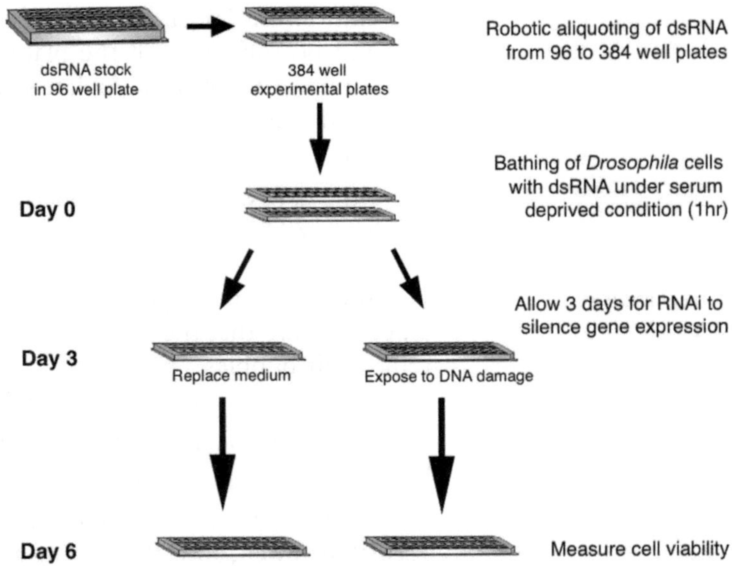

Fig. 2. Work flow for high-throughput screening with dsRNA library. Robotics are used to realiquot dsRNA from 96-well stock plates into 384-well experimental plates. *Drosophila* cells are aliquoted into the plates in the absence of serum to promote uptake of the dsRNA; serum-containing medium is then added after 1 h. Depending on the experiment, different timings can be used; here we incubate the cells for 3 days to achieve robust knockdown of target proteins. At that time either medium is exchanged for medium with or without DNA-damaging agent. Following 3 further days of incubation cell viability is then tested.

genome are available for screening from *Drosophila* RNAi Screening Center (DRSC), Harvard University (4), and commercial sources such as Invitrogen (formerly from Ambion) and Thermo Scientific (formerly from Open Biosystems). Initially the genomic screen for DNA damage is performed using the full library of dsRNA to target the knockdown of known and predicted mRNA encoded by the entire genome of *Drosophila*, and then a validation experiment is performed to confirm results obtained from the genomic screening experiment. Validation experiments are performed using a different set of dsRNAs which target a different region of mRNA than that targeted by the dsRNAs used in the genomic screening library. The experimental procedure for genomic screening and validation experiment is same, but the experimental design for validation experiment is more stringent because the objective of validation is to obtain high-quality data by reducing the rates of false positivity or false negativity from the genomic screening experiment.

1. dsRNAs used for screening or validation experiments—supplied in a 96-well plate—are aliquoted into 384-well plates for conducting the experiments with *Drosophila* cells.
   (a) *Aliquoting dsRNA for the genomic screen*: About 22,915 individual dsRNAs are present in the original DRSC library

(version 1), and these stocks are maintained in 96-well plates, consisting of 92 unique dsRNA and 4 controls per plate. For the screening experiment, dsRNAs from four different 96-well plates are combined to fill one 384-well plate consisting of one dsRNA/well. Thus the entire DRSC library (22,915 individual dsRNAs) is represented on sixty-three 384-well plates. Each well (excepting empty control wells) contains, on average, 0.05 μg (0.01–0.43 μg) of dsRNA. The experimental plates with mapping for dsRNA in each plate for genomic screening are readily available from DRSC, while the commercial sources provide the dsRNA stock plates using a robotic liquid handling system dsRNA that can be aliquoted into 384-well plates. The limitation of this experimental plate format is the single representation of a unique dsRNA per 384-well plate such that replicates (duplicates of control or treatment with DNA-damaging agents) have to be performed in separate plates. This format can provide a source for interplate variations. It is therefore recommended to conduct duplicate experiments simultaneously or use the same batch of experimental plates for the duplicate experiments.

(b) *Aliquoting dsRNA for validation experiment*: For validation experiments, a validation library could be either purchased from DRSC or synthesized by the end users in their own laboratory using primer sequences provided by DRSC and the method described in the section for HTP synthesis of dsRNA library. For the validation experiments, it is highly recommended that each dsRNA be represented in quadruplicates within a 384-well plate, and that 10 % of the wells of a plate be dedicated to control dsRNA. Also by using a standardized concentration of dsRNA, 0.4 μg/well, maximal RNAi efficiency can be achieved with the potential of eliminating false negative results as a consequence of incomplete target knockdown. The control dsRNAs are useful to normalize data and improve the robustness of validation analysis, further discussed in Subheading 3.4.2, and for additional reading refer to (5). The control dsRNAs suggested to use in validation experiments are the following: (a) control dsRNA (non-targeting dsRNA, such as dsRNA against luciferase, with 40 wells of a 384-well plate containing this dsRNA (see Note 11)); (b) positive control for dsRNA uptake (dsRNA against anti-apoptotic gene Thread (Th), which results in loss of cell viability in itself, in quadruplicates); (c) positive control for DNA damage (dsRNA against a gene known to result in loss of cell viability only in the presence of DNA damage, but not in itself, in quadruplicates); and (d) for background correction include high dose of a damaging agent that would result in 100 % cell death, in quadruplicates. A suggested distributed pattern for aliquoting the dsRNA in quadruplicates is discussed in ref. 5 with the main intension being to offset edge effects, reduce interplate variations, and enhance the statistical reliability of the readout from the plate. The pattern described in ref. 5 is achievable only with certain types of robotic liquid handling systems; in the absence of availability of such instruments, users may define their own patterns and consider the points discussed here and in ref. 5.

2. Experimental plates should be equalized to room temperature and centrifuged at $110 \times g$ for 1 min before proceeding to next step. If the plates are sealed, carefully peel the wrap and cover the plate with the lid.
3. (See Note 12) Harvest *Drosophila* cells such as Kc_{167} from a sub-confluent culture grown in a tissue culture flask (such as T175), transfer the cell suspension into a 50 mL conical tube, pellet down the cells, and resuspend in an appropriate volume of serum-free Schneider's medium. Count the cells, adjust the concentration of cells to 1.2×10^6/mL, using serum-free Schneider's medium, and prepare the cell suspension in bulk for the required number of plates.
4. Dispense 10 μL (1.2×10^4 cells) of cell suspension to each well of 63 different 384-well plates, using a liquid dispensing system (similar to Wellmate, Matrix Technologies).
5. Centrifuge the plate at $110 \times g$ for 1 min at room temperature.
6. Incubate for 1 h at 24 °C (or room temperature) to allow phagocytic uptake of the dsRNA, then dispense 20 μL of serum-containing medium (15 % heat-inactivated FBS), centrifuge the plate at $110 \times g$ for 1 min at room temperature, and incubate the plates at 24 °C for a further 72 h. Since the medium present in the wells located at the edges of the plate tends to evaporate during the incubation, it is recommended to stack the plates in a loosely covered plastic container that has moist paper towels at the bottom and overlay few moist paper towels on the top of the plates that are above on the stack (see Note 13).

3.3. Induction of DNA Damage

Incubating the 384-well plates consisting *Drosophila* cells and dsRNA for 3 days should result in an efficient RNAi-mediated knockdown of the target mRNA and presumably depletion of the corresponding protein within the cell. This procedure in effect generates a library of *Drosophila* cells with one gene of the genome being silenced in each well by utilizing the RNAi technology mediated by a library of dsRNA. Applying DNA damage to these cells allows the investigation of which proteins are essential for cell survival following damage exposure. Using this approach a genome-wide impact on cell survival could be analyzed using different types and different concentration of DNA-damaging agents to compare and contrast the mode of cellular recovery after such damages. Choosing a dose for DNA-damaging agent to use in the RNAi screening experiment depends upon the screeners' interest to measure decrease or increase (resistance) in viability in the absence of certain genes and in the context of the DNA-damaging agent. For example, to measure a decrease in cell viability following RNAi, screeners must choose a dose that would produce minimum but measurable effect on viability in the absence of any RNAi. Such a dose should then allow identification and quantification of further decrease in viability when the DNA-damaging agent is combined with an effective RNAi. In our genomic screening experiments we chose to use a concentration of DNA-damaging agent that would result in 70 % viability (in the absence of any RNAi) so that we could identify genes essential for cell survival following DNA damage in a robust manner (2). If the screen is intended to measure resistance, higher dose with stronger effect on viability could be selected for the screening experiment.

1. Remove the plates from the incubator, and centrifuge at 110 × *g* for 1 min at room temperature.
2. Prepare a stock of medium in bulk, with or without appropriate concentration of DNA-damaging agents sufficient for the number of plates to be used in the experiment (for example, 0.004 % (w/v) MMS in Schneider's medium consisting of 10 % heat-inactivated FBS for inducing DNA damage).
3. Aspirate the medium from the 384-well experimental plates carefully using a 16- or 24-channel wand (V & P Scientifics), under a low vacuum suction. Do not let the plate to dry out for long; proceed to next step as quickly as possible (see Note 14).
4. Dispense 40 μL/well of fresh (10 % serum containing) medium, with or without DNA-damaging agents, using a liquid dispensing system.
5. Centrifuge the plate at 110 × *g* for 1 min at room temperature.
6. Incubate the plates at 24 °C for an additional 72 h in the humid container as described before.

3.4. Assessment of Cell Viability

Following the knockdown of intended mRNA targets with RNAi and application of DNA damage, genes essential for cell survival could be surveyed on a genome level by using a cell viability assay. The cell viability assay described in this section quantifies the amount of live cells present based on the intracellular levels of ATP, detected by using luciferin as a substrate and luciferase as an enzyme that utilizes ATP to produce luminescence. Therefore, the amount of luminescence detected is directly proportional to the amount of cells present in each well in these experiments. Using a commercially available assay (CellTiter-Glo, Promega) an RNAi-based genome-wide screen was originally performed to detect the genes essential for cell survival in *Drosophila* (3). This same method was then adapted to identify genes essential for cell survival following DNA damage (2). Using Celltiter-Glo provides a measure of viable cells in terms of luminosity values. These values are then used to compare damage-treated cells with untreated cells and the effect of RNAi-mediated knockdown on the loss of viability after DNA damage is determined. Therefore, in this section the experimental procedure and different methods to analyze the results are discussed.

3.4.1. Experimental Procedure

1. Following an additional incubation for 72 h, remove the plates from the incubator, and centrifuge at $110 \times g$ for 1 min at room temperature.
2. Aspirate the medium from the 384-well experimental plates carefully using a 16- or 24-channel wand (V & P Scientifics), under a low vacuum suction. Do not let the plate to dry out for long; proceed to next step as quickly as possible (see Note 15).
3. Dispense 40 μL/well of CellTiter-Glo reagent using a liquid dispensing system (see Note 16).
4. Centrifuge the plate at $110 \times g$ for 1 min at room temperature and proceed to read using a microplate reader capable of detecting luminescence (such as, Spectramax M5 plate reader, Molecular Devices).
5. Collect and map the results to identifiers by referring to the dsRNA used in these experiments to corresponding genes and proceed to data analysis.

3.4.2. Data Analysis

1. Results obtained from the genomic screen

   Multiple methods are available for normalizing the data obtained from genomic screen; however we have tested the available methods and observed that a combination of background subtraction and scaling followed by quantile normalization to generate better quality data worked better than any single procedure (5). Therefore, the approaches for data normalization are discussed below.

   Background subtraction and scaling: Background luminescence for each plate is calculated based on the median value of the plate and the value obtained for the well containing the lowest value. For example, the DRSC screening library has included a dsRNA targeting Thread (Th; CG12284), and an anti-apoptotic gene is included in all experimental plate, as a positive control for RNAi. Silencing the expression of Th results in loss of cell viability and therefore, for background subtraction, the luminescence value of the well containing Th could be subtracted out from the value of each other well in the plate. Then the data could be scaled as follows. For untreated plates, the median value of each plate could be assumed to be equivalent to 100 % viability and the data should be scaled accordingly, as described in ref. 5. For damage-treated plates, a prior knowledge based on an estimate from cell viability (in the absence of any RNAi) with the amount of damaging agent used in the experiment should be used for scaling the data. For example, exposure with 0.004 % alkylating agent MMS results in 65 % viability and therefore, the median value of the plate treated with MMS could be assumed to be equivalent to 65 % viability of untreated cells.

Quantile normalization: Quantile normalization (6) is based on comparing quantiles between data sets, with no special allowance for outliers. Each of the 63 plate types following background subtraction and scaling could be quantile normalized between replicates of untreated and between replicates of damage treatment. Following quantile normalization, "M," the moderated *t*-statistic in $\log_2$ fold change, could be calculated for each dsRNA, by which the genes could be ranked.

2. For results obtained from the validation experiments

Validation plates (described in Subheading 3.2) consist of 40 wells of non-targeting dsRNA against luciferase (Luc) to provide a control for the presence of dsRNA and 4 wells containing a high concentration of DNA-damaging agent (such as MMS (0.12 %, w/v) resulting in 99 % cell death) to determine the background luminescence in the absence of any viable cells. Luminescence values for each well could be scaled similar to the screen data, except the value for high concentration of DNA-damaging agent should be adjusted to 0 %. For the data analysis background subtracted well values should be normalized by dividing with median of scaled Luc values, and a *T-test* (two tailed, Type II) could be performed between normalized quadruplicates to determine if there is a difference between control and treatment. Since it is common in RNAi-based screens that removal of certain genes could result in decreased viability without any treatment, to account for reduced cell number under these circumstances and to determine if further reduction in cell viability with damage treatment is significant, a second *T*-test (two tailed, Type II) could be performed with Percent Control Viability (*PCV*) for each dsRNA targeting an mRNA against the *PCV* of dsRNA targeting Luc. To determine PCV, the data from normalized results could be scaled to 100 % based on values for Luc with untreated set, and to an appropriate percentage (based on previous estimate) for damage-treated set. PCV could then be determined by comparing the normalized values for any gene with scaled percentage for Luc within untreated or damage-treated gene set. Statistically significant hits from these analyses could be selected to compare with results obtained from genomic screening. Values obtained based on PCV along with statistics could be used to rank hits to select candidate genes for follow-up studies, or the entire results could be combined to analyze for an interpretation to provide systems-level understanding on damage response.

4. Notes

1. To facilitate simultaneous PCR amplification of different genes in the HTP format, it is essential to maintain uniform conditions for all PCR reactions. For example, using working stock of primers prepared from a standardized concentration of primer stock will ensure uniform primer-to-template ratio in all PCR reactions. Therefore, it is recommended that forward or reverse primers be custom synthesized to a scale of 25 nmol, and then adjusted separately to yield 10 μM stock when resuspended in a volume of 400 μL.
2. *Caution*: Ethidium bromide is a carcinogen, so handle and dispose according to local regulations.
3. Genomic DNA could be used instead of reverse transcribed cDNA for gene-specific PCR amplification; however doing so introduces intronic sequences into the dsRNA and might increase off-target effects with RNAi. But if the cDNA for a particular gene of interest could not be amplified when the mRNA is used as a starting material then genomic DNA could be used as a substitute. In order to perform the reactions in HTP scale, it is necessary to prepare reagents at least in 10 % excess than needed so that reagent shortage while dispensing is avoided. For example, preparing reagents for 105 PCR reactions will ensure equal dispensing of PCR reagents to a 96-well plate.
4. *Caution*: UV exposure from transilluminator is harmful to eye and skin, so use appropriate personal protective equipment while handing the gel for this step.
5. Alternatively, Qiagen gel isolation kit could be used to isolate the individual PCR products by using 96 individual microfuge tubes and following manufacturer's recommendations; however such isolation of PCR products using individual tubes will be time consuming and laborious.
6. Making thin slice of gel would weigh approximately about 200 mg for all PCR products, if Subcell GT cell (Biorad) is used with a 20-well comb that is 0.75 mm thick.

7. The volume of the buffer can be adjusted based on the weight of the gel; accordingly it is necessary to adjust the volumes of other buffers to be used in the subsequent steps.
8. Wash buffer provided in the Invitrogen kit might require addition of ethanol.
9. Drying for extended period of time is undesirable for the elution of PCR products from the filter plate, if few wells were incompletely dried; keep the remaining wells hydrated by adding extra water or seal the plate with parafilm or saran wrap to reestablish the vacuum, and punch holes to selectively drain the incompletely dried wells.
10. Using nuclease-free water for quantification of dsRNA by this method has been observed to yield discordant results; therefore, it is highly recommended to use Tris buffer for the quantification of dsRNA instead of nuclease-free water.
11. dsRNA against GFP has off-target effects in *Drosophila*.
12. Important notes before proceeding with this step: The amount of *Drosophila* cells required for this step depends upon the number of experimental plates used for the screening or validation experiment. User may have to plan in advance and scale-up the numbers of flask to yield appropriate amounts of cells. In order to improve the efficiency of dsRNA uptake by phagocytosis avoid passaging or exchanging growth medium on the previous day of this step.
13. Using sterile water to moisten paper towels will reduce contamination issues.
14. *Drosophila* cell line such as *Kc167* is semi-adherent, so in order to avoid losing cells by vacuum suction, insert the tip of the wand along one corner of the well, gently slide it down towards the bottom of the well, and lift it up quickly.
15. Follow the precautions described in step 3, under Subheading 2.3.
16. Prepare Celltiter-Glo reagent in bulk for the required numbers of plates. The Celltiter-Glo reagent is supplied in 2× concentration and therefore, should be diluted 1:1 with PBS, prior to use.

Acknowledgement

This work was supported by NIH grants K22-ES12264 and R15ES019128 as well as a Voelcker Fund Young Investigator Award from the Max and Minnie Tomerlin Voelcker Fund to AJRB. We also thank Aparna Gorthi for her constructive criticism and editing of the manuscript.

References

1. Mathey-Prevot B, Perrimon N (2006) Drosophila genome-wide RNAi screens: are they delivering the promise? Cold Spring Harb Symp Quant Biol 71:141–148
2. Ravi D, Wiles AM, Bhavani S, Ruan J, Leder P, Bishop AJ (2009) A network of conserved damage survival pathways revealed by a genomic RNAi screen. PLoS Genet 5:e1000527
3. Boutros M, Kiger AA, Armknecht S, Kerr K, Hild M, Koch B, Haas SA, Paro R, Perrimon N (2004) Genome-wide RNAi analysis of growth and viability in Drosophila cells. Science 303:832–835
4. Ramadan N, Flockhart I, Booker M, Perrimon N, Mathey-Prevot B (2007) Design and implementation of high-throughput RNAi screens in cultured Drosophila cells. Nat Protoc 2:2245–2264
5. Wiles AM, Ravi D, Bhavani S, Bishop AJ (2008) An analysis of normalization methods for Drosophila RNAi genomic screens and development of a robust validation scheme. J Biomol Screen 13:777–784
6. Irizarry RA, Bolstad BM, Collin F, Cope LM, Hobbs B, Speed TP (2003) Summaries of Affymetrix GeneChip probe level data. Nucleic Acids Res 31:e15

Chapter 3

Isolating Genes Involved with Genotoxic Drug Response in the Nematode *Caenorhabditis elegans* Using Genome-Wide RNAi Screening

Lone Vedel Schøler, Tine Hørning Møller, Steffen Nørgaard, Lotte Vestergaard, and Anders Olsen

Abstract

The soil nematode *Caenorhabditis elegans* has become a popular genetic model organism used to study a broad range of complex biological processes, including development, aging, apoptosis, and DNA damage responses. Many genetic tools and tricks have been developed in *C. elegans* including knock down of gene expression via RNA interference (RNAi). In *C. elegans* RNAi can effectively be administrated via feeding the nematodes bacteria expressing double-stranded RNA targeting the gene of interest. Several commercial *C. elegans* RNAi libraries are available and hence gene inactivation using RNAi can relatively easily be performed in a genome-wide fashion. In this chapter we give a protocol for using genome-wide RNAi screening to identify genes involved with the response to genotoxic stress.

Key words: *C. elegans*, DNA damage, Whole genome RNAi screening, Genotoxic compounds

1. Introduction

1.1. *Caenorhabditis elegans*: A Popular Genetic Model Organism

The small soil nematode *Caenorhabditis elegans* (*C. elegans*) is a popular model organism for studying a variety of biological processes, including checkpoint functions, DNA damage and repair and apoptosis. The somatic cells of the adult nematode are postmitotic but in the germ line and during development, mitotic, meiotic, and embryonic cell divisions can be studied (1). The field of *C. elegans* genetics has changed dramatically since the last edition of this book with the introduction of forward genetic approaches such as RNA interference (RNAi), establishment of systematic

Lotte Bjergbæk (ed.), *DNA Repair Protocols*, Methods in Molecular Biology, vol. 920,
DOI 10.1007/978-1-61779-998-3_3, © Springer Science+Business Media New York 2012

Table 1
Commonly used DNA damaging agents

Compound	Mechanism	Type of DNA damage	Reference
Ionizing radiation		ss/ds breaks	(16)
ENU	Alkylation	ds breaks	(17)
EMS	Alkylation	G/C→A/T transitions	(3)
HU	dNTP depletion/stalled replication forks	ss/ds breaks	(18)
Paraquat	Superoxide formation	Oxidation	(19)
TMP/UV		Deletion	(20)

gene knock out consortia, and strategies for targeted gene inactivation.

Ethyl methanesulfonate (EMS) has been widely used to mutagenize *C. elegans* in classical reverse genetic screens in order to investigate gene function. The recently developed protocols for using next-generation sequencing for identification of point mutations (2) will undoubtedly keep this approach popular. The protocol for EMS mutagenesis published in the previous edition of this book (3), is still an excellent point of reference for researchers interested in EMS screening. Other mutagens have also successfully been used in *C. elegans* (Table 1) and since these are all DNA damaging agents they can also be used as tools to understand the genetics of the DNA damage response and DNA repair processes.

This chapter focuses on using genome-wide RNAi screening to identify genes with altered sensitivity to DNA damaging agents. While our protocol describes the use of the ribonucleotide reductase (RNR) inhibitor hydroxyurea (HU), it can in principle be adapted to include any DNA damaging agent (see Note 1). Various transgenic strains have been constructed to study DNA damage and apoptosis in *C. elegans* and protocols for using some of these will briefly be introduced. We conclude the chapter with useful links for the further characterization of the identified genes.

1.2. General Principles of RNAi Screens

The RNAi screen protocol involves the five basic steps illustrated in Fig. 1 and is described in detail in the following sections. Compared to classical mutagenesis the RNAi screen approach has the major advantage that the identities of the inactivated genes are immediately known and consequently no efforts need to be devoted to mapping and identifying mutations. Importantly, by using RNAi it is also possible to study reduced function of essential genes that

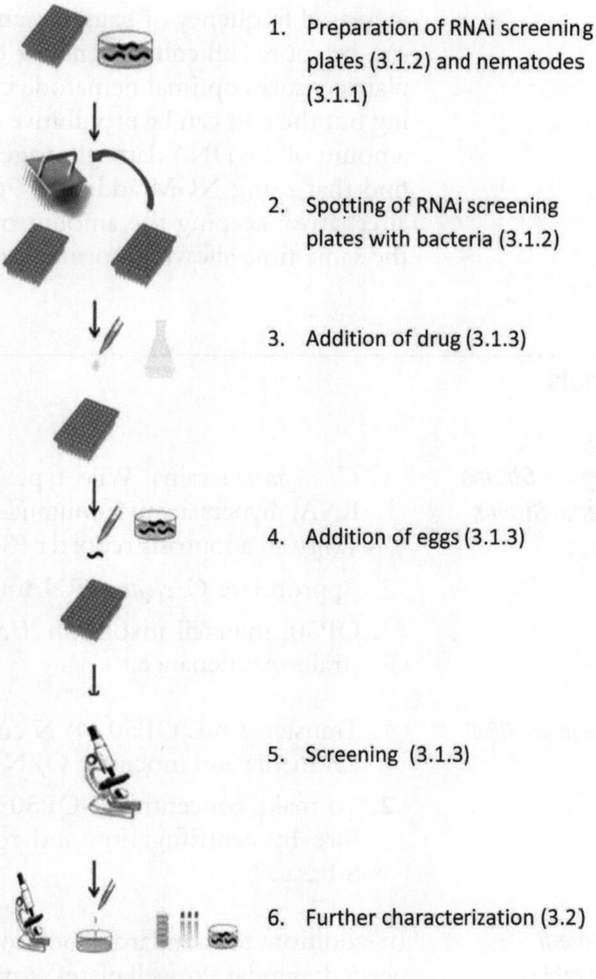

Fig. 1. The basic steps in an RNAi screen.

cause lethality when inactivated by mutation. However, there are also certain drawbacks that should be considered: (i) the RNAi clone may target more than one gene if the genes share sequence homology, (ii) not all tissues are equally susceptible to RNAi, (iii) the screen will not be saturated since the libraries do not offer full coverage of the entire genome, (iv) not all bacterial clones will grow up, and (v) sometimes phenotypes caused by RNAi are less reproducible than those arising from mutations. Thus, it is often worth ordering a mutant from the *Caenorhabditis* Genetics Center (CGC) if available for further characterization of the gene function.

Whole genome RNAi screens have been successfully performed using liquid cultures in 96-well format (4), normal NGM plates (5), or NGM added to 96-well plates (6). The advantages of using liquid cultures are reduced labor cost and requirement of fewer materials compared to the other approaches. The drawbacks are difficulties in culturing the nematodes without slight starvation, increased frequency of bagging, and finally the phenotypic scoring can be more difficult depending on the endpoint. Using regular plates ensures optimal nematode culture conditions and easy scoring but the cost can be prohibitive due to the need for much higher amount of the DNA damaging agent and general consumables. We find that using NGM added to 96-well plates offers an attractive alternative; keeping the amount of compound needed low and at the same time allowing normal growth of the nematodes.

2. Materials

2.1. *C. elegans* Strains and Bacterial Strains (See Note 2)

1. *C. elegans* strains: Wild-type N2, NL2099 *rrf-3(pk1426)* an RNAi hypersensitive mutant strain (7), MD701 P_{lim-7}, *ced-1::gfp* an apoptosis reporter (8).
2. Appropriate *C. elegans* RNAi feeding libraries.
3. OP50, an uracil auxotroph *E. coli*, as food for normal *C. elegans* strain maintenance.

2.1.1. Concentrated OP50

1. Transfer 2 mL OP50 O/N culture to 2 L nonselective sterile LB media and inoculate O/N at 37 °C, shaking.
2. To make concentrated OP50 pellet the 2 L OP50 O/N culture by centrifugation and resuspend final pellet in 15 mL S-basal.

2.2. Equipment and Disposables

In addition to standard laboratory equipment the following is needed: regular 96-well plates, 2 mL 96-well plates, 50 and 90 mm Petri dishes, air tight boxes, platinum wire worm pick, silicoated 1.5 mL microcentrifuge tubes, silicoated p100 pipette tips, pin tool replicator microcentrifuge, centrifuges for 15 and 500 mL tubes, heating block with magnetic stirring, dissection microscope, and fluorescence microscope equipped with minimum 40× objective.

2.3. Reagents and Solutions

2.3.1. Reagents

1 mM Levamisole solution, LB medium, LB agar, 3 % formaldehyde/0.1 M K_2HPO_4 (pH 7.2), 1× phosphate buffered saline (PBS), PBT (0.1 % Tween 20 in 1× PBS), VECTASHIELD® Mounting Medium (hard set with DAPI), agarose, 100 % methanol, ampicillin, isopropyl β-D-1-thiogalactopyranoside (IPTG), agar, NaCl, and peptone..

2.3.2. Solutions

1. *1 M KPO4 buffer*: Mix 108.3 g KH_2PO_4 and 35.6 g K_2HPO_4 and adjust to 1 L with milli-Q H_2O. Verify pH 6.0, autoclave, and store at RT.
2. *S-basal w/o cholesterol*: Mix 5.8 g NaCl, 50 mL 1 M KPO_4 buffer and adjust to 1 L with milli-Q H_2O. Autoclave and store at RT.

3. *Hydroxyurea solution*: Prepare a fresh stock solution before each experiment by dissolving 184 mg HU in 3 mL S-basal. From the stock a dilution can be made in S-basal to obtain the desired concentration.
4. *Hypochlorite solution*: Mix 40 mL milli-Q H_2O, 7.5 mL NaOCl (12 %), 2.5 mL 5 M NaOH.

2.4. Plates

2.4.1. Regular NGM Plates

1. Mix: 17 g agar, 3 g NaCl, and 2.5 g peptone and add up to 1 L of milli-Q H_2O.
2. Autoclave.
3. Cool liquid NGM to 55 °C (use of magnetic stirring can help prevent the agar from setting).
4. Add: 25 mL 1 M KPO_4 buffer, 1 mL 1 M $CaCl_2$, 1 mL 1 M $MgSO_4$, and 1 mL 5 mg/mL cholesterol in ethanol and mix thoroughly.
5. Pour liquid NGM into Petri dishes of desired sizes using sterile techniques and let dry O/N at RT.
6. Spot plates with O/N OP50 culture using a transfer pipette—mix before adding a few drops in the middle of the plates.
7. Allow to dry O/N at RT before using the plates.

Use 50 mm plates for maintenance and 90 mm for mass cultures or 96-well plates for screening. NGM plates should be stored in an air tight box to prevent water evaporation. Plates can be stored at RT or 4 °C to prolong durability.

2.4.2. RNAi Screening Plates

Follow the protocol for regular NGM plates but under step 4 also add ampicillin and IPTG to a final concentration of 100 µg/mL and 1 mM, respectively.

Prepare 96-well RNAi screening plates by adding 200 µL NGM for RNAi to each well and let dry O/N at RT. It is important to have the same amount of NGM in each well to ensure consistent drug concentration if the drug is applied to the bacterial lawn (see Note 1). RNAi plates should be used within 2 weeks. Fungicide can be added to the plates to prevent fungal contamination.

3. Methods

Standard *C. elegans* culturing should be performed as described (9).

3.1. Isolating Mutants Hypersensitive/Resistant to Hydroxyurea

HU is a chemotherapeutic drug that induces replication fork stalling during S-phase, by depleting the cells for dNTP pools. HU induces DNA damage in a dose-dependent manner (10). To identify genes, causing hypersensitivity to HU when inactivated, adjust the

HU concentration (via dilution series) so that normal development in WT nematodes is allowed. Hypersensitive worms will arrest development at an early larval stage and will be significantly smaller in size compared to control nematodes or develop into sick looking animals with gross developmental defects. To identify genes causing resistance to HU when inactivated, adjust the HU concentration so that WT nematodes just arrest development. Genes causing HU resistance will allow development into adult worms and will be larger in size compared to control nematodes. We find that using this experimental protocol *cep-1* (p53) or *cid-1* mutants/inactivation can be used as positive control for HU resistance and *chk-1* can be used as positive control for HU sensitivity (6, 11).

A large number of eggs is required to perform the RNAi screen and these can be obtained by synchronization via hypochlorite treatment of gravid adult worms in mass cultures. To prevent starvation of the mass cultures the nematodes must be fed with concentrated OP50 bacteria (see 2.1.1).

3.1.1. Nematode Mass Culturing and Synchronization (Harvesting of Eggs) (Step 1)

1. Prepare five maintenance plates with nematodes of desired genotype (40 egg laying adults on each). Leave for 3 days.
2. Spot 2 mL concentrated OP50 culture on large NGM plates and leave to dry.
3. Wash of nematodes from maintenance plates (from 1) in S-basal. Remove supernatant and spot nematodes onto the large NGM plates spotted with concentrated OP50. Expand the population by letting them grow for 1–2 generations. The generation time depends on which strain is used. For wild-type N2 worms 3 days are required to complete development from egg to egg-laying adult. The plates should contain as many gravid worms as possible without starving them.
4. Wash gravid worms off the large plates in S-basal and transfer them to a 15 mL falcon tube.
5. Pellet worms by centrifugation at 1,800 rfc for 30 s. Wash with S-basal and repeat until supernatant is cleared of bacteria.
6. Add 10 mL hypochlorite solution to each falcon tube and shake/vortex until worms break open and eggs are released—observe progress under a stereo microscope (it can be necessary to repeat the step twice in order to dissolve all adult worms).
7. Add S-basal and do a final vortexing.
8. Pellet the released eggs by centrifugation at 3,300 rfc for 30 s.
9. Wash eggs five times in S-basal.
10. Determine number of eggs by mixing thoroughly and count number of eggs in, e.g., 1 μL. Eggs can be used for screening (see 3.1.3) or for maintaining mass cultures.

3.1.2. Spotting of RNAi Screening Plates (Step 2)

1. To culture the RNAi feeding bacteria in 96-well format, add 1 mL LB medium with ampicillin (final concentration 100 μg/mL) to each well in a 2 mL 96-well plate.
2. Defrost HT115 RNAi library plate on ice.
3. Using a 96 pin replicator inoculate bacterial cultures from the HT115 RNAi library plate to the 2 mL 96-well plate.
4. Grow cultures O/N at 37 °C (shaking) before spotting 20 μL in each well of the RNAi screening plates—include control wells with empty vector.
5. Add glycerol to the remaining bacteria at a final concentration of 10 % and store plate at −80 °C for back up.
6. Allow spotted NGM plates to dry for 2 days at RT before use. Make sure all wells are equally dry.

3.1.3. Screening for Potential DNA Damage-Sensitive "Mutants" (Steps 3–5)

1. Spot the HU solution onto the bacterial lawn of the RNAi screening plates and let dry for 1 h (adjust volume up to 20 μL to obtain the desired concentration) (see note 3).
2. Add 30 eggs to each well of the prepared RNAi screening plate—remember to mix egg solution carefully before each transfer.
3. Incubate at 20 °C for 4 days to let the eggs develop.
4. Determine the level of sensitivity/resistance by scoring growth and development of nematodes in each well. Ensure that appropriate negative and positive controls behave as expected.

3.2. Further Characterization of Genes Involved in DNA Damage Responses (Step 6)

When the RNAi clones causing the desired phenotype have been identified it is time to start a more detailed characterization of the DNA damage-associated phenotypes (see Note 4). Many approaches exist for analyzing the functions of identified genes in regard to a possible role in checkpoint function and DNA damage response (1, 12). Protocols for some of the most popular methods are given in the following and a brief overview of additional methods is collected in Table 2.

3.2.1. Detection of Apoptosis Using the P_{lim-7}, ced-1::gfp Reporter Strain (Modified from Ref. 8)

DNA damage induces apoptosis in the germ line of *C. elegans* (13). CED-1 is a transmembrane protein expressed in engulfing cells. It clusters at the surface of the phagocytic cup during the removal of apoptotic cells, and thus the CED-1::GFP reporter strain labels apoptotic cells in the process of being engulfed (8). The *lim-7* promoter ensures expression exclusively in the sheet cells surrounding the germ cells.

1. Culture the worms to adulthood on the desired RNAi bacteria. Growth at 25 °C results in a stronger GFP signal of the CED-1::GFP reporter.
2. Mount the worms in 15 μL S-basal on a 2 % agarose pad (optional add one drop of levamisole to immobilize the worms and skip heat immobilization (step 4)).

Table 2
Methods for gene characterization

Method	Detection	References
Embryonic lethality	Genomic instability results in high frequencies of embryonic lethality	(21)
Male frequency	Nondisjunction of the X chromosomes results in higher male frequencies	(22)
DIC	Apoptotic cells in the germ line can be seen as reflective, disc-like objects	(23, 24)
ced-1::gfp	CED-1::GFP accumulates around phagosomes and apoptotic cells appear as green rings in the arch of the germ line	(8)
SYTO12	SYTO12 binds to and stains DNA in apoptotic cells	(25)
TUNEL assay	Labels free DNA ends with modified nucleotides detectable with fluorescence	(26)
DAPI staining	Staining of DNA in all cells Number of germ cells Morphology of germ cells	(15, 16, 27–29)
Cell cycle arrest	Decrease in number and enlargement of germ cells due to DNA damage	(17)
hus-1::gfp Localization	HUS-1 accumulates at distinct foci in the nucleus after DNA damage	(22)
RAD-51 immunostaining	RAD-51 localizes to DSBs in the DNA	(28)
qPCR of *egl-1* and *ced-13*	Transcription of *egl-1* and *ced-13* is upregulated in response to DNA damage	(30)
Mutator assays	Determination of the number of spontaneous mutations	(31)

3. Apply a cover slip and seal with nail polish.
4. Immobilize the worms by placing the slide on a heating block at 60 °C for 10 s.
5. Observe the worms using a fluorescent microscope. Apoptotic cells appear as green rings in the arch of the germ line.

3.2.2. Examination of Germ line and Germ Cell Morphology and Germ Cell Number Using DAPI Staining of Whole Animals (Modified from Ref. 12)

The fluorescent dye 4′,6-diamidino-2-phenylindole (DAPI) binds to A-T rich sequences in the DNA (14) and can be used to visualize DNA in whole worms.

1. Transfer 30–50 worms into 1.5 mL microcentrifuge tubes containing 500 μL S-basal. Remove S-basal and fix worms for 15 min at RT in 1 mL 3 % formaldehyde/0.1 M K_2HPO_4 (pH 7.2). Allow worms to settle by gravity between each resuspension.
2. Wash worms in 1 mL PBT. Remove supernatant.
3. Postfix worms in 1 mL 100 % methanol (−20 °C) for 5 min.
4. Wash worms two times in 1 mL PBT. Remove most of supernatant, leave a small amount of liquid to allow for easier transfer of the worms.
5. Mount worms on a glass slide in one drop of VECTASHIELD® Mounting Medium and apply cover slide. Leave for 15–20 min to harden.
6. Observe worms under a fluorescent microscope. The worms can be scored for different phenotypes like germ cell morphology, number of germ cells, developmental stage of germ cells, and chromosome morphology (see Table 2).

3.2.3. Immunostaining of Gonads (Modified from Ref. 15)

Immunostaining can be used to determine if the localization of DNA repair proteins is altered following inactivation of the gene of interest (see Note 5). To reduce background noise dissected gonads can be used.

1. Transfer 15–20 worms onto a glass slide containing 50 μL S-basal + one drop of 1 mM levamisole.
2. Cut off the head at the pharynx by using a syringe. At least one arm of the germ line will pop out along with the intestine.
3. Transfer worms to a silicoated 1.5 mL microcentrifuge tube containing 500 μL 1× PBS by using a P100 with silicoated tips (see Note 6).
4. Remove PBS and fix the worms in 1 mL of 3 % formaldehyde/0.1 M K_2HPO_4 (pH 7.2). Incubate 15 min at RT. Remove supernatant.
5. Wash the worms in 1 mL PBT. Remove supernatant.
6. Postfix the worms in 1 mL 100 % methanol (−20 °C) for 5 min.
7. Wash the worms two times in 1 mL PBT. Remove supernatant.
8. Dilute the primary antibody in 1 mg/mL BSA in PBT.
9. Add 100 μL antibody solution to the worms and incubate O/N at 4 °C, shaking.
10. Remove antibody and wash the worms 4 × 5 min in 500 μL PBT. Additional washes might be necessary if too much background staining is observed.

11. Dilute the secondary antibody in 1 mg/mL BSA in PBT.
12. Add 100 μL secondary antibody solution to the worms and incubate 4 h at RT, shaking.
13. Remove antibody and wash the worms 3×5 min in 500 μL PBT. Additional washes might be necessary.
14. Mount the worms on a glass slides in one drop of VECTASHIELD® Mounting Medium and apply a cover slide. Leave for 15–20 min to harden. Alternatively, if DAPI staining is not needed, mount the worms in 15 μL S-basal on a 2 % agarose pad and cover with a glass slide and seal with nail polish.

4. Notes

1. Drug delivery in *C. elegans* can be problematic if the compounds are not water soluble, as organic solvents might affect the worm. Drug delivery in EtOH and DMSO can be performed at low concentrations (0.5–1 %).
2. Two commercial RNAi libraries (and sub-libraries of these) are available (Open Biosystems and Geneservice, Ltd.). These libraries target approximately 11,000 and 16,800 open reading frames, respectively. Since some of the ORFs in the libraries are overlapping the entire genome is not covered. *C. elegans* strains and OP50 can be obtained from the CGC at the University of Minnesota, Twin Cities (http://www.cbs.umn.edu/CGC/). *C. elegans* mutants with increased/decreased sensitivity to RNAi in various tissues can also be obtained from the CGC.
3. *C. elegans* can be treated with compounds by casting it directly into the nematode growth media, adding it to nematode culture dish seeded with bacterial food or by soaking the nematode directly in the compound. Casting the compound directly in NGM ensures an even drug concentration throughout the plate but may affect growth of the bacteria as well and hence the efficiency of the RNAi.
4. It should be noted that not all genes identified in the screen will be involved in DNA repair since for example checkpoint defective and drug uptake/transport-defective "mutants" will also be isolated in this type of screen.
5. This protocol can be used to stain any desired protein in the gonads, whether involved in DNA damage or not.
6. The use of silicoated tubes and tips are not necessary, but they minimize the loss of worms while working, as worms do not stick as easily to silicoated surfaces.

References

1. van den Heuvel S (2005) The *C. elegans* cell cycle: overview of molecules and mechanisms. Methods Mol Biol 296:51–67
2. Sarin S, Prabhu S, O'Meara MM, Pe'er I, Hobert O (2008) *Caenorhabditis elegans* mutant allele identification by whole-genome sequencing. Nat Methods 5:865–867
3. Hartman PS, Ishii N (1999) Isolating mutants of the nematode *Caenorhabditis elegans* that are hypersensitive to DNA-damaging agents. Methods Mol Biol 113:11–16
4. Nollen EA, Garcia SM, van Haaften G, Kim S, Chavez A, Morimoto RI, Plasterk RH (2004) Genome-wide RNA interference screen identifies previously undescribed regulators of polyglutamine aggregation. Proc Natl Acad Sci U S A 101:6403–6408
5. Hansen M, Hsu AL, Dillin A, Kenyon C (2005) New genes tied to endocrine, metabolic, and dietary regulation of lifespan from a *Caenorhabditis elegans* genomic RNAi screen. PLoS Genet 1:119–128
6. Olsen Lab, L. L. U. R.
7. Simmer F, Moorman C, van der Linden AM, Kuijk E, van den Berghe PV, Kamath RS, Fraser AG, Ahringer J, Plasterk RH (2003) Genome-wide RNAi of *C. elegans* using the hypersensitive rrf-3 strain reveals novel gene functions. PLoS Biol 1:E12
8. Zhou Z, Hartwieg E, Horvitz HR (2001) CED-1 is a transmembrane receptor that mediates cell corpse engulfment in *C. elegans*. Cell 104:43–56
9. Stiernagle T (2006) Maintenance of *C. elegans*. WormBook 1–11
10. Hakansson P, Hofer A, Thelander L (2006) Regulation of mammalian ribonucleotide reduction and dNTP pools after DNA damage and in resting cells. J Biol Chem 281:7834–7841
11. Olsen A, Vantipalli MC, Lithgow GJ (2006) Checkpoint proteins control survival of the postmitotic cells in *Caenorhabditis elegans*. Science 312:1381–1385
12. Gartner A, MacQueen AJ, Villeneuve AM (2004) Methods for analyzing checkpoint responses in *Caenorhabditis elegans*. Methods Mol Biol 280:257–274
13. Gartner A, Milstein S, Ahmed S, Hodgkin J, Hengartner MO (2000) A conserved checkpoint pathway mediates DNA damage-induced apoptosis and cell cycle arrest in *C. elegans*. Mol Cell 5:435–443
14. Kapuscinski J (1995) DAPI: a DNA-specific fluorescent probe. Biotech Histochem 70: 220–233
15. http://www.genetics.wustl.edu/tslab/
16. Sakashita T, Takanami T, Yanase S, Hamada N, Suzuki M, Kimura T, Kobayashi Y, Ishii N, Higashitani A (2010) Radiation biology of *Caenorhabditis elegans*: germ cell response, aging and behavior. J Radiat Res (Tokyo) 51:107–121
17. De Stasio EA, Dorman S (2001) Optimization of ENU mutagenesis of *Caenorhabditis elegans*. Mutat Res 495:81–88
18. Timson J (1975) Hydroxyurea. Mutat Res 32:115–132
19. Cocheme HM, Murphy MP (2008) Complex I is the major site of mitochondrial superoxide production by paraquat. J Biol Chem 283:1786–1798
20. Gengyo-Ando K, Mitani S (2000) Character_ ization of mutations induced by ethyl methanesulfonate, UV, and trimethylpsoralen in the nematode *Caenorhabditis elegans*. Biochem Biophys Res Commun 269:64–69
21. Song A, Labella S, Korneeva NL, Keiper BD, Aamodt EJ, Zetka M, Rhoads RE (2010) A *C. elegans* eIF4E-family member upregulates translation at elevated temperatures of mRNAs encoding MSH-5 and other meiotic crossover proteins. J Cell Sci 123:2228–2237
22. Hofmann ER, Milstein S, Boulton SJ, Ye M, Hofmann JJ, Stergiou L, Gartner A, Vidal M, Hengartner MO (2002) *Caenorhabditis elegans* HUS-1 is a DNA damage checkpoint protein required for genome stability and EGL-1-mediated apoptosis. Curr Biol 12: 1908–1918
23. Sulston JE, Horvitz HR (1977) Post-embryonic cell lineages of the nematode, *Caenorhabditis elegans*. Dev Biol 56:110–156
24. Lu N, Yu X, He X, Zhou Z (2009) Detecting apoptotic cells and monitoring their clearance in the nematode *Caenorhabditis elegans*. Methods Mol Biol 559:357–370
25. Gumienny TL, Lambie E, Hartwieg E, Horvitz HR, Hengartner MO (1999) Genetic control of programmed cell death in the *Caenorhabditis elegans* hermaphrodite germline. Development 126:1011–1022
26. Wu YC, Stanfield GM, Horvitz HR (2000) NUC-1, a *Caenorhabditis elegans* DNase II homolog, functions in an intermediate step of DNA degradation during apoptosis. Genes Dev 14:536–548
27. Parusel CT, Kritikou EA, Hengartner MO, Krek W, Gotta M (2006) URI-1 is required for DNA stability in *C. elegans*. Development 133:621–629

28. Alpi A, Pasierbek P, Gartner A, Loidl J (2003) Genetic and cytological characterization of the recombination protein RAD-51 in *Caenorhabditis elegans*. Chromosoma 112:6–16
29. Dorsett M, Schedl T (2009) A role for dynein in the inhibition of germ cell proliferative fate. Mol Cell Biol 29:6128–6139
30. Schumacher B, Hanazawa M, Lee MH, Nayak S, Volkmann K, Hofmann ER, Hengartner M, Schedl T, Gartner A (2005) Translational repression of *C. elegans* p53 by GLD-1 regulates DNA damage-induced apoptosis. Cell 120:357–368
31. Harris J, Lowden M, Clejan I, Tzoneva M, Thomas JH, Hodgkin J, Ahmed S (2006) Mutator phenotype of *Caenorhabditis elegans* DNA damage checkpoint mutants. Genetics 174:601–616

Chapter 4

Establishment of the DNA Repair-Defective Mutants in DT40 Cells

Masamichi Ishiai, Emi Uchida, and Minoru Takata

Abstract

The chicken B cell line DT40 has been widely used as a model system for reverse genetics studies in higher eukaryotes, because of its advantages including efficient gene targeting and ease of chromosome manipulation. Although the genetic approach using the RNA interference technique has become the standard method particularly in human cells, DT40 still remains a powerful tool to investigate the regulation and function of genes and proteins in a vertebrate system, because of feasibility of easy, rapid, and clear genetic experiments. The use of DT40 cells for DNA repair research has several advantages. In addition to canonical assays for DNA repair, such as measurement of the sensitivities toward DNA damage reagents, it is possible to measure homologous recombination and translesion synthesis activities using activation-induced deaminase (AID)-induced diversification of the immunoglobulin locus. In this chapter, we would describe a detailed protocol for gene disruption experiments in DT40 cells.

Key words: DT40, DNA repair, Reverse genetics, Gene disruption, Conditional targeting

1. Introduction

The chicken DT40 B cell line was established from Avian Leukosis Virus-induced lymphoma in the Bursa of Fabricius (1). DT40 cells display the high ratio of targeted to random DNA integration, which occurs at essentially all loci with efficiencies that are orders of magnitude higher than those observed in mammalian cells (2). Targeted integration of DNA constructs by homologous recombination (HR) enables us to inactivate a gene of interest (knock-out). Alternatively, we can modify a specific aspect of a gene function by the introduction of a more subtle mutation (knock-in). These approaches are essentially the same as those in the yeast *S. cerevisiae*, or in the murine embryonic stem (ES) cells to produce mutant mouse strain.

Lotte Bjergbæk (ed.), *DNA Repair Protocols*, Methods in Molecular Biology, vol. 920,
DOI 10.1007/978-1-61779-998-3_4, © Springer Science+Business Media New York 2012

Recently, genetic approaches using RNA interference techniques ("knock-down" experiments) have been widely used for higher eukaryotes including *C. elegans*, *Drosophila*, and human cells. However, it always remains a possibility that the reduction in the target mRNA/protein expression level might be insufficient or the off-target effects may occur. DT40 knockout cell lines are much easier to make compared to mouse ES cells. Therefore, gene targeting experiments in DT40 cells still remain a powerful option to investigate the regulation and function of gene and protein in vertebrate cells. This is particularly true if the species (chicken) and cell types (lymphocytes) are not a major concern.

DT40 has a number of advantages as a model system. First of all, the cells grow very rapidly and therefore its handling is easy. Second, the gene targeting procedure has been established very well. We describe our protocol in this chapter. Third, a number of mutant cell lines have already been created and they constitute a useful panel. Forth, highly sophisticated experiments are possible such as creation of double, conditional, or knock-in mutants. Fifth, DT40 cell has a relatively invariant character in both karyotype and phenotype even during extended period of cell culture. Therefore, data from DT40 cells could be directly compared mutants by mutants. In contrast, cells from human patients or knock-out mice from different genetic backgrounds are more difficult to compare. Sixth, DT40 cell is also a good source for biochemical analyses, because of the ease of large-scale culture with the stable characters under the same genetic background.

The use of DT40 cells for DNA repair research has some advantages. In addition to canonical assays of DNA repair, such as measurements of the sensitivities to DNA damage reagents, several unique DNA repair assays are available. First, the homologous recombination (HR) and translesion synthesis (TLS) activities can be measured using the intrinsic property of DT40 that diversifies its immunoglobulin (Ig) antigen specificity (3). This is depending on expression of activation-induced deaminase (AID), transcription of the Ig gene, and HR and TLS mechanisms (3, 4). The efficiency can be measured by simple subculture and FACS analysis of the reexpression of the surface Ig. Second, the gene targeting efficiency can be measured by transfection of a targeting construct (2). Targeting and nontargeting events can be determined by Southern blotting or PCR analysis. Third, the repair efficiency of the chromosomal double strand break (DSB) in an artificially integrated recombination substrate at a specific locus can be measured (5). DSB is introduced by plasmid-based expression of the rare cutting enzyme I-*Sce*I (6).

As in other systems, there are a few drawbacks in DT40. Since they are chicken cells, availability of the reagents can be a problem. In particular, antibodies against human or mouse proteins often do not cross-react with the chicken counterparts. Knock-in of an epitope tag including a green fluorescent protein (GFP) could overcome this problem. Another potential problem is that chicken genome sequence data are far from complete, even though the draft of chicken genome had been published in 2004 (7). This may cause a trouble in designing targeting vector or comparative genomic hybridization array. In addition, there might be some chance of missing information in the database if you attempt to identify proteins by mass spectrometry. Finally, we have to take into account the absence of functional p53 in DT40, which induces apoptosis and cell cycle arrest upon DNA damage.

2. Materials

1. Maintenance of DT40 cell line. Culture medium is RPMI 1640 with L-glutamate, supplemented with 10 % fetal calf serum (FCS), 1 % chicken serum, 2 mM L-glutamate, and 50 μM β-mercaptoethanol. Addition of penicillin and streptomycin is optional. FCS is preheated at 56 °C for 30 min to prevent complement activation. Chicken serum need not be heat-inactivated.
2. Gene pulser II Electroporator (Biorad).
3. Antibiotics used for selection of DT40 transfectants (Table 1).
4. Standard set up for molecular biology lab: Reagents for plasmid construction, PCR, and Southern blotting.

3. Methods

3.1. DT40 Cell Culture

Optimal growth conditions for DT40 cells are in complete RPMI 1640 medium (see step 1 of Subheading 2) at 39.5 °C with 5 % CO_2. The doubling time is about 8 h in this condition (see Note 1). For long-term storage, cells suspended in FCS containing 10 % DMSO could be frozen at −80 °C or in liquid nitrogen. For short-term storage of clones awaiting Southern blotting, we mix one volume of cultured cells with the same volume of FCS containing 20 % DMSO in an 1.5 mL tube and put them directly into deep freezer.

3.2. How to Design a Gene Targeting Construct

You should consider the following points.

1. How to obtain chicken DNA sequences. Prior to design a gene targeting vector, you need to identify the chicken cDNA and genomic sequences of the gene of your interest in the database, for example, at National Center for Biotechnology Information (NCBI) Web site (http://www.ncbi.nlm.nih.gov/) (see Notes 2 and 3).

Table 1
The marker genes and reagents used for the selection of transfected DT40 cells

Name of the selection marker gene	Abbreviation of the gene	Reagent	Supplier	Code	Stock solution	Final concentration
Blasticidin S deaminase[a]	*bsr*	Blasticidin S hydrochloride	Funakoshi	KK-400	10 mg/mL H_2O	25 μg/mL
Histidinol dehydrogenase[a]	*hisD*	L-Histidinol dihydrochloride	SIGMA	H6647-5G	50 mg/mL H_2O	1 mg/mL
Puromycin *N*-acetyltransferase[a]	*puro*	Puromycin dihydrochloride	SIGMA	P8833-25MG	1 mg/mL H_2O	0.5 μg/mL
Zeocin resistance gene	*zeo/Sh ble*	Zeocin	InvivoGen	ant-zn-1	100 mg/mL[b]	1 mg/mL[c]
Hygromycin B phosphotransferase[a]	*hygro/hph*	Hygromycin B	Invitrogen	10687-010	50 mg/mL[b]	2.5 mg/mL[d]
Neomycin phosphotransferase II[a]	*neo*	G418 disulfate	nacalai tesque	16513-26	50 mg/mL[b]	2 mg/mL

[a]These selection gene cassettes are available from our lab upon request
[b]These are supplied as solution
[c]In DNA double strand break repair mutant cells, use 1/2–1/4 concentration
[d]Add 1 M HEPES 1 mL/40 mL culture to neutralize

2. Components of the vector. Gene targeting constructs contain a selection drug maker cassette flanked by genomic sequences at both sides (often called left and right or upstream and downstream arms), and a vector backbone. Any cloning vector can be used for this purpose. In DT40 cells, the gene targeting efficiency is very high, thus you do not need negative selection.
3. How you would like to disrupt the gene of your interest? Upon gene targeting event, the selection gene cassette is inserted into or replaced by the gene of interest. The size of the deleted genome is determined by the position of left and right arms within the genome. To fully inactivate a gene, it is theoretically best to replace the complete coding sequences with the cassette. However, this is not always easily achievable because a large distance between the arms may drastically decrease the ratio of targeted to random integration, though more than 20 kb of the gene locus has been deleted (8). As a compromise, we usually design a vector with the deletion size not exceeding 10 kb, and try to delete exon(s) encoding a critical domain of the protein in interest. We also place the boundary of the arm within an exon to ensure disruption of the coding sequence. Introducing an in-frame stop codon might help.
4. The length of the arms. This would greatly affect the efficiency of gene targeting, and generally speaking the longer the better. However, compared with mouse ES cells system that requires more than 10 kb of the genomic sequences, shorter genomic sequences have been used in DT40 cells: 2–5 kb arms usually resulted in successful targeting. Our minimal arm length that worked was 0.8 kb (9). We usually design our constructs in such a way that the size of individual left and right arm of are more than 1.5 kb, the combined size of the arms is more than 3 kb (preferably 4–5 kb) and <10 kb.
5. All of our selection drug marker cassette are flanked by *Bam*HI sites and can be conventionally cloned into *Bam*HI (or *Bgl*II) sites located between the left and light arms. It is best to use *Bam*HI, since it is sometimes necessary to swap the cassette with another one. A number of the selection marker gene cassettes are available upon request (the marker genes and drugs are summarized in Table 1). At the moment we have *hisD*, *bsr*, *puro*, and *neo* cassettes flanked with loxP sequences, and they are useful for selection marker recycling to make double or triple knockout. Our first choice is the *hisD* and *bsr* cassettes, since the drugs kill cells rapidly, and are not expensive. Puromycin selection is also very efficient; however, we tend to save it for re-expression experiments.
6. The targeting vector should normally be linearized before transfection, and it is best to consider which enzyme site to use while designing the vector. We often use *Pvu*I or *Sca*I that are

located in the *bla* (ampicillin resistant) gene of the vector backbone, or a restriction enzyme, which cut a unique site in the multicloning sites flanking the either arm. The linearization seems to increase the number of the transfectants obtained after the selection.

7. Screening strategy. We normally screen transfected clones for targeting events with Southern blotting. You should consider the enzyme appropriate for screening and DNA fragment used for probe. The probe should not hybridize with the arms. An alternative screening method is described in Note 4.
8. DT40 is diploid except for chromosome 2, which is in trisomy. If the gene of your interest is on chromosome 2, unfortunately you will have to delete three alleles with three serial transfections with three different selection drugs. On the other hand, since DT40 was derived from a female chicken, and it has only one chromosome Z (chicken sex chromosome). Human chromosome 9 and chicken chromosome Z are in synteny (10), thus a gene in human chromosome 9 is expected to be deleted in a single gene targeting event in DT40 (11, 12) (see Note 5).

3.3. Construction of the Gene Targeting Vector

We describe an example for the construction using pBluescript vector.

1. Select the restriction sites for cloning both arms, which are not present in the arm sequences. In the example shown in Fig. 1, we use *Sal*I-*Bam*HI sites for left arm and *Bam*HI-*Not*I sites for light arm.
2. Synthesize PCR primers incorporating the restriction sites at the ends (Fig. 1a).
3. Amplify the target arms by long-range PCR (e.g., Takara LA-Taq) using designed primers and genomic DNA from DT40 cells as template. This will produce arm sequence isogenic to at least one allele of DT40 cells and would enhance the targeting efficiency. LA-Taq may create a number of mutations in the arms, but this seems not to affect targeting efficiency.
4. Cloning the left arm into *Sal*I-*Bam*HI sites and right arm into *Bam*HI-*Not*I sites of pBluescript (Fig. 1b). Confirm the arms by sequencing, for example, using M13 forward and/or reverse primers, and by restriction digestion.
5. Finally, the drug marker cassette is inserted into *Bam*HI site (Fig. 1c). Check the orientation of the drug marker cassette by restriction enzyme digestion and/or sequencing. We prefer the drug resistance cassette placed in reverse orientation to the direction of transcription of the gene. It seems good practice to save the vector with another orientation as well, since sometimes the vector with one orientation shows better targeting efficiency compared to the vector with the other.

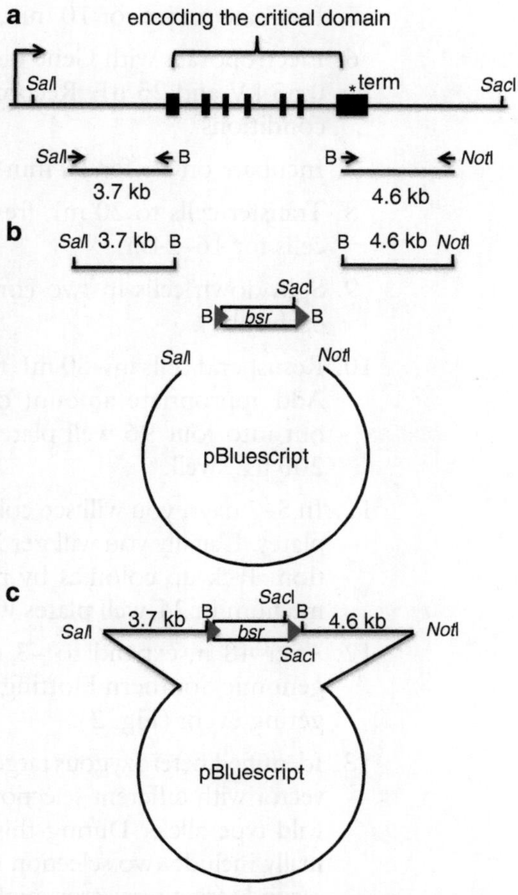

Fig. 1. Construction of the targeting vector. (**a**) Schematic representation of part of the "gene of interest" allele. For simplification, only one allele is shown. Primers were designed according to DNA sequence of the gene with proper restriction enzyme sites. The fragments of the target vector are amplified by PCR. *Black boxes* indicate the positions of exons. **term*, termination codon, *B*, *BamHI* site. (**b**) Prepared target arm fragments, a drug marker cassette, and the vector (pBluescript) digested with proper restriction enzyme. They are ligated in sequential steps. *Triangles* indicate loxP sequences. (**c**) The targeting vector. See details in the text.

3.4. Transfection in DT40

The DT40 cells are readily transfected with DNA by electroporation.

1. Linearize 30 μg plasmid DNA (targeting vector) with a restriction enzyme in sufficient volume of buffer (e.g., 200 μL and 60–90 U of enzyme), then precipitate with ethanol and sodium acetate and dry up. Extraction of DNA with phenol/chloroform is not essential.

2. Spin down $0.5–1 \times 10^7$ cells. Use cells in log phase. Aspirate medium.

3. Resuspend cells in 0.5 mL PBS and transfer to a cuvette with 4 mm gap (Biorad or BTX).

4. Dissolve linearized DNA in 20 μL of PBS, and add to the cuvette. Mix well. Avoid forming bubbles.

5. Incubate on ice for 10 min.
6. Electroporate with Gene pulser II (Biorad) at the condition of 0.55 kV and 25 μF. Record time constant each time to verify conditions.
7. Incubate on ice for 10 min.
8. Transfer cells to 20 mL fresh medium in 10 cm dish. Culture cells for 16–24 h.
9. Spin down cells in two conical 50 mL tubes (~10 mL cells/ each tube).
10. Resuspend cells in ~80 mL fresh medium (~40 mL/each tube). Add appropriate amount of selection drugs. Mix well. Plate out into four 96-well plates using an 8-channel pipetter. Put 200 μL/well.
11. In 5–7 days, you will see colonies growing at the bottom of the plates. Usually you will get 20–100 colonies from one transfection. Pick up colonies by pipette and transfer them to 1 mL medium in 24-well plates when they are <2 mm in diameter.
12. After 48 h, expand to ~3 mL culture scale and analyze with genomic Southern blotting to identify clones with correct targeting event (Fig. 2).
13. Identified heterozygous targeted clone is transfected again using a vector with different selection drug cassette to disrupt remaining wild-type allele. During this second selection process, we normally include two selection drugs to reduce the chance that the second targeting vector might replace the integrated first one.

3.5. Design for Conditional Gene Knockout

Mutants homozygous for genes essential for cell survival or proliferation cannot be produced by the above standard targeted gene disruption. After targeting the first allele by standard methods, further manipulations are required to express the gene product (by a "rescue construct") that can be shut-off conditionally. Then the second allele should be disrupted. For this purpose, several methods have been used successfully to render DT40 cells conditionally null for essential genes. Each of these methods requires a specific set of plasmids. For details, please see the references indicated.

1. Tet-off system. A conditionally repressible gene is introduced through random integration and the "transgene" is expected to keep the null cell alive by expressing a rescue construct. This rescue construct can be inactivated using the tetracycline repressible promoter (13).
2. Cre-loxP system. The Cre recombinase-loxP-mediated deletion system can be used to turn off expression (14). An expression cassette might be placed between two loxP sequences in the rescue construct. Alternatively, the loxP sequences can be knocked-in into the gene of interest at appropriate positions flanking important exons. In this case, the expression is driven

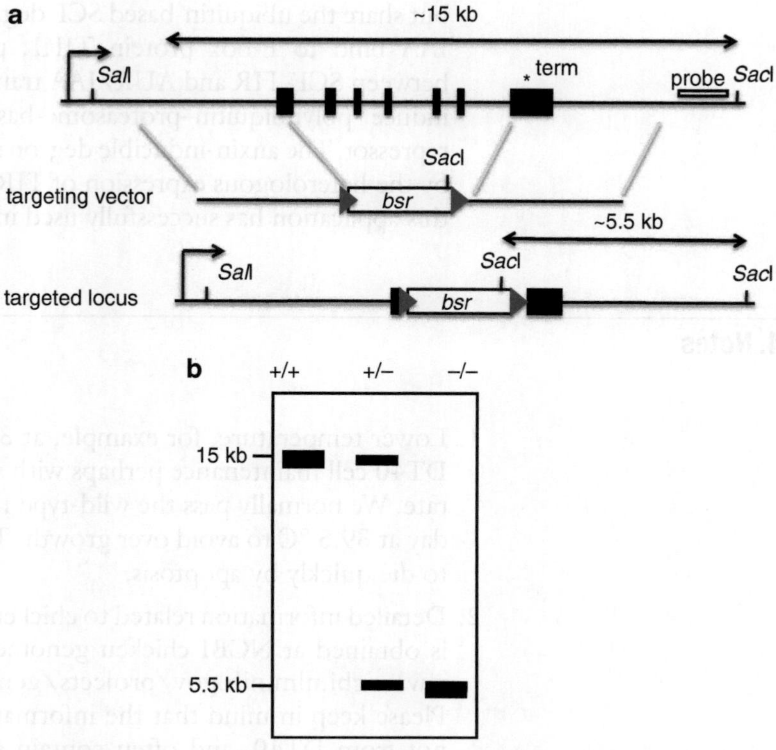

Fig. 2. Gene targeting of the "gene of interest" loci. (**a**) Schematic representation of the "gene of interest," the gene targeting construct, and the configuration of targeted allele are shown. (**b**) Schematic results of Southern blot analysis. In this case, genomic DNA from cells with indicated genotypes are digested with *SacI*, and probed with the flanking DNA fragment shown as *white box*.

by the endogenous promoter, which should provide better expression control. The expression is shut off by Cre enzyme fused with the ligand-binding domain of estrogen receptor (MerCreMer) (15) upon activation of tamoxifen addition. The efficiency of the Cre-mediated excision was surprisingly high (close to 100 %), making this option attractive.

3. A temperature sensitive mutant transgene. This is a possibility since it was used once for CENP-C deletion (16).

4. Degron system. Recently, two protein knock-down approaches to rapidly deplete the protein from a rescue construct have been reported. Both methods use a domain to induce degradation (i.e., degron), which is fused to the protein of interest. One is the application of temperature-sensitive degron system using a degradation pathway based on the N-end rule, which has been widely used in yeast (17). Another is the auxin-inducible degron system (18). Indol-3-acetic acid (IAA), a plant hormone auxin (AUX), directly induces rapid degradation of the AUX/IAA transcription repressors by a specific form of the SCF (Skp1, Cullin, and F-box) E3 ubiquitin ligase complexes. Other eukaryotes including vertebrate cells lack the auxin response

but share the ubiquitin-based SCF degradation pathway. AUX/IAA bind to F-box protein TIR1, promote the interaction between SCF-TIR and AUX/IAA transcription repressors, and induce polyubiquitin–proteasome-based degradation of the repressor. The auxin-inducible degron system can be introduced by the heterologous expression of TIR1 in non-plant cells, and this application has successfully used in DT40 cells (18).

4. Notes

1. Lower temperature, for example, at 37 °C, is also suitable for DT40 cell maintenance perhaps with some decrease in growth rate. We normally pass the wild-type DT40 cells at every other day at 39.5 °C to avoid over growth. The overgrown cells tend to die quickly by apoptosis.
2. Detailed information related to chicken (*Gallus gallus*) genome is obtained at NCBI chicken genome resource page (http://www.ncbi.nlm.nih.gov/projects/genome/guide/chicken/). Please keep in mind that the information in the databases are not from DT40, and often contain some differences due to polymorphisms. Do not trust the database too much.
3. If you cannot find any chicken cDNA sequence information in the database, even not in the EST collections, it is worth trying cross hybridization (i.e., Southern blotting of chicken genomic DNA with human or mouse probe) at reduced stringency (19). When you obtain signals, go on to screening the cDNA and/or genomic libraries with that condition.
4. Genomic PCR-based screening is also reported (20). In this case, at least one primer should be located outside of the region used in the gene targeting vector.
5. We have noticed that four transfections were required for disruption of *cbl* gene locus (21).

Acknowledgements

The protocols resented here is the combinations of those developed through the work of our previous and current colleagues. The authors' work was supported in part by Grants-in aid from the Ministry of Education, Science, Sports, and Culture of Japan (MI and MT). Financial support were also provided by the Uehara Memorial Foundation (MT), Takeda Foundation (MT), the Ichiro Kanehara Foundation (MI), and the Mochida Memorial Foundation for Medical and Pharmaceutical Research (MI).

References

1. Baba TW, Giroir BP, Humphries EH (1985) Cell lines derived from avian lymphomas exhibit two distinct phenotypes. Virology 144: 139–151
2. Buerstedde JM, Takeda S (1991) Increased ratio of targeted to random integration after transfection of chicken B cell lines. Cell 67: 179–188
3. Sale JE, Calandrini DM, Takata M, Takeda S, Neuberger MS (2001) Ablation of XRCC2/3 transforms immunoglobulin V gene conversion into somatic hypermutation. Nature 412: 921–926
4. Okada T, Sonoda E, Yamashita YM, Koyoshi S, Tateishi S, Yamaizumi M, Takata M, Ogawa O, Takeda S (2002) Involvement of vertebrate polκ in Rad18-independent postreplication repair of UV damage. J Biol Chem 277:48690–48695
5. Fukushima T, Takata M, Morrison C, Araki R, Fujimori A, Abe M, Tatsumi K, Jasin M, Dhar PK, Sonoda E, Chiba T, Takeda S (2001) Genetic analysis of the DNA-dependent protein kinase reveals an inhibitory role of Ku in late S-G2 phase DNA double-strand break repair. J Biol Chem 276:44413–44418
6. Johnson RD, Liu N, Jasin M (1999) Mammalian XRCC2 promotes the repair of DNA double-strand breaks by homologous recombination. Nature 401:397–399
7. Consortium, I. C. G. S (2004) Sequence and comparative analysis of the chicken genome provide unique perspectives on vertebrate evolution. Nature 432:695–716
8. Arakawa H, Saribasak H, Buerstedde JM (2004) Activation-induced cytidine deaminase initiates immunoglobulin gene conversion and hypermutation by a common intermediate. PLoS Biol 2:E179
9. Inabe K, Ishiai M, Scharenberg AM, Freshney N, Downward J, Kurosaki T (2002) Vav3 modulates B cell receptor responses by regulating phosphoinositide 3-kinase activation. J Exp Med 195:189–200
10. Nanda I, Shan Z, Schartl M, Burt DW, Koehler M, Nothwang H, Grutzner F, Paton IR, Windsor D, Dunn I, Engel W, Staeheli P, Mizuno S, Haaf T, Schmid M (1999) 300 million years of conserved synteny between chicken Z and human chromosome 9. Nat Genet 21:258–259
11. Yamamoto K, Ishiai M, Matsushita N, Arakawa H, Lamerdin JE, Buerstedde JM, Tanimoto M, Harada M, Thompson LH, Takata M (2003) Fanconi anemia FANCG protein in mitigating radiation- and enzyme-induced DNA double-strand breaks by homologous recombination in vertebrate cells. Mol Cell Biol 23:5421–5430
12. Hirano S, Yamamoto K, Ishiai M, Yamazoe M, Seki M, Matsushita N, Ohzeki M, Yamashita YM, Arakawa H, Buerstedde JM, Enomoto T, Takeda S, Thompson LH, Takata M (2005) Functional relationships of FANCC to homologous recombination, translesion synthesis, and BLM. EMBO J 24:418–427
13. Lemaire R, Prasad J, Kashima T, Gustafson J, Manley JL, Lafyatis R (2002) Stability of a PKCI-1-related mRNA is controlled by the splicing factor ASF/SF2: a novel function for SR proteins. Genes Dev 16:594–607
14. Sauer B, Henderson N (1988) Site-specific DNA recombination in mammalian cells by the Cre recombinase of bacteriophage P1. Proc Natl Acad Sci U S A 85:5166–5170
15. Zhang Y, Wienands J, Zurn C, Reth M (1998) Induction of the antigen receptor expression on B lymphocytes results in rapid competence for signaling of SLP-65 and Syk. EMBO J 17: 7304–7310
16. Fukagawa T, Regnier V, Ikemura T (2001) Creation and characterization of temperature-sensitive CENP-C mutants in vertebrate cells. Nucleic Acids Res 29:3796–3803
17. Su X, Bernal JA, Venkitaraman AR (2008) Cell-cycle coordination between DNA replication and recombination revealed by a vertebrate N-end rule degron-Rad51. Nat Struct Mol Biol 15:1049–1058
18. Nishimura K, Fukagawa T, Takisawa H, Kakimoto T, Kanemaki M (2009) An auxin-based degron system for the rapid depletion of proteins in nonplant cells. Nat Methods 6: 917–922
19. Ausubel FM, Brent R, Kingston RE, Moore DD, Seidman JG, Smith JA, Struhl K (1999) Current protocols in molecular biology. John Wiley and Sons, Inc., Hoboken, New Jersey
20. Wang X, Takenaka K, Takeda S (2010) PTIP promotes DNA double-strand break repair through homologous recombination. Genes Cells 15:243–254
21. Yasuda T, Maeda A, Kurosaki M, Tezuka T, Hironaka K, Yamamoto T, Kurosaki T (2000) Cbl suppresses B cell receptor-mediated phospholipase C (PLC)-γ2 activation by regulating B cell linker protein-PLC-γ2 binding. J Exp Med 191:641–650

Part II

Assessing DNA Damage and Repair

Chapter 5

Electrophoretic Mobility Shift Assays for Protein–DNA Complexes Involved in DNA Repair*

Chun Tsai, Vaughn Smider, Byung Joon Hwang, and Gilbert Chu

Abstract

The electrophoretic mobility shift assay (EMSA) can be used to study proteins that bind to DNA structures created by DNA-damaging agents. UV-damaged DNA-binding protein (UV-DDB), which is involved in nucleotide excision repair, binds to DNA damaged by ultraviolet radiation or the anticancer drug cisplatin. Ku, XRCC4/Ligase IV, and DNA–PKcs, which are involved in the repair of DNA double-strand breaks by nonhomologous end joining, assemble in complexes at DNA ends. This chapter will describe several EMSA protocols for detecting different DNA repair protein–DNA complexes. To obtain additional information, one can apply variations of the EMSA, which include the reverse EMSA to detect binding of ^{35}S-labeled protein to damaged DNA, and the antibody supershift assay to detect the presence of a specific protein in the protein–DNA complex.

Key words: UV-DDB, Global genomic repair, Nucleotide excision repair, Ku, DNA-PKcs, XRCC4/Ligase IV, DNA-PK, Nonhomologous end joining, Double-strand break repair, V(D)J recombination

1. Introduction

DNA repair pathways must include proteins that recognize and bind to DNA lesions. The search for such proteins has been facilitated by the use of electrophoretic mobility shift assays (EMSAs), which were first used to detect transcription factors that bind to specific DNA sequences (1, 2). To study DNA repair, EMSAs have been adapted to detect proteins that bind to specific DNA lesions or structural features.

In an EMSA, non-denaturing polyacrylamide gel electrophoresis resolves protein–DNA complexes. A radiolabeled DNA probe contains the DNA lesion or DNA structure of interest. The protein–DNA

* Authors Chun Tsai, Vaughn Smider, Byung Joon Hwang shared equally in this work.

Lotte Bjergbæk (ed.), *DNA Repair Protocols*, Methods in Molecular Biology, vol. 920,
DOI 10.1007/978-1-61779-998-3_5, © Springer Science+Business Media New York 2012

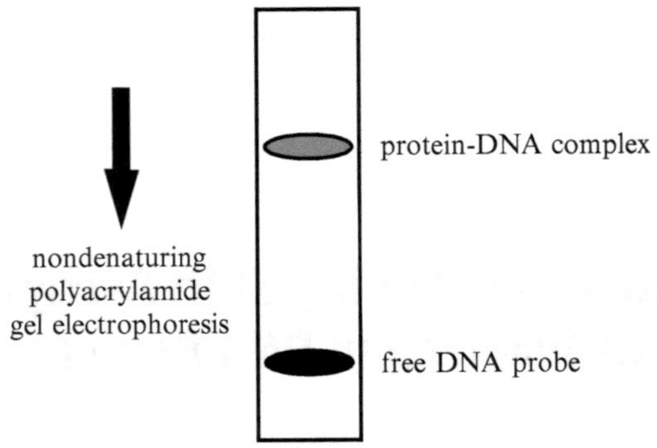

DNA repair protein	*UV-DDB*	*Ku*
structure bound by protein	bulky DNA adduct	DNA end
radiolabeled DNA probe	UV-damaged DNA	linear DNA
unlabeled DNA competitor	undamaged DNA	supercoiled DNA

Fig. 1. EMSA for studying UV-DDB and Ku. UV-damaged DNA probe for UV-DDB or linear DNA probe for Ku may be incubated with cell extracts in the presence of the competitor DNA to block nonspecific binding. After incubation, the protein–DNA complex is resolved from the free DNA probe by non-denaturing gel electrophoresis.

complexes assemble by incubation of the DNA probe with either crude protein extract or purified proteins (Fig. 1).

EMSA experiments must address the specificity of protein binding. With crude extracts, extraneous proteins may bind to the DNA probe, rather than to the DNA lesion. With purified proteins, some DNA repair proteins may have a weaker but detectable affinity for undamaged regions of the DNA. To eliminate nonspecific binding events, the DNA probe can be mixed with a molar excess of undamaged competitor DNA.

For any EMSA, several general principles ensure proper detection of the protein–DNA complex. The procedure for extracting proteins from the cell nuclei must be optimized for the target protein. A cocktail of protease inhibitors must be present, particularly for high molecular weight proteins that may be vulnerable to degradation. The salt concentration of the extraction buffer must be high enough to dissociate the target protein from genomic DNA, but not so high that interfering components are also extracted. The electrophoresis conditions must be optimized to obtain a well-defined mobility shift: in particular, the salt concentration of the electrophoresis buffer should permit stable formation of the specific protein–DNA complex while minimizing the formation of nonspecific complexes. Use of a mini-gel apparatus permits resolution of the desired complexes in about 30 min; end labeling of the probe DNA with ^{32}P or ^{33}P permits detection of the complexes in a few hours. In our laboratory, gels are poured in lots of 13 and stored for up to 3 weeks at 4 °C so that experiments can be done with minimal setup time.

Commercially prepared gels have provided acceptable results for many experiments. However, in some cases involving the detection of large protein–DNA complexes, we have been unable to replicate earlier results even though the commercially prepared gels were ostensibly manufactured with the same reagents in the same concentrations. In such cases, we were able to replicate our results only after preparing the gels in our own laboratory.

Once a complex is detected, its specificity for the DNA probe should be verified. If the mobility shift of the DNA probe is due to formation of a protein–DNA complex rather than binding by a nonprotein molecule, complex formation should be sensitive to protease digestion. If the protein binds specifically to damaged DNA, a complex of the same mobility should not form with undamaged probe DNA. Alternatively, unlabeled competitor DNA with and without damage can be compared for its ability to compete away the labeled complex. The spectrum of DNA structures recognized by the protein can be determined by testing different competitor DNAs.

Of course, the discovery of a protein that binds to damaged DNA does not prove that it is involved in DNA repair. To establish biological significance, we have found it fruitful to screen mutant cell lines for abnormalities in binding activity. Successful screening depends on procedures described below for rapidly making cell extracts from small numbers of cells.

EMSAs have been used to study a number of DNA repair proteins. In this chapter, we describe our protocols for studying protein complexes involved in nucleotide excision repair and nonhomologous end joining. For nucleotide excision repair, UV-DDB recognizes UV-induced damage in both transcribed and nontranscribed DNA, thus facilitating global genomic repair (3). UV-DDB consists of two subunits, DDB1 and DDB2. For nonhomologous end joining, Ku binds DNA ends and subsequently recruits other components of the nonhomologous end joining repair complex, including DNA–PKcs, XRCC4/Ligase IV, and XLF (Cernunnos) (4).

In the case of UV-DDB, the probe DNA was a linear DNA fragment damaged by exposure to UV-radiation, and the competitor DNA consisted of unlabeled linear DNA that was left undamaged. A complex with a reasonably well-defined mobility was detected even though UV-radiation introduced DNA lesions throughout the length of the DNA probe (5). This phenomenon was previously reported for the binding of a sequence-specific DNA-binding protein to a series of DNA probes containing the binding sequence at different sites. In that case, the mobility of the protein–DNA complex was relatively insensitive to the position of the binding sequence (6). The specificity of UV-DDB was established by showing that the binding activity was competed away by the addition of unlabeled UV-damaged competitor DNA to the binding reaction (Fig. 2a). UV-DDB was then purified (7) and partially sequenced in order to isolate a cDNA encoding a polypeptide of the expected molecular

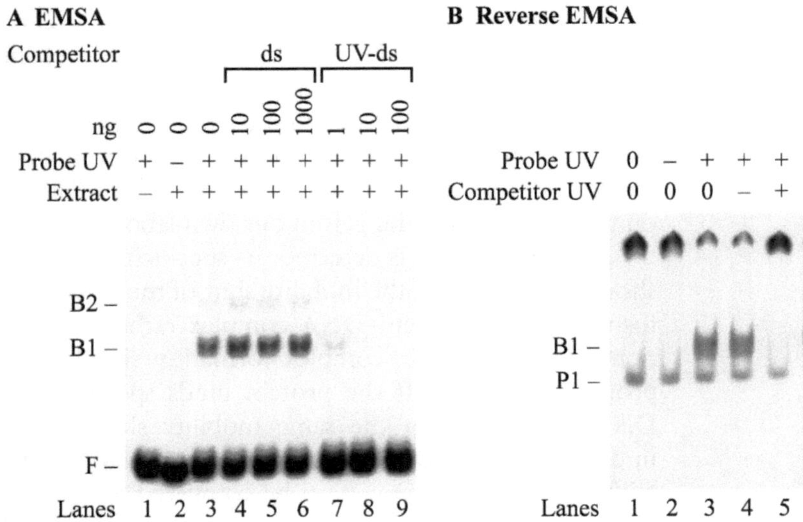

Fig. 2. EMSA and reverse EMSA for UV-DDB. (**a**) EMSA. Crude protein extracts from HeLa cells were incubated with f148 probe DNA in the presence of different amounts of unlabeled competitor DNA. The probe DNA was UV irradiated (*lanes 1* and *3–9*) or left intact (*lane 2*). Protein extract was omitted (*lane 1*) or added (*lanes 2–9*). Control binding reactions omitted competitor DNA (*lanes 1–3*). Competition was carried out with intact double-stranded DNA (ds) (*lanes 4–6*), UV-irradiated double-stranded DNA (UV-ds) (*lanes 7–9*). (**b**) Reverse EMSA. The fraction of ^{35}S-labeled DDB1 protein that bound to a UV-DNA cellulose column was assayed in a binding reaction with unlabeled f148 DNA. Probe DNA was omitted (*lane 1*), added as undamaged DNA (*lane 2*), or added as UV-damaged DNA (*lanes 3–5*). Unlabeled plasmid competitor DNA was either omitted (*lanes 1–3*), added as undamaged DNA (*lane 4*), or added as UV-damaged DNA (*lane 5*). Some of the ^{35}S-labeled DDB1 protein is retained in the well of the gel (P2) because it was purified from a UV-DNA cellulose column, so that some high molecular weight UV-damaged DNA co-eluted as a complex with ^{35}S-labeled DDB1 protein.

weight. To demonstrate that this polypeptide bound specifically to UV-damaged DNA, the cDNA was transcribed and translated in vitro so that the protein product was labeled with ^{35}S-labeled methionine (8). A reverse EMSA was then used to show that the mobility of the labeled protein was specifically shifted by the addition of damaged DNA, but not be undamaged DNA (Fig. 2b).

To determine the biological relevance of UV-DDB, we screened mutant cell lines hypersensitive to UV radiation. A role in nucleotide excision repair was supported by the discovery that UV-DDB was absent in a subset of xeroderma pigmentosum group E cells, which are defective in this repair pathway (5). Furthermore, UV-DDB also recognized competitor DNA carrying intrastrand DNA crosslinks induced by the anticancer drug cisplatin, and levels of UV-DDB were increased in extracts from cells selected for resistance to cisplatin (9). Biological relevance was established by showing that microinjection of XPE cells with purified UV-DDB restored DNA repair to the cells (10). More specifically, the ectopic expression of UV-DDB in hamster cells enhanced the global genomic repair of cyclobutane dimers (11). Global genomic repair refers to the removal of DNA lesions from nontranscribed regions of the genome and from the nontranscribed strand in transcribed regions of the genome.

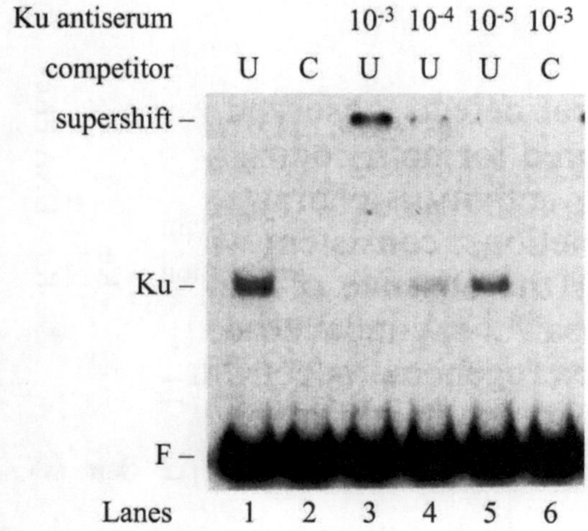

Fig. 3. EMSA and antibody supershift for Ku. Nuclear extract from the hamster cell line AA8 was incubated in the presence of radiolabeled f148 probe and uncut plasmid (U) DNA competitor in lane 1, or cut plasmid (C) in lane 2. The Ku antiserum was included in lanes 3–5 in dilutions ranging from 10^{-3} to 10^{-5} in the presence of uncut plasmid (U). The supershifted complex is specific for DNA ends as shown in lane 6, where cut plasmid (C) is the competitor.

In the case of Ku, the probe was an undamaged linear DNA fragment. Supercoiled plasmid DNA served as competitor for nonspecific DNA-binding proteins, so that binding to the linear DNA probe was specific for the DNA ends (Fig. 3). This assay system detected a protein–DNA complex, which we originally denoted as DNA end-binding (DEB) factor (12). The specificity for DNA ends was established by showing competition for the protein–DNA complex by unlabeled competitor plasmid DNA cleaved with any one of several different restriction enzymes, producing DNA ends with 5′ or 3′ overhanging, or blunt, ends. The identification of DEB factor as Ku protein was first established by a useful adjunct to the EMSA, in which incubation of the binding reactions with anti-Ku antibodies produced a supershift of the original protein–DNA complex (13).

To determine the biological relevance of DEB factor, we screened a series of mutant cell lines hypersensitive to ionizing radiation, reasoning that a DEB protein might be involved in the repair of the double-strand breaks produced by ionizing radiation. This hypothesis was confirmed by the discovery that DEB factor was absent in three cell lines from X-ray complementation group 5, which is defective in double-strand break repair (13). Strikingly, these cells were also defective in V(D)J recombination, the pathway that generates immunological diversity by cleaving and rearranging the immunoglobulin genes in B cells and the T cell receptor genes in T cells. Biological relevance was established by showing that transfection of the mutant cells with a cDNA expression vector for the 86 kDa subunit of Ku restored both ionizing radiation

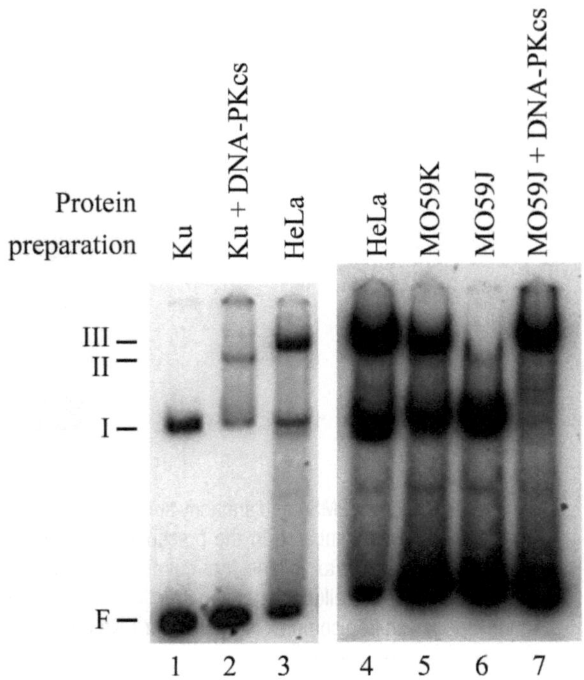

Fig. 4. EMSA for complex of DNA–PKcs and Ku bound to DNA. Radiolabeled f32 probe was incubated with purified protein consisting of 11 ng of Ku (*lane 1*), or 11 ng of Ku plus 35 ng of DNA–PKcs (*lane 2*). In addition, the probe was incubated with 500 ng of nuclear extract prepared from HeLa cells (*lanes 3* and *4*), M059K cells derived from a human glioma (*lane 5*), or M059J cells lacking DNA–PKcs derived from the same glioma (*lanes 7* and *8*). In lane 8, the M059J extract was supplemented with 50 ng of purified DNA–PKcs. The gel shows migration of the free probe (F) and protein–DNA complexes containing Ku (I), Ku and DNA–PKcs (II), and Ku, DNA–PKcs, and an unidentified factor (III).

resistance and V(D)J recombination (14, 15), establishing a role for Ku in nonhomologous end joining.

Upon binding to DNA ends, Ku is capable of translocating inward to allow binding of additional Ku molecules. Ku also recruits the DNA-dependent protein kinase catalytic subunit (DNA–PKcs) to the DNA ends, thus forming a higher-order complex with DNA (Fig. 4). (DNA-PK consists of its catalytic subunit DNA–PKcs and the heterodimeric Ku protein.) To detect binding by DNA–PKcs, we modified the EMSA by shortening the length of the radiolabeled DNA probe to 32 base pairs and eliminating competitor DNA. Indeed, higher-order complexes were detected with either purified proteins or cell extracts. Interestingly, cell extracts generate a higher-order complex with a slightly lower mobility than seen with purified proteins (compare lanes 2 and 3 in Fig. 4). Supershift assays with antibodies demonstrate that this complex contains both Ku and DNA–PKcs (data not shown). Extracts from M059J cells, which lack DNA–PKcs, do not support the formation of the higher-order complex, but supplementation of the mutant extracts with purified DNA–PKcs restores the higher-order complex (compare

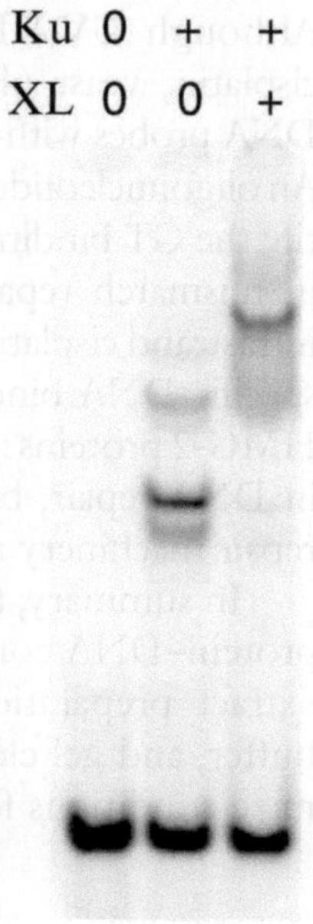

Fig. 5. EMSA for complex of Ku and XRCC4/Ligase IV bound to DNA. Probe DNA f298 labeled with ^{33}P was incubated in the absence of protein (*lane 1*), together with recombinant Ku alone (*lane 2*), or together with both recombinant Ku and recombinant XRCC4/Ligase IV (*lane 3*). The mobility of the protein–DNA complex was not due to DNA ligation, because the reaction conditions permitted only negligible ligation, as confirmed by gel electrophoresis (data not shown). The ^{33}P labeling produced sharper bands, allowing for greater resolution compared to the ^{32}P labeling (Figs. 2–4).

lanes 6 and 7 in Fig. 4). Finally, DNA–PKcs is capable of binding to DNA ends in the absence of Ku. Indeed, when the proteins are incubated with a short 32 bp oligonucleotide in the absence of competitor DNA, the EMSA is capable of detecting complexes containing DNA–PKcs. These complexes include DNA bound to DNA–PKcs, or Ku in a complex with DNA–PKcs (16).

In addition to recruiting DNA–PKcs, Ku also recruits XRCC4/Ligase IV to DNA ends (17). Ku and XRCC4/Ligase IV form a complex on DNA that migrates with a slower mobility than Ku alone (Fig. 5). Furthermore, antibodies against XRCC4 or Ku supershift the complex to an even higher position in the gel (17), indicating that Ku and XRCC4 are present in the complex. XRCC4 and Ligase IV form a stable dimeric complex under the conditions used for the binding reaction, suggesting that the protein–DNA complex contains all three proteins, Ku, XRCC4, and Ligase IV.

EMSAs have been used to identify other structure-specific proteins. The randomly damaged UV-irradiated DNA probe, which was used to detect UV-DDB in human extracts, failed to detect a corresponding factor in yeast extracts, but instead detected photolyase, which repairs UV-induced cyclobutane pyrimidine dimers by photoreactivation (18). Like UV-DDB, yeast photolyase recognized both UV-damaged and cisplatin crosslinked DNA.

Although UV-DDB expression was associated with resistance to cisplatin, yeast photolyase conferred sensitivity to cisplatin (19). DNA probes with other structures have detected additional proteins. An oligonucleotide containing a single GT mismatch was used to identify the GT binding protein that has now been shown to be involved in mismatch repair (20). An oligonucleotide containing a single intrastrand cisplatin crosslink was used to search extracts for damage-specific DNA-binding proteins that proved to be the HMG-1 and HMG-2 proteins (21, 22). In this case, the proteins were not involved in DNA repair, but rather interfered with the nucleotide excision repair machinery and conferred sensitivity to cisplatin (23).

In summary, this chapter describes EMSAs for studying several protein–DNA complexes. Variations in probe length and labeling, extract preparation, protein and DNA concentrations, binding buffer, and gel electrophoresis conditions reflect the need to optimize conditions for each protein–DNA complex.

2. Materials

2.1. Probe DNA

2.1.1. Preparation of f148 and f32 Probe

1. f148: pRSVcat plasmid.
2. f32: oligonucleotides f32-1, 5′-GTAGTTGTACCCGATGGTGGACGCGTGC-3′ and f32-2, 5′-GGCCGCACGCGTCCATCGGGTACAACTAC3′.
3. Hind III and Pvu II: 10 units/μL (New England Biolabs, Beverly, MA).
4. Agarose.
5. 1× TBE: 89 mM Tris–HCl, pH 8.0, 89 mM borate, and 2 mM EDTA.
6. Ethidium bromide solution: 2 μg/mL, dissolved in distilled water and filtered through 0.4 μm filter.
7. Long-wavelength UV light source (366 nm): Model UVL-56 (Ultra-violet products, Inc., San Gabriel, CA).
8. IBI-electroeluter: Model 46000 (IBI-Kodak, Rochester, NY).
9. Polyethylene tubing: I.D. 1.6 mm.
10. Salt solution: 10 μL of 0.5 % bromophenol blue and 1 mL of 10 M ammonium acetate.
11. 18-gauge needle.
12. Razor blade.
13. Isobutanol.
14. Ethanol, 100 %.
15. Ethanol, 80 %.
16. TE: 10 mM Tris–HCl, pH 8.0, and 1 mM EDTA.

2.1.2. Preparation of f298 Probe

1. f298: any plasmid containing the Amp® gene for the PCR template; forward primer 5′-CATACCAAACGACGAGCGTG AC-3′ and reverse primer 5′-ATCCATAGTTGCCTGACT CCCC-3′.
2. Herculase II Fusion DNA Polymerase (Agilent, Santa Clara, CA).
3. 5× Herculase II reaction buffer.
4. dNTPs.
5. Thermal cycler for PCR amplification.
6. QIAquick PCR purification Kit (Qiagen, Valencia, CA).

2.1.3. Labeling of Probe DNA (Klenow Method)

1. 10× Klenow buffer: 100 mM Tris–HCl, pH 7.5, 50 mM $MgCl_2$, and 75 mM dithiothreitol.
2. 10 mM dATP, dGTP, and dTTP.
3. α-^{32}P-dCTP: 3,000 Ci/mmol, 10 mCi/mL (GE Healthcare Life Sciences, Piscataway, NJ).
4. DNA polymerase I, Klenow fragment: 5 units/μL (New England Biolabs, Beverly, MA).

2.1.4. Labeling of f148 Probe (Exonuclease III/ Klenow Method)

1. Items 1–4 of Subheading 2.1.3.
2. Exonuclease III: 100 units/μL, diluted in TE (New England Biolabs, Beverly, MA).

2.1.5. Labeling of f298 Probe (T4 Polynucleotide Kinase Method)

1. 10× T4 polynucleotide kinase reaction buffer: 700 mM Tris–HCl, pH 7.6, 100 mM $MgCl_2$, 50 mM Dithiothreitol (DTT).
2. [γ-^{33}P]ATP: 3,000 Ci/mmol, 10 mCi/mL (Perkin Elmer, Waltham, MA). See note 4 for comments on the use of ^{33}P to label the DNA probe.
3. T4 polynucleotide kinase: 10 U/μL (New England Biolabs, Beverly, MA).
4. Nucleotide-removal kit (Qiagen, Valencia, CA).

2.2. Cell Extracts

2.2.1. Whole Cell Extract

1. Phosphate-buffered saline: 1 g/L D-glucose, 36 mg/L sodium pyruvate, 36 mg/L calcium phosphate, and 36 mg/L magnesium phosphate.
2. Lysis buffer: 700 mM NaCl, 1 mM EGTA, 1 mM EDTA, 10 mM β-glycerol-phosphate, 2 mM $MgCl_2$, 10 mM KCl, 1 mM sodium vanadate, 1 mM PMSF, 1 mM dithiothreitol, 0.1 % NP-40, 10 μg/mL of each pepstatin, leupeptin, and aprotinin.
3. Bradford solution (BioRad, Hercules, CA).

2.2.2. Cytoplasmic and Nuclear Extract

1. Phosphate-buffered saline: 1 g/L D-glucose, 36 mg/L sodium pyruvate, 36 mg/L calcium phosphate, and 36 mg/L magnesium phosphate.

2. Buffer A without NP-40: 10 mM Hepes-KOH, pH 7.4, 1.5 mM $MgCl_2$, 10 mM KCl, 1 mM dithiothreitol, and 1 mM EDTA.
3. Buffer A with NP-40: 10 mM Hepes-KOH, pH 7.4, 1.5 mM $MgCl_2$, 10 mM KCl, 1 mM dithiothreitol, 1 mM EDTA, and 1.0 % (w/v) NP-40.
4. Buffer C: 20 mM Hepes-KOH, pH 7.4, 20 % glycerol (v/v), 500 mM NaCl, 1.5 mM $MgCl_2$, 0.2 mM EDTA, and 0.5 mM dithiothreitol.
5. 100× Protease inhibitor cocktail: 100 mM PMSF, 10 mg/mL pepstatin A, 10 mg/mL leupeptin, 10 mg/mL aprotinin.

2.2.3. Non-denaturing Polyacrylamide Gel

1. Mini-gel plates: 0.75 mm thickness, 80×100 mm (Hoefer, Holliston, MA).
2. Gel caster: SE215 Mighty-Small multiple gel caster (Hoefer, Holliston, MA).
3. Acrylamide-bisacrylamide: 29:1, w/w (Molecular biology grade, Sigma-Aldrich, St. Louis, MO).
4. 5× TGE buffer: 50 mM Tris–HCl, pH 8.5, 380 mM glycine, and 2 mM EDTA (see Note 1).
5. Ammonium persulfate: 10 % of fresh solution.
6. TEMED.
7. Gel combs: 10 or 15 lane combs with 0.75 mm thickness (Hoefer, Holliston, MA).

2.3. EMSA for Studying UV-DDB

2.3.1. UV-Damaged f148 DNA

1. Germicidal lamp: G15T8 (General Electric, Cleveland, OH).
2. Radiometer/photometer: Model IL1350 (International light Inc., Newburyport, MA).

2.3.2. Competitor DNA to Mask Nonspecific Binding

1. Poly(dI-dC) (GE Healthcare Life Sciences, Piscataway, NJ).
2. Salmon sperm DNA: Sheared by passing through 18-gauge needle with pressure.
3. Linear plasmid DNA: pRSVcat plasmid digested with Pvu II and extracted with phenol/chloroform, and precipitated in ethanol.

2.3.3. Competitor DNA to Test Binding Specificity

1. Supercoiled pRSVcat plasmid.
2. AP treatment buffer: 0.01 M sodium citrate, 0.1 M NaCl, pH 5.0.
3. Pt treatment buffer: 3 mM NaCl, 1 mM NaH_2PO_4, pH 7.4.
4. *Cis*-diamminedichloroplatinum (*cis*-DDP, Sigma, St. Louis, MO): Stock solution was dissolved in Pt treatment buffer and stored at −20 °C in the dark.

5. Ethanol, 100 %.
6. *N*-Methyl-*N*′-nitro-*N*-nitrosoguanidine (MNNG, Sigma) was dissolved to 15 mM in ethanol and stored at −20 °C.
7. MNNG treatment buffer: 10 mM Tris–HCl, pH 7.5 and 50 mM NaCl.

2.3.4. UV-DDB Binding Reaction

1. 5× binding buffer: 12 mM Hepes-KOH, pH 7.9, 60 mM KCl, 5 mM $MgCl_2$, 4 mM Tris, 0.6 mM EDTA, 1 mM dithiothreitol, and 12 % glycerol (v/v).
2. BSA: 10 mg/mL.
3. Plasmid DNA: pRSVcat, 1 mg/mL.
4. Poly(dI-dC)/salmon sperm DNA: 1/0.5 mg/mL.
5. Whole cell or nuclear extract: 0.1–0.5 mg/mL.
6. Loading dye: 0.025 % bromophenol blue in 5× binding buffer.
7. Whatman 3MM paper (GE Healthcare Life Sciences, Piscataway, NJ).
8. X-ray film: Kodak XAR-5 (Sigma-Aldrich, St. Louis, MO).

2.4. Variation on EMSA: Reverse Electrophoretic Mobility Shift Assay

2.4.1. In Vitro Synthesis of ^{35}S-Labeled DDB1

1. DDB1 plasmid: cloned into pCDNA3 vector (Invitrogen, Carlsbad, CA).
2. Rabbit reticulocyte lysate (Promega, Madison, WI).
3. ^{35}S-labeled methionine: 1,000 Ci/mmol, 50 mCi/mL (GE Healthcare Life Sciences, Piscataway, NJ).
4. T7 RNA polymerase buffer: 40 mM Tris–HCl, pH 7.9, 6 mM $MgCl_2$, 2 mM spermidine, 10 mM dithiothreitol, 0.5 mM ATP, 0.5 mM CTP, 0.5 mM GTP, and 0.5 mM UTP.
5. T7 RNA polymerase: 10 units/μL (Promega, Madison, WI).
6. Amino acid mixture minus methionine: 1 mM (Promega, Madison, WI).
7. RNasin ribonuclease inhibitor: 40 units/μL (Promega, Madison, WI).
8. RNase-free water: treated with 0.1 % DEPC at 37 °C overnight and autoclaved at 120 °C for 20 min.

2.4.2. Purification of Active ^{35}S-Labeled DDB1

1. UV-DNA cellulose resin: prepared as described previously (7).
2. 200 μL pipetman tip.
3. Buffer D: 10 mM Hepes-KOH, pH 7.9, 2 mM EDTA, 2 mM dithiothreitol, 0.01 % NP-40 (v/v).

2.4.3. Reverse Electrophoretic Mobility Shift Assay

1. 5× binding buffer: 12 mM Hepes-KOH, pH 7.9, 60 mM KCl, 5 mM $MgCl_2$, 4 mM Tris, 0.6 mM EDTA, 1 mM dithiothreitol, and 12 % glycerol (v/v).
2. BSA: 10 mg/mL.

3. Plasmid DNA: pRSVcat, 1 mg/mL.
4. Poly(dI-dC)/salmon sperm DNA: 1/0.5 mg/mL.
5. ^{35}S-labeled DDB1.
6. Loading dye: 0.025 % bromophenol blue in 5× binding buffer.
7. 50 % methanol and 10 % acetic acid.
8. Amersham Amplify™ Fluorographic Reagent (GE Healthcare Life Sciences, Piscataway, NJ).
9. Whatman 3MM paper (GE Healthcare Life Sciences, Piscataway, NJ).
10. Baby powder.
11. Whatman 3MM paper (GE Healthcare Life Sciences, Piscataway, NJ).
12. X-ray film: Kodak XAR-5 (Sigma-Aldrich, St. Louis, MO).

2.5. EMSA for Studying Nonhomologous End Joining Proteins

2.5.1. Probe DNA

1. Materials for purifying and radiolabeling the f148 and f32 probe are described in Subheading 2.1.

2.5.2. Competitor DNA to Mask Nonspecific Binding

1. Supercoiled pRSVcat plasmid: purified by CsCl centrifugation (24).

2.5.3. Competitor DNA to Test Binding Specificity

1. Poly(dA) or poly(dT) (GE Healthcare Life Sciences, Piscataway, NJ).
2. Double-stranded linear DNA: unlabeled f148 DNA.
3. Single-stranded circular DNA: phage M13 DNA (New England BioLabs, Beverly, MA).
4. Supercoiled pRSVcat plasmid: purified by CsCl centrifugation (24).

2.5.4. Ku Binding Reaction

1. 5× binding buffer: 12 mM Hepes-KOH, pH 7.9, 60 mM KCl, 5 mM $MgCl_2$, 4 mM Tris, 0.6 mM EDTA, 1 mM dithiothreitol, and 12 % glycerol (v/v).
2. Supercoiled plasmid DNA: pRSVcat, 1 mg/mL.
3. Protein extract: 0.1–0.5 mg/mL.
4. Loading dye: 0.025 % bromophenol blue in 5× binding buffer.
5. Polyacrylamide gel: 4 %.
6. Whatman 3 MM paper (GE Healthcare Life Sciences, Piscataway, NJ).
7. X-ray film: Kodak XAR-5 (Sigma-Aldrich, St. Louis, MO).

2.5.5. Ku Plus DNA–PKcs Binding Reaction

1. 5× binding buffer: 50 mM Tris–HCl, pH 7.8, 5 mM EDTA, 500 mM NaCl, 25 % glycerol, 1 mM dithiothreitol.
2. Protein extract: 0.1–5.0 mg/mL.
3. Loading dye: 0.025 % bromophenol blue in 1× binding buffer.
4. Polyacrylamide gel: 4 %.
5. Whatman 3 MM paper (GE Healthcare Life Sciences, Piscataway, NJ).
6. X-ray film: Kodak XAR-5 (Sigma-Aldrich, St. Louis, MO).

2.5.6. Ku Plus XRCC4/Ligase IV Binding Reaction

1. 2× binding buffer: 50 mM Tris–HCl, pH 7.5, 150 mM NaCl, 145 mM KCl, 10 mM Mg_2Cl, 4 mM DTT, 100 μg/mL BSA, 0.05 % Triton X-100, 10 % (v/v) Glycerol, 0.2 mM EDTA and 10 % (w/v) polyethylene glycol (MW > 8,000 Da).
2. Recombinant Ku and XRCC4/Ligase IV. For methods to express and purify the recombinant proteins, see ref. 4.
3. 5 % Polyacrylamide TBE gel (Bio-Rad, Hercules, CA).
4. Whatman 3MM paper (GE Healthcare Life Sciences, Piscataway, NJ).
5. Storage phosphor screen (GE Healthcare Life Sciences, Piscataway, NJ).
6. Typhoon 9400 imager (GE Healthcare Life Sciences, Piscataway, NJ).

2.5.7. Variation on EMSA: Antibody Supershift

1. Anti-Ku antibody: HT serum from a human autoimmune patient with polymyositis-scleroderma overlap syndrome (13). Supershifts can also utilize antibodies against other nonhomologous end joining proteins, such as XRCC4, to characterize the components of a protein–DNA complex (17).
2. 1 % BSA.

3. Methods

3.1. Probe DNA

The 148 bp DNA fragment (f148) was isolated from bacterial chloramphenicol acetyltransferase gene.

3.1.1. Preparation of f148 Probe

1. Digest pRSVcat plasmid with Hind III and Pvu II restriction enzymes.
2. Separate the digested DNA fragments on a 1.5 % agarose gel.
3. Stain the gel with ethidium bromide solution and visualize the 148 bp DNA fragment (f148) with long-wavelength UV light.

Long-wavelength UV was used instead of short-wavelength UV to prevent damage to the DNA.

4. Excise the band containing the f148 fragment with a razor blade.
5. Elute the f148 DNA fragment from the gel slice by using an IBI-electroeluter, as described below.
6. Soak the gel slice containing f148 in 0.2× TBE buffer for 5–10 min.
7. Preclear the electroeluter chamber by electrophoresis in 0.2× TBE for 20 min at 150 V. Carefully remove air bubbles in the chamber.
8. Place the gel slice in a circular receptacle. Surround the gel slice with 0.2× TBE buffer but do not cover the gel slice. Keep valve up.
9. Flush V-channel with 0.2× TBE buffer by using 18-gauge needle on 1 mL syringe with polyethylene tubing to protect electroeluter device.
10. Underlay 125 μL of salt solution into the V-channel.
11. Close the cover to move the valve to the intermediate position.
12. Run the electroeluter at 150 V; watch for the migration of the DNA band out of the gel slice with the handheld long-wavelength UV light.
13. Stop the electroelution when all of the DNA fragments have left the gel slice (approximately 10–20 min).
14. Push the valve to the lowest position.
15. Withdraw the contents in the V-channel and rinse the channel with 100 μL of the salt solution.
16. Extract the eluate twice with 400 μL of isobutanol, which will remove the contaminated dyes and reduce the volume.
17. Measure the final volume of the eluate and add two volumes of 100 % ethanol and precipitate at −20 °C for 2 h.
18. Spin the ethanol precipitate at 13,000 × *g* for 20 min at 4 °C.
19. Wash the pellet with cold 80 % ethanol solution.
20. Suspend the pellet in TE solution and measure the concentration.

3.1.2. Preparation of f32 Probe

1. Mix 300 μL of 5 ng/μL of f32-1 with 300 μL 5 ng/μL f32-2 in TE buffer (materials are described in Subheading 2.1.1).
2. Heat the mixture of oligonucleotides to 100 °C for 2 min.
3. Anneal the oligonucleotides by turning off the heat source and allowing the mixture to cool to room temperature.

3.1.3. Preparation of f298 Probe

1. Set up reaction mixture at room temperature as follows:

(a) 5× Herculase II reaction buffer	10 μL
(b) dNTPs (25 mM each)	0.5 μL
(c) Template DNA	10 ng
(d) Forward primer (10 μM)	1.25 μL
(e) Reverse primer (10 μM)	1.25 μL
(f) Nuclease-free water	to 49.5 μL

2. Add 0.5 μL of Herculase II fusion DNA polymerase.
3. Amplify the DNA by PCR for 35 cycles, with annealing temperature 60 °C and extension time of 20 s.
4. To verify successful amplification of the probe, run 1 μL of the reaction on a 1 % agarose gel prior to purification.
5. Purify the PCR product using Qiaquick PCR purification kit (Qiagen, Valencia, CA)

3.1.4. Labeling of f148 or f32 Probe (Klenow Method) (See Note 2)

1. Set up reaction mixture at room temperature as follows:

(a) 10× Klenow buffer	1 μL
(b) f148 or f32 DNA (5 ng/μL)	4 μL
(c) 10 mM dATP	0.5 μL
(d) 10 mM dGTP	0.5 μL
(e) 10 mM dTTP	0.5 μL
(f) α-^{32}P-dCTP (10 μCi/μL)	1 μL
(g) Klenow (5 units/μL)	1 μL
(h) Distilled water	to 10 μL

2. Incubate at room temperature for 20 min.
3. Inactivate the Klenow by incubating the reaction mixture at 65 °C for 10 min. This step may be omitted if step 4 is performed immediately.
4. Purify the labeled f148 with nucleotide-removal kit (Qiagen, Valencia, CA).

3.1.5. Labeling of f148 Probe (Exonuclease III/ Klenow Method) (See Note 3)

1. Set up exonuclease III digestion reaction as follows:

(a) 10× Klenow buffer	2 μL
(b) f148 DNA fragment (5 ng/μL)	4 μL
(c) Exonuclease III (0.2 units/μL, diluted in TE)	1 μL
(d) Distilled water	to 10 μL

2. Incubate at room temperature for 10 min.
3. Inactivate the exonuclease III by incubating the reaction mixture for 10 min at 65 °C
4. Add the following to the cooled reaction mixture at room temperature:

(a) 10 mM dATP	1 μL
(b) 10 mM dGTP	1 μL
(c) 10 mM dTTP	1 μL
(d) α-^{32}P-dCTP (10 μCi/μL)	1 μL
(e) Klenow (5 units/μL)	1 μL
(f) Distilled water	to 20 μL

5. Incubate at 37 °C for 30 min.
6. Inactivate the Klenow by incubating the reaction mixture at 65 °C for 10 min. This step may be omitted if step 7 is performed immediately.
7. Purify the labeled f148 with nucleotide-removal kit (Qiagen, Valencia, CA).

3.1.6. Labeling of f298 Probe (T4 Polynucleotide Kinase Method) (See Note 4)

1. Set up reaction mixture at room temperature as follows:

(a) 10× kinase buffer	2 μL
(b) f298 DNA	10 pmole
(c) γ-^{33}P-dATP (3,000 Ci/mmole, 10 mCi/mL)	20 pmole (6 μL)
(d) Nuclease-free water	to 19 μL

2. Heat the mixture to 70 °C for 5 min and put on ice.
3. Add 1 μL of T4 polynucleotide kinase and incubate at 37 °C for 30 min.
4. Purify the labeled f298 DNA with nucleotide-removal kit (Qiagen, Valencia, CA).

3.2. Preparation of Cell Extracts

3.2.1. Whole Cell Extract

1. Harvest 2×10^6 cells from culture dishes in 1 mL of ice-cold phosphate-buffered saline.
2. Pellet the cells by centrifugation for 1 min at 13,000 × *g*.
3. Resuspend the pellet in 30 μL of the lysis buffer.
4. Incubate the lysates at 4 °C for 30 min with gentle shaking.
5. Centrifuge the lysates at 13,000 × *g* for 30 min at 4 °C.
6. Save the supernatant at −80 °C (see Note 5).
7. Measure the protein concentration by a modification of the Bradford method (25).

3.2.2. Cytoplasmic and Nuclear Extract

1. Harvest 2×10^6 adherent cells from culture dishes in 1 mL of ice-cold phosphate-buffered saline.
2. Wash the cells once with 500 μL of 1× phosphate-buffered saline.
3. Add 10 μL protease inhibitor cocktail to 1 mL buffer A without NP-40. Wash the cells once with 500 μL of Buffer A without NP-40.
4. Add 1 μL protease inhibitor cocktail to 100 μL buffer A with NP-40. Resuspend the pellets in 20 μL of Buffer A with NP-40.
5. Incubate the suspensions at 4 °C for 10 min with gentle shaking.
6. Spin down the lysates at 13,000 × *g* for 5 s at 4 °C.
7. Save the supernatant as cytoplasmic extract at −80 °C (see Note 5).
8. Wash the pellet in 300 μL buffer A without NP-40 by gentle pipetting. Microfuge at 13,000× *g* for 5 s.
9. Add 10 μL of protease inhibitor cocktail to 1 mL buffer C. Resuspend the pellet from step 8 in 75 μL of Buffer C (see Notes 6 and 7).
10. Incubate the suspensions at 4 °C for 20 min with gentle shaking.
11. Spin down the nuclear lysates at 13,000 × *g* for 15 min at 4 °C.
12. Save the supernatant as nuclear extract at −80 °C (see Note 5).
13. Measure the protein concentration by the method of Bradford (25)

3.3. Preparation of Non-denaturing Gel

1. Assemble 13 sets of 0.75 mm thick mini-gels in a gel caster.
2. Prepare 4 % polyacrylamide gel solution as follows (see Note 8):

(a) Acrylamide-bisacrylamide (29:1)	14 mL
(b) 5× TGE buffer	20 mL
(c) Distilled water	65 mL
(d) Ammonium persulfate (10 %)	1 mL
(e) TEMED	120 μL

3. Gently mix the gel solution and immediately pour to the top of the assembled gel caster with caution to avoid bubbles.
4. Tap the gel caster gently several times to remove the bubbles in the gel caster.
5. Insert the comb into the top of each gel.
6. Leave at room temperature for at least 4 h.
7. Disassemble the gel caster and wrap each gel with Saran Wrap and store at 4 °C until use. Be careful to remove residual polyacrylamide, which adheres to the outside of the individual gels, since this can interfere with efficient electrophoresis (see Note 9).

3.4. EMSA for Studying UV-DDB

3.4.1. UV Irradiation of the f148 Probe

The labeled f148 was damaged by UV radiation at a DNA concentration of 0.2 μg/mL with a germicidal lamp at a flux of 10.4 J/m^2/s for total doses of 100–5,000 J/m^2 (see Notes 10 and 11).

3.4.2. Competitor DNA Against Nonspecific Binding

Crude cellular extracts have nonspecific DNA-binding proteins including DEB proteins (see Note 12). Thus, excess unlabeled linear double-stranded DNA was included in the reaction mixture. The amount of the competitor DNA was determined empirically because different extracts require different amounts of competitor DNA. A mixture of poly(dI-dC) and salmon sperm DNA (2:1, w/w) was generally used.

3.4.3. Competitor DNA to Test Specificity of Binding

The binding specificity for UV-damaged DNA and single-stranded DNA was measured by competition assay. The single-stranded competitor DNA was prepared by heating linear double-stranded DNA at 100 °C for 5 min and then cooling rapidly in ice water. UV-damaged single-stranded DNA was prepared by exposing the single-stranded DNA to UV radiation.

The chemically damaged competitor DNA was prepared as described previously (26) (see Note 13). Apurinic DNA was prepared by incubating DNA (300 μM DNA phosphate) in AP treatment buffer at 70 °C for different times. Approximately one purine base is released every 4 min under these reaction conditions. Cisplatin damaged DNA was prepared by incubating supercoiled plasmid (300 μM DNA phosphate) with *cis*-DDP at 37 °C for 12–18 h in the dark. DNA was purified by ethanol precipitation. MNNG was incubated with DNA (300 μM DNA phosphate) for 24 h at 37 °C in MNNG treatment buffer. DNA was purified by ethanol precipitation.

3.4.4. Binding Reaction

1. Assemble the reaction mixture as follows:

(a) 5× binding buffer	2 μL
(b) UV-^{32}P-f148 probe	0.2 or 0.02 ng
(c) BSA (10 mg/mL)	0.3 μL
(d) Plasmid DNA (1 mg/mL)	2 μL
(e) Poly(dI-dC)/salmon sperm DNA (1.0/0.5 mg/mL)	0.5–2 μL
(f) Distilled water	to 8 μL
(g) Whole cell or nuclear extract (0.1–0.5 mg/mL)	2 μL

2. Add whole cell or nuclear extract last.
3. Incubate the reaction mixture at room temperature for 30 min.
4. Add 2 μL of loading dye to the reaction mixture and gently mix together.

5. Resolve the protein–DNA complexes by non-denaturing gel electrophoresis at 10 V/cm at room temperature.
6. Dry the gel on Whatman 3 MM paper and expose to X-ray film at −80 °C (Fig. 2a).

3.5. Variation on EMSA: Reverse Electrophoretic Mobility Shift Assay

Specific binding of UV-DDB to UV-damaged DNA probe was also visualized by a reverse mobility shift assay with labeled UV-DDB and unlabeled UV-f148 probe (Fig. 2b).

3.5.1. In Vitro Synthesis of ^{35}S-Labeled DDB1

UV-DDB consists of a 125 kDa subunit, DDB1, and a 48 kDa subunit, DDB2. Labeled DDB1 protein was synthesized by transcribing the DDB1 cDNA from a T7 promoter with T7 RNA polymerase and translating the mRNA in a rabbit reticulocyte lysate in the presence of (^{35}S) methionine.

1. Assemble the in vitro transcription/translation reaction mixture as follows:

(a) pCDNA3 (DDB1) plasmid (1 mg/mL)	1 μL
(b) Rabbit reticulocyte lysate	25 μL
(c) (^{35}S) methionine	4 μL
(d) T7 RNA polymerase buffer	2 μL
(e) T7 RNA polymerase	1 μL
(f) Amino acid mixture minus methionine	1 μL
(g) RNasin ribonuclease inhibitor	1 μL
(h) RNase-free water	to 50 μL

2. Incubate at 30 °C for 120 min.
3. Store the in vitro translation products at −80 °C.

3.5.2. Purification of Active ^{35}S-Labeled DDB1 Protein

Not all of the DDB1 protein synthesized in vitro is active in binding UV-damaged DNA. Thus, the active fraction was purified through UV-DNA cellulose affinity chromatography.

1. Pack 10 μL of UV-DNA cellulose resin in a 200 μL Pipetman tip and equilibrate the column with Buffer D.
2. Load the in vitro translated ^{35}S-labeled DDB1 proteins onto the column (50 μL of the reticulocyte lysate).
3. Wash the column extensively with 600 μL of Buffer D containing 200 mM NaCl.
4. Elute the bound ^{35}S-labeled DDB1 protein with Buffer D containing 1 M NaCl.
5. Store the fractions at −80 °C.

3.5.3. Reverse Electrophoretic Mobility Shift Assay

1. Assemble the reaction mixture as follows:

(a) 5× binding buffer	2 μL
(b) UV-f148 probe	1 ng
(c) BSA (10 mg/mL)	0.3 μL
(d) Plasmid DNA (1 mg/mL)	2 μL
(e) Poly (dI-dC)/salmon sperm DNA (1/0.5 mg/mL)	0.5 μL
(f) Distilled water	to 8 μL
(g) ^{35}S-labeled DDB1	2 μL

2. Incubate the reaction at room temperature for 30 min.
3. Add 2 μL of loading dye to the reaction mixture and gently mix together.
4. Resolve the protein–DNA complex by non-denaturing gel electrophoresis at 10 V/cm at room temperature.
5. Fix the gel in 50 % methanol and 10 % acetic acid for 2 h.
6. Soak the gel in the Amersham Amplify™ Fluorographic Reagent for 30 min.
7. Dry the gel on Whatman 3 MM paper.
8. Treat the sticky surface of the dried gel with baby powder, which allows direct contact of the gel with the X-ray film.
9. Expose the gel to X-ray film at −80 °C.

3.6. EMSA for Studying Ku

3.6.1. Probe DNA

Prepare the f148 probe as described in Subheading 3.1.1 and radiolabel the probe as described in Subheading 3.1.4 (see Notes 14 and 15).

3.6.2. Competitor DNA Against Nonspecific Binding

Crude nuclear extracts have proteins that may bind nonspecifically to the labeled f148 probe without binding to the DNA ends. To minimize the effects of these proteins on the EMSA, supercoiled plasmid is included in the binding reaction. Because the supercoiled pRSVcat plasmid contains the f148 fragment but has no free ends, it is a good nonspecific competitor (see Note 16). In general, human cell extracts require 2 μg of nonspecific competitor in a 10 μL reaction. Rodent cell extracts, however, require only 50–100 ng of nonspecific competitor.

3.6.3. Competitor DNA to Test Binding to Different Types of DNA Ends

Competition assays are performed using single-stranded linear DNA (poly(dA) or poly(dT)), double-stranded linear DNA (unlabelled f148), single-stranded circular DNA (phage M13), or supercoiled double-stranded DNA (pRSVcat). These competitor DNAs are included in the binding reaction in concentrations ranging from 0.2 to 200 ng. Competitors containing free DNA ends (unlabelled f148) or DNA containing single-stranded to double-stranded DNA transitions (phage M13) compete for Ku binding activity.

3.6.4. Binding Reaction

1. Prepare the reaction mixture as follows:

(a)	5× binding buffer	2 μL
(b)	f148 probe	0.2 ng
(c)	Supercoiled plasmid	2 μg (human extracts) or 50–100 ng (rodent extracts)
(d)	Distilled water	to 8 μL
(e)	Nuclear extract	0.5 μg

2. Incubate the reaction for 5 min at room temperature.
3. Add 2 μL of loading dye to the reaction and mix by gentle pipetting.
4. Resolve the protein–DNA complexes by non-denaturing gel electrophoresis at 10 V/cm at room temperature.
5. Dry the gel on 3 MM Whatman paper and expose to X-ray film at −80 °C.

3.6.5. Antibody Supershift

Anti-Ku antibodies may be added to the binding reaction to supershift the Ku/DNA complex. Serial dilutions of the antibodies are made in 1 % BSA and added to the reaction mix above, prior to adding the protein extract (Fig. 3).

3.7. EMSA for Studying Ku and DNA–PKcs Bound to DNA

3.7.1. Probe DNA

Prepare the f32 probe as described in Subheading 3.1.2 and radiolabel the probe as described in Subheading 3.1.4.

3.7.2. Competitor DNA

We have found that shortening the length of the DNA probe allows a complex containing DNA–PKcs to form on DNA ends in the absence of competitor DNA (16) (see Note 17). Additionally, certain competitor DNAs, including poly dA and supercoiled plasmid, generate a supershift of the DNA–PKcs-containing complex. This may be due to the ability of DNA–PKcs to synapse two DNA molecules (27).

3.7.3. Binding Reaction (See Notes 18 and 19)

1. Prepare the reaction mixture as follows:

(a)	5× binding buffer	2 μL
(b)	f32 probe	0.2 ng
(c)	Distilled water	to 8 μL
(d)	Nuclear extract	(human) 0.5 μg (rodent) 2.0 μg
(e)	NaCl	to a final concentration of 200 mM (human extracts) 400 mM (rodent extracts)

2. Incubate the reaction for 5 min at room temperature.
3. Add 2 μL of loading dye to the reaction and mix by gentle pipetting.
4. Resolve the protein–DNA complexes by non-denaturing gel electrophoresis at 10 V/cm at room temperature.
5. Dry the gel on 3 MM Whatman paper and expose to X-ray film at −80 °C.

3.7.4. Antibody Supershift

Anti-DNA–PKcs antibodies may be added to the binding reaction to supershift or disrupt the DNA–PKcs containing complexes. Serial dilutions of the antibodies are made in 1 % BSA and added to the reaction mix above, prior to adding the protein extract.

3.8. EMSA for Studying Ku and XRCC4/Ligase IV Bound to DNA

3.8.1. Probe DNA

Prepare the f298 probe as described in Subheading 3.1.3 and radiolabel the probe as described in Subheading 3.1.6.

3.8.2. Binding Reaction

1. Prepare the reaction mixture as follows:

(a) 2× binding buffer	5 μL
(b) ^{33}P-labeled f298 probe	20 fmole
(c) Recombinant Ku	10 fmole
(d) Recombinant XRCC4/Ligase IV	40 fmole
(e) Nuclease-free water	to 10 μL

The concentrations of purified proteins must be optimized for each protein–DNA complex under study. The concentrations in this assay are suggestions for a starting point.

2. Incubate the reaction at 37 °C for 30 min.
3. Pre-run the gel for 1.5 h in 0.5× TBE buffer at 60 V.
4. Load at least one well with 0.01 % bromophenol blue in 1× binding buffer to provide a tracking dye. No dyes are included in the binding reactions.
5. Load binding reactions and run the 5 % polyacrylamide gel in fresh and chilled 0.5× TBE buffer at 60 V for 3 h.
6. Remove the gel from the cassette and attach it to 3MM Whatman paper. Cover the gel with Saran wrap. To dry the gel, apply vacuum with heat for 30 min, and then turn off heat to let cool for 30 min.
7. Expose the gel to a storage phosphor screen. Image with Typhoon scanner.

4. Notes

1. The high-salt buffers TGE and TBE were superior to low-salt buffers in these EMSAs because the high-salt buffers reduced nonspecific binding more efficiently.
2. When labeling the DNA probe with Klenow fragment, the radiolabeled nucleotide used to fill in the 5′ overhang of the DNA should be complementary to an internal base of the 5′ overhang. Inefficient labeling may occur if the radiolabeled nucleotide is complementary to the last base of the 5′ overhang because of 3′ to 5′ exonuclease activity in the Klenow fragment.
3. The exonuclease III/Klenow method increases the specific activity of the probe (Subheading 3.1.5). This can be helpful in detecting UV-DDB in rodent cells, where levels are substantially lower than in human cells.
4. We have used both ^{32}P and ^{33}P to prepare radiolabeled DNA probes. Choice of ^{32}P provides greater sensitivity, while choice of ^{33}P provides greater resolution of the bands in the EMSA. For the analysis of very large protein–DNA complexes, which migrate only small distances during gel electrophoresis, we prefer to use ^{33}P-labeled DNA probes.
5. Extracts may be stored at 4 °C for 1–3 days, but should be stored at −80 °C for longer periods of time. Extracts should be stored in aliquots to minimize repeated freeze–thaw steps. Thawing of the extracts should be done on ice to minimize protease activity, which may occur at higher temperatures.
6. The extraction procedure should be optimized by testing different salt concentrations in buffer C or changing the NP-40 concentration.
7. The presence of NP-40 in the extraction buffer (buffer C) can inhibit extraction of some proteins. Washing the nuclei following the treatment with buffer A containing NP-40 removes the NP-40 and overcomes this effect (Subheading 3.2.2, step 8). Additionally, larger proteins may be cleaved by proteases to a greater extent than smaller proteins, making it critical that protease inhibitors be present at all steps in the extraction procedure following the lysis in buffer A.
8. Lowering the acrylamide concentration and decreasing the length of the DNA probe can facilitate the detection of larger protein–DNA complexes by permitting the migration of these complexes into the gel. Examples include the complexes of Ku-DNA–PKcs (Fig. 4) and Ku-XRCC4/Ligase IV (Fig. 5) that assemble on DNA ends. In such cases, we have tested a range of acrylamide concentrations and acrylamide to bisacrylamide ratios in order to find conditions that best stabilize the

protein–DNA complex, while permitting sufficient migration of the complex into the gel. Commercially prepared gels may vary from lot to lot. To ensure reproducibility, we recommend casting one's own gels.

9. Conduction leaks in the gel can lead to aberrant migration of the free probe and protein–DNA complexes. Cleaning areas around the gel and gel box where salt buildup can occur ameliorates the problem.
10. In detecting UV-DDB, increasing the UV doses to the probe DNA will increase the number of lesions per DNA molecule. Higher-order complexes corresponding to multiple UV-DDB binding events can be visualized at these higher doses (7).
11. When cyclobutane pyrimidine dimers were removed from a UV-damaged DNA probe by treatment with purified photolyase, the binding was decreased compared to the untreated UV-damaged DNA probe. This suggests that UV-DDB recognizes at least some cyclobutane dimers (7).
12. Because these EMSAs rely on the structure-specific binding properties of the DNA repair proteins, it is important that the competitor DNA be free of contaminating structures that might compete for binding activity.
13. UV-DDB recognizes DNA damaged by several agents including UV irradiation, but direct binding was observed only to the UV-damaged double-stranded DNA probe. The binding of UV-DDB to apurinic sites, cisplatin adducts, nitrogen mustard adducts, and single-stranded DNA was detected by showing that these forms of damaged DNA could compete with UV-damaged DNA for binding to UV-DDB (7, 26). UV-DDB may have low affinities for these forms of DNA, so that the competition assay may be more effective than the direct binding assay in detecting binding activity.
14. Increasing the length of the radiolabeled DNA probe can be utilized to detect multiple Ku-binding events. This is possible because Ku is able to bind DNA and translocate along the molecule. Thus, longer DNA lengths permit more Ku molecules to bind and produce a "ladder" pattern in the EMSA (12).
15. Decreasing the length of the DNA probe can allow a decrease in the amount of competitor DNA used in the EMSA for Ku.
16. The competitor used in EMSA for Ku should be supercoiled DNA that is free of contaminating nicked or linear DNA. Ku binds strongly to free DNA ends and has been reported to bind nicks, so competitors with these structures will obscure Ku-binding activity.
17. There should not be competitor DNA in EMSAs to detect DNA–PKcs. Additionally, the probe length should be short.

We have found 32 base pairs to be optimal in EMSAs to detect DNA–PKcs.

18. DNA–PKcs levels are lower in rodent extracts than human extracts, thus the concentration of extract used in an EMSA should be increased from 0.5 μg (human extract) to 2.0 μg (rodent extract).
19. Since DNA–PKcs EMSAs do not utilize competitor DNA, we have tested the effect of different salt concentrations on nonspecific probe binding and DNA–PKcs complex formation. DNA–PKcs complexes can be detected with NaCl concentrations as high as 500 mM, with higher salt decreasing nonspecific binding. For rodent extracts, which use more protein extracts, we recommend increasing the NaCl concentration to 400 mM.

References

1. Fried M, Crothers DM (1981) Equilibrium and kinetics of lac repressor-operator interactions by polyacrylamide gel electrophoresis. Nucleic Acids Res 9:6505–6525
2. Garner MM, Revzin A (1981) A gel electrophoresis method for quantifying the binding of proteins to specific DNA regions: application to the components of the E. coli lactose operon regulatory system. Nucleic Acids Res 9: 3047–3059
3. Tang J, Chu G (2002) Xeroderma pigmentosum complementation group E and UV-damaged DNA-binding protein. DNA Repair 1:601–616
4. Tsai CJ, Kim SA, Chu G (2007) Cernunnos/XLF promotes the ligation of mismatched and noncohesive DNA ends. Proc Natl Acad Sci USA 104:7851–7856
5. Chu G, Chang E (1988) Xeroderma pigmentosum group E cells lack a nuclear factor that binds to damaged DNA. Science 242:564–567
6. Singh H et al (1986) A nuclear factor that binds to a conserved sequence motif in transcriptional control elements of immunoglobulin genes. Nature 319:154–158
7. Hwang BJ, Chu G (1993) Purification and characterization of a protein that binds to damaged DNA. Biochemistry 32:1657–1666
8. Hwang BJ, Liao J, Chu G (1996) Isolation of a cDNA encoding a UV-damaged DNA binding factor defective in xeroderma pigmentosum group E cells. Mutat Res 362:105–117
9. Chu G, Chang E (1990) Cisplatin-resistant cells express increased levels of a factor that recognizes damaged DNA. Proc Natl Acad Sci USA 87:3324–3327
10. Keeney S et al (1994) Correction of the DNA repair defect in xeroderma pigmentosum group E by injection of a DNA damage-binding protein. Proc Natl Acad Sci USA 91:4053–4056
11. Tang J et al (2000) Xeroderma pigmentosum p48 gene enhances global genomic repair and suppresses UV-induced mutagenesis. Mol Cell 5:737–744
12. Rathmell WK, Chu G (1994) A DNA end-binding factor involved in double-strand break repair and V(D)J recombination. Mol Cell Biol 14:4741–4748
13. Rathmell WK, Chu G (1994) Involvement of the Ku autoantigen in the cellular response to DNA double-strand breaks. Proc Natl Acad Sci USA 91:7623–7627
14. Smider V et al (1994) Restoration of X-ray resistance and V(D)J recombination in mutant cells by Ku cDNA. Science 266:288–291
15. Taccioli GE et al (1994) Ku80: product of the XRCC5 gene and its role in DNA repair and V(D)J recombination. Science 265:1442–1445
16. Hammarsten O, Chu G (1998) DNA-dependent protein kinase: DNA binding and activation in the absence of Ku. Proc Natl Acad Sci USA 95:525–530
17. Nick McElhinny SA et al (2000) Ku recruits the XRCC4-ligase IV complex to DNA ends. Mol Cell Biol 20:2996–3003
18. Patterson M, Chu G (1989) Evidence that xeroderma pigmentosum cells from complementation group E are deficient in a homolog of yeast photolyase. Mol Cell Biol 9:5105–5112
19. Fox M, Feldman B, Chu G (1994) A novel role for DNA photolyase: binding to drug-induced DNA damage is associated with enhanced cytotoxicity in yeast. Mol Cell Biol 14:8071–8077
20. Jiricny J (1994) Colon cancer and DNA repair: have mismatches met their match? Trends Genet 10:164–168

21. Donahue BA et al (1990) Characterization of a DNA damage-recognition protein from mammalian cells that binds specifically to intrastrand d(GpG) and d(ApG) DNA adducts of the anticancer drug cisplatin. Biochemistry 29: 5872–5880
22. Toney J et al (1989) Isolation of cDNAs encoding a human protein that binds selectively to DNA modified by the anticancer drug cis-diamminedichloroplatinum(II). Proc Natl Acad Sci USA 86:8328–8332
23. Brown SJ, Kellet PJ, Lippard SJ (1993) Ixr1, a yeast protein that binds to platinated DNA and confers sensitivity to cisplatin. Science 261: 603–605
24. Sambrook J, Fritsch E, Maniatis T (1989) Molecular cloning: a laboratory manual. Cold Spring Harbor Laboratory Press, Cold Spring Harbor, NY
25. Bradford MM (1976) A rapid and sensitive method for the quantitation of microgram quantities of protein utilizing the principle of protein-dye binding. Anal Biochem 72:248–254
26. Payne A, Chu G (1994) Xeroderma pigmentosum group E binding factor recognizes a broad spectrum of DNA damage. Mutat Res 310: 89–102
27. DeFazio L et al (2002) Synapsis of DNA ends by the DNA-dependent protein kinase. EMBO J 21:3192–3200

Chapter 6

The Comet Assay: A Sensitive Genotoxicity Test for the Detection of DNA Damage and Repair

Günter Speit and Andreas Rothfuss

Abstract

The comet assay (single-cell gel electrophoresis) is a simple and sensitive method for studying DNA damage and repair. In this microgel electrophoresis technique, a small number of cells suspended in a thin agarose gel on a microscope slide is lysed, electrophoresed, and stained with a fluorescent DNA-binding dye. Cells with increased DNA damage display increased migration of chromosomal DNA from the nucleus towards the anode, which resembles the shape of a comet. The assay has manifold applications in fundamental research for DNA damage and repair, in genotoxicity testing of novel chemicals and pharmaceuticals, environmental biomonitoring, and human population monitoring. This chapter describes a standard protocol of the alkaline comet assay and points to some useful modifications.

Key words: Alkaline comet assay, Alkali-labile sites, Biomonitoring, Crosslinks, DNA-strand breaks, Excision repair, Genotoxicity testing, Single-cell gel electrophoresis

1. Introduction

The comet assay (single-cell gel electrophoresis) is a useful technique for studying DNA damage and repair with manifold applications. In this microgel electrophoresis technique, a small number of cells suspended in a thin agarose gel on a microscope slide is lysed, electrophoresed, and stained with a fluorescent DNA-binding dye. Cells with increased DNA damage display increased migration of chromosomal DNA upon electrophoresis from the nucleus towards the anode, which resembles the shape of a comet (Fig. 1). In its alkaline version, which is mainly used, DNA-strand breaks and alkali-labile sites become apparent, and the extent of DNA migration correlates with the amount of DNA damage in the cell. The comet assay combines the simplicity of biochemical techniques for detecting DNA single-strand breaks and/or alkali-labile sites with the single cell approach typical of cytogenetic assays. The advantages of the

Lotte Bjergbæk (ed.), *DNA Repair Protocols*, Methods in Molecular Biology, vol. 920,
DOI 10.1007/978-1-61779-998-3_6, © Springer Science+Business Media New York 2012

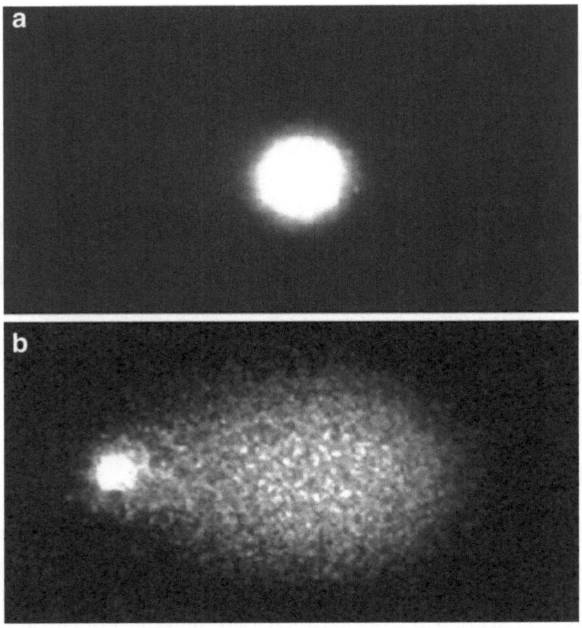

Fig. 1. Photomicrographs of human lymphocytes in the comet assay. (**a**) Untreated cell (control). (**b**) Cell exhibiting increased DNA migration after mutagen treatment.

comet assay include its simple and rapid performance, its sensitivity for detecting DNA damage, the analysis of data at the level of the individual cell, the use of extremely small cell samples, and the usability of virtually any eukaryote cell population. Apart from image analysis, which greatly facilitates and enhances the possibilities of comet measurements, the cost of performing the assay is extremely low. The comet assay has already been used in many studies to assess DNA damage and repair induced by various agents in a variety of cells in vitro and in vivo (1). The test has widespread applications in genotoxicity testing in vitro and in vivo (2–5), DNA damage and repair studies (6), environmental biomonitoring (7, 8), human population monitoring (9, 10), reproductive toxicology (11), and radiobiology (12).

The alkaline version (pH >13) of the comet assay introduced by Singh and coworkers (13) detects a broad spectrum of DNA lesions, that is, DNA single- and double-strand breaks and alkali-labile sites. Modified versions of the assay introduced by Olive (14) involved lysis in alkaline buffer followed by electrophoresis at either neutral or mild alkaline (pH 12.1) conditions to detect DNA double-strand breaks or single-strand breaks, respectively. However, since the majority of genotoxic agents induce much more single-strand breaks and alkali-labile sites than double-strand breaks, the alkaline version (pH >13) of the comet assay has been identified to show the highest sensitivity for detecting induced DNA damage and has been recommended for genotoxicity testing (2). Important improvements of the test procedure were introduced by Klaude and coworkers (15). The use of agarose-precoated slides in combination with drying of gels and fixation of the comets led to a further simplification and a much better handling of the test. The comet assay is especially suited for studies with a high number of samples since it can be performed in a high-throughput fashion and analysis of slides can be automated (16–18).

1.1. Detection of DNA Damage

A broad spectrum of DNA-damaging agents increases DNA migration in the comet assay such as ionizing radiation, hydrogen peroxide and other radical-forming chemicals, alkylating agents, polycyclic aromatic hydrocarbons (PHAs) and other adduct-forming chemicals, radiomimetic chemicals, various metals, or UV-irradiation. In principle, the alkaline version of the comet assay detects all kinds of directly induced DNA single-strand breaks and any lesion that can be transformed into a single-strand break under alkaline conditions (i.e., alkali-labile sites).

In addition to directly induced strand breakage, processes which introduce single-strand nicks in the DNA, such as incision during excision repair processes, are also detectable. In some cases (e.g., UV, PAHs) the contribution of excision repair to the induced DNA effects in the comet assay seems to be of major importance (19). Some specific classes of DNA base damage can be detected with the comet assay in conjunction with lesion-specific endonucleases (6). These enzymes, applied to the slides for a short time after lysis, nick DNA at sites of specific base alterations and the resulting single-strand breaks can be quantified in the comet assay. Using this modification of the comet assay, oxidized DNA bases have been detected with high sensitivity with the help of endonuclease III, formamidopyrimidine-DNA-glycosylase (FPG) or cell extracts in in vitro tests and in samples obtained from human studies (6, 20–22).

Crosslinks (DNA-DNA or DNA-protein) as induced by chemicals, such as nitrogen mustard, cisplatin, cyclophosphamide or formaldehyde may cause problems in the standard protocol. The induction of crosslinks reduces the ability of the DNA to migrate in the agarose gel by stabilizing chromosomal DNA (23, 24). Crosslinks can be detected by adjusting the duration of unwinding and/or electrophoresis to such an extent that control cells exhibit significant DNA migration. A lower extent in DNA migration in treated samples compared to untreated controls would then indicate an induction of crosslinks (25). Another possibility is to induce DNA migration with a second strand-breaking agent (e.g., ionizing radiation, methyl methanesulfonate) after exposure towards the (assumed) crosslinking agent and performing the comet assay immediately thereafter. A crosslinking effect is then determined as reduced migration in comparison with the effect of the strand-breaking agent alone (23, 24). Post-treatment of samples with Proteinase K allows to distinguish between DNA-DNA and DNA-protein cross-links (24).

1.2. Measuring DNA Repair

A widely used approach for determining DNA repair is to monitor a time-dependent removal of lesions (i.e., the decrease in DNA migration) after treatment with a DNA-damaging agent. The comet assay has been successfully used to follow the rejoining of strand breaks induced by ionizing radiation or reactive oxygen species (26, 27) as well as the repair of various kinds of DNA damage induced by chemical mutagens (28, 29). A useful extension of repair studies includes the additional use of lesion-specific enzymes (6), or cell extracts (30, 31). Thereby, the repair of specific types of DNA lesions can be followed and, due to its high sensitivity, this approach enables the analysis of very low ("physiological") levels of DNA damage (32). A common alternative approach is the use of repair inhibitors or repair-deficient cells. Incubation of cells with inhibitors of DNA- (repair-) synthesis, such as hydroxyurea, cytosine arabinoside, or aphidicolin leads to an accumulation of incomplete repair sites as DNA breaks (19, 33). Mutant cell lines either with a specific defect in a repair pathway (e.g., xeroderma pigmentosum) or with a hypersensitivity towards specific DNA damaging agents (e.g., various mutant rodent cell lines) are well suited to elucidate DNA repair pathways and the biological consequences of disturbed DNA repair or to evaluate the repair competence of cells (19, 34–36). While the standard version of the comet assay provides information on DNA damage and repair in the whole genome of a cell, the introduction of a combination of the comet assay with fluorescence in situ hybridization (FISH) additionally allows to measure DNA damage and repair in specific genomic regions (37–39).

The purpose of this protocol is to provide information on the application of the alkaline comet assay for the investigation of DNA damage and repair in mammalian cells in vitro. For establishing the method, we recommend to start with experiments using blood samples and the induction of DNA damage by a standard mutagen (e.g., methyl methane sulfonate, MMS). The method described here is based on a protocol established by R. Tice according to the original work of Singh et al. (13) and includes the modifications introduced by Klaude and coworkers (15). An outline of the protocol is diagrammed in Fig. 2.

2. Materials

1. Microscope slides (with frosted end).
2. Coverslips (24 × 60 mm).
3. Normal melting-point agarose.
4. Low melting point (LMP) agarose.
5. Horizontal gel electrophoresis unit.

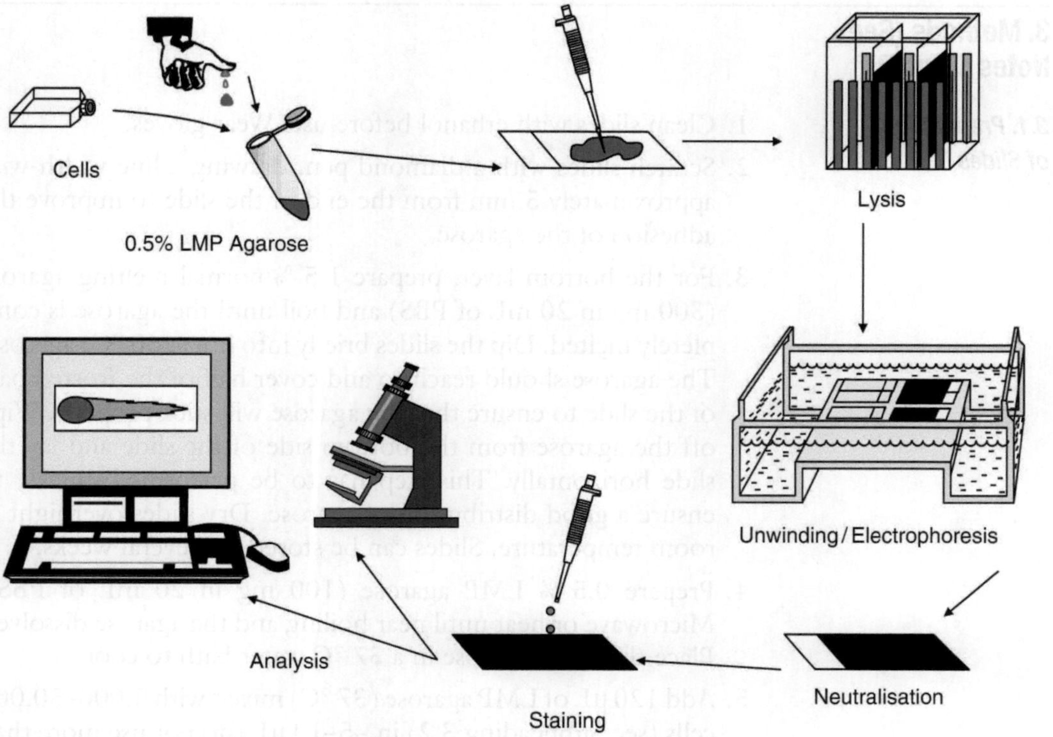

Fig. 2. Scheme for the performance of the comet assay.

6. Fluorescence microscope equipped with an excitation filter of 515–560 nm and a barrier filter of 590 nm.
7. Phosphate-buffered saline (PBS) (without Ca^{2+} and Mg^{2+}).
8. Lysing solution (1 L): 2.5 M NaCl, 100 mM EDTA, 10 mM Tris (set pH to 10.0 with ~7 g solid NaOH). Store at room temperature. Final lysing solution (100 mL, made fresh): add 1 mL of Triton X-100 and 10 mL of DMSO to 89 mL of lysing solution, and then refrigerate (4 °C) for 60 min before use.
9. Electrophoresis buffer: 300 mM NaOH/1 mM EDTA. Prepare from stock solutions of 10 N NaOH (200 g/500 mL distilled H_2O), 200 mM EDTA (14.89 g/200 mL of dH_2O, pH 10.0). Store at room temperature. For 1× buffer, mix 45 mL NaOH, 7.5 mL of EDTA, and add water to 1,500 mL (total volume needed depends on gel box capacity). Mix well and store at 4 °C. Make fresh before each run.
10. Neutralization buffer: 0.4 M Tris–HCl, pH 7.5. Store at room temperature.
11. Ethidium bromide staining solution: 10× stock: 200 μg/mL. Store at room temperature. For 1× stock (20 μg/mL), mix 1 mL with 9 mL dH_2O and filter. *Caution*: Ethidium bromide is a mutagen. Handle with care.

3. Methods (See Notes 1 and 2)

3.1. Preparation of Slides

1. Clean slides with ethanol before use. Wear gloves.
2. Scratch slides with a diamond pen, drawing a line width-wise approximately 5 mm from the end of the slide to improve the adhesion of the agarose.
3. For the bottom layer, prepare 1.5 % normal melting agarose (300 mg in 20 mL of PBS) and boil until the agarose is completely melted. Dip the slides briefly into hot (>60 °C) agarose. The agarose should reach to and cover half of the frosted part of the slide to ensure that the agarose will stick properly. Wipe off the agarose from the bottom side of the slide and lay the slide horizontally. This step has to be performed quickly to ensure a good distribution of agarose. Dry slides overnight at room temperature. Slides can be stored for several weeks.
4. Prepare 0.5 % LMP agarose (100 mg in 20 mL of PBS). Microwave or heat until near boiling and the agarose dissolves. Place the LMP agarose in a 37 °C water bath to cool.
5. Add 120 μL of LMP agarose (37 °C) mixed with 5,000–50,000 cells (see Subheading 3.2) in ~5–10 μL (do not use more than 10 μL). Add coverslip, and place the slide in a refrigerator for ~2 min (until the agarose layer hardens). Using ~10,000 cells results in ~1 cell per microscope field (250× magnification). From this step until the end of electrophoresis, direct light irradiation should be avoided to prevent additional DNA damage.
6. Gently slip off the coverslip and slowly lower slide into cold, freshly made lysing solution. Protect from light, and place at 4 °C for a minimum of 1 h. Slides may be stored for extended periods of time in cold lysing solution (but generally not longer than 4 week). If precipitation of the lysing solution is observed, slides should be rinsed carefully with distilled water before electrophoresis.

3.2. Preparation of Cells (See Notes 3 and 4)

1. Whole blood: Mix ~5 μL whole blood with 120 μL of LMP agarose, and layer onto the slide.
2. Isolated lymphocytes: Add 4 mL of whole blood to a tube with 4 mL prewarmed (37 °C) Ficoll. Centrifuge for 25 min at ~320×*g*. Carefully remove the lymphocytes and resuspend them in 8 mL RPMI 1640 medium. Centrifuge again for 10 min at ~180×*g*. Remove the supernatant and repeat the washing step. Incubate the cells for 30 min at 37 °C. Centrifuge for 10 min at ~130×*g*, discard the supernatant and resuspend the pellet in 375 μL of RPMI 1640 medium. Count the cells and adjust to 1,500 cells/μL. Mix 10 μL of the suspension with 120 μL LMP agarose and layer onto the slide.

3. Cell cultures.
   (a) Monolayer cultures: Gently trypsinize the cells (for approx. 2 min with 0.15 % trypsin, stop by adding serum or complete cell culture medium) to yield approximately 1×10^6 cells/mL. Add 10 μL of cell suspension to 120 μL LMP agarose, and layer onto the slide.
   (b) Suspension cultures: Add ~15,000 cells in 10 μL (or smaller volume) to 120 μL of LMP agarose and layer onto the slide.

3.3. Electrophoresis and Staining (See Notes 5–7)

1. After at least 1 h at 4 °C, gently remove the slides from the lysing solution.
2. Place the slides in the gel box near the anode (+) end, positioning them as close together as possible.
3. Fill the buffer reservoirs with electrophoresis buffer (4 °C) until the slides are completely covered (avoid bubbles over the agarose). Perform the electrophoresis in an ice bath (4 °C).
4. Let slides sit in the alkaline buffer for 20–60 min to allow unwinding of the DNA and the expression of alkali-labile damage. For most experiments with cultured cells, 20 min are recommended.
5. Turn on power supply to 25 V (~0.8–1.5 V/cm, depending on gel box size) and adjust current to 300 mA by slowly raising or lowering the buffer level. Depending on the purpose of the study and on the extent of migration in control samples, allow the electrophoresis to run for 20–40 min. For most experiments, 20 min is recommended.
6. Turn off the power. Gently lift the slides from the buffer and place on a staining tray. Coat the slides with drops of neutralization buffer, and let sit for at least 5 min. Repeat two more times.
7. Drain the slides, rinse carefully with distilled water, and let them dry (inclined) at room temperature. Slides can be stored for a longer time before staining. To stain, rinse the slides briefly in distilled water, add 30 μL 1× ethidium bromide staining solution, and cover with a coverslip. Antifade can be used to prevent slides from drying or fading out if necessary, i.e., when automated analysis is used.

Slides should be stained one by one and evaluated immediately. It is possible to rinse stained (evaluated) slides in distilled water, remove the coverslip, let the slides dry and stain them at a later time point for reevaluation.

3.4. Evaluation of DNA Effects (See Note 8)

For visualization of DNA damage, observations are made of ethidium bromide-stained DNA at 250× (or 400×) magnification using a fluorescence microscope. Generally, 100 randomly selected cells per sample are analyzed. In principle, evaluation can be done in four different ways:

1. Image analysis systems are used to quantify DNA damage. Parameters such as tail intensity (percentage DNA in the tail), tail moment, and tail length are commonly used. It is important to note that the some parameters (e.g., tail moment) may be calculated differently among image analysis systems. For the purpose of inter-laboratory comparison of DNA damage parameters, tail intensity is probably the most suited.
2. Automated systems have been established, which search for comets and carry out the analysis with minimal human intervention.
3. Cells are scored visually according to tail size into five classes (from undamaged, 0, to maximally damaged, 4). Thus the total score for 100 comets can range from 0 (all undamaged) to 400 (all maximally damaged).
4. Cells are analyzed using a calibrated scale in the ocular lens of the microscope. For each cell, the image length (diameter of the nucleus plus migrated DNA) is measured in microns, and the mean is calculated. Alternatively, the length of the comet (or comet tail) can be measured on a photomicrograph. These measurements are very laborious and may only give limited information because the tail length saturates at higher levels of damage.

For the statistical analysis of comet assay data, a variety of parametric and nonparametric statistical methods are used. The most appropriate means of statistical analysis depends on the kind of study and has to take into account the various sources of assay variability. For a powerful statistical analysis of in vitro test data, appropriate replication and repeat experiments have to be performed (2, 40, 41). For example, the median DNA migration of 50 cells per sample and the mean of 2–3 samples per data point may be determined. Also, the mean from repeat experiments can be determined. The use of the median should be preferred over the average since a normal size distribution is usually not observed. Analyses are mainly based on changes in group mean response but attention should also be paid to the distribution among cells which often provides additional important information. Recommendations for appropriate statistical analyses of comet assay data have been published (40, 41).

4. Notes

1. Many technical variables have been used including the concentration and amount of LMP agarose, the composition of the lysing solution and the lysis time, the alkaline unwinding, the electrophoresis buffer, electrophoretic conditions, and

DNA-specific dyes for staining. Some of these variables may affect the sensitivity of the test. To allow for a comparison obtained in different laboratories and for a critical evaluation of data, it is absolutely necessary to clearly describe the technical details of the method employed.

2. The simplicity of the comet assay combined with the need of only low number of cells per sample enables the conduct of in vitro studies with high efficiency. Therefore, the comet assay can be used in a high-throughput fashion (17). Furthermore, the introduction of automated image analysis systems for comet assay slides further can speed up test performance (18).
3. Many other cell types have been used and it is an advantage of the comet assay that virtually any eukaryote cell population is amenable to analysis. The comet assay is particularly suited for the investigation of organ- or tissue-specific genotoxic effects in vivo (2–5), the only requirement being the preparation of an intact single cell suspension.
4. For the demonstration of a positive effect, mix 200 μL heparinized whole blood with 50 μL of a 2.5×10^{-4} M methyl methanesulfonate (MMS) solution (final concentration: 5×10^{-5} M), incubate for 1 h at 37 °C and then use 10 μL for the test.
5. For each cell type, the method should be adjusted scientifically to obtain valid and reproducible results. It is important to define the optimal time for alkaline treatment and electrophoresis. It is recommended that the conditions must be such that the DNA from the control cells exhibit, on the average, some migration. This effect ensures sensitivity and enables an evaluation of intralaboratory experiment-to-experiment variability (5).
6. The temperature during alkaline treatment and electrophoresis significantly influences the amount of DNA migration (42). It is necessary to establish stable and reproducible conditions and it may be useful to use a cooled electrophoresis unit or to place the electrophoresis unit in a jar filled with ice or in a cooled room.
7. If specific types of base damage should be determined by using lesion-specific endonucleases or cell extracts, the standard protocol has to be modified in the following way: after at least 1 h at 4 °C, gently remove slides from the lysing solution and wash three times in enzyme buffer. Drain slides and cover with 200 μL of either buffer or enzyme in buffer. Seal with a coverslip and incubate for 30 min at 37 °C. Remove coverslip, rinse slides with PBS and place them on the electrophoretic box (6, 20–22).
8. It is strongly recommended to include some measure of cytotoxicity into any study since increased DNA migration may also occur due to non-genotoxic cell killing. However, such an effect may depend on the cell type used. While no increased DNA migration had been observed in human leukocytes (43) or cell

lines such as V79 (43, 44) and L5178Y (45), TK-6 cells showed increased DNA migration after treatment with non-genotoxic cytotoxins when viability in treated cultures fell below 75 % (46). Therefore, acute cytotoxic effects should be determined by Trypan blue exclusion measurements or fluorochrome-mediated viability tests. Furthermore, individual dead or dying cells may be identified by their specific microscopical image, i.e., necrotic or apoptotic cells may result in comets with small or nonexistent head and large, diffuse tails (47). These cells are commonly called "hedgehogs," "ghost cells," "clouds," or "non-detectable cell nuclei (NDCN)." Such cells have been detected after treatment with cytotoxic, non-genotoxic agents (44, 46, 48). However, since these microscopic images are also seen after treatment with high doses of radiation or high concentrations of strong mutagens, such comets are not uniquely diagnostic for apoptosis/necrosis (49, 50). For the evaluation of genotoxic effects, it is recommended to record these cells but to exclude them from image analysis under the principle that they represent dead cells.

References

1. Dhawan A, Bajpayee M, Parmar D (2009) Comet assay: a reliable tool for the assessment of DNA damage in different models. Cell Biol Toxicol 25:5–32
2. Tice RR, Agurell E, Anderson D, Burlinson B, Hartmann A, Kobayashi H, Miyamae Y, Rojas E, Ryu J-C, Sasaki YF (2000) The single cell gel/comet assay: guidelines for in vitro and in vivo genetic toxicology testing. Environ Mol Mutagen 35:206–221
3. Hartmann A, Agurell E, Beevers C, Brendler-Schwaab S, Burlinson B, Clay P, Collins A, Smith A, Speit G, Thybaud V, Tice RR (2003) Recommendations for conducting the in vivo alkaline Comet assay. Mutagenesis 18:45–51
4. Brendler-Schwaab S, Hartmann A, Pfuhler S, Speit G (2005) The in vivo comet assay: use and status in genotoxicity testing. Mutagenesis 20:245–254
5. Burlinson B, Tice RR, Speit G, Brendler-Schwaab SY, Collins AR et al (2007) Fourth International Workshop on Genotoxicity Testing: results of the in vivo Comet assay workgroup. Mutat Res 627:31–35
6. Collins AR (2009) Investigating oxidative DNA damage and its repair using the comet assay. Mutat Res 681:24–32
7. Jah AN (2008) Ecotoxicological applications and significance of the comet assay. Mutagenesis 23:207–221
8. Frenzilli G, Nigro M, Lyons BP (2009) The Comet assay for the evaluation of genotoxic impact in aquatic environments. Mutat Res 681:80–92
9. Dusinska M, Collins AR (2008) The comet assay in human biomonitoring: gene-environment interactions. Mutagenesis 23:191–205
10. Valverde M, Rojas E (2009) Environmental and occupational biomonitoring using the Comet assay. Mutat Res 681:93–109
11. Baumgartner A, Cemeli E, Anderson D (2009) The comet assay in male reproductive toxicology. Cell Biol Toxicol 25:81–98
12. Olive PL (2009) Impact of the comet assay in radiobiology. Mutat Res 681:13–23
13. Singh NP, McCoy MT, Tice RR, Schneider EL (1988) A simple technique for quantification of low levels of DNA damage in individual cells. Exp Cell Res 175:184–191
14. Olive PL (1989) Cell proliferation as a requirement for development of contact effect in Chinese hamster V79 spheroids. Radiat Res 117:79–92
15. Klaude M, Erikson S, Nygren J, Ahnström G (1996) The comet assay: mechanisms and technical considerations. Mutat Res 363:89–96
16. Frieauff W, Hartmann A, Suter W (2001) Automatic analysis of slides processed in the comet assay. Mutagenesis 16:133–137

17. Wood DK, Weingeist DM, Bhatia SN, Engelward BP (2010) Single cell trapping and DNA damage analysis using microwell arrays. Proc Natl Acad Sci U S A 107:10008–10013
18. Stang A, Brendamour M, Schunck C, Witte I (2010) Automated analysis of DNA damage in the high-throughput version of the comet assay. Mutat Res 698:1–5
19. Speit G, Hartmann A (1995) The contribution of excision repair to the DNA-effects seen in the alkaline single cell gel test (comet assay). Mutagenesis 10:555–559
20. Collins AR, Duthie SJ, Dobson VL (1993) Direct enzymic detection of endogenous oxidative base damage in human lymphocyte DNA. Carcinogenesis 14:1733–1735
21. Dennog C, Hartmann A, Frey G, Speit G (1996) Detection of DNA damage after hyperbaric oxygen (HBO) therapy. Mutagenesis 11:605–609
22. Speit G, Schütz P, Bonzheim I, Trenz K, Hoffmann H (2004) Sensitivity of the FPG protein towards alkylation damage in the comet assay. Toxicol Lett 146:151–158
23. Pfuhler S, Wolf HU (1996) Detection of DNA-crosslinking agents with the alkaline comet assay. Environ Mol Mutagen 27:196–201
24. Merk O, Speit G (1999) Detection of crosslinks with the comet assay in relationship to genotoxicity and cytotoxicity. Environ Mol Mutagen 33:167–172
25. Fuscoe JC, Afshari AJ, George MH, DeAngelo AB, Tice RR, Salman T, Allen JW (1996) In vivo genotoxicity of dichloroacetic acid: evaluation with the mouse peripheral blood micronucleus assay and the single cell gel assay. Environ Mol Mutagen 27:1–9
26. Collins AR, Ai-guo A, Duthie SJ (1995) The kinetics of repair of oxidative DNA damage (strand breaks and oxidised pyrimidines) in human cells. Mutat Res 336:69–77
27. Collins AR, Dusinska M, Horvathova E, Munro E, Savio M, Stetina R (2001) Inter-individual differences in repair of DNA base oxidation, measured in vitro with the comet assay. Mutagenesis 16:297–301
28. Hartmann A, Speit G (1996) The effect of arsenic and cadmium on the persistence of mutagen-induced DNA lesions in human cells. Environ Mol Mutagen 27:98–104
29. Hartmann A, Speit G (1995) Genotoxic effects of chemicals in the single cell gel (SCG) test with human blood cells in relation to the induction of sister chromatid exchanges (SCE). Mutat Res 346:49–56
30. Langie S, Knaapen AD, Brauers K, van Berlo D, van Schooten F-J, Godschalk WL (2006) Development and validation of a modified comet assay to phenotypically assess nucleotide excision repair. Mutagenesis 21:153–158
31. Gaivao I, Piasek A, Brevik A, Shaposhnikov S, Collins AR (2009) Comet assay-based methods for measuring DNA repair in vitro; estimates of inter- and intra-individual variation. Cell Biol Toxicol 25:45–52
32. Collins AR, Harrington V, Drew J, Melvin R (2003) Nutritional modulation of DNA repair in a human intervention study. Carcinogenesis 24:511–515
33. Gedik CM, Ewen SWB, Collins AR (1992) Single-cell gel electrophoresis applied to the analysis of UV-C damage and its repair in human cells. Int J Radiat Biol 62:313–320
34. Green MHL, Lowe JE, Harcourt SA, Akinluyi P, Rowe T, Cole J, Anstey AV, Arlett CF (1992) UV-C sensitivity of unstimulated and stimulated human lymphocytes from normal and xeroderma pigmentosum donors in the comet assay: a potential diagnostic technique. Mutat Res 273:137–144
35. Helbig R, Speit G (1997) DNA effects in repair-deficient V79 Chinese hamster cells studied with the comet assay. Mutat Res 377:279–286
36. Tebbs RS, Flannery ML, Meneses JJ, Hartmann A, Tucker JD, Thompson LH, Cleaver JE, Pedersen RA (1999) Requirement for the Xrcc1 DNA base excision repair gene during early mouse development. Dev Biol 208:513–529
37. Shaposhnikov S, Frengen E, Collins AR (2009) Increasing resolution of the comet assay using fluorescent in situ hybridization—a review. Mutagenesis 24:383–389
38. Glei M, Hovhannisyan G, Pool-Zobel BL (2009) Use of Comet-FISH in the study of DNA damage and repair: review. Mutat Res 681:33–43
39. Spivak G, Cox RA, Hanawalt PC (2009) New applications of the Comet assay: Comet-FISH and transcription-coupled DNA repair. Mutat Res 681:44–50
40. Lovell DP, Omori T (2008) Statistical issues in the use of the comet assay. Mutagenesis 23:171–182
41. Wiklund SJ, Agurell E (2003) Aspects of design and statistical analysis in the Comet assay. Mutagenesis 18:167–175
42. Speit G, Trenz K, Schütz P, Rothfuss A, Merk O (1999) The influence of temperature during alkaline treatment and electrophoresis on results obtained with the comet assay. Toxicol Lett 110:73–78
43. Hartmann A, Speit G (1997) The contribution of cytotoxicity to effects seen in the alkaline comet assay. Toxicol Lett 90:183–188

44. Hartmann A, Kiskinis E, Fjaellman A, Suter W (2001) Influence of cytotoxicity and compound precipitation on test results in the alkaline comet assay. Mutat Res 497:199–212
45. Kiskinis E, Suter W, Hartmann A (2002) High-throughput comet assay using 96-well plates. Mutagenesis 17:37–43
46. Henderson L, Wolfreys A, Fedyk J, Bourner C, Windebank S (1998) The ability of the comet assay to discriminate between gentoxins and cytotoxins. Mutagenesis 13:89–94
47. Olive PL, Banath JP (1995) Sizing highly fragmented DNA in individual apoptotic cells using the comet assay and a DNA crosslinking agent. Exp Cell Res 221:19–26
48. Kiffe M, Christen P, Arni P (2003) Characterization of cytotoxic and genotoxic effects of different compounds in CHO K5 cells with the comet assay (single-cell gel electrophoresis assay). Mutat Res 537: 151–168
49. Meintieres S, Nesslany F, Pallardy M, Marzin D (2003) Detection of ghost cells in the standard alkaline comet assay is not a good measure of apoptosis. Environ Mol Mutagen 41:260–269
50. Rundell MS, Wagner ED, Plewa MJ (2003) The comet assay: genotoxic damage or nuclear fragmentation? Environ Mol Mutagen 42:61–67

Chapter 7

Detection of DNA Damage by Comet Fluorescence In Situ Hybridization

Wiebke Schlörmann and Michael Glei

Abstract

Comet fluorescence in situ hybridization (Comet-FISH) is a useful method to detect overall and region-specific DNA damage in individual cells. Two well-established methods are combined, the Comet assay (single cell gel electrophoresis) and fluorescence in situ hybridization (FISH). The Comet assay is the method of choice for the detection of DNA damage. With the alkaline version the influence of specific substances such as water pollutants or ingredients of food on individual cells can be easily measured. The Comet assay involves the embedding of cells in agarose on microscopic slides, lysis of cells, and separation of DNA via electrophoresis. Damaged DNA migrates from the nucleus (head of the comet) forming a tail. The percentage of DNA in the tail correlates with the degree of DNA strand breaks (DNA damage). The combination of FISH with the Comet assay uses labeled probes which hybridize specifically to selected DNA sequences. This allows the detection of specific DNA damage or repair capacity in single cells. Here we present exemplarily the Comet-FISH method by detection of DNA damage using hydrogen peroxide as a genotoxic model substrate.

Key words: Comet-FISH, Comet Assay, FISH, DNA damage, Hybridization

1. Introduction

1.1. General Aspects of the Comet-FISH Technique

The Comet fluorescence in situ hybridization (Comet-FISH) technique is a combination of two well-established methods. The first step of this method is the Comet assay (single cell gel electrophoresis) which was described primarily by Östling and Johanson (1) in its neutral version. Single cells are embedded in agarose onto microscopic slides and after cell lysis an electrophoresis step separates damaged DNA from undamaged DNA. Damaged DNA migrates towards the anode forming a tail of a comet while the intact DNA in the nucleus represents its head. So the Comet assay is a relatively simple and fast method that allows separating

Lotte Bjergbæk (ed.), *DNA Repair Protocols*, Methods in Molecular Biology, vol. 920,
DOI 10.1007/978-1-61779-998-3_7, © Springer Science+Business Media New York 2012

fragmented from non-fragmented DNA and quantification of DNA damage by genotoxic compounds and also its repair. It is supposed that the neutral version of this method mainly detects double strand breaks whereas its alkaline variant described by Singh et al. (2) also detects single strand breaks and alkalilabile sites, though this is discussed controversially (3). The combination of the Comet assay with the fluorescence in situ hybridization (FISH) was first described by Santos et al. (4). The combination of this two techniques helps to detect specifically labeled DNA sequences of interest, including whole chromosomes. This modification of the Comet assay is a useful tool to detect overall and region-specific DNA damage and repair in individual cells. The Comet-FISH technique inserts a hybridization step with specific fluorescent labeled probes to particular DNA sequences after unwinding and electrophoresis. This enables the detection of specific gene sequences or telomeres. The combination of both techniques can also be used for detection of site-specific breaks in DNA regions which are relevant for the development of different diseases. DNA sequence-modifications as well as the distribution of DNA damage and repair in the complete genome can be detected at the level of individual cells. Comet-FISH also allows the assignment of the probed sequences to the damaged or undamaged part of the comet (tail or head, respectively). In contrast the Comet assay alone comprises only overall DNA damage.

For more information about the Comet-FISH technique and its modifications for different applications see Glei et al. (5). Schaeferhenrich et al. (6), Knoebel et al. (7, 8), Glei et al. (9), and Shaposhnikov et al. (10, 11) also provide information about evaluation and interpretation of the Comet-FISH results.

1.2. Application of the Comet-FISH Technique

The classical alkaline Comet assay is a useful and simple method to detect DNA damage in individual cells. The genotoxic effects of specific test compounds can be analyzed and the use of lesion-specific endonucleases makes it possible to analyze different kinds of DNA damage. Furthermore, the capacity of DNA repair can be studied in single cells. The combination of the Comet assay with FISH allows the detection of DNA damage or repair at an even higher level of resolution since fluorescent labeled probes are used which are specific for particular DNA sequences. With the Comet-FISH method further insights in mechanisms of cancer development and chemoprevention can be studied. This technique enables the evaluation on whether the damage and repair are occurring within the vicinity of the probed genes of interest (5). Several probes are described for the use of Comet-FISH (10) and the size of the probes represent a clear limitation for the detection of actually gene-specific effects. The limited resolution of the Comet assay is approximately 10–800 kb using the standard conditions and fragments smaller than 10 kb might get lost by this technique (12). Probes targeting at least 10 kb or longer are necessary for FISH detection of gene-specific loci. The most probes which are commercially available rank from 30 to 100 kb in size and are an assembly of approximately 300 bases covering unique DNA sequences, spanning the target gene (13). Comet-FISH however offers a unique possibility to study the sensitivity of genes toward genotoxic compounds (14, 15) by quantification of migrated DNA fragments.

Two fluorescence signals of a particular gene in the head of a comet indicate that the gene is in an undamaged region of DNA. If one spot or several spots appear in the tail of a comet, this indicates that a break or breaks has/have occurred in the proximity of the probed gene (Fig. 1). The presence of a scaffold- or matrix-associated region in or near the gene (16) can be supposed if certain parts of genes remain in the head even when there is a DNA break nearby. This was observed for example for the dihydrofolate reductase gene (DHFR) from Chinese hamster ovary (CHO) cells upon treatment with H_2O_2 or a photosensitizer plus light (17).

The classical FISH protocol has been adjusted to the specific experimental requirements of DNA embedded in agarose. Kelvey-Martin et al. (15) used thermal co-denaturation of probe and DNA (74 °C, 5 min); however, Rapp et al. (18) suggested chemical denaturation to separate the target strand DNA. The parameters for denaturation must be chosen in such a way that the gel matrix is not damaged. Also conditions for further steps such as washing, signal amplification, and microscopical analysis need to be optimized for the Comet-FISH protocol. Additionally, optimized DNA probes can be used to ameliorate the hybridization efficiency, which should be always determined (6). Such probes need to have a higher DNA concentration and a reduced size compared to the probes used for conventional FISH on metaphase chromosomes (18). Padlock probes for example are very stable as well as specific and have the advantage that the reactions can be performed at 37 °C which avoids a gel matrix damage (10). For example, Rapp et al. (18) described two versions of Comet-FISH, one based on the alkaline version, and the other on the neutral version of the Comet assay. Though the Comet-FISH method is a relative rapid and simple method to study DNA damage and its repair throughout the genome and in defined chromosomal regions (19) on the level of individual cells, there are only a few further developments described. Shaposhnikov et al. (20) for example presented new applications of Comet-FISH by increasing the output and sample size with a 12-gel slide format. They also reviewed methodological features of the Comet-FISH technique (10, 11). Beyond that, Hovhannisyan (21) described not only the combination of FISH with the Comet assay, but also with the micronucleus test.

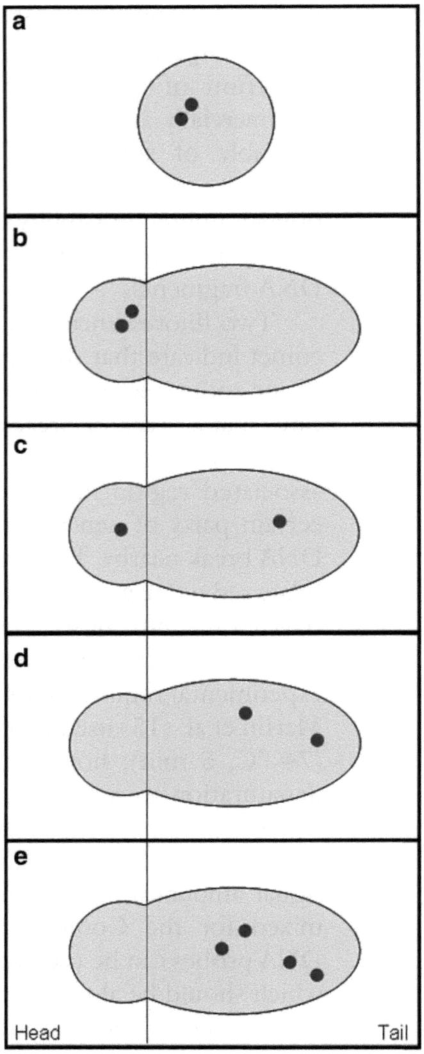

Fig. 1. Schematic overview of possible breakage pattern and resulting signal distribution after Comet-FISH: a) undamaged DNA; b) damaged DNA without a break in the vicinity of the specifically labeled DNA sequence of interest; c, d) damaged DNA with one or two signals clustered in the tail; e) damaged DNA with more than two signals resulting from strand breaks in the vicinity of the specifically labeled DNA of interest clustered in the tail.

2. Materials

2.1. Pre-coating of Slides

1. Glass slides.
2. Glass cover slips.
3. Phosphate buffered saline (PBS): 8.0 g/L NaCl, 0.2 g/L KCl, 0.2 g/L KH_2HPO_4, 1.15 g/L $Na_2HPO_4 \times 2H_2O$, pH 7.3. Store at 4 °C.

4. Normal Melting Agarose (NMA; 0.5 %): 150 mg NMA in 30 ml PBS (microwave). Store at 4 °C (2–3 weeks, see Note 1).
5. Water quench.
6. Heating plate.
7. Ice, icebox.

2.2. Treatment of Cells with Test Compounds (e.g., Genotoxic Model Chemicals Like H_2O_2)

1. H_2O_2 (see Note 2).
2. Pipettors and tips.
3. Vortex mixer.
4. Tubes.

2.3. Comet Assay (Detection of DNA Damage)

1. Precoated slides from Subheading 2.1.
2. Low Melting Agarose (LMA; 0.7 %): 210 mg LMA in 30 ml PBS (microwave). Store at 4 °C (see Note 1).
3. Glass cover slips.
4. Lysis stock solution: 146.1 g/L NaCl, 37.2 g/L Na_2EDTA, 1.2 g/L Tris-Base, 8 g/L NaOH, 10 g/L *N*-lauroyl sarcosin sodium salt, pH 10. Store at room temperature.
5. Lysis working solution: 1 % Triton-X, 10 % DMSO, 89 % lysis stock solution. Store at 4 °C, prepare freshly.
6. Electrophoresis chamber (e.g., Renner GmbH, Dannstadt, Germany).
7. Electrophoresis stock solution: 400 g/L NaOH, 74.4 g/L Na_2EDTA.
8. Electrophoresis working solution: 60 ml NaOH stock solution, 10 ml Na_2EDTA stock solution, 1,930 ml distilled water, store at 4 °C, prepare freshly.
9. Neutralization buffer: 4.2 M Tris–HCl, 0.08 M Tris-base, pH 7.2.
10. TE-buffer: 10 mM Tris–HCl, 1 mM EDTA, pH 8. Store at room temperature.
11. Antifade buffer: 2.5 g DABCO, 50 ml TE-buffer, 50 ml glycerol, store at 4 °C, protect from light.
12. SYBR green [1:10,000]: 1 μl SYBR®-Green, 9,999 μl antifade buffer, store at 4 °C, protect from light.
13. Ethanol (100 %).

2.4. FISH: Preparation of Slides

1. 0.5 M NaOH.
2. PBS.
3. Ethanol (70, 80, 95, 100 %).
4. Hybrisol VI (Oncor, Gaithersburg, UK).
5. Plastic cover slips.

2.5. FISH: Hybridization

1. Digoxigenin labeled DNA probes (e.g., Oncor, Gaithersburg, UK).
2. Hybrisol VI (Oncor, Gaithersburg, UK).
3. Water quench.
4. Plastic cover slips.
5. Hybridization chamber.
6. Saline sodium citrate (2× SSC): 0.3 M NaCl, 0.03 M sodium citrate, pH 7.2.
7. 1× Phosphate buffered detergent (PBD) (Oncor, Gaithersburg, UK).

2.6. Detection of Digoxigenin-Labeled Probes

1. Anti-digoxigenin-AP Fab-fragments (Roche, Mannheim, Germany).
2. HNPP-Fluorescence Detection Kit (Roche, Mannheim, Germany).
3. SYBR®-Green (1 μl/10 ml).
4. Fluorescence microscope (e.g., ZEISS Axiovert M100, Carl Zeiss Jena GmbH, Jena, Germany).
5. Fluorescence filter (green and red, e.g., ZEISS filter No. 09 and 15).
6. Digital CCD camera and imaging software (e.g., MicroMAX, BFI OPTILAS GmbH; Visitron Systems GmbH, Puchheim, Germany).
7. Image analysis system (e.g., Kinetic Imaging, Liverpool, UK or Perceptive Instruments Suffolk, UK).

3. Method

3.1. Pre-coating of Slides

1. Dissolve 0.5 % NMA at 60 °C (water quench, see Note 1).
2. Drop 50 μl of 0.5 % NMA on one side of each slide and distribute NMA with another slide by dragging it to the opposite side.
3. Dry slides at ~60 °C (heating plate).
4. Coat slides with another 200 μl of 0.5 % NMA (see Note 3).
5. Cover slides with cover slips rapidly to allow a homogeneous disposition of NMA.
6. Cool slides for 10 min on an icebox (see Note 4).
7. Store slides at 4 °C (see Note 5).

3.2. Treatment of Cells with Test Compounds (e.g., Genotoxic Model Chemicals Like H_2O_2)

1. Dilute, e.g., H_2O_2 in PBS to favored concentrations, vortex.
2. Incubate cell suspension, e.g., HT29 cells (2×10^6 cells/ml) with H_2O_2 solution for 5 min at 4 °C.
3. Remove H_2O_2 solution by centrifugation at $400 \times g$ for 5 min.
4. Discard the supernatant.

3.3. Comet Assay (Determination of DNA Damage)

1. Dissolve 0.7 % LMA at 40 °C (water quench, see Note 1) and resuspend cell pellet in a volume of 50 μl.
2. Distribute 50 μl of the cell suspension onto pre-coated slides, cover them with cover slips, and allow agarose to solidify for 10 min on an icebox.
3. Remove cover slips and immerse slides in lysis solution for at least 60 min at 4 °C (see Note 6).
4. Place slides into an electrophoresis chamber containing alkaline electrophoresis buffer and incubate them for 20 min.
5. Carry out electrophoresis at 1.25 V/cm and 300 mA for 20 min by adjusting the total volume of electrophoresis buffer.
6. Remove slides from the electrophoresis chamber and wash them three times for 5 min in neutralization buffer.
7. Stain slides with SYBR Green (1 μl/10 ml, 30 μl per slide) and determine DNA damage or continue with step 8.
8. Dehydrate slides in abs. EtOH for at least 3 days.

3.4. FISH: Preparation of Slides

1. Rehydrate slides in dd H_2O for 10 min.
2. Denature DNA in 0.5 M NaOH for 30 min.
3. Carry out neutralization in PBS for 1 min.
4. Dehydrate slides in an ascending EtOH series: 70, 80, 95 % each for 5 min.
5. Dry slides at room temperature.
6. Drop 30 μl Hybrisol VI onto each slide and spread it with a plastic cover slip.

3.5. FISH: Hybridization

1. Prepare the hybridization mixture containing 10 μl digoxigenin labeled probe (commercially acquired) and 20 μl Hybrisol VI.
2. Denature hybridization mixture at 37 °C for 5 min (water quench) (see Note 7).
3. Apply hybridization mixture onto one half of each slide and Hybrisol VI as negative control on the other half and cover it with a plastic cover slip.
4. Incubate slides at 37 °C in hybridization chambers for 24–72 h.
5. Wash slides in 2× SSC at 72 °C for 5 min.
6. Wash slides in 1× PBD for 5 min at room temperature.

3.6. Detection of Digoxigenin-Labeled Probes

1. For detection of digoxigenin-labeled probes, use anti-digoxigenin-AP-Fab fragments and 2-hydroxy-3-naphthoicacid-20-phenylanilide-phosphate (HNPP) fluorescent detection set (Roche, Mannheim, Germany) according to the manufacturer's instructions.
2. Use SYBR-Green (1 µl/10 ml; 30 ml per slide) to counterstain the HNPP-detected probes (see Note 8).
3. Evaluate slides with a fluorescence microscope (e.g., ZEISS Axiovert M100, Carl Zeiss Jena GmbH, Jena, Germany), equipped with filters to score images stained with SYBR Green and HNPP or rather HNP/TR (see manufacturer's instructions) (green and red, respectively) (see Notes 9–11).
4. For capturing images, use a Digital CCD camera (e.g., MicroMAX, BFI OPTILAS GmbH, Puchheim, Germany equipped with the Meta View Imaging Software, Visitron Systems GmbH, Puchheim, Germany).

3.7. Evaluation

1. Determine hybridization efficiency (6) (see Note 12).
2. Categorize comets into four degrees of damage ranging from undamaged images (class 1) to severely damaged images (class 4) (22) and/or by determination of the tail intensity with an image analysis system (e.g., Perceptive Instruments, Suffolk, UK)
3. For the further Comet-FISH evaluation, only use cells with two fluorescent spots and record the position of the signals in the comet head or comet tail (Fig. 1).
4. Record the percentage of damaged cells (belonging to comet classes 2–4), in which it is possible to find a migration of at least one signal into the comet tail as the parameter of migration.
5. Score about 100 cells per slide (see Notes 11 and 12).

4. Notes

1. NMA and LMA are dissolved by heating in a microwave and kept fluid in a water quench. NMA and LMA can be stored for 2–3 weeks at 4 °C.
2. H_2O_2 can be used in a favored concentration (usually 75 µM).
3. Pre-coating of the slides is more homogeneous when slides stay on the heating plate at lower temperature (~30 °C) while the second agarose coat is applied.
4. A metal box filled with ice is useful to cool down pre-coated slides. Cooling causes a more homogeneous coating of the slides and less background signals.

5. The slides can be stored for maximal 1 week at 4 °C in a humidified chamber.
6. The slides can also be lysed over night at 4 °C.
7. Denaturation procedure depends on DNA probe used, for self made dig-labeled samples denature for 5 min at 72 °C, 4 min on ice, 30 min at 37 °C and mix 5 μl sample with 10 μl Hybrisol VI (7).
8. If the fluorescence signals are too weak dilute SYBR Green 1:1,000 instead of 1:10,000.
9. In the case that the hybridization signals are too weak reduce SSC concentration to improve the stringency of the washing steps and wash for a longer period of time.
10. Peroxides may induce a fading of SYBR Green. So it could be necessary to use higher concentrations of DABCO Antifade and/or to wash the slides before staining.
11. If there are too many background signals which disturb evaluation, rehydrate slides for a longer period of time before staining.
12. Improvement of the hybridization efficiency might be achieved by alterations in the hybridization temperature and duration as well as by varying the temperature and stringency of the washing steps (15 min in ddH_2O).

References

1. Őstling O, Johanson KJ (1984) Microelectrophoretic study of radiation-induced DNA damages in individual mammalian cells. Biochem Biophys Res Commun 123:291–298
2. Singh NP, McCoy MT, Tice RR, Schneider EL (1988) A simple technique for quantitation of low levels of DNA damage in individual cells. Exp Cell Res 175:184–191
3. Collins AR, Oscoz AA, Brunborg G, Gaivao I, Giovannelli L, Kruszewski M, Smith CC, Stetina R (2008) The comet assay: topical issues. Mutagenesis 23:143–151
4. Santos SJ, Singh NP, Natarajan AT (1997) Fluorescence in situ hybridization with comets. Exp Cell Res 232:407–411
5. Glei M, Hovhannisyan G, Pool-Zobel BL (2009) Use of Comet-FISH in the study of DNA damage and repair: review. Mutat Res 681:33–43
6. Schaeferhenrich A, Beyer-Sehlmeyer G, Festag G, Kuechler A, Haag N, Weise A, Liehr T, Claussen U, Marian B, Sendt W, Scheele J, Pool-Zobel BL (2003) Human adenoma cells are highly susceptible to the genotoxic action of 4-hydroxy-2-nonenal. Mutat Res 526: 19–32
7. Knobel Y, Glei M, Weise A, Osswald K, Schaferhenrich A, Richter KK, Claussen U, Pool-Zobel BL (2006) Uranyl nitrilotriacetate, a stabilized salt of uranium, is genotoxic in nontransformed human colon cells and in the human colon adenoma cell line LT97. Toxicol Sci 93:286–297
8. Knobel Y, Weise A, Glei M, Sendt W, Claussen U, Pool-Zobel BL (2007) Ferric iron is genotoxic in non-transformed and preneoplastic human colon cells. Food Chem Toxicol 45:804–811
9. Glei M, Schaeferhenrich A, Claussen U, Kuechler A, Liehr T, Weise A, Marian B, Sendt W, Pool-Zobel BL (2007) Comet fluorescence in situ hybridization analysis for oxidative stress-induced DNA damage in colon cancer relevant genes. Toxicol Sci 96:279–284
10. Shaposhnikov S, Frengen E, Collins AR (2009) Increasing the resolution of the comet assay using fluorescent in situ hybridization–a review. Mutagenesis 24:383–389

11. Shaposhnikov S, Thomsen PD, Collins AR (2011) Combining fluorescent in situ hybridization with the comet assay for targeted examination of DNA damage and repair. Methods Mol Biol 682:115–132
12. Rapp A, Bock C, Dittmar H, Greulich KO (2000) UV-A breakage sensitivity of human chromosomes as measured by COMET-FISH depends on gene density and not on the chromosome size. J Photochem Photobiol B 56:109–117
13. Ooi FA (2001) Oncogene amplification detection by fluorescence in situ hybridization (FISH). Acta Histochem Cytochem 34:391–397
14. Bock C, Rapp A, Dittmar H, Monajembashi S, Greulich KO (1999) Localisation of specific sequences and DNA single strand breaks in individual UV-A irradiated human lymphocytes by COMET FISH. Prog Biomed Optics 3568:207–217
15. Kelvey-Martin VJ, Ho ET, McKeown SR, Johnston SR, McCarthy PJ, Rajab NF, Downes CS (1998) Emerging applications of the single cell gel electrophoresis (Comet) assay. I. Management of invasive transitional cell human bladder carcinoma. II. Fluorescent in situ hybridization Comets for the identification of damaged and repaired DNA sequences in individual cells. Mutagenesis 13:1–8
16. Anderson D, Yu TW, Browne MA (1997) The use of the same image analysis system to detect genetic damage in human lymphocytes treated with doxorubicin in the Comet and fluorescence in situ hybridisation (FISH) assays. Mutat Res 390:69–77
17. Horvathova E, Dusinska M, Shaposhnikov S, Collins AR (2004) DNA damage and repair measured in different genomic regions using the comet assay with fluorescent in situ hybridization. Mutagenesis 19:269–276
18. Rapp A, Hausmann M, Greulich KO (2005) The comet-FISH technique: a tool for detection of specific DNA damage and repair. Methods Mol Biol 291:107–119
19. Spivak G (2010) The Comet-FISH assay for the analysis of DNA damage and repair. Methods Mol Biol 659:129–145
20. Shaposhnikov S, Azqueta A, Henriksson S, Meier S, Gaivao I, Huskisson NH, Smart A, Brunborg G, Nilsson M, Collins AR (2010) Twelve-gel slide format optimised for comet assay and fluorescent in situ hybridisation. Toxicol Lett 195:31–34
21. Hovhannisyan GG (2010) Fluorescence in situ hybridization in combination with the comet assay and micronucleus test in genetic toxicology. Mol Cytogenet 3:17
22. Wollowski I, Ji ST, Bakalinsky AT, Neudecker C, Pool-Zobel BL (1999) Bacteria used for the production of yogurt inactivate carcinogens and prevent DNA damage in the colon of rats. J Nutr 129:77–82

Chapter 8

Quantitative DNA Damage and Repair Measurement with the Yeast Comet Assay

Rui Oliveira and Björn Johansson

Abstract

The yeast comet assay is a fast, sensitive, and inexpensive technique to measure oxidative DNA damage, DNA damage repair, and the genotoxic or protective effects of chemicals. The main advantage over the comet assay using cells of higher organisms is the genetic tractability and ease of cultivation of yeast. A drawback is the lower DNA content of the cells as well as the need for cell wall digestion prior to electrophoresis. Here, we describe in detail a recently developed protocol that permits sensitive and reproducible measurement of DNA damage and DNA repair using *Saccharomyces cerevisiae* as model system. The combination of this assay with yeast mutants affected in genome maintenance and the wide selection of available yeast molecular biology tools can contribute to illuminate fundamental mechanisms of DNA damage, repair, and activity of DNA protective compounds.

Key words: Comet assay, *Saccharomyces cerevisiae*, DNA damage, Oxidative stress, Genome integrity, DNA damage repair

1. Introduction

The comet assay, also known as single-cell gel electrophoresis, is an assay for detection of DNA damage in single cells. Single-strand breaks are indirectly detected through the relaxation of the negative supercoiling of chromosomal DNA. The relaxation is detected by higher migration rates of relaxed DNA under a briefly applied low-intensity electric field (1). In practical terms, cells are mixed with liquid low-melting agarose (LMA), spread on the surface of a microscopy slide in order to obtain cells embedded in a thin agarose gel. Cells are lysed and DNA is forced to migrate out of the nucleoid by electrophoresis. The DNA is stained with a fluorescent dye and each cell is observed using fluorescence microscopy. Each agarose-trapped cell will look like a comet with a head constituted

Lotte Bjergbæk (ed.), *DNA Repair Protocols*, Methods in Molecular Biology, vol. 920,
DOI 10.1007/978-1-61779-998-3_8, © Springer Science+Business Media New York 2012

by the nucleoid and a tail of migrated genomic DNA. The length of the tail and/or the amount of DNA in the tail is proportional to the number of DNA strand breaks.

The DNA bases may be damaged by oxidation, leaving the sugar phosphate backbone intact, so oxidative DNA damage does not necessarily directly result in single-strand breaks detectable by the comet assay. The most common oxidized base formed upon reaction of reactive oxygen species with DNA is 7,8-dihydro-8-oxoguanine (2). This kind of oxidative damage often leads to abasic sites repaired by the nucleotide or base excision repair pathway (3). These repair pathways introduce a single-strand break as an intermediate product that can be detected by the comet assay as an indirect indication of oxidative DNA base damage.

Oxidative attack to the deoxyribose can cause break in N-glycosyl bonds leading to the formation of abasic sites, strand breaks (4), or oxidation at C1′, C2′, or C4′. Abasic sites and oxidation at C1′, C2′, and C4′ can lead to the so-called alkali-labile sites that under alkaline conditions spontaneously undergo hydrolysis of the phosphodiester bond at C3′ forming a single-strand break (5). When the comet assay is performed under alkaline conditions, alkali-labile sites can be detected in addition to single-strand breaks.

Double-strand breaks further increase electrophoretic mobility over single-strand breaks due to fragmentation of DNA. Hence, the comet assay is suitable for the assessment of damage caused by virtually all DNA-damaging agents that cause oxidative damage and DNA breaks (single or double). Östling and Johanson (6) implemented this technique the first time with pH-neutral conditions and Singh et al. (7) adapted the comet assay to alkaline conditions with improved sensitivity.

Simplicity and low reagent cost have greatly contributed to the popularity of this method. Applications include human genotoxicology studies, assessment of genome integrity in sperm for *in vitro* fertilization, and monitorization of environmental pollution by DNA-damaging agents in ecotoxicology. The possibility of analysis of DNA damage at cell level and the robustness provided by the statistical analysis of high number of comets in each sample have contributed to the applicability of the method.

Fundamental studies on mechanisms of DNA damage and repair usually are supported by results obtained with other techniques, involving *in vitro* approaches. In *in vitro* systems, reactions leading to DNA damage and repair are performed without the "native intracellular environment." In addition, these techniques do not provide information of damage at single-cell level, so heterogeneity of response by a population of cells cannot be analyzed. Here we present a protocol of the comet assay applied to the model organism *Saccharomyces cerevisiae*, which can widen the range of applications of this assay to fundamental studies and applied research in mechanistic studies of activity of DNA-protecting

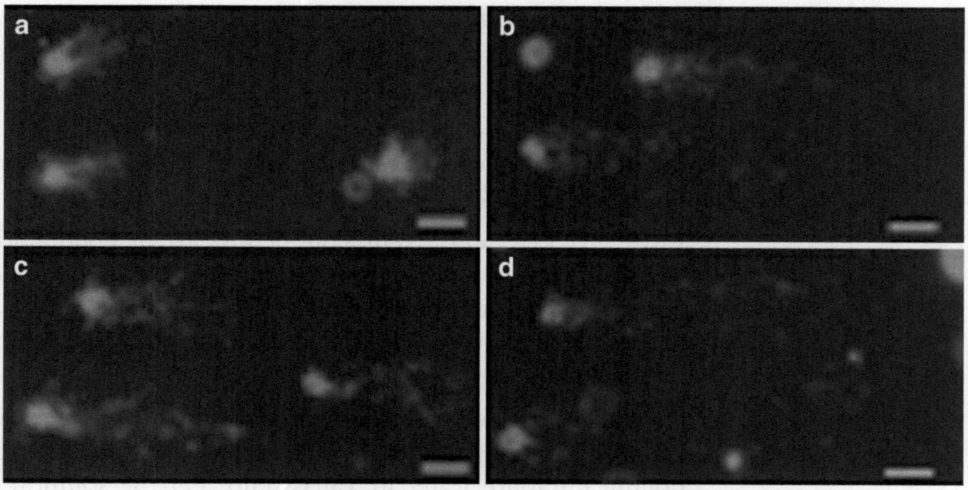

Fig. 1. Photomicrographs of yeast comets from nonexposed cells to the DNA-damaging agent (**a**), exposed to 300 μM (**b**), 500 μM (**c**), and 1 mM (**d**) $KMnO_4$. For visualization, DNA was stained with GelRed™ and observed at ×400 magnification. Bar = 10 μm.

compounds. The use of yeast cells as model can contribute for this given the genetic amenability to create mutants and other available molecular tools.

The use of lower eukaryotes and prokaryotes in the comet assay has been hampered by the low cellular DNA content that causes irregular comet tail shapes, increasing the difficulty in the implementation of quantification systems. Our improvements in the protocol of the alkaline version with lower pH and milder chemical conditions, such as lower concentration of detergent in cell lysis and increased concentration of LMA as electrophoretic matrix, allowed us to obtain reproducible and quantifiable results. Therefore, we were able to obtain a dose–response effect (Fig. 1) with different genotoxic compounds and to study kinetics of DNA repair in wild-type cells and mutant strains affected in DNA repair mechanisms (8, 9).

2. Materials

1. YPD: Mix 1 % (w/v) yeast extract, 2 % (w/v) peptone, and 2 % (w/v) glucose with an appropriate volume of deionized water, stir to dissolve, and autoclave at 120 °C, 1 atm for 20 min.
2. Zymolyase buffer: Dissolve 2 mg Zymolyase (ImmunO™—20 T) in 1 mL S buffer (see composition below) (see Note 1) and 50 mM β-mercaptoethanol. Prepare fresh solution just before use.

3. S buffer: 1 M sorbitol, 25 mM KH_2PO_4 in ultrapure water and adjust to pH 6.5 with NaOH. Autoclave at 120 °C, 1 atm for 20 min.
4. 10 mM H_2O_2: 1 M solution in S buffer and store at −20 °C in the dark. Before each experiment dilute stock solution serially to 10 mM with S buffer.
5. LMA: 1.5 % (w/v) LMA in S buffer, dissolve by boiling briefly, let solidify at room temperature, and store at 4 °C. Before each experiment warm up at 90–100 °C to liquefy, incubate at 35 °C in a water bath, and use only after stabilization at this temperature (see Note 2).
6. Glass slides covered with agarose: Dissolve 0.5 % (w/v) agarose (multipurpose agarose with normal melting temperature) in boiling ultrapure water in a glass cup. Let the solution cool down to 50–55 °C, dip glass slides in the melted agarose, and let drain excess agarose. Keep glass slides in horizontal position, clean the lower surface with absorbent paper, and let stand on the laboratory bench until cooled down. These slides can be prepared in advance and stored at 4 °C for several weeks or months (see Note 3).
7. Coverslips: Use 22 × 22 mm, 0.13–0.17 mm thick coverslips. Each glass slide can accommodate two samples with these coverslips.
8. Lysing buffer: 30 mM NaOH, 1 M NaCl, 0.05 % (w/v) lauroylsarcosine, 50 mM EDTA, and 10 mM Tris–HCl, pH 10. This buffer should be prepared just before use by mixing appropriate volumes of stock solutions of 300 mM NaOH, 5 M NaCl, 500 mM EDTA, and 100 mM Tris–HCl. Stock solutions may be stored at 4 °C at least for several weeks. Lauroylsarcosine should be dissolved in the end and the solution should be gently stirred in order to avoid foam formation. Keep at 4 °C until use.
9. Electrophoresis buffer: 30 mM NaOH, 10 mM EDTA, and 10 mM Tris–HCl, pH 10. This buffer should be prepared just before use with the same stock solutions used for lysing buffer. Keep at 4 °C until use.
10. Neutralization buffer: 10 mM Tris–HCl, pH 7.4. This buffer can be prepared in advance and stored at 4 °C.
11. Ethanol 76 % and 96 %: Mix appropriate volumes of ethanol and deionized water to obtain 76 % and 96 %v/v. Store at 4 °C.
12. GelRed™: Dilute GelRed™ Nucleic Acid Stain (10,000× in water) (Biotium, Inc, Hayward, CA, USA) with ultrapure water to obtain 1× solution. Store in the dark at room temperature (see Note 4).
13. Comet Score™ v1.5: This is a freeware that can be downloaded at http://autocomet.com/products_cometscore.php (see Note 5).

3. Methods

1. Prepare a pre-inoculum of 5 mL YPD with yeast cells from a fresh (1–4 days old) stock culture on solid YPD medium. Incubate overnight at 30 °C, 200 rpm, and 1:4–1:10 air:liquid ratio.
2. Dilute the pre-inoculum with fresh YPD medium to $OD_{600} = 0.1$. Make 25 mL final volume in a 100 mL Erlenmeyer flask and incubate at 30 °C, 200 rpm, until $OD_{600} = 0.4–0.8$ (see Note 6).
3. Harvest cells by centrifugation at 5,000 × *g*, 4 °C for 2 min and wash twice with the same volume of deionized water at 4 °C.
4. Collect 1 mL suspension by centrifugation at 5,000 × *g*, 4 °C for 2 min, resuspend in Zymolyase (ImmunO™—20 T) buffer, and incubate at 30 °C, 200 rpm, for 30 min in order to obtain spheroplasts (see Note 7).
5. Collect spheroplasts by centrifugation at 5,000 × *g*, 4 °C for 2 min, wash with the same volume of ice-cold S buffer, resuspend in 1 mL 10 mM H_2O_2 (in S buffer) (see Note 8), and incubate for 20 min at 4 °C (see Notes 9–11).
6. Collect spheroplasts by centrifugation at 5,000 × *g*, 4 °C for 2 min, and wash once with the same volume of ice-cold S buffer.
7. After the second centrifugation resuspend carefully the pellet in the same volume of LMA 1.5 % (w/v in S buffer) at 35 °C (see Note 12).
8. Immediately spread 40 μL of each sample onto glass slides pre-coated with 0.5 % (w/v) agarose and cover with a coverslip (see Note 13). Place glass slides on ice to solidify the agarose and to preserve spheroplasts until the subsequent step.
9. Remove the coverslip from the microgels (see Note 14), submerge glass slides in lysing buffer, and incubate at 4 °C for 20 min.
10. Remove the glass slides, drain the excess of lysing buffer, submerge in electrophoresis buffer, and incubate at 4 °C for 20 min.
11. Place the slides in the electrophoresis tank filled with electrophoresis buffer at 4 °C and apply an electric field of 0.7 V/cm for 10 min (see Note 15).
12. Neutralize the microgels by incubation in neutralization buffer at room temperature for 10 min.
13. Drain excess neutralization buffer and incubate samples in ethanol 76 % (v/v) and subsequently in 96 % (v/v), both for 10 min and at room temperature.

14. Let the microgels dry at room temperature, add 10 μL GelRed™, cover with a coverslip, and analyze comets by fluorescence microscopy with ×400 magnification. If microgels are not for immediate observation, store glass slides without dying at 4 °C (see Note 16).
15. Record the tail length of 50 random comets (see Notes 17 and 18) from different parts of the microgel using the freeware Comet Score™ v1.5.

4. Notes

1. Whenever in suspension, spheroplasts are maintained in S buffer containing 1 M sorbitol. Sorbitol acts as osmotic stabilizer of cells deprived of cell wall, which become sensitive to lysis when exposed to hypo-osmotic conditions.
2. Ensure that LMA is, at most, at 35 °C in order to minimize heat shock of spheroplasts.
3. Use frosted glass slides and use the frosted edge to hold them.

   Use a pencil to identify samples throughout manipulations. Do not use water-resistant pens, as subsequent steps of incubations with ethanol will erase any identification.
4. Higher sensitivity is achieved with this fluorescent dye when compared with ethidium bromide. Other advantages include low toxicity, high stability, and safe disposal in the drain. We have also tried SYBR Green to stain DNA with high sensitivity; however photobleaching was higher than in the case of GelRed™.
5. Commercial automatic comet scoring software systems are available for download in the Internet, which are useful for rapid analysis of samples. We have not tried these systems due to the possible erroneous identification of yeast comet tails due to their irregular shape. However, if used, we advise to validate analyses by manually scoring representative samples with Comet Score™ v1.5.
6. This will ensure that the culture has grown for at least two generations. Therefore, young cells at mid-log phase will constitute the majority of the cell population. If cells are collected at higher optical density (provided that the culture is in log phase), adjust OD_{600} to 0.4 in subsequent wash steps as this optical density corresponds to an optimal cell density for observation of microgels without overlapping comets.

7. Include as much as 1 mL aliquots as the number of treatments predicted to assay. For instance, to assay four different toxicant concentrations, collect 4 × 1 mL. You may also scale down the volume of each aliquot because only 40 μL will be transferred to the glass slide. However higher volume than 40 μL is advised in order to facilitate manipulations.

8. For negative control replace H_2O_2 solution by S buffer, which is used as the solvent of the toxicant. From this step on, maintain spheroplasts at 4 °C in order to inhibit DNA repair mechanisms.

9. To analyze antigenotoxicity of compounds, additional incubation steps with these compounds (prepared in S buffer) can be performed before incubation with the toxicant, simultaneously with the toxicant or after toxicant treatment. Compounds should be prepared in an osmotic-protecting vehicle in order to avoid lysis of spheroplasts. The use of 1 M sorbitol, contained in S buffer, ensures this protection. For nonsimultaneous incubations a washing step with ice-cold S buffer is introduced between treatments.

10. For DNA repair analysis several aliquots of the cell suspension are subjected to the same toxicant treatment, washed with ice-cold S buffer, and incubated at 30 °C in S buffer. After incubation of each sample for a given period of time (0, 5, 10 min …), procedure proceeds to step 7.

11. Alternatively, toxicant treatment can be performed on LMA-embedded spheroplasts. In this case, jump to step 7 and, after step 9, cover the microgel with 150–200 μL 10 mM H_2O_2. This incubation should be made at 4 °C; subsequently the reagent should be drained and glass slides dipped in the subsequent reagent. This alternative procedure can also be used for antigenotoxicity and DNA repair analyses.

12. This step should be as quickly as possible so that exposure of cells to 35 °C is minimized.

13. Press gently on top of the coverslip in order to uniformly distribute LMA-embedded spheroplasts and to create a thin microgel. This way, when observing under the microscope, comet structures will be placed at nearly the same focusing distance, which facilitates observation.

14. Press the coverslip gently with a finger and then move sideways in a single movement only hard enough to displace the coverslip without affecting the microgel.

15. Due to the short running time of electrophoresis, it is sufficient to use prechilled electrophoresis buffer and run the electrophoresis at room temperature without affecting significantly the temperature of samples.

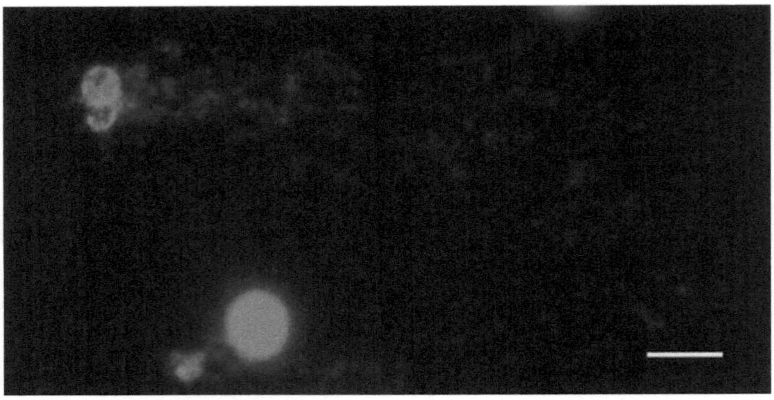

Fig. 2. Photomicrograph of yeast comets from cells exposed to 1 mM $KMnO_4$. For visualization, DNA was stained with GelRed™ and observed at ×400 magnification. Note the upper comet with two "heads" typical of budding cells and the lower cell with undigested cell wall. Bar = 10 μm.

16. Samples can be stored at 4 °C for weeks or months. Fluorescence becomes lower after a few days if fluorescent dye has been added to the microgels. In this case, gently remove the coverslip to avoid damaging the microgel and add new fluorescent dye.
17. We have been recording data from 20 comets without significant variation of results when compared with data from 50 comets.
18. A small fraction of cells are resistant to cell wall enzymatic degradation. These cells are easily distinguishable from comets without DNA tail by their well-defined borders (nucleoids of comets have always irregular edges) (Fig. 2) and intense homogeneous staining. Comets formed from budding cells are easily identified by the "double-headed" comet structure (two nucleoids) (Fig. 2) and should be excluded from analysis.

References

1. Shaposhnikov SA, Salenko VB, Brunborg G et al (2008) Single-cell gel electrophoresis (the comet assay): loops or fragments? Electrophoresis 29:3005–3012
2. Barnes DE, Lindahl T (2004) Repair and genetic consequences of endogenous DNA base damage in mammalian cells. Annu Rev Genet 38:445–476
3. Gellon L, Barbey R, van der Kemp PA et al (2001) Synergism between base excision repair, mediated by the DNA glycosylases Ntgl and Ntg2, and nucleotide excision repair in the removal of oxidatively damaged DNA bases in *Saccharomyces cerevisiae*. Mol Genet Genomics 265:1087–1096
4. Mello-Filho AC, Meneghini R (1984) *In vivo* formation of single-strand breaks in DNA by hydrogen peroxide is mediated by the Haber-Weiss reaction. Biochim Biophys Acta 781: 56–63
5. Povirk LF, Steighner RJ (1989) Oxidized apurinic/apyrimidinic sites formed in DNA by oxidative mutagens. Mutat Res 214:13–22
6. Ostling O, Johanson KJ (1984) Microelectrophoretic study of radiation-induced DNA damages in individual mammalian cells. Biochem Biophys Res Commun 123:291–298
7. Singh NP, McCoy MT, Tice RR et al (1988) A simple technique for quantitation of low levels

of DNA damage in individual cells. Exp Cell Res 175:184–191

8. Azevedo F, Marques F, Fokt H et al (2011) Measuring oxidative DNA damage and DNA repair using the yeast comet assay. Yeast 28:55–61

9. Marques F, Azevedo F, Johansson B et al (2011) Stimulation of DNA repair in *Saccharomyces cerevisiae* by *Ginkgo biloba* leaf extract. Food Chem Toxicol. doi: 10.1016/j.fct.2011.03.020 49:1361–1366

Chapter 9

Analysis of DNA Damage and Repair in Nuclear and Mitochondrial DNA of Animal Cells Using Quantitative PCR

Amy M. Furda, Amanda Smith Bess, Joel N. Meyer, and Bennett Van Houten

Abstract

This chapter was written as a guide to using the long-amplicon quantitative PCR (QPCR) assay for the measurement of DNA damage in mammalian as well as nonmammalian species such as *Caenorhabditis elegans* (nematodes), *Drosophila melanogaster* (fruit flies), and two species of fish (*Fundulus heteroclitus* and *Danio rerio*). Since its development in the early 1990s (Kalinowski et al., Nucleic Acids Res 20:3485–3494, 1992; Salazar and Van Houten, Mutat Res 385:139–149, 1997; Yakes and Van Houten, Proc Natl Acad Sci USA 94:514–519, 1997), the QPCR assay has been widely used to measure DNA damage and repair kinetics in nuclear and mitochondrial genomes after genotoxin exposure (Yakes and Van Houten, Proc Natl Acad Sci USA 94:514–519, 1997; Santos et al., J Biol Chem 278:1728–1734, 2003; Mandavilli et al., Mol Brain Res 133:215–223, 2005). One of the main strengths of the assay is that the labor-intensive and artifact-generating step of mitochondrial isolation is not needed for the accurate measurement of mitochondrial DNA copy number and damage. Below we present the advantages and limitations of using QPCR to assay DNA damage in animal cells and provide a detailed protocol of the QPCR assay that integrates its usage in newly developed animal systems.

Key words: QPCR, DNA damage, DNA repair, Mitochondria, Mitochondrial DNA, mtDNA damage, mtDNA repair, Oxidative stress, Nucleotide excision repair, Base excision repair

1. Introduction

1.1. Principle of the Assay

The idea that many kinds of DNA lesions are able to block DNA polymerase progression is the basis for the quantitative PCR (QPCR) assay (1). The QPCR assay requires preparing samples with equal amounts of DNA so that the only variable influencing fragment amplification is the number of polymerase-blocking DNA lesions such as single-strand breaks, abasic sites, and bulky adducts (1–3). In this way, one can compare lesion frequencies of samples based on amplification alone, in which higher amplification

Lotte Bjergbæk (ed.), *DNA Repair Protocols*, Methods in Molecular Biology, vol. 920,
DOI 10.1007/978-1-61779-998-3_9, © Springer Science+Business Media New York 2012

corresponds with a lower lesion frequency (4, 5). By assuming a Poisson distribution of lesions, the amplification of treated samples is compared to amplification of mock-treated samples to calculate relative lesion frequency, expressed in lesions per kilobase. As the damage is repaired, amplification is restored and thus, the QPCR assay is useful in measuring the kinetics of DNA repair in both nuclear and mitochondrial genomes after treatment with a DNA damaging agent (6–11).

1.2. Advantages of the Assay

Advantages of QPCR in measuring DNA damage and repair include its high sensitivity, robust function despite low amounts of DNA (nanograms), and ability to measure DNA damage and repair in specific genes based on primer design. In addition, the DNA extraction procedure for QPCR allows for both nuclear and mitochondrial genomes to be extracted in the same sample without the need for mitochondrial isolation. Because long fragments of DNA (10–15 kb) are amplified, QPCR is highly sensitive to biologically relevant low levels of lesions (approximately 1 lesion per 10^5 bases) (2, 12). Because QPCR relies on PCR amplification, the assay can be performed on DNA samples containing as little as 1–2 ng of genomic DNA, and we have even performed this assay on the equivalent of a single nematode (<1,000 cells, and <1 ng DNA) (13). Thus, the QPCR assay is less demanding on starting material than Southern blots or HPLC, which require at least 10–50 μg of DNA for analysis.

Careful adherence to the rules of QPCR primer design (see Subheading 3.5) should allow for any PCR-amplifiable gene or DNA region to be studied with the QPCR technique. In this way, the QPCR assay can allow for a direct comparison of DNA damage sensitivity and repair rate between two genes that are hypothesized to be differentially damaged and repaired. Finally, the DNA extraction procedure allows for the simultaneous extraction of both nuclear and mitochondrial genomes in a single sample and thus, damage sensitivity and repair rates in nuclear and mitochondrial DNA (mtDNA) can be compared in one biological sample. In fact, the QPCR assay has been utilized in a wide variety of cells and tissues to compare nuclear DNA (nDNA) damage and repair to mtDNA damage and repair after treatment with various genotoxins (14–22).

1.3. Limitations of the Assay

The QPCR assay has four main limitations. First, the QPCR assay can only detect lesions that stall or halt DNA polymerase progression, which may include the 8-hydroxydeoxyguanosine (8-OHdG) lesion, one of the most-studied lesions generated during oxidative stress (23, 24). Second, although QPCR is able to detect lesions that stall or stop the polymerase, the specific type of lesion cannot be deduced from the QPCR assay alone. An additional concern is that the QPCR assay is not able to detect region-specific damage in the nDNA that is not within the amplification region of the primer set. Thus, the results can be skewed if the DNA damaging agent specifically targets the amplified region of the genome, or specifically targets a non-amplified region. One study has shown that oxidative lesions seem to occur within promoter regions of specific genes during aging (25). Of note, limitations similar to the preceding three are common to many other available methods for measuring DNA damage. The last potential issue has only recently been discovered in samples obtained using automated DNA extraction methods (QIAcube), for which it appears that mtDNA emerges from the extraction mostly in supercoiled form, and we have found it difficult to amplify this DNA due to poor primer access following the denaturing step (supercoiled covalently closed mtDNA rapidly reanneals). However, this issue with mtDNA amplification can be surmounted with use of restriction enzymes to linearize the mtDNA in a region outside of the amplified area. For instance, the HaeII restriction enzyme cuts the mouse mtDNA at a single site outside of the amplification region. The use of restriction enzymes with QIAcube-extracted DNA greatly enhances mtDNA amplification in both mock-treated and treated samples (see Note 1).

1.4. Newer Applications of the Assay

Historically used in mammalian models or systems (e.g., cell culture), the QPCR assay has recently been adapted to nonmammalian laboratory animals including *Caenorhabditis elegans* (26), *Drosophila melanogaster* (27), *Danio rerio* (27), *Trypanosoma cruzi* (28), and *Oryzias latipes* (Meyer, unpublished). These species complement mammalian models with attributes such as short generation times, large numbers of offspring, transparency of embryos and/or adults, ease and cost-effectiveness of maintenance, and powerful genetic tools. In one case (*C. elegans*), the assay has been further improved to permit measurement of DNA damage in single nematodes comprising between 30 and ~1,000 cells (13), greatly reducing the number of animals that need to be cultured and permitting analysis of interindividual variability in response. The assay has also been adapted to the ecological sentinel species *Fundulus heteroclitus* (22) and *Gambusia holbrooki* (Meyer, unpublished), allowing analysis of DNA damage in wild populations inhabiting contaminated sites. For example, Jung et al. (22, 29) found that a population of *F. heteroclitus* inhabiting a creosote-contaminated estuary exhibited elevated levels of mitochondrial and nuclear DNA damage compared to fish from a reference site. In addition to these new nonmammalian species, the QPCR assay was recently used to determine if exogenously introduced ligase constructs could complement ligase III function in the repair of oxidatively damaged mtDNA (30), to measure DNA damage in white blood cells from patients suffering from the disease Friedreich's ataxia

(19), to elucidate the role of tyrosyl-DNA phosphodiesterase (TDP1) in mtDNA repair (31), and to measure mtDNA damage in models of age-related macular degeneration (32, 33), and cataracts (34).

2. Materials (See Notes 2 and 3 for Reagent Storage Conditions)

2.1. DNA Sample Extraction

2.1.1. Manual Method of Extraction (Qiagen Genomic-Tip Protocol) (See Note 4)

1. Cell pellets of 1×10^6 cells/pellet or tissue sample.
2. QIAGEN Genomic-tips: 20/G for small pellets and tissues <20 mg (QIAGEN), 100/G for tissue samples <100 mg (QIAGEN).
3. Associated buffers with Genomic-tip kit (QIAGEN) (Buffers G2, QBT, QC, and QF, RNase A, and Proteinase K).
4. Isopropanol.
5. Cold 70 % ethanol in H_2O.
6. 1× Tris–EDTA 10 mM, 1 mM, pH 8.0.
7. 15 mL conical tubes (four per sample, BD Falcon, BD Biosciences, Bedford, MA).
8. 2.2 mL microcentrifuge tubes (two per sample).
9. Water bath that can be set to 50 °C.
10. Refrigerated microcentrifuge.

2.1.2. Automated Method of Extraction (QIAcube) (See Note 4)

1. Cell pellets of 1×10^6 cells/pellet or tissue sample <25 mg.
2. 1× PBS.
3. QIAcube (QIAGEN).
4. QIAcube materials (pipette tips, 2 mL screw-cap sample tubes, rotor adapters, 1.5 mL elution tubes) (QIAcube starter kit).
5. For human samples, associated buffers and spin columns with the QIAamp DNA mini kit (QIAGEN), and for animal samples, associated buffers and spin columns with the DNeasy blood and tissue kit (QIAGEN).
6. Ethanol, 200 proof.

2.2. Pre-PCR DNA Quantitation and Post-PCR Analysis

1. PicoGreen dsDNA quantitation reagent (Molecular Probes).
2. 1× Tris–EDTA 10 mM, 1 mM, pH 8.0.
3. Lambda (λ)/HindIII DNA (Gibco) to generate a standard curve.
4. A fluorescence reader capable of measuring fluorescence with 485 nm excitation and 528 nm emission (BioTek Synergy 2 Multi-Mode microplate reader, BioTek Instruments), black-bottomed 96-well plates (Costar, Corning Inc., Corning, NY),

single-channel p10 and p200 pipettes, multichannel pipette, and pipet basin.

5. 15 mL conical tube (four per sample, BD Falcon, BD Biosciences, Bedford, MA).
6. 0.5 mL autoclaved microcentrifuge tubes (to dilute DNA for PicoGreen reads).

2.3. Restriction Digest

1. 225 ng purified DNA.
2. Restriction enzyme (mouse: HaeII; human: PvuII; rat: XhoI) and associated components (10× NEBuffer, 100× BSA).
3. ddH_2O.
4. 0.2 mL PCR tubes (USA Scientific).
5. Thermocycler or another method of incubating samples at 37 °C/56 °C.

2.4. QPCR

1. GeneAmp XL PCR Kit (Applied Biosystems).
2. Bovine Serum Albumin (Roche, cat. no. 10-711-454-001).
3. dNTPs (GeneAmp dNTP Mix with dTTP, Applied Biosystems) (see Note 5).
4. Primer oligos.
5. 0.2 mL PCR tubes (USA Scientific).
6. A 96-well format thermocycler (Biometra Professional Standard Thermocycler 96, Biometra).

2.5. Dedicated Equipment and Workstations (See Notes 6 and 7)

1. PCR enclosure with UV sterilization and, if possible, an air purifier (Labconco).
2. Distinct, dedicated workstations (preferably in separate laboratories) for (a) DNA isolation; (b) setting up the PCR reactions; and (c) post-PCR analysis.
3. Pipettes and tips exclusively used for QPCR setup kept apart from pipettes and tips used for pre- and post-QPCR DNA analysis.

3. Methods

3.1. DNA Extraction

3.1.1. Manual DNA Extraction

1. To extract DNA from tissues that are very tough, or from organisms with cuticles or exoskeletons, first grind the sample in liquid nitrogen (27). For softer tissues such as brain and liver, use a motorized pellet pestle to homogenize the sample.
2. For a 1×10^6 cell pellet (or tissue samples below 15–20 mg), follow the 20/G Genomic-tip protocol for tissues (see Note 8). For larger amounts of tissue (>20 mg), follow the

100/G Genomic-tip protocol. For the purposes of this chapter, the 20/G Genomic-tip protocol will be explained in further detail.

3. Add 2 mL of Buffer G2 and 4 μL RNase A in a labeled 15 mL conical tube and vortex it briefly.
4. Add 100 μL Proteinase K to each tube and vortex it until pellet is completely resuspended (see Note 9).
5. Incubate sample in 50 °C water bath for 2 h and ensure that lysate is clear after incubation. At the same time, prewarm Buffer QF in 50 °C water bath.
6. While sample is incubating, set up one QIAGEN Genomic-tip 20/G column per sample over a prelabeled 15 mL conical tube and equilibrate the column with 2 mL of QBT buffer.
7. After incubation, vortex sample for 10 s and apply it to equilibrated column (see Note 9). Allow it to flow freely.
8. Move column to fresh 15 mL conical tube and wash column with 3 × 1 mL wash buffer QC.
9. Move column to fresh 15 mL conical tube and elute DNA with 1 mL elution buffer QF; repeat this one additional time over a fresh 15 mL tube.
10. Precipitate DNA by adding 1 mL eluted DNA + 700 μL isopropanol to a labeled 2.2 mL microcentrifuge tube and inverting tube at least 25 times. Each DNA sample will now have two 2.2 mL tubes from the two elutions. After precipitating DNA, place tubes at −80 °C overnight.
11. The following day, thaw sample and centrifuge it at ~15,000 × *g* for 60 min at 4 °C.
12. Discard the supernatant and resuspend the pellet in 1 mL cold 70 % ethanol.
13. Centrifuge sample at ~15,000 × *g* for 45 min at 4 °C.
14. Discard the supernatant and air-dry samples by inverting opened tubes and placing on paper towel for 10 min. Use clean pipette tip to remove droplets from tube until a dry white pellet is all that can be seen in the tube. Add 30–100 μL TE to the tube and combine the pellets in the two 2.2 mL tubes for each sample by taking the TE containing the resuspended pellet and moving it to the sample's other tube and ensuring that the second pellet is in the TE. Pipet up and down slowly and carefully to prevent DNA breakage.

3.1.2. Automated DNA Extraction

1. Begin appropriate program in QIAcube for human (QIAamp DNA Mini Kit, blood or body fluid, standard lysis) or animal (DNeasy Blood & Tissue Kit, animal blood or cells, standard lysis). The QIAcube is designed for tissue samples below 25 mg and pellets equal to or below 5×10^6 cells.

2. Follow the on-screen instructions to set up QIAcube for DNA extraction (add reagents to bottles in correct order, label elution tubes, prepare columns and elution tubes in rotor adapters and place in centrifuge, set up tube of proteinase K, load pipette tips).
3. Resuspend cell pellet in 200 μL PBS (human samples) or 100 μL PBS (animal samples) and transfer to a labeled 2 mL screw-cap tube.
4. Place screw-tap tubes in appropriate positions in the QIAcube sample rack and start the protocol.
5. After the protocol is finished, save labeled 1.5 mL elution tubes and place samples at 4 °C.

3.2. Restriction Digest for Mitochondrial Analysis of DNA from Automated Extraction (See Note 1)

1. Total volume of all digest components is 50 μL.
2. Calculate volume of DNA to load into each digest. Optimal DNA amount for restriction digest is 225 ng. All samples should have same amount of DNA for restriction digest (225 ng).
3. To calculate amount of ddH_2O to be used in the digest, subtract the volume of DNA to be used in the digest from 39 μL. For example, if 10 μL DNA will be used for the digest, 29 μL ddH_2O will be used. The individual volumes of DNA and ddH_2O are expected to vary between samples, but the total volume of ddH_2O and DNA will always be 39 μL.
4. NEBuffer stock is 10× and is diluted 10× in the digest mix (5 μL).
5. BSA stock is 100× and is first diluted to 10× in ddH_2O and then diluted 10× in the digest mix (5 μL).
6. Add ddH_2O to a labeled 0.2 mL PCR tube, followed by DNA, 5 μL 10× appropriate NEBuffer, and 5 μL 10× BSA.
7. Add appropriate restriction enzyme (20 units HaeII for mouse, 10 units PvuII for human, and 20 units XhoI for rat) and place samples in thermocycler. Set thermocycler for 37 °C for 60 min for mouse DNA, 37 °C for 120 min for human DNA, and 56 °C for 60 min for rat DNA. After digest, immediately place samples on ice or at 4 °C. Further purification is not necessary prior to QPCR.

3.3. DNA Extraction via Lysis from a Small Number of C. elegans Nematodes

1. Between 1 and 10 nematodes can be lysed at a ratio of 1 worm per 10 μL of lysis buffer for early-stage larvae to 1 worm per 20–30 μL for gravid adults.
2. Prepare lysis buffer containing 1× rTth XL PCR buffer and 1 mg/mL proteinase K (30 % 3.3× rTth XL PCR buffer, 65 % deionized, sterile H_2O, and 5 % 20 mg/mL proteinase K).

3. Aliquot lysis buffer into PCR tubes and place tubes on ice or in a chilled 96-well plate.
4. While observing the PCR tube with a dissecting microscope, use a platinum wire pick or an eyelash to pick nematode(s) into the PCR tube.
5. Immediately after picking the nematode(s) into lysis buffer, place the PCR tube on dry ice or directly at −80 °C. Do not allow the nematode(s) to sit in lysis buffer unfrozen for more than 5 min.
6. Place all samples at −80 °C for at least 10 min to disrupt the nematode cuticle. Store samples at −80 °C until lysed.
7. To lyse samples, heat to 65 °C for 5 min, vortex at high speed for 5 s, heat for an additional 55 min at 65 °C, heat to 95 °C for 15 min, and cool to 4 °C.

3.4. Quantitation of DNA Template (See Note 10)

1. For manually extracted samples, dilute the DNA 1:20 or 1:30 in 1× TE, depending on the size of cell pellet. For samples with automated extraction, dilute the DNA 1:10 in 1× TE because the initial DNA yield is lower.
2. Pipet 95 μL 1× TE into wells of a black-bottomed 96-well plate in duplicate for standards and samples.
3. Pipet 5 μL of standards into wells in duplicate (0 ng/μL, 1.25 ng/μL, 2.5 ng/μL, 5 ng/μL, 10 ng/μL, and 20 ng/μL). A standard of 40 ng/μL can also be used for the manually extracted DNA, which tends to have high initial yields.
4. Pipet 5 μL of diluted DNA samples into wells in duplicate.
5. Thaw PicoGreen reagent in the dark at room temperature and then make up PicoGreen solution by adding 5 μL PicoGreen reagent to 1 mL 1× TE.
6. In subdued lighting, pipet 100 μL PicoGreen solution into each well, cover plate, and place in the dark for 10 min to allow for color development.
7. Set up fluorescence reader by setting excitation filter at 485 and emission filter at 528 nM. Add a 20 s shaking step before read step. For the Biotek Synergy 5, the sensitivity limit is 75 and shaking of the plate is set at level 3 for 20 s.
8. After a 10 min incubation, place plate in fluorescence reader and read sample concentrations. If DNA concentration is higher than the highest standard, dilute DNA further and reread sample (see Note 11).
9. For manually extracted samples, calculate volume of sample and 1× TE needed to make a 10 ng/μL concentration of DNA. Once the 10 ng/μL samples are made, load into plate and do PicoGreen read.

10. After acquiring the actual concentrations for the 10 ng/μL samples, calculate the volume of sample and 1× TE needed to further dilute to a 3 ng/μL concentration of DNA required for QPCR (see Note 12). If samples are more dilute we have found that you can run the QPCR assay using less DNA (~1 ng/μL), as long as ALL the samples are diluted equally. It may be necessary to add an additional one to two PCR cycles to increase the signal.
11. For samples extracted with the QIAcube, restriction digest is required for maximum product amplification. Using the initial read of diluted DNA, back-calculate to determine the original DNA concentration, and determine the volume of DNA needed for 225 ng total for the restriction digest.

3.5. Primer Design and Optimization (See Note 13)

3.5.1. Design Long (~10–15 kb) and Short (~200 bp) Primers for New Genomic Targets or Species

1. Search GenBank to select genomic sequence to be amplified.
2. If applicable, paste the genomic sequence into DNA Duster (http://users.soe.ucsc.edu/~kent/dnaDust/dnadust.html) to remove numbers and spaces.
3. Copy the genomic sequence into primer design software such as Primer 3.
4. Set primer specifications:
   (a) Primer length of 20–25 nt
   (b) Tm of 68–70 °C for long product and 63–65 °C for short product (or 60 °C if you anticipate utilizing real-time PCR)
   (c) GC content of 40–60 %
   (d) Desired product length (~10–15 kb for long, ~200 bp for short) (it is possible to design primers that amplify a <10 kb region, but decreased amplicon size results in decreased sensitivity)
   (e) Return at least three primer options
5. Select at least three forward and three reverse primers.
6. Order primers with standard desalting.
7. Store in −20 °C until use.

3.5.2. Selection of Optimal Primers

1. Spin down lyophilized primers to pull down the pellet.
2. Resuspend in sterile 0.1× TE (1 mM Tris + 0.1 mM EDTA) to 540 μM.
3. Vortex briefly.
4. Store stock primer solution at −20 °C until use. Large volumes may be aliquoted (50 μL or less) in sterile 1.5 mL DNase- and RNase-free tubes prior to freezing to avoid multiple freeze–thaw cycles.

5. Dilute an aliquot to 10 μM from 540 μM with 0.1× TE. Store 10 μM working solution at 4 °C.
6. Run PCR as described below with each primer combination at 63 °C and 68 °C annealing temperature for short and long primers, respectively. Annealing temperature optimization is described in Subheading 3.5.3.
7. Run all PCR products on an agarose gel (1 % and 2 % for long and short products, respectively).
8. Pick the primer combination that results in a bright and unique band of the expected size. Very minor nontarget products may be eliminated with optimization of magnesium concentration and annealing temperature (see Subheading 3.5.3).
9. Confirm the identity of products of expected size by sequencing or restriction digest followed by gel electrophoresis of cut products.

3.5.3. Optimizing Primer Conditions

1. After picking the optimal primer pair for each product, repeat QPCR at a range of annealing temperatures and magnesium concentrations (Table 1) (see Note 14). Typically a range of 4–6 °C around the designed annealing temperature and 1–1.4 mM (final concentration) $MgO(Ac)_2$ works well. This process is expedited with the use of a thermocycler with gradient capacity.
2. Run products on an agarose gel and, again, select the conditions that yield a bright band of expected size with no secondary products.

3.6. Standard QPCR Conditions for Animal Cells

Using the XL PCR kit mentioned in the materials section, the PCR reactions are prepared as follows:

3.6.1. PCR Reaction

1. To a 0.2 mL PCR tube, add 15 ng of DNA (total).
2. Controls needed include a PCR tube containing 1× TE instead of DNA ("no template" control) and a PCR tube containing 50 % DNA amount (DNA diluted 1:1 first).
3. When several samples are being run, a master mix can be made for each primer set, which consists of the following components, added in this order:
   (a) Sterile Sigma W3500 water (9.6 μL water per reaction when DNA volume is ~5 μL and 8.6 μL water per reaction when DNA volume is ~6 μL)
   (b) 1× buffer (15 μL 3.3× buffer per reaction)
   (c) 100 ng/μL final concentration of BSA (5 μL of 1 mg/mL stock solution of BSA in ddH_2O per reaction)
   (d) 200 μM final concentration of dNTPs (4 μL of 2.5 mM/each nt stock solution per reaction) (see Note 5)

Table 1
Gene targets and primer pairs for QPCR

Human		
13.5 kb fragment from the 5′ flanking region near the β-globin gene, accession number J00179		
48510	5′-CGA GTA AGA GAC CAT TGT GGC AG-3′	Sense
62007	5′-GCA CTG GCT TAG GAG TTG GAC T-3′	Antisense
12.2 kb region of the DNA polymerase gene β, accession number L11607		
2372	5′-CAT GTC ACC ACT GGA CTC TGC AC-3′	Sense
3927	5′-CCT GGA GTA GGA ACA AAA ATT GCT G-3′	Antisense
10.4 kb fragment encompassing exons 2–5 of HPRT gene, accession number J00205		
14577	5′-TGG GAT TAC ACG TGT GAA CCA ACC-3′	Sense
24997	5′-GCT CTA CCC TGT CCT CTA CCG TCC-3′	Antisense
8.9 kb mitochondria fragment, accession number J01415		
5999	5′-TCT AAG CCT CCT TAT TCG AGC CGA-3′	Sense
14841	5′-TTT CAT CAT GCG GAG ATG TTG GAT GG-3′	Antisense
221 bp mitochondria fragment		
14620	5′-CCC CAC AAA CCC CAT TAC TAA ACC CA-3′	Sense
14841	5′-TTT CAT CAT GCG GAG ATG TTG GAT GG-3′	Antisense
Mouse		
8.7 kb fragment of the β-globin gene, accession number X14061		
21582	5′-TTG AGA CTG TGA TTG GCA ATG CCT-3′	Sense
30345	5′-CCT TTA ATG CCC ATC CCG GAC T-3′	Antisense
6.6 kb fragment of the DNA polymerase gene β, accession number AA79582		
Chr8, 23735019	5′-TAT CTC TCT TCC TCT TCA CTT CTC CCC tgg-3′	Sense
Chr8, 23741702	5′-CGT GAT GCC GCC GTT GAG GGT CTC CTG-3′	Antisense
10 kb mitochondria fragment		
3278	5′-GCC AGC CTG ACC CAT AGC CAT AAT AT-3′	Sense
13337	5′-GAG AGA TTT TAT GGG TGT AAT GCG G-3′	Antisense
117 bp mitochondria fragment		
13597	5′-CCC AGC TAC TAC CAT CAT TCA AGT-3′	Sense
13688	5′-GAT GGT TTG GGA GAT TGG TTG ATG T-3′	Antisense
Rat		
12.5 kb fragment from the clusterin (TRPM-2) gene, accession number M64733		
5781	5′-AGA CGG GTG AGA CAG CTG CAC CTT TTC-3′	Sense
18314	5′-CGA GAG CAT CAA GTG CAG GCA TTA GAG-3′	Antisense
13.4 kb mitochondria fragment		
13559	5′-AAA ATC CCC GCA AAC AAT GAC CAC CC-3′	Sense
10633	5′-GGC AAT TAA GAG TGG GAT GGA GCC AA-3′	Antisense
211 bp mitochondria fragment		
14678	5′-CCT CCC ATT CAT TAT CGC CGC CCT TGC-3′	Sense
14885	5′-GTC TGG GTC TCC TAG TAG GTC TGG GAA-3′	Antisense

(continued)

Table 1
(continued)

Caenorhabditis elegans[a]		
13.7 kb fragment from the polymerase epsilon gene		
	5′-AGT CGT TGA ACG CAG TGG TGT CAT-3′	Sense
	5′-CAG TCT TTC TTC GAC GCA TTC AAC G-3′	Antisense
9.3 kb fragment from the unc-2 gene		
unc2fl	5′-TGG CTG GAA CGA ACC GAA CCA T-3′	Sense
unc2rl	5′-GGC GGT TGT GGA GTG TGG GAA G-3′	Antisense
225 bp nuclear fragment		
1334	5′-TCC CGT CTA TTG CAG GTC TTT CCA-3′	Sense
1536	5′-GAC GCG CAC GAT ATC TCG ATT TTC-3′	Antisense
10.9 kb mitochondrial fragment		
1528	5′-CCA TCA ATT GCC CAA AGG GGA GT-3′	Sense
12442	5′-TGT CCT CAA GGC TAC CAC CTT CTT CA-3′	Antisense
195 bp mitochondrial fragment		
12272	5′-CAC ACC GGT GAG GTC TTT GGT TC-3′	Sense
12442	5′-TGT CCT CAA GGC TAC CAC CTT CTT CA-3′	Antisense
Drosophila melanogaster		
11.5 kb fragment of the β tubulin gene		
	5′-GTA TTC CTG CGC CAG GAG GAT CG-3′	Sense
	5′-CAG ATG CTG GAG CTG CCT TTG GA-3′	Antisense
10.3 kb fragment of the β tubulin gene		
	5′-GAG GAG CCT TGC GAA CAA CAG CA-3′	Sense
	5′-CAA TGA CAG CTG CGC CTC GAG AT-3′	Antisense
152 bp nuclear fragment		
	5′-CGA GGG ATA CCT GTG AGC AGC TT-3′	Sense
	5′-GTC ACT TCT TGT GCT GCC ATC GT-3′	Antisense
14.2 kb mitochondrial fragment		
	5′-GCC GCT CCT TTC CAT TTT TGA TTT CC-3′	Sense
	5′-TGC CAG CAG TCG CGG TTA TAC CA-3′	Antisense
151 bp mitochondrial fragment		
	5′-GCT CCT GAT ATA GCA TTC CCA CGA-3′	Sense
	5′-CAT GAG CAA TTC CAG CGG ATA AA-3′	Antisense
Fundulus heteroclitus (Atlantic killifish)		
11.5 kb fragment of the CFTR gene, accession number AY028263		
	5′-CAG CCG CCC GCA AAT TCT CA-3′	Sense
	5′-CAG AAT GCG GGC CTT GCT GA-3′	Antisense
234 bp nuclear fragment		
	5′-GCC GCT GCC TTC ATT GCT GT-3′	Sense
	5′-ATG AGC TGG GTG TGC GCT GA-3′	Antisense
9.4 kb mitochondria fragment, accession number CN984995		
	5′-TTG CAC CAA GAG TTT TTG GTT CCT AAG ACC-3′	Sense
	5′-GAT GTT GGA TCA GGA CAT CCC AAT GGT GCA-3′	Antisense
264 bp mitochondria fragment		
	5′-ATC TGC ATG GCC AAC GCC TA-3′	Sense
	5′-GGC GGT GCC AGT TTC CTT TT-3′	Antisense

(continued)

Table 1 (continued)

Danio rerio (zebrafish)		
10.7 kb fragment of the AHR2 gene		
	5′-AGA GCG CGA TTG CTG GAT TCA C-3′	Sense
	5′-GTC CTT GCA GGT TGG CAA ATG G-3′	Antisense
233 bp nuclear fragment		
	5′-ATG GGC TGG GCG ATA AAA TTG G-3′	Sense
	5′-ACA TGT GCA TGT CGC TCC CAA A-3′	Antisense
10.3 kb mitochondria fragment		
	5′-TTA AAG CCC CGA ATC CAG GTG AGC-3′	Sense
	5′-GAG ATG TTC TCG GGT GTG GGA TGG-3′	Antisense
198 bp mitochondria fragment		
	5′-CAA ACA CAA GCC TCG CCT GTT TAC-3′	Sense
	5′-CAC TGA CTT GAT GGG GGA GAC AGT-3′	Antisense

[a]Primers and PCR conditions for additional nuclear targets are listed in Meyer et al. (2007)

(e) 1.2 mM final concentration of Mg^{2+} (2.4 μL of 25 mM $MgO(Ac)_2$ per reaction) (see Note 15)

(f) 0.4 μM of each of the two primers (2 μL of 10 μM primer working solution per primer per reaction)

Note: The volumes noted above are per reaction tube. Always make at least two reaction tubes more master mix than needed, or more if a large number of samples will be run.

4. Set up conditions for PCR machine, explained in further detail below (Subheading 3.6.3).
5. Begin the PCR reaction by a "hot start." Bring the reaction mixture to 75 °C prior to addition of enzyme (1 unit/reaction, dilute 0.5 μL of the polymerase in 4.5 μL of sterile water for each reaction (see Note 16)) and subsequent cycling.
6. After adding enzyme, close tubes and press continue to run the remainder of PCR program.
7. For *C. elegans* extracts from low number of nematodes, the assay is set up in a similar manner, but the DNA concentration is not known, as a fixed number of nematodes is used (see Subheading 3.3 under Methods).

3.6.2. Cycle Number and Optimization

1. Perform cycle tests to determine quantitative conditions for the gene of interest by using a non-damaged sample (15 ng DNA) and a 50 % control containing half of the amount of the non-damaged template (7.5 ng DNA). Include a "no template" control (explained above in Subheading 3.6.2) as well.

2. The experimental setup is appropriate if the 50 % control shows a 50 % reduction of the amplification signal (see Note 17) when quantifying PCR products.

3.6.3. Quantitation of PCR Products

Protocol is similar to that of Subheading 3.2 in terms of fluorescence reader protocol, 10 min incubation step, makeup of PicoGreen solution, and pipetted amounts of DNA, 1× TE, and PicoGreen solution for each well; however, there are some notable differences:

1. The DNA is not diluted, and is taken directly from the PCR tube.
2. Measurements are performed in triplicate wells.
3. No standards are needed, although they can be useful to compare samples from one plate to another if the plate reader settings vary from run to run.
4. Read blank-corrected fluorescence values. The blank is the "no template" control PCR product, explained in Subheading 3.6.2.

3.7. Data Analysis

3.7.1. Calculation of Lesion Frequency

1. Open Microsoft Excel.
2. Put blank-corrected fluorescence values acquired from fluorescence reader in a column.
3. Average the non-treated (undamaged) samples' blank-corrected fluorescence values.
4. Divide each sample's blank-corrected fluorescence by the average blank-corrected fluorescence value to get a ratio.
5. Take the negative natural log (-ln) of the ratio to determine the lesion frequency per fragment. We normalize this lesion frequency to number of lesions/10 kb.
6. The lesion frequencies *from at least two separate QPCR runs* are calculated by combining the average lesion frequency for each biological sample (we use at least two treated and two control samples) with the average lesion frequency.

3.7.2. Normalization to mtDNA Copy Number

1. Open Microsoft Excel.
2. Put blank-corrected fluorescence values for the small mitochondrial amplicon in a column after the column for large mitochondrial blank-corrected fluorescence values.
3. Average all small mitochondrial blank-corrected fluorescence values.
4. Divide each sample's small mitochondrial blank-corrected fluorescence values by the average of all samples to get a ratio.
5. Divide the large mitochondrial blank-corrected fluorescence value by its corresponding small mitochondrial blank-corrected

fluorescence ratio to get a "normalized" large mitochondrial blank-corrected fluorescence value.

6. Average the non-treated (undamaged) samples' "normalized" large mitochondrial blank-corrected fluorescence values.
7. Divide each sample's "normalized" large mitochondrial blank-corrected fluorescence values by the average non-treated blank-corrected fluorescence value to get a ratio.
8. Take the negative natural log (-ln) of the ratio to determine the lesion frequency per fragment. Now the lesion frequencies for mtDNA damage are normalized to mtDNA copy number.

3.8. Statistical Analysis

3.8.1. Quality Control

1. Ensure that 50 % controls are adequate (generally between 40 % and 60 %) (see Note 17).
2. Compare blank-corrected fluorescence values for replicate QPCR reactions by correlation analysis. If the correlation is poor, perform a third PCR. Comparison of the three values in the context of the correlation provides the basis for removal of outliers.

3.8.2. Data Compilation and Assessment

1. Combine lesion values for QPCR replicates (we do at least two separate QPCR runs). These are considered analytical replicates and do not contribute to the "n" of the experiment. However, combining two or more replicates is important in minimizing technical noise between QPCR runs.
2. Test normality of the data (e.g., with the Kolmogorov–Smirnov Normality test). In our experience, DNA damage data has always been normally distributed. If this were not the case, graphical presentations (e.g., box-and-whisker plots) and statistical analyses (e.g., Kruskal Wallis test) should be substituted for those described below.

3.8.3. Data Presentation

1. Present the data graphically as mean ± standard error (unless the goal is an understanding of population distribution of DNA damage levels, in which case standard deviations would be more appropriate).
2. For repair studies, it may be clearer to present the data using a "percent removal" format.

3.8.4. Other Statistical Considerations

1. Error values are generated for control samples by comparing each individual control to the average of all controls.
2. If only two samples are being compared, a *t*-test or 1-factor ANOVA can be used to test for statistical significance.
3. If more than two samples are being compared, ANOVA should be used.

4. If there is more than one variable (e.g., dose, chemical), then an initial (global) multifactor ANOVA should be used, and post hoc comparisons of subsets of the data should only be carried out if warranted by significant global ANOVA results.

4. Notes

1. As noted above, if DNA is extracted using an automated system employing the QIAcube, linearization of the mtDNA is necessary to get an accurate level of mtDNA. This is accomplished using digestion of the samples with a restriction enzyme (discussed in further detail in Subheading 2.3 in materials and Subheading 3.2 in methods).
2. The extraction reagents and buffers for manual and automated extraction, including 1× TE, 1× PBS, and isopropanol, are stored at room temperature. DNA and diluted primers at a concentration for QPCR are kept at 4 °C. The GeneAmp XL PCR kit and its reagents, the BSA stock and working stocks, the dNTP stocks and working stocks, restriction enzymes and associated NEbuffer and BSA, PicoGreen reagent, diluted (λ) DNA standards, and primer stocks are stored at −20 °C. Initial DNA extracts can be kept at 4 °C temporarily but should be moved to −80 °C for long-term storage. Cell pellets are stored at −80 °C. Ensure that all components kept at −20 °C and −80 °C are thawed completely before use.
3. The diluted (λ) DNA standard is kept at −20 °C and the diluted DNA samples are kept at 4 °C. For long-term storage, we routinely store the concentrated samples at −20 °C and try to avoid many cycles of freeze/thaw.
4. DNA template integrity is essential for the reliable amplification of long PCR targets (35). Although various kits are commercially available for DNA isolation, procedures that involve phenol extraction should be avoided due to potential introduction of artifactual DNA oxidation. The manual and automated procedures we use for DNA extraction (QIAGEN) give rise to templates of relatively high molecular weight and highly reproducible yield.
5. A solution of 10 mM of total dNTPs is prepared (2.5 mM of each nucleotide) and is stored as 100 μL aliquots in −20 °C to minimize degradation. Higher misincorporation frequency for the enzyme and reduction in effective magnesium concentration can occur if dNTPs exceed 200 μM in the QPCR reaction.
6. In order to avoid carryover of PCR products into the next subsequent reactions, separate workstations must be set up,

preferably in different laboratory spaces. We house our PCR thermocyclers in a separate room and set up all PCR reactions in a dedicated UV-equipped hood. Once the PCR run is underway, the tubes are never again opened in that room.

7. In addition to high-quality reagents, the most important factor for the success of QPCR is the diligent avoidance of sample cross-contamination with PCR products. We use sterile technique for all steps. The constant use of disposable gloves when handling samples and reagents is essential to avoid the introduction of nucleases, foreign DNA, or other contaminants that can cause degradation of the template or inhibition of the polymerase during cycling. Additionally, we have found that it is extremely important to have distinct, dedicated workstations for different steps of the procedure, preferably in physically separate laboratories. It is of extreme importance not to open the PCR tubes after the last cycle in the same laboratory where the reactions were set up. Small DNA quantities can volatilize and contaminate other reactions, particularly if the tube is still hot, and completed reactions contain very high numbers of PCR products. The inclusion of a blank sample (where no DNA is added) helps to assure that no contamination has occurred with spurious DNA or PCR products. This sample should give no DNA band, if checked on gel, nor high fluorescence signal (as gauged by PicoGreen).
8. When using the manual Genomic-tip protocol, the tissue protocol is used irrespective of whether tissue or cells are being studied, since the protocol for DNA extraction of cultured cells involves isolation of nuclei and hence loss of mtDNA.
9. When extracting DNA manually, vortex the samples well prior to lysis and again before adding them to the columns. This vortexing does not affect the subsequent amplification of the DNA.
10. The accuracy of the data obtained with the PicoGreen assay is comparable to or can exceed the reproducibility that is accomplished with ^{32}P radiolabeled nucleotides (Chen and Van Houten, unpublished observation) followed by subsequent agarose gel electrophoresis.
11. If samples are still highly concentrated after the first dilution (i.e., well above 10 ng/μL), we recommend an additional round of quantitation. This ensures accuracy of the concentration of the final 3 ng/μL solution.
12. The success of QPCR is absolutely dependent upon the accurate quantitation of the DNA present in the samples (7).
13. To simply compare DNA damage between genomes, a single long nuclear and mitochondrial product is sufficient. However, to detect differential damage or repair (potentially in coding and noncoding regions), primers can be designed for multiple

gene targets. Organism-specific Web sites (e.g., WormBase for *C. elegans*) can be rich sources of information about relative expression levels of different genes. A single short product is sufficient since this is used for copy number normalization and does not reflect lesion frequency. Avoid highly repetitive regions of the nuclear genome. Gels illustrating primer analysis and other aspects of primer design troubleshooting are presented in Meyer (36). If the full genome sequence of the species is known, it is best to use in silico PCR to check for possible amplification of additional targets. This could in some cases escape detection even by restriction digest, especially in the case of duplicated genes. Any primer design software can be used as long as it allows for primer/product conditions (i.e., annealing temperature, primer length, product length, GC content) to be specified. Primer 3 has been a reliable source of good primers for QPCR and this free Web program can be found at http://frodo.wi.mit.edu/primer3/. Primers can be suspended in sterile 0.1× TE, 1× TE, or deionized water. The use of 0.1× TE rather than 1× TE is a precautionary measure to reduce the amount of EDTA such that it does not affect PCR through chelation of Mg^{2+}. Aliquots of the stock primer solution can be made after suspension and before freezing in order to avoid multiple freeze/thaw cycles. If frozen, primer stocks should be completely thawed prior to use. Once a 540 μM primer aliquot is diluted to a working concentration of 10 μM it is kept at 4 °C and is stable for several months. Primers should be renewed from time to time to maintain amplification. When first developing the assay in your laboratory it is essential to assess the PCR product of your primer set by agarose gel electrophoresis to verify the size of the product and to assure that no other spurious products are generated.

14. Another concern when performing QPCR is finding the optimal thermal conditions for amplification of your target gene. As mentioned before, QPCR in our laboratory is routinely performed using hot start, which produces cleaner PCR products because it prevents nonspecific annealing of primers to each other, as well as to template, before enzyme addition. Keep in mind that the melting temperature of the primers and the annealing temperature used in the PCR determine how stably and specifically the primers hybridize to the DNA template. Thus, it is important to check this parameter with suitable software beforehand, and annealing temperatures must be experimentally optimized. Table 2 shows the most favorable conditions for animal DNA amplifications currently used in our laboratory.
15. The optimal magnesium concentration must be determined for each set of primers and template. The *rTth* polymerase is extremely sensitive to magnesium; we advise that amplification

Table 2
PCR conditions for animal targets

Target	Amplicon size	Cycles	Temp (°C)	Primer concentration (μm, each)	$MgO(Ac)_2$ concentration (mM)
Human					
β-globin	13.5 kb	27	64	0.4	1.3
Polymerase β	12.2 kb	26	64	0.4	1.2
HPRT	10.4 kb	29	64	0.4	1.3
Large Mito	8.9 kb	19	64	0.4	1.2
Small Mito	221 bp	19	60	0.4	1.1
Mouse					
β-globin	8.7 kb	25	65	0.4	1.1
Polymerase β	6.6 kb	25	64	0.4	1.2
Long Mito	10 kb	20	64	0.4	1.2
Short Mito	117 bp	18	60	0.4	1.1
Rat					
Clusterin	12.5 kb	28	65	0.4	1.0
Long Mito	13.4 kb	20	65	0.4	1.2
Short Mito	211 bp	20	60	0.4	1.1
C. elegans					
Polymerase ε	13.7 kb	21	68	0.4	1.2
unc-2	9.3 kb	20	68	0.4	1.2
Polymerase ε	225 bp	23	63	0.4	1.2
Long Mito	10.9 kb	16	66	0.4	1.2
Short Mito	195 bp	18	63	0.3	1.2
D. melanogaster					
β tubulin	11.5 kb	22	67	0.4	1.2
β tubulin	10.3 kb	22	67	0.4	1.2
Short nuclear	152 bp	24	65	0.4	1.2
Long Mito	14.2 kb	17	66	0.4	1.2
Short Mito	151 bp	19	61	0.4	1.2
F. heteroclitus (Atlantic killifish)					
CFTR	11.5 kb	24	68	0.4	1.1
CFTR	234 bp	24	62	0.4	1.2
Long Mito	9.4 kb	16	65	0.4	1.2
Short Mito	264 bp	24	62	0.4	1.2
D. rerio (zebrafish)					
AHR2	10.7 kb	24	69	0.4	1.2
AHR2	233 bp	27	60	0.4	1.2
Long Mito	10.3 kb	19	68	0.4	1.2
Short Mito	198 bp	21	62	0.4	1.2

Temperature is the annealing temperature and is based on the melting temperature of the oligonucleotide pair

of the fragment of interest be evaluated using varying quantities of Mg^{2+}, starting from 0.9 mM and increasing by 0.1 increments.

16. Increasing amounts of the thermostable polymerase beyond 2.5 units per reaction can increase the production of nonspecific amplification products.
17. The usefulness of the QPCR assay for the detection of DNA damage requires that amplification yields be directly proportional to the starting amount of template. These conditions must be met by keeping the PCR in the exponential phase. During each set of amplifications we routinely amplify a control sample in which only 50 % of the template is added to the QPCR. It is best to dilute control samples to 50 % and add 5 μL of this dilution, rather than pipetting 2.5 μL into the reaction tube. Depending on the DNA quality and the products being amplified, relative amplification ranging from 40 to 60 % is considered acceptable. Any experiments that are outside this range are not satisfactory and the entire set of reactions is discarded. It may be necessary to re-optimize the PCR by varying the number of cycles to establish a linear response to increasing template concentrations from 1.25 to 30 ng. Once the optimal number of cycles is identified, always run this 50 % control as a quality control.

References

1. Ponti M, Forrow SM, Souhami RL, D'Incalci M, Hartley JA (1991) Measurement of the sequence specificity of covalent DNA modification by antineoplastic agents using Taq DNA polymerase. Nucleic Acids Res 19:2929–2933
2. Ayala-Torres S, Chen Y, Svoboda T, Rosenblatt J, Van Houten B (2000) Analysis of gene-specific DNA damage and repair using quantitative polymerase chain reaction. Methods 22:135–147
3. Santos JH, Meyer JN, Mandavilli BS, Van Houten B (2006) Quantitative PCR-based measurement of nuclear and mitochondrial DNA damage and repair in mammalian cells. Methods Mol Biol 314:183–199
4. Kalinowski DP, Illenye S, Van Houten B (1992) Analysis of DNA damage and repair in murine leukemia L1210 cells using a quantitative polymerase chain reaction assay. Nucleic Acids Res 20:3485–3494
5. Jennerwein MM, Eastman A (1991) A polymerase chain reaction-based method to detect cisplatin adducts in specific genes. Nucleic Acids Res 19:6209–6214
6. Salazar JJ, Van Houten B (1997) Preferential mitochondrial DNA injury caused by glucose oxidase as a steady generator of hydrogen peroxide in human fibroblasts. Mutat Res 385:139–149
7. Yakes FM, Van Houten B (1997) Mitochondrial DNA damage is more extensive and persists longer than nuclear DNA damage in human cells following oxidative stress. Proc Natl Acad Sci USA 94:514–519
8. Chen KH, Yakes FM, Srivastava DK, Singhal RK, Sobol RW, Horton JK, Van Houten B, Wilson SH (1998) Up-regulation of base excision repair correlates with enhanced protection against a DNA damaging agent in mouse cell lines. Nucleic Acids Res 26:2001–2007
9. Ballinger SW, Patterson C, Yan CN, Doan R, Burow DL, Young CG, Yakes FM, Van Houten B, Ballinger CA, Freeman BA et al (2000) Hydrogen peroxide- and peroxynitrite-induced mitochondrial DNA damage and dysfunction in vascular endothelial and smooth muscle cells. Circ Res 86:960–966
10. Mandavilli BS, Ali SF, Van Houten B (2000) DNA damage in brain mitochondria caused by aging and MPTP treatment. Brain Res 885:45–52
11. Ballinger SW, Van Houten B, Jin GF, Conklin CA, Godley BF (1999) Hydrogen peroxide

causes significant mitochondrial damage in human RPE cells. Exp Eye Res 68:765–772

12. Van Houten B, Chen Y, Nicklas JA, Rainville IR, O'Neill JP (1998) Development of long PCR techniques to analyze deletion mutations of the human hprt gene. Mutat Res 403: 171–175
13. Boyd WA, Crocker TL, Rodriguez AM, Leung MC, Lehmann DW, Freedman JH, Van Houten B, Meyer JN (2010) Nucleotide excision repair genes are expressed at low levels and are not detectably inducible in Caenorhabditis elegans somatic tissues, but their function is required for normal adult life after UVC exposure. Mutat Res 683:57–67
14. Wang AL, Lukas TJ, Yuan M, Du N, Tso MO, Neufeld AH (2009) Autophagy and exosomes in the aged retinal pigment epithelium: possible relevance to drusen formation and age-related macular degeneration. PLoS One 4:e4160
15. Wang AL, Lukas TJ, Yuan M, Neufeld AH (2008) Increased mitochondrial DNA damage and down-regulation of DNA repair enzymes in aged rodent retinal pigment epithelium and choroid. Mol Vis 14:644–651
16. Acevedo-Torres K, Berrios L, Rosario N, Dufault V, Skatchkov S, Eaton MJ, Torres-Ramos CA, Ayala-Torres S (2009) Mitochondrial DNA damage is a hallmark of chemically induced and the R6/2 transgenic model of Huntington's disease. DNA Repair (Amst) 8:126–136
17. Bonner M, Kmiec EB (2009) DNA breakage associated with targeted gene alteration directed by DNA oligonucleotides. Mutat Res 669: 85–94
18. Khurana RN, Parikh JG, Saraswathy S, Wu GS, Rao NA (2008) Mitochondrial oxidative DNA damage in experimental autoimmune uveitis. Invest Ophthalmol Vis Sci 49:3299–3304
19. Haugen AC, Di Prospero NA, Parker JS, Fannin RD, Chou J, Meyer JN, Halweg C, Collins JB, Durr A, Fischbeck K et al (2010) Altered gene expression and DNA damage in peripheral blood cells from Friedreich's ataxia patients: cellular model of pathology. PLoS Genet 6:e1000812
20. Chatterjee A, Mambo E, Zhang Y, Deweese T, Sidransky D (2006) Targeting of mutant hogg1 in mammalian mitochondria and nucleus: effect on cellular survival upon oxidative stress. BMC Cancer 6:235
21. Duxin JP, Dao B, Martinsson P, Rajala N, Guittat L, Campbell JL, Spelbrink JN, Stewart SA (2009) Human Dna2 is a nuclear and mitochondrial DNA maintenance protein. Mol Cell Biol 29:4274–4282
22. Jung D, Cho Y, Collins LB, Swenberg JA, Di Giulio RT (2009) Effects of benzo(a)pyrene on mitochondrial and nuclear DNA damage in Atlantic killifish (Fundulus heteroclitus) from a creosote-contaminated and reference site. Aquat Toxicol 95:44–51
23. Wu LL, Chiou CC, Chang PY, Wu JT (2004) Urinary 8-OHdG: a marker of oxidative stress to DNA and a risk factor for cancer, atherosclerosis and diabetics. Clin Chim Acta 339:1–9
24. Belousova EA, Rechkunova NI, Lavrik OI (2006) Thermostable DNA polymerases can perform translesion synthesis using 8-oxoguanine and tetrahydrofuran-containing DNA templates. Biochim Biophys Acta 1764:97–104
25. Lu T, Pan Y, Kao SY, Li C, Kohane I, Chan J, Yankner BA (2004) Gene regulation and DNA damage in the ageing human brain. Nature 429:883–891
26. Meyer JN, Boyd WA, Azzam GA, Haugen AC, Freedman JH, Van Houten B (2007) Decline of nucleotide excision repair capacity in aging Caenorhabditis elegans. Genome Biol 8:R70
27. Hunter SE, Jung D, Di Giulio RT, Meyer JN (2010) The QPCR assay for analysis of mitochondrial DNA damage, repair, and relative copy number. Methods 51:444–451
28. Cabrera G, Barria C, Fernandez C, Sepulveda S, Valenzuela L, Kemmerling U, Galanti N (2011) DNA repair BER pathway inhibition increases cell death caused by oxidative DNA damage in Trypanosoma cruzi. J Cell Biochem 112:2189–2199
29. Jung D, Cho Y, Meyer JN, Di Giulio RT (2009) The long amplicon quantitative PCR for DNA damage assay as a sensitive method of assessing DNA damage in the environmental model, Atlantic killifish (Fundulus heteroclitus). Comp Biochem Physiol C Toxicol Pharmacol 149:182–186
30. Simsek D, Furda A, Gao Y, Artus J, Brunet E, Hadjantonakis A, Van Houten B, Shuman S, McKinnon PJ, Jasin M (2011) Crucial role for DNA ligase III in mitochondria but not in Xrcc1-dependent repair. Nature 471:245–248
31. Das BB, Dexheimer TS, Maddali K, Pommier Y (2010) Role of tyrosyl-DNA phosphodiesterase (TDP1) in mitochondria. Proc Natl Acad Sci USA 107:19790–19795
32. Karunadharma PP, Nordgaard CL, Olsen TW, Ferrington DA (2010) Mitochondrial DNA damage as a potential mechanism for age-related macular degeneration. Invest Ophthalmol Vis Sci 51:5470–5479.
33. Lin H, Xu H, Liang FQ, Hao L, Gupta P, Havey AN, Boulton ME, Godley BF (2011) Mitochondrial DNA damage and repair in RPE

associated with aging and age-related macular degeneration. Invest Opthalmol Vis Sci 52:3521–3529

34. Zhang Y, Zhang L, Zhang L, Bai J, Ge H, Liu P (2010) Expression changes in DNA repair enzymes and mitochondrial DNA damage in aging rat lens. Mol Vis 16:1754–1763
35. Cheng S, Chen Y, Monforte JA, Higuchi R, Van Houten B (1995) Template integrity is essential for PCR amplification of 20- to 30-kb sequences from genomic DNA. PCR Methods Appl 4:294–298
36. Meyer JN (2010) QPCR: a tool for analysis of mitochondrial and nuclear DNA damage in ecotoxicology. Ecotoxicology 19: 804–811
37. Santos JH, Hunakova L, Chen Y, Bortner C, Van Houten B (2003) Cell sorting experiments link persistent mitochondrial DNA damage with loss of mitochondrial membrane potential and apoptotic cell death. J Biol Chem 278:1728–1734
38. Mandavilli BS, Boldogh I, Van Houten B (2005) 3-nitroproprionic acid-induced hydrogen peroxide, mitochondrial DNA damage, and cell death are attenuated by Bcl-2 overexpression in PC12 cells. Mol Brain Res 133:215–223

Part III

Repair of Incorrect or Damaged DNA Bases

Chapter 10

In Vitro DNA Mismatch Repair in Human Cells

Liya Gu, Charles M. Ensor, and Guo-Min Li

Abstract

The in vitro DNA mismatch repair (MMR) assay is a very useful technique for studying the functions and the mechanisms of the MMR system in genome maintenance. This assay has been effectively used to evaluate MMR proficiency in various tumor cells and to identify the majority of the protein components required for MMR. The procedure for setting up and performing the MMR assay involves mismatch substrate preparation, cell extract preparation, and the repair assay. In this chapter, we describe the detailed methods for this functional in vitro assay.

Key words: DNA mismatch repair, *HeLa-S3* nuclear extract, Mismatched DNA substrate, Column chromatography, Electrophoresis

1. Introduction

DNA mismatch repair (MMR) plays a very important role in the maintenance of genome integrity. A deficiency in MMR causes genomic instability and a predisposition to certain types of cancer, including hereditary nonpolyposis colorectal cancer (HNPCC) (1–3). It is now clear that MMR maintains genome stability by correcting mismatches generated during DNA replication, suppressing nonhomologous recombination, and mediating DNA damage-induced cell cycle arrest and apoptosis (2).

The MMR system was initially identified in *Escherichia coli*, and involves the following protein components: MutS, MutL, MutH, DNA helicase II (MutU/UvrD), four exonucleases (ExoI, ExoVII, ExoX, and RecJ), single-stranded DNA-binding protein (SSB), DNA polymerase III holoenzyme, and DNA ligase (4). MMR is initiated by binding of the MutS protein to a mismatch, which recruits MutL and activates the MutH endonuclease activity to cleave a hemimethylated dGATC sequence several hundred base

Lotte Bjergbæk (ed.), *DNA Repair Protocols*, Methods in Molecular Biology, vol. 920,
DOI 10.1007/978-1-61779-998-3_10, © Springer Science+Business Media New York 2012

pairs away from the mismatch. The strand break is used by one of the exonucleases to remove the mismatch in a manner dependent on MutS, MutL, and UvrD. The gap is then filled by DNA polymerase III holoenzyme in concerted action with SSB (4).

Like *E. coli* MMR, the human system is nick-directed (5), and requires protein components homologous to those indentified in *E. coli.* These human components include MutSα (or MutSβ), MutLα, proliferating cellular nuclear antigen (PCNA), exonuclease 1 (EXO1), replication protein A (RPA), replication factor C (RFC), high-mobility group box 1 protein (HMGB1), DNA polymerase δ, and DNA ligase I (6).

It is worth mentioning that the identification and characterization of many MMR components in both *E. coli* and human cells relied on a functional in vitro MMR assay that was developed almost three decades ago (7). In this chapter, we describe a detailed and modified protocol to perform the in vitro MMR assay, which includes the preparation of nicked DNA substrate from derivatives of M13mp18 phages and nuclear extract from *HeLa-S3.*

2. Materials

Prepare all of the solutions in double-distilled water (ddH$_2$O) and store at room temperature unless indicated otherwise.

2.1. Preparation of Mismatched Substrates

1. M13mp18-UKY1 and M13mp18-UKY2 phage stocks: These phages were derived from M13mp18 (7,249 bp, New England Biolabs, Boston, MA, USA) by cloning a G/C- (M13mp18-UKY1) or an A/T- (M13mp18-UKY2) containing oligonucleotide duplex into the BamHI–SpHI sites of M13mp18 (8), which resulted in two M13mp18 derivatives that differ by a base pair at the same position. Thus, a G/T or an A/C mismatch can be generated if M13mp18-UKY1 double-stranded DNA (ds-DNA) hybridizes with M13mp18-UKY2 single-stranded DNA (ss-DNA) or M13mp18-UKY2 ds-DNA hybridizes with M13mp18-UKY1 ss-DNA, respectively. The size of M13mp18-UKY1 or UKY2 is 7,164 bp. There is an NsiI site in M13mp18-UKY1 and an XhoI site in M13mp18-UKY2. These restriction sites are lost when the G/T or A/C mismatch is created by the cross-hybridization of the DNA from the two phage (see below). In addition, M13mp18-UKY1 and UKY2 also contain other unique restriction sites, including PstI (255 bp 3′ to the mismatch) and BglI (172 bp 5′ to the mismatch), which can be used for linearization of dsDNA to provide a 3′ or 5′ nick for the heteroduplex (see Fig. 1).
2. *E. coli* strain XL1-Blue (*recA1 endA1 gryA96 thi hsdR17 supE44 relA1 lac [F proAB lacIqZØM15 Tn109Tetr]*) (Agilent, Santa Clara, CA, USA).

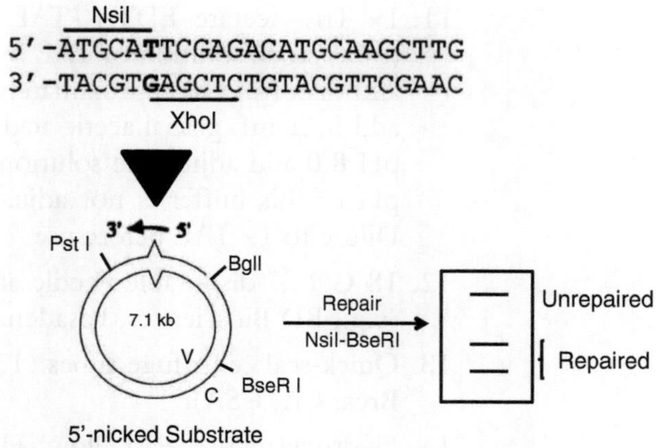

Fig. 1. Schematic diagram of 5′ nick-directed MMR. The circular G-T heteroduplex (7.1 kbp in size) containing a 5′ nick (at the BglI site) 255 bp away from the mismatch was constructed, as described in Subheading 2. Similarly, a 3′-nicked substrate can be generated by digesting the ds-DNA with PstI, which creates a nick 3′ to the G-T mismatch. After incubation with nuclear extracts or a purified MMR system, the DNA sample is double-digested with NsiI (the repair scoring enzyme) and BseRI, followed by agarose gel electrophoresis. The unrepaired substrate is not digested by NsiI and appears as the larger band in the gel while the repaired product results in two smaller fragments upon restriction endonuclease digestion.

3. DNA preparation Solution I: 0.9 % glucose, 10 mM EDTA, pH 8.0, Tris–HCl, pH 8.0. Autoclave and store at 4 °C.
4. DNA preparation Solution II: 0.2 M NaOH, 1 % SDS. This solution has to be freshly prepared.
5. DNA preparation Solution III: For 100 mL, mix 60 mL of 5 M potassium acetate and 11.5 mL of glacial acetic acid. Add ddH_2O to a final volume of 100 mL (see Note 1).
6. TE buffer: 10 mM Tris–HCl, pH 8.0 and 1 mM EDTA, pH 8.0. Autoclave.
7. TE-buffered phenol: Melt a bottle of phenol in a 65 °C water bath. Add an equal volume of 0.5 M Tris–HCl, pH 8.0 to the phenol and mix well. Allow the phases to separate (it takes approximately 5 min). Aspirate off the top layer and discard. Repeat this step several times by adding an equal volume of 0.1 M Tris–HCl, pH 8.0 until the aqueous phase reaches a pH of ~7.8 (measure with pH paper), followed by extraction twice with TE buffer, pH 8.0. Aliquot the TE-buffered phenol into 50 mL brown bottles and add ~1 cm layer of TE buffer over the phenol. The TE buffer-saturated phenol can be stored at 4 °C for up to 6 months (see Note 2).
8. Agarose (e.g., Sigma/St. Louis, MO, USA).
9. 6× loading buffer: 40 % sucrose, 0.25 % bromophenol blue, and 0.25 % xylene cyanol. Autoclave and store at 4 °C.
10. Ethidium bromine (EtBr): 10 mg/mL in water (e.g., Affymetrix/USB, Santa Clara, CA, USA).

11. 1× Tris–Acetate–EDTA (TAE) buffer: Make a concentrated (50×) stock solution of TAE by weighing out 242 g Tris base and dissolving in approximately 750 mL of ddH_2O. Carefully add 57.1 mL glacial acetic acid and 100 mL of 0.5 M EDTA, pH 8.0 and adjust the solution to a final volume of 1 L. The pH of this buffer is not adjusted and should be about 8.5. Dilute to 1× TAE before use.
12. 18 G 1½″ disposable needle and a 10 mL disposable syringe (e.g., BD Biosciences, Pasadena, CA, USA).
13. Quick-seal centrifuge tubes (13 mL) (e.g., Beckman Coulter, Brea, CA, USA).
14. Hydroxylapatite, Fast Flow (HAP) (e.g., Calbiochem/EMD, Gibbstown, NJ, USA).
15. Liquid chromatography column, 2.5 × 10 cm (e.g., Sigma/St. Louis, MO, USA).
16. Sodium Phosphate buffers (Na-PB) for HAP column: 0.03 M, 0.16 M, 0.42 M, and 1 M Na-PBs are needed for the HAP column. Make 1 L of 0.03 M Na_2PO_4 and 0.03 M $NaHPO_4$ separately. Mix 700 mL of 0.03 M Na_2PO_4 with 700 mL of 0.03 M $NaHPO_4$ and adjust pH to 6.9 by adding some of the remaining 0.03 M Na_2PO_4 or $NaHPO_4$ solution. Make 0.16 M, 0.42 M, and 1 M Na-PB in the same way. Autoclave and store at room temperature (RT).
17. Exonuclease V (ExoV) (e.g., Affymetrix/USB, Santa Clara, CA, USA).
18. Sephacryl™ S-500 resin (S-500) (e.g., GE healthcare, Piscataway, NJ, USA).
19. Kontes Flex column, 1 × 50 cm (e.g., Fisher Scientific, Pittsburg, PA, USA).
20. S-500 column buffer: 0.3 M NaCl in TE buffer. Autoclave.
21. Restriction Enzymes and 10× buffers: BglI and BglI 10× buffer. PstI and PstI 10× buffer (e.g., New England Biolab/ Boston, MA, USA).

2.2. Preparation of HeLa-S3 Nuclear Extract

1. Washing buffer: 20 mM Hepes–KOH, pH 7.5, 5 mM KCl, 0.5 mM $MgCl_2$, and 0.2 M sucrose. Autoclave and store at 4 °C (see Note 3).
2. Hypotonic buffer: 20 mM Hepes–KOH, pH 7.5, 5 mM KCl, and 0.5 mM $MgCl_2$. Autoclave and store at 4 °C (see Note 3).
3. Extraction buffer: 50 mM Hepes–KOH, pH 7.5, 10 % sucrose. Autoclave and store in 4 °C (see Note 3).
4. Dialysis buffer: 25 mM Hepes–KOH, pH 7.5, 0.1 mM EDTA, pH 8.0, and 50 mM KCl. Autoclave and store at 4 °C (see Note 4).

5. 10× protease inhibitor cocktail: 100 mM phenylmethylsulfonyl fluoride (PMSF), 5 μg/mL leupeptin, 5 μg/mL pepstatin, and 191.5 mM benzamidine. Dissolve 1.75 g PMSF in 95 mL of 100 % ethanol. Add 0.5 mL of 10 mg/mL pepstatin (in methanol), 0.5 mL of 10 mg/mL leupeptin (in ddH_2O), and 3 g benzamidine. Adjust the volume to 100 mL with 100 % ethanol and store at −20 °C.
6. 100 mM Dithiothreitol (DTT): Weigh 15.4 g DTT and dissolve it in ddH_2O. Adjust the volume to 100 mL. Store as 1-mL aliquots at −20 °C.
7. Dounce homogenizers, 15 mL and 50 mL (e.g., Bellco, Vinland, NJ, USA).
8. Quick Start Bradford Protein Assay solution (e.g., Bio-Rad, Hercules, CA, USA).

2.3. In Vitro DNA Mismatch Repair Assay

1. 10× cocktail buffer for MMR assay: 200 mM Tris–HCl, pH 7.6, 500 μg/mL bovine serum albumin (BSA), 1 mM each of the four dNTPs (dATP, dCTP, dGTP, and dTTP), 10 mM glutathione (reduced form), 50 mM $MgCl_2$, 15 mM ATP, pH 7.0 (see Note 5).
2. Proteinase K stock solution (3 mg/mL): Dissolve proteinase K in 10 mM Tris–HCl, pH 8.0, 20 mM $CaCl_2$, and 50 % glycerol. Store at −20 °C.
3. Proteinase K digestion buffer: Mix 67 μL of 10 % SDS, 50 μL of 0.5 M EDTA, pH 8.0, and 30 μL of 3 mg/mL proteinase K stock solution. Add ddH_2O to a final volume of 1 mL. This solution has to be freshly prepared.
4. Restriction enzymes: XhoI, NsiI, BseRI, and 10× digestion buffer for each enzyme (e.g., New England Biolab/Boston, MA, USA).
5. RNase I (e.g., Affymetrix/USB, Santa Clara, CA, USA).

3. Methods

3.1. Preparation of Mismatched Substrates

1. Preparation of phage stock: In duplicate 13-mL culture tubes, inoculate 3 mL 2×YT with 200 μL of overnight culture of bacterial strain XL1-Blue. Inoculate the cells with a single plaque of either M13mp18-UKY1 or M13mp18-UKY2. Grow the cultures at 37 °C for 6–7 h in an incubator with 250 rpm shaking. Spin the cells down at 12,000 × *g* for 10 min at 4 °C and save the supernatant (phage stock).
2. Amplification of phage stock: In duplicate 250-mL flasks, inoculate 50 mL of 2×YT with 1 mL of overnight culture of

XL1-Blue and grow the cells at 37 °C × 250 rpm to an optical density of 0.3 at 600 nm (OD_{600}). Add 0.5 mL phage stock obtained above and culture the flasks for 7 h at 37 °C × 250 rpm. Collect the supernatants after centrifugation at 12,000 × *g* for 10 min at 4 °C. Perform a phage titer assay as described (9). The phage titer should be ~5×10^{11} PFU/mL.

3. ds-DNA preparation: In duplicate 3-L flasks, inoculate 1.5 L of 2×YT with 10 mL of overnight culture of XL1-Blue and grow the cells at 37 °C × 250 rpm. Measure the OD_{600} of the cells every 30 min. When the cell density reaches 0.3 O.D. (~5×10^{8} cells/mL and 7.5×10^{11} total cells), add 15 mL of M13mp18-UKY1 or M13mp18-UKY2 phage stock (7.5×10^{12} PFU) to each flask and grow the cells for 7 h at 37 °C × 250 rpm. The ratio of phage to XL1-Blue cells is ~10:1. Place the flasks in ice water for 20 min and then spin the cells down at 5,900 × *g* for 20 min at 4 °C. Save the supernatant which contains phage for ss-DNA preparation. Resuspend the cell pellet from the 3 L culture in 60 mL of Solution I with 5 mg/mL of lysozyme (see Note 6) and incubate at RT for 10 min. Add 120 mL of Solution II and gently mix (see Note 7). Incubate the mixture on ice for 10 min. Add 90 mL of ice-cold Solution III and mix well. Keep it on ice for 10 min and then spin at 17,600 × *g* for 60 min at 4 °C. Collect the supernatant by passing it through four layers of cheesecloth. Add 0.6 vol. (~150 mL) of iso-propanol to the supernatant, mix very well, and let it stand at RT for 30 min. Spin at 17,600 × *g* for 30 min at RT (see Note 8). Collect the pellet and wash once with 70 % ethanol. Remove as much ethanol as possible and dry the pellet under a vacuum for 5–10 min. Resuspend the pellet in 20 mL of TE (pH 8.0) (see Note 9). Weigh the solution on a balance and record the grams of DNA solution. For each gram of DNA solution, add 1.05 g of CsCl and 50 μL of 10 mg/mL EtBr (see Note 10). Divide the DNA solution into two 13-mL Quick-seal centrifuge tubes and spin at 192,000 × *g* for 18 h at RT. Collect the band of supercoiled DNA with an 18 G 1½″ disposable needle and a 10 mL disposable syringe (see Note 11). Remove the EtBr by extraction with n-butanol. Add an equal volume of n-butanol, vortex, and allow the phases to separate. Remove the upper organic phase. Repeat the extraction until no more red color remains. Dialyze the DNA three times against 1 L TE buffer, pH 8.0 (see Note 12). Measure the ds-DNA concentration at A_{260}. An optical density of 1 corresponds to a concentration of 50 μg/mL for double-stranded DNA.

4. ss-DNA preparation: Spin the M13mp18-UKY1 or M13mp18-UKY2 supernatant obtained from the ds-DNA preparation at 5,900 × *g* for 30 min at 4 °C and transfer the supernatant to a

3 L flask. Measure the volume of supernatant and add 36 g of NaCl and 50 g of PEG-3500 per 1 L of supernatant to the flask. Stir at RT for 1 h. Spin at 5,900 × *g* for 30 min at 4 °C and collect the phage pellet. Resuspend the pellet with 22 mL of 10 mM Tris–HCl, pH 8.0 and shake in a 37 °C water bath for 2 h to get the phage particles resuspended completely. Spin the solution at 12,000 × *g* for 10 min at 4 °C. Save the supernatant and discard the pellet. Weigh the solution on a balance and record the grams of phage solution. Add 0.4348 g of CsCl for each 1 g of phage solution and mix until the CsCl completely dissolves. Divide the solution into two 13 mL Quick-seal centrifuge tubes and spin at 192,000 × *g* for 18 h at RT. Collect the white phage band using an 18G 1½″ needle and dialyze the phage three times against 1 L TE buffer, pH 8.0 (see Note 12). Extract the phage four times with an equal volume of TE-buffered phenol to release the ss-DNA from the phage. Dialyze the ss-DNA three times against 1 L TE buffer, pH 8.0 (see Note 12). Measure the ss-DNA concentration at A_{260}. An optical density of 1 corresponds to a concentration of 37 μg/mL for ss-DNA.

5. Linearization of ds-DNA: Mix 1.5 mg of M13mp18-UKY1 or M13mp18-UKY2 ds-DNA with 0.5 mL of appropriate 10× digestion buffer and adjust the volume to 5 mL with ddH_2O. Add 500 U of BglI (for 5′-nicked substrate) or 500 U of PstI (for 3′-nicked substrate) to the DNA and incubate at 37 °C for 2–3 h. Check the completion of digestion by agarose gel electrophoresis. Mix 100–200 ng digested or undigested DNA with 6× gel loading buffer and load it on a 1 % agarose gel containing 1× TAE. Run the gel at 25 mA until the dye front has reached the bottom of the gel. DNA bands are visualized by UV illumination after staining the gel in 0.5 μg/mL of EtBr solution. Extract the linear ds-DNA once with TE-buffered phenol and precipitate the DNA using ethanol. Dry the DNA in a speed vacuum.

6. Heterodulex preparation: The annealing buffer contains 50 mM Tris–HCl (pH 7.6), 1 mM EDTA, and 10 mM NaCl. Dissolve the linear M13mp18-UKY1 or M13mp18-UKY2 ds-DNA in 2 mL of TE buffer and transfer it to a 50 mL tube. Add 10 mg of M13mp18-UKY2 ss-DNA to M13mp18-UKY1 ds-DNA (GC mismatch), or M13mp18-UKY1 ssDNA to M13mp18-UKY2 ds-DNA (AT mismatch). Then add 2.96 mL of 0.5 M Tris–HCl, pH 7.6, 60 μL of EDTA, pH 8.0, and 100 μL of 3 M NaCl and adjust the volume to 30 mL by adding ddH_2O (this makes the annealing buffer containing the DNA). Mix well and then add 900 μL of fresh-made 10 M NaOH and incubate at RT for 5 min to denature the ds-DNA. Add 3 mL of 2.9 M acetic acid, 1.35 mL of 3 M KCl, and 3.72 mL of

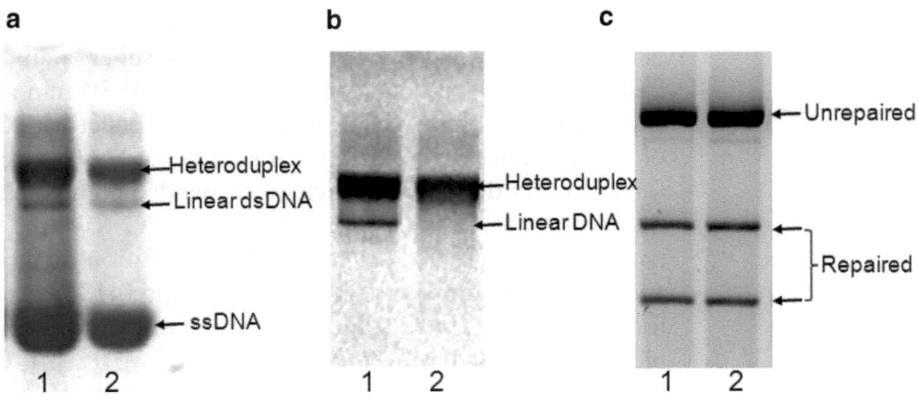

Fig. 2. Preparation of mismatched DNA substrate. (**a**) Crude heteroduplex after hybridizing a linearized ds-DNA with tenfold of an ss-DNA, followed by agarose gel electrophoresis. *Lane 1*, 3′-nicked substrate; *lane 2*, 5′-nicked substrate. (**b**) Heteroduplex after the HAP column without (*lane 1*) or with (*lane 2*) ExoV digestion for 2 h at 37 °C and separated by agarose gel electrophoresis. (**c**) Repair gel. 100 ng of the heteroduplex DNA was incubated with 50 μg of *HeLa-S3* nuclear extract for 10 min at 37 °C. After phenol extraction and ethanol precipitation, the DNA sample was digested with NsiI and separated on an agarose gel. *Lane 1*, 3′-nicked substrate; *lane 2*, 5′-nicked substrate.

1 M Na-PB, pH 6.9 to the tube. Incubate the mixture at 65 °C for 30 min and then slowly cool down to 37 °C over a period of about 2 h. Incubate at 37 °C for 30 min and then store on ice. Check the heteroduplex product by running it on a 1 % agarose gel as described above and use supercoiled ds-DNA (from step 3) (see Panel a, Fig. 2, and Note 13).

7. HAP column chromatography: The HAP column is used for removal of excess ss-DNA. Weigh 8 g of HAP and wash three to four times with 0.03 M Na-PB, pH 6.9. Pour the HAP into a 2×10 cm column and wash the column with 40 mL of 0.03 M Na-PB, pH 6.9. Dilute the heteroduplex threefold with ddH_2O and load slowly onto the column. Wash the column with 40 mL of 0.03 M Na-PB, pH 6.9 and then 80 mL of 0.16 M Na-PB, pH 6.9. The ss-DNA is eluted in the 0.16 M Na-PB wash. Elute the heteroduplex with 40 mL of 0.42 M Na-PB and collect 1 mL fractions. The heteroduplex is eluted with linear ds-DNA. Run a 1 % agarose gel to locate the heteroduplex by loading 5 μL from each fraction and use supercoiled ds-DNA, linear ds-DNA, and ss-DNA as controls. Combine the tubes containing the heteroduplex and dialyze three times against 1 L TE, pH 8.0 (see Note 12).

8. ExoV digestion: ExoV is used for removal of linear ds-DNA from the heteroduplex DNA. The ExoV digestion buffer contains 66.7 mM of glycine, pH 9.4, 5 mM MgCl, 8.3 mM 2-mercaptoethanol, and 0.5 mM ATP. Remove 10–20 μL of the heteroduplex DNA from the tube as a control before adding ExoV. Then add 2 mL of 1 M glycine, pH 9.4, 150 μL of 1 M MgCl, 17 μL of 14.4 M 2-mercaptoethanol, and 150 μL

of 100 mM ATP (this makes the ExoV digestion buffer). After adjusting the volume to 30 mL with ddH_2O, add 100 U of ExoV and incubate at 37 °C for 2–3 h. Check the completion of digestion by running a 1 % agarose gel by comparing with the undigested control sample. Extract the heteroduplex with TE-buffered phenol and concentrate the sample to 1 mL by butanol extraction (see Panel b, Fig. 2, and Note 14).

9. S-500 gel filtration chromatography: S-500 chromatography is used for removal of nucleotides after ExoV digestion. S-500 is washed and then autoclaved for 10 min in TE buffer, pH 8.0 with 0.3 M NaCl. Pack a 1 × 50 cm column with 30 mL of S-500 and wash the column with 60 mL of the same buffer. Load 1 mL of concentrated DNA sample to the column and elute the DNA with the same buffer (see Note 15). Collect 1 mL fractions and locate the DNA by running fractions on a 1 % agarose gel. Combine the tubes containing the DNA heteroduplex and dialyze three times against 1 L TE, pH 8.0 (see Note 12). The heteroduplex is the final product. Measure the DNA concentration at A_{260}. Aliquot and store in −80 °C (see Note 16).

3.2. Preparation of Hela-S3 Nuclear Extract (See Note 17)

All of the steps are at 4 °C or on ice.

1. Harvest of *HeLa-S3*: *HeLa-S3* cells are grown in suspension with RPMI 1640 containing 5 % FBS and antibiotics (penicillin, 100 U/mL; streptomycin, 100 U/mL). Harvest 6 L of cells when the cell density reaches 1×10^6/mL (10–12 g cells). Cool the cells on ice for 30 min and spin at 2,600 × *g* for 8 min. Resuspend the cell pellet in 120 mL of washing buffer (20 mL/L medium). Check for cell viability by mixing 2 μL of cell suspension with 2 μL of trypan blue dye and observe the cells under a microscope. Dead cells pick up the stain and appear deep blue. Spin the cells at 2,000 × *g* for 5 min. Pour the liquid out completely and weigh the cell pellet.
2. Preparation of cell nuclei: Resuspend the washed cells in hypotonic buffer (2.78 mL/g of original cell pellet) and let the cells swell on ice for 10 min. Lyse the cells by douncing on ice. Begin with ten strokes and check the completion of lysis using trypan blue dye. Nuclei appear light blue and round in shape.
3. Extraction of nuclear proteins: Pellet the nuclei at 2,000 × *g* for 5 min and resuspend the nuclear pellet in cold extraction buffer (1.5 mL/g of original cell pellet). Measure the volume and add 0.031 vol. of 5 M NaCl solution. Rotate the mixture on a Labquake shaker or equivalent for 60 min. Spin the mixture at 14,600 × *g* for 20 min to remove cell debris and collect the supernatant. Measure the volume and slowly add 0.42 g ammonium sulfate per 1 mL of supernatant with stirring on ice.

Continue the stirring on ice for 10–15 min until the salt completely dissolves. Pellet the nuclear protein by spinning at 14,600 × *g* for 20 min. Remove as much of the liquid as possible (see Note 18). Resuspend the pellet in dialysis buffer using ~50 μL/g of original cell pellet and make the pellet into a pipettable slurry. Transfer the slurry to a dialysis bag and monitor the course of dialysis by measuring the conductivity by removing 10 μL aliquots from the dialysis bag and diluting to 4.0 mL with ddH_2O every 20–30 min. Stop dialysis when the conductivity of the diluted aliquot is ~50 μS/cm. The salt concentration of the nuclear extract is about 100 mM.

4. Aliquot and storage nuclear extract: Clarify the nuclear extract by centrifugation in a microcentrifuge at 11,200 × *g* for 15 min. Transfer the supernatant to a new tube and mix very well. Aliquot the nuclear extract at 20–30 μl/tube and store the aliquots at −80 °C. Measure the protein concentration by using the Quick Start Bradford Protein Assay solution by using BSA as a standard. The concentration of *HeLa-S3* nuclear extract is ~20 mg/mL.

3.3. In Vitro DNA Mismatch Repair Assay

1. Reaction assembly: Assemble MMR reactions using 100 ng of DNA heteroduplex, 2 μL 10× cocktail buffer, and 50 μg of *Hela-S3* nuclear extract. Use 0.5 M KCl to adjust the salt concentration to a final concentration of 110 mM and add ddH_2O to a final volume of 20 μL (see Note 19). Mix well and incubate at 37 °C for 15–20 min.
2. Proteinase K digestion: Terminate the reactions by adding 30 μL of proteinase K digestion buffer and incubating at 37 °C for 20 min.
3. Phenol extraction: Add 50 μL of TE-buffered phenol to each reaction, vortex, and spin for 2 min at RT. Transfer the aqueous phase to a new tube and then back-extract the phenol with 70 μL of TE. Spin for 2 min and combine the aqueous phases. Add 100 μL of TE-buffered phenol to the aqueous tube, vortex, and spin for 5 min. Transfer the aqueous phase to a new tube (see Note 20). The volume of the aqueous phase is about 100 μL.
4. Ethanol precipitation: Add 11 μL of 3 M NaOAC and 270 μL of cold 100 % ethanol to each tube, mix well, and store at −80 °C for 15 min. Spin in a microcentrifuge at 11,000 rpm for 15 min at 4 °C and carefully remove the liquid. Carefully wash the pellet with 500 μL cold 70 % ethanol. Dry the DNA in a speed vacuum.
5. Restriction digestion: Dissolve the DNA in 17.5 μL of ddH_2O and add 2 μL of appropriate 10× digestion buffer and 0.5 μL of RNase I (see Note 21). Mix well and add 2 U of BseRI (scoring enzyme) and NsiI (for the G/T mismatch substrate)

or BseRI and XhoI (for the A/C mismatch substrate). Incubate the samples at 37 °C for 2–3 h (see Note 22).

6. Detection of MMR products: Add 4 μL of 6× loading dye to each sample tube and run on a 1 % agarose gel as described before (see Panel c, Fig. 2).

4. Notes

1. It is best if the solution is freshly made and kept on ice for 1–2 h before use.
2. Phenol is volatile and caustic. Prepare the phenol in a fume hood and wear gloves and eye protection.
3. Add 1/10 volume of 10× protease inhibitor cocktail and 1/100 volume of 100 mM DTT to the buffer before use.
4. Add 0.3 g/L of DTT and 1/10 volume of 10× protease inhibitor cocktail to the dialysis buffer before use.
5. For each 1 mL of 10× cocktail buffer for the MMR assay, mix all of the components except $MgCl_2$ and add ddH_2O to a final volume of 950 μL. Make 19-μL aliquots and store at −20 °C. Before use add 1 μL of 1 M $MgCl_2$ to each tube and mix well.
6. Dissolve 0.3 g of lysozyme in 60 mL of Solution I first and then add half of the solution to the cell pellet. Resuspend the cell pellet completely by using a cell scraper and then add the other half of the solution to the resuspended cells.
7. Mix the cells gently. Vigorous mixing will shear the bacterial genomic DNA into small fragments which could end up being co-purified with the phage DNA.
8. Both the rotor and the centrifuge should be at RT before centrifugation.
9. Do not dry the ds-DNA too long under vacuum. Incubate the DNA at 37 °C for 1 h with gentle shaking to make the DNA completely dissolved. If there is some insoluble material in the DNA solution, spin the DNA solution and remove the pellet before adding CsCl.
10. EtBr is a mutagen and should be handled with care.
11. Two bands will appear after the centrifugation. The top band is nicked DNA and the lower band is supercoiled DNA. Collect the lower band carefully.
12. We use a dialysis membrane with a 6–8,000 MWCO. Dialysis against each change of buffer is for a minimum of 2 h at 4 °C.

The sample is usually dialyzed against the third buffer change overnight.

13. Three bands will appear on the agarose gel. The slowest band is the heteroduplex DNA. The middle band is linear ds-DNA and the fastest moving band is ss-DNA.
14. Use a 1:1 (v/v) ratio of 2-butanol:DNA solution to concentrate the DNA. Repeat the extractions until the desired volume is reached. The DNA may appear yellow in color when the volume is ~1 mL.
15. For better separation do not load more than 1 mL of DNA onto the column. Repeat the S-500 column step if it is necessary. The column can be used many times after washing.
16. Check the final products by NsiI or XhoI enzyme digestion. Both the G/T and A/T substrates cannot be cut by NsiI or XhoI. Use M13mp18-UKY1 or UKY2 ds-DNA as a control.
17. The protocol can be used for any other mammalian cell lines. For a better nuclear extract preparation, start with 5 g or more of cells.
18. There is usually some white, flocculent material that does not pellet well which can be discarded with the supernatant.
19. The salt concentrations of each component added to the assay must be considered when determining the final salt concentration of 110 mM.
20. Avoid taking any of the interphase which may contain a trace amount of phenol. Phenol in the DNA will interfere in the enzyme digestion.
21. Heat RNase I at 75 °C for 10 min to inactivate any DNase contamination before use. Nuclear extract contains a lot of RNAs. Repaired DNA bands will appear more clearly on the agarose gel if RNase I is added to the digestion buffer.
22. The repaired products are 3 and 4.1 kbp. The repair efficiency is ~30–40 %.

Acknowledgement

The work was supported in part by NIH grants CA104333 (to L. Gu) and GM089684 (to G-M Li).

References

1. Kolodner RD, Alani E (1994) Mismatch repair and cancer susceptibility. Curr Opin Biotechnol 5:585–594
2. Li GM (2008) Mechanisms and functions of DNA mismatch repair. Cell Res 18:85–98
3. Modrich P (1994) Mismatch repair, genetic stability, and cancer. Science 266:1959–1960
4. Modrich P (1991) Mechanisms and biological effects of mismatch repair. Annu Rev Genet 25:229–253
5. Holmes J Jr, Clark S, Modrich P (1990) Strand-specific mismatch correction in nuclear extracts of human and Drosophila melanogaster cell lines. Proc Natl Acad Sci USA 87:5837–5841
6. Zhang Y, Yuan F, Presnell SR, Tian K, Gao Y, Tomkinson AE, Gu L, Li GM (2005) Reconstitution of 5′-directed human mismatch repair in a purified system. Cell 122:693–705
7. Lu A-L, Clark S, Modrich P (1983) Methyl-directed repair of DNA base pair mismatches in vitro. Proc Natl Acad Sci USA 80:4639–4643
8. Guo S, Presnell SR, Yuan F, Zhang Y, Gu L, Li GM (2004) Differential requirement for proliferating cell nuclear antigen in 5′ and 3′ nick-directed excision in human mismatch repair. J Biol Chem 279:16912–16917
9. Ausubel FM, Brent R, Kingston RE, Moore DD, Seidman JG, Smith JA, Struhl K (1995) Current protocols in molecular biology. John Wiley & Sons, New York

Chapter 11

Eukaryotic DNA Mismatch Repair In Vitro

Fenghua Yuan, Limin Song, Fengsong Liu, Liya Gu, and Yanbin Zhang

Abstract

Mismatch repair corrects biosynthetic errors generated during DNA replication. Mismatch repair deficiency causes a mutator phenotype and directly underlies hereditary nonpolyposis colorectal cancer and some sporadic cancers. Because of remarkably high conservation of the mismatch repair machinery between the budding yeast (*Saccharomyces cerevisiae*) and humans, the study of mismatch repair in yeast has provided tremendous insights into the mechanisms of this repair pathway in humans. Here we describe a set of practical protocols for how to prepare the yeast and HeLa cell-free nuclear extracts and site-specific DNA mismatch substrates, and how to carry out the in vitro mismatch repair assay. We validated the yeast cell-free system by the mismatch repair deficient strain (Δ*msh2*) and the complementation assay with purified yeast MutSα.

Key words: In vitro DNA mismatch repair, MMR, Cancer, Cell-free nuclear extract, Yeast, *Saccharomyces cerevisiae*, Human

1. Introduction

Mismatch repair (MMR) is an important maintenance mechanism for genome stability, whose defects lead to hereditary nonpolyposis colorectal cancer (HNPCC) and a variety of other sporadic cancers (1–5). MMR machinery corrects DNA replication-associated biosynthetic errors (base–base mismatches and small insertion/deletion loops) that escaped from the proofreading activity of the replicative DNA polymerases. Mismatch repair also affects many other cellular functions including DNA damage signaling, apoptosis, mitotic and meiotic recombination, class-switch recombination, somatic hypermutation, and triplet expansion (3, 6, 7).

MMR is best characterized in *Escherichia coli*. The *E. coli* MMR pathway has been reconstituted in vitro with 11 proteins including MutH, MutS, MutL, four DNA exonucleases, SSB, helicase II, DNA polymerase III holoenzyme, and DNA ligase (8, 9).

Lotte Bjergbæk (ed.), *DNA Repair Protocols*, Methods in Molecular Biology, vol. 920,
DOI 10.1007/978-1-61779-998-3_11, © Springer Science+Business Media New York 2012

Eukaryotic MMR pathway is similar to *E. coli* MMR in terms of repair mechanism and key components. For example, both eukaryotic and bacterial MMR are strand-specific and bidirectional. Eukaryotic MutS- and MutL-like activities are highly homologous to their bacterial counterparts. Many components of the eukaryotic MMR machinery have been identified including homologs of MutS, MutL, exonucleases (EXO1), RPA, DNA polymerase δ, RFC, and PCNA. However, MMR in eukaryotes seems to be more complicated. Although there is only one form of MutS or MutL in *E. coli*, at least three forms of MutS (MutSα or MSH2-MSH6 heterodimer, MutSβ or MSH2-MSH3, and MutSγ or MSH4-MSH5) and three forms of MutL homologs (MutLα, MutLβ, and MutLγ) have been found in eukaryotic cells. Eukaryotic MMR activities that play roles similar to *E. coli* MutH and helicase II have not yet been identified (3, 10, 11). Additionally, more DNA nucleases, besides EXO1, may also be involved in eukaryotic MMR, because it has been observed that EXO1 null mutants in yeast and mice confer only a partial MMR defect in vivo and multiple nucleases are required in *E. coli* MMR (8, 12, 13).

In order to facilitate the identification of novel MMR factors in eukaryotes, we developed a cell-free extract based in vitro MMR protocol using the budding yeast (*Saccharomyces cerevisiae*) cells. Favorable factors contributing to the use of yeast to study the MMR mechanism include the following: (1) the remarkably high conservation of the MMR machinery between the budding yeast and humans, (2) the easy knockout manipulation, (3) the availability of whole genome deletion strains, and (4) the relatively low cost for large quantities of yeast cells and for setting up the system. The cell-free repair system has long been proved to be a powerful tool in DNA repair studies (14–19).

2. Materials

Prepare all solutions using ultrapure water and analytical grade reagents. In many failed cases, quality of dH_2O had been identified as one of the major problems. Prepare and store all reagents at room temperature unless indicated otherwise. Diligently follow all waste disposal regulations when disposing of waste materials. All chemicals were purchased from Sigma-Aldrich unless indicated otherwise.

2.1. For Yeast Cell-Free Nuclear Extract

1. YPD: 2 % Bacto-Peptone, 1 % yeast extract, and 2 % dextrose.
2. ED solution: 0.1 M EDTA-KOH, pH 8.0, 10 mM dithiothreitol (DTT).
3. YPD/S: 2 % Bacto-Peptone, 1 % yeast extract, 2 % dextrose, 1 M sorbitol.

4. Ficoll buffer: 18 % Ficoll 400, 10 mM Tris–HCl, pH 7.5, 20 mM KCl, 2 mM EDTA, 0.125 mM spermidine, 0.05 mM spermine, 1 % thiodiglycol, 3 mM DTT, and a cocktail of protease inhibitors (18).
5. Extraction buffer: 100 mM Tris–acetate, pH 7.9, 50 mM potassium acetate, 10 mM $MgSO_4$, 20 % glycerol, 2 mM EDTA, 3 mM DTT, and a cocktail of protease inhibitors.
6. Dialysis buffer: 20 mM HEPES, pH 7.6, 10 mM $MgSO_4$, 10 mM EGTA, 3 mM DTT, 20 % glycerol, and a cocktail of protease inhibitors.
7. Strains: Yeast wild-type strain BY4741 (MATa his3Δ1 leu2Δ0 met15Δ0 ura3Δ0) and Δ*msh2* deletion strain (MATa his3Δ1 leu2Δ0 met15Δ0 ura3Δ0 Kan^r) (Invitrogen).
8. Yeast lytic enzyme in power was from MP Biomedicals LLC.

2.2. For HeLa Cell-Free Nuclear Extract

1. Wash buffer: 20 mM HEPES-KOH (pH 7.5), 5 mM KCl, 0.5 mM $MgCl_2$, 0.2 M Sucrose. Add the following reagents immediately before use to the cold buffer: 0.5 mM DTT and a cocktail of protease inhibitors.
2. Hypotonic buffer: 20 mM HEPES-KOH (pH 7.5), 5 mM KCl, 0.5 mM $MgCl_2$. Add the following reagents immediately before use to the cold buffer: 0.5 mM DTT and a cocktail of protease inhibitors.
3. Extraction buffer: 50 mM HEPES-KOH (pH 7.5), 10 % Sucrose. Add the following reagents immediately before use to the cold buffer: 0.5 mM DTT and a cocktail of protease inhibitors.
4. Dialysis buffer: 25 mM HEPES-KOH (pH 7.6), 50 mM KCl, 0.1 mM EDTA. Add 2 mM DTT and a cocktail of protease inhibitors immediately before use.

2.3. For Substrate Preparation

1. 2× YT (1 L): 16 g tryptone, 10 g yeast extract, 5 g NaCl, pH 7.0.
2. Solution I: 0.9 % glucose, 10 mM EDTA, 25 mM Tris–HCl (pH 8.0).
3. Solution II: 0.2 M NaOH, 1 % SDS (should be made before use).
4. Solution III: 3 M potassium acetate, 11.5 % glacial acetic acid. For 100 mL, mix 60 mL of 5 M potassium acetate (pH 5.3), 11.5 mL of glacial acetic acid, and 28.5 mL of ddH_2O.
5. TE buffer: 10 mM Tris–HCl (pH 8.0), 1 mM EDTA.
6. Strains and virus: f1MR1 (GC) and f1MR3 (AT) viral stocks with only one base difference from each other, M13mp18-GC and M13mp18-AT viral stocks, and *E. coli* strain XL1 Blue (recA1 endA1 gyrA96 thi-1 hsdR17 supE44 relA1 lac [F′ proAB lacIqZΔM15 Tn10 (Tetr)]) that is resistant to tetracycline.

7. Hydroxyapatite (HAP) resin was Bio-Gel HTP Gel from Bio-Rad. 50-mL and 5-mL empty columns were from VWR international. Sephacryl S-300 resin and K9/60 empty column were from GE Healthcare. Plasmid-Safe ATP-dependent DNase was purchased from Epicentre Biotechnologies. All restriction enzymes were from New England Biolabs. dNTPs and ATP were purchased from Invitrogen.

3. Methods

3.1. Preparation of Yeast Cell-Free Nuclear Extract

The protocol for the preparation of yeast nuclear extract has been well documented previously (18, 20–22), but only recently we have described the first protocol for preparing the yeast nuclear extract that supports mismatch repair assay in vitro (23).

1. Cell growth: Inoculate 3 × 1 mL of overnight yeast culture to 3 × 1 L of YPD in 4-L flasks (3 flasks) at 5:00 pm. Overnight culture at 30 °C with shaking (200 rpm) should have a yield of OD_{600} at ~3 the second day at 9:00 am (see Note 1). Harvest cells by centrifugation at 2,700 × *g* for 10 min at 20 °C in a Sorvall SLA-3000 rotor (500-mL bottle). Combine the cell pellets into one by resuspending cells in dH_2O and centrifugate at 2,700 × *g* for another 10 min at 20 °C. After decanting the supernatant, weigh the pellet to obtain the net weight of cells. A typical yield of ~30 g is expected from 3 L of culture.
2. Preparation of spheroplasts: Resuspend the cell pellet in ED solution at a ratio of 10 mL/g of cells. Incubate at 30 °C with shaking (100 rpm) for 20 min. Collect cells by centrifugation at 2,700 × *g* for 10 min and completely remove any liquid anywhere in the bottle by Kimwipes because ED solution is a strong inhibitor of yeast lytic enzyme in the latter steps. Add YPD/S at 1 mL/g of cells and 15 mL of 2 M sorbitol and resuspend the pellet. Add 50 mg of yeast lytic enzyme (MP Biochemicals, LLC, 70,000 U/g) and incubate at 30 °C with shaking (100 rpm) for about 1 h to convert cells to spheroplasts (18). Every 15 min, check the progress of spheroplasting by mixing 4 μL of cells with 20 μL of 1 % SDS and observing the number of cell ghosts under microscope. Digestion is complete when approximately 80 % of cells burst into cell debris in water. Additional yeast lytic enzyme and longer incubation time may be required if less than 80 % of spheroplasts is obtained within 1 h.
3. Preparation of nuclei: Keep everything cold on ice from this step. Add 300 mL of ice-cold YPD/S to stop the enzyme digestion. Pelletize spheroplasts by centrifugation at 2,700 × *g*

at 4 °C for 10 min. Wash the spheroplasts twice by resuspending the pellet in 300 mL of YPD/S solution and centrifugation. The resuspension of spheroplasts is easier to do by adding 50 mL of YPD/S first and using a spatula. The washed pellet is then resuspended in 300 mL of cold 1 M sorbitol and pelletized again by centrifugation. Resuspend spheroplasts in 100 mL of Ficoll buffer. Break spheroplasts by a 40-mL glass Dounce homogenizer with a tight-fitting type B pestle for three to five strokes on ice. Spin the lysate in a Sorvall SS-34 rotor at 3,800 × *g* for 10 min and transfer the supernatant to new tubes. It does not matter if the loose pellet also transfers. Repeat spinning the supernatant three more times until the slimy loose pellet disappears and only trace amount of pellet should be formed. Harvest nuclei by centrifugation at 20,000 × *g* for 30 min. The nuclei pellet can be stored at −80 °C overnight.

4. Extraction of nuclei: Resuspend the nuclei pellet in 15 mL of extraction buffer. Measure the volume, transfer to a cold beaker with a stirring bar, and then slowly add neutralized 4 M $(NH_4)_2SO_4$ to a final concentration of 0.9 M while stirring gently. Incubate on ice for 30 min with stirring (see Note 2). Spin the suspension at 170,000 × *g* in an ultracentrifuge 50Ti rotor (Beckman) at 4 °C for 1 h. Transfer the supernatant to a cold beaker with a 5-mL pipette carefully. Avoid disturbing the pellet and do not take the white floating material. Measure the volume of the supernatant again and add 0.35 g powdered solid $(NH_4)_2SO_4$ per mL supernatant gradually over a period of 30 min with stirring. Add 10 μL of 1 N NaOH/g of $(NH_4)_2SO_4$ added to neutralize the solution and continue stirring on ice for another 30 min. Transfer to a 50-mL tube for the Sorvall SS-34 rotor and spin at 14,500 × *g* for 20 min at 4 °C. Remove the supernatant by dumping, spin at 14,500 × *g* again for 5 min, and carefully remove all supernatant as much as possible with a Pasteur pipette.
5. Dialysis: Resuspend the pellet in 300 μL of the dialysis buffer. Dialyze the protein suspension overnight at 4 °C against the dialysis buffer with an extra 75 mM $(NH_4)_2SO_4$. Clarify the protein extract by centrifugation at 14,500 × *g* in an SS-34 rotor. Measure the protein and salt concentrations of the supernatant, aliquot, and store in −80 °C (see Note 3).

3.2. Preparation of HeLa Nuclear Extract

1. Preparation of HeLa nuclei: Grow HeLa cells in suspension. At least 3 L of HeLa suspension culture is needed for extract preparation. Chill the cell culture on ice for 30 min. The temperature should fall to less than 11 °C. Transfer the cells into 1-L centrifuge bottles and pelletize in a Sorvall RC-3B rotor at 3,000 × *g* for 8 min. The supernatant should be sanitized by Chlorox bleach before disposal. Resuspend the cell pellet in wash buffer (20 mL/L culture) and recover by centrifugation

at 3,300 × *g* for 5 min. Record the wet cell weight. Resuspend the washed pellet in hypotonic buffer (5 mL/L culture or 2.78 mL/g cell pellet). Let cells swell on ice for 10 min. Check for cell viability by mixing the cell suspension with an equal volume of 0.4 % trypan blue and observe the cells under a microscope. Dead cells take up the stain and appear blue. Under normal condition, most HeLa cells (>90 %) are supposed to be viable.

2. Extraction of nuclei: Break HeLa cells on ice by a Dounce homogenizer. Begin with ten strokes and check the progress of the lysis by observing the ratio of free intact nuclei and intact cells through trypan blue staining of a small aliquot. When ~80 % of the cells are broken (stained by trypan blue), pellet the nuclei in a SS 34 rotor at 4 °C (2,000 × *g*) for 5 min. Resuspend the nuclei pellet in cold extraction buffer (2.5 mL/L culture or 1.39 mL/g nuclei pellet). Measure the volume of the resuspended nuclei and add 0.031 volumes of 5 M NaCl stock solution (see Note 2). Mix the mixture on a rotator for 60 min at 4 °C. Pellet and discard the debris using a SS 34 rotor at 4 °C (14,500 × *g*) for 20 min. Measure the volume of the supernatant and transfer it to a small glass beaker with a stirring bar. Slowly add 0.42 g powdered ammonium sulfate/mL supernatant over a period of 20 min with slow stirring on ice. Keep stirring for an additional 20 min on ice. Pellet the proteins by centrifugation for 20 min in a SS 34 rotor at 4 °C (15,800 × *g*). Remove the supernatant as much as possible. Usually there will be some white flocculant material that does not pellet well. Discard it with the supernatant.
3. Dialysis: Resuspend the pellet in the dialysis buffer (note the difference in pH and DTT concentration) using ~90 μL/L original cell culture. The goal is to make the pellet pipettable by adding the least possible amount of buffer. Transfer the slurry to a dialysis bag (molecular weight cutoff: 6–8,000). Closely monitor the course of dialysis by measuring the conductivity of diluted aliquots (10 μL into 4 mL of ddH_2O). Stop dialysis when the conductivity is approximately 50 μS/cm. Clarify the extract by centrifugation in an Eppendorf tube at 22,000 × *g* for 15 min at 4 °C. Aliquot the nuclear extract in aliquots of 15–30 μL (one drop from a 9″ Pasteur pipette). Freeze the aliquots by liquid nitrogen and store at −80 °C (see Note 3).

3.3. Preparation of Mismatch Substrates

Eukaryotic MMR is bidirectional and specific to the newly synthesized daughter strand. A strand discontinuity such as nick and gap can direct strand-specific MMR and is required for repair activity in vitro. The strand discontinuities that target MMR to the newly synthesized daughter strand are of important biological relevance, because only the moiety on the newly synthesized strand, not on the parental strand, of a mismatch is faulty (3, 4, 24). Our method to generate mismatch substrates follows the protocol reported previously with modifications (14, 25, 26). Here, we describe a protocol for preparation of the gapped mismatch substrates that dramatically reduces DNA ligation prior to repair.

1. Preparation of virus stocks: To two 250-mL flasks containing 10 mL 2× YT medium each, add 100 μL of overnight culture of XL1 Blue and inoculate a single phage plaque of either phage A (e.g., M13mp18-GC) or phage B (e.g., M13mp18-AT). Incubate at 37 °C with shaking (250 rpm) for 6 h. Then spin the culture at 11,000 × *g* with a Sorvall SS-34 rotor and save the supernatants as phage stocks. The phage stocks should have a titer of ~5×10^{11} phages/mL and can be stored at 4 °C for years. The phage stocks need to be amplified before going to the large scale: To two 250-mL flasks containing 50 mL of 2× YT medium, add 0.5 mL of overnight culture of XL1 Blue. Incubate at 37 °C with shaking until OD_{600} reaches 0.2 (~90 min). Add 0.5 mL of each phage stock to each of the two flasks. Incubate another 6 h at 37 °C with shaking. Collect the supernatants again after spinning with the SS-34 rotor. The titer of the phages should be ~2×10^{12}/mL. These phage suspensions can be stored at 4 °C for a week before going to the next step. Use the pellet to miniprep the phage dsDNA for verification.
2. Virus amplification and DNA preparation: At 10:00 pm, to each of the four 6-L flasks containing 1.5 L 2× YT medium, add 10 mL of overnight culture of XL1 Blue. Incubate at 37 °C with 250 rpm shaking. Retrieve 1 mL of sample every 30 min for OD_{600} measurement. When OD_{600} reaches ~0.3 (about 90 min, ~5×10^{8} cells/mL), add 8 mL of phage A to each of two flasks with 1.5 L cultures and 8 mL of phage B to each of the other two flasks. Incubate at 37 °C with shaking for another 7 h (from 12:00 am to 7:00 am) (see Note 4). Cultures were placed on ice for about 20 min to cool down followed by centrifuging at 2,700 × *g* in a Sorvall SLA-3000 rotor for 10 min. Both the pellets and the supernatant should be saved for DNA preparation. The cell pellets (from 3 L culture) were resuspended by vortex in 60 mL Solution I with 5 mg/mL lysozyme. Incubate the suspension at room temperature for 10 min and then on ice for 10 min. Add 120 mL fresh made Solution II, gently mix well, incubate on ice for 10 min. Add 90 mL ice cold Solution III, mix well, incubate on ice for 10 min. Spin in a SS-34 rotor at 11,000 × *g* at 4 °C for 60 min. Collect the supernatant by four layers of cheesecloth. Add 0.6 vol. of isopropanol to the supernatant, shake and vortex vigorously several times, and let it stand at room temperature for 30 min. Spin at 11,000 × *g* for 30 min at room temperature. Discard the supernatant and wash the pellet with room temperature 70 % ethanol. Briefly spin and decant ethanol as much as possible. Dry under vacuum for about 5 min. Add 20 mL

1×TE to the GSA bottles and gently shake the bottles for 1–2 h at room temperature to dissolve the pellets. Transfer the dissolved DNA to a 50 mL Falcon tube and wash with an additional 2 mL TE (total 22 mL for 3 L culture). To every 1.0 g of DNA solution, add 1.05 g CsCl, 50 μL 10 mg/mL ethidium bromide and then centrifuge at 160,000×*g* in a Beckman NVT 65 rotor for 18 h at 20 °C. Collect the band of supercoiled DNA. Remove ethidium bromide with n-butanol and dialyze against 3×1.0 L TE buffer (pH 8.0). Determine DNA concentration at A_{260}.

Phage supernatant obtained from dsDNA preparation was spun again at 3,500×*g* for 20 min at 4 °C. Transfer the phage suspension to a 4 L flask. To each Liter of solution, add 36 g NaCl, 50 g PEG. Stir at room temperature for 1 h. Spin at 3,500×*g* for 30 min at 4 °C. Resuspend phage in 15 mL 10 mM Tris–HCl (pH 8.0), then wash bottles with 5 mL Tris–HCl twice. Shake in 37 °C water bath for 0.5 h to dissolve phage particles. Spin at 11,000×*g* in a SS34 rotor for 10 min at 4 °C, discard pellet. To 1.0 g phage solution, add 0.4348 g CsCl, centrifuge at 160,000×*g* in a NVT 65 rotor for 16 h. Collect the fluffy band (usually the top band), dialyze against 3×1.0 L TE (pH 8.0), extract with phenol three times followed by ether extraction three times. Incubate the extracted DNA at 37 °C for 10 min in order to remove ether. Dialyze against 3×1.0 L TE buffer (pH 8.0) in order to remove CsCl. Measure DNA concentration at A_{260}.

3. Enzyme digestion: To prepare the 5′ gapped G-T mismatch substrate, *Sau*96I and DrdI restriction enzymes were used to digest f1MR3 (GC) dsDNA in order to generate a 36-base gap (Fig. 1a). Similarly, PstI and SalI restriction enzymes were used to digest M13mp18-GC dsDNA in order to generate the 3′ gapped (26-base) G-T mismatch substrate (Fig. 1b). To prepare nicked mismatch substrate, use only *Sau*96I to digest M13mp18-GC or M13mp18-AT dsDNA. Because the procedure to prepare both 5′ and 3′ substrates is basically the same, we use preparation of the 3′ gapped substrate based on M13 as example. A typical reaction is conducted in 8 mL of reaction volume that contains 1.5 mg of dsDNA and 600 units of each PstI and SalI. Incubate at 37 °C for 2 h and check the completion of digestion by retrieving 1 μL of DNA sample for agarose gel analysis. Additional enzymes may be needed in case of incomplete digestion. To remove the restriction enzymes, phenol extraction (once), ether extraction (twice), and ethanol precipitation are required. Resuspend the pellets in 2 mL of TE buffer (pH 8.0).

4. Heteroduplex formation: Set aside 133 μL (~0.1 mg) of PstI-SalI digested M13mp18-GC dsDNA for the homoduplex control substrate. To the remaining 1.4 mg of dsDNA, add

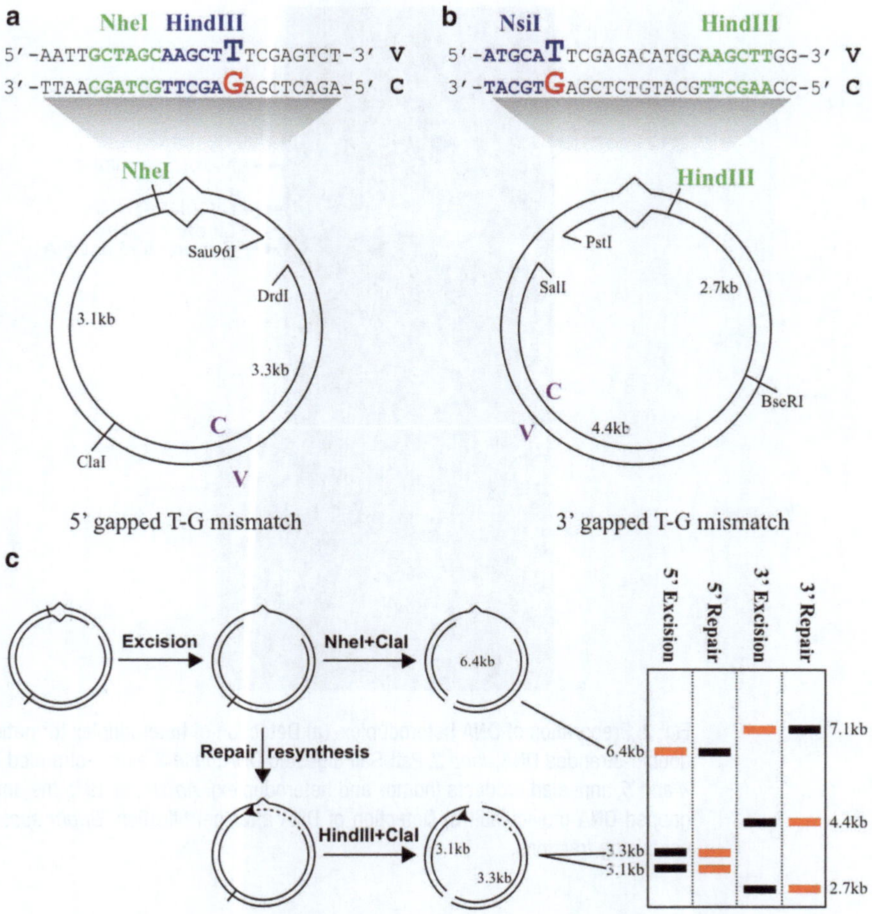

Fig. 1. 5′ and 3′ gapped mismatch substrates. (**a**) The plasmid-based 5′ gapped T-G mismatch substrate. This substrate is based on the f1MR1 phage as described previously (14, 15). The sequence detail around the mismatch is shown on the *top*. T-G mismatch is highlighted in *bigger fonts*. The mismatch excision scoring enzyme NheI are shown in *green* (*Gray* and *bold* if colorless) and the mismatch repair scoring enzyme HindIII is shown in *blue* (*Dark* and *bold* if colorless). Sau96I and DrdI are the enzymes for creation of the gap. ClaI is used as the reference enzyme to create two pieces of DNA fragments. The numbers in kb are the sizes of DNA fragments after digestion. V and C indicate the viral (continuous) strand and the complementary (discontinued, gapped) strand respectively. (**b**) 3′ gapped T-G mismatch substrate. It is based on the M13 phage as described previously (14). The sequence detail around the mismatch is shown on the *top*. T-G mismatch is highlighted in *bigger fonts*. The 3′ mismatch excision scoring enzyme HindIII are shown in *green* (*Gray* and *bold* if colorless) and the 3′ mismatch repair scoring enzyme NsiI is shown in *blue* (*Dark* and *bold* if colorless). SalI and PstI are the enzymes for creation of the gap. BseRI is used as the reference enzyme to create two pieces of DNA fragments. The numbers in kb are the sizes of DNA fragments after digestion. V and C indicate the viral (continuous) strand and the complementary (discontinued, gapped) strand respectively. (**c**) Schematic representation of the 5′ and 3′ gap-directed mismatch excision and repair on agarose gel. *Orange* (*gray* if colorless) bars highlight what to observe in an indicated reaction. For mismatch excision, successful removal of mismatches will lead to a single-stranded area around the mismatch and therefore destroy the recognition sites of excision scoring enzymes, resulting in a bigger band after enzyme digestion (6.4 kb for the 5′ gapped substrate or 7.1 kb for the 3′ gapped substrate). For mismatch repair, successful correction of mismatches will lead to the recovery of the recognition sites of repair scoring enzymes, resulting in two smaller bands (3.3 kb and 3.1 kb for the 5′ gapped substrate or 4.4 kb and 2.7 kb for the 3′ gapped substrate). This figure is reproduced from Yuan et al. (23) with permission from the publisher.

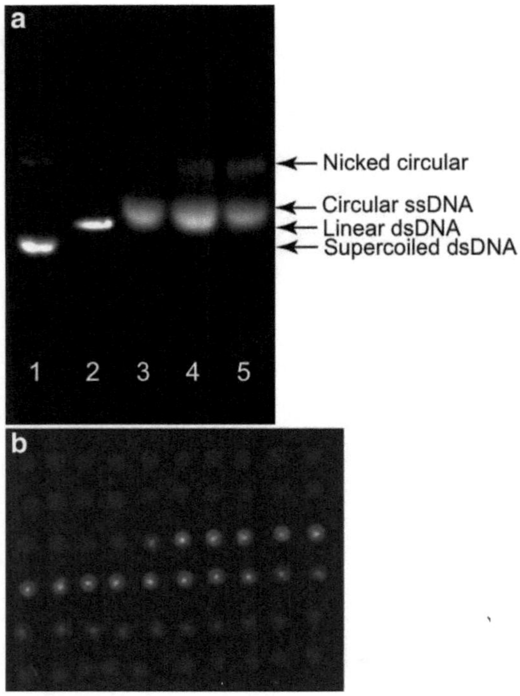

Fig. 2. Preparation of DNA heteroduplex. (**a**) Detection of heteroduplex formation. *Lane 1*, double-stranded DNA; *lane 2*, PstI-SalI digested DNA; *lane 3*, single-stranded DNA; *lanes 4* and *5*, annealed products (homo- and heteroduplex). *Arrow* points to the annealed and gapped DNA molecules. (**b**) Detection of DNA after gel filtration. *Bright spots* are DNA-containing fractions.

7 mg of M13mp18-AT closed circular ssDNA (20 mL if 0.35 mg/mL) that makes the ratio of ssDNA:dsDNA 10:1 in molecules; Add 3 mL of 0.5 M Tris–HCl (pH 7.5), 60 μL of 0.5 M EDTA, pH 8.0, 100 μL of 3 M NaCl. Add dH_2O to the final volume of 30 mL. The final concentrations for Tris, EDTA, and NaCl are 50 mM, 1 mM, and 10 mM respectively. In this well mixed mixture, add 900 μL of 10 N NaOH and incubate at room temperature for 5 min; add 3 mL of 2.9 N acetic acid (1:6 diluted glacial acetic acid), 1.35 mL of 3 M KCl, and 3.72 mL of 1 M potassium phosphate buffer, pH 7.4 and incubate at 65 °C for 30 min in a water bath. Slowly cool down the water bath to 37 °C (~3 h). Incubate at 37 °C for an additional 30 min and store on ice. Retrieve 3 μL of the mix and check the annealing product by running an agarose gel. The slowest moving nicked-circular position is where the gapped G-T mismatch substrate is (Fig. 2a). The homoduplex control DNA substrate is prepared simply by scaling down the annealing reaction to 2 mL and annealing the PstI-SalI digested M13mp18-GC dsDNA with M13mp18-GC ssDNA.

5. Hydroxyapatite (HAP) column purification: Weigh 14 g (for purification of heteroduplex mismatch substrates) and 5 g

(for homoduplexes) of hydroxyapatite and suspend them in 0.03 M sodium phosphate buffer (pH 6.9, buffer made by mixing monobasic and dibasic sodium phosphate without additional components). After three more times of wash with the buffer, load the suspended hydroxyapatite into empty columns (50-mL for the heteroduplex, 5-mL for the homoduplex). After washing the columns with 2 column-volume of the phosphate buffer, load the 3× ddH_2O-diluted DNA samples onto the columns. Wash the columns with 2 column-volume of the 0.03 M sodium phosphate buffer and 4 column-volume of 0.16 M sodium phosphate buffer (pH 6.9). The ssDNA should be eluted at this step. Elute the annealed dsDNA with 2 column-volume of 0.42 M sodium phosphate buffer (pH 6.9) by proper fractionation (i.e., 3 mL for the 50-mL column and 0.5 mL for the 5 mL column). Use nicked circular DNA as control to verify the successful elution of the annealed and gapped mismatch substrates. The pooled samples were dialyzed against TE buffer overnight.

6. Removal of the trace amount of linear dsDNA. In the gapped circular mismatch substrates after HAP column purification, there are still trace amount of contaminated linear dsDNA. The Plasmid-Safe ATP-dependent DNase (Epicentre Biotechnologies) is specialized in removing these contaminants but has no activity on the circular dsDNA. In the dialyzed HAP fractions (~35 mL), add 4.2 mL 10× reaction buffer and ATP to 1 mM. Add 420 U of the DNase (42 μL). Incubate at 37 °C for 2 h and check for the completion of digestion by running a 1 % agarose gel. The mixture was dialyzed against TE buffer for 3 h with a buffer change in the middle. Extract the mixture with phenol once and extract with butanol until the volume reduced to ~1–2 mL.

7. Gel filtration purification. This step further removes contaminants in the substrates. Pack a K9/60 empty column (GE Healthcare) with 40 mL of Sephacryl S-300 resin. After washing the column with 2 column-volume of TE buffer (pH 7.6) containing 0.3 M NaCl, the concentrated DNA sample from DNase digestion was loaded onto the S-300 column. For better separation, only less than 1 mL of DNA sample was applied each time. Multiple running is required when the sample volume is more than 1 mL. Resolve the column with the same NaCl containing TE buffer and collect fractions at 1 mL/tube. Locate the DNA by mixing 1 μL of fractions with 7 μL of ethidium bromide (1 μg/mL) on Saran plastic wrap and exposing to UV light (Fig. 2b). Check the DNA-containing fractions by 1 % agarose gel and pool the DNA-containing fractions (see Note 5) and dialyze against TE buffer to remove NaCl. The dialyzed DNA will be the final product and should be aliquoted to proper size (i.e., 200 μL/tube) and stored in –80 °C. The DNA substrates can be stored in 4 °C for several weeks.

3.4. Mismatch Excision and Repair Assays

Gap-directed mismatch excision and repair are assayed essentially as described previously for the nick-directed excision and repair (14, 15) in a 20-μL reaction mixture containing 100 ng of mismatched or homoduplex DNA, 30 μg of yeast nuclear extract (see Note 6), 10 mM Tris–HCl (pH 7.6), 5 mM $MgCl_2$, 110 mM KCl, and 1.5 mM ATP. 100 μM dNTPs is required for the repair assay, while no dNTPs should be included in the excision assay (see Note 7). Purified yeast MSH2-MSH6 (MutSα, see Note 8) is added as indicated in Fig. 3. After incubation at 30 °C for 15 min, reactions are terminated by adding 40 μL of fresh-made proteinase K solution (0.67 % SDS, 25 mM EDTA, and 0.3 mg/mL proteinase K) and incubation at 37 °C for at least 30 min. After proteinase K digestion, add TE buffer to bring the volume of mixture to 100 μL and extract it with phenol twice before ethanol precipitation. Ethanol precipitation works best by adding 1/10 volume of 3 M sodium acetate, 2.5 volume of 100 % ethanol, and −80 °C for exactly 13 min (see Note 9). The pellet after centrifugation is visible (because of non-DNA materials) and should be subjected to vigorous wash by 70 % ethanol at room temperature for 10 min. Spin, decant the 70 % ethanol, and dry the pellet in a speed vac. Resuspend the dried pellet in 10 μL dH_2O and digest it with the scoring and reference enzymes (in a 20 μL mixture) as indicated in Fig. 1a, b. The digested products are separated on 1 % agarose gels, and DNA bands were visualized by UV illumination in the presence of ethidium bromide (see Note 10 for some preliminary results).

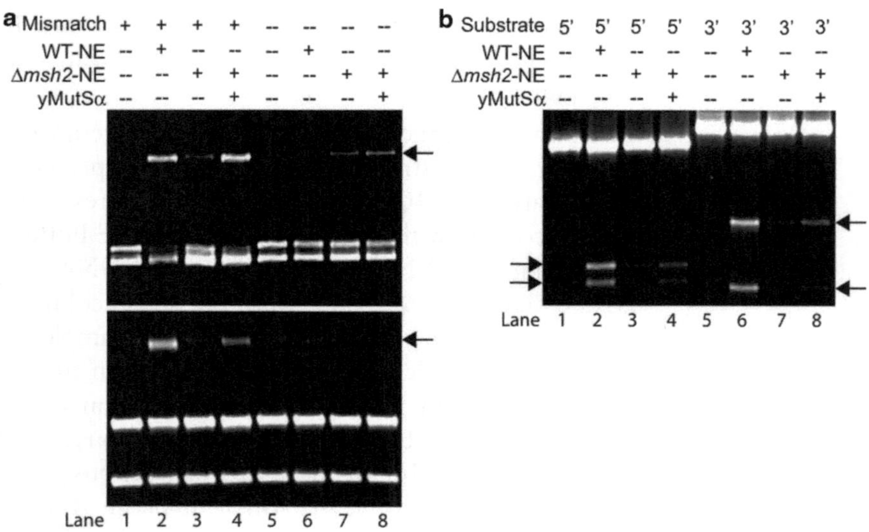

Fig. 3. Mismatch excision and repair activities in yeast nuclear extracts. (**a**) Mismatch excision on the 5′ and 3′ gapped mismatch and homoduplex substrates. WT-NE, wild type nuclear extract; Δ*msh2*-NE, MSH2 deletion nuclear extract; yMutSα, purified yeast MSH2-MSH6 (20 nM). *Arrows* point to the excision products. *Lanes 1* and *5* show the substrates only. (**b**) Mismatch repair on the 5′ and 3′ gapped mismatch substrates. *Arrows* point to the repair products. *Lanes 1* and *5* show the substrates only. This figure is reproduced from Yuan et al. (23) with permission from the publisher.

4. Notes

1. A yeast culture with OD_{600} over 3 may experience hard time when being converted to spheroplasts in the latter steps.
2. The extraction of nuclei should be very carefully performed to reduce contamination from nonspecific nucleases and yet to obtain enough nuclear proteins for repair meanwhile.
3. Cell-free nuclear extract may be stored in small working aliquots. Repeated freeze–thaw leads to the instability of proteins.
4. Less ssDNA will be recovered when incubated for more than 7 h.
5. Balance the amount of DNA and its purity before combining all DNA-containing fractions. Do not include fractions with limited DNA.
6. The procedures for mismatch excision and repair assays using HeLa nuclear extract is the same as the yeast extract except that 50 μg of HeLa nuclear extract should be used.
7. The conductivity of the distilled water should be lower than 2.0 μS/cm. Quality of the distilled water is critical for mismatch excision and repair assay. It is best to prepare all solutions involved in mismatch excision and repair assay fresh. If not, solution stocks should be tested before use.
8. The purified yeast MutSα is available at the authors' lab for research purpose only.
9. Samples should be placed at −80 °C for exactly 13 min, and centrifuged for 13 min at 4 °C. Longer time in any situations may result in more precipitated salt in samples that will affect subsequent enzyme digestion.
10. As shown in Fig. 3a, the wild-type yeast nuclear extract has robust excision activities on both 5′ and 3′ gapped mismatch substrates, but not on the homoduplex control DNA (Fig. 3a, compare lane 2 with 6). The 5′ and 3′ repair activities of the wild-type extract are also clear as indicated in Fig. 3b, lanes 2 and 6. This activity is comparable to that of human HeLa nuclear extract (25), considering that less protein is used for the yeast extract. We noticed that more extract does not help increase the repair activity because the contaminated nucleases nonspecifically degraded the DNA substrate if more extract is used. To further prove the validity of the cell-free repair system, we included an extract from the mismatch repair deficient strain Δ*msh2*. As expected, the Δ*msh2* deletion extract lost the mismatch specific excision and repair activities on both 5′ and 3′ gapped substrates (Fig. 3a, b, compare lane 3 with 2, Fig. 3b compare lane 7 with 6). Msh2 deletion extract shows some nonspecific excision on the 5′ gapped T-G substrate (Fig. 3a, compare lanes

7 and 3 with 6 on the 5′ T-G substrate), which can be explained by that MutSα increases mismatch excision and repair specificity (10, 27). Very importantly, the complementation experiment with purified yeast MutSα protein shows the restoration of both mismatch-specific excision and repair activities to the deletion extract (Fig. 3a, b lanes 4, Fig. 3b lane 8).

Acknowledgment

This work was supported by a NIH grant R01HL105631 (to Y.Z.), a New Investigator Research grant (to Y.Z.) from Florida Biomedical Research Program, and a grant #31101669 from National Science Foundation of China (to F.L.).

References

1. Harfe BD, Jinks-Robertson S (2000) Annu Rev Genet 34:359–399
2. Kolodner RD, Marsischky GT (1999) Curr Opin Genet Dev 9:89–96
3. Kunkel TA, Erie DA (2005) Annu Rev Biochem 74:681–710
4. Modrich P, Lahue R (1996) Annu Rev Biochem 65:101–133
5. Schofield MJ, Hsieh P (2003) Annu Rev Microbiol 57:579–608
6. Jiricny J (2006) Nat Rev Mol Cell Biol 7:335–346
7. Li GM (2008) Cell Res 18:85–98
8. Burdett V, Baitinger C, Viswanathan M, Lovett ST, Modrich P (2001) Proc Natl Acad Sci USA 98:6765–6770
9. Lahue RS, Au KG, Modrich P (1989) Science 245:160–164
10. Zhang Y, Yuan F, Presnell SR, Tian K, Gao Y, Tomkinson AE, Gu L, Li GM (2005) Cell 122:693–705
11. Song L, Yuan F, Zhang Y (2010) IUBMB Life 62:548–553
12. Amin NS, Nguyen MN, Oh S, Kolodner RD (2001) Mol Cell Biol 21:5142–5155
13. Wei K, Clark AB, Wong E, Kane MF, Mazur DJ, Parris T, Kolas NK, Russell R, Hou H Jr, Kneitz B, Yang G, Kunkel TA, Kolodner RD, Cohen PE, Edelmann W (2003) Genes Dev 17:603–614
14. Guo S, Presnell SR, Yuan F, Zhang Y, Gu L, Li GM (2004) J Biol Chem 279:16912–16917
15. Holmes J Jr, Clark S, Modrich P (1990) Proc Natl Acad Sci USA 87:5837–5841
16. Thomas DC, Roberts JD, Kunkel TA (1991) J Biol Chem 266:3744–3751
17. Wang H, Hays JB (2000) Mol Biotechnol 15:97–104
18. Wang Z (2006) Methods Mol Biol 313:317–331
19. Wang Z, Wu X, Friedberg EC (1993) Proc Natl Acad Sci USA 90:4907–4911
20. Muster-Nassal C, Kolodner R (1986) Proc Natl Acad Sci USA 83:7618–7622
21. Ponticelli AS, Struhl K (1990) Mol Cell Biol 10:2832–2839
22. Verdier JM, Stalder R, Roberge M, Amati B, Sentenac A, Gasser SM (1990) Nucleic Acids Res 18:7033–7039
23. Yuan F, Lai F, Gu L, Zhou W, El Hokayem J, Zhang Y (2009) Methods 48:14–18
24. Larson ED, Nickens D, Drummond JT (2002) Nucleic Acids Res 30:E14
25. Parsons R, Li GM, Longley MJ, Fang WH, Papadopoulos N, Jen J, de la Chapelle A, Kinzler KW, Vogelstein B, Modrich P (1993) Cell 75:1227–1236
26. Su SS, Lahue RS, Au KG, Modrich P (1988) J Biol Chem 263:6829–6835
27. Genschel J, Modrich P (2003) Mol Cell 12:1077–1086

Chapter 12

Immuno-Slot Blot Assay for Detection of UVR-Mediated DNA Damage

Mahsa Karbaschi, Nigel J. Brady, Mark D. Evans, and Marcus S. Cooke

Abstract

Solar ultraviolet radiation (UVR), through the formation of DNA photolesions, is the primary cause of most skin cancers. A better understanding of the mechanisms of UVR-induced DNA damage may help prevent skin cancer and this may be achieved using methods to quantify DNA damage. The immuno-slot blot (ISB) method is routinely used for detection and quantification of any heat- and alkali-stable DNA adducts for which a sufficiently specific monoclonal antibody is available. The main steps in ISB are fragmentation and denaturation of the DNA, immobilization of DNA to a nitrocellulose filter, incubation with primary antibody against a specific DNA adduct, incubation with an enzyme-linked secondary antibody and finally chemiluminescence detection and quantification of the DNA adducts.

Key words: Reactive oxygen species, Cyclobutane pyrimidine dimers, Ultraviolet radiation, Nonelectrophoretic transfer, Immuno-slot blot, Immunoassay, Monoclonal antibodies, Nitrocellulose membrane

1. Introduction

Solar ultraviolet radiation (UVR) is reaching the Earth's surface. DNA is one of the main targets for UVR to induce damage in different organisms ranging from bacteria to humans (1). Over the last few decades, there has been a significant and growing interest in understanding the mechanisms of various effects of UVR on DNA. The UVR-induced DNA damage depends on the wavelength of UVR (2). UVB (280–320 nm) absorption by DNA results in direct dimerization of adjacent pyrimidine bases which results in different forms of DNA lesions such as cyclobutane-pyrimidine dimers (CPDs), (6-4) photoproducts (6-4PPs), and their Dewar valence isomers (3, 4). Cellular absorption of UVR also produces reactive oxygen species (ROS), which result in

Lotte Bjergbæk (ed.), *DNA Repair Protocols*, Methods in Molecular Biology, vol. 920,
DOI 10.1007/978-1-61779-998-3_12, © Springer Science+Business Media New York 2012

oxidatively generated modification of DNA, e.g., 8-oxo-7,8-dihydro-2′-deoxyguanosine (5). UVA radiation excites endogenous chromophores which either transfer energy to DNA (type I photosensitization) or transfer energy to oxygen, producing singlet oxygen (type II photosensitization) (6–8). In both cases, guanine is the major target of such photoreactions. UVA may also produce hydroxyl radicals through the photosensitized production of superoxide anions (4). Because of high reactivity and low specificity of hydroxyl radicals, this may produce a wide range of DNA damage. UVA has also been reported to induce thymine dimer (T<>T) (9–11), but the mechanism of CPD induction by UVA is not known (5). UVR also leads to the induction of lipid peroxidation and subsequently damage to DNA (12) via induction of reactive carbonyl species such as malondialdehyde. Consequently, reaction of malondialdehyde with deoxyguanosine, can produce malondialdehyde-deoxyguanosine (M_1dG), a mutagenic adduct (13, 14). M_1dG is not just formed from lipid peroxidation, but also by hydrogen abstraction from the deoxyribose moiety, most likely by ROS.

The sensitive detection and quantification of DNA modifications is an essential requirement for analysis of specific DNA base modifications and cellular DNA repair activity. Whilst a number of related techniques have been developed for this purpose, each appears to have some limitations. Here we describe a protocol for immuno-slot blot (ISB).

ISB is a highly sensitive assay which is applied for detection and quantification of DNA adducts that are produced by carcinogens, mutagens, or chemotherapeutic agents. This method is used in risk estimation of exposure to environmental carcinogens, molecular dosimetry and on cancer patients receiving medications with alkylating cytostatic drugs (15). This assay, which was first developed by Muller et al. (16), is designed for the detection of adducts for which a specific monoclonal antibody (which does not have a cross-activity with other normal or altered DNA nucleobases) is available. This method is limited to the study of DNA adducts that are stable during heat and alkali denaturation of the doubled stranded DNA, during the procedural steps in the assay.

Briefly, the ISB assay comprises fragmentation of the DNA containing modified bases by sonication and heat denaturation at alkaline pH. The denatured DNA is then immobilized on a nitrocellulose membrane using apparatus with slot shaped wells under vacuum. After drying, the membrane is incubated with primary antibody against the DNA adduct under investigation. Then, an enzyme-linked secondary antibody raised against the primary antibody is added. Bound antibodies are then detected by using chemiluminescence reagents which fog photographic film resulting in dark bands, whose intensity is proportional to the antibody bound to the DNA adduct and hence damage present. The DNA adducts are detected and quantified by densitometric evaluation of the bands.

In comparison to other routinely used techniques ISB has some advantages. Some of these advantages over other procedures used for the detection of specific DNA lesions are as follows:

- The ISB assay was developed predominantly to detect and quantify a small level of DNA adducts in a small amount of DNA. Only 3.5 μg of DNA is required for a triplicate determination of modified DNA nucleobases. Therefore, this method is applicable to small biopsy samples, DNA restriction fragments, or certain subfractions of chromatin (17). This compares well to the sensitivity of other routinely used methods including radioimmunoassay (17).
- Minimal processing of DNA is required with ISB (18). No enzymatic digestion is required which is the main advantage of ISB over methods such as the enzyme-modified comet assay. Full accessibility of the enzymes to all DNA adducts in the nucleoid with the modified comet assay may also be of concern.
- With the use of appropriate antibodies against specific DNA adducts, a high sensitivity for detection of femtomole to picomole amounts of modified nucleosides in the DNA is possible (15, 17). This exceeds the sensitivity of other routinely used methods for detection and quantification of DNA adducts.
- Since ISB is not competitive, the modified nucleobases are detected without the use of labeled molecules such as radioactive isotopes which is a notable advantage over techniques such as radioimmunoassay.
- The ISB results are reproducible, whereas the reproducibility of competitive ELISA has often been a problem, partly because of variations in the absorption of the antigens to plastic surfaces of the plates (17).

In this chapter, we introduce a semiquantitative ISB method for the detection of T<>T on the DNA (Fig. 1) and a quantitative ISB method for the detection of M_1dG adduct.

2. Materials

All reagents are used as received unless otherwise specified. All solutions are prepared using ultrapure water (≥18 MΩ resistance) or autoclaved HPLC grade water. Prepare and store all the buffers at room temperature unless mentioned otherwise.

2.1. Preparation of Buffers and Reagents

1. 10 mM potassium phosphate, pH 7.0 (KP Buffer 1): Dissolve 348.36 mg of K_2HPO_4 (MW. 174.18) in 200 mL of water (10 mM). Adjust pH to 7.0 with 2 M phosphoric acid.

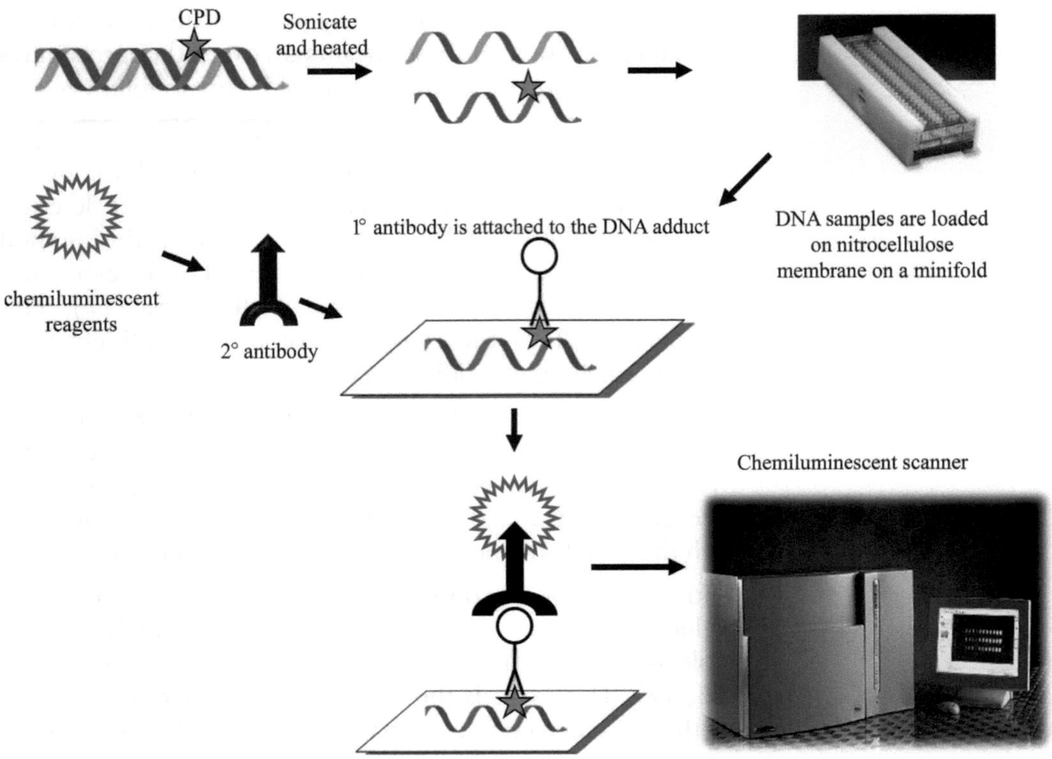

Fig. 1. Schematic representation of the immuno-slot blot detection of CPD on DNA.

2. 30 mM potassium phosphate buffer, (KP Buffer 2):
   Buffer should be made with 0.0088 M of K_2HPO_4 (Sigma) and 0.0035 M of KH_2PO_4 (Sigma). To make 1 L of this buffer, you need 1.53 g of K_2HPO_4 and 0.48 g of KH_2PO_4, both dissolved in 1 L of water.
3. Phosphate Buffered Saline, (PBS):
   Make 0.01 M PBS by dissolving 1 PBS tablet (Sigma) in 200 mL of water according to the manufacturer's instructions.
4. PBS plus 0.1 % (v/v) Tween-20 (PBS-T):
   For 2 L of PBS-T, add 2 mL of Tween-20 to 1,998 mL of PBS and mix using a magnetic stirrer for 30 min.
5. 1 M ammonium acetate
   Make 1 M ammonium acetate (MW. 77.08) by dissolving 77.08 g ammonium acetate (Fisher Scientific) in 1 L of water.
6. 2 M ammonium acetate:
   Make 1 M ammonium acetate (MW. 77) by dissolving 154.16 g ammonium acetate (Fisher Scientific) in 1 L of water.
7. 5 μg/mL Propidium iodide, (PI)
   Dilute 250 μg of PI (Sigma) in 50 mL of PBS (see Note 1).

2.2. Immuno-Slot Blot Components

1. Gel-blotting papers (GB 002, Schleicher and Schuell, Anderman and Company Ltd., Kingston-Upon-Thames, UK).
2. Immuno-Slot blot apparatus (Schleicher and Schuell) (see Note 2).
3. Cellulose nitrate filter (0.1 μM, BA79, Schleicher and Schuell).
4. Blocking solution, 5 %w/v dry powder low fat milk in PBS-T.
5. SuperSignal ultra chemiluminescent substrate (Pierce, Illinois, USA).
6. ECL-hyperfilm (Amersham International plc.).
7. Aspirator.

2.3. Antibodies

The following section describes the antibodies for the detection of malondialdehyde-2′-deoxyguanosine (M1dG) and cyclobutane thymine-thymine dimer (T<>T):

1. The stock solution of the primary antibody (mouse monoclonal D101A antibody) for M_1dG (Dako, Denmark) with 0.3 mg/mL concentration should be diluted 1:6 (to a final concentration of 50 ng/μL) with PBS prior to use, and from this aliquots should be prepared and stored at −80 °C until further use (see Note 3).
2. Stock solution of primary antithymine dimer, mouse monoclonal, IgG_1 (0.5 mg/mL purified monoclonal antibody in PBS with protein stabilizer and 0.1 % sodium azide (Kamiya Biomedical, Seattle, USA)) should be diluted 1 in 10 in PBS to a final concentration of 50 μg/mL. This stock should be aliquoted and stored at −80 °C until further use (see Note 3).
3. Secondary antibody (for both M_1dG and T<>T assays), goat anti-mouse immunoglobulin horseradish peroxidase conjugated, (Dako A/S, Denmark) which should be used as received (1 g/L) and stored at 4 °C.

3. Methods

3.1. Immuno-Slot Blot Determination of T<>T in DNA Samples

3.1.1. Preparation of Samples

1. Extract the DNA from cells or tissues of interest and measure the concentration of the extracted DNA. For example, DNA can be extracted from human skin keratinocytes using AllPrep DNA/RNA/protein mini kit (Qiagen, UK). The DNA is eluted in DNase/RNase-free water. Determine DNA concentration by spectrophotometry.
2. Pipette the appropriate volume corresponding to 3.5 μg of DNA and make the volume up to 100 μL with KP buffer 1. Vortex-mix and centrifuge at 20,000 × *g* for 1 min (see Note 4).

3. Sonicate the samples for 20 min in a sonicating water bath.
4. Add 150 μL of PBS and vortex-mix.
5. Heat to 100 °C for 5 min and then cool on ice for at least 10 min (see Note 5).
6. Vortex-mix and centrifuge at 20,000 × g in a microfuge for 2 min. Then add 250 μL of 2 M ammonium acetate, vortex-mix, and centrifuge again at 20,000 × g for 1 min.
7. Pipette 142.9 μL for 1 μg of DNA onto the Nitrocellulose filter.

3.1.2. Immuno-Slot Blot

The slot blot apparatus used here is a 24 × 3 well format. This allows for the analysis of nine standards (to construct the calibration line) and 15 samples in triplicate.

1. Two gel-blotting papers are wetted in 1 M ammonium acetate for 10 s before being put onto the immuno-slot blot apparatus (see Note 6).
2. Cut the cellulose nitrate filter to size, using a gel-blotting paper template. A V-shaped nick (no longer than 2–3 mm for each arm of the "V") should be cut out of the top-left corner to indicate the orientation of the nitrocellulose filter. The filter must then be wetted in distilled water followed by 1 M ammonium acetate for 10 s before being placed on top of the gel-blotting papers. Assemble the immuno-slot blot apparatus together ensuring that there are no air bubbles between the filter and blotting paper by placing two plastic clamps on each side and subsequently connecting to an aspirator (Fig. 2).

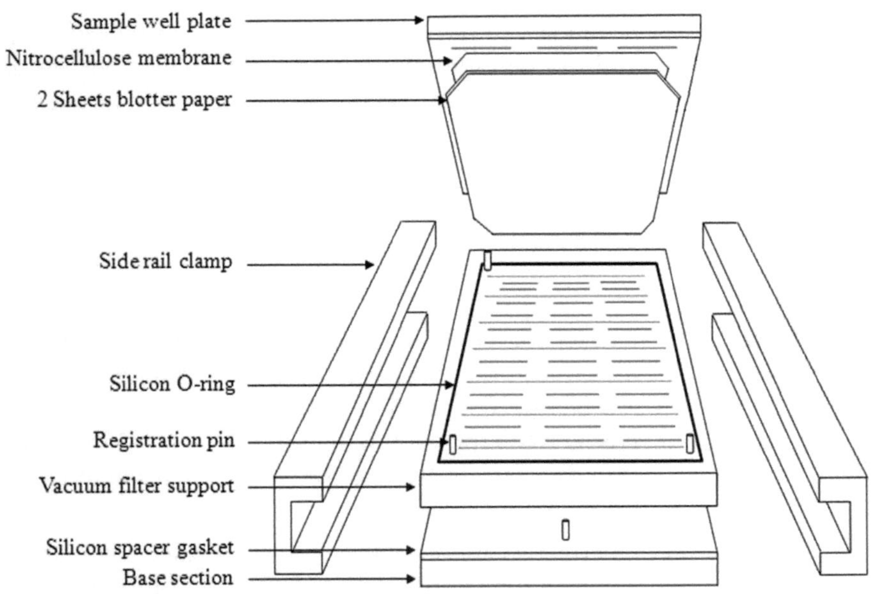

Fig. 2. Assembly instruction for the Minifold II device based upon the manufacturer's diagram with the product insert.

3. Pipette an appropriate volume that contains 1 μg of DNA for each standard and sample into each well of the immuno-slot blot apparatus while the aspirator is running at −3.5 in Hg pressure.
4. Allow the wells to run dry and then add 200 μL of 1 M ammonium acetate to each well and again allow the wells to run dry, and then leave for 20 min.
5. Carefully place the nitrocellulose membrane between filter paper and heat at 80 °C for 1.5 h in the oven.
6. After heating, place the nitrocellulose filter in a plastic tray slightly bigger than the size of your membrane and bathe in 100 mL 5 % (w/v in PBS-T) nonfat milk solution (5.0 g, Marvel, Nestle Ltd.) and allow it to rock gently at room temperature (20 °C) for 1 h. Then wash the filter twice for 5 min each to remove the milk solution, adding 50 mL of PBS-T and allowing to rock at room temperature.
7. After blotting the nitrocellulose membrane in 5 % (w/v in PBS) milk powder, dilute the primary anti-T<>T antibody in PBS-T plus 0.5 % milk powder to a final concentration of 6.25 ng/mL. Next, incubate the filter with the primary anti-T<>T in a plastic tray slightly bigger than the size of the NC membrane and allow it to rock at room temperature for 1 h.
8. Wash the filter with PBS-T (50 mL) by gentle rocking in a tray for 1 min followed by a further two 5 min washes with PBS-T.
9. Dilute the secondary antibody (goat anti-mouse immunoglobulin horseradish peroxidase conjugated) 1 in 4,000 in PBS-T plus 0.5 % milk powder to a final concentration of 0.25 g/mL. Then allow the filter to rock at room temperature for 2 h.
10. Wash the filter with PBS-T (50 mL) by gentle rocking in a tray for 15 min followed by a further two 5 min washes with PBS-T. Be careful not to let the filter dry out after the final wash (see Note 7).
11. Just before using the SuperSignal ultra chemiluminescent substrate prepare a working solution by mixing equal volumes (4 mL of each) of the ultraluminol/enhancer solution with the ultrastable peroxide solution.
12. Bathe the filter in the chemiluminescence reagents for 5 min, gently blot on a paper towel or filter paper, and then cover in cling film and expose to the ECL-hyperfilm.
13. Determine the level of the adduct in each DNA sample by obtaining a chemiluminescent image of the nitrocellulose filter using and image acquisition system, e.g., the ChemiGenius^{2} image acquisition system (Syngene, Cambridge, UK) (see Notes 8 and 9) (Fig. 3).

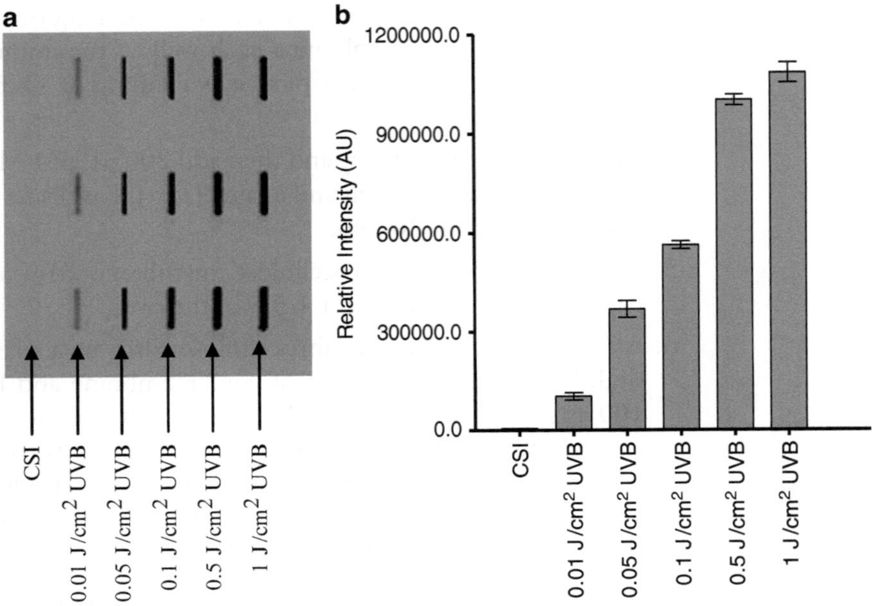

Fig. 3. (**a**) Immuno-slot blots obtained using antithymine dimer to detect the amount of T<>T in CT-DNA exposed to different doses of UVB; (**b**) Densitometric evaluation of the slot areas on the slot blot. *CSI* represents control sham irradiated samples.

3.2. Immuno-Slot Blot Determination of M_1dG in DNA Samples

A series of standards should be prepared, along with preparation of the samples, before performing the immuno-slot blot assay.

3.2.1. Preparation of MDA-Treated Calf Thymus DNA Samples

1. Prepare 2 mg/ mL of Calf Thymus DNA (CT-DNA) in ultrapure water.
2. Pipette out 1 mL of 2 mg/mL of CT-DNA and dry using speed vac or by freeze drying.
3. Make up 10 mL of 100 mM malondialdehyde (MDA) solution by dissolving 0.071 g MDA in 10 mL of KP buffer 2.
4. Make up 2 mL of 1 mM MDA solution by carrying out a 1 in 100 dilution of the 100 mM MDA solution. This is done by pipetting 20 μL of 100 mM MDA solution into 1,980 μL KP buffer 2. Prepare a range of MDA solution according to Table 1.
5. Incubate each MDA solution (1 mL) with 2 mg of CT-DNA for 24 h at 37 °C in a water bath incubator, and then vortex the samples.
6. After 24 h, add 100 μL of 3 M sodium acetate, followed by, either 700 μL of room-temperature isopropanol or 2 mL of ice-cold ethanol to the tubes (see Note 10). Mix the tubes by gentle inversion and centrifuge at 20,000 × *g* for 5 min. Discard the supernatant, add 1 mL of 70 % ethanol, and then centrifuge at 20,000 × *g* for 5 min.

Table 1
Preparation of MDA solution for generation of MDA-treated CT-DNA standard

Vial number	Concentration of MDA (μM)	Volume of MDA (1 mM)	KP buffer 1 (μL)
1	0	0 μL	1,000
2	1	1 μL	999
3	10	10 μL	990
4	50	50 μL	950
5	100	100 μL	900
6	250	250 μL	750
7	500	500 μL	500
8	1,000	1,000 μL	0

7. Again, discard the supernatant, add 1 mL of absolute ethanol, and then centrifuge at 20,000 × *g* for 5 min. Discard the supernatant and allow the samples to air-dry for 5 min.
8. Dissolve the pellet in 500 μL of HPLC grade water by leaving on a rotating wheel for 4–6 h.
9. Quantitate the concentration of the DNA in the solution in number 7. Take an aliquot of this solution equivalent to 2 μg of DNA and acid hydrolysis using 0.1 M formic acid at 70 °C for 1 h.
10. Determine the level of M1dG in each sample using HPLC fluorescence detection. For example use the method reported by Sing et al. (19).

3.2.2. Preparation of M_1dG Calibration Curve

1. Dilute CT-DNA treated with MDA to give a series of standards with 5 μg of CT-DNA in a volume of 50 μL. Vortex-mix and centrifuge at 20,000 × *g* for 1 min. Example dilutions are given in Table 2 using a MDA-modified CT-DNA standard which contains 10 fmol/μg of DNA.
2. Pipette 35 μL (3.5 μg) of each standard into an Eppendorf tube, and then add 65 μL of KP buffer 2.
3. Vortex-mix and centrifuge at 20,000 × *g* for 1 min.
4. Each standard should then be sonicated in a sonicating water bath for 20 min, this allows for the DNA to be broken up into approximately 100 base pair long fragments.
5. Then add 150 μL of PBS and vortex-mix.
6. Heat to 100 °C for 5 min and then cool on ice for at least 10 min. Heating and rapid cooling denatures the DNA and prevents reannealing (see Note 5).

Table 2
Preparation of M_1dG calibration curve (see Note 11)

Vial number	Control calf thymus DNA (100 μg/mL)	MDA-treated calf thymus DNA (100 μg/mL)	M1dG (fmol/μg)
1	80 μL	0 μL	0
2	48 μL	2 μL	0.4
3	45 μL	5 μL	1.0
4	40 μL	10 μL	2.0
5	35 μL	15 μL	3.0
6	25 μL	25 μL	5.0
7	20 μL	30 μL	6.0
8	10 μL	40 μL	8.0
9	0 μL	50 μL	10.0

7. Vortex-mix and centrifuge at 20,000 × g for 2 min and then add 250 μL of 2 M ammonium acetate, vortex-mix, and centrifuge again at 14,000 × g for 1 min.
8. Pipette 142.9 μL of each standard which will contain 1 μg of DNA onto the nitrocellulose filter.

3.2.3. Preparation of Genomic DNA Samples

Prepare the genomic DNA samples for determination of M_1dG by following the procedure described above in Subheading 3.1.1 under Subheading 3.1.

3.2.4. Immuno-Slot Blot

In order to determine the induction of M_1dG on the DNA using immuno-slot blot, follow the steps in Subheading 3.1.2 under Subheading 3.1, with the exception of step 7 (see Note 10). Step 7 is replaced by the step below which describes the use of anti-T<>T antibody instead of the anti-M_1dG antibody:

After blotting the nitrocellulose membrane in 5 % (w/v in PBS) milk powder, Dilute the M_1dG antibody in 5 % (w/v in PBS-T) nonfat milk powder to a final concentration of 6.25 ng/mL. Incubate the filter with the primary antibody for M_1dG in a plastic tray slightly bigger than the size of the membrane and allow it to rock at room temperature for 2 h and then 4 °C overnight.

The level of the adduct in each sample should be corrected for the amount of DNA bound to the filter, determined by propidium iodide staining (see below).

3.2.5. Correction of DNA Bound to Filter by Propidium Iodide Staining

1. Following chemiluminescence step (above), bathe the nitrocellulose filter in 50 mL of PBS with gentle rocking for 10 min to wash the membrane.

2. Incubate the nitrocellulose filter with 250 μg of propidium iodide (dilute 50 μL PI from the stock in 50 mL of PBS) with gentle rocking for 3 h in a plastic tray dedicated for PI, covered with aluminum foil.
3. Wash the nitrocellulose filter with 50 mL of PBS by gentle rocking for 1 h.
4. Place the filter between two acetate sheets and capture the PI-luminescence bands. The correction factor for varying levels of bound DNA is introduced by relating the PI intensity of the samples to the average PI intensity of the standards.

4. Notes

1. Propidium iodide is a suspected carcinogen and great care must be taken when handling.
2. The number of slots depends on preference and is not critical to the assay. For example, we use a Minifold II 72 well slot-blot System (Schleicher and Schuell, via Whatman, UK).
3. Primary antibody must be aliquoted before storage and use as freeze–thaw will damage the antibody and reduce its efficacy.
4. The DNA samples and standards must be cooled in ice immediately. Heating causes denaturation of the DNA, while rapid cooling prevents renaturation. Slow cooling at room temperature may result in renaturation and double-stranded DNA, therefore inhibiting the binding efficacy of the antibody.
5. For the samples, if there are varying volumes of DNA to achieve 3.5 μg, it may be more convenient to speed-vac the samples to dryness and redissolve all the DNA samples in 100 μL of KP buffer.
6. It is important to note that gloves should be worn at this stage especially when handling the nitrocellulose filter.
7. The care must be taken to avoid drying of the nitrocellulose filter at all the steps.
8. Here we used Syngene machine by which the frame width fitting the slot areas is adjusted via software (Gene tools), and the light intensities (relative units) per slot area are displayed. The signals are then incorporated to reduce the background fluctuations made by electronic noise or the light source.
9. The immuno-slot blot method for determination of the level of the T<>T in each DNA sample, does not require to be corrected for the amount of DNA bound to the filter by propidium iodide staining.

10. It is probably a little easier to mix the isopropanol/aqueous solution in the 15 mL falcon tube, hence the addition of 2 mL of ethanol.
11. This table gives an example of how to prepare a range of MDA-modified DNA standard, using a MDA-modified CT-DNA standard which contains 10 fmol M_1dG/μg DNA. Control unmodified CT-DNA is used as a carrier for the modified DNA.

Acknowledgment

The authors would like to acknowledge Prof. Peter B. Farmer (University of Leicester, UK) for use of immuno-slot blot equipment and laboratory space, Dr. Rajinder Singh (University of Leicester, UK) for advice, and Prof. Lawrence Marnett (Vanderbilt University, USA) for supplying the M_1dG antibody.

References

1. Sinha RP, Hader DP (2002) UV-induced DNA damage and repair: a review. Photochem Photobiol Sci 1:225–236
2. Ravanat JL, Douki T, Cadet J (2001) Direct and indirect effects of UV radiation on DNA and its components. J Photochem Photobiol B 63:88–102
3. Cadet J, Sage E, Douki T (2005) Ultraviolet radiation-mediated damage to cellular DNA. Mutat Res 571:3–17
4. Douki T, Reynaud-Angelin A, Cadet J, Sage E (2003) Bipyrimidine photoproducts rather than oxidative lesions are the main type of DNA damage involved in the genotoxic effect of solar UVA radiation. Biochemistry 42:9221–9226
5. Cooke MS, Duarte TL, Cooper D, Chen J, Nandagopal S, Evans MD (2008) Combination of azathioprine and UVA irradiation is a major source of cellular 8-oxo-7,8-dihydro-2′-deoxyguanosine. DNA Repair (Amst) 7:1982–1989
6. Cadet J, Berger M, Douki T, Ravanat JL (1997) Oxidative damage to DNA: formation, measurement, and biological significance. Rev Physiol Biochem Pharmacol 131:1–87
7. Coohill TP, Peak MJ, Peak JG (1987) The effects of the ultraviolet wavelengths of radiation present in sunlight on human cells in vitro. Photochem Photobiol 46:1043–1050
8. Douki T, Perdiz D, Grof P, Kuluncsics Z, Moustacchi E, Cadet J, Sage E (1999) Oxidation of guanine in cellular DNA by solar UV radiation: biological role. Photochem Photobiol 70:184–190
9. Mouret S, Baudouin C, Charveron M, Favier A, Cadet J, Douki T (2006) Cyclobutane pyrimidine dimers are predominant DNA lesions in whole human skin exposed to UVA radiation. Proc Natl Acad Sci USA 103: 13765–13770
10. Kvam E, Tyrrell RM (1997) Induction of oxidative DNA base damage in human skin cells by UV and near visible radiation. Carcinogenesis 18:2379–2384
11. Tyrrell RM, Reeve VE (2006) Potential protection of skin by acute UVA irradiation – from cellular to animal models. Prog Biophys Mol Biol 92:86–91
12. Poli G, Dianzani MU, Cheeseman KH, Slater TF, Lang J, Esterbauer H (1985) Separation and characterization of the aldehydic products of lipid peroxidation stimulated by carbon tetrachloride or ADP-iron in isolated rat hepatocytes and rat liver microsomal suspensions. Biochem J 227:629–638
13. Everett SM, Singh R, Leuratti C, White KL, Neville P, Greenwood D, Marnett LJ, Schorah CJ, Forman D, Shuker D, Axon AT (2001) Levels of malondialdehyde-deoxyguanosine in

the gastric mucosa: relationship with lipid peroxidation, ascorbic acid, and Helicobacter pylori. Cancer Epidemiol Biomarkers Prev 10:369–376

14. Dennis KJ, Shibamoto T (1989) Production of malonaldehyde from squalene, a major skin surface lipid, during UV-irradiation. Photochem Photobiol 49:711–716
15. Adamkiewicz J, Eberle G, Huh N, Nehls P, Rajewsky MF (1985) Quantitation and visualization of alkyl deoxynucleosides in the DNA of mammalian cells by monoclonal antibodies. Environ Health Perspect 62:49–55
16. Muller R, Adamkiewicz J, Rajewsky MF (1982) Immunological detection and quantification of carcinogen-modified DNA components. IARC Sci Publ 463–479
17. Nehls P, Adamkiewicz J, Rajewsky MF (1984) Immuno-slot-blot: a highly sensitive immunoassay for the quantitation of carcinogen-modified nucleosides in DNA. J Cancer Res Clin Oncol 108:23–29
18. Ludeke BI (1992) An immuno-slot-blot assay for detection and quantitation of alkyldeoxyguanosines in DNA. In: Manson MM (ed.) Immunochemical Protocols. Humana Press, Totowa, NJ, p 307
19. Singh R, Leuratti C, Josyula S, Sipowicz MA, Diwan BA, Kasprzak KS, Schut HA, Marnett LJ, Anderson LM, Shuker DE (2001) Lobe-specific increases in malondialdehyde DNA adduct formation in the livers of mice following infection with Helicobacter hepaticus. Carcinogenesis 22:1281–1287

Chapter 13

Quantification of DNA Photoproducts in Mammalian Cell DNA Using Radioimmunoassay

Thomas R. Berton and David L. Mitchell

Abstract

Over the past 25 years, the use of polyclonal and monoclonal antibodies to quantify DNA damage has burgeoned. Immunoassays offer distinct advantages over other analytical procedures currently used to measure DNA damage including adaptability, sensitivity, and selectivity. This combination of attributes allows for the development of powerful analytical techniques to visualize and quantify specific types of DNA damage in cells, tissue, and organisms exposed to subtoxic levels of xenobiotics with distinct advantages over the other procedures in the analysis of DNA damage in human and environmental samples. Radioimmunoassay (RIA) is readily applied to a variety of biological materials and has typically been used to measure DNA damage in cell and organ cultures, tissue sections and biopsies, buccal cells, bone marrow aspirates, peripheral blood lymphocytes, and urine. Here we describe the use of a very sensitive RIA for the specific quantitation of cyclobutane dimers and (6-4) photoproducts in DNA extracted from mammalian cells and tissues.

Key words: Cyclobutane dimer, (6-4) Photoproduct, Polyclonal antibody, Radioimmunoassay, Ultraviolet light

1. Introduction

Radioimmunoassay is a competitive binding assay between an unlabeled and a radiolabeled antigen for binding to antibody raised against that antigen. Yalow and Berson (1) received a Nobel Prize in Medicine for its development. For detailed theory and troubleshooting of RIA refer to Harlow and Lane (2) or Chard (3). We have adapted this technique to the measurement of specific DNA photoproducts in the DNA of UV-irradiated cells (4–6). The following description is given for quantification of cyclobutane pyrimidine dimers (CPDs) and pyriminidine(6-4)pyrimidinone photodimers [(6-4)PDs] in DNA using RIA. For convenience, the radiolabeled antigen is referred to as the "probe" and the unlabeled

Lotte Bjergbæk (ed.), *DNA Repair Protocols*, Methods in Molecular Biology, vol. 920,
DOI 10.1007/978-1-61779-998-3_13, © Springer Science+Business Media New York 2012

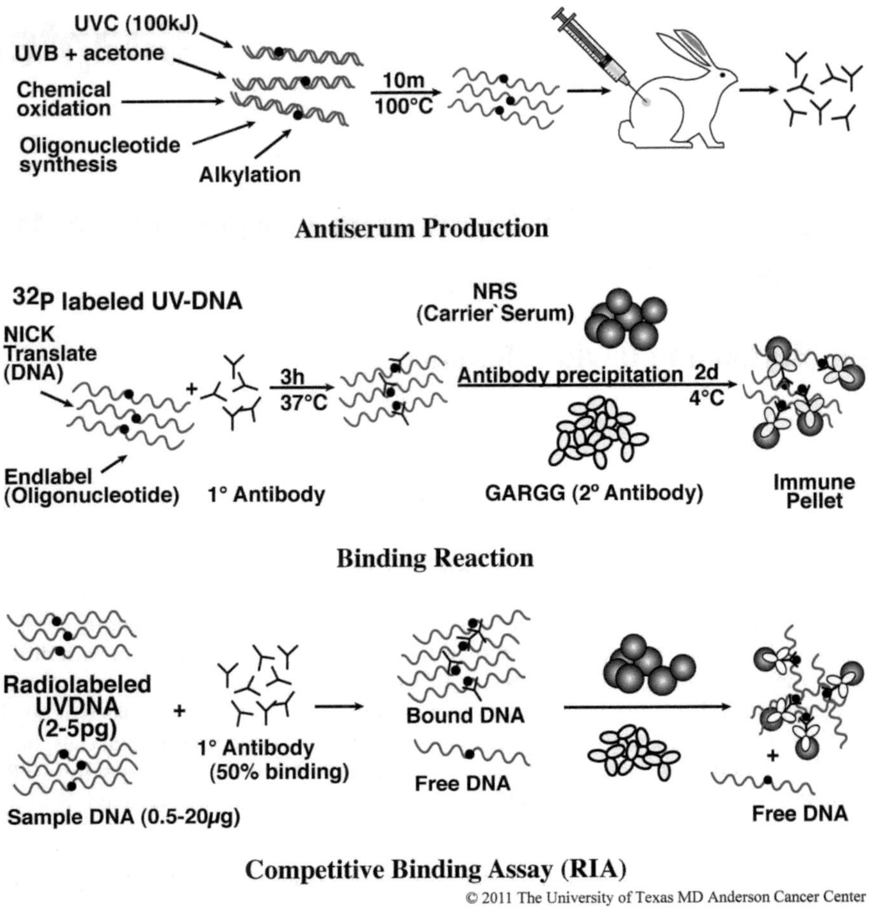

Fig. 1. Diagram of RIA protocol. (*Top*) Antibodies are raised against a specific type of DNA lesion, e.g., CPD. A variety of methods and treatments can be used to damage or modify bases in DNA for use as an immunogen. (*Middle*) Binding activity of antiserum is characterized. (*Bottom*) RIA is used to measure lesion levels in sample DNA.

competitor as "sample" or "standard." The amount of radiolabeled antigen bound to antibody is determined by separating the antigen–antibody complex from free antigen by secondary antibody by immunoprecipitation (Fig. 1). The amount of radioactivity in the antigen–antibody complex in the presence of known amounts of competitor (i.e., standards) is used to quantify the amount of unknown sample present in the reaction. The sensitivity of the RIA is determined by the affinity of the antibody and specific activity of the radiolabeled antigen. Using high affinity antibody and probe labeled to a high specific activity, the reaction can be limited to such an extent that extremely low levels of damage in sample DNA can be quantified. This particular procedure has resulted from 25 to 30 years research and has proven to be a reliable and facile technique for measuring DNA damage and repair endpoints. This is not to say that variations on this basic procedure will not be as productive or useful in DNA damage and repair studies.

2. Materials

Prepare all solutions using ultrapure water and analytical grade reagents. Kits should be stored according to the manufactures specifications. Buffers and secondary antibodies should be stored at 4 °C. Radioactive reagents should be stored and disposed of according to institutional regulations.

1. H_2O (HPLC or Millipore-filtered).
2. Salmon testes (or calf thymus) DNA.
3. Acetone.
4. Methylated bovine serum albumin (see Note 1).
5. DNA nick-translation kit.
6. ^{32}P-Labeled deoxynucleotide triphosphates (dNTPs).
7. TE buffer: 10 mM Tris–HCl, pH 8.0, 1 mM EDTA.
8. ^{14}C-labeled thymidine deoxyribonucleoside ([^{14}C]TdR).
9. Lysis buffer A: 10 mM Tris–HCl, pH 8.0, 1 mM EDTA, 0.5 % sodium dodecylsulfate (SDS), and 0.3 mg/mL Proteinase K (see Note 2).
10. Lysis buffer B: 10 mM Tris, pH 8.0, 1 mM EDTA, 0.5 % SDS, 100 μg/mL DNase-free RNase A.
11. Chloroform–isoamyl alcohol (24:1).
12. Tris-saturated phenol (pH > 7.6).
13. 10× TES: 100 mM Tris, pH 8.0, 10 mM EDTA, 1.5 M NaCl.
14. Gelatin (Type B: bovine skin).
15. RIA buffer: 1× TE + 0.15 % gelatin (see Note 3).
16. Normal rabbit serum (EMD4Biosciences NS01L); stored frozen in 200 μL aliquots (see Note 4).
17. Goat anti-rabbit IgG (EMD4Biosciences), stored frozen in 0.5 mL aliquots (see Note 5).
18. Tissue Solubilizer (NCS-II from GE Healthcare) supplemented with 10 % (v:v) H_2O.
19. Scintillation cocktail (e.g., ScintiSafe from Fisher) containing 1 mL/L acetic acid (see Note 6).
20. Gentra Puregene Tissue Kit (QIAGEN).

3. Methods

3.1. Preparation of Immunogen

1. Commercial DNA (e.g., salmon testes or calf thymus from Sigma) is diluted to 1 mg/mL in 10 mL sterile H_2O (as determined by optical density at 260 nm) (see Note 7).

2. Diluted UV-irradiate double-stranded (or single-stranded) DNA using one of the following protocols:
   (a) Prepare the immunogen for anti-CPD sera by irradiating diluted DNA in 10 % acetone (7) with ~75 kJ/m^2 UV-B light in a glass 100-mm plate (see Note 8).
   (b) Prepare the immunogen for anti-(6-4)PD sera by irradiating DNA with 60 kJ/m^2 UV-C light (see Note 9).
3. Heat-denature UV-irradiated DNA at 100 °C for 10 min.
4. Electrostatically couple the UV-irradiated DNA (heat denatured or not; see Note 9) to methylated bovine serum albumin (MBSA) (see Note 10) (8).

3.2. Immunization Schedule

1. Initially, inject subcutaneously four New Zealand White female rabbits (see Note 11) at ten sites (100 μL each) with 0.5 mL immunogen mixed with an equal volume of Freund's Complete Adjuvant (final concentration of 1.0 mg/mL UV-DNA).
2. Subsequently, inject rabbits at 2-week intervals except that Freund's Incomplete rather than Complete Adjuvant is mixed with 0.5 mL immunogen giving a final concentration of 0.4 mg/mL UV-DNA.
3. At 10–12 days following the second injection, draw 1 mL serum and evaluate the binding affinity using immunoprecipitation (see Subheading 3.3).
4. Continue immunization at 2-week intervals until sufficient binding activity is attained at which time antisera (60–80 mL) are drawn from the animal using heart puncture (see Note 12).
5. Antisera are dispensed into 1-mL aliquots and stored at −20 °C (see Note 13).

3.3. Determination of Antiserum Binding Using Immunoprecipitation (Fig. 1)

1. Nick-translate DNA (0.1 μg) with ^{32}P-dCTP and/or ^{32}P-TTP to give a specific activity of at least 5×10^8–10^9 cpm/μg (see Note 14). A typical reaction includes the following:
   (a) 2 μL 10× buffer (from kit).
   (b) 2 μL dATP [for poly(dA):poly(dT)]; or 2 μL each dATP and dGTP (for DNA).
   (c) 0.5 μL poly(dA):poly(dT) or *C. perfringens* DNA (diluted to 20 μg/100 μL).
   (d) 12.5 μL ^{32}P-TTP (10 mCi/mL, NEN or GE Healthcare).
   (e) 3–4 μL enzyme mix (from kit).
   (f) Incubate for 30–45 min at 15 °C.
   (g) Separate radiolabeled ligand from free dNTPs using a Nick Column (Pharmacia) equilibrated and eluted with 1× TE buffer.

2. Irradiate ^{32}P-labeled probe with 30 kJ/m^2 UV-C light (see Notes 9 and 15).
3. Restore volume (due to evaporation) with H_2O and dilute 2,500- to 5,000-fold in RIA buffer (yielding 2.5–5.0 pg probe in 50 μL buffer) (see Note 16).
4. Add 1 mL RIA buffer to duplicate 12-mm disposable culture tubes (see Note 17).
5. Add 50 μL antiserum diluted in RIA buffer at half-log increments from 1/1,000 to 1/1,000,000 (dilution prior to dispensing). Duplicate tubes without antiserum are dispensed to determine background.
6. Add 50 μL diluted ^{32}P-labeled probe (from step 3) and vortex well.
7. Incubate 3–4 h with gentle rotation (optional) in a 37 °C dry incubator.
8. Separately, add 50 μL normal rabbit serum diluted 1/40 in RIA buffer and 50 μL goat anti-rabbit IgG diluted 1/20 and vortex-mix well.
9. Incubate at 4 °C for 2 day until immune pellet (translucence) develops.
10. Centrifuge tubes at ~2,500 × *g* (3,500–4,000 rpm) for 30–45 min.
11. Decant supernatant, invert tubes onto absorbent paper in test tube rack, and drain for 5–10 min (see Note 18).
12. Swab lip of test tube with cotton-tipped applicator wrapped in tissue (to remove any accumulated liquid).
13. Add 100 μL NCS Tissue Solubilizer supplemented with 10 % H_2O and incubate at 37 °C (or room temperature) with rotation until immune pellet is completely dissolved (see Note 19).
14. Add 2 mL scintillation cocktail (e.g., ScintiSafe from Fisher) supplemented with 1 mL/L acetic acid and vortex.
15. Either place tube directly in 20 mL scintillation vial or decant sample into 20 mL scintillation vial and wash 2× with 4 mL additional scintillation cocktail.
16. Count ^{32}P using liquid scintillation counter.

3.4. Treatment and Isolation of Cultured Mammalian Cell DNA (See Note 20)

1. Plate 2.5–3 × 10^6 cells in 7 × 100-mm plates (duplicate or triplicate plates can be used) with medium (α-MEM) containing 0.005–0.01 mCi/mL [^{14}C]TdR 2 days prior to irradiation (see Note 21).
2. For a DNA repair experiment, irradiate all but one plate (-UV control) with 10–20 J/m^2 UV-C light or UV-B equivalent (see Note 22). The perimeter of the plates should be swabbed with a cotton-tipped applicator to remove cells that would otherwise

be shielded from the irradiation. The medium is poured off and the plates are washed once with 1× TES (or PBS).

3. Harvest one irradiated plate (with duplicate) at the time of irradiation by scraping with a rubber policeman into a 15 mL polypropylene centrifuge tube (trypsinization can be used to life adherent cells from the plate). Additional plates should be harvested at, e.g., 1.5, 3, 6, 24, and 48 h post irradiation for repair studies.
4. Centrifuge at 150×*g* (500–1,000 rpm) for 5 min to pellet cells; decant buffer.
5. Add 4 mL Lysis Buffer A or B, mix vigorously, and incubate overnight at 37 °C or 2–3 h at 60 °C.
6. Extract with 4 mL Sevag (chloroform–isoamyl alcohol, 24:1 v/v) and transfer aqueous phase to a 30 mL Corex tube.
7. Add 0.4 volume (1.6 mL) 5 M sodium acetate (see Note 23) and 2.5 volumes (14 mL) ice-cold absolute ethanol. Place in freezer overnight.
8. Centrifuge sample at 12,000×*g* in a SS-34 rotor (10,000 rpm) at 0 °C for 20 min to pellet DNA. Decant supernatant away from side containing pellet.
9. Wash the pellet with 5–10 mL 70 % ice-cold ethanol.
10. Invert and partially dry for 30–60 min at 23 °C and resuspend in 1.5 mL sterile H_2O or TE buffer. Allow several hours with periodic vortexing for pellet to completely resuspend. Drying the pellet too long will make resuspension difficult.
11. Determine DNA concentration using absorption or spectrofluorometry (see Note 21).
12. After heat denaturation count 20–50 μL to determine the level of [^{14}C]DNA (if cells were prelabeled).
13. Place in refrigerator at 4 °C for short term or in −20 °C freezer for long-term storage.

3.5. Competitive Binding Assay (RIA)

The RIA is simply the basic immunoprecipitation reaction outlined in Subheading 3.3 into which a standard or sample DNA has been added to compete with the radiolabeled probe for antibody binding (Fig. 1). Hence, the procedure is exactly the same as that used for immunoprecipitation with the following additions/modifications:

1. A single dilution of antiserum is used. This dilution is determined from immunoprecipitation analyses of binding activity (Subheading 3.3) and should yield 30–60 % of the radiolabeled probe in the immune pellet.
2. For quantification of CPDs or (6-4)PDs a dose response of heat-denatured salmon testes (Sigma) UV-irradiated DNA is used to generate a standard curve. We routinely use doses of 3,

a

	A	B	C	D	E	F	G	H	I	J	K	
1	Tube #	Sample	UV Dose	ul/ (x) ug	ug/ml	cpm	cpm-bkg	%inhibition	CPM/mb			
2	1	-Ab										
3							=AVERAGE(F2:F3)					
4	=A2+1	+Ab					=F4-G$3					
5							=F5-G$3	=AVERAGE(G4:G5)				
6	=A4+1	Standard	3	**=x/E6*1000**	x		=F6-G$3	=(1-($G6/H$5))*100		=LOG(C6*8.1)	**=RSQ(H6:H13,J6:J13)**	Corr. Coef
7			3				=F7-G$3	=(1-($G7/H$5))*100		=LOG(C7*8.1)		
8	=A6+1	Standard	10	**=x/E8*1000**	x		=F8-G$3	=(1-($G8/H$5))*100		=LOG(C8*8.1)	**=SLOPE(H6:H13,J6:J13)**	Slope
9			10				=F9-G$3	=(1-($G9/H$5))*100		=LOG(C9*8.1)		
10	=A8+1	Standard	30	**=x/E10*1000**	x		=F10-G$3	=(1-($G10/H$5))*100		=LOG(C10*8.1)	**=INTERCEPT(H6:H13,J6:J13)**	Y-intercept
11			30				=F11-G$3	=(1-($G11/H$5))*100		=LOG(C11*8.1)		
12	=A10+1	Standard	100	**=x/E12*1000**	x		=F12-G$3	=(1-($G12/H$5))*100		=LOG(C12*8.1)		
13			100				=F13-G$3	=(1-($G13/H$5))*100		=LOG(C13*8.1)		
14	=A12+1	Sample	X	**=x/E14*1000**	x		=F14-G$3	=(1-($G14/H$5))*100	=10^(($H14-$K$10)/$K$8)			
15							=F15-G$3	=(1-($G15/H$5))*100	=10^(($H15-$K$10)/$K$8)			
16	=A14+1	Sample	Y	**=x/E16*1000**	x		=F16-G$3	=(1-($G16/H$5))*100	=10^(($H16-$K$10)/$K$8)			
17							=F17-G$3	=(1-($G17/H$5))*100	=10^(($H17-$K$10)/$K$8)			

b

	A	B	C	D	E	F	G	H	I	J	K	L	M
1	Tube #						cpm	cpm-bkg	%inhibition				
2	1	-Ab											
3								=AVERAGE(G2:G3)					
4	=A2+1	+Ab						=G4-H$3					
5								=G5-H$3	=AVERAGE(H4:H5)				
6		Cell Line	Treatment	Time Point	ul/(x)cpm	cpm/20 ul				% PD remaining			
7	=A4+1		-UV	0h	(y)ul			=G7-$H$3					
8								=G8-$H$3	=AVERAGE(H7:H8)				
9	=A7+1		+UV	0h	=E13*0.1			=G9-$H$3	=(1-($H9/I$5))*100	10	=LOG(J9)	**=RSQ(I9:I14,K9:K14)**	Corr. Coef
10								=G10-$H$3	=(1-($H10/I$5))*100	10	=LOG(J10)		
11	=A9+1		+UV	0h	=E15*0.1			=G11-$H$3	=(1-($H11/I$5))*100	30	=LOG(J11)	**=SLOPE(I9:I14,K9:K14)**	Slope
12								=G12-$H$3	=(1-($H12/I$5))*100	30	=LOG(J12)		
13	=A11+1		+UV	0h	(y)ul			=G13-$H$3	=(1-($H13/I$5))*100	100	=LOG(J13)	**=INTERCEPT(I9:I14,K9:K14)**	Y-intercept
14								=G14-$H$3	=(1-($H14/I$5))*100	100	=LOG(J14)		
15	=A13+1		+UV	1.5h	(y)ul			=G15-$H$3	=(1-($H15/I$5))*100	=10^(($I15-$L$13)/$L$11)			
16								=G16-$H$3	=(1-($H16/I$5))*100	=10^(($I16-$L$13)/$L$11)			
17	=A15+1		+UV	3 h	(y)ul			=G17-$H$3	=(1-($H17/I$5))*100	=10^(($I17-$L$13)/$L$11)			
18								=G18-$H$3	=(1-($H18/I$5))*100	=10^(($I18-$L$13)/$L$11)			
19	=A17+1		+UV	6 h	(y)ul			=G19-$H$3	=(1-($H19/I$5))*100	=10^(($I19-$L$13)/$L$11)			
20								=G20-$H$3	=(1-($H20/I$5))*100	=10^(($I20-$L$13)/$L$11)			
21	=A19+1		+UV	24 h	(y)ul			=G21-$H$3	=(1-($H21/I$5))*100	=10^(($I21-$L$13)/$L$11)			
22								=G22-$H$3	=(1-($H22/I$5))*100	=10^(($I22-$L$13)/$L$11)			

Fig. 2. Microsoft Excel spreadsheets showing calculations used in RIA experiments. (**a**) Formulae used for quantifying CPDs. (**b**) Formulae for quantifying relative levels of photoproducts (PDs).

10, 30, 100, and 300 J/m^2 as our standard curve and assay the same amount of standard as sample DNA (see Note 24). The yield of CPDs and (6-4)PDs from this UV source is independently determined using an alternative bioanalytical technique (e.g., HPLC). Hence, in the spreadsheet (Fig. 2a) a value of 8.1 CPDs/mb/J/m^2 was used in column "J." This value will vary depending on the UV source (see Notes 9 and 22). When relative, rather than exact, amounts of CPDs or (6-4)PDs are adequate for experimental purposes (as in repair kinetics experiments) the sample harvested at the time of irradiation is titrated in half-log increments to determine the optimal amount required for assay.

3. Unlabeled competitor mammalian DNA, radioactive ligand, and diluted antibody are incubated together for 3 h at 37 °C with gentle rotation (optional). (As above, it is prudent to perform a preliminary titration of sample DNA to determine the amount required for adequate inhibition in the RIA). The total volume and concentration of sample DNA added can vary within certain limitations (see Note 25).

3.6. Data Analysis

1. A sample Excel spreadsheet for quantification of CPDs or (6-4) PDs is shown in Fig. 2. Formulae for quantifying CPDs are shown in Fig. 2a. An identical spreadsheet can be used to quantify (6-4)PDs.

2. A sample Excel spreadsheet for quantification of *relative* photoproducts (PDs) remaining at specific times post UV-irradiation (e.g., in a DNA repair experiment) is shown in (Fig. 2b).

4. Notes

1. Methylated BSA can be frozen and thawed ad infinitum. MBSA is added dropwise with a Pasteur pipet (–50 μL/drop) until the UV-DNA is significantly opaque (i.e., until further addition of mBSA does not change the cloudiness of the solution).
2. Lysis buffer A is used for "crude" extractions in which the DNA has been prelabeled with [^{14}C]TdR and the RIA is set up to determine *relative* amounts of photoproduct remaining in a DNA repair experiment (as shown in Fig. 2b). In such an experiment actual quantification of damage is not required since the 0 h sample is titrated to serve as a standard curve. In this case it is only necessary to analyze equivalent amounts of ^{14}C. Lysis buffer B is used when numbers of photoproducts per megabase DNA are required. In this case the concentration of the DNA sample is critical and care must be taken to assure accurate quantification. A more standard DNA isolation procedure is called for which includes lysis in the presence of RNase, followed by Proteinase K digestion, organic extractions with equal volumes of phenol, phenol:Sevag (1:1), and Sevag, and precipitation with 2 volumes ethanol in the presence of 0.4 volume 5 M ammonium acetate. After precipitation and washing (with 70 % ethanol) duplex DNA is quantified using absorbance at 260 nm (assuming the $A_{260/280}$ is > 1.7) or spectrofluorometry using a DNA specific (Hoescht of DAPI) dye. From Fig. 2a, b it is evident that equivalent amount of standard and sample are required to determine photoproduct concentrations.
3. RIA buffer consists of 1× TES to which 0.15 % gelatin (w/v) has been added to reduce nonspecific binding. The gelatin is heated into solution using a hotplate magnetic stirrer (not microwave) and heated to precisely 39–40 °C. Overheating (by as much as 1 °C) will result in prohibitive background! The cause of this is unknown. This is the first step that should be checked when troubleshooting poor or erratic binding and high background.
4. Normal rabbit serum (NRS) from EMD4Biosciences has been titrated and we have found that a 1/40 dilution is optimal for immune pellet formation. Obviously other sources are readily available; however, we suggest that titrations be performed using the described immunoprecipitation protocol to determine optimal dilution.

5. Goat anti-rabbit IgG can be purchased from EMD4Biosciences in bulk or in smaller aliquots. The bulk product requires a greater concentration than the individual 5 mL aliquots and we suggest, as above, that the optimal dilution be determined using the immunoprecipitation protocol.
6. Tissue solubilizer is very basic and will chemoluminesce in scintillation cocktail resulting in very high counts. This effect will naturally dissipate within 12–18 h but can be immediately quenched by the addition of acetic acid in the cocktail.
7. Commercial DNA does not usually require repurification. However, the purity should be checked using the $A_{260/280}$ with values >1.7 acceptable.
8. The UV-B source consists of four Westinghouse FS20 sunlamps filtered through cellulose acetate (Kodacel from Kodak) with a wavelength cutoff of 290 nm (9). Dosimetry is determined with an appropriate photometer/radiometer (e.g., IL1400 photometer coupled to a SCS 280 probe). At a distance of ~10 cm the fluence rate is ~5 J/m^2/s, hence, exposure times of ~4 h are required for adequate CPD induction. DNA diluted in photosensitizing solvent should not be irradiated in plastic. The DNA is extensively dialyzed post irradiation to remove any acetone.
9. The UV-C source consists of a bank of 5 Philips Sterilamp G8T5 bulbs emitting predominantly 254 nm light. At a distance of ~20 cm the fluence rate is ~14 J/m^2/s and at this fluence rate the average duration of exposure is ~2 h. This high UV-C dose results in extensive fragmentation and denaturation of the DNA, precluding heat denaturation prior to MBSA coupling (see next Note).
10. Methylated BSA is added drop-wise (~50 μL each drop) until the UV-irradiated DNA solution turns cloudy.
11. We have found that individual rabbits have very different immune responses, hence, we recommend that at least four animals be used for raising anti-UV DNA antibodies until an antiserum with high sensitivity is obtained (e.g., 50 % inhibition in the RIA using 1 μg of denatured DNA containing ~10 CPDs or (6-4)PDs per megabase.).
12. An alternate strategy has yielded high titer antisera. Briefly, if an adequate titer (and sensitivity) is not reached after several bi-weekly injections, a 4–6 week hiatus followed by a single injection with 100 μg/100 μL UV-DNA may yielded a more robust immune response.
13. Repeated freezing and thawing of antisera is to be strictly avoided since this can severely reduce binding activity.
14. Both CPD and (6-4)PD frequencies are greatest in nucleic acid substrates containing a high A:T content. Hence, optimal

substrates for the radiolabeled probe include *Clostridium perfringens* DNA (70 % A:T) as well as the homopolymer poly(dA):poly(dT). The higher the specific activity the less probe can be used (see Note 16). Therefore, DNA or polynucleotide can be labeled with >1 nucleoside triphosphate to increase specific activity.

15. Facile irradiation of small volumes of DNA can be achieved using a 25-mm plate or 24-well culture plate in which a depression has been made in a parafilm covering. By drilling a small hole in the bottom of the well air is released to prevent wrinkling.

16. The amount of probe added to the RIA determines its sensitivity. It is essential to use 10 pg or less and have enough cpm in the assay to yield useful binding (and inhibition) data. Hence, if a 5,000 dilution of probe leaves < 500 cpm in 50 μL, a greater concentration must be used. Good assay conditions should be limited to at least 500 cpm/50 μL (added to reaction) at a probe dilution not to exceed 1/1,250.

17. We use 12 mm culture tubes (from Fisher or Baxter) that have colored labels. This helps separate the components of the RIA (i.e., binding conditions, standard curve and sample groups) for more facile visual recognition and less error.

18. Pellets can be inverted for 15–30 min to allow drainage but care should be taken that pellets do not slide down the face of the tube. A second wash is optional and may reduce background.

19. It is extremely important that the immune pellet be *completely* solubilized but not allowed to dry. Partial solubility will result in reduced and variable counts in duplicate samples.

20. To isolate solid tissue from either fresh, frozen, or fixed and paraffin embedded tissue follow the protocol directions for the Cell Lysis, RNAse Treatment, and Protein Precipitation from the Gentra Puregene Tissue Kit. After protein precipitation we have found that pipetting the DNA-containing supernatant into a 1.5 mL microfuge tube improves yield rather than pouring off the supernatant. After protein precipitation continue with the protocol directions for DNA Precipitation and DNA Hydration.

21. Prelabeling with [^{14}C]TdR is optional. Radiolabeled sample DNA allows for the facile dilution to equivalent amounts of DNA for assay (assuming the specific activities of the samples are the same). For repair experiments, using equivalent ^{14}C counts assures that only DNA at the time of irradiation (excluding any nascent DNA) is assayed, thus avoiding any artifactual repair due to cell proliferation and DNA dilution. Unlabeled DNA can be used for DNA damage measurements as long as the concentration of DNA is confidently known and any

"proliferation effects" are accounted for. We measure DNA concentrations after sample denaturation and immediately prior to assay using an Oligreen fluorescent reagent (Molecular Probes, Eugene, OR). The amount of [^{14}C]TdR added to the cells and the duration of the prelabeling depends on the doubling time of the particular cells being studied. Our experience with transformed human fibroblasts and Chinese hamster ovary cells (with doubling times of 24 h) has shown that 2-day incubation with label results in specific activities of 1,000–3,000 cpm/μg DNA.

22. We have found that unfiltered FS20 sunlamps induce 1/10 the amount of CPDs in DNA as UV-C irradiation. FS20 sunlamps filtered through cellulose acetate induce 1/100 the amount of damage as that produced by UV-C irradiation.
23. 5 M sodium acetate should be filter-sterilized and stored at 4 °C. This buffer should be checked prior to use for growth of contaminating organisms (cloudiness).
24. The damage frequencies used in the standard curves (Fig. 2a, b) are determined from independent, nonimmunological measures of CPDs and (6-4)PDs. For our light source the rate of CPD induction was determined using a supercoiled plasmid assay and T4 endonuclease V; (6-4)PDs were determined using the same assay and photoinduced alkali-labile site (PALS) analysis (10).
25. Sample volumes <100 μL do not significantly affect the reaction conditions (e.g., total binding). Sample volumes >100 μL can be used; however, the total reaction volume should be increased accordingly (i.e., doubled). We have found that sample amounts up to 20 μg yield reasonable data.

References

1. Yalow RS, Berson SA (1959) Assay of plasma insulin in human subjects by immunological methods. Nature 184(Suppl 21):1648–1649
2. Harlow E, Lane D (1988) Antibodies: a laboratory manual. Cold Spring Harbor Laboratory, Cold Spring Harbor, NY
3. Chard T (1990) An introduction to radioimmunoassay and related techniques, 4th rev edn. Elsevier, Amsterdam, NY
4. Mitchell DL, Clarkson JM (1981) The development of a radioimmunoassay for the detection of photoproducts in mammalian cell DNA. Biochim Biophys Acta 655:54–60
5. Mitchell DL (1996) Radioimmunoassay of DNA damaged by ultraviolet light. In: Pfeifer G (ed) Technologies for detection of DNA damage and mutations. Plenum, New York, pp 73–85
6. Jeffrey WH, Mitchell DL (2001) Measurement of UVB induced DNA damage in marine planktonic communities. In: Paul JH (ed) Methods in marine microbiology. Academic, New York, pp 469–488
7. Lamola AA, Yamane T (1967) Sensitized photodimerization of thymine in DNA. Proc Natl Acad Sci USA 58:443–446
8. Plescia OJ, Braun W, Palczuk NC (1964) Production of antibodies to denatured deoxyribonucleic acid (DNA). Proc Natl Acad Sci USA 52:279–285
9. Rosenstein BS (1984) Photoreactivation of ICR 2A frog cells exposed to solar UV wavelengths. Photochem Photobiol 40:207–213
10. Mitchell DL, Brash DE, Nairn RS (1990) Rapid repair kinetics of pyrimidine(6-4)pyrimidone photoproducts in human cells are due to excision rather than conformational change. Nucleic Acids Res 18:963–971

Chapter 14

Measuring the Formation and Repair of UV Damage at the DNA Sequence Level by Ligation-Mediated PCR

Ahmad Besaratinia and Gerd P. Pfeifer

Abstract

The formation and repair of DNA damage at specific locations in the genome is modulated by DNA sequence context, by DNA cytosine-5 methylation patterns, by the transcriptional status of the locus and by proteins associated with the DNA. The only method currently available to allow precise sequence mapping of DNA lesions in mammalian cells is the ligation-mediated polymerase chain reaction (LM-PCR) technique. We provide an update on technical details of LM-PCR. LM-PCR can be used, for example, for mapping of ultraviolet (UV) light-induced DNA photoproducts such as cyclobutane pyrimidine dimers.

Key words: Ligation-mediated PCR, UV damage, Cyclobutane pyrimidine dimers, (6-4) Photoproducts

1. Introduction

The most abundant and well-characterized lesions formed upon irradiation of DNA with ultraviolet (UV) light are the cyclobutane pyrimidine dimers (CPDs) and the pyrimidine (6-4) pyrimidone photoproducts [(6-4) photoproducts; (6-4)PPs] (1). CPDs are several times more frequent than (6-4)PPs (2). Both photoproducts can cause mutations, but the CPD is the more mutagenic lesion in mammalian cells (3). CPDs persist much longer in mammalian DNA than (6-4)PPs owing to faster removal of (6-4)PPs (4, 5). CPDs are subject to a specialized transcription-coupled repair pathway (6, 7), which removes these lesions selectively from the template strand of genes transcribed by RNA polymerase II.

Nucleotide excision repair plays an important role in preventing UV-induced skin cancer. Cells from patients suffering from xeroderma pigmentosum (XP) are hypersensitive to UV light (8). The incidence of skin cancer in certain XP patients is increased by several thousand-fold relative to the normal population (9) and

Lotte Bjergbæk (ed.), *DNA Repair Protocols*, Methods in Molecular Biology, vol. 920,
DOI 10.1007/978-1-61779-998-3_14, © Springer Science+Business Media New York 2012

this probably is a consequence of a severe deficiency in repair of UV photolesions. In previous work, we have developed a technique, based on the ligation-mediated PCR (LM-PCR) reaction, which can be used to analyze the formation and repair of UV photoproducts along specific human genes at the DNA sequence level (10–22). LM-PCR methods for the detection of (6-4)PPs (12) and CPDs (13) are available. LM-PCR provides a sufficient level of sensitivity when physiologically relevant UV doses (equivalent to 10–20 J/m^2 of UVC) are used for irradiation, and repair of CPDs can be measured reliably at these doses (10, 11, 14, 17–21, 23).

The ability of LM-PCR to detect DNA adducts depends on the specific conversion of adducts into strand breaks with a 5′-phosphate group. (6-4)PPs and their Dewar isomers can be converted by heating UV-irradiated DNA in piperidine (24). CPDs can be mapped at the DNA sequence level by cleavage with specific enzymes such as T4 endonuclease V (25, 26). T4 endonuclease V cleaves the glycosidic bond of the 5′ base in a pyrimidine dimer and also cleaves the sugar phosphate backbone between the two dimerized pyrimidines. The digestion products still contain a dimerized pyrimidine base at the cleavage site. We determined that these fragments could be amplified efficiently by LM-PCR after photoreversal of the cyclobutane ring with *Escherichia coli* photolyase to result in a normal base on a 5′ terminal sugar-phosphate (13).

This updated LM-PCR protocol describes the sequential steps employed for the detection of these DNA lesions, at the level of nucleotide resolution, in the genome of eukaryotic or prokaryotic cells (Fig. 1). The updated protocol utilizes a computerized fluorescence-based labeling and detection method (LI-COR system; LI-COR; Lincoln, NE), which offers great advantage over the earlier versions of this protocol that included laborious and radioactivity-based steps, such as gel transfer, electroblot, ^{32}P-containing probe synthesis, hybridization, and autoradiography. The sequential order of the steps involved in this updated protocol is as follows: (1) conversion of DNA lesions to ligatable single-stranded DNA breaks using chemical or enzymatic treatments; (2) primer extension towards the single-stranded DNA breaks; (3) ligation of the extended fragments; (4) PCR amplification of the ligation products; (5) fluorescent-labeling of the PCR-amplified products; and (6) simultaneous gel electrophoresis and fluorescence detection of the labeled products. The updated protocol is a 2-day procedure, although it may also be shortened to 1 day by replacing the overnight ligation step with a 2-h ligation step. However, since the ligation efficiency depends upon the activity of ligase, we recommend the overnight ligation, which in our hands has shown high reproducibility. The protocol has been standardized and extensively validated in our laboratory for applications in DNA-lesion footprinting of various carcinogens in mammalian genomes. In the following sections, we describe the technical aspects of this updated

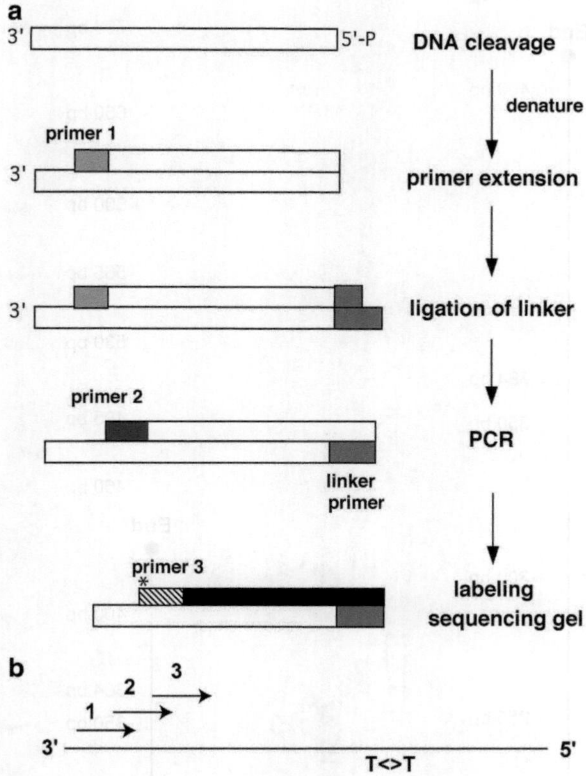

Fig. 1. Outline of the LM-PCR procedure. (**a**) General scheme showing the LM-PCR approach for detection of strand breaks or DNA damage sites. DNA containing strand breaks introduced at the sites of UV damage is used in a primer extension reaction (with primer 1), followed by ligation of a linker, PCR (with primer 2) and a labeling step (using infrared dye-labeled primer 3). (**b**) Primer arrangement. The relative orientation of primers 1, 2, and 3 relative to the DNA template containing a pyrimidine dimer (T<>T) is shown.

protocol and specify its detailed steps. We highlight the applications of this protocol for footprinting of sunlight-induced DNA lesions with special focus on ultraviolet radiation-derived photodimers, including CPDs and (6-4)PPs (see Fig. 2 for an example).

2. Materials

2.1. Cleavage of DNA at Sites of UV Photodamage

1. 1 M Piperidine, freshly prepared.
2. 10×T4 endonuclease V buffer: 500 mM Tris–HCl, pH 7.6, 500 mM NaCl, 10 mM EDTA, 10 mM dithiothreitol (DTT),1 mg/mL bovine serum albumin (BSA).
3. T4 endonuclease V. This enzyme is commercially available, for example, from Epicentre Technologies (Madison, WI), or from New England Biolabs (Ipswich, MA).

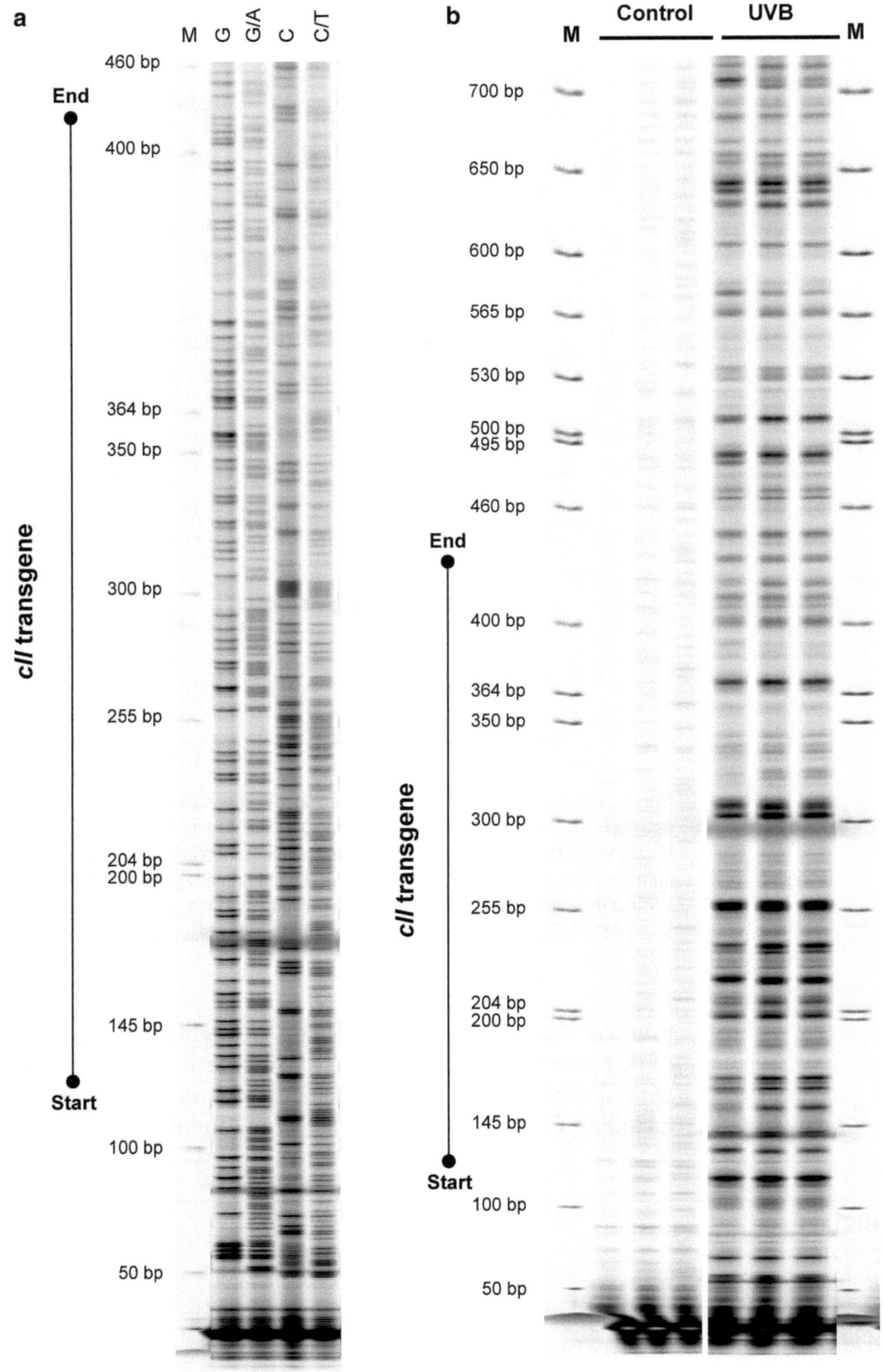

Fig. 2. LM-PCR of the *cII* transgene gene in mouse cells. (**a**) Genomic DNA of transgenic Big Blue® mouse embryonic fibroblasts was subjected to standard Maxam and Gilbert chemical reactions, and subsequently DNA footprinting of the *cII* transgene was performed using our updated LM-PCR protocol, as described in the text. Individual Maxam/Gilbert sequencing ladders are: "G", "G+A", "C", and "C+T". (**b**) Transgenic Big Blue® mouse embryonic fibroblasts were irradiated with ultraviolet light B (UVB) to produce CPDs. Control DNA was not irradiated. The cellular DNA was extracted and subsequently subjected to T4 endonuclease V cleavage and CPD photolyase reactivation, followed by LM-PCR to detect CPDs. M = Molecular size marker (IRDye® 700 Sizing Standard).

4. *E. coli* photolyase. This enzyme was kindly provided by Dr. A. Sancar (University of North Carolina at Chapel Hill) and can commercially be obtained from Trevigen (Gaithersburg, MD).
5. Two 360 nm black lights (Sylvania 15 W F15T8).
6. TE buffer: 10 mM Tris–HCl, pH 7.6, 1 mM EDTA.

2.2. Estimation of Lesion Frequency by Alkaline Agarose Gels

1. Agarose.
2. 50 mM NaCl, 4 mM EDTA.
3. Running buffer: 30 mM NaOH, 2 mM EDTA.
4. Loading dye: 50 % glycerol, 1 M NaOH, 0.05 % bromocresol green.
5. 0.1 M Tris–HCl, pH 7.5.
6. Ethidium bromide (1 μg/mL).

2.3. Ligation-Mediated PCR

2.3.1. Primer Extension Using a Biotinylated Primer 1

Recipe 1 (Primer extension mix)

Component	1× (μL)
H_2O	14.4
10× Vent Buffer	3.0
100 mM $MgSO_4$	1.2
25 mM dNTP mix	0.3
20 μM Primer 1 (Biotinylated)	0.1
Mix thoroughly by pipetting before adding the enzyme	–
Vent $^{(exo-)}$ (2 Units/μL)	1.0
Total	20.0

Vent $^{(exo-)}$, 10× Vent Buffer, and $MgSO_4$ (New England Biolabs (NEB); Ipswich, MA); Deoxynucleoside Triphosphate Set (Roche Diagnostics; Indianapolis, IN); add equal volumes of dATP, dCTP, dGTP, and dTTP (100 mM each) from this set to make the 25 mM dNTP.

Biotinylated primers can be synthesized in-house or alternatively purchased from various companies, such as Integrated DNA Technologies, Inc. (IDT; San Diego, CA).

2.3.2. Preparation of Magnetic Beads

1. Streptavidin-coupled magnetic beads (Dynal Biotech ASA; Oslo, Norway).
2. 2× magnetic bead wash buffer: 2 M NaCl, 10 mM Tris–HCl, pH 7.5, and 1 mM EDTA, pH 8.0.

2.3.3. Overnight Ligation

Recipe 2 (Ligation mix)

Component	1× (μL)
H_2O	7.95
1 M Tris–HCl, pH 7.5	1.50
1 M $MgCl_2$	0.30
1 M DTT	0.30
100 mM ATP	0.30
10 mg/mL BSA	0.15
20 μM LM-PCR Linker	3.00
Mix thoroughly by pipetting before adding the enzyme	–
T4 DNA Ligase (3 Units/μl)	1.50
Total	15.00

ATP = Adenosine-5-triphosphate, lithium salt 100 mM, pH 7.0 (Roche Diagnostics).

BSA = Bovine serum albumin (NEB).

DTT = Dithiothreitol (Sigma-Aldrich Inc.; Saint Louis, MO).

T4 DNA Ligase (Promega; Madison, WI).

The LM-PCR linker is prepared in 250 mM Tris–HCl, pH 7.7, by annealing a 25-mer oligonucleotide (5′ –GCGGTGACCCGG GAGATCTGAATTC) to an 11-mer (5′ –GAATTCAGATC) (final concentration of both oligonucleotides: 20 pmol/μL). This mixture is heated to 95 °C for 3 min, and subsequently cooled down to 4 °C over a time period of at least 4 h. The LM-PCR linker is aliquoted to working solution of 100–200 μL each, and stored in non-defrost –20 °C freezers.

2.3.4. PCR Amplification

Recipe 3 (PCR Amplification Mix)

Component	1× (μL)
H_2O	18.5
Q Solution	5.0
5× Taq buffer	10.0
25 mM $MgCl_2$	4.0
25 mM dNTP	0.5
20 μM Primer 2	0.5
20 μM LP25	0.5
Mix thoroughly by pipetting before adding the enzyme	–
AmpliTaq DNA polymerase (5 Units/μl)	1.0
Total	40.0

5× Taq buffer = 200 mM NaCl, 50 mM Tris–HCl, pH 8.9, and 0.05 % (w/v) gelatin.

AmpliTaq DNA polymerase and 25 mM $MgCl_2$ (Applied Biosystems; Foster City, CA).

LP25 = This is a 25-mer universal linker primer with the following sequence: 5′-GCGGTGACCCGGGAGATCTGAATTC-3′.

Q Solution (Qiagen; Valencia, CA).

Primer 2 is the second gene-specific primer, which is downstream of primer 1 but may overlap a few bases with primer 1 (see Fig. 1b).

2.3.5. Labeling

Recipe 4 (Labeling Mix)

Component	1× (μL)
H_2O	0.79
Q Solution	0.30
5× Taq buffer	0.60
25 mM $MgCl_2$	0.18
25 mM dNTP	0.03
1 μM Primer 3 (IR-Dye® 700/800)	1.00
Mix thoroughly by pipetting before adding the enzyme	–
AmpliTaq DNA polymerase (5 Units/μL)	0.10
Total	3.00

Fluorescence infrared dye-labeled primers (IR-Dye® 700/800 Primer 3) can be ordered from various companies, such as LI-COR Biosciences (LI-COR; Lincoln, NE) and Integrated DNA Technologies, Inc. (IDT; San Diego, CA). Primer 3 is 3′ (downstream) of primer 2 but can overlap a few bases with primer 2 (Fig. 1b).

3. Methods

3.1. Cleavage of DNA at Sites of UV Photodamage

3.1.1. (6-4) Photoproducts

DNA from irradiated cells is isolated by standard methods, for example by phenol–chloroform extraction. To obtain DNA fragments with a 5′ phosphate group at the positions of (6-4)PPs, DNA is heated in 1 M piperidine (12). This will destroy the photolesion and create strand breaks with 5′ phosphate groups since the sugar residue at the 3′-base of the (6-4)PP is cleaved by beta-elimination.

1. Dissolve 10–50 μg of UV-irradiated DNA in 100 μL of 1 M piperidine.
2. Heat the DNA at 90 °C for 30 min in a heat block (use lid locks to prevent tubes from popping). Cool samples briefly on ice after heating.

3. Add 10 μL of 3 M sodium acetate, pH 5.2 and 2.5 volumes of ethanol. Put on dry ice for 20 min.
4. Spin at 14,000 rpm (10,000×*g*) in a microcentrifuge for 15 min.
5. Wash twice with 1 mL of 75 % ethanol.
6. Remove traces of remaining piperidine by drying the sample overnight in a vacuum concentrator. Dissolve DNA in TE buffer to a concentration of approximately 0.5–1 μg/μL.
7. Determine the frequency of (6-4)PPs by separating 1 μg of the DNA on a 1.5 % alkaline agarose gel along with appropriate size markers (see Subheading 3.2).

3.1.2. Cyclobutane Pyrimidine Dimers

DNA is first incubated with T4 endonuclease V and then with *E. coli* photolyase to create fragments with 5′-phosphate groups and ligatable ends (13).

1. The UV-irradiated DNA (about 10 μg in 50 μL) is mixed with 10 μL of 10× T4 endonuclease V buffer and a saturating amount of T4 endonuclease V in a final volume of 100 μL. Saturating amounts of T4 endonuclease V activity can be determined by incubating UVC-irradiated (20 J/m^2) genomic DNA with various enzyme dilutions and separating the cleavage products on alkaline agarose gels (see Subheading 3.2). Incubate at 37 °C for 1 h.
2. Add dithiothreitol to a final concentration of 10 mM. Add 5 μg of *E. coli* photolyase under yellow light.
3. Irradiate the samples in 1.5 mL tubes from two 360 nm UVA-emitting black lights (Sylvania 15W F15T8) filtered through 0.5 cm thick window glass for 1 h at room temperature at a distance of 3 cm.
4. Extract once with phenol–chloroform.
5. Precipitate the DNA by adding one tenth volume of 3 M sodium acetate, pH 5.2 and 2.5 volumes of ethanol. Leave on dry ice for 20 min. Centrifuge samples for 10 min at 14,000 rpm (10,000×*g*) in microcentrifuge at 4 °C.
6. Wash pellets with 1 mL of 75 % ethanol and air-dry.
7. Dissolve DNA in TE buffer to a concentration of about 0.5–1 μg/μL.
8. Determine the frequency of CPDs by running 1 μg of the samples on a 1.5 % alkaline agarose gel.

3.2. Estimation of Cleavage Frequency by Alkaline Agarose Gels

The approximate size of the fragments obtained after cleavage of UV-irradiated DNA is determined on an alkaline 1.5 % agarose gel.

1. Prepare a 1.5 % alkaline agarose gel by suspending agarose in 50 mM NaCl, 4 mM EDTA, and microwaving. Pour the gel.
2. After the gel solidifies, soak it in running buffer for at least 2 h.

3. Dilute the DNA sample with one volume of loading dye. Incubate for 15 min at room temperature. Load the samples.
4. Run the gel at 40 V for 3–4 h.
5. Neutralize the gel by soaking for 60 min in 500 mL of 0.1 M Tris–HCl, pH 7.5.
6. Stain with ethidium bromide (1 μg/mL) for 30 min.
7. Destain in water for 30 min.

3.3. Ligation-Mediated PCR

3.3.1. Primer Extension Using a Biotinylated Primer 1

1. Prepare the Primer extension mix (see Subheading 2.3, Recipe 1) based on the number of samples to be analyzed (we recommend an additional 10 % extra mix per preparation).
2. Add 20 μL of the primer extension mix to 10 μL of DNA digest (0.5–1.0 μg) in a siliconized 0.65-mL microcentrifuge tube.
3. Mix by pipetting (three to four times).
4. Overlay with 20 μL of mineral oil (see Notes 1 and 2).
5. Place the samples in a pre-programmed thermocycler (paused at step 1) and resume the run. The thermocycler is set up for the following program:
   (a) 1st step: 95 °C for 3 min;
   (b) 2nd step: *Tm* of primer 1 (or up to 5 °C higher) for 5 min;
   (c) 3rd step: 72 °C for 10 min; and
   (d) 4th step: 4 °C for cooling.

Our choice of preference for Vent $^{(exo-)}$ DNA polymerase in the primer extension reaction of LM-PCR is due to its lack of 3′ exonuclease activity, which degrades single-stranded DNA templates/oligonucleotide primers. The use of other DNA polymerases for the extension reaction of LM-PCR should be tested by the investigator(s), and verified for optimal performance.

3.3.2. Binding to Streptavidin-Coupled Magnetic Beads

1. Gently swirl the bottle containing the streptavidin-coupled magnetic beads (Dynal Biotech ASA; Oslo, Norway) to fully resuspend the beads.
2. Aliquot 20 μL of the beads per sample in a siliconized 0.65-mL microcentrifuge tube.
3. Place the microcentrifuge tubes containing the beads in a magnetic particle concentrator (MPC) (Dynal Biotech ASA), and allow for magnetic separation to occur (this may take approximately 1 min).
4. Discard the supernatant, and wash the beads twice with 50 μL of 2× magnetic bead wash buffer (see step 3 above).
5. Resuspend the beads in 30 μL of 2× wash buffer and transfer the resuspended beads to the microcentrifuge tubes containing the primer extension products (prepared in Subheading 3.3.1).

6. Immobilize the primer-extension products to the beads by rotating the mixture at room temperature for 15–60 min.
7. Pulse-spin the microcentrifuge tubes (at low speed for ~2 s), place them in an MCP, and allow for magnetic separation to occur.
8. Discard the supernatant, and wash the beads twice with 50 μL of 2× wash buffer.
9. Resuspend the beads in 15 μL of 0.1× TE buffer, pH 7.5 (10 mM Tris–HCl, 1 mM EDTA, pH 7.5) (see Note 3).

3.3.3. Overnight Ligation

1. Prepare the Ligation mix (see Recipe 2 in Subheading 2.3.3) based on the number of samples to be analyzed (we recommend an additional 10 % extra mix per preparation) (see Note 4).
2. Add 15 μL of Ligation mix to each sample.
3. Mix by pipetting (three to four times).
4. Overlay with 20 μL of mineral oil (see Note 1).
5. Place the samples in a thermocycler, and incubate overnight at 17 °C.

3.3.4. PCR Amplification

1. Pulse-spin the microcentrifuge tubes containing the ligation products (samples prepared in Subheading 3.3.3, and place them in an MCP).
2. Discard the supernatant, and wash the beads three times with 100 μL of 1× TE buffer, pH 8.0 (10 mM Tris–HCl, 1 mM EDTA, pH 8.0).
3. Resuspend the beads in 10 μL of 0.1× TE buffer, pH 8.0 (1 mM Tris–HCl, 0.1 mM EDTA, pH 8.0) (see Note 5).
4. Prepare the PCR amplification mix (see Subheading 2.3.4, Recipe 3) based on the number of samples to be analyzed (we recommend an additional 10 % extra mix per preparation).
5. Add 40 μL of the PCR amplification mix to each sample [ligation product] (see, step 3 above).
6. Mix by pipetting (three to four times).
7. Overlay with 20 μL of mineral oil (see Note 1).
8. Place the samples in a pre-programmed thermocycler (paused at cycle 1), and resume the run. The thermocycler is programmed for the following set up:
   (a) 1st step: 95 °C for 2 min;
   (b) 2nd step: *Tm* of primer 2 (or 1 °C lower) for 2 min;
   (c) 3rd step: 72 °C for 3 min;
   (d) 4th step: 20 cycles of (95 °C for 45 s, *Tm* of primer 2 (or 1 °C lower) for 2 min, and 72 °C for 3 min);

(e) 5th step: 95 °C for 45 s;
(f) 6th step: *Tm* of primer 2 (or 1 °C lower) for 2 min;
(g) 7th step: 72 °C for 10 min; and
(h) 8th step: cool at 4 °C.

As a well-characterized recombinant thermostable polymerase, AmpliTaq DNA polymerase is our polymerase of choice for both PCR amplification and labeling reactions of LM-PCR. The use of other DNA polymerases (or a combination thereof) for the amplification and/or labeling reactions of LM-PCR should be tested by the investigator(s), and verified for optimal performance.

3.3.5. Labeling

1. Pulse-spin the microcentrifuge tubes (prepared in Subheading 3.3.4), and place them in an MCP, and allow for magnetic separation to occur.
2. Aspirate 10 μL of the supernatant (without touching the beads), and transfer it to a new microcentrifuge tube on ice (see Note 6).
3. Prepare the Labeling mix (see Subheading 2.3.5, Recipe 4) based on the number of samples to be analyzed (we recommend an additional 10 % extra mix per preparation).
4. Add 3 μL of the Labeling mix to each sample [PCR product] (see step 2 above).
5. Mix by pipetting (three to four times).
6. Overlay with 20 μL of mineral oil (see Note 1).
7. Place the samples in a pre-programmed thermocycler (paused at cycle 1), and resume the run. The thermocycler is programmed for the following set up:
   (a) 1st step: 95 °C for 2 min;
   (b) 2nd step: *Tm* of primer 3 (or up 5 °C higher) for 2 min;
   (c) 3rd step: 72 °C for 3 min;
   (d) 4th step: three to six cycles of (95 °C for 45 s, *Tm* of primer 3 (or up 5 °C higher) for 2 min, and 72 °C for 3 min). More cycles lead to lower signal to noise ratio, i.e., higher background and more nonspecific bands.
   (e) 5th step: 95 °C for 45 s;
   (f) 6th step: *Tm* of primer 3 (or up 5 °C higher) for 2 min;
   (g) 7th step: 72 °C for 10 min; and
   (h) 8th step: cool at 4 °C.

3.3.6. Detection

The IRDye® 700/800 fluorescence-labeled products (prepared in Subheading 3.3.5) are run on a polyacrylamide–urea gel electrophoresis system coupled to a computerized DNA sequencer (e.g., Long Read IR 4200 DNA Sequencing system (LI-COR)). The sequencer is equipped with a real-time IRDye® 700/800-laser (dual) detector and data acquisition software, which enable simultaneous scanning of the sequencing gel during the electrophoresis run. A typical run consists of the following steps:

1. Prepare a 5–8 % LI-COR sequencing gel, as instructed by the manufacturer (LI-COR).
2. Pre-run the gel for 20 min on an LI-COR DNA sequencer.
3. Preparation of samples for loading onto the gel: Add 4 μL of the IR^2 Stop solution (LI-COR) to 13 μL of the IR-Dye® 700/800 labeled products (prepared in Subheading 3.3.5, and denature at 95 °C for 2 min).
4. Cool the samples on ice for approximately 5 min.
5. Load 1.5–2.0 μL of samples onto each well of the pre-run gel.
6. Run the gel, as instructed by the manufacturer (LI-COR).

It is highly recommended that Maxam and Gilbert chemical reactions be prepared from the genomic sequence of interest, and run in parallel to the samples (see Fig. 2a for an example of such sequencing lanes). Also, it is helpful to include the IRDye® 700/800 Sizing Standards (LI-COR), which contain labeled DNA fragments with different lengths, in all runs (Fig. 2) (see Note 7). The inclusion of Maxam and Gilbert reactions and appropriate size markers in the sequencing run will help locate the exact position of each base in the sequence ladders from all samples. During the electrophoresis run, a solid-state laser diode excites the infrared dye present in the labeled DNA fragments as they migrate past the detector window. Simultaneously, a focusing fluorescence microscope containing a solid-state silicon avalanche photodiode scans back and forth across the width of the sequencing gel, and collects the data in real time. The raw data are processed and analyzed as the electrophoresis run progresses. The output image data are saved as TIFF files, and can be retrieved at any time during the electrophoresis run. Quantification of the image data can be achieved by specialized image analysis softwares, e.g., Gene ImagIR (Scanalytics Inc., Rockville, MD).

4. Notes

1. Overlaying the samples with mineral oil is not needed if a thermocycler with hot-bonnet heated lid is used for the primer extension reaction. In such case, skip step 4, pre-start the thermocycler run, and pause at 95 °C (1st step) to allow the hot bonnet to equilibrate at this temperature.

2. Handling and processing of all samples in Subheading 3.3.1 should be done on ice.
3. Handling and processing of all samples in Subheading 3.3.2 can be done at room temperature.
4. Handling and processing of all samples in Subheading 3.3.3 should be done on ice.
5. Handling and processing of all samples in Subheading 3.3.4 should be done on ice.
6. Preserve the leftover PCR product at 4 °C (Do not freeze!). We have successfully used the leftover PCR products, which were stored at 4 °C for several weeks, for subsequent labeling reactions.
7. Both the IR-Dye® 700/800 fluorescence primers and the sizing standards are light sensitive and should be handled under dimmed or yellow light.

Acknowledgement

This work was supported by a grant from the National Institute of Environmental Health Sciences (ES06070) to G.P.P.

References

1. Pfeifer GP, You YH, Besaratinia A (2005) Mutations induced by ultraviolet light. Mutat Res 571:19–31
2. Yoon J-H, Lee C-S, O'Connor T, Yasui A, Pfeifer GP (2000) The DNA damage spectrum produced by simulated sunlight. J Mol Biol 299:681–693
3. You YH, Lee DH, Yoon JH, Nakajima S, Yasui A, Pfeifer GP (2001) Cyclobutane pyrimidine dimers are responsible for the vast majority of mutations induced by UVB irradiation in mammalian cells. J Biol Chem 276:44688–44694
4. Mitchell DL, Nairn RS (1989) The biology of the (6-4) photoproduct. Photochem Photobiol 49:805–819
5. Mitchell DL, Fernandez AA (2011) Different types of DNA damage play different roles in the etiology of sunlight-induced melanoma. Pigment Cell Melanoma Res 24:119–124
6. Mellon I, Spivak G, Hanawalt PC (1987) Selective removal of transcription blocking DNA damage from the transcribed strand of the mammalian DHFR gene. Cell 51:241–249
7. Bohr VA, Smith CA, Okumoto DS, Hanawalt PC (1985) DNA repair in an active gene: removal of pyrimidine dimers from the DHFR gene of CHO cells is much more efficient than in the genome overall. Cell 40:359–369
8. Cleaver JE (1968) Defective repair replication of DNA in xeroderma pigmentosum. Nature 218:652–656
9. Hanawalt PC, Sarasin A (1986) Cancer-prone hereditary diseases with DNA processing abnormalities. Trends Genet 2:124–129
10. Dammann R, Pfeifer GP (1997) Lack of gene- and strand-specific DNA repair in RNA polymerase III transcribed human tRNA genes. Mol Cell Biol 17:219–229
11. Gao S, Drouin R, Holmquist GP (1994) DNA repair rates mapped along the human PGK-1 gene at nucleotide resolution. Science 263: 1438–1440
12. Pfeifer GP, Drouin R, Riggs AD, Holmquist GP (1991) In vivo mapping of a DNA adduct at nucleotide resolution: detection of pyrimidine (6-4) pyrimidone photoproducts by ligation-mediated polymerase chain reaction. Proc Natl Acad Sci USA 88:1374–1378
13. Pfeifer GP, Drouin R, Riggs AD, Holmquist GP (1992) Binding of transcription factors creates hot spots for UV photoproducts in vivo. Mol Cell Biol 12:1798–1804

14. Tornaletti S, Pfeifer GP (1994) Slow repair of pyrimidine dimers at p53 mutation hotspots in skin cancer. Science 263:1436–1438
15. Tornaletti S, Pfeifer GP (1996) Ligation-mediated PCR for analysis of UV damage. In: Pfeifer GP (ed) Technologies for detection of DNA damage and mutations. Plenum, New York, pp 199–209
16. Törmänen VT, Pfeifer GP (1992) Mapping of UV photoproducts within *ras* protooncogenes in UV-irradiated cells: correlation with mutations in human skin cancer. Oncogene 7:1729–1736
17. Tu Y, Tornaletti S, Pfeifer GP (1996) DNA repair domains within a human gene: selective repair of sequences near the transcription initiation site. EMBO J 15:675–683
18. Tommasi S, Oxyzoglou AB, Pfeifer GP (2000) Cell cycle-independent removal of UV-induced pyrimidine dimers from the promoter and the transcription initiation domain of the human CDC2 gene. Nucleic Acids Res 28:3991–3998
19. Hu W, Feng Z, Chasin LA, Tang MS (2002) Transcription-coupled and transcription-independent repair of cyclobutane pyrimidine dimers in the dihydrofolate reductase gene. J Biol Chem 277:38305–38310
20. Zhu Q, Wani MA, El-Mahdy M, Wani AA (2000) Decreased DNA repair efficiency by loss or disruption of p53 function preferentially affects removal of cyclobutane pyrimidine dimers from non-transcribed strand and slow repair sites in transcribed strand. J Biol Chem 275:11492–11497
21. Tu Y, Bates S, Pfeifer GP (1997) Sequence-specific and domain-specific DNA repair in xeroderma pigmentosum and Cockayne syndrome cells. J Biol Chem 272:20747–20755
22. Hendriks G, Calleja F, Besaratinia A, Vrieling H, Pfeifer GP, Mullenders LH, Jansen JG, de Wind N (2010) Transcription-dependent cytosine deamination is a novel mechanism in ultraviolet light-induced mutagenesis. Curr Biol 20:170–175
23. Besaratinia A, Kim SI, Pfeifer GP (2008) Rapid repair of UVA-induced oxidized purines and persistence of UVB-induced dipyrimidine lesions determine the mutagenicity of sunlight in mouse cells. FASEB J 22:2379–2392
24. Lippke JA, Gordon LK, Brash DE, Haseltine WA (1981) Distribution of UV light-induced damage in a defined sequence of human DNA: detection of alkaline-sensitive lesions at pyrimidine nucleoside-cytidine sequences. Proc Natl Acad Sci USA 78:3388–3392
25. Gordon LK, Haseltine WA (1980) Comparison of the cleavage of pyrimidine dimers by the bacteriophage T4 and Micrococcus luteus UV-specific endonucleases. J Biol Chem 255: 12047–12050
26. Radany EH, Friedberg EC (1980) A pyrimidine dimer-DNA glycosylase activity associated with the v gene product of bacterophage T4. Nature 286:182–185

Chapter 15

Construction of Plasmids Containing Site-Specific DNA Interstrand Cross-Links for Biochemical and Cell Biological Studies

Milica Enoiu, The Vinh Ho, David T. Long, Johannes C. Walter, and Orlando D. Schärer

Abstract

Plasmids containing a site-specific DNA interstrand cross-link (ICL) are invaluable tools for the investigation of ICL repair pathways at the biochemical and cellular level. We describe a procedure for preparation of plasmid DNA substrates containing a single ICL at a specific site. The procedure is versatile, leads to reliable yields of pure DNA substrate, and is suitable for the incorporation of virtually any type of DNA lesion into plasmids.

Key words: Plasmid DNA substrate, Interstrand cross-link, Site-specific damage, BbsI restriction enzyme, Gel filtration, Ligation

1. Introduction

DNA interstrand cross-links (ICLs) are among the most cytotoxic DNA lesions, covalently linking together both strands of a DNA duplex. ICLs are formed by endogenous metabolites and by cross-linking agents used in antitumor therapy (1, 2). Despite their biological and clinical importance, the cellular responses to ICLs, particularly the mechanisms by which they are repaired, are not yet understood in detail. The investigation of ICL repair has been hampered by the heterogeneity of DNA lesions produced by treatment of cells with cross-linking agents, which yield mostly mono-adducts and intrastrand cross-links (a covalent linkage between two bases on the same DNA strand) (3). The recent development

Lotte Bjergbæk (ed.), *DNA Repair Protocols*, Methods in Molecular Biology, vol. 920,
DOI 10.1007/978-1-61779-998-3_15, © Springer Science+Business Media New York 2012

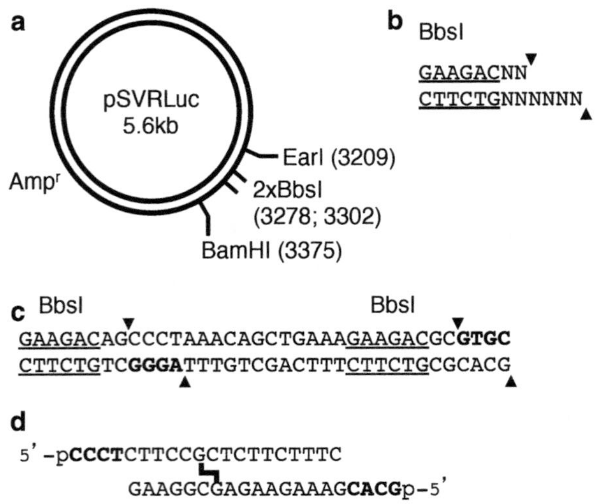

Fig. 1. Strategy for the construction of the site-specific, modified plasmid. (**a**) pSVRLuc plasmid containing 2 BbsI sites in tandem. (**b**) The recognition sequence (*underlined*) and cleavage site of the BbsI restriction enzyme. (**c**) The sequence of the tandem BbsI sites in the pSVRLuc plasmid. (**d**) The sequence of the cisplatin ICL-containing duplex oligonucleotide. The 5′ phosphorylated four nucleotide overhangs are shown in *bold*.

of methods to generate oligonucleotides containing site-specific ICLs and their incorporation into plasmids has been extremely useful for the dissection of different ICL repair pathways using biochemical (4–6) and cell biological methods (7–10). However, the preparation of such substrates is technically difficult, generally resulting in low yields or nonhomogenous substrates that contain varying amounts of impurities. We report a procedure for preparing ICL-containing plasmids with high purity and yield that overcomes these limitations.

The challenge of incorporating duplex oligonucletides into plasmids in vitro lies in the low efficiency of the ligation reaction, risk of incorporating multiple copies of the oligonucleotide or linearized plasmid, and controlling the orientation of incorporation. Our method addresses these issues by inserting two BbsI sites in tandem within a plasmid (Fig. 1a). BbsI is a type IIS restriction enzyme that cleaves outside of a non-palindromic recognition sequence (Fig. 1b). After treatment with BbsI, the plasmid forms linear fragments (a vector and the excised fragment) with two different non-palindromic cohesive ends (Fig. 1a, c). Following linearization with BbsI, the plasmid vector is separated from the excised fragment by gel filtration, and the 4-nucleotide 5′ ends are ligated to the complimentary 5′ phosphorylated ends of an oligonucleotide insert containing a single ICL (Fig. 1d). We have found that the best yield of closed circular plasmid is obtained when very low concentrations of linearized plasmid vector (0.5–2 nM) and slight excess of the oligonucleotide insert (1.5–3×) are used. Under these conditions, the formation of multimeric species and the ligation of separate insert molecules at each end of the linear vector are reduced. Finally, the covalently closed circular plasmid is purified by ethidium bromide cesium chloride density gradient centrifugation. This design ensures the incorporation of a single molecule of the damage-containing oligonucleotide duplex in a controlled orientation into a closed circular plasmid. The non-palindromic sticky ends preclude intramolecular and intermolecular ligation amongst separate molecules of plasmid vector or insert, and are a key factor for the specificity of the ligation procedure.

We have successfully used this method for the incorporation of different ICLs (cisplatin or nitrogen mustard-like), as well as intrastrand cross-links (cisplatin) into plasmids. The substrates obtained have been used, in their circular form, for in vitro investigations of replication-associated ICL repair (4, 5) and the repair of a transcription-blocking ICL in mammalian cells (using a reporter gene reactivation assay), or as a linear fragment for an in vitro transcription assay (Enoiu M, Jiricny J, Schärer OD, manuscript in revision). Herein, we describe the construction of the plasmid substrate used for in vitro replication assays in *Xenopus laevis* egg extracts (described in the accompanying Chapter 16 by Knipscheer et al.). Our procedure is versatile and can be used for any plasmid engineered to contain the tandem BbsI sites, as well as for the incorporation of duplex oligonucleotides containing virtually any type of DNA modification.

2. Materials

2.1. DNA Components

1. Plasmid DNA: pSVRLuc (5.6 kb) (Maxiprep) (see Note 1). This DNA and detailed sequence information is available upon request from the authors.
2. Oligonucleotides 5′ phosphorylated, for the preparation of the DNA duplex containing a cisplatin ICL (see Note 2).

Upper strand sequence: 5′pCCCTCTTCC**G**CTCTTCTTTC; bottom strand sequence: 5′pGCACGAAAGAAGA**G**CGGAAG (the cross-linked bases are indicated in bold) (Fig. 1d) (see Note 3). Dissolve the oligonucleotides to 2 mM in H_2O.

2.2. Reagents and Buffers

Use ultrapure water (18 MΩ) for the preparation of solutions and in all reactions. All the reagents are of analytical grade.

1. Cisplatin: To obtain a 4 mM cisplatin solution, freshly dissolve 6 mg cisplatin in 5 mL H_2O. Vortex until completely dissolved. Protect from light.
2. 15.2 mM $AgNO_3$:12.9 mg $AgNO_3$ (MW: 169.9) in 5 mL H_2O.

3. 20 mM $NaClO_4$, pH 5.2: 281 mg $NaClO_4 \cdot H_2O$ (MW: 140.46) in 90 mL H_2O. Adjust the pH to 5.2 with 1 % acetic acid and the volume to 100 mL. Dilute with an equal volume of H_2O to obtain 10 mM $NaClO_4$.
4. 100 mM $NaClO_4$: 1.4 g $NaClO_4 \cdot H_2O$ in 100 mL H_2O.
5. 0.5 M NaCl: 2.92 g NaCl (MW: 58.44) in 100 mL H_2O.
6. Mobile phases for anion exchange chromatography:

   A1: 10 mM Tris–HCl, pH 7.4: 1.21 g Tris in 800 mL H_2O, adjust the pH to 7.4 with HCl and the volume to 1 L. Filter and degas.

   B1: 1 M NaCl, 10 mM Tris–HCl, pH 7.4: 1.21 g Tris and 58.4 g NaCl in 800 mL H_2O, adjust the pH to 7.4 with HCl and the volume to 1 L. Filter and degas.

   A2: 10 mM NaOH: 0.4 g NaOH in 1 L H_2O. Filter and degas.

   B2: 1 M NaCl, 10 mM NaOH: 0.4 g NaOH and 58.4 g NaCl in 1 L H_2O. Filter and degas.
7. BbsI (NEB R0539L) 5,000 U/mL (see Note 4).
8. Analytical restriction enzymes: EarI, BamHI.
9. T4 DNA Ligase (NEB M0202L) 400,000 U/mL.
10. 10× Ligase buffer: 500 mM Tris–HCl, pH 7.5, 100 mM $MgCl_2$, 10 mM DTT, 1 mM ATP, 25 μg/mL bovine serum albumin (BSA). Make 1 mL aliquots and store at −20 °C.
11. NEBuffer 2 (10×).
12. BSA 10 mg/mL (100×): Dilute 1/10 in H_2O to obtain a 10× stock.
13. 10 mM Tris–HCl, pH 8.0.
14. TE buffer: 10 mM Tris–HCl pH 8.0, 1 mM EDTA. Filter and degas for use as a gel filtration buffer.
15. DNA ladders: 1 kb DNA Ladder (NEB N3232) and 100 bp DNA Ladder (NEB 3231). Store at −20 °C.
16. 1× Tris–acetate buffer (TAE): 40 mM Tris–acetate, 1 mM EDTA, pH 8.0. Prepare a 50× stock solution from 242 g Tris base, 57.1 mL glacial acetic acid, 100 mL 0.5 M EDTA, H_2O to 1 L (pH 8.0).
17. 1× Tris–borate buffer (TBE): 89 mM Tris–borate, 2 mM EDTA, pH 8.0. Prepare a 5× stock solution from 54 g Tris base, 27.5 g boric acid, 0.5 M EDTA, H_2O to 1 L (pH 8.0).
18. Gel loading buffer (6×): 0.25 % Bromophenol blue, 0.25 % xylene cyanol FF, 30 % glycerol in water. Store at 4 °C.
19. Acrylamide/Bisacrylamide 29:1. Store at 4 °C.
20. Agarose, ultrapure.

21. 10% Ammonium persulfate. To 1 g ammonium persulfate, add H_2O to 10 mL. Store at 4 °C for up to 4 weeks.
22. *N*,*N*,*N′*,*N′*-Tetramethylethylenediamine (TEMED). Store at 4 °C.
23. Phenol:chloroform:isoamyl alcohol (25:24:1). Store at 4 °C.
24. 1-Butanol.
25. Ethanol.
26. Chloroform.
27. Ethidium bromide 10 mg/mL (Sigma E 8751).
28. 3 M sodium acetate pH 5.2: 40.8 g of sodium acetate·$3H_2O$ in 80 mL of H_2O and set pH to 5.2 with 3 M acetic acid.
29. Cesium chloride.

2.3. Equipment and Consumables

1. Fast protein liquid chromatography (FPLC) system (e.g., ÄKTA FPLC from GE Healthcare).
2. Mono Q 5/50 GL column (GE Healthcare).
3. Gel filtration column HiLoad Superdex 75 prep grade 16/60. Store in 20 % ethanol at 4 °C.
4. UV spectrophotometer.
5. Ultracentrifuge (Sorvall Ultra Pro 80), Vertical rotor 65 V13.
6. PA Ultracrimp Tubes 6 mL (Sorvall 792730).
7. Ultracrimp sealing tool (Sorvall 79263).
8. High-speed centrifuge (Sorvall Evolution RC), Sorvall SS-34 fixed angle rotor.
9. Oak Ridge centrifuge tube (30 mL and 50 mL).
10. Bench-top centrifuge (Eppendorf 5804R), Swing-bucket rotor A-4-44.
11. Microcentrifuge (Eppendorf 5417R).
12. UV lamp (360 nm wavelength).
13. UV transilluminator.
14. Agarose gel electrophoresis system.
15. Bio-Rad Mini-PROTEAN 3 Electrophoresis system.
16. Water bath.
17. Thermomixer compact (Eppendorf).
18. Amicon Ultra-15 3K (centrifugal filter device).
19. Amicon Ultra-15 30K (centrifugal filter device).
20. Amicon Ultra-4 30K (centrifugal filter device).
21. 0.47 μm Millipore filter.
22. 21-gauge needle.

23. 18-gauge needle attached to 1 mL syringe.
24. Qiagen Plasmid Maxi Kit. Kit includes QIAGEN-tip 500 columns as well as QBT and QC and QF buffers.

3. Methods

3.1. Preparation of the DNA Duplex Containing a Cisplatin ICL

This protocol is based on the procedure established by Hofr and Brabec (11).

1. Generation of monoaqua species from cisplatin:
   Cisplatin can react with DNA only after a chloride ligand is replaced by H_2O (see Note 5). The activation of cisplatin to monoaquamonochloro cisplatin is performed by precipitating chloride in a reaction with 0.95 molar equivalents of $AgNO_3$ in 10 mM $NaClO_4$, pH 5.2.

   Mix in an Eppendorf tube 250 μL 4 mM cisplatin, 187.5 μL H_2O, 500 μL 20 mM $NaClO_4$, pH 5.2, and 62.5 μL 15.2 mM $AgNO_3$. Incubate for 24 h at 37 °C in the dark. Filter through a 0.47 μm Millipore filter to remove the AgCl precipitate. Keep the solution on ice and use it immediately for the next step.
2. Platination of the single-stranded oligonucleotide to a cisplatin monoadduct: The single-stranded oligonucleotide (upper strand sequence, *see* Subheading 2) is reacted with monoaquamonochloro cisplatin at a cisplatin-to-DNA ratio of 3:1 in 10 mM $NaClO_4$, pH 5.2.

   Mix in an Eppendorf tube 45 μL 10 mM $NaClO_4$ pH 5.2, 30 μL of 1 mM monoaquamonochloro cisplatin, and 5 μL of 2 mM oligonucleotide (10 nmol DNA) (see Note 6). Incubate for 12 min at 37 °C. Stop the reaction by adding 20 μL of 0.5 M NaCl (0.1 M final concentration). Keep on ice until next step.
3. Purification of the cisplatin monoadduct containing oligonucleotide: The cisplatin monoadduct is separated from the unreacted oligonucleotide and overplatinated by-products by anion exchange FPLC using a Mono Q column.

   Connect the Mono Q 5/50 GL column to the ÄKTA purifier. Wash the column with H_2O and then equilibrate with 0.1 M NaCl and 10 mM Tris–HCl, pH 7.4 (10 % buffer B1, 90 % buffer A1).

   Dilute the DNA sample to 1 mL with 0.1 M NaCl and 10 mM Tris–HCl, pH 7.4 and inject it using a 1 mL syringe into the 2 mL sample loop. Start a three-step gradient: 10–37 % buffer B1 in 5 column volumes, 37–47 % buffer B1 in 40 column volumes, and 47–100 % buffer B1 in 1 column volume at a flow rate of 2 mL/min. Re-equilibrate to 10 % buffer B1. Collect 0.5 mL fractions during the second gradient step.

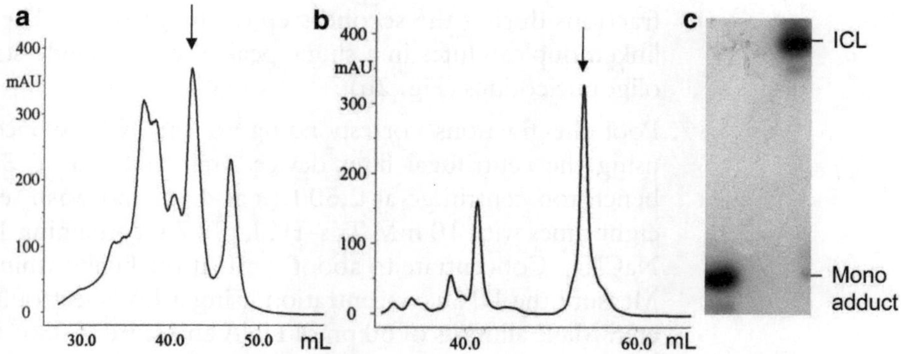

Fig. 2. Purification of the cisplatin monoadduct and cisplatin ICL DNA. (**a**) Typical anion exchange chromatogram showing separation of the cisplatin monoadduct DNA (indicated by *arrow*) from the unreacted oligonucleotide (*last peak*) and the overplatinated by-products (*early eluting peaks*). (**b**) Typical anion exchange chromatogram showing separation of the cisplatin ICL (indicated by *arrow*) from the single-stranded oligonucleotides. (**c**) Analysis of radiolabeled (^{32}P) cisplatin ICL and cisplatin monoadduct DNA in denaturing polyacrylamide gel.

The cisplatin monoadduct elutes at about 0.4 M NaCl before the unreacted oligonucleotide (Fig. 2a). Pool the fractions corresponding to the monoadduct and measure the DNA concentration by UV (the contribution of cisplatin to the absorbance is ignored). The yield of the cisplatin monoadduct oligonucleotide is typically around 15–20 %.

4. Annealing of the cisplatin monoadduct containing oligonucleotide to the complementary strand: Add $MgCl_2$ to the solution containing the cisplatin monoadduct oligonucleotide to a final concentration of 2 mM. Add 1.05 molar equivalents of the complementary oligonucleotide (bottom strand sequence, see Subheading 2). Mix and incubate for 24 h at 25 °C (protect from light).
5. Cross-linking reaction: The formation of the ICL is performed in 100 mM $NaClO_4$. Desalt and exchange the buffer of the DNA sample by extensive washing with 100 mM $NaClO_4$ in a centrifugal filter device Amicon Ultra-15 3K at 4 °C in a benchtop centrifuge (Eppendorf 5804R) at 2,500 × *g* at 4 °C (see Note 7). Repeat the washing step seven to eight times. Stop the last washing step when the volume of the sample is about 1 mL. Incubate for 48 h in the dark at 37 °C for the cross-linking reaction.
6. Purification of the cross-linked duplex: The cross-linked DNA is separated from the single-stranded oligonucleotides on the Mono Q column in an alkaline gradient. Equilibrate the Mono Q column with 0.1 M NaCl and 10 mM NaOH (10 % mobile phase B2, 90 % mobile phase A2). Dilute the DNA sample to 2 mL with 0.1 M NaCl and 10 mM NaOH. Inject the DNA sample into the 2 mL sample loop. Start a three-step gradient: 10–55 % B2 in 7 column volumes; 55–70 % B2 in 40 column volumes; and 55–100 % B2 in 1 column volume. Collect 0.5 mL

fractions during the second step of the gradient. The cross-linked duplex elutes in a single peak after the single-stranded oligonucleotides (Fig. 2b).

7. Pool the fractions corresponding to the ICL. Concentrate using the centrifugal filter device Amicon Ultra-15 3K in a bench-top centrifuge at 2,500×*g* at 4 °C and wash seven to eight times with 10 mM Tris–HCl, pH 7.4 containing 10 mM $NaClO_4$. Concentrate to about 1 mL at the final washing step. Measure the DNA concentration using a UV spectrophotometer. Make aliquots of 60 pmol DNA and store at −80 °C. The typical yield of the ICL duplex is 4–5 % with respect to the single-stranded oligonucleotide used in the first platination reaction (see Note 8).

3.2. BbsI Digestion of the Plasmid DNA

1. Set up a BbsI digest by mixing in an Eppendorf tube 300 μg plasmid DNA (see Note 9), 30 μL BbsI (150 U), and 60 μL 10× NEB2, and add H_2O to 600 μL total volume. Mix well (without vortexing) and incubate at 37 °C overnight in a Thermomixer.
2. Check BbsI digestion on an agarose gel. Prepare a 0.8 % mini agarose gel in 1×TAE; add ethidium bromide to a final concentration of 0.5 μg/mL and allow to polymerize completely. Mix 0.5 μL of the BbsI digestion mix (250 ng DNA) with 4.5 μL H_2O and 1 μL 6× gel loading dye. Load digested sample on the agarose gel alongside a sample of the undigested plasmid (250 ng) and 1 kb DNA ladder. Run the gel at 5 V/cm until the bromophenol blue and xylene cyanol are separated by about 2 cm and take a photograph. The BbsI-digested plasmid should be linearized (Fig. 3a) (see Note 10).
3. Check by restriction analysis whether the BbsI digest was complete (see Note 11).

   Place 2 μL of the BbsI digest mix (1 μg DNA) into each of the two Eppendorf tubes. Mix the first sample with 0.5 μL EarI (10 U), 1.8 μL 10× NEBuffer 2, and 15.7 μL H_2O. Mix the second sample with 0.5 μL BamHI (10 U), 1.8 μL 10× NEBuffer 2, 2 μL 10× BSA, and 13.7 μL H_2O. Incubate both tubes for 2 h at 37 °C.
4. Prepare a 12 % non-denaturing polyacrylamide gel using the BioRad Miniprotean 3 gel casting system. For a minigel with 1.5 mm spacers, prepare 10 mL of polyacrylamide gel: 3 mL 40 % Acrylamide (Acrylamide/Bisacrylamide 29:1), 2 mL 5×TBE, 4.93 mL H_2O, 70 μL 10 % Ammonium persulfate, and 3.5 μL TEMED. Pour the gel and insert a comb with ten slots.
5. Add 4 μL 6× gel-loading buffer (without xylene cyanol) to the EarI and BamHI digests and load on the gel alongside the 100 bp DNA marker. Load 1× gel loading dye (with both

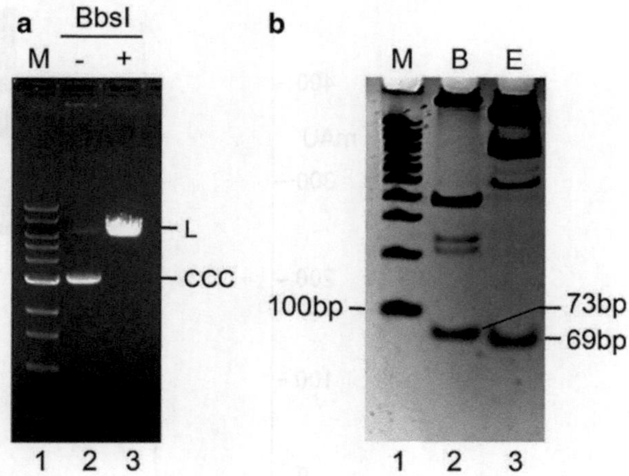

Fig. 3. BbsI digest of the pSVRLuc plasmid. (**a**) Analyzing BbsI digestion by agarose gel. Lane 1: 1 kb DNA Ladder, lane 2: undigested plasmid, lane 3: BbsI-digested plasmid. *CCC* covalently closed circular form, *L* linear plasmid. (**b**) Restriction analysis of the BbsI-digested plasmid. The fragments released after BamHI (73 bp, lane 2) and EarI (69 bp, lane 3) digests were analyzed by 12 % non-denaturing PAGE (*inverted mode picture*). Lane 1: 100 bp DNA Ladder.

marker dyes) in one slot to estimate the migration of the DNA fragments during the run (see Note 12). Run the gel in 1×TBE buffer at 1–8 V/cm until the bromophenol blue runs out of the gel.

6. Perform ethidium bromide post-staining by submerging the gel in 1×TBE containing 0.5 μg/mL ethidium bromide for 30–45 min shaking at room temperature. Wrap the gel in Saran Wrap and photograph with an ultraviolet transilluminator (see Note 13).

   Only the expected fragments of 69 and 73 base pairs should be detected (Fig. 3b). If any residual fragments containing 24 extra base pairs are detected, add 1 μL BbsI (in 20 μL 1× NEB2) and continue the digest until completion (check again by EarI and BamHI restriction analysis). Do not continue to the next step until complete BbsI digestion is ensured (see Note 14).

3.3. Purification of the Digested Plasmid

1. Deproteinize the BbsI-digested sample by extraction with phenol–chloroform–isoamyl alcohol. Add an equal volume of phenol–chloroform–isoamyl alcohol (25:24:1) to the BbsI reaction, vortex contents of the tube until an emulsion forms, and centrifuge at 12,000 ×*g* for 15–20 s in a microfuge at room temperature. Transfer the aqueous phase (containing the DNA) to a fresh tube and repeat the extraction with phenol–chloroform–isoamyl alcohol. Extract once more with an equal volume of chloroform and save the aqueous phase.
2. Connect the HiLoad Superdex 75 prep grade 16/60 gel filtration column to the FPLC system at 4 °C. Wash the column with

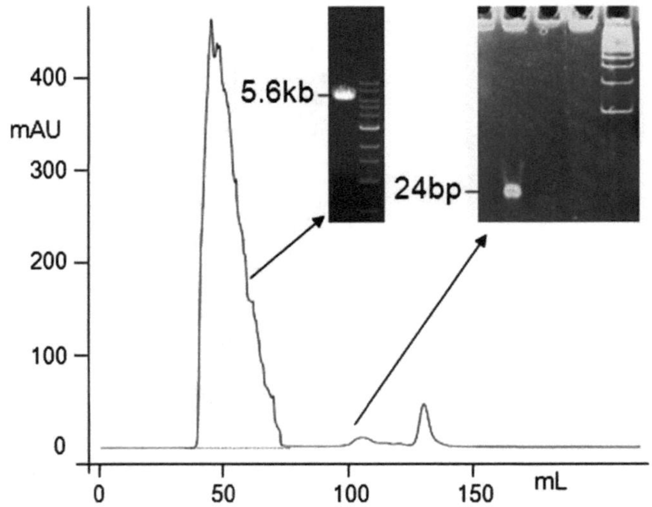

Fig. 4. Purification of the BbsI-digested pSVRLuc plasmid. A typical gel filtration profile shows the elution of the linearized plasmid after 1 column volume and of the excised fragment after 2 column volumes. The insets show the DNA corresponding to the peaks, analyzed by 1 % agarose gel (linearized plasmid) and by 8 % non-denaturing PAGE (24 bp oligonucleotide).

1 column volume H_2O at a flow rate of 1 mL/min (see Note 15), and then equilibrate with 2–3 column volumes of TE buffer (see Note 16).

3. Program the FPLC run as follows: flow rate 1 mL/min, UV detection: 260 nm, sample injection size of 1 mL, start of fractionation after 1 column volume, collecting 2 mL fractions, and stop the run after 3 column volumes.
4. Pre-warm the digested DNA sample at 37 °C, then load into a 1 mL syringe, and inject it into the sample loop (1 mL). Start the FPLC run.
5. The plasmid vector elutes first (Fig. 4). The excised DNA fragment is eluted later (usually after 2 column volumes) and is hardly seen in the chromatogram (Fig. 4) (see Note 17). The fractions containing the plasmid DNA are further analyzed using a UV spectrophotometer (260 nm) to accurately determine the DNA concentration (this allows for an estimation of the subsequent precipitation efficiency). Pool the relevant fractions and measure the volume.
6. Concentrate the DNA by ethanol precipitation. Transfer the diluted DNA solution to an Oak Ridge centrifuge tube (30 mL or 50 mL capacity, depending on the final volume obtained after addition of ethanol; see below). Add 1/10th volume of 3 M sodium acetate to adjust the salt concentration to 300 mM. Add 2 volumes of ice-cold ethanol, mix well, and store at –20 °C for 1 h or overnight.

7. Recover the DNA by centrifugation at 20,000 × *g* for 30 min at 4 °C. Carefully remove the supernatant without disturbing the pellet, and then wash the pellet with 10 mL 70 % ethanol. Centrifuge at 20,000 × *g* for 10 min at 4 °C. After removing the supernatant, air-dry the pellet until the last traces of fluid have evaporated (do not overdry the pellet).
8. Dissolve the DNA pellet in 0.5–1 mL 10 mM Tris–HCl pH 8.0 (warm up at 30 °C to help dissolve). Transfer the DNA to an Eppendorf tube.
9. Measure the DNA concentration using a UV spectrophotometer (260 nm). The yield of purified DNA is typically ~80 % of the BbsI-digested DNA loaded onto the column. The purified DNA can be stored at −20 °C.

3.4. Ligation of the ICL-Containing Oligonucleotide to the Linearized Plasmid

1. Test the efficiency of the ligation in small-scale reactions. In general, we obtain good yields of the closed circular plasmid by using 0.5–1 nM plasmid vector and a 1.5× molar excess of the oligonucleotide DNA insert. However, since the efficiency of the ligation reaction might vary between different plasmid preparations, it is important to test the optimal amounts of DNA in small-scale reactions using various combinations of vector to insert ratios and concentrations (see Note 18). Set up ligation reactions containing 0.5, 1, and 2 nM plasmid vector, and 1.5 and 3× molar excess of the insert for each concentration of plasmid. Add 4 μL 10× Ligase buffer, 0.2 μL T4 DNA ligase, and H_2O to 40 μL. Perform a control reaction in parallel without insert whenever a new preparation of the digested plasmid is used. Mix well and incubate overnight at 16 °C.
2. Load the same amount of DNA from each of the small-scale ligation reactions in a 0.8 % agarose gel and compare the yield of the closed circular DNA (Fig. 5a). In the control reaction without insert, only the linear form of the plasmid should be detected.
3. Scale up the ligation reaction using the DNA concentrations that give the best yield in the small-scale test reactions. The preparative ligation reaction can be set up with a volume of 20–50 mL depending on the desired yield of closed circular plasmid (see Note 19). For a typical ligation reaction (1 nM vector and 1.5× insert), mix in a 50 mL Falcon tube: 40 pmol pSVRLuc plasmid vector (148 μg), 60 pmol oligonucleotide insert, 4 mL 10× Ligase buffer, 10 μL T4 DNA ligase, and H_2O to 40 mL. Mix by inverting the tube and incubate overnight at 16 °C in a water bath.
4. Take a reaction sample corresponding to 100 ng DNA and analyze on a 0.8 % agarose gel to check the efficiency of the preparative ligation reaction (Fig. 5b). Load an aliquot of the

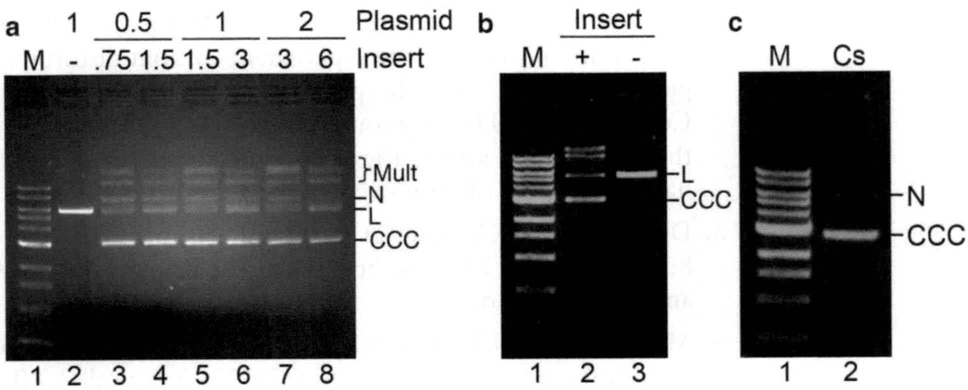

Fig. 5. Ligation of the ICL-containing oligonucleotide into the BbsI-digested plasmid vector. (**a**) Small-scale ligation reactions. The concentrations of the plasmid DNA and insert are indicated in [nM]. Lane 1: 1 kb DNA Ladder, lane 2: control ligation reaction without insert. *CCC* covalently closed circular form, *L* linear plasmid, *N* nicked plasmid, *Mult*: multimers. The same amount of total DNA was loaded for each ligation reaction. (**b**) Large-scale ligation reaction. A typical ligation reaction is shown in lane 2. Lane 1: 1 kb DNA Ladder, lane 3: control ligation reaction without insert. (**c**) Covalently closed circular plasmid containing the interstrand cross-link isolated from cesium chloride (Cs) gradient. A small amount of nicked plasmid may be formed during the purification procedure.

linearized plasmid vector without insert for comparison. The covalently closed circular plasmid should represent at least 20 % of the total DNA (typically ranging from 20 to 50 %).

5. Concentrate the DNA from the ligation reaction using one of the procedures below, depending on the volume of the ligation reaction.
   (a) For a ligation reaction volume up to 50 mL, concentrate the DNA solution to 0.7 mL using the centrifugal filter device Amicon Ultra-15 30K in a bench-top centrifuge at 4,000 × *g* at 4 °C (see Note 20).
   (b) For a ligation reaction volume higher than 50 mL, concentrate the DNA using a QIAGEN-tip 500 column from a Qiagen Plasmid Maxi Prep kit. Pre-equilibrate the column with 10 mL of buffer QBT, apply the ligation reaction, and then wash the column with 30 mL of QC buffer. Elute with QF buffer (1.25 M NaCl, 50 mM Tris–HCl pH 8.5), collecting 1 mL fractions. Measure the DNA concentration of each fraction using a UV spectrophotometer, pool the concentrated fractions, and dialyze against TE.

3.5. Purification of the Closed Circular Plasmid

1. The concentrated ligation solution (obtained in step 5(a) of the Subheading 3.4) is mixed with a cesium chloride ethidium bromide solution or alternatively, cesium chloride is dissolved directly into the DNA solution (obtained in step 5(b) of the Subheading 3.4).

(a) Dissolve 12 g cesium chloride in 10.7 mL 10 mM Tris–HCl, pH 8.0, 1 mM EDTA buffer in a Falcon tube. Cover the tube with aluminum foil and add 80 μL 10 mg/mL Ethidium Bromide. Mix well. Load the 0.7 mL concentrated DNA sample into the 6 mL ultracentrifuge tube. Add the same volume of TE buffer to the balance tube. In the dark, fill the tubes with the CsCl/Ethidium bromide solution.

(b) For every 1 mL of concentrated ligation solution, mix with 1 g of cesium chloride in a Falcon tube. In a separate Falcon tube, dissolve 15 g cesium chloride with 15 mL TE buffer. Load the CsCl-ligation solution into a 6 mL ultracentrifuge tube wrapped with aluminum foil. Add 30 μL 10 mg/mL ethidium bromide to the ultracentrifuge tube. Fill the remainder of the ultracentrifuge tube with the CsCl–TE solution. Fill a second ultracentrifuge tube with the CsCl–TE solution to serve as a balance.

2. Seal the tubes using the Ultracrimp sealing tool. Balance the tubes with the adaptors for Stepsaver 65V13 ultracentrifuge rotor (using different combinations of the adaptors) so that the difference in weight is less than 10 mg.
3. Centrifuge at 60,000 rpm (about 340,000 × *g*) for 16 h at 4 °C (see Note 21). Set the break off from 20,000 rpm (about 37,000 × *g*) to 0.
4. Stop the ultracentrifuge run and wait until the rotor has come to a complete stop. This can take up to 1 h.
5. Take out the tube from the ultracentrifuge rotor carefully (without disturbing the gradient). In a darkroom, place the tube on a holder and visualize the bands using a UV lamp at 360 nm (try to minimize the exposure to UV light). Two bands of DNA should be visible: the upper band consists of linear and nicked DNA, and the lower band consists of closed circular DNA. Pierce the top of the tube with a 21-gauge needle, which serves as a vent (heating up the needle with a lighter can facilitate piercing). Pierce the tube with an 18-gauge needle attached to a 1 mL syringe just below the lower DNA band (make sure that the bevel of the needle is facing upwards toward the DNA band), and carefully extract the closed circular DNA. Avoid contamination with the upper DNA band.
6. Dilute the collected DNA to 2 mL with TE buffer in a 15 mL Falcon tube wrapped with aluminum foil.
7. Add an equal volume of water-saturated butanol to extract the ethidium bromide. Mix the two phases by inverting several times.
8. Centrifuge the mixture at 450 × *g* in a bench-top centrifuge for 2 min.

9. Remove the upper phase (butanol containing the ethidium bromide), and repeat the extraction (steps 7 and 8) four to six times to remove any remaining ethidium bromide.
10. Remove the CsCl by spin dialysis through an Amicon Ultra-4 30K centrifugal filter device. Concentrate to about 100 μL by spinning at 4,000 × *g*, 4 °C in the bench-top centrifuge.

    Add 3.9 mL TE buffer and spin to concentrate as above. Repeat this buffer exchange step four more times. Perform a last buffer exchange with 10 mM Tris–HCl pH 7.5 and stop the spin when the volume of the DNA sample is about 200 μL.
11. Transfer the DNA sample to an Eppendorf tube. Measure the concentration using a UV spectrophotometer and analyze the DNA by agarose gel electrophoresis (Fig. 5b) (see Note 22). The final yield of the closed circular plasmid varies between 20 and 50 % of the linear vector DNA included in the ligation reaction. In general, we obtain 30–50 μg of closed circular plasmid from a 40 mL ligation reaction.
12. Dilute the DNA to 10 ng/μL with 10 mM Tris–HCl, pH 7.5 containing 10 mM $NaClO_4$ (10 mM $NaClO_4$ is added for cisplatin ICL-containing plasmids, but not for other types of ICLs). Make single-use aliquots and store at −80 °C.

4. Notes

1. The plasmid purity is important for the efficiency of the BbsI digest. We use the Qiagen Maxiprep kit and pay attention to completely dry out residual ethanol, since traces of ethanol inhibit BbsI.
2. We have also incorporated duplexes containing nitrogen mustard-like ICLs (12, 13) into the same plasmid using identical procedures with very similar yields and efficiency.
3. Use 5′ phosphorylated, highly pure oligonucleotides. We use chemically phosphorylated and HPLC- or PAGE-purified oligonucleotides. Oligonucleotides phosphorylated enzymatically (T4 Polynucleotide kinase) can also be used, but since enzymatic phosphorylation is less efficient, the yield of closed circular plasmid after ligation may be significantly decreased.
4. For BbsI, storage at −80 °C is recommended for periods longer than 30 days. We make aliquots (30 μL aliquot for digestion of 300 μg DNA) and store them at −80 °C.
5. We have found that some batches of cisplatin are extremely reactive and do not need activation prior to the cisplatination

reaction. We recommend testing each new batch of the reagent in small-scale platination reactions (using 2 nmol DNA) in order to find out whether the activation of cisplatin is needed or not.

6. The platination reaction can be scaled up to 20 nmol single-stranded oligonucleotide (keep the final DNA concentration to 125 μM). For optimal resolution during the anion exchange chromatography step, it is important not to load more than 20 nmol DNA on the Mono Q column. For the preparation of higher amounts of cisplatin monoadduct DNA, make several reactions and inject them separately on the Mono Q column.
7. The desalting and buffer exchange step can be also done by dialysis against 0.1 M $NaClO_4$, using a Spectra/Por Biotech RC MWCO 3.5K membrane, 4 h at 4 °C.
8. The presence of the cisplatin ICL within the DNA duplex can be further confirmed by radioactive labeling (^{32}P) in a phosphate exchange reaction and analysis of the size by denaturing PAGE (Fig. 2c).
9. The BbsI digest reaction can be scaled up to 500 μg DNA using 0.5 U BbsI/μg DNA.
10. This step may be skipped since it does not discern whether BbsI cleavage occurred at one site or at both sites. The next step (restriction analysis of the digested DNA) is absolutely needed and sufficient to make sure that BbsI digest was complete.
11. The choice of the analytical restriction enzyme depends on the particular plasmid. Use two restriction enzymes that cut close to the BbsI sites and generate fragments of 50–300 bp on the BbsI-linearized plasmid. This size ensures that incompletely digested BbsI DNA fragments (+24 bp) can easily be detected.

    For the pSVRLuc, EarI digest generates a 69 bp fragment (83 bp if BbsI digest is incomplete), and BamHI generates a 73 bp fragment (97 bp if BbsI digest is incomplete). BamHI and EarI are not unique sites in pSVRLuc, but the fragments generated by cleavage at the other sites are larger and do not interfere with the detection of the fragments of interest.
12. In a 12 % nondenaturing polyacrylamide gel, bromophenol blue runs at ~20 bp and xylene cyanol runs at ~70 bp DNA fragments.
13. Choose a UV exposure long enough to make sure that besides the expected fragments, no fragments containing an additional 24 base pairs, generated from an incomplete BbsI digest, can be detected (even very faint bands). Visualizing the picture in inverted mode often confers a better sensitivity.
14. In general, BbsI digestion goes to completion, and the presence of two BbsI sites might contribute to the activity of the

enzyme, as known for some type IIS enzymes. However, it is very important to make sure that the digest goes to completion, since partially digested DNA can re-ligate, leading to contamination of the ICL-containing plasmid with uncross-linked parental DNA.

15. The gel filtration column has to be free of any contaminants in order to avoid the degradation of single-stranded overhangs. Prior to use, perform a rigorous cleaning of the column by washing at a flow rate of 0.8 mL/min with 4 column volumes of 1 M NaOH, 4 column volumes of H_2O, then 0.5 column volumes of 30 % isopropanol, followed by 2 column volumes of H_2O.
16. The equilibration of the gel filtration column with the running buffer can be done overnight at a flow rate of 0.2 mL/min.
17. If the excised fragment is not detected in the FPLC profile, fractions eluting after 2 column volumes can be pulled, ethanol precipitated, and analyzed by 10–20 % native PAGE with ethidium bromide post-staining (Fig. 4). The detection of the short DNA fragment will ensure that it has been efficiently removed by gel filtration and is particularly important when performing the procedure for the first time.
18. It is important to accurately determine the concentration of both the digested plasmid vector and the oligonucleotide insert by UV absorbance, since the efficiency of the ligation reaction is critically dependent on DNA concentration.
19. The ligation reaction can be scaled up to a volume higher than 50 mL if necessary (for the preparation of larger amounts of substrate, or for combinations of plasmid vectors and inserts that give lower yields of closed circular form than usual). However, it is important to stay well below the maximum binding capacity of the QIAGEN-tip 500 column (500 μg).
20. In order to decrease the time needed for the concentration of the DNA sample after the preparative ligation reaction, deproteinization by phenol–chloroform–isoamyl alcohol extraction (2×), followed by chloroform extraction (1×), and finally butanol extraction can be performed before loading into the Amicon Ultra-15 30K filter device. We have found that removal of the Ligase and BSA from the ligation reaction significantly increases the flow rate.
21. The ultracentrifugation can also be done for more than 16 h.
22. The quality of the ICL-containing plasmid preparation can be further analyzed using several methods. A fragment containing the ICL can be cut out from the plasmid, radiolabeled using a phosphate exchange reaction, and analyzed in denaturing polyacrylamide gel. Alternatively, use the restriction digest analysis at the unique SapI site that is inactivated by the cisplatin ICL.

Acknowledgements

This work was supported by the New York State Office of Science and Technology and Academic Research NYSTAR (C040069), the Association for International Cancer Research (AICR, Nr. 00-225), the Swiss Cancer League (OCS-01413-080-2003), the National Institutes of Health (GM080454 and CA092584 to O.D.S. and GM62267, HL098316, and GM80676 to J.C.W.), and the American Cancer Society (postdoctoral fellowship PF-10-146-01-DMC to D.T.L.).

References

1. Noll DM, Mason TM, Miller PS (2006) Formation and repair of interstrand cross-links in DNA. Chem Rev 106:277–301
2. Schärer OD (2005) DNA interstrand cross-links: natural and drug-induced DNA adducts that induce unique cellular responses. Chembiochem 6:27–32
3. Guainazzi A, Schärer OD (2010) Using synthetic DNA interstrand crosslinks to elucidate repair pathways and identify new therapeutic targets for cancer chemotherapy. Cell Mol Life Sci 67:3683–3697
4. Knipscheer P, Räschle M, Smogorzewska A, Enoiu M, Ho TV, Schärer OD, Elledge SJ, Walter JC (2009) The fanconi anemia pathway promotes replication-dependent DNA interstrand cross-link repair. Science 326:1698–1701
5. Räschle M, Knipsheer P, Enoiu M, Angelov T, Sun J, Griffith JD, Ellenberger TE, Schärer OD, Walter JC (2008) Mechanism of replication-coupled DNA interstrand crosslink repair. Cell 134:969–980
6. Zhang N, Lu X, Zhang X, Peterson CA, Legerski RJ (2002) hMutSbeta is required for the recognition and uncoupling of psoralen interstrand cross-links in vitro. Mol Cell Biol 22:2388–2397
7. Shen X, Do H, Li Y, Chung WH, Tomasz M, de Winter JP, Xia B, Elledge SJ, Wang W, Li L (2009) Recruitment of fanconi anemia and breast cancer proteins to DNA damage sites is differentially governed by replication. Mol Cell 35:716–723
8. Shen X, Jun S, O'Neal LE, Sonoda E, Bemark M, Sale JE, Li L (2006) REV3 and REV1 play major roles in recombination-independent repair of DNA interstrand cross-links mediated by monoubiquitinated proliferating cell nuclear antigen (PCNA). J Biol Chem 281:13869–13872
9. Wang X, Peterson CA, Zheng H, Nairn RS, Legerski RJ, Li L (2001) Involvement of nucleotide excision repair in a recombination-independent and error-prone pathway of DNA interstrand cross-link repair. Mol Cell Biol 21:713–720
10. Zheng H, Wang X, Warren AJ, Legerski RJ, Nairn RS, Hamilton JW, Li L (2003) Nucleotide excision repair- and polymerase eta-mediated error-prone removal of mitomycin C interstrand cross-links. Mol Cell Biol 23:754–761
11. Hofr C, Brabec V (2001) Thermal and thermodynamic properties of duplex DNA containing site-specific interstrand cross-link of antitumor cisplatin or its clinically ineffective trans isomer. J Biol Chem 276:9655–9661
12. Angelov T, Guainazzi A, Schärer OD (2009) Generation of DNA interstrand cross-links by post-synthetic reductive amination. Org Lett 11:661–664
13. Guainazzi A, Campbell AJ, Angelov T, Simmerling C, Schärer OD (2010) Synthesis and molecular modeling of a nitrogen mustard DNA interstrand crosslink. Chemistry 16:12100–12103

Chapter 16

Replication-Coupled DNA Interstrand Cross-Link Repair in Xenopus Egg Extracts

Puck Knipscheer, Markus Räschle, Orlando D. Schärer, and Johannes C. Walter

Abstract

Interstrand cross-links (ICL) are one of the most hazardous types of DNA damage as they form a roadblock to all processes that involve strand separation. Repair of these lesions involves several different DNA repair pathways, but the molecular mechanism is unclear. Here we describe a system that allows the examination of ICL repair, via a physiological mechanism, in vitro. This system, which uses Xenopus egg extracts in combination with a DNA template that contains a site-specific ICL, represents a unique tool to study the molecular mechanism of ICL repair.

Key words: DNA repair, Interstrand cross-link, Xenopus egg extract, DNA replication, Cross-linked plasmid, Cell-free system

1. Introduction

DNA interstrand cross-links (ICLs) are toxic DNA lesions whose repair is poorly understood. ICL repair takes place primarily during S phase of the cell cycle, although there may be some repair activity in G1 phase as well. Genetic studies have identified several different DNA repair pathways that likely participate in ICL repair (1). However, the mechanistic details underlying this reaction remain unresolved. Progress has long been hampered by the lack of a system that recapitulates ICL repair in vitro. We have recently developed a system that supports ICL repair in cell-free extracts, which allows the examination of various steps in the repair process (2). In this system, ICL repair is coupled to DNA replication, indicating that it represents the primary S phase-specific repair pathway (2–4).

Lotte Bjergbæk (ed.), *DNA Repair Protocols*, Methods in Molecular Biology, vol. 920,
DOI 10.1007/978-1-61779-998-3_16, © Springer Science+Business Media New York 2012

To recapitulate replication-coupled ICL repair, we make use of *Xenopus laevis* egg extracts that support efficient DNA replication of plasmid DNA templates (5). This is achieved by first incubating the DNA in a High-Speed Supernatant (HSS) of egg cytoplasm, which leads to pre-RC assembly via sequential recruitment of the origin recognition complex (ORC), Cdt1, Cdc6, and MCM2–7 to DNA. The subsequent addition of a highly concentrated NucleoPlasmic Extract (NPE) promotes initiation of a single round of DNA replication (6). For our studies of repair, a 5.6 kb plasmid DNA template that contains a sequence specific cisplatin ICL (pICL) is added to the extract system. Details on how to make such a cross-linked plasmid can be found in Chapter 15 of this book.

Incubation of pICL in HSS/NPE leads to the following series of events (2). First, two DNA replication forks converge on the lesion with their leading strands stalled between 20 and 40 nt from the lesion (Fig. 1a, i). Next, one of the forks resumes synthesis and pauses again one nucleotide before the cross-linked nucleotide (Fig. 1a, ii). Subsequently, endonucleolytic incisions on either side of the cross-link "unhook" the ICL (Fig. 1a, iii), creating a DNA double strand break (DSB) in one sister and leaving a monoadduct

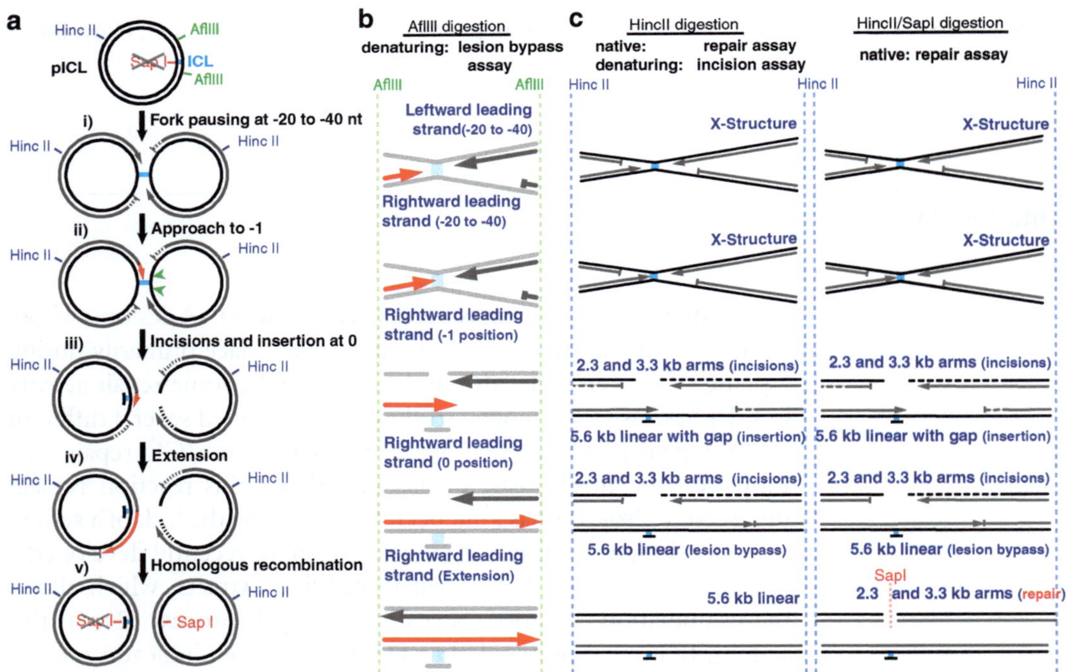

Fig. 1. Schematic Overview of DNA repair intermediates and products generated by the various assays described in this chapter. (**a**) Cartoon of pICL showing the restriction sites and intermediates formed during repair. (**b**) DNA products analyzed on sequencing gels in the lesion bypass assay (Subheading 3.3). *Dark gray* and *red* strands are visible on the gel in Fig. 2. Lesion bypass of the rightward moving fork (*red* strands) can be followed at single nucleotide resolution (see Fig. 2). (**c**) Products analyzed under denaturing and native conditions in the incision and repair assays respectively (Subheadings 3.4 and 3.5).

in the other. The lesion is bypassed by insertion of a nucleotide across from the adducted base (Fig. 1a, iii), followed by extension of the strand well beyond the ICL (Fig. 1a, iv). The final steps in repair involve homologous recombination to repair the dsDNA break in the broken sister chromatid (Fig. 1a, v; (7)). In cells, the monoadduct in the other sister is likely removed via excision repair, but this reaction is inefficient in our extract system. Ultimately, 5–25 % of the replicated pICL is fully repaired, as measured by regeneration of a SapI restriction site that coincides with the cross-link in the parental plasmid. ICL repair in this system is fully dependent on, and directly coupled to, active DNA replication (2, 4).

This system is ideally suited to study the various steps in ICL repair and the roles of specific proteins in the repair process. Immunodepletion of specific factors from the extracts provides valuable insights into the function of each factor. Using this approach, we previously showed that the translesion DNA polymerase ζ is involved in the extension step during translesion synthesis (2). Furthermore, we demonstrated that the Fanconi anemia protein complex FANCD2-FANCI is required for ICL repair and that its depletion abrogates the incision and lesion bypass steps (4). In addition to immunodepletions, the extracts can also be easily manipulated by using specific inhibitors. For example, addition of a peptide derived from BRCA2 that inhibits Rad51 function showed that homologous recombination is a late step in ICL repair that acts downstream of the Fanconi anemia pathway (7). Together with the ability to monitor specific ICL repair intermediates, these approaches make the system a powerful means to decipher the molecular mechanism of ICL repair.

This chapter describes several assays that examine specific steps in ICL repair. We first describe how to set up the ICL repair reaction using HSS/NPE and pICL. Detailed protocols on how to make HSS and NPE extracts can be found elsewhere (6). Then, we describe how to use the purified DNA repair intermediates to examine the lesion bypass reaction in detail, making use of high resolution sequencing gels. We explain how to analyze the dual incision step that unhooks the ICL. Finally, we outline an assay that allows for the determination of the repair efficiency of the reaction.

2. Materials

2.1. Performing Repair of pICL in Egg Extract

1. [α-P^{32}]-dATP (3,000 Ci/mmol). Take the necessary radiation safety training and use standard precautions when working with this material.

2. 1 M PC (Phosphocreatine disodium salt) (Sigma): 10 mM sodium phosphate, pH 7.0, store 50 μL aliquots at −20 °C.
3. 0.2 M ATP (Adenosine 5′-triphosphate disodium salt hydrate) (Sigma): dissolve in sterile water, adjust the pH to 7.0 with 10 M NaOH using pH indicator strips. Store 50 μL aliquots at −20 °C.
4. 5 mg/mL CPK (Creatine phosphokinase)(Sigma): 50 mM NaCl, 50 % glycerol and 10 mM HEPES-KOH, pH 7.5, store 250 μL aliquots at −20 °C. These aliquots are stable for 2–6 months.
5. ATP regeneration mix: Combine 10 μL 1 M PC (see item 2), 5 μL 0.2 M ATP (see item 3), and 0.5 μL 5 mg/mL CPK (see item 4) immediately before use. Store on ice.
6. 1 M DTT (Dithiothreitol)(Bio-Rad): dissolve in sterile water, store 20 μL aliquots at −20 °C.
7. 10× ELB-salts: 25 mM $MgCl_2$, 500 mM KCl, 100 mM HEPES-KOH, pH 7.7. Filter-sterilize and store at 4 °C.
8. 1 M sucrose: dissolve in water, filter-sterilize, and keep at 4 °C. Discard this stock solution after 1–2 months or when visibly contaminated.
9. ELB-sucrose: 1× ELB salts (from a 10× stock, see item 7), 0.25 M sucrose (from a 1 M stock, see item 8). Make fresh once a week.
10. 0.5 mg/mL Nocodazole (Sigma): Dissolve in DMSO, store 50 μL aliquots at −20 °C.
11. 75 ng/μL pICL in TE buffer. A detailed protocol for pICL preparation is described in Chapter 15 of this book.
12. 75 ng/μL pControl in TE buffer (plasmid with the same sequence as pICL but not containing a cross-link).
13. 75 ng/μL pQuant in TE buffer (see Note 1).
14. High Speed Supernatant (HSS). Preparation of HSS is described in detail in ref. 6. Extract needs to be quality tested before use.
15. Nucleoplasmic extract (NPE). Preparation of NPE is described in detail in ref. 6. Extract needs to be quality tested before use.
16. 0.5 mL Safe-Lock tubes (Eppendorf).
17. Stop solution I: 8 mM EDTA, 0.13 % phosphoric acid, 10 % ficoll, 5 % SDS, 0.1 % bromophenol blue in 80 mM Tris–HCl, pH 8.0.
18. Stop solution II: 0.5 % SDS, 25 mM EDTA pH 8.0 in 50 mM Tris–HCl pH 7.5.

19. 20 mg/mL Proteinase K (Roche): Dissolve in milliQ water, store 50 μL aliquots at −20 °C.
20. 2 mg/mL RNase, heat inactivated to remove background DNase activity.

2.2. Purification of pICL Repair Intermediates

1. 10 mM Tris–HCl, pH 8.0.
2. Phenol–Chloroform 1:1: Mix equal volumes of water-saturated phenol, pH 8.0 (USB) (include some of the water layer) and Chloroform (Fisher). Wait until the water and phenol–chloroform layers have completely separated. Store the solution protected from light at 4 °C for up to 2 months. Do not keep on ice as the solution becomes turbid.
3. 0.5 mL Safe-Lock tubes.
4. 3 M Sodium acetate, pH 5.5. Dissolve in milliQ water and adjust pH with glacial acetic acid.
5. 20 mg/mL glycogen from mussels (Roche). Store in 20 μL aliquots at −20 °C.
6. Ice-cold 100 % Ethanol. Store at −20 °C.
7. Ice-cold 70 % Ethanol. Store at −20 °C.

2.3. Detailed Analysis of Lesion Bypass Products on Sequencing Gels

1. AflIII restriction enzyme, 500 units/mL (NEB). BSA solution and buffer 3 supplied with the enzyme.
2. Formamide loading buffer: Gel loading buffer II for denaturing PAGE (Ambion): 95 % Formamide, 18 mM EDTA, and 0.025 % SDS, 0.025 % Xylene Cyanol, and 0.025 % Bromophenol Blue.
3. 10 μM Primer for sequencing ladder (see Note 2).
4. [γ-P^{32}]-ATP (3,000 Ci/mmol).
5. T4 polynucleotide kinase (PNK), 10,000 units/mL (NEB).
6. 10× PNK buffer (NEB).
7. Thermo Sequenase Cycle Sequencing Kit (USB/Affymetrix).
8. pControl (250 ng/μL).
9. Glass plates, large and small (see Note 3).
10. Spacers: sequencing gel spaces, 0.4 mm thick (GibcoBRL/Labrepco).
11. Comb: shark tooth comb, 0.4 mm thick (GibcoBRL/Labrepco).
12. Clips: large binder clips (Staples).
13. Windex window cleaner.
14. Silanizing reagent: Safety Coat Silanizing Reagent, 250 mL (VWR).

15. Gel sealing tape (GibcoBRL/Labrepco).
16. 40 % acrylamide: RapidGel-XL™ 40 % Liquid Acrylamide Stock Solution (USB).
17. GTG buffer: 20× Glycerol Tolerant Gel (GTG) Buffer (USB).
18. Running buffer: 0.8× GTG buffer in water.
19. Urea (Fisher Scientific).
20. 10 % Ammonium persulfate (APS) in water, store at 4 °C.
21. TEMED: *N*,*N*,*N*′,*N*′-tetramethylethylenediamine (Bio-Rad).
22. Needles, 18 gauge.
23. S2 sequencing apparatus (GibcoBRL/Labrepco).
24. High voltage power supply.
25. Gel loading tips, flat.
26. DEAE (Diethylaminoethyl) paper: DE81 ion exchange paper (Whatman).
27. Whatman paper: Whatman 3 MM chromatography paper (Whatman).

2.4. Timing of Dual Incisions

1. HincII restriction enzyme (NEB). BSA solution and buffer 4 supplied with the enzyme.
2. 3 M Sodium acetate, pH 5.5. Dissolve in milliQ water and adjust pH with glacial acetic acid. Keep at room temperature.
3. 20 mg/mL glycogen from mussels (Roche). Store in 20 μL aliquots at −20 °C.
4. Ice-cold 100 % Ethanol. Store at −20 °C.
5. Ice-cold 70 % Ethanol. Store at −20 °C.
6. Alkaline loading buffer: 50 mM NaOH, 1 mM EDTA (pH 8.0), 2.5 % Ficoll-400, 0.05 % Bromocresol green.
7. Alkaline running buffer: 50 mM NaOH, 1 mM EDTA (pH 8.0).
8. 0.25 M HCl: dilute concentrated HCl to 0.25 M in water.
9. NaOH/NaCl: 0.5 M NaOH and 1.5 M NaCl in water.
10. Whatman paper: Whatman 3 MM chromatography paper (Whatman).
11. Membrane: Neutral nylon membrane (Perkin Elmer).
12. 10× SSC: 3 M NaCl, 0.3 M NaCitrate, pH 7.0.
13. 0.5 M NaCl and 0.5 M Tris–HCl, pH 7.5 in water.
14. UV cross-linker: UV Stratalinker (stratagene, model 2400).
15. Random priming labeling kit (GE, Megaprime DNA Labeling System).
16. 0.5 M EDTA pH 8.0.

17. G-50 columns (Illustra MicroSpin G-50 Columns, GE).
18. Hybridization mix: Ultrahyb (Ambion).
19. Hybridization oven.
20. 2× SSC/0.1 % SDS and 0.1× SCC/0.1 % SDS. Made from a 20× SSC stock and a 10 % SDS stock in water.

2.5. Determining ICL Repair Efficiency

1. HincII and SapI restriction enzymes (NEB). BSA solution and buffer 4 supplied with the enzymes.
2. 10× DNA loading buffer: 50 % glycerol, 2× TBE buffer, 25 mM EDTA and 0.1 % Bromophenol Blue.
3. DEAE (Diethylaminoethyl) paper: DE81 ion exchange paper (Whatman).
4. Whatman paper: Whatman 3 MM chromatography paper (Whatman).
5. Phosphorimager.
6. Quantification software: we use Quantity One software from Bio-Rad, the basic version can be downloaded for free from the Bio-Rad Web site.

3. Methods

3.1. Performing Replication/Repair Reaction of pICL in HSS/NPE

Preparing an ICL repair reaction is very similar to a DNA replication reaction (described in detail in ref. 6), the main differences being that repair involves use of pICL as a template and requires longer incubation times. Starting from a master replication reaction, samples are taken over a time course of up to 4 h. The following protocol is designed to yield enough repair intermediates to analyze ten time points using the assays described in this chapter. It is freely scalable, allowing further analyses of the isolated repair intermediates (e.g., by native 2D electrophoresis, electron microscopy, or ChIP). As ICL repair in this system is dependent on efficient DNA replication, it is important to take samples to check the replication efficiency. In addition, samples can be taken from the reaction to analyze post-translational protein modifications during repair using Western blotting (see Note 4).

Unless otherwise stated, all incubations are performed at 20–22 °C.

1. Prepare the following solutions just before you start the reaction: ATP regeneration mix, 0.5 M DTT, ELB-sucrose (see Note 5).
2. Thaw a 33 μL HSS aliquot, keep on ice after thawing. Add 1 μL of the ATP regeneration mix and 0.2 μL of 0.5 mg/mL nocodazole (see Note 6). Mix by pipetting up and down a few times.

3. Transfer 30 μL of HSS-mix to a fresh tube and add pICL and pQuant to final concentrations of 6 ng/μL and 0.3 ng/μL, respectively. This will result in a pICL concentration of 2 ng/μL, and pQuant of 0.1 ng/μL in the HSS/NPE mixture (see Notes 1 and 7). Avoid diluting the extract more than 20 %. Mix by gently pipetting up and down two times (see Note 8). Transfer the HSS-mix to 20–22 °C for 20 min prior to addition of NPE (see Note 9).
4. Thaw 40 μL of NPE and keep it on ice after thawing. Add 2.4 μL ATP regeneration mix, 1.0 μL 0.5 M DTT and 2 μL of [α-P^{32}]-dATP (see Notes 10 and 11). Add 32 μL ELB-sucrose to dilute the NPE (see Note 12). Mix by gently pipetting two to three times. Incubate the NPE mix at 20–22 °C for 15 min prior to addition to HSS.
5. Add 72 μL of NPE mix to the 36 μL of HSS mix to start the replication/repair reaction.
6. At each time point take 9 μL from the reaction mix and add 90 μL stop solution II in a 0.5 mL Safe-Lock tube (see Note 13). From this sample, DNA repair intermediates will be extracted for further analysis. At the same time take an additional 1 μL from the reaction mix and add it to 5 μL stop solution I. This sample will be run on a native agarose gel to assess replication efficiency.
7. When all samples are collected, add 7 μL RNase to the samples that will be extracted and incubate at 37 °C for 30 min. Then, add 3 μL Proteinase K and incubate for at least 1 h at 37 °C, or overnight at room temperature. These samples are now ready to be extracted.
8. To the samples that will be used to determine the replication efficiency, add 1 μL proteinase K. Incubate for at least 1 h at 37 °C, or overnight at room temperature and proceed as described in ref. 6.

3.2. Purification of pICL Repair Intermediates

1. Briefly spin samples from Subheading 3.1, step 7 to collect possible condensation on the lid.
2. Dilute each sample with 40 μL of 10 mM Tris–HCl, pH 8.0 (see Note 14).
3. Add 150 μL Phenol–Chloroform 1:1. Close the tubes firmly and mix by carefully inverting the tubes 10–15 times (do not vortex).
4. Centrifuge samples at 16,000 ×*g* for 4 min at room temperature. During this step, prepare a new series of Safe-Lock tubes.
5. Extract the DNA, which is in the upper aqueous phase, and add it to a new tube. Be conservative to prevent contamination with the phenol–chloroform layer. It is important to get equal

amounts from each sample. We usually take ~140 μL of the upper layer.

6. To the extracted DNA, add an equal volume of chloroform and mix by inverting the tubes 10–15 times. Centrifuge at 16,000 × *g* for 4 min at room temperature.
7. Prepare a new series of tubes. In this case we do not use the Safe-Lock tubes as they are made of a thick layer of plastic that prevents clear vision of the pellet after precipitation.
8. To each tube add: 12 μL (1/10th of the volume) 3 M Sodium acetate, pH 5.5 and 1 μL glycogen.
9. Extract the aqueous phase from the chloroform extraction tubes. We usually remove ~120 μL, and add it to the tubes containing sodium acetate and glycogen.
10. Add 280 μL ice-cold 100 % ethanol, close tubes firmly, and mix by carefully inverting 15 times. Incubate on ice for 15 min.
11. Centrifuge samples at 16,000 × *g* for 30 min at 4 °C.
12. Aspirate the supernatant carefully (see Note 15) and add 250 μL ice-cold 70 % ethanol to wash the pellet.
13. Centrifuge at 16,000 × *g* for another 5 min at 4 °C and aspirate the supernatant (see Note 15). Try to get all the fluid off.
14. Air-dry the pellets for no longer than 10 min (see Note 16).
15. Resolubilize the pellet in 10 μL 10 mM Tris–HCl, pH 8.0. Incubate at room temperature for at least half an hour and mix by tapping the tube. Do not vortex or pipet up and down. Store at 4 °C until further use (see Note 17).

3.3. Detailed Analysis of Lesion Bypass Products on Sequencing Gels

Lesion bypass in ICL repair is a stepwise process (Fig. 1a, and (2)) that can be followed over time using high-resolution sequencing gels. We make use of two AflIII sites on pICL situated 149 nt to the left and 540 nt to the right of the ICL (Fig. 1a). Digestion of repair intermediates with AflIII generates products assignable to leading and lagging strand products (see Note 18) (Figs. 1b and 2). This way it is possible to show the pausing of the leading strand at the −20 to −40 position (Figs. 1b and 2, lane 1), the approach to the −1 position (Figs. 1b and 2, lanes 2–4), the insertion of a nucleotide across from the damaged base (Figs. 1b and 2, lanes 4 and 5), and the extension of the leading strand beyond the lesion (Figs. 1b and 2, lane 5–10) (see Note 19). This is a powerful assay to monitor various steps of lesion bypass under specific conditions (e.g., addition of a specific inhibitor or immunodepletion of a specific protein). The procedure is described in four steps: sample preparation (Subheading 3.3.1), generating the sequencing ladder (Subheading 3.3.2), pouring the sequencing gel (Subheading 3.3.3), and running the sequencing gel (Subheading 3.3.4).

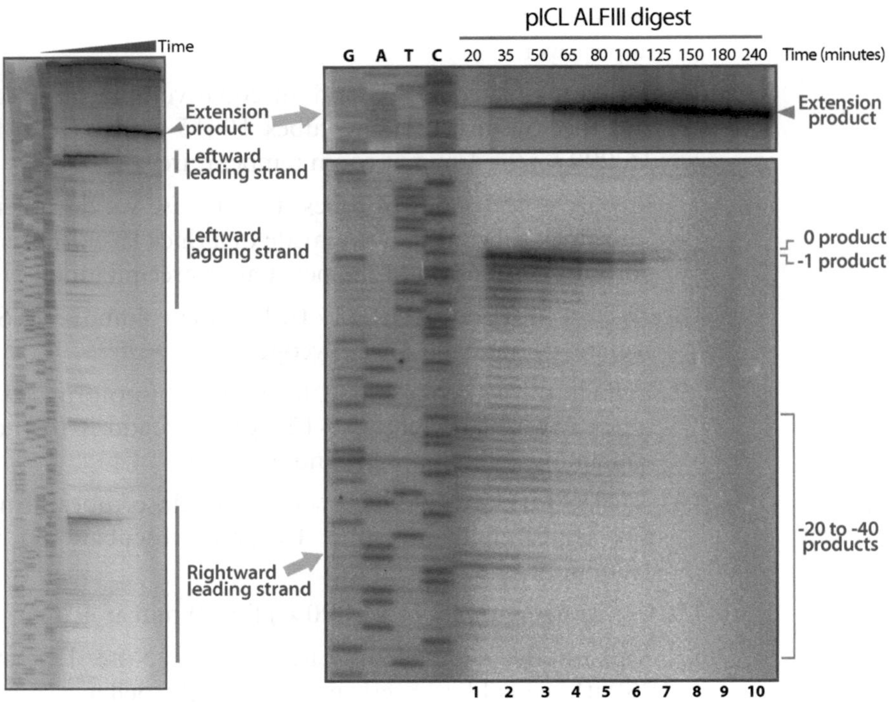

Fig. 2. Analysis of lesion bypass during ICL repair (Subheading 3.3). AflIII digested repair intermediates are separated on a sequencing gel. Leading and lagging strands of the leftward moving fork, leading strands of the rightward moving fork, and lesion bypass product ("Extension product") are indicated on the *left panel*. *Right panel* shows an enlargement of the rightward moving leading strand on which −20 to −40, −1, 0, and extension products are indicated.

3.3.1. Sample Preparation

1. Digest 3 μL of the purified repair intermediates (Subheading 3.2, step 15) with 0.4 μL AflIII (2 units) in a maximal volume of 5 μL (see Note 20).
2. Incubate the reaction at 37 °C for 3 h.
3. Briefly spin the samples and add 2.5 μL formamide loading buffer to each sample. The samples are ready for use or can be stored at −80 °C.

3.3.2. Preparing the Sequencing Ladder

1. Prepare a 10 μM solution of a suitable primer (see Note 2).
2. Setup an end-labeling reaction by mixing 1 μL of this primer with 3.5 μL [γ-P^{32}]-ATP, 0.5 μL 10× PNK buffer, and 0.2 μL T4 polynucleotide kinase.
3. Incubate at 37 °C for 10 min.
4. Incubate at 95 °C for 2 min to inactivate the PNK (see Note 21).
5. During these incubations, prepare the sequencing reactions. We use the Thermo Sequenase Cycle Sequencing Kit. Take four PCR tubes and add 4 μL of the ddGTP, ddATP, ddTTP, and ddCTP termination mixes provided by the kit to each tube.

6. Make a master stock of the reaction mix by adding 2 μL 10× reaction buffer provided by the kit, 1 μL of the end-labeled primer from step 4, and 0.7 μL pControl (at 250 ng/μL) to 11.8 μL milliQ water. Mix and add 2 μL of sequenase provided by the kit just prior to starting the reaction.
7. Add 4 μL of the reaction mix to each tube of termination mixes, mix, and run the following protocol in a PCR machine: (1) 95 °C for 30 s, (2) 60 °C for 30 s, (3) 72 °C for 90 s, (4) GoTo step 2 for 50 times, (5) Hold at 4 °C.
8. When the reaction is finished, add 4 μL formamide loading buffer to each sample and store at −80 °C until use.

3.3.3. Pouring the Sequencing Gel

1. Thoroughly clean the inside of the inner and outer glass plates with Windex window cleaner, rinse with water and dry. Rinse the plates with 100 % ethanol and air-dry. Mark the outside of the plate and treat the inside with care (see Note 22).
2. Silanize the smaller glass plate by applying a few milliliters of silanizing reagent on a tissue and spreading it out on the glass plate. Rub thoroughly over the entire surface. Leave the plate to dry for about half an hour. Repeat step 1 (see Notes 23 and 24).
3. Assemble the gel sandwich with the spacers. Make sure that the plates are well aligned and the inside of the plates is on the inside of the sandwich. Add a clip to each side to fix the sandwich.
4. Tape the bottom of the gel sandwich and the bottom half of the sides with sequencing gel tape (see Notes 25 and 26). Take care not to wrinkle the tape, as this will cause leaking. Add four additional clips to each side of the gel sandwich.
5. Prepare the gel mix by adding 14 mL 40 % acrylamide, 3.2 mL GTG buffer, and 38.4 g Urea to 34.8 mL milliQ water (see Notes 27 and 28).
6. Filter the gel mix through a 22 μM filter using a vacuum filter setup.
7. Degas the mix by leaving it under vacuum for about 10 min.
8. Add 480 μL 10 % APS and 40 μL TEMED, and fill a 50 mL syringe with the gel mix. Add a needle to the syringe and immediately start injecting the mix into the gap between the gel plates. Do this by keeping the gel sandwich vertically and slightly tilted to one side. Inject gel mix from one side (see Note 29).
9. When the space between the plates is almost filled, lay down the gel horizontally. If enough of the gel mix was added, the entire gel space will now fill up and excess gel mix will come out at the top. Immediately insert the flat side of the comb in

between the glass plates, 4–5 mm deep. Add one or two clips to the top of the sandwich.

10. Leave the gel for at least 2 h to complete polymerization (see Note 30).

3.3.4. Running the Sequencing Gel

1. Remove the clips and the tape from the gel sandwich.
2. Rinse the sides of the gel sandwich to remove precipitated urea.
3. Mount the gel into the running apparatus.
4. Add running buffer to the upper compartment and verify that the gel properly seals this compartment.
5. Carefully remove the comb and thoroughly rinse the slot with running buffer using a syringe with a needle (see Note 31).
6. Insert the comb back between the glass plates, this time with the teeth pointing down. At this stage, do not let the comb touch the gel surface but leave about 1 mm in between (see Note 32).
7. Add running buffer to the lower buffer compartment and pre-run the gel at 60 W for 40 min to 2 h. The gel should warm up to approximately 50 °C during this period.
8. Just prior to running, prepare the samples and the sequencing ladder by incubating them at 75 °C for 2 min. Snap cool them on ice–water and keep them on ice until use.
9. After prerunning, the comb can be pushed into the gel until the teeth are about 1 mm deep to create the wells for the samples. Choose an area of the gel with high quality wells (a flat bottom surface) and carefully rinse the wells with running buffer using a syringe and needle.
10. Using ultrathin flat gel loading tips, load 1–4 μL of each sample (see Note 33). It is important to load the samples as a tight band at the bottom of each well. This will improve the resolution of the bands (see Note 34).
11. Run the gel at 60 W for about 2 h and 20 min until the Xylene Cyanol dye has run 30 cm from the top of the gel (see Note 35).
12. Start preheating the gel dryer at 80 °C 10 min before the electrophoresis is finished. Cut a piece of DEAE paper and a piece of Whatman paper at the size of the gel.
13. Remove the gel sandwich from the apparatus and rinse the bottom with water to reduce radioactive contamination.
14. Lay down the gel sandwich on a piece of paper and carefully lift the upper (small) glass plate by wedging a piece of plastic/metal in between the plates (see Note 36). As the small glass

plate is silanized, the gel should stick to the large plate. Try to remove the small glass plate in one movement, if it retouches the gel, the gel will likely stick to the plate and break.

15. Carefully lay down the DEAE paper (see Note 37) on the gel starting on one side and moving to the other by rolling a plastic pipet over it. Try to prevent wrinkles and bubbles. Put the Whatman paper on top of the DEAE paper.
16. Lift up both papers starting from one corner of the gel and peel off the entire gel. It should stick well to the paper.
17. Immediately cover the gel with a piece of saran wrap, try to avoid wrinkles, and dry the gel for at least 45 min in a gel dryer at 80 °C.
18. Expose the dried gel to a phosphor screen for 16–72 h and scan the screen using a phosphorimager (for an example see Fig. 2).

3.4. Timing of Dual Incisions

During ICL repair, the cross-link is released from one of the DNA strands by dual incisions on either side of the lesion. To analyze these incisions we use denaturing gel electrophoresis of HincII-digested repair intermediates (see Fig. 1c, left panel) (4). Before any incisions have taken place, the parental DNA migrates in gels as a large X-shaped molecule because the two strands are connected by the ICL (Fig. 3, lane 4). Dual incisions result in the collapse of this large X-shaped molecule into a linear molecule and two arm fragments (Fig. 3, lane 8–14, for size markers lane 1 and 2). The arm fragments disappear over time due to resection and homologous recombination (Fig. 3, lanes 10–14, (4)). As the parental strands are not radioactively labeled, we use Southern blotting for their detection. The 2.3 and 3.3 kb arm fragments visible at early time points (Fig. 3, lane 5–7) are leading and lagging strand products of replication. The procedure is described in three steps: preparing the samples (Subheading 3.4.1), running and transferring the alkaline gel (Subheading 3.4.2), and hybridization and detection (Subheading 3.4.3).

3.4.1. Preparing the Samples

1. Digest 1 μL of the purified repair intermediates (Subheading 3.2, step 15) with 0.4 μL HincII (4 units) in a volume of 20 μL. Incubate samples at 37 °C for 3 h (see Notes 38 and 39).
2. Precipitate the DNA by adding 2 μL 3 M sodium acetate, pH 5.5, 0.2 μL 0.5 M EDTA, pH 8.0, and 55 μL ice-cold 100 % ethanol. Mix by carefully inverting 10–15 times and incubate on ice for 15 min.
3. Centrifuge samples at 16,000 × g for 30 min at 4 °C.
4. Aspirate the supernatant carefully (see Note 40) and add 100 μL ice-cold 70 % ethanol to the pellet.

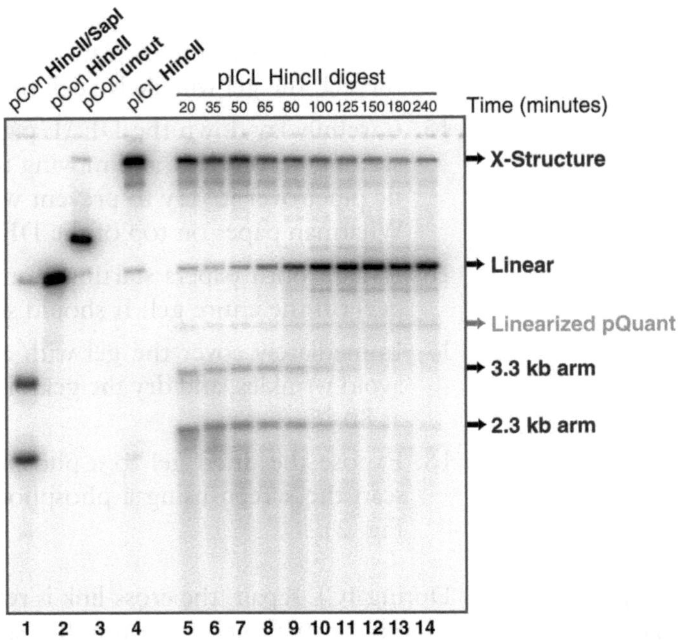

Fig. 3. Examination of dual incisions during ICL repair. HincII-digested repair intermediates are separated on a denaturing gel and visualized using Southern blotting (*lane 5–14*) (Subheading 3.4). Undigested and HincII and/or SapI-digested pControl and pICL are shown in *lane 1–4* and serve as size markers. X-structures, linears, and 2.3/3.3 arm fragments are indicated. pQuant is visible due to some background reactivity with the probe.

5. Centrifuge at 16,000 × *g* for another 5 min at 4 °C and aspirate the supernatant (see Note 40).
6. Air-dry the pellets for about 10 min at room temperature.
7. Resolubilize the pellet in 10 μL alkaline loading buffer.

3.4.2. Running and Transferring the Alkaline Gel

1. Poor a large (20 × 30 cm) 0.9 % agarose gel in milliQ water. Use a comb with thin (1 mm) slots (see Note 41). Let it solidify completely. Incubate the gel in 2.5 L alkaline running buffer half an hour before starting the run.
2. Load the samples and start the run at 0.8 V/cm (cm referring to the distance between the electrodes). Run until the samples have just entered the gel (this will take around 40 min).
3. Put a glass plate on top of the gel covering at least the area from the wells to halfway the gel. Run for another 18 h at 0.8 V/cm (until the dye front is halfway the gel).
4. Cut the gel just above the dye front and incubate the top half for 7 min in 1 L 0.25 M HCl while slowly shaking.
5. Rinse gel with water and incubate for 30 min in 1 L NaOH/NaCl.

6. Transfer the DNA to the membrane by making the following stack on top of a tray:
   (a) 1 Large piece of Whatman paper (prewet in 10× SCC)
   (b) Gel
   (c) Membrane (prewet in milliQ followed by 10× SCC)
   (d) 6 Gel-sized Whatman papers (prewet in 10× SCC)
   (e) A big stack of paper towels
   (f) A glass plate or tray
   (g) A weight (1–2 kg)

   The stack is placed in a larger tray that contains 10× SSC to a level just below the tray the stack is assembled on. The large piece of Whatman paper hangs over the sides of the inner tray into the buffer. After covering the membrane with the Whatman papers air bubbles are removed by rolling with a pipet.
7. Transfer for 18 h.
8. Disassemble the stack and incubate the membrane in 250 mL 0.5 M Tris–HCl, pH 7.5, 0.5 M NaCl for 15 min.
9. Air-dry the membrane.
10. Cross-link the DNA to the membrane using the "Autocrosslink" settings of the UV cross-linker (output is 120,000 μJ/cm^2).

3.4.3. Hybridization and Detection

1. The probe is generated using the random prime labeling kit (GE). pControl, linearized by digestion with HincII, is used as a template. For this, 50 ng of linearized pControl in 21 μL milliQ water is added to 5 μL of the random primers provided by the kit.
2. Incubate at 98 °C for 5 min. Quickly spin down and add: 5 μL 10× buffer (kit), 4 μL dCTP, dGTP and dTTP each (kit), 5 μL [α-P^{32}]-dATP, and 2 μL Klenow (kit).
3. Incubate at 37 °C for 1 h and stop the reaction by adding 2 μL 0.5 M EDTA.
4. Remove the unincorporated [α-P^{32}]-dATP by spinning through a G-50 column according to manufacturer's protocol.
5. Preheat the hybridization mix at 68 °C until the precipitation has dissolved.
6. Prehybridize the blot for at least 30 min at 42 °C in 22 mL hybridization mix.
7. Incubate the probe at 98 °C for 3 min and snap cool on water/ice slurry for 2 min. Add the probe to the hybridization mix and incubate the blot overnight at 42 °C in a hybridization oven.
8. Wash the membrane twice with prewarmed (42 °C) 2× SSC/0.1 % SDS for 5 min. Then, wash the blot twice with

prewarmed 0.1× SSC/0.1 % SDS for 15 min. Air-dry the membrane and expose to a phosphorimager screen for 4–24 h. Scan the screen using a phosphorimager.

3.5. Determining ICL Repair Efficiency

The recognition site of the restriction endonuclease SapI overlaps with the cross-link in pICL (Fig. 1a). Because the cross-link prevents cleavage by SapI (2), repair of pICL can be monitored by the regeneration of the SapI site. Regeneration of the SapI site is determined by dual digestion with both SapI and HincII (HincII cuts only once in pICL) (Fig. 1c) and quantification of the 2.3 and 3.3 kb dual digestion products. A complication is that even in unrepaired molecules, any incisions near the ICL (during the repair process, (2)) would result in 2.3 and 3.3 kb fragments after HincII digestion (Fig. 1a, c). To identify the background level of these products, we digest an equal volume of the repair intermediates with HincII only and subtract these products from the HincII/SapI digestion products (a detailed description of how this is done is given below). This assay determines the percentage of the input pICL that is repaired "error free" (no errors at the SapI site). We first describe how to run the repair gel (Subheading 3.5.1) followed by an explanation how to determine the repair efficiency from this gel (Subheading 3.5.2).

3.5.1. Running the pICL Repair Gel

1. Digest 1 μL of the purified repair intermediates (Subheading 3.2, step 15) with 0.4 μL HincII (4 units), and 1 μL with 0.4 μL HincII and 0.6 μL SapI (1.2 units), in a volume of 20 μL. Incubate samples at 37 °C for 3 h (see Notes 42 and 43).
2. Briefly spin down the samples. Add 5 μL DNA loading buffer to each sample and load the entire sample on a large (20 × 30 cm) 0.8 % agarose gel in 1× TBE. Run the gel ~3 h at 4.5 V/cm (cm referring to the distance between the electrodes), until the dye front is half way the gel (see Note 44).
3. Cut the gel just above the dye front. Discard the bottom half of the gel containing nonincorporated nucleotides in an appropriate radioactive waste container.
4. Dry the gel in a stack containing, from top to bottom: saran wrap, approximately 20 paper towels, a piece of Whatman paper, a piece of DEAE paper, the gel, and again DEAE paper, Whatman paper, paper towels, and saran wrap. Dry the gel by applying a weight of about 1 kg on top of the stack for approximately 1 h. Dispose of the paper towels, taking care to check whether they are radioactive.
5. Finish drying by placing the gel surrounded by the DEAE and Whatman paper in a gel dryer. Cover with saran wrap and vacuum dry at 80 °C for 30–60 min.
6. Wrap the dried gel in saran wrap and expose it to a phosphor screen for 2–18 h. Scan the screen using a phosphorimager.

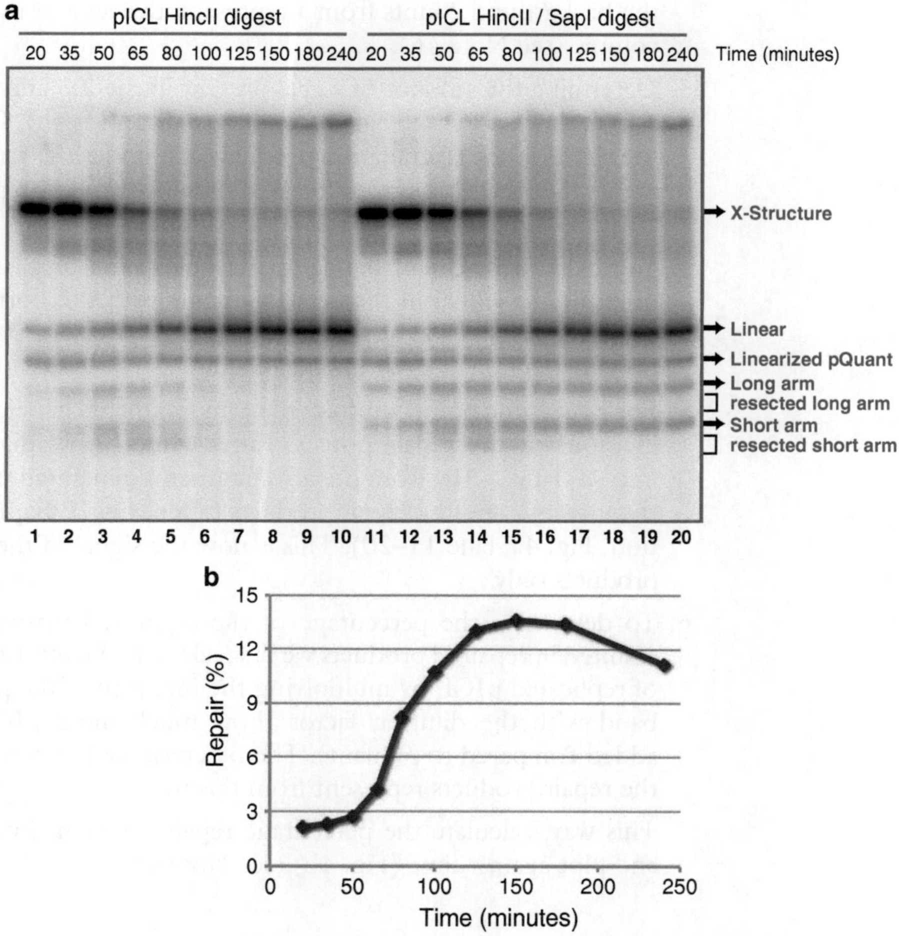

Fig. 4. Determination of ICL repair efficiency. (**a**) Analysis of HincII (*lane 1–10*) and HincII/SapI (*lane 11–20*)-digested repair intermediates on a native agarose gel (Subheading 3.5). (**b**) Calculated repair efficiency plotted against time.

3.5.2. Interpreting the Repair Gel and Determining Repair Efficiency

1. An example of a repair gel is given in Fig. 4a. After HincII digestion (Fig. 4a, lane 1–10), the major species at early time points is the large X-shaped structure that is formed when replication forks converge at the cross-link (Fig. 1a, c, (2)). At later time points, dual incisions take place that unhook the cross-link from one strand, resulting in the formation of a linear molecule and two arm fragments (Figs. 1c and 4a lane 3–6). The arm fragments are degraded over time (2). The HincII/SapI digested samples (Fig. 4a, lane 11–20) show the same species but in addition show a buildup of 2.3 and 3.3 kb arm fragments at later time points which represent the repair products (Figs. 1c and 4a lanes 15–20). All the lanes on the repair gel show the linearized pQuant, which will be used for normalization (also see Note 45).
2. First, using quantification software, determine the signal present in the linearized pQuant bands of each lane (Fig. 4a). Subtract

the background counts from a representative area of the same volume (see Note 46).

3. Determine the amount of signal present in the 2.3 and 3.3 kb arm fragments in each lane (draw a box around both). Include the signal in the degradation products just under the arm fragments. Subtract background signal obtained from a representative background area of the same volume.
4. Normalize all pQuant bands to the one in the first lane and multiply the signal in the arm fragments with the normalization factor. This will adjust for unequal recovery of products upon extraction and/or unequal loading.
5. For every time point, subtract the arm fragments generated by incisions in the repair process (after HincII-only digestion, Fig. 4a, lane 1–10) from the arm fragments generated by both the incisions and the repair products (after HincII/SapI digestion, Fig. 4a, lane 11–20). This is now the signal of the repair products only.
6. To determine the percentage of the replicated plasmid that resulted in repaired products we first calculate the total amount of replicated pICL by multiplying the intensity of the pQuant band with the dilution factor (how much more pICL was added compared to pQuant). Then, determine the percentage the repair products represent from this total.
7. This way, calculate the percentage repair for each time point and plot against time (Fig. 4b) (see Note 47).

4. Notes

1. A low amount of pQuant is added to the repair reaction to determine repair efficiency (see Subheading 3.5.2). Generally, the concentration of pQuant should be similar to the expected concentration of repair products. Therefore, if the concentration of pICL is 2 ng/μL, and 10 % repair is expected, 0.2 ng/μL pQuant should be added. pQuant can be any 3–4 kb plasmid with little homology to pICL (to prevent adding a potential donor sequence for homologous recombination). It should be cut once with HincII and not with SapI. We use pCDF-DUET (Novagen) that we modified to contain only one HincII restriction site. For accurate quantification it is important that linearized pQuant can be separated from any fragments of pICL on native agarose gels (Subheading 3.5 and Fig. 4). Digestion of pQuant with AflIII should not generate any short fragments (<700 bp) that might obscure bands in the lesion bypass assay (Subheading 3.3 and Fig. 2).

2. Choose a sequencing primer that starts at the digestion site used in Subheading 3.3.1. For the example described here, samples are digested with AflIII and the gel is run under conditions that allow nucleotide resolution of fragments between ~80 and ~160 nt. One of the AflIII sites is located at a distance of ~150 nt from the ICL and the primer starts exactly at this restriction site. This way, the sequence and size of the products present in the sequencing ladder correspond to those of the newly synthesized (and therefore radioactively labeled) leading strands of the repair intermediates.

3. The size of the glass plates depends on the sequencing apparatus. We use a model S2 sequencing apparatus (GibcoBRL/Labrepco). The large plate is 42 × 33 cm and the small plate 39 × 33 cm.

4. It can also be useful to take samples for western blot analysis at this stage. These samples can be used to examine the activation of the Fanconi anemia pathway by monitoring the ubiquitylation of FANCD2 and FANCI (2, 8). Another useful marker to study is the phosphorylated form of Chk1 (2), which is used as a measure of activation of the DNA damage checkpoint kinase ATR (also see Note 10).

5. Use ELB-sucrose for dilution of stock solutions (DTT, plasmids etc.).

6. In principle, a low speed supernatant (LSS) extract can replace HSS in this procedure. This, however, only works if plasmid DNA is used as a template. If chromatin DNA is used as a template, nuclei formation and DNA replication initiates before NPE addition and synchronization is lost.

7. The plasmid concentration in a replication/repair reaction generally should not be lower than ~0.3 ng/μL since this will not support efficient DNA replication (9). Optimal concentration is between 0.5 and 10 ng/μL. For a repair reaction we use a concentration of pICL between 1 and 3 ng/μL.

8. When using pICL, do not vortex and try to avoid extensive pipetting, preferably mix by stirring with the pipet tip in the solution (if pipetting up and down is needed, do it carefully) to make sure that no additional damage is added to the potentially fragile cross-linked template.

9. This deviates from the DNA replication protocol as described in ref. 6 where a 30 min incubation is described. In a repair reaction this time is reduced to avoid potential processing of the ICL in HSS.

10. [α-P^{32}]-dATP can be added to either HSS or NPE reaction mixes. Since incorporation of nucleotides only starts a few minutes after adding NPE to HSS, [α-P^{32}]-dATP can also be added immediately after combining HSS and NPE. This is

useful for taking nonradioactive samples of the reaction for analysis by Western blotting.

11. The amount of [α-P^{32}]-dATP that is added to a repair reaction depends on several variables including what assays will be performed with the samples. A sequencing gel requires more [α-P^{32}]-dATP incorporation than a repair assay. Second, the efficiency of replication. In case of low anticipated replication efficiency (for example after extensive depletion) addition of more [α-P^{32}]-dATP may be necessary to increase the radioactivity of the samples. Finally it depends on how fresh the batch of [α-P^{32}]-dATP is. The amount of fresh [α-P^{32}]-dATP we use in this example is enough to perform all assays described in this chapter using samples from a single repair reaction.
12. NPE often requires dilution with ELB-sucrose to allow for rapid and complete replication (6). The optimal degree of dilution has to first be established for every extract.
13. It is important that the repair reaction samples are collected in Safe-Lock tubes. This prevents radioactive contamination during subsequent phenol–chloroform extraction.
14. We found that if samples are not diluted at this step a crystalline white pellet forms at the ethanol precipitation step that inhibits digestion of the samples by restriction endonucleases.
15. The pellet after ethanol precipitation and wash steps is very small and often does not stick well to the bottom of the tube. Take extreme care during aspiration, as it is very easy to lose the pellet. We usually process no more than four samples at a time and leave the rest spinning.
16. Do not overdry the pellets and do not use a vacuum dryer. This will cause problems in dissolving the pellet, likely due to the fact that repair intermediates contain complex structures.
17. These purified repair intermediate samples are used for the assays described below. We have used these samples up to 10 days after generating them but we recommend using the samples as soon as possible for maximal signal intensity ($T_{1/2}$ for ^{32}P is 14.3 days).
18. The method we used to distinguish between leading and lagging strand products is described in supplementary figure 8 of (2).
19. Note that the glycerol and salt concentrations as well as the sample volume loaded on the gel can affect the migration of DNA molecules on denaturing polyacrylamide gels. Therefore we recommend loading equal volumes of samples and sequencing ladder. In case the migration of the DNA intermediates differs from the expected, the sequence ladder samples can be precipitated and taken up in the same solution as the repair intermediate samples.

20. It is important to keep the volume of the digestion reaction small to be able to load sufficient repair intermediates for detection. We use AflIII to digest our products because this results in fragments of suitable sizes for the analysis on sequencing gels. In case a plasmid with a different sequence is used, choose an appropriate restriction enzyme.
21. At this stage, unincorporated nucleotides can be removed by spinning the sample through a G-50 column. This will reduce radioactive contamination during the following steps.
22. Only touch the silanized surface of the glass plate with soft tissues.
23. If the gel sticks to both plates upon separation causing it to rupture, it sometimes helps to silanize the large glass plate as well as the small one.
24. Silanization of the glass plates does not need to be done before every experiment. The plates can be used up to ten times before resilanization is required.
25. We use proper gel sealing tape for this. We have tried many types of more general tape but none were adequate (for example the gel-mix dissolves the glue of the tape, or the tape is very difficult to remove from the glass plates after use).
26. A rubber casting-clamp is available for sealing the gel sandwich, which can be used instead of tape.
27. Heating of the acrylamide mix in a microwave oven for 15 s helps to dissolve urea.
28. The percentage of the gel can be varied depending on the size of the product generated by digestion. In this case we use a 7 % gel to get nucleotide resolution of fragments between ~80 and ~160 nt. A 5 % acrylamide gel allows examination of products up to 550 nt at nucleotide resolution.
29. Occasionally, bubbles appear on the side of the gel during pouring. This is usually a sign that the glass plates need to be resilanized. If the samples are loaded at a distance of a centimeter or more from the bubbles, they will not affect the migration of the samples.
30. Alternatively it can be left overnight to polymerize, cover the top of the gel with wet tissues to prevent it from drying out.
31. Carefully inspect the surface of the gel made by the comb. Mark areas where the surface is not perfectly straight. Because this surface forms the bottom of the sample slots it should be as straight as possible to ensure high resolution.
32. If the teeth of the comb are inserted into the gel during the prerunning of the gel it is very difficult to seal the wells properly for sample loading.

33. Try to load the samples as quickly as possible after stopping the prerun, the gel should not cool down too much. Load similar volumes of the repair samples and sequencing ladder samples (see Note 19).
34. We have found that the last sample of a series often shows less tight bands compared to the previous ones. We solve this by adding two or more sequencing ladder samples after the last sample.
35. Running time depends on the percentage of the gel and the size of the products. Here we examine a ~150 nt product by running it on a 7 % gel for 2.2 h at 60 W. However, a 500 nt product can be examined at nucleotide resolution on a 5 % gel by running it for ~5.5 h at 55 W.
36. If there are air bubbles on one side of the gel, open the gel from the other side as they can cause the gel to break.
37. It's critical to use positively charged DEAE paper as it will help to retain small DNA molecules.
38. For the incision assay it is preferable to prepare samples without [α-P^{32}]-dATP incorporation. This way, double detection of the signal from the nascent strands is prevented (once due to body labeling and once due to the hybridization with the radioactive probe in the Southern blot).
39. Suitable markers for the Southern blot can be generated by digestion of similar amounts of pICL and pControl with HincII, and pControl with HincII/SapI.
40. Because there is very little DNA in the samples and no glycogen was added, the pellet is difficult to see. The pellet does seem to be firmly stuck to the side of the tube as we have never lost a pellet at this stage (all tubes should be put into the centrifuge in a known orientation, so that one knows where to expect the pellet after centrifugation).
41. We found it is important to use a comb with thin (1 mm) slots, as it increases the resolution of the gel.
42. We sometimes observe problems with SapI digestion. To keep SapI active during the entire digestion period we add 0.3 μL per sample at the beginning and another 0.3 μL SapI after 2 h.
43. In some cases, contaminants in the purified DNA samples inhibit proper digestion. From the gel it is not always obvious whether they have been properly digested as one will always find a certain degree of heterogeneous, large products regardless of the quality of digestion. Therefore it is best to have an independent digestion control. For this, set up a replication/repair reaction with pControl in parallel with pICL, which will result in radioactive products that generate the same 2.3 and 3.3 kb fragments after HincII/SapI digestion. Alternatively, it is possible to

include a nonradioactive sample of pControl during the extraction procedure. Digestion of the extracted pControl with HincII/SapI and detection of the products on an agarose gel stained with Sybr gold allows verification of digestion.

44. Running a large gel is required to properly separate the repair products from linearized pQuant. In addition, it is important to stop the gel before unincorporated radioactive nucleotides migrate off the gel (when the dye front it halfway down the gel). This helps to decrease background radioactivity on the gel and reduces radioactive waste.

45. In both the HincII and the HincII/SapI digested samples, at later time points, a pronounced high molecular weight smear is often observed. This high molecular weight smear likely consists of partially degraded X-shape structures and repair intermediates formed by homologous recombination ((4, 7).

46. The pQuant band may overlap with considerable background smearing that is not equal in each lane of the gel. If this is the case, we choose the background signal for each lane separately in an area close to the pQuant band.

47. If there is a small fraction of non-cross-linked plasmid in the pICL prep, these products will show up as "repair products" at the earliest time points (in the example in Fig. 1 this represents 2 %). The actual repair products are formed only after about 1 h and show a sharp increase at that time (2). To determine the true repair efficiency, the former should be subtracted from the latter.

References

1. Legerski RJ (2010) Repair of DNA interstrand cross-links during S phase of the mammalian cell cycle. Environ Mol Mutagen 51: 540–551
2. Raschle M, Knipscheer P, Enoiu M, Angelov T, Sun J, Griffith JD, Ellenberger TE, Scharer OD, Walter JC (2008) Mechanism of replication-coupled DNA interstrand crosslink repair. Cell 134:969–980
3. Akkari YM, Bateman RL, Reifsteck CA, Olson SB, Grompe M (2000) DNA replication is required to elicit cellular responses to psoralen-induced DNA interstrand cross-links. Mol Cell Biol 20:8283–8289
4. Knipscheer P, Raschle M, Smogorzewska A, Enoiu M, Ho TV, Scharer OD, Elledge SJ, Walter JC (2009) The Fanconi anemia pathway promotes replication-dependent DNA interstrand cross-link repair. Science 326: 1698–1701
5. Walter J, Sun L, Newport J (1998) Regulated chromosomal DNA replication in the absence of a nucleus. Mol Cell 1:519–529
6. Lebofsky R, Takahashi T, Walter JC (2009) DNA replication in nucleus-free Xenopus egg extracts. Methods Mol Biol 521:229–252
7. Long DT, Räschle M, Joukov V, Walter JC (2011) Mechanism of RAD51-dependent DNA interstrand cross-link repair. Science 333(6038):84–87
8. Sobeck A, Stone S, Costanzo V, de Graaf B, Reuter T, de Winter J, Wallisch M, Akkari Y, Olson S, Wang W, Joenje H, Christian JL, Lupardus PJ, Cimprich KA, Gautier J, Hoatlin ME (2006) Fanconi anemia proteins are required to prevent accumulation of replication-associated DNA double-strand breaks. Mol Cell Biol 26:425–437
9. Lebofsky R, van Oijen AM, Walter JC (2011) DNA is a co-factor for its own replication in Xenopus egg extracts. Nucleic Acids Res 39:545–555

Chapter 17

In Vitro Base Excision Repair Using Mammalian Cell Extracts

Jason L. Parsons and Grigory L. Dianov

Abstract

Base excision repair (BER) is an essential cellular mechanism that maintains genome stability by repairing DNA damage, such as DNA base lesions, base loss (AP sites) and single strand breaks, generated through endogenous metabolism or via exogenous mutagens. Therefore, in vitro BER assays are important for our understanding of the mechanism of cellular response to mutagens and may also reveal important information about the development of several DNA repair-related human diseases, such as cancer, and aging. Here, we describe the preparation and use of mammalian cell extracts in in vitro BER assays using both oligonucleotide and closed circular DNA substrates containing site-specific DNA lesions, in combination with denaturing acrylamide gel electrophoresis and phosphor imaging analysis.

Key words: Base excision repair, DNA repair assays, Cell extracts, AP endonuclease, DNA polymerase, DNA ligase

1. Introduction

Base excision repair (BER) is an essential cellular process required for genome stability and a deficiency or misregulation of BER has been linked to premature aging, increased rate of mutagenesis and the development of human diseases, such as cancer. BER is initiated by damage-specific DNA glycosylases that hydrolyze the N-glycosylic bond linking the damaged DNA base to the sugar phosphate backbone, thus causing excision of the damaged base. The arising apurinic/apymidinic site (AP site) is recognized by AP endonuclease-1 (APE1) that cleaves the phosphodiester bond 5′ to the AP site, generating a DNA single strand break (SSB) with 5′-deoxyribose phosphate (dRP) and 3′-hydroxyl (OH) ends. The 5′-dRP residue is removed by DNA polymerase β (Pol β) that also adds the correct new nucleotide into the arising one nucleotide gap and the DNA ends are finally sealed by XRCC1-DNA ligase

Lotte Bjergbæk (ed.), *DNA Repair Protocols*, Methods in Molecular Biology, vol. 920,
DOI 10.1007/978-1-61779-998-3_17, © Springer Science+Business Media New York 2012

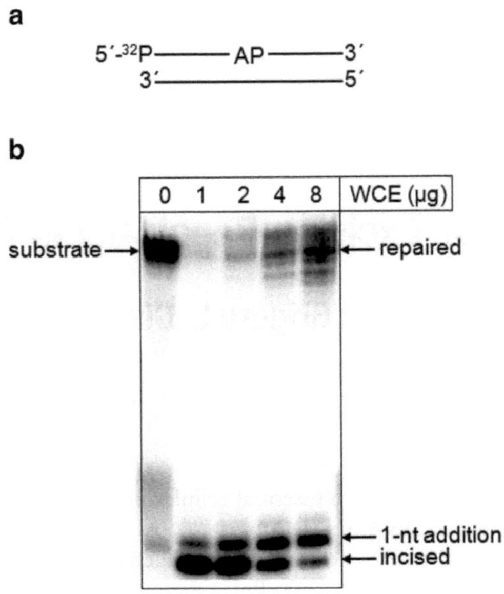

Fig. 1. Repair of an AP site containing duplex oligonucleotide substrate using human WCE. (**a**) Schematic representation of a 5′-γ-^{32}P-labeled AP site containing duplex oligonucleotide substrate. (**b**) A 5′-γ-^{32}P-labeled AP site containing substrate (300 fmol) was incubated with HeLa WCE (0–8 μg) for 20 min at 37 °C. The products of the reaction were separated by 20 % denaturing acrylamide gel electrophoresis and analyzed by phosphor imaging. Following incubation of the AP site containing substrate with increasing amounts of WCE, the three steps of BER can be visualized. Firstly, the initial substrate is incised by APE1 to yield a smaller DNA fragment, secondly Pol β generates a one nucleotide (1-nt) addition product into the incised AP site and finally the nick is sealed by DNA ligase that generates the full length repaired product.

IIIα complex (1). This is commonly referred to as the short patch BER pathway by which the majority (~80 %) of repair is achieved. If the 5′-dRP is resistant to the AP lyase activity of Pol β, then a switch to DNA polymerase δ/ε (Pol δ/ε) occurs that allows incorporation of two to eight more nucleotides into the repair gap, creating a 5′-flap structure that is excised by flap endonuclease-1 (FEN-1) in a proliferating cell nuclear antigen (PCNA)-dependent manner. The remaining nick in the DNA backbone is then sealed by DNA ligase I (Lig I) and this completes the long patch BER pathway (2, 3).

Studying BER is important for our understanding of how cells respond to DNA damage induced endogenously through oxidative metabolism or via exogenous mutagens and also to examine the development of human diseases, such as cancer, and aging that have been associated with deficiencies in BER. To achieve this, the simplest approach is the generation and use of whole cell extracts (WCE) prepared from mammalian cell culture lines or tissues, in combination with in vitro BER assays using duplex oligonucleotide or closed circular DNA substrates containing a single site-specific lesion, the repair of which can be subsequently analyzed by denaturing acrylamide gel electrophoresis. To generate DNA repair substrates of

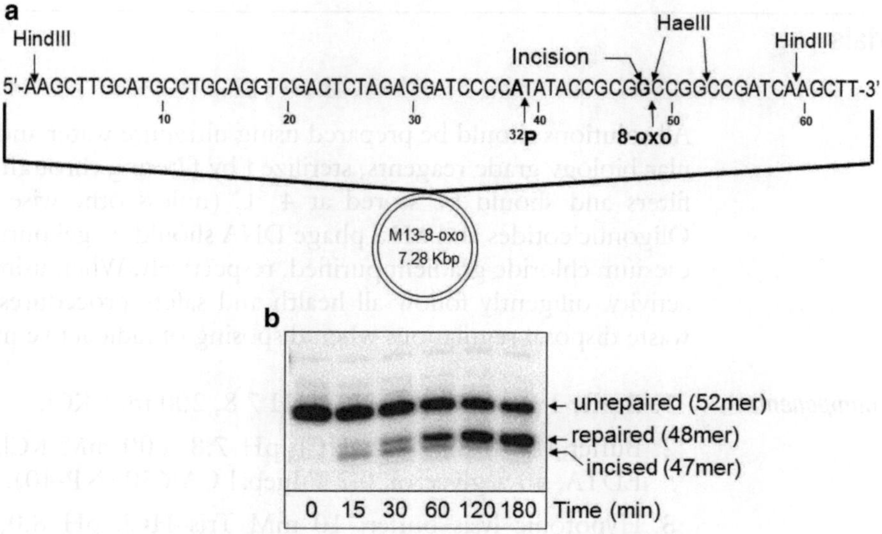

Fig. 2. Repair of an 8-oxoguanine-containing closed circular DNA substrate using HeLa WCE. (**a**) Schematic representation of a closed circular DNA substrate containing a site-specific 8-oxoguanine lesion. The sites of cleavage by the restriction enzymes *Hin*d III and *Hae* III and the position of the ^{32}P label are shown. The site of incision by APE1 subsequent to DNA glycosylase processing of 8-oxoguanine (8-oxo) is indicated with an *arrow*. (**b**) Time-dependent repair of 8-oxoguanine by HeLa WCE. Reactions containing 40 fmol of 8-oxoguanine containing substrate and 100 μg of WCE were incubated at 37 °C for the indicated time prior to isolation of the substrate DNA, followed by digestion with *Hin*d III and *Hae* III. Reaction products were separated by 10 % denaturing acrylamide gel electrophoresis and analyzed by phosphor imaging. The 8-oxoguanine substrate has two *Hae* III restriction sites, although one of them is blocked by 8-oxoguanine and therefore simultaneous cleavage with *Hae* III and *Hind* III generates a 52-mer labeled product. Following incubation with the WCE, 8-oxoguanine is excised and the arising AP site is incised by APE1 to generate a 47-mer incised product. The incised product is subsequently repaired by Pol β and DNA ligase that restores the *Hae* III restriction site and therefore this is revealed as a 48-mer product following simultaneous digestion with *Hae* III and *Hind* III. The figure has been adapted and modified from ref. 6.

interest, oligonucleotides containing a specific lesion at a predetermined position should be synthesized or purchased from a commercially available source. The optimal length of the oligonucleotide is about 40 bases, with the specific lesion located centrally, and the sequence should be carefully chosen so as not to contain any hairpins, repeats, and not to form any primer dimers. Both duplex oligonucleotides and closed circular DNA substrates can be used for DNA repair assays; however, closed circular molecules are more stable following incubation in the WCE and are the preferred substrate to study long patch repair enzymology. Single lesion closed circular DNA substrates should also contain appropriate DNA restriction sites in the vicinity of the lesion that will allow further detailed analysis of DNA repair. Typical oligonucleotide and closed circular DNA substrates are presented in Figs. 1a and 2a, correspondingly.

In this chapter, we cover methods for preparation of DNA repair substrates and WCE, and describe in vitro BER assays for the measurement of DNA glycosylase, APE1, DNA polymerase, and DNA ligase activities, as well as measuring the complete BER process, in combination with denaturing acrylamide gel electrophoresis and phosphor imaging analysis.

2. Materials

All solutions should be prepared using ultrapure water and molecular biology grade reagents, sterilized by filtering through 0.2 μm filters and should be stored at 4 °C (unless otherwise stated). Oligonucleotides and M13 phage DNA should be gel purified and caesium chloride gradient purified, respectively. When using radioactivity, diligently follow all health and safety procedures and all waste disposal regulations when disposing of radioactive material.

2.1. WCE Components

1. Buffer I: 10 mM Tris–HCl, pH 7.8, 200 mM KCl.
2. Buffer II: 10 mM Tris–HCl, pH 7.8, 600 mM KCl, 2 mM EDTA, 40 % glycerol, 0.2 % Igepal CA-630 (NP-40).
3. Hypotonic lysis buffer: 10 mM Tris–HCl, pH 8.0, 1 mM EDTA.
4. Sucrose–glycerol buffer: 50 mM Tris–HCl, pH 8.0, 10 mM $MgCl_2$, 25 % sucrose, 50 % glycerol (see Note 1).
5. Protease inhibitors: 1 mg/ml chymostatin in DMSO, 1 mg/ml pepstatin in ethanol, 1 mg/ml aprotinin in water, 1 mg/ml leupeptin in water, 0.25 M phenylmethylsulfonyl fluoride (PMSF) in isopropanol (see Note 2).
6. Reducing agent: 1 M dithiothreitol (DTT) in water and stored at −20 °C.
7. Ammonium sulfate: Saturated ammonium sulfate dissolved in water and pH 7.0 with ammonium hydroxide.
8. Sodium hydroxide: 1 M sodium hydroxide.
9. Dialysis buffer: 25 mM HEPES-KOH, pH 7.9, 100 mM KCl, 12 mM $MgCl_2$, 1 mM EDTA, 17 % glycerol.
10. Fine glass homogenizers.
11. Ultracentrifuge: Centrifuge capable of achieving speeds of up to 100,000 × *g*.

2.2. Oligonucleotide Labeling Components

1. TE buffer: 10 mM Tris–HCl, pH 8.0, 1 mM EDTA.
2. T4-PNK: T4-polynucleotide kinase: (10 U/μl; New England Biolabs, Hitchin, UK) stored at −20 °C.
3. T4-PNK buffer (10×): 70 mM Tris–HCl, pH 7.6, 10 mM $MgCl_2$, 5 mM DTT. Aliquot and store at −20 °C.
4. γ-^{32}P-ATP: γ-^{32}P-Adenosine 5′-triphosphate (10 μCi/μl, 6,000 Ci/mmol; Perkin Elmer, Cambridge, UK) (see Note 3).

5. Purification columns: Micro Bio-spin P-6 columns (Biorad Laboratories, Hemel Hempstead, UK).
6. Speed vac concentrator.

2.3. Closed Circular DNA Components

1. Single-stranded M13 phage DNA: Streak a master culture of JM109 bacterial cells onto a minimal agar plate and incubate for 24–36 h at 37 °C. Inoculate 50 ml LB medium with a single colony in a 250 ml flask and agitate for 6–8 h at 37 °C at 250 rpm. Cool the plating culture to 0 °C and this can then be stored at 4 °C for up to 1 week. Add 2.5 ml plating bacteria to a 15 ml tube, add 0.5 ml of bacteriophage stock, carefully mix, and allow the tube to stand for 5 min at room temperature. Dilute the infected cells into 250 ml LB medium prewarmed to 37 °C in a 2 l flask and incubate for 5 h at 37 °C in a rotary shaker at 250 rpm. Centrifuge the cultures at 10,000 × *g* for 20 min at 4 °C, remove the supernatant, and add 50 ml cold buffer containing 30 % polyethylene glycol (PEG-6000) and 3 M NaCl. Invert the solution several times, incubate on ice for 60 min with regular inversions and centrifuge at 10,000 × *g* for 20 min at 4 °C. Carefully remove all of the supernatant and resuspend the pellet in 25 ml buffer containing 10 mM Tris–HCl (pH 8.0), 1 mM EDTA, and 0.05 % SDS. Add proteinase K to 50 μg/ml and incubate at 56 °C for 60 min. Add 12.5 ml chloroform:isoamyl alcohol (24:1), mix and centrifuge at 3,000 × *g* for 5 min. Remove the upper aqueous layer and repeat the chloroform:isoamyl alcohol extraction. Following centrifugation, add one tenth volume of 3 M NaCl and 2 volumes of ethanol and precipitate the DNA overnight at −20 °C. Centrifuge at 40,000 × *g* for 30 min at 4 °C, remove the supernatant, and allow the pellet to air-dry for 10 min. Resuspend the pellet in 1 ml TE buffer, measure the DNA concentration, aliquot, and store at −20 °C.
2. Annealing buffer (10×): 0.5 M Tris–HCl, pH 8.0, 0.2 M KCl, 70 mM $MgCl_2$, 1 mM EDTA.
3. DNA polymerase buffer (5×): 50 mM Tris–HCl, pH 8.0, 0.25 M NaCl, 50 mM $MgCl_2$, 5 mM DTT.
4. dNTP (10 mM): From 100 mM stocks of dATP, dCTP, dGTP and dTTP, mix 10 μl of each dNTP into a total volume of 100 μl water. Aliquot and store at −20 °C.
5. ATP (20×): 40 mM ATP in water. Filter sterilize and store at −20 °C.
6. T4 gene 32 protein: (10 mg/ml; New England Biolabs, Hitchin, UK) stored at −20 °C.
7. T4 DNA ligase: (20 U/μl; New England Biolabs, Hitchin, UK) stored at −20 °C.

8. T4 DNA polymerase: (3 U/μl; New England Biolabs, Hitchin, UK) stored at −20 °C.
9. CsCl: Caesium chloride.
10. EtBr: 10 mg/ml ethidium bromide.
11. Ultracentrifuge: Centrifuge capable of achieving speeds of up to 300,000 × *g*.
12. OptiSeal tubes: 11.2 ml tubes (Beckman-Coulter, High Wycombe, UK).
13. Amicon Ultra-30: Centrifugal filter units with molecular weight cut-off of 30 kDa (Millipore, Watford, UK).
14. TE buffer: 10 mM Tris–HCl, pH 8.0, 1 mM EDTA.

2.4. BER Assay Components

1. BER buffer (20×): 0.8 M HEPES-KOH, pH 7.8, 100 mM $MgCl_2$, 2 mM EDTA, 10 mM DTT. Aliquot and store at −20 °C.
2. ATP (20×): 40 mM ATP in water. Filter sterilize, aliquot, and store at −20 °C.
3. dNTP (20×): From 100 mM stocks of dATP, dCTP, dGTP, and dTTP, mix 4 μl of each dNTP into a total volume of 1 ml water. Aliquot and store at −20 °C.
4. PC (20×): 0.5 M phosphocreatine in water. Filter sterilize, aliquot, and store at −20 °C.
5. PCK (25×): Dissolve 5 mg creatine phosphokinase in 1 ml 10 mM glycine, pH 9.0, and then add 1 ml glycerol. Aliquot and store at −20 °C.
6. NAD (20×): Dissolve 3.3 mg β-nicotinamide adenine dinucleotide in 1 ml water. Aliquot and store at −20 °C (see Note 4).
7. Carrier DNA (25×): 1 mg/ml single-stranded 30-mer oligonucleotide (see Note 5).
8. Dialysis buffer: 25 mM HEPES-KOH, pH 7.9, 100 mM KCl, 12 mM $MgCl_2$, 1 mM EDTA, 17 % glycerol, 2 mM DTT. Aliquot and store at −20 °C.
9. Purification columns: Micro Bio-spin P-6/P-30 columns (Biorad Laboratories, Hemel Hempstead, UK).
10. Sodium borohydride: 0.5 M sodium borohydride in water prepared just before use (see Note 6).
11. 0.5 M EDTA: 0.5 M ethylenediaminetetraacetic acid in water and stored at room temperature.
12. 10 % SDS: 10 % sodium dodecyl sulfate in water and stored at room temperature.
13. Proteinase K: 10 mg/ml proteinase K stored at −20 °C.
14. Phenol/chloroform/isoamyl alcohol: 25:24:1 mixture.
15. Speed vac concentrator.

2.5. Denaturing Acrylamide Gel Electrophoresis Components

1. TBE buffer (10×): 890 mM Tris-borate, pH 8.3, 20 mM EDTA.
2. 20 % denaturing acrylamide gel: Dissolve 480 g urea (8 M) in 500 ml 40 % acrylamide (19:1 acrylamide:Bis-acrylamide) and 100 ml 10× TBE. Make volume up to 1 l with water, filter sterilize, and store at room temperature (see Note 7).
3. Ammonium persulfate: 10 % solution in water.
4. TEMED: *N,N,N′,N′*-tetramethylethylenediamine (Biorad Laboratories, Hemel Hempstead, UK) stored at 4 °C.
5. FA dye: 95 % formamide containing 0.02 % xylene cyanol and 0.02 % bromophenol blue, aliquoted, and stored at −20 °C.
6. Acrylamide gel electrophoresis apparatus: Model SA sequencing gel apparatus with glass plates, spacers, and combs (Biometra, Goettingen, Germany).
7. Gel sealing tape and steel clips.

3. Methods

3.1. Harvesting of Cell Cultures

1. For adherent cells growing on 10 cm dishes, wash the plates with 5 ml cold PBS and remove PBS using a vacuum pump.
2. Add 5 ml PBS and scrape off cells using a cell scraper.
3. Wash each plate with a further 5 ml PBS and combine to the cell suspension.
4. Centrifuge the cells at 350 × *g* at 4 °C for 5 min in a 15 ml tube.
5. Remove supernatant using a vacuum pump and resuspend the cell pellet in 1 ml cold PBS and transfer to a 1.5 ml microcentrifuge tube.
6. Centrifuge the cells at 350 × *g* at 4 °C for 5 min and remove liquid.
7. Freeze the cell pellet at −80 °C for at least 1 h (see Note 8).

3.2. Preparation of Whole Cell Extracts (Method 1, Adapted and Modified from Ref. 4)

1. Add protease inhibitors (1 μl of 1 mg/ml chemostatin, pepstatin, aprotinin, leupeptin, and 4 μl of 0.25 mM PMSF) and DTT (1 μl 1 M DTT) to 1 ml Buffer I and 1 ml Buffer II before use (see Note 9).
2. Measure a packed cell volume (PCV) and resuspend the cells in the same volume of Buffer I (see Note 10).
3. Add 2 volumes of Buffer II and mix the cell suspension thoroughly.
4. Rotate the cell suspension at 4 °C for 30 min.
5. Centrifuge at 100,000 × *g* for 20 min and collect supernatant into a fresh 1.5 ml microcentrifuge tube.

6. Dialyze the cell extract against dialysis buffer overnight at 4 °C.
7. Collect dialysate, centrifuge at 16,000 × *g* for 10 min at 4 °C, and collect supernatant.
8. Measure the protein concentration of the WCE, aliquot, and store aliquots at −80 °C.

3.3. Preparation of Whole Cell Extracts (Method 2, Adapted and Modified from Ref. 5)

1. Precool glassware (beakers, homogenizer, and centrifuge tubes) to 4 °C before use and preferably perform all subsequent steps in a cold room (4 °C).
2. Add protease inhibitors (1 μl of 1 mg/ml chemostatin, pepstatin, aprotinin, leupeptin, and 4 μl of 0.25 mM PMSF) and DTT (1 μl 1 M DTT) to hypotonic lysis buffer and sucrose–glycerol buffer just before use (see Note 9).
3. Measure a PCV and resuspend the cells in four times the volume (4× PCV) of hypotonic lysis buffer containing protease inhibitors and leave on ice for 20 min (see Note 11).
4. Homogenize the cells in a glass homogenizer on ice using approximately 30 strokes.
5. Transfer the cell lysate to a glass beaker and stir slowly whilst adding dropwise 4 volumes (4× PCV) of sucrose-glycerol buffer containing protease inhibitors.
6. Add 1 volume (1× PCV) of saturated ammonium sulfate (pH 7.0) dropwise to the cell lysate whilst stirring and the viscosity will increase.
7. Leave the beaker on ice and gently stir briefly every 5 min for 30 min.
8. Transfer the cell suspension to clean centrifuge tubes and centrifuge at 200,000 × *g* for 3 h at 4 °C (see Note 12).
9. Carefully remove the supernatant into a clean tube using a Pasteur pipette (see Note 13).
10. Measure the volume of the supernatant and add 0.33 g/ml of finely ground ammonium sulfate gently and slowly to the supernatant whilst stirring.
11. Add 10 μl 1 M sodium hydroxide per gram of ammonium sulfate and stir for 30 min.
12. Centrifuge the suspension at 75,000 × *g* for 20 min at 4 °C.
13. Discard the supernatant and resuspend the protein pellet in 1 volume (1× PCV) of dialysis buffer containing 1 mM DTT (see Note 14).
14. Dialyze the WCE against 1 l of dialysis buffer containing 1 mM DTT overnight at 4 °C.
15. Centrifuge the WCE at 16,000 × *g* for 10 min at 4 °C.

16. Collect the supernatant, measure the protein concentration, aliquot, and store at −80 °C.

3.4. Labeling of Oligonucleotides for Duplex Oligonucleotide Substrates

1. Mix 30 pmol of single-stranded oligonucleotide with 2 μl 10× T4-PNK buffer and make volume up to 17.5 μl with water.
2. Add 2 μl γ-^{32}P-dATP (10 μCi/μl) to the tube followed by 0.5 μl T4-PNK (10 U/μl), mix by vortexing, briefly centrifuge, and incubate for 1 h at 37 °C.
3. Centrifuge a Biospin P-6 column for 1,000 × *g* for 2 min.
4. Add 500 μl TE buffer to the center of the column and spin at 1,000 × *g* for 2 min, discard flow-through, and repeat.
5. Make labeling reaction volume up to 50 μl with TE buffer and add to the center of the prewashed Biospin P-6 column and spin at 1,000 × *g* for 4 min.
6. Make volume of labeled oligonucleotide up to 100 μl with TE buffer.

3.5. Preparation of 5′-Labeled Duplex Oligonucleotide Substrates

1. To generate a duplex oligonucleotide for the measurement of DNA glycosylase, AP site incision activity or for the analysis of full BER repair activity, mix 50 μl of γ-^{32}P-labeled single-stranded oligonucleotide containing the site-specific modification with 30 pmol of complementary single stranded oligonucleotide.
2. To generate a duplex oligonucleotide for the measurement of DNA polymerase or DNA ligase activity, mix 50 μl of γ-^{32}P-labeled single-stranded oligonucleotide with 22.5 pmol of the corresponding 3′-downstream oligonucleotide and 30 pmol of the complementary single-stranded oligonucleotide.
3. Add 4 μl 5 M sodium chloride and make volume up to 100 μl with TE buffer.
4. Incubate the oligonucleotides at 95 °C for 5 min and slow cool to room temperature (see Note 15).

3.6. Labeling of Oligonucleotides for Single Lesion Closed Circular DNA Substrates

1. Mix 30 pmol of a single-stranded oligonucleotide containing a site-specific modification with 2 μl 10× T4-PNK buffer and make volume up to 8 μl with water.
2. Add 10 μl γ-^{32}P-dATP (10 μCi/μl) to the tube followed by 2 μl T4-PNK (10 U/μl), mix by vortexing, briefly centrifuge, and incubate for 1 h at 37 °C.
3. Centrifuge a Biospin P-6 column for 1,000 × *g* for 2 min.
4. Add 500 μl TE buffer to the center of the column, spin at 1,000 × *g* for 2 min, discard flow-through, and repeat.
5. Add the labeling reaction to the center of the prewashed Biospin P-6 column and spin at 1,000 × *g* for 4 min.
6. Use a speed vac to dry the 5′-labeled oligonucleotide.

3.7. Preparation of 5′-Labeled Single Lesion Plasmid Substrates

1. Mix the dried 5′-labeled single-stranded oligonucleotide containing a site-specific modification (30 pmol) with 12.5 pmol single-stranded M13 phage DNA, 10 μl of 10× annealing buffer, and make volume up to 100 μl with water.
2. Incubate at 80 °C for 5 min and slow cool to room temperature (see Note 15).
3. Mix 100 μl of the annealed oligonucleotide/M13 phage DNA with 50 μl 5× DNA polymerase buffer, 12.5 μl 10 mM dNTP, 2.5 μl 20× ATP, 100 μg T4 gene 32 protein, 35 U T4 DNA ligase, 40 U T4 DNA polymerase, and make volume up to 250 μl with water.
4. Incubate the reaction for 4 h at 37 °C.
5. In a 15 ml tube, mix 8.33 g of CsCl, 250 μl of the labeled closed circular DNA from step 4 and 8.22 ml of water.
6. Invert the tube until the CsCl has dissolved, add 0.56 ml of EtBr (see Note 16).
7. Transfer the mix to 11.2 ml OptiSeal tubes and centrifuge at 300,000 × *g* for 16 h at 20 °C.
8. Remove the tubes from the centrifuge, carefully insert an 18-gauge needle into the bottom side of the tube and collect fractions of five drops in size into 1.5 ml tubes.
9. Locate the position of the DNA by analysis of 2 μl of each fraction on a 1 % agarose gel containing EtBr.
10. Pool fractions containing the 5′-labeled closed circular DNA, add an equal volume of water saturated butanol, invert the tube several tubes, and centrifuge at 16,000 × *g* for 1 min.
11. Remove and discard the upper organic layer and repeat the extraction a further three to five times until all the EtBr has been extracted from the DNA.
12. Dilute the DNA up to 4 ml with TE buffer, concentrate the sample using Amicon-30 filter units and repeat a further three times.
13. In the final step, concentrate to a volume of 300 μl and store at −20 °C.

3.8. Analysis of DNA Glycosylase Activity Using Duplex Oligonucleotide Substrates

1. Prepare Mix A containing 1 μl (300 fmol) of a 5′-γ-^{32}P-labeled duplex oligonucleotide substrate containing a site-specific modification (e.g., Uracil or 8-oxoguanine), 0.5 μl 20× BER buffer, 0.5 μl 20× PC, 0.2 μl 25× PCK, and 0.2 μl 25× carrier DNA to a total volume of 5 μl with water.
2. Prepare Mix B containing WCE (up to 10 μg) in dialysis buffer up to a volume of 5 μl.
3. Mix 5 μl Mix A and 5 μl Mix B and incubate at 37 °C for up to 20 min.

4. Add 10 μl FA dye to stop the reaction.
5. Analyze products of the reaction by denaturing acrylamide gel electrophoresis and phosphor imaging.

3.9. Analysis of DNA Glycosylase Activity Using a Single Lesion Closed Circular DNA Substrate

1. Prepare Mix A containing 1 μl (40 fmol) of a γ-^{32}P-labeled closed circular DNA substrate, 2.5 μl 20× BER buffer, 2.5 μl 20× PC, 1 μl 25× PCK, and 1 μl 25× carrier DNA to a total volume of 25 μl with water.
2. Prepare Mix B containing WCE (up to 100 μg) in dialysis buffer up to a volume of 25 μl.
3. Mix 25 μl Mix A and 25 μl Mix B and incubate at 37 °C for up to 3 h.
4. Add 2 μl of 0.5 M EDTA, 2 μl 10 % SDS and 2 μl proteinase K and incubate for 30 min at 37 °C.
5. Meanwhile, centrifuge a Biospin P-6 column for 1,000 × *g* for 2 min.
6. Add 500 μl TE buffer to the center of the column and spin at 1,000 × *g* for 2 min and repeat.
7. Add 50 μl of phenol/chloroform/isoamyl alcohol to the reactions, mix thoroughly, and centrifuge at 16,000 × *g* for 1 min.
8. Add the top aqueous layer to the center of the prewashed Biospin P-6 column and spin at 1,000 × *g* for 4 min.
9. Dry the samples in a speed vac for approximately 45 min at 45 °C.
10. Resuspend the sample in 10 μl of the appropriate restriction enzyme digestion buffer, add the restriction enzyme (10–40 U), and digest for 2 h at 37 °C.
11. Add 10 μl FA dye.
12. Analyze products of the reaction by denaturing acrylamide gel electrophoresis and phosphor imaging.

3.10. Analysis of AP Site Incision Activity Using Duplex Oligonucleotide Substrates

1. Incubate 10 μl (3 pmol; for ten reactions) of the 5′-γ-^{32}P-labeled duplex oligonucleotide substrate containing a site-specific uracil residue with 2 pmol uracil DNA glycosylase in 1.25 μl 20× BER buffer to a total volume of 25 μl with water at 37 °C for 30 min (see Note 17).
2. Centrifuge a Biospin P-30 column for 1,000 × *g* for 2 min.
3. Add 500 μl TE buffer to the center of the column and spin at 1,000 × *g* for 2 min and repeat.
4. Add the reaction mix to the center of the prewashed Biospin P-30 column and spin at 1,000 × *g* for 4 min.
5. Check that the volume of the 5′-γ-^{32}P-labeled AP site containing duplex oligonucleotide substrate is approximately 26 μl.

6. Prepare Mix A containing 2.6 μl of the 5′-γ-^{32}P-labeled AP site containing duplex oligonucleotide substrate, 0.5 μl 20× BER buffer, 0.5 μl 20× ATP, 0.5 μl 20× NAD, 0.5 μl 20× PC, 0.2 μl 25× PCK, and 0.2 μl 25× carrier DNA to a total volume of 5 μl with water.
7. Prepare Mix B containing WCE (0–10 μg) in dialysis buffer up to a volume of 5 μl.
8. Mix 5 μl Mix A and 5 μl Mix B and incubate at 37 °C for up to 20 min.
9. Add 2 μl 0.5 M sodium borohydride to the reactions and incubate on ice for 10 min (see Note 18).
10. Add 12 μl FA dye to stop the reaction.
11. Analyze products of the reaction by denaturing acrylamide gel electrophoresis and phosphor imaging.

3.11. Analysis of AP Site Incision Activity Using a Single Lesion Closed Circular DNA Substrate

1. Incubate 10 μl (400 fmol; for ten reactions) of the γ-^{32}P-labeled closed circular DNA substrate containing a site-specific uracil residue with 1 pmol uracil DNA glycosylase in 1.25 μl 20× BER buffer to a total volume of 25 μl with water at 37 °C for 30 min (see Note 17).
2. Centrifuge a Biospin P-30 column for 1,000 × *g* for 2 min.
3. Add 500 μl TE buffer to the center of the column and spin at 1,000 × *g* for 2 min and repeat.
4. Add the reaction mix to the center of the prewashed Biospin P-30 column and spin at 1,000 × *g* for 4 min.
5. Make the volume of the γ-^{32}P-labeled AP site containing closed circular DNA substrate up to 30 μl.
6. Prepare Mix A containing 3 μl of the 5′-γ-^{32}P-labeled AP site containing duplex oligonucleotide substrate, 2.5 μl 20× BER buffer, 2.5 μl 20× ATP, 2.5 μl 20× NAD, 2.5 μl 20× PC, 1 μl 25× PCK, and 1 μl 25× carrier DNA to a total volume of 25 μl with water.
7. Prepare Mix B containing WCE (0–100 μg) in dialysis buffer up to a volume of 25 μl.
8. Mix 25 μl Mix A and 25 μl Mix B and incubate at 37 °C for up to 3 h.
9. Add 2 μl of 0.5 M EDTA, 2 μl 10 % SDS, 2 μl proteinase K, and incubate for 30 min at 37 °C.
10. Add 50 μl of phenol/chloroform/isoamyl alcohol to the reactions, mix thoroughly, and centrifuge at 16,000 × *g* for 1 min.
11. Remove the top aqueous layer, add 10 μl 0.5 M sodium borohydride, and incubate for 10 min on ice (see Note 18).
12. Meanwhile, centrifuge a Biospin P-6 column for 1,000 × *g* for 2 min.

13. Add 500 μl TE buffer to the center of the column, spin at 1,000 × *g* for 2 min, and repeat.
14. Add the sodium borohydride-treated DNA to the center of the prewashed Biospin P-6 column and spin at 1,000 × *g* for 4 min.
15. Dry the samples in a speed vac for approximately 45 min at 45 °C.
16. Resuspend the sample in 10 μl of the appropriate restriction enzyme digestion buffer, add the restriction enzyme (10–40 U), and digest for 2 h at 37 °C.
17. Add 10 μl FA dye.
18. Analyze products of the reaction by denaturing acrylamide gel electrophoresis and phosphor imaging.

3.12. Analysis of DNA Polymerase Activity Using Duplex Oligonucleotide Substrates

1. Prepare Mix A containing 1 μl (300 fmol) of the 5′-γ-^{32}P-labeled one nucleotide gap containing duplex oligonucleotide substrate, 0.5 μl 20× BER buffer, 0.5 μl 20× dNTP, 0.5 μl 20× NAD, 0.5 μl 20× PC, 0.2 μl 25× PCK, and 0.2 μl 25× carrier DNA to a total volume of 5 μl with water.
2. Prepare Mix B containing WCE (0–10 μg) in dialysis buffer up to a volume of 5 μl.
3. Mix 5 μl Mix A and 5 μl Mix B and incubate at 37 °C for 20 min.
4. Add 10 μl FA dye to stop the reaction.
5. Analyze products of the reaction by denaturing acrylamide gel electrophoresis and phosphor imaging.

3.13. Analysis of DNA Ligase Activity Using Duplex Oligonucleotide Substrates

1. Prepare Mix A containing 1 μl (300 fmol) of the 5′-γ-^{32}P-labeled SSB containing duplex oligonucleotide substrate, 0.5 μl 20× BER buffer, 0.5 μl 20× ATP, 0.5 μl 20× NAD, 0.5 μl 20× PC, 0.2 μl 25× PCK, and 0.2 μl 25× carrier DNA to a total volume of 5 μl with water.
2. Prepare Mix B containing WCE (0–10 μg) in dialysis buffer up to a volume of 5 μl.
3. Mix 5 μl Mix A and 5 μl Mix B and incubate at 37 °C for 20 min.
4. Add 10 μl FA dye to stop the reaction.
5. Analyze products of the reaction by denaturing acrylamide gel electrophoresis and phosphor imaging.

3.14. Analysis of Full BER Activity Using Duplex Oligonucleotide Substrates

1. Prepare Mix A containing 1 μl (300 fmol) of the 5′-γ-^{32}P-labeled duplex oligonucleotide substrate containing a site-specific modification, 0.5 μl 20× BER buffer, 0.5 μl 20× ATP, 0.5 μl 20× dNTP, 0.5 μl 20× NAD, 0.5 μl 20× PC, 0.2 μl 25× PCK,

and 0.2 μl 25× carrier DNA to a total volume of 5 μl with water.

2. Prepare Mix B containing WCE (0–10 μg) in dialysis buffer up to a volume of 5 μl.
3. Mix 5 μl Mix A and 5 μl Mix B and incubate at 37 °C for 20 min.
4. Add 2 μl 0.5 M sodium borohydride to the reactions and incubate on ice for 10 min (see Note 18).
5. Add 12 μl FA dye to stop the reaction.
6. Analyze products of the reaction by denaturing acrylamide gel electrophoresis and phosphor imaging (*Typical result is shown in* Fig. 1b).

3.15. Analysis of Full BER Activity Using a Single Lesion Closed Circular DNA Substrate

1. Prepare Mix A containing 1 μl (40 fmol) of the γ-^{32}P-labeled closed circular DNA substrate, 2.5 μl 20× BER Buffer, 2.5 μl 20× ATP, 2.5 μl 20× NAD, 2.5 μl 20× dNTP, 2.5 μl 20× PC, 1 μl 25× PCK, and 1 μl 25× carrier DNA to a total volume of 25 μl with water.
2. Prepare Mix B containing WCE (0–100 μg) in dialysis buffer up to a volume of 25 μl.
3. Mix 25 μl Mix A and 25 μl Mix B and incubate at 37 °C for up to 3 h.
4. Add 2 μl of 0.5 M EDTA, 2 μl 10 % SDS and 2 μl proteinase K and incubate for 30 min at 37 °C.
5. Meanwhile, centrifuge a Biospin P-6 column for 1,000 × *g* for 2 min.
6. Add 500 μl TE buffer to the center of the column and spin at 1,000 × *g* for 2 min and repeat.
7. Add 50 μl of phenol/chloroform/isoamyl alcohol to the reactions, mix thoroughly, and centrifuge at 16,000 × *g* for 1 min.
8. Remove the top aqueous layer, add 10 μl 0.5 M sodium borohydride, and incubate for 10 min on ice (see Note 18).
9. Add the sodium borohydride treated DNA to the center of the prewashed Biospin P-6 column and spin at 1,000 × *g* for 4 min.
10. Dry the samples in a speed vac for approximately 45 min at 45 °C.
11. Resuspend the sample in 10 μl of the appropriate restriction enzyme digestion buffer, add the restriction enzyme (10–40 U), and digest for 2 h at 37 °C.
12. Add 10 μl FA dye.
13. Analyze products of the reaction by denaturing acrylamide gel electrophoresis and phosphor imaging. (*Typical result is shown in* Fig. 2b).

3.16. Denaturing Acrylamide Gel Electrophoresis Analysis

1. Clean glass plates and spacers thoroughly with 70 % ethanol before use and allow to dry.
2. Assemble the glass plates together with the spacers at the sides of the plates, seal the bottom and sides of the plates with gel sealing tape and clamp together using steel clips.
3. Pour 40 ml acrylamide into a glass beaker, add 280 μl APS, and mix briefly.
4. Add 14 μl TEMED, mix and pour the gel between the glass plates (see Note 19).
5. Remove air bubbles from the gel and insert a comb (see Note 20).
6. Leave the acrylamide to polymerize for approximately 30 min.
7. Carefully remove the comb, remove the steel clamps and gel sealing tape and insert the plates into the electrophoresis tank, ensuring that the smaller plate is on the inner surface of the tank.
8. Clamp the plates to the tank tightly, close the valve to stop the reservoir emptying, and fill the reservoir with 1× TBE to the top of the glass plates (see Note 21).
9. Fill the lower chamber with enough 1× TBE to cover the plate ends and wash wells with the buffer thoroughly using a syringe and a needle (see Note 22).
10. Close the upper and lower lids, connect to a power supply, and run the gel for at least 30 min at a constant 42 W to prewarm the gel and also to remove any impurities.
11. Heat samples to 95 °C for 5 min and in the meantime, stop the power supply to the gel, and open the upper lid.
12. Wash wells thoroughly again with the buffer using a syringe and a needle (see Note 22).
13. Briefly centrifuge the samples to spin the contents to the bottom of the tube and carefully load 10 μl of the sample using a gel loading tip.
14. Repeat loading all of the samples, close the upper lid, reconnect the power supply, and run the gel for 1.5–2 h at a constant 42 W.

3.17. Phosphor Imaging

1. Empty the upper chamber of the electrophoresis tank reservoir by opening the valve and allowing the buffer to drain.
2. Remove the glass plates containing the gel from the electrophoresis tank and separate the glass plates.
3. Remove the gel from the attached glass plate using Saran wrap (see Note 23).

4. Place in an autoradiography cassette and add a storage phosphor screen on top of the gel before closing the cassette.
5. Place the cassette at 4 °C overnight.
6. Scan the storage phosphor screen using a phosphor imager.

4. Notes

1. Dissolve the sucrose into the solution containing the Tris–HCl, $MgCl_2$, and made up with water by gently heating the mixture. Then once cooled, add the glycerol and mix.
2. Prepare protease inhibitors for storage at −20 °C and use a fresh aliquot every month depending on the number of freeze–thaw cycles.
3. Use the freshest radioactivity possible to ensure clear and sharp bands following separation of the radiolabeled products by gel electrophoresis and analysis by phosphor imaging.
4. Prepare a 1 ml stock of 5 mM NAD in water, aliquot, and use the aliquots within 1 week.
5. Use any sequence of single-stranded oligonucleotide (~30-mer) to minimize exonucleases within the extract from degrading the radiolabeled duplex oligonucleotide substrate.
6. Weigh a small amount of $NaBH_4$ in a 1.5 ml microcentrifuge tube, add the appropriate volume of water just before use, and discard once used.
7. Prepare 1 l solutions of acrylamide containing urea by gently heating the solution whilst stirring, filter sterilize, and keep bottle wrapped in aluminum foil to protect from sunlight and minimize decomposition. Generally the solution should only be kept for approximately 6–12 months before disposal.
8. Cells can be stored indefinitely at −80 °C but should be left for at least 1 h prior to cell extract preparation to promote cell lysis.
9. Buffers containing protease inhibitors should only be used on the day of use and fresh buffers should be prepared otherwise.
10. If the cell pellet volume is less than 50 μl, then the volume of Buffer I to be used should be kept at 50 μl (and consequently 100 μl Buffer II) so that efficient mixing is maintained.
11. It is recommended that the volume of cell pellet used for this method should be no lower than 0.5 ml.

12. Centrifuge tubes should be washed with water, followed by 70 % ethanol, and then dried before storing at 4 °C ready for use.
13. A layer of fat will be present on the top of the solution and should be removed using a Pasteur pipette. The cell extract can then be removed but be careful not to remove the cell debris or the viscous DNA that are present at the bottom of the tube.
14. Ensure that the cell extract pellet is thoroughly resuspended in buffer and use a larger volume of buffer if required.
15. For annealing, heat the sample to the required temperature and then either remove the block from the heater or switch off the heat block and allow to slow cool to room temperature.
16. Any steps using EtBr should be performed in a dark room using a red safety light.
17. An AP site containing substrate should be prepared just before use, should not be stored and therefore should be prepared daily. To ensure the 100 % efficiency of uracil DNA glycosylase activity, AP site formation should be routinely checked by piperidine treatment that cleaves AP sites causing strand breaks. This is achieved by incubating the AP site substrate with 10 % piperidine at 90 °C for 15 min, the sample dried in a speed vac and resuspended in formamide loading dye prior to analysis by acrylamide gel electrophoresis and phosphor imaging.
18. Treatment with sodium borohydride is required to stabilize the uncleaved AP site, since this will degrade in the heat generated during the electrophoresis step creating a smear.
19. Start by pouring the gel between the glass plates to one corner whilst angling the plates at 45° to the vertical and resting the one bottom corner on the bench. As the plates are filled with the gel, gradually tilt so that both bottom edges of the plates are touching the bench and once completely filled, lay horizontally.
20. Use a spare spacer to remove the air bubbles by inserting between the glass plates and drawing the bubble upwards towards the top to release the air. Insert a comb between the glass plates making sure not to push the comb too far that it cannot be removed when the acrylamide is set.
21. If the buffer leaks between the glass plates and the gel tank, disassemble the apparatus, use gel seal (GE Healthcare, Little Chalfont, UK) to lubricate the rubber seals prior to reassembling the apparatus.
22. Also use a spacer cut to the size of one well, to insert into the wells containing buffer and carefully move up and down to remove any excess acrylamide.

23. Place Saran wrap over the exposed gel being careful to remove any air bubbles, then flip over the gel so that it is face down on the bench. Carefully peel away the Saran wrap containing the gel from the glass plate and once removed, fold the edges of the Saran wrap to seal the gel.

References

1. Dianov GL, Parsons JL (2007) Co-ordination of DNA single strand break repair. DNA Repair (Amst) 6:454–460
2. Frosina G, Fortini P, Rossi O, Carrozzino F, Raspaglio G, Cox LS, Lane DP, Abbondandolo A, Dogliotti E (1996) Two pathways for base excision repair in mammalian cells. J Biol Chem 271:9573–9578
3. Podlutsky AJ, Dianova II, Podust VN, Bohr VA, Dianov GL (2001) Human DNA polymerase beta initiates DNA synthesis during long- patch repair of reduced AP sites in DNA. EMBO J 20:1477–1482
4. Tanaka M, Lai JS, Herr W (1992) Promoter-selective activation domains in Oct-1 and Oct-2 direct differential activation of an Snrna and messenger-RNA promoter. Cell 68: 755–767
5. Manley JL, Fire A, Samuels M, Sharp PA (1983) In vitro transcription: whole-cell extract. Methods Enzymol 101:568–582
6. Allinson SL, Dianova II, Dianov GL (2001) DNA polymerase beta is the major dRP lyase involved in repair of oxidative base lesions in DNA by mammalian cell extracts. EMBO J 20: 6919–6926

Chapter 18

Using Arabidopsis Cell Extracts to Monitor Repair of DNA Base Damage In Vitro

Dolores Córdoba-Cañero, Teresa Roldán-Arjona, and Rafael R. Ariza

Abstract

Base excision repair (BER) is a major pathway for the removal of endogenous and exogenous DNA damage. This repair mechanism is initiated by DNA glycosylases that excise the altered base, and continues through alternative routes that culminate in DNA resynthesis and ligation. In contrast to the information available for microbes and animals, our knowledge about this important DNA repair pathway in plants is very limited, partially due to a lack of biochemical approaches. Here we describe an in vitro assay to monitor BER in cell-free extracts from the model plant *Arabidopsis thaliana*. The assay uses labeled DNA substrates containing a single damaged base within a restriction site, and allows detection of fully repaired molecules as well as DNA repair intermediates. The method is easily applied to measure the repair activity of purified proteins and can be successfully used in combination with the extensive array of biological resources available for *Arabidopsis*.

Key words: In vitro DNA repair assay, Plant cell extracts, Denaturing polyacrylamide gel electrophoresis, Oligonucleotide, Uracil, AP site, DNA glycosylase, Base excision repair

1. Introduction

Cells have evolved sophisticated DNA repair mechanisms to avoid the deleterious consequences of DNA damage. Prominent among such mechanisms is the base excision repair (BER) pathway, that plays a critical role in cellular protection against most common forms of non-bulky DNA damage, including alkylated or oxidized bases, inappropriate bases arisen during replication or spontaneous deamination, and sites of base loss (1–3).

BER is a multistep process initiated by enzymes termed DNA glycosylases that catalyze hydrolysis of the N-glycosidic bond between the target base and deoxyribose (4, 5). Cells contain different DNA glycosylases, and each type removes a specific base

Lotte Bjergbæk (ed.), *DNA Repair Protocols*, Methods in Molecular Biology, vol. 920,
DOI 10.1007/978-1-61779-998-3_18, © Springer Science+Business Media New York 2012

alteration or a limited spectrum of structurally related lesions (1, 6). The excision of the DNA damage as a free base is a hallmark of BER, and generates an abasic (apurinic/apyrimidinic, AP) site that must be processed in order to restore the structure and integrity of DNA. AP sites may be incised either by AP endonucleases or by the AP lyase activity associated with the so-called bifunctional DNA glycosylases/lyases, leaving 5′- or 3′-blocked ends, respectively (7). Additional enzymatic activities convert these terminal ends into 5′-P and 3′-OH termini (7, 8). The ensuing gap filling is carried out by a DNA polymerase and may take place either by insertion of a single nucleotide (SN-BER) or by DNA synthesis involving several nucleotides (long-patch repair, LP-BER) (2). Finally, continuity of the processed strand is restored by a DNA ligase. The BER pathway was initially elucidated in bacteria, and subsequent studies have shown that its main features are conserved in eukaryotes (9). Nevertheless, accumulating evidence indicates that a number of differences exist between different species, particularly during post-excision events (10).

Plants comprise an important group of organisms in which research on DNA repair, and particularly BER, has lagged far behind the rapid progress made in bacteria, yeast, and mammals (9). Much of the knowledge accumulated during the past 20 years about BER in microbes and animals has been obtained through the development of in vitro repair assays that allow biochemical studies with cell-free extracts, and combining their results with genetic analyses (11). However, until recently there were no reliable in vitro assays for specifically detecting BER in plant cells. In this chapter we describe a specific assay to monitor repair of damaged bases in cell-free extracts from the model plant *Arabidopsis thaliana* (12, 13). To develop such a plant BER assay, we have chosen as a convenient model lesion uracil, a ubiquitous DNA alteration that arises by spontaneous deamination of cytosine or from erroneous incorporation of dUMP during replication (14).

The Arabidopsis BER assay is performed by incubating whole cell extracts from *Arabidopsis* plants, normally 15-days old seedlings, with a labeled 51-mer duplex oligonucleotide containing a single uracil residue within an *Hpa*II restriction site. The presence of U instead of C at this location makes the DNA substrate resistant to *Hpa*II cleavage. The repair of the lesion, that involves uracil excision and repair synthesis to insert a cytosine, is detected as conversion of the duplex DNA to a form susceptible to digestion. Thus, repair is visualized by the emergence a 21-nt labeled fragment following denaturing polyacrylamide gel electrophoresis (PAGE) (Fig. 1).

This chapter contains detailed methods to prepare DNA substrate, obtain Arabidopsis whole-cell extracts, setting up repair reactions, and resolve repair products by denaturing PAGE. The methods described here can be easily adapted to study BER of

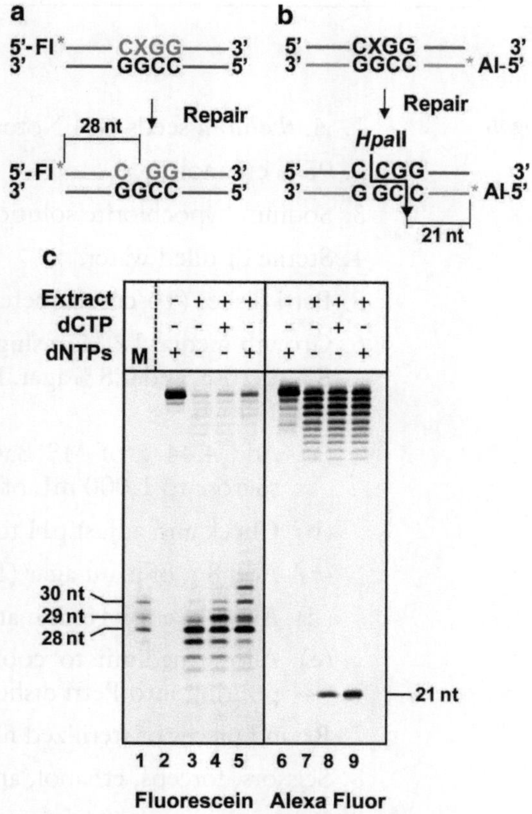

Fig. 1. Schematic diagram of molecules used as DNA substrates. Double-stranded oligonucleotides contain a lesion (X) at an *Hpa*II site on the upper strand. Fluorescein (Fl) (**a**) or Alexa Fluor (Al) (**b**) at 5′ end of the upper or lower strand, respectively, are indicated by an *asterisk*. (**c**) Arabidopsis cell extract (70 μg) was incubated with duplex DNA that contained a U residue in the upper strand. The DNA substrate was labeled either at the 5′ end of the upper strand with Fluorescein (*lanes 2–5*) or at 5′ end of the lower strand with Alexa Fluor (*lanes 6–9*). *Lane 1* contains DNA size markers. Reactions were incubated at 30 °C for 3 h in a reaction mixture containing either dCTP or all four dNTPs, as indicated. Reaction products were separated in a 12 % denaturing polyacrylamide gel either before (*lanes 2–5*) or after (*lanes 6–9*) *Hpa*II digestion. Fluorescein- (*lanes 2–5*) and Alexa Fluor-labeled fragments (*lanes 6–9*) were detected by fluorescence scanning.

different base alterations. Thus, we have successfully used DNA substrates containing the AP site synthetic analogue tetrahydrofuran (THF) instead of uracil (12). In addition to study cell-free extracts, the assay may also be used for measuring the activity of purified repair proteins and for analyzing their capacity to complement repair-defective cell extracts obtained from Arabidopsis mutants (13). It is also possible to supplement the repair reaction mixture with additional factors such as antibodies or inhibitors, in order to further characterize proteins involved in plant BER (12). The use of such a versatile repair assay in a model system for which a vast array of genomic tools and biological resources is available (http://arabidopsis.org/) may help to undertake a comprehensive biochemical and genetic analysis of BER in plants.

2. Materials

2.1. Plant Growth

1. *A. thaliana* seeds (see Note 1).
2. 95 % ethanol.
3. Sodium hypochlorite solution: 50 % (v/v) commercial bleach.
4. Sterile distilled water.
5. Petri dishes (10-cm diameter).
6. Growth media: 1× Murashige and Skoog (MS) Basal Medium, 3 % sucrose, and 0.8 % agar. Preparation of 1× MS agar medium is as follows:
   (a) Add 4.44 g of MS Basal Medium (Sigma) and 30 g of sucrose to 1,000 mL of distilled water and stir to dissolve.
   (b) Check and adjust pH to 5.8 with 1 M KOH.
   (c) Add 8 g of plant agar (Duchefa).
   (d) Autoclave for 15 min at 15 psi, 121 °C.
   (e) Allow medium to cool to approximately 50 °C before pouring into Petri dishes.
7. Round pieces of sterilized filter paper (10-cm diameter).
8. Scissors, forceps, ethanol, and burner.
9. Balance.
10. Aluminum foil.
11. Liquid nitrogen.
12. Laminar-flow hood.

2.2. Plant Cell Extract Preparation

1. Liquid nitrogen, mortar, and pestle.
2. Spatula.
3. Extraction buffer: 25 mM Hepes KOH, pH 7.8, 100 mM KCl, 5 mM $MgCl_2$, 250 mM sucrose, 1 mM DTT, 10 % glycerol, 1 μL/mL protease inhibitor cocktail (Sigma-Aldrich). Freshly prepared (see Note 2).
4. Microcentrifuge and 2 mL microcentrifuge tubes.
5. Nylon mesh (20 μm).
6. Mini-spin columns and scalpel.
7. Dialysis tubing and clamps.
8. Dialysis buffer: 25 mM Hepes KOH 7.8, 100 mM KCl, 2 mM DTT, 17 % glycerol.
9. Beaker or flask, magnetic stirring bar, and magnetic stirring plate.

Table 1
DNA sequence of oligonucleotides used as substrates

Name	DNA sequence[a]	Strand	Label[b]
Fl-UGF	TCACGGGATCAATGTGTTCTTTCAGCTCUGGT CACGCTGACCAGGAATACC	Upper	Fl at 5′
Fl-APGF	TCACGGGATCAATGTGTTCTTTCAGCTCFGGTC ACGCTGACCAGGAATACC	Upper	Fl at 5′
UGF	TCACGGGATCAATGTGTTCTTTCAGCTCUGGTC ACGCTGACCAGGAATACC	Upper	–
CGR	AGTGCCCTAGTTACACAAGAAAGTCGAGGCCAG TGCGACTGGTCCTTATGG	Lower	–
Al-CGR	AGTGCCCTAGTTACACAAGAAAGTCGAGGCCAG TGCGACTGGTCCTTATGG	Lower	Al at 5′

[a]F = AP site analog (tetrahydrofuran)
[b]Fl = fluorescein; Al = alexa fluor 647

2.3. DNA Repair Reactions

2.3.1. DNA Substrate Preparation

DNA substrates are prepared by annealing oligonucleotides commercially synthesized according to our specifications (Operon, IDT, Fisher Scientific). Table 1 shows the sequence of some oligonucleotides that we routinely employ in our laboratory. Those used for the damaged version of the upper strand contain either U or the abasic spacer THF within an *Hpa*II recognition site (see Note 3). To detect intermediates arising during DNA repair we prepare DNA substrates with a 5′-end labeled upper strand annealed to a non-labeled complementary strand. To monitor full-repair, a non-labeled upper strand is annealed to a 5′-end labeled lower strand (Fig. 1b; see Note 4).

1. TE buffer: 10 mM Tris–HCl, 1 mM EDTA pH 8.0. Sterilize by autoclaving.
2. Labeled oligonucleotide: stock solution at 100 μM and working solution at 5 μM in TE. Store at −20 °C.
3. Non-labeled oligonucleotide: stock solution at 100 μM and working solution at 10 μM in TE. Store at −20 °C.
4. Annealing buffer: 10 mM Tris–HCl pH 8.0, 20 mM NaCl. Prepared from sterile stock solutions.
5. Water bath, tube rack, and glass beaker.
6. Thermoblock.

2.3.2. DNA Repair Reaction

1. DNA duplex (see Subheading 3.3.1 below).
2. 10× DNA repair buffer: 450 mM HEPES-KOH, pH 7.8, 700 mM KCl, 50 mM $MgCl_2$, 10 mM DTT, 4 mM EDTA, 360 μg BSA, 10 mM NAD, 2 % glycerol.
3. 2 mM dCTP (Roche). Store in small aliquots at −20 °C.

4. 2 mM deoxyribonucleoside triphosphate (dNTP) mixture (Roche): 2 mM dCTP, 2 mM dATP, 2 mM dGTP, and 2 mM dTTP. Store in small aliquots at −20 °C.
5. 100 mM ATP (Sigma). Store in small aliquots at −20 °C.
6. 50 mM NAD (Sigma). Store in small aliquots at −20 °C.
7. 220 mM phosphocreatine (Sigma). Store in small aliquots at −20 °C.
8. Creatine phosphokinase (from rabbit muscle, Sigma). Dissolve at a concentration of 5 mg/mL and store at −20 °C. Prepare working solution at 5 μg/mL in H_2O and store in small aliquots at −20 °C.
9. Proteinase K (Sigma). Dissolve at a concentration of 20 mg/mL in H_2O. Store in small aliquots at −20 °C.
10. K buffer: 20 mM EDTA, 0.6 % SDS, and 0.5 mg/mL proteinase K. Prepare at least 30 min before use and incubate at 37 °C to dissolve.
11. TE buffer: 10 mM Tris–HCl, 1 mM EDTA pH 8.0. Prepared from sterile stock solutions.
12. Phenol:chloroform:isoamyl alcohol (25:24:1) (Sigma).
13. Glycogen. Dissolve at a concentration of 2 mg/mL in H_2O. Store in small aliquots at −20 °C.
14. Glycogen solution: 0.3 mM NaCl, 16 mg/mL glycogen. Prepare at least 30 min before use and keep on ice.
15. 70 % and absolute ethanol (−20 °C).
16. *Hpa*II restriction endonuclease (10 U/μL) and 10× NEBuffer 1 (New England Biolabs).
17. 10× TBE: 0.9 M Trizma base, 0.9 boric acid, 20 mM EDTA pH 8.0.
18. 90 % formamide in 1× TBE.
19. Thermoblock.

2.4. Gel Electrophoresis

1. Glass plates (40 × 20 cm, H × W).
2. Detergent and paper towels.
3. Spacers and comb, 0.2 mm thick.
4. Box sealing tape and binder clips.
5. TBE 10×: 0.9 M Trizma base, 0.9 M boric acid, 20 mM EDTA pH 8.0.
6. Ammonium persulfate (APS) (Bio-Rad): 10 % solution in H_2O. Prepare immediately before use.
7. TEMED (Sigma).
8. Glass graduated pipette (25 mL) and manual or automatic pipettor.

9. Denaturing polyacrylamide gel solution (12 %) containing 7 M urea: 25.22 g urea, 18 mL of 40 % Acrylamide/Bis Solution 19:1, 6 mL TBE 10×, 17.5 mL H_2O.
10. Aluminum heat dispersion plate.
11. Vertical electrophoresis system and high voltage power supply.
12. Formamide dye mix (FDM): 80 % formamide, 1 mg/mL bromophenol blue, 10 mM EDTA pH 8.0.
13. Fluor imaging gel scanning equipment (see Note 5).

3. Methods

3.1. Plant Growth

3.1.1. Seeds Sterilization

Work in sterile conditions and carry all steps in a laminar-flow hood.

1. Surface sterilize 10–20 mg seeds (see Note 6) in an Eppendorf tube by adding 1 mL of 95 % ethanol and vortex gently for 1 min. Allow seeds to sediment to the bottom of the tube (10–30 s) and carefully remove ethanol.
2. Add 1 mL of 50 % (v/v) bleach. Vortex gently for 10–15 min and allow seeds to sediment. Remove the bleach and wash the seeds at least four times with 1 mL of sterile distilled water.
3. Leave the seeds in water (50 μL of water per 1 mg of sterilized seeds) and incubate at 4 °C in the dark for 2 days to break dormancy and achieve uniform and efficient germination.

3.1.2. Planting of Seeds

Work in sterile conditions and carry all steps in a laminar-flow hood.

1. Using sterile forceps put a round piece of filter paper onto the surface of the MS medium agar plates (see Note 7).
2. Plate about 50 seeds on each plate. Take a 50 μL aliquot of sterile seeds with a micropipette and distribute the seeds evenly over the filter surface (see Notes 8 and 9).
3. Seal the plates with parafilm.
4. Transfer plates to a growth chamber at 23 °C and long-day conditions (16 h light/8 h dark).
5. Incubate for 14–15 days or until the seedlings develop four true leaves (see Note 10). At this point, each plate should yield 0.5–1.0 g fresh weight plant material.

3.2. Plant Cell Extract Preparation

1. Harvest the seedlings carefully with forceps.
2. Wrap 0.5–1.0 g of seedlings with aluminum foil and freeze them immediately in liquid nitrogen (see Note 11).

3. Precool mortar and pestle and add liquid nitrogen. Add the frozen seedlings and grind the plant material to a fine powder, adding more liquid nitrogen if necessary (see Note 12).
4. Transfer the powder to a 2 mL microcentrifuge tube and resuspend the homogenized plant material in 3 volumes (w/v) of ice-cold extraction buffer (see Note 13).
5. Incubate on ice for 1 h. Occasionally, mix the sample by inverting the tube several times.
6. Centrifuge at 16,000 × *g* for 1 h at 4 °C using a microfuge.
7. Carefully transfer the supernatant to a new 2 mL centrifuge tube (see Note 14).
8. Place the supernatant in a bottomless minispin column fitted into 2 mL tube with a piece of 20 μm nylon mesh between the two tubes (Fig. 2; see Note 15). Spin for 10 s. Remove the minispin column from the tube and keep the filtered homogenate on ice.
9. Estimate how much dialysis tubing membrane you will need to hold your sample and cut a piece that has an extra couple of inches (see Note 16).
10. Wash your dialysis tubing membrane with distilled water. Never let your dialysis tubing to dry out once it has been wetted.
11. Remove the water carefully from the tubing. Clip the bottom of the tubing with a dialysis clamp. Using a micropipette, carefully transfer the filtered homogenate (obtained in step 8, above) into the dialysis tubing, clip the top tightly, and place the tubing in a beaker or flask containing dialysis buffer precooled at 4 °C (see Note 17).

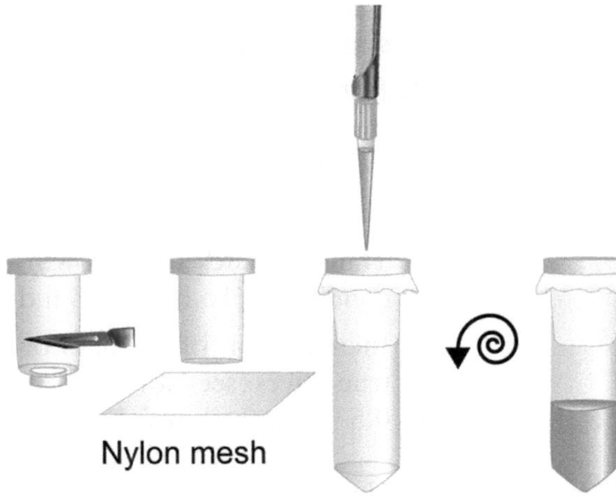

Fig. 2. Filtration of the plant tissue homogenate. The homogenate is placed inside a bottomless minispin column inserted into a microcentrifuge tube and filtered through the nylon mesh by centrifugation.

12. Place the beaker or flask on a magnetic stirring plate and stir at 4 °C during 12–14 h.
13. After the dialysis, carefully remove the dialysate from the tubing into a microcentrifuge tube.
14. Determine protein concentration by Bradford assay with BSA standards and store in small aliquots at −80 °C, until needed (see Note 18). The protein concentration is usually around 2–4 μg/μL.

3.3. DNA Repair Reactions

3.3.1. DNA Substrate Preparation

1. Preheat a thermoblock and a water bath at 95 °C.
2. Mix in an Eppendorf tube 0.4 μL of the 5 μM solution of the labeled oligonucleotide, 0.4 μL of the 10 μM solution of the non-labeled oligonucleotide and 1.2 μL of annealing buffer (the volumes are provided as per repair reaction and should be scaled up for experiments involving multiple samples; make enough premix for one or two more reactions than needed).
3. Heat the mix at 95 °C for 5 min in the thermoblock (see Note 19).
4. Quickly put the tube in a rack inside of a glass beaker, containing enough preheated water to cover just the bottom of the tube.
5. Let the sample slowly cool until water reaches room temperature (25 °C approximately).
6. Use immediately or store at 4 °C until needed (see Note 20).

3.3.2. Setting Up Repair Reactions

Perform all operations on ice.

1. Unfreeze plant cell extracts slowly on ice.
2. Prepare a premix with the following reagents (the volumes are provided as per reaction and should be scaled up for experiments involving multiple samples; make enough premix for one or two more reactions than needed): 5 μL of 10× DNA repair buffer, 5 μL of 220 mM phosphocreatine, 1 μL of 100 mM ATP, 1 μL of 50 mM NAD, 0.5 μL of 5 μg/mL creatine phosphokinase, 2 μL of DNA substrate (prepared as described in Subheading 3.3.1 above), and 15 μL of H_2O to adjust the volume per reaction to 29.5 μL.
3. Prepare 40–80 μg of plant cell extract and dialysis buffer to a final total volume of 20 μL in an Eppendorf tube (see Note 21). Add 29.5 μL premix to each tube and 0.5 μL of the appropriate deoxynucleotide solution (2 mM dCTP or 2 mM dNTP mixture).
4. Briefly mix and centrifuge the tubes.
5. Incubate at 30 °C for 3 h (see Note 22).
6. Stop the reaction by adding 10 μL of K buffer and incubating at 37 °C for 30 min.

7. Add 60 μL of TE buffer and 120 μL of phenol:chloroform: isoamyl alcohol (25:24:1). Vortex vigorously during 10–15 s, or until the sample develops a white color.
8. Centrifuge 1 min at 16,000 × *g* using a microfuge.
9. Carefully remove the supernatant and transfer to a new Eppendorf tube containing 8 μL of glycogen solution.
10. Add 360 μL of cold absolute ethanol and incubate at −20 °C for at least 30 min (see Note 23).
11. Centrifuge 15 min at 16,000 × *g* using a microfuge and discard the supernatant.
12. Add 200 μL of cold 70 % ethanol, centrifuge 5 min at 16,000 × *g*, and discard the supernatant.
13. Dry the pellet in a Speed-Vac 5 min or by inverting the tubes over a paper towel and allow to air dry for several minutes.
14. To monitoring full-repair, proceed to *Hpa*II digestion (see Subheading 3.4 below).
15. To detect DNA repair intermediates, resuspend the pellet in 10 μL of 90 % formamide (see Note 24) and keep samples on ice until ready to perform denaturing PAGE (Subheading 3.5 below).

3.4. HpaII Digestion

1. Make a digestion premix containing 4 μL of H_2O, 0.5 μL of 10× NEBuffer 1 and 0.5 μL (5 U) of *Hpa*II restriction endonuclease per repair reaction.
2. Resuspend the precipitated DNA (Subheading 3.3.2, step 13 above) in 5 μL of the digestion premix.
3. Incubate at 37 °C for 1 h.
4. Stop the reaction by adding 5 μL of 90 % formamide. Keep samples on ice until ready to perform denaturing PAGE (see Subheading 3.5 below).

3.5. Denaturing Polyacrylamide Gel Electrophoresis

3.5.1. Gel Preparation

1. Prepare the denaturing polyacrylamide gel solution and stir to dissolve urea.
2. Clean glass plates and spacers thoroughly with a nonabrasive detergent and rinse thoroughly (see Note 25).
3. Dry the plates carefully with paper towels.
4. Put the 0.4 mm spacers between the glass plates and seal them with box sealing tape (see Notes 26 and 27). Place the glass plate sandwich in a vertical position.
5. Add 600 μL APS 10 % and 60 μL TEMED to the gel solution, mix gently and immediately pour the gel using a 25 mL pipette and a manual or automatic pipettor, with caution to avoid bubbles (see Notes 28 and 29).

6. Immediately insert the comb into the gel, and clamp it in place using binder clips. Make sure that no acrylamide solution is leaking from the gel assembly.
7. Allow the acrylamide to polymerize for at least 1 h at room temperature (see Note 30).

3.5.2. Gel Electrophoresis

1. Remove sealing tape and clamps. Wash gel assembly with detergent and rinse. Carefully slide off the comb from the gel (see Note 31).
2. Place the gel assembly in the electrophoresis apparatus and secure it with binder clips on the sides (see Note 32). Place the aluminum heat dispersion plate and fix to the gel with binder clips.
3. Fill the upper and lower buffer reservoirs with electrophoresis buffer (1× TBE).
4. Rinse the wells thoroughly with a syringe filled with 1× TBE to wash away urea that has diffused into the wells.
5. Connect the electrodes to a power supply and allow the gel to pre-run at 1,000–1,500 constant voltage (V) until the gel temperature reaches 42–45 °C (see Notes 33 and 34).
6. Denature DNA samples by heating at 95 °C during 5 min. Spin and transfer to ice to prevent re-annealing.
7. Rinse again the wells thoroughly with a syringe filled with 1× TBE to wash away urea that has diffused into the wells (see Note 35).
8. Load 10 μL of each sample into adjacent wells. In a separate well, load 10 μL of FDM to monitor the progress of electrophoresis (see Notes 36 and 37).
9. Run the gel at 1,300–1,500 V (constant voltage) during approximately 3 h or until the FDM is within 5–10 cm of the bottom of the gel (see Note 38).
10. Turn off the power supply, disconnect the leads, discard the electrophoresis buffer from the reservoirs, and remove the gel assembly from the electrophoresis equipment.
11. Release the binder clips and remove the gel sandwich from the sequencing apparatus. Wash glass plates with a nonabrasive detergent, rinse thoroughly with water and dry carefully with paper towels.
12. Place the gel sandwich onto the Fluor Imager. Scan the gel with the appropriate laser and analyze the image (see Note 5).

3.6. Analysis of Results

Typical results obtained with the Arabidopsis BER assay are shown in Fig. 1c. The right-half of the gel (lanes 6–9) shows the results obtained with a duplex DNA substrate labeled at the lower strand, which allows detection of fully repaired molecules. Incubation with the cell extract converted the uracil-containing DNA to a form susceptible to HpaII digestion, which generates a labeled 21-mer fragment (lanes 8 and 9). Repair is dependent on the presence of dNTPs in the reaction mixture. The repair observed with dCTP as the only deoxyribonucleotide reflects the SN-BER subpathway (lane 8). We usually find that uracil repair is more efficient when all four deoxynucleotides are present in the reaction mixture (lane 9). Fragments shorter than 51 nt detected in the upper part of the gel are the result of exonuclease activity in extracts, since they are not detected when incubations are performed in the absence of Mg^{2+}. The left half of the gel (lanes 2–5) shows the results obtained with a duplex DNA substrate labeled at the upper strand, which allows detection of DNA repair intermediates. In the presence of dCTP as the only deoxynucleotide, the major DNA repair intermediate detected is a 29-nt fragment, corresponding to a short-patch DNA synthesis that involves the insertion of one dCMP in the repair gap (lane 4). In the presence of all four dNTPs, several upper-strand reaction intermediates are detected that co-migrate with 29-, 30-, and 31-nt fragments (lane 5). These results indicate that up to three nucleotides are inserted during LP-BER.

4. Notes

1. Any *A. thaliana* may be used. Different natural accessions (or ecotypes) are available from seeds stock centers. We routinely use *A. thaliana Col* ("Columbia"), the ecotype most commonly used by Arabidopsis researchers, but we have also obtained repair-proficient extracts from the *Ws* (Wassilewskija) ecotype.
2. The presence of sucrose in this buffer increases contamination risk. Extraction buffer can be prepared in advance without protease inhibitor and stored during 1 or 2 days at 4 °C. Add protease inhibitor immediately before use.
3. Other lesions and other restriction sites may be used.
4. We use Fluorescein and Alexa Fluor 647 as fluorescent labels for the 5′ ends of the upper and lower strands, respectively, but other types of DNA labeling may be used.
5. We routinely use an FLA5100 Fluor Imager (Fujifilm) but other equipments should yield equivalent results. Fluorescein and Alexa Fluor 647 are detected at 473 and 635 nm, respectively.
6. One Arabidopsis seed weighs approx 0.02 mg. Growing about 50 seeds (1 mg of seeds approximately) per plate will yield 0.5–1 g of fresh weight plant material. About 1.5 mL of plant cell extract is obtained using 0.5 g of seedlings (total protein concentration is usually 2–4 μg/μL).

7. It is possible planting seeds using top agar, but it will make more difficult to collect seedlings. If nevertheless you prefer this method, prepare top agar (0.8 % agar in distilled water) and dispense into 3 mL aliquots in sterile tubes placed in a preheated thermoblock at 42 °C. Add a 50-μL aliquot of seeds per tube, mix quickly, pour onto the dishes and swirl to distribute seeds.
8. Cut a short piece of the micropipette tip with sterile scissors to seeds to flow.
9. Try to place the seeds as evenly spaced as possible. If the seeds are not well distributed all over the plate, you can move them with a sterile micropipette tip.
10. In our experience, use of older plants results in plant cell extracts less active and concentrated.
11. This tissue can be stored at −80 °C until needed. In our experience, tissue stored up to 1 year in these conditions yields fully active extracts.
12. Never let the tissue to unfreeze; if liquid nitrogen evaporates, quickly add more and keep grinding until the powder has a flour-like appearance.
13. While grinding the tissue, precool a spatula in liquid nitrogen and use it to transfer the powder into 2 mL microcentrifuge tubes placed on ice. Do it quickly to prevent unfreezing. Distribute the powder obtained from 0.5 g of seedlings in two tubes and add 750 μL of ice-cold extraction buffer to each one.
14. Pipet slowly to avoid taking cellular debris.
15. With a scalpel cut the bottom half of a used minispin column. Put a piece of nylon mesh between the column and a new 2 mL microcentrifuge tube. Place the supernatant in the minispin column, close its cap, and quickly centrifuge for a few seconds. This method allows filtration of small volumes and avoids loss of cell extract by capillarity (Fig. 2).
16. Follow manufacturer's directions. Always handle dialysis tubing with gloves.
17. We recommend performing this step in a cold room. Use a volume of dialysis buffer at least 200 times that of the sample.
18. Unfrozen aliquots can be stored at −20 °C, although it is possible that extracts lose some repair activity after multiple rounds of freezing and thawing. Therefore, aliquoting into small fractions is highly recommended.
19. The use of a thermoblock instead of a waterbath avoids possible water entry when tubes occasionally popping open during incubation.

20. Annealed DNA substrates can be prepared and stored at 4 °C 1 day before use.
21. Do not forget to prepare a negative control with no plant cell extract.
22. When performing kinetic analyses, we incubate reactions over several time points, usually spanning a period from 1 to 6 h.
23. If required, samples can be left overnight at this point.
24. Samples resuspended in 90 % formamide can be stored at –20 °C for up to several weeks.
25. Wear gloves and avoid touching the inside surface of the plates, which must be free of grease spots to prevent air bubbles formation when pouring the gel.
26. Assemble the glass plate sandwich over two separate level surfaces (for example, two micropipette tip boxes), to facilitate the next step. Put two binder clips along each side of the glass sandwich to fix the gel assembly.
27. Seal the bottom first, and then the two sides of the gel assembly. Apply the sealing tape as smoothly as possible to avoid forming air channels or bubbles along the edges of the glass plates. Rub the tape firmly onto the glass with towel paper to eliminate air channels and ensure a liquid-tight seal. Take particular care with the bottom corners of the plates, as these are the places where leaks often occur. An extra band of tape around the bottom of the plates can help to prevent leaks.
28. Maintain a constant flow to reduce the formation of bubbles. If some bubble appears, you can try to remove it with a spacer.
29. Again place two clamps along each side near the middle and bottom of the glass plate sandwich. This will help to maintain uniform gel thickness while pouring the gel and help to avoid the gel coming out. It is important that the clamps be placed over the side spacers *only*. Clamping unsupported glass will distort the thickness of the gel.
30. Polymerization may be extended overnight. Short polymerization times could result in distorted wells.
31. Wash the plates and the wells with water under the faucet to remove rest of acrylamide.
32. The notched plate should face inward toward the buffer reservoir.
33. We use a high voltage power supply with temperature probe. This is particularly useful to minimize variation from one gel run to the next. If you do not have a temperature probe, you can use thermometer strips to monitor gel temperature.

34. Excessive power will cause the gel to overheat and crack the glass plates.
35. It is important to wash away urea that has diffused into the wells and may adversely affect running of samples.
36. We avoid adding FDM to the samples since the dye may interfere with fluorescence detection. To facilitate loading of colorless samples, we mark the wells in the glass plate with a permanent marker.
37. Choose your appropriate marker dye taking into account the expected sample sizes.
38. Keep gel temperature constant at about 42–45 °C. Running time will depend on the DNA size and on products size differences. Run the gel until the marker dyes have migrated the desired distance.

Acknowledgements

We thank members of our laboratory for helpful discussion and advice. This work was supported by the Spanish Ministry of Science and Innovation and the European Regional Development Fund (grant BFU2010-18838), and by the Junta de Andalucía (grant P07-CVI-02770).

References

1. Zharkov DO (2008) Base excision DNA repair. Cell Mol Life Sci 65:1544–1565
2. Fortini P, Dogliotti E (2007) Base damage and single-strand break repair: mechanisms and functional significance of short- and long-patch repair subpathways. DNA Repair (Amst) 6:398–409
3. Barnes DE, Lindahl T (2004) Repair and genetic consequences of endogenous DNA base damage in mammalian cells. Annu Rev Genet 38:445–476
4. Stivers JT, Jiang YL (2003) A mechanistic perspective on the chemistry of DNA repair glycosylases. Chem Rev 103:2729–2759
5. Zharkov DO, Grollman AP (2005) The DNA trackwalkers: principles of lesion search and recognition by DNA glycosylases. Mutat Res 577:24–54
6. Huffman JL et al (2005) DNA base damage recognition and removal: new twists and grooves. Mutat Res 577:55–76
7. Levin JD, Demple B (1990) Analysis of class II (hydrolytic) and class I (beta-lyase) apurinic/apyrimidinic endonucleases with a synthetic DNA substrate. Nucleic Acids Res 18:5069–5075
8. Wiederhold L et al (2004) AP endonuclease-independent DNA base excision repair in human cells. Mol Cell 15:209–220
9. Friedberg EC et al (2006) DNA repair and mutagenesis. ASM Press, Washington, D.C
10. Kelley MR et al (2003) Disparity between DNA base excision repair in yeast and mammals: translational implications. Cancer Res 63:549–554
11. Dianov GL (2003) Monitoring base excision repair by *in vitro* assays. Toxicology 193:35–41
12. Córdoba-Cañero D et al (2009) Single-nucleotide and long-patch base excision repair of DNA damage in plants. Plant J 60:716–728
13. Córdoba-Cañero D et al (2010) Arabidopsis uracil DNA glycosylase (UNG) is required for base excision repair of uracil and increases plant sensitivity to 5-fluorouracil. J Biol Chem 285:7475–7483
14. Kavli B et al (2007) Uracil in DNA—general mutagen, but normal intermediate in acquired immunity. DNA Repair (Amst) 6:505–516

Chapter 19

In Vitro Monitoring of Base Excision Repair in *Saccharomyces cerevisiae*

Simone Moertl and Fred Ahne

Abstract

Base excision repair (BER) is an important mechanism to maintain genomic stability. Here we offer a set of protocols to quantitatively analyze BER capacity in whole cell-free yeast extracts. Cell-free yeast extracts were obtained by a French press procedure and repair capacities were measured by using oligonucleotide substrates. Repair products were separated by polyacrylamide gel electrophoresis and detected by autoradiography. These set of methods allow the analysis of different kinds of base damage and of individual mechanistic steps within BER. We used these protocols to investigate a new role of the DNA double strand break repair protein XRS1 in BER (1).

Key words: Base excision repair assay, Incision assay, Gap filling assay, Whole cell-free extracts, Radioactive labeling, Oligonucleotide-based assays, Denaturing polyacrylamide gel electrophoresis

1. Introduction

BER is a DNA damage repair pathway, which removes small, non-helix-distorting base lesions resulting from oxidation, alkylation, and deamination or inappropriate bases like uracil or 8-oxo-guanine. Defects in BER can lead to malignancies and age-related degenerative diseases in mammals (2). BER is a complex pathway system, consisting of several subpathways and numerous enzymes with partially redundant functions. Like in mammals, there are also two main BER subpathways in *Saccharomyces cerevisiae*, differing in the amount of nucleotides incorporated during repair. In the short-patch repair pathway only one single molecule is built into the DNA, while long-patch repair incorporates several new nucleotides. Both pathways start with the removal of the inappropriate base by specific DNA-glycosylases, resulting in an AP site. In the major BER pathway in *S. cerevisiae*, the AP site is then incised by Apn1 or Apn2 creating a 5′-deoxyribophosphate end. Rad27 catalyzes the removal

Lotte Bjergbæk (ed.), *DNA Repair Protocols*, Methods in Molecular Biology, vol. 920, DOI 10.1007/978-1-61779-998-3_19, © Springer Science+Business Media New York 2012

of the 5′-blocked end by extending the AP sites to gaps of up to 5 nt. New DNA synthesis and ligation through Cdc9 completes BER. In the alternative BER subpathway, the damaged base is processed by an enzyme with combined N-glycosylase/AP-lyase activities, namely Ogg1, Ntg1, or Ntg2. These enzymes produce 3′-α,β-unsaturated aldehydic ends, which are difficult to extend by DNA polymerases. Genetic studies suggest that their removal is catalyzed by the AP-endonucleases Apn1 and Apn2. Finally, the gap is filled by polymerase activity and the break is reconstituted by ligase Cdc9 (reviewed in refs. 3, 4).

Here we describe a cell-free system in which damage-specific base excision repair (BER) synthesis can be demonstrated in vitro with extracts from the yeast *S. cerevisiae*. This way of biochemical analysis of BER offers several advantages:

- *S. cerevisiae* can be easily and economically cultivated. This is important because the biochemical analysis of BER requires relatively high protein amounts.
- Fast and efficient methods for the construction of strains with mutated or completely deleted genes are available for yeast. As BER consists of several subpathways and as there are enzymes with partially overlapping function the investigation of strains mutated in multiple genes is desirable.
- The French press procedure yields highly active whole cell-free extracts within 3 h working time and a dialysis step overnight. Therefore laborious preparations of nuclear extracts including ultracentrifugation are not necessary.
- Due to the high availability of oligonucleotides including nucleotides with various modifications these methods allow the repair analysis of a broad spectrum of base damages.

2. Materials

1. YEPD medium: 2 % dextrose; 1 % yeast extract, and 2 % peptone, used as complete growth medium for yeast.
2. Potassium phosphate buffer: 25 mM potassium phosphate, pH 8.0.
3. Extraction buffer: 100 mM Tris–HCl, pH 8.0, 10 mM β-mercaptoethanol, and a protease inhibitor cocktail (e.g., Sigma) in an appropriate concentration.
4. Dialysis buffer: 10 mM Tris–HCl, pH 7.5; 10 mM β-mercaptoethanol.
5. Substrates for BER assays: Table 1 shows nucleotide sequences and melting points of all used oligonucleotides.

Table 1
Oligonucleotides for BER assays

	Sequence	T_m (°C)
BER1	5′-GCC CTG CAG GTC GAU TCT AGA GGA TCC CCG GGT AC-3′	81.0
BER2	5′-GTA CCC GGG GAT CCT CTA GAG TCG ACC TGC AGG GC-3′	82.0
BER3	5′-GCC CTG CAG GTC GAG* TCT AGA GGA TCC CCG GGT AC-3′	81.0
BER4	5′-GTA CCC GGG GAT CCT CTA GAC TCG ACC TGC AGG GC-3′	82.0
BER5	5′-GCC CTG CAG GTC GAC-3′	52.0
BER6	5′-AGA GGA TCC CCG GGT AC-3′	56.0

G*: 8-oxo-G

6. BER assay buffer (5×): 250 mM HEPES, pH 7.5, 2.5 mM EDTA, 10 mM DTT, 100 mM KCl, 20 mM ATP, 25 mM phosphocreatine, 2.5 mM NAD, 0.5 mM ddTTP. Store at −20 °C in 50 μL aliquots. Add 100 μg/mL phosphocreatine kinase to the diluted buffer (freshly).
7. Incision buffer: 50 mM HEPES, pH 7.7, 7.5 mM $MgCl_2$, 0.5 mM EDTA, 1 mM dithiothreitol, 20 mM phosphocreatine, 2 % glycerol, 100 μg/mL bovine serum albumin.
8. Gap filling buffer: 50 mM Tris–HCl, pH 7.4, 50 mM KCl, 1 mM dithiothreitol, 5 mM $MgCl_2$, 5 % glycerol.
9. 18 % polyacrylamide gel solution: 36 mL 30 % acrylamide/bisacrylamide (30 %, 37.5:1), 28.8 g urea, 10.9 mL BER-TBE, 0.6 mL AMPS, 0.03 mL TEMED.

   24 % polyacrylamide gel solution: 36 mL acrylamide/bisacrylamide (40 %, 19:1), 28.8 g urea, 10.9 mL BER-TBE, 0.6 mL AMPS, 0.03 mL TEMED.
10. BER-TBE: 5.34 mL 1 M Tris–HCl pH 8.8, 5.4 mL 1 M boric acid, 0.24 mL 0.5 M EDTA.
11. DNA gel loading buffer: 95 % formamide, 20 mM EDTA, 0.02 % bromophenol blue, 0.02 % xylene cyanol.

3. Methods

3.1. Preparation of Cell-Free Extracts

Whole cell-free yeast extracts are prepared by a French press procedure according to Jha et al. 1993 (5).

1. Inoculate YEPD medium (50 mL) with a single colony and grow the culture overnight to stationary growth phase at 30 °C.

2. Dilute the pre-culture to a volume of 500 mL with YEPD medium and grow under vigorous shaking to a cell density of 5×10^6–7×10^6 cells/mL (early logarithmic growth phase) (see Note 1).
3. Keep cells on ice for 20 min, harvest by centrifugation (4,500 × *g*, 15 min) and wash twice with potassium phosphate buffer.
4. Next, resuspend the cell pellet in 5 mL ice-cold extraction buffer and add 1 mM protease inhibition mix per 20 g of cell pellet.
5. Crush cells in a prechilled French Press Cell (e.g., Aminco), applying an internal pressure of 20,000 psi (see Note 2).
6. Adjust the crude extract to pH 8.0 by adding 0.1 M NaOH and remove cell debris by centrifugation at 21,000 × *g* in an SS34 rotor for 40 min at 4 °C. Carefully remove the supernatant without demolishing the pellet and keep on ice.
7. Precipitate proteins by stepwise addition of ammonium sulfate to a final concentration of 0.35 g/mL for at least 20 min on ice, collect by centrifugation (21,000 × *g*, 30 min), and dissolve in 3 mL dialysis buffer.
8. Dialyse overnight in the same buffer.
9. Centrifuge extract for 10 min (4,500 × *g*) to remove insoluble particles. The crude extract can either be directly used or stored up to 6 month at −80 °C. Protein concentrations are determined according to Bradford using bovine serum albumin as standard.

3.2. Preparation of Assay Substrates

The incision assay and the gap filling assays require oligonucleotide labeling at the 5′-end with [γ-^{32}P]ATP prior to annealing.

3.2.1. End Labeling of Oligonucleotides

1. Mix 0.1 nM oligonucleotide (BER1, incision assay or BER5, gap filling assay) with 1/10 vol 10× reaction buffer A (NEB Biolabs), 2 nM ATP, 10 mCi/mL [γ-^{32}P]ATP, and 10 U T4 polynucleotide kinase (NEB Biolabs) in a volume of 10 μL.
2. Incubate the mixture at 37 °C for 20 min and heat to 75 °C for 10 min.
3. Finally, purify the oligonucleotides using G-25 Sephdex micro-columns (Roche) to remove unincorporated [γ-^{32}P]ATP. To do this, resuspend the matrix in the column buffer by inversion. Remove excess buffer and pack the column bed by centrifugation for 1 min at 1,000 × *g*. Apply the sample to the center of the column (see Note 3). Following sample application, centrifuge the column for 4 min at 1,000 × *g* and collect the sample in a microcentrifuge tube. Discard the column in an appropriate waste receptacle.

3.2.2. Annealing of Oligonucleotides

Annealing is done by mixing equimolar amounts of the oligonucleotides diluted in PCR buffer. Mixtures are incubated for 10 min at 95 °C following slow cooling down to room temperature. For the standard BER assay, monitoring the removal of uracil, oligonucleotides BERl and BER2 are annealed. For the investigation of incision activity 5′-labeled BERl is combined with BER2. To analyze the processing of 8-oxoG, a substrate resulting from annealing of BER4 and BER3 is used. To measure gap filling activity the 14 nt ssDNA, 5′-end radioactive labeled oligonucleotide BER5 and oligonucleotide BER6 are annealed to BER2.

3.3. BER Assays

3.3.1. Standard BER Assay

To measure the BER capacity of cell-free extracts an in vitro assay adapted from Harrigan et al. is done (6). During the repair process damaged bases (uracil (BERl) or 8-oxo-guanin (BER3)) are removed from the substrate oligonucleotide and [α-^{32}P]dCTP ([α-^{32}P]dGTP in the case of the 8-oxoG oligo) is incorporated opposite G (or C). Before ligation this intermediate has a size of 15 nt; after ligation the original length of 35 nt is restored. To distinguish between short patch repair, where only 1 nt is substituted, and long patch repair, were a number of oligonucleotides are replaced, ddTTP is added to the reaction mixture to abort elongation. This allows the detection of long-patch repair products with a size of 16 nt (Fig. 1).

1. For each BER assay, dilute 20 μg of protein in 1/5 Vol 5× BER buffer, 250 nM oligonucleotide substrate, 10 mM $MgCl_2$ and 2.2 μM [α-^{32}P]dCTP (2.2 μM [α-^{32}P]dGTP for the 8-oxoG oligonucleotide) (final volume 20 μL).
2. Incubate samples at 37 °C for 2–20 min.
3. Stop reactions by adding 1 μL of 100 mM EDTA and heat to 72 °C for 5 min. Remove unincorporated [α-^{32}P]dCTP with G-25 Sephadex minicolumns (Subheading 3.2.1). Samples can be immediately incubated with loading buffer and analyzed by gel electrophoresis or they can be stored at −20 °C.

3.3.2. Incision Assay

To analyze recognition of the damaged base and incision of the phosphate backbone by lyase or endonuclease activity an in vitro incision assay is performed on the basis of Wang et al.(7) (Fig. 2).

1. For substrate preparation: Radioactive label oligonucleotide BERl at the 5′-end and anneal to oligonucleotide BER2.
2. Incubate 250 nM oligonucleotide substrate ([γ-^{32}P]dCTP-BERl + BER2) in 10 μL incision reaction buffer with 1 μg cell-free protein extract for 2–15 min at 23 °C.
3. Abort the reactions by adding 5 mM EDTA and heat to 65 °C for 20 min. Ethanol precipitate repair products and solubilize in DNA loading dye.

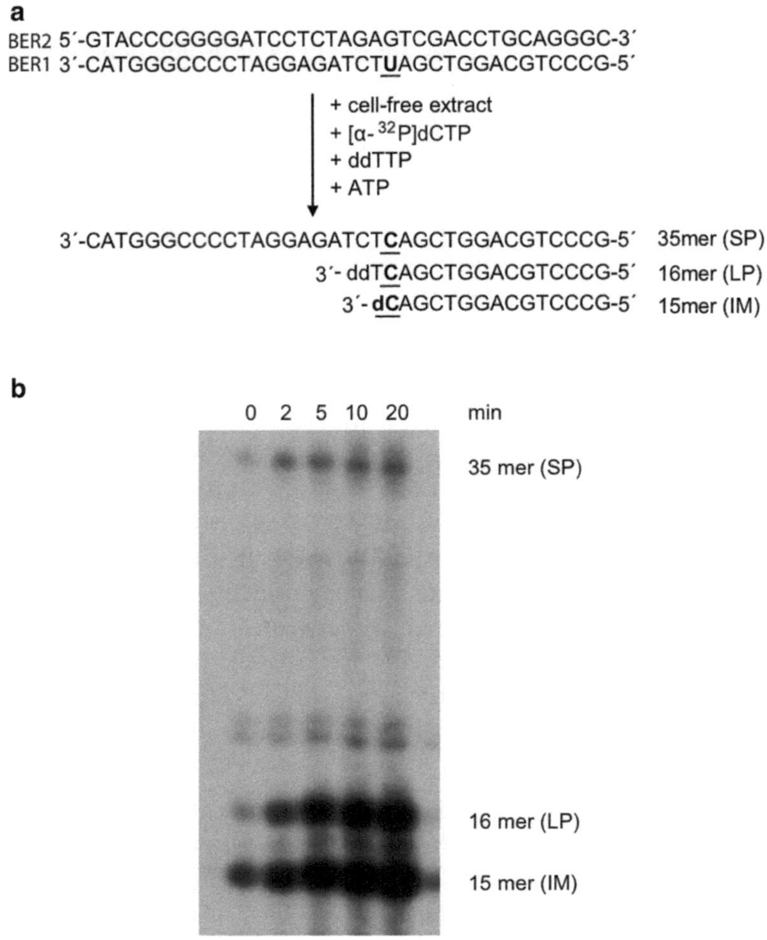

Fig. 1. Quantification of BER using a uracil-containing oligonucleotide substrate and whole cell-free extracts (**a**) principle of the in vitro assay, long patch repair (LP) is visible through the 16mer repair product, short patch repair (SP) leads to the restoration of the original 35mer and the repair intermediate (IM) is visible as 15mer (**b**): in vitro BER repair capacity of the *S. cerevisiae* wildtype strain MKP-0.

3.3.3. Gap Filling Assay

After elimination of the damaged base and AP site processing a single strand gap (one to several nucleotides) has to be filled by polymerase activity. To analyze this step we employ a special assay in order to monitor strand elongation for gap filling adapted from Wong et al. (8). We use a 35 bp double-stranded oligonucleotide with a 4 nt gap in one strand (Fig. 3).

1. For substrate preparation: Radioactive label oligonucleotide BER5 at the 5′-end and anneal with oligonucleotides BER6 and BER2.
2. Mix 40 μg cell-free extract with 10 nM oligonucleotide substrate (BER2 + [γ-^{32}P]BER5 + BER6) and gap filling buffer. Add 2.5 mM dCTP and 2.5 mM dTTP.

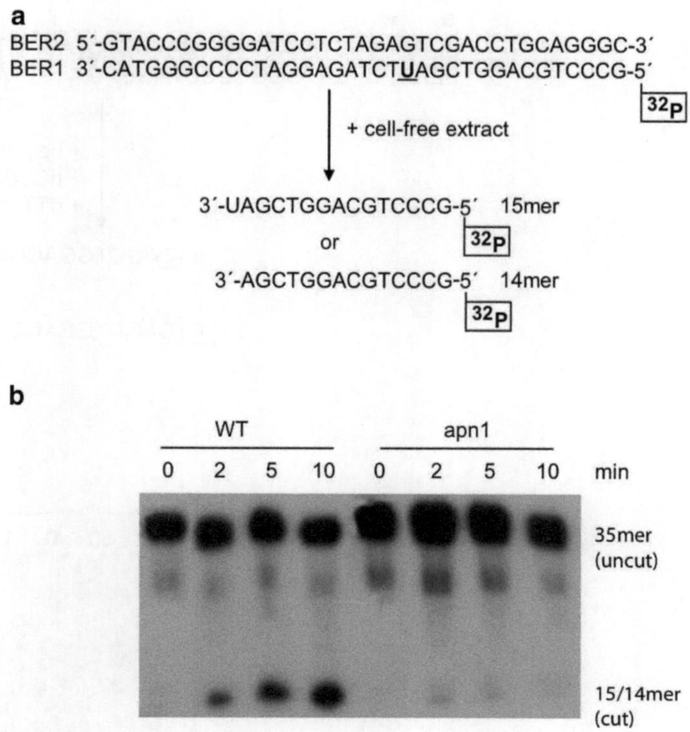

Fig. 2. Incision assay using a uracil-containing oligonucleotide. (**a**) Generation of an AP site and incision activity can be monitored through the appearance of cut products of the labeled substrate. (**b**) Incision assay with whole cell extracts derived from wildtype cells and from a *apn1* mutant that is deficient in backbone incision.

3. Stop reactions by the addition of 5 mM EDTA and mix with 10 µL loading dye.

3.4. Resolution of BER Assay Products by SDS Polyacrylamide Gel electrophoresis

Polyacrylamide gels are especially useful for the separation of small DNA fragments up to the discrimination at the 1 nt level. For the separation of BER repair products, denaturing polyacrylamide gels are used. We employ 18 % gels for the analysis of incision products, while 24 % gels are used for the BER assay and the gap filling assay. To fasten the gel casting procedure the polyacrylamide solution containing acrylamide/bisacrylamide, urea, boric acid, and EDTA can be prepared in advance and stored at 4 °C.

1. Add 0.1 % AMPS and 0.005 % TEMED to the polyacrylamide solution to start polymerization. Apply the mixture directly to the gel plate sandwich (24 cm length, e.g., CBS Scientific). For complete polymerization keep the gel at 4 °C overnight.
2. Place the gel in the electrophoresis unit and pour freshly prepared buffer in the upper and lower buffer chamber. Rinse the gel slots with electrophoresis buffer before running, and pre-run for 30 min at 600 V (see Note 4).

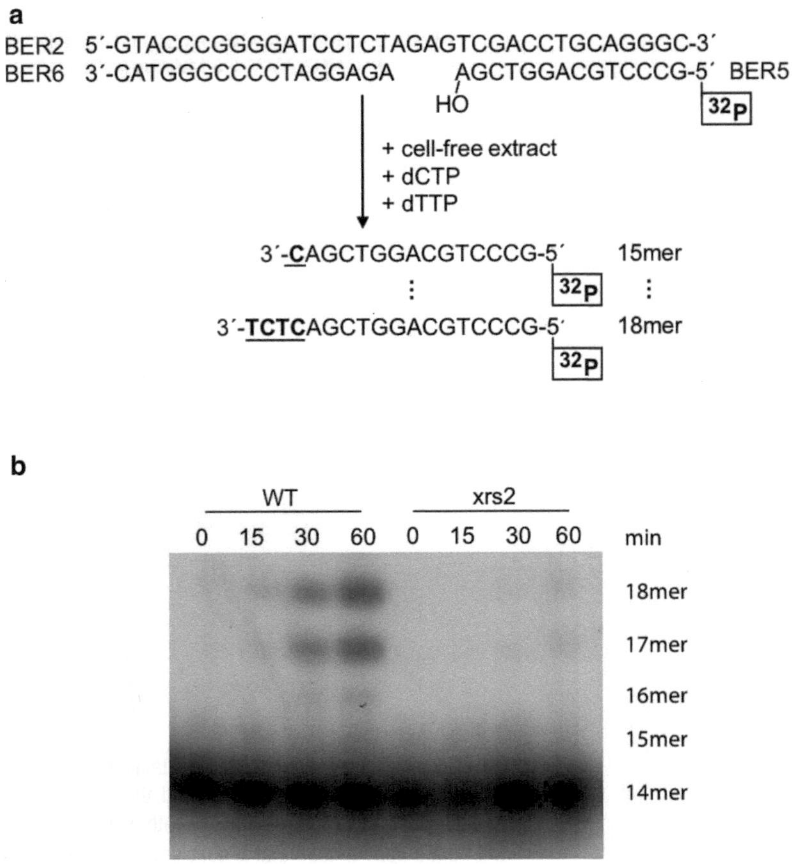

Fig. 3. Four nt gap filling assay; (**a**) gap filling activity is visible through the extension of the 5′-labeled substrate. (**b**) Gap filling activity of whole cell extract derived from the wildtype strain MKP-0.

3. Mix identical volumes of 2× loading buffer and BER assay samples, boil at 95 °C and apply to the gel slots. Conduct the electrophoretic separation for 1.5 h to 4 h at 600 V.

3.5. Visualization and Quantification

Repair products are detected by autoradiography and quantified by digital imaging.

4. Notes

1. Usage of cells in early log phase (5×10^6 cells/mL) is critical for successful extract preparation. Using higher cell densities results in poor protein yield or reduced activity.
2. French press cylinders should be cooled to –20 °C, we also suggest meanwhile cooling when more than one extract is prepared.

3. For successful removal of unincorporated radioactive nucleotides the reaction mixture must be pipetted directly on the matrix

4. Before applying the samples on the polyacrylamide gel carefully rinse the gel slots, otherwise samples start at different levels. Furthermore all slots should be loaded at least with loading dye to avoid "smiling". Running parameters should not be increased because this results in poor resolution.

References

1. Steininger S, Ahne F, Winkler K et al (2010) A novel function for the Mre11-Rad50-Xrs2 complex in base excision repair. Nucleic Acids Res 38:1853–1865
2. Friedberg EC (2003) DNA damage and repair. Nature 421:436–440
3. Boiteux S, Guillet M (2004) Abasic sites in DNA: repair and biological consequences in Saccharomyces cerevisiae. DNA Repair (Amst) 3:1–12
4. Alseth I, Eide L, Pirovano M et al (1999) The Saccharomyces cerevisiae homologues of endonuclease III from Escherichia coli, Ntg1 and Ntg2, are both required for efficient repair of spontaneous and induced oxidative DNA damage in yeast. Mol Cell Biol 19: 3779–3787
5. Jha B, Ahne F, Kistler M et al (1998) A rapid method to monitor repair and mis-repair of DNA double-strand breaks by using cell extracts of the yeast Saccharomyces cerevisiae. Curr Genet 33:1–3
6. Harrigan JA, Wilson DM 3rd, Prasad R et al (2006) The Werner syndrome protein operates in base excision repair and cooperates with DNA polymerase beta. Nucleic Acids Res 34:745–754
7. Wang Z, Wu X, Friedberg EC (1997) Molecular mechanism of base excision repair of uracil-containing DNA in yeast cell-free extracts. J Biol Chem 272:24064–24071
8. Wong HK, Kim D, Hogue BA et al (2005) DNA damage levels and biochemical repair capacities associated with XRCC1 deficiency. Biochemistry 44:14335–14343

Chapter 20

Mitochondrial Base Excision Repair Assays

Ricardo Gredilla and Tinna Stevnsner

Abstract

Mitochondrial DNA (mtDNA) is constantly exposed to oxidative injury. Due to its location close to the main site of reactive oxygen species, the inner mitochondrial membrane, mtDNA is more susceptible than nuclear DNA to oxidative damage. The accumulation of DNA damage is thought to be particularly deleterious in post-mitotic cells, including neurons, and to play a critical role in the aging process and in a variety of diseases. Thus, efficient mtDNA repair is important for the maintenance of genomic integrity and a healthy life. The base excision repair (BER) mechanism was the first to be described in mitochondria, and consequently it is the best known. This chapter outlines protocols for isolating mitochondria from mammalian cells in culture and from rodent tissues including liver and brain. It also covers the isolation of synaptic mitochondria. BER takes place in four distinct steps, and protocols describing in vitro assays for measuring these enzymatic steps in lysates of isolated mitochondria are included.

Key words: Mitochondria, mtDNA, DNA repair, Base excision repair

1. Introduction

DNA is constantly exposed to endogenous and exogenous agents that generate DNA lesions and induce DNA instability. The main endogenous genotoxic agents within the cells are the mitochondrial reactive oxygen species (ROS). They generate a large number of DNA lesions, including oxidized DNA bases, abasic sites, and single- and double-strand breaks. Many of these ROS-induced DNA lesions show mutagenic or cytotoxic effects due to mispairing of bases, which may give rise to mutations upon DNA replication (1, 2). Due to its particular location, close to the mitochondrial free radical generation sites, the mitochondrial genome is intensely exposed to ROS damage. It is well established that mitochondrial DNA (mtDNA) contains higher steady-state levels of oxidative DNA lesions compared to nuclear DNA (3–5). Avoidance of mtDNA damage accumulation is critical for cellular survival, because it can generate blockage of DNA replication and transcription,

Lotte Bjergbæk (ed.), *DNA Repair Protocols*, Methods in Molecular Biology, vol. 920,
DOI 10.1007/978-1-61779-998-3_20, © Springer Science+Business Media New York 2012

leading to genomic instability. Genomic instability and mtDNA mutations are associated with hereditary diseases (6), age-related diseases such as Parkinson's disease (7), and aging itself (8–10).

In order to maintain genomic integrity, different DNA repair mechanisms exist and the specific pathway employed depends, in part, upon the type of DNA damage that is being repaired. These mechanisms take place both in nuclei and mitochondria, although not all the different pathways have been described in both compartments. Nowadays, only the base excision repair (BER) and the mismatch repair (MMR) pathways have been described to occur in mammalian mitochondria (11). Lesions such as alkylation, deamination, or oxidation products are repaired by the BER pathway (12) while MMR removes mismatches in the DNA (13). Since DNA oxidation products are mainly generated by ROS in mitochondria, efficient BER is a critical cellular response to ROS-induced DNA damage to avoid mtDNA mutation accumulation and its deleterious effects.

Briefly, the BER pathway takes place in four distinct steps, both in nuclei and mitochondria. The pathway is initiated by a specific DNA glycosylase, which recognizes the modified or inappropriate base and cleaves the N-glycosidic bond, creating an abasic site. The glycosylases can be divided into mono-functional and bi-functional glycosylases. In addition to the cleavage activity towards the N-glycosylic bond, the bi-functional glycosylases also possess apurinic/apyrimidinic (AP) lyase activity, which allows the cleavage of the DNA backbone and generation of a 3′-deoxyribose phosphate (dRP) at the AP site. The resulting AP site is then processed by an AP endonuclease, which creates a strand break with a 3′-hydroxyl end and a 5′-(dRP) residue. In mitochondria repair then proceeds with the filling of the single nucleotide gap by the mtDNA polymerase, pol γ. Besides the polymerase activity, polymerase γ (pol γ) possesses 3′–5′ exonuclease and 5′dRP lyase activities. Thus, when mtBER is initiated by a monofunctional DNA glycosylase, the 5′-dRP moiety generated when the AP endonuclease cleaves the strand break can be removed by the 5′dRP lyase function of pol γ. Finally, the resulting nick is sealed by a DNA ligase.

Although the mechanisms in principle are similar in the nucleus and the mitochondrion, the latter possesses an independent BER machinery, the components of which are encoded by nuclear genes (14). Most of the enzymes contributing to the mitochondrial pathway are isoforms of nuclear enzymes, either by alternative splicing or alternative transcription initiation sites. Thus, when assaying the BER activity of the mitochondrial fraction it is critical to ensure the absence of nuclear protein contamination.

In the following sections of this chapter we describe the techniques and basic protocols required for isolation of mitochondria from various sources and for measuring accurate and specific BER activities.

2. Materials

2.1. Isolation of Mitochondrial Fraction

2.1.1. Rodent Tissues

1. MSHE buffer: 210 mM mannitol, 70 mM sucrose, 10 mM HEPES, 1 mM EDTA, 1 mM EGTA, pH 7.4.
2. STE buffer: 32 mM sucrose, 10 mM Tris–HCl, 1 mM EDTA, pH 7.4.
3. Subtilisin A protease.
4. Protease inhibitors: spermine, spermidine, phenylmethylsulfonyl fluoride (PMSF), ditihiothreitol (DTT), protease inhibitor cocktail set III (Calbiochem).
5. Mitochondrial buffer: 20 mM HEPES, 1 mM EDTA, 2 mM DTT, 5 % glycerol, pH 7.4.
6. Glass-glass homogenizer.

2.1.2. Cultured Cells

1. PBS buffer: 140 mM NaCl, 8 mM Na_2HPO_4, 3 mM KCl, 1 mM KH_2PO_4.
2. MSHE buffer (see Subheading 2.1.1).
3. Protease inhibitors.
4. Mitochondrial buffer.
5. Percoll.
6. Glass-glass homogenizer.

2.2. Assesing the Presence of Mitochondrial and Nuclear Proteins

1. Specific primary antibodies against mitochondrial proteins such voltage-dependent anion-selective channel protein 1 (VDAC-1 (Abcam)) or cytochrome c oxidase subunit II (Santa Cruz).
2. Specific primary antibodies for detection of nuclear proteins such as lamin B (Santa Cruz), lamin A/C (Santa Cruz), proliferating cell nuclear antigen (PCNA (Santa Cruz)) or CDC47 (Thermo Scientific).
3. HRP-linked secondary antibodies.
4. GeneGnome (Syngene) equipment or equivalent for visualizing and quantifying Western blots.

2.3. 5′-Labeling and Duplexing of Specific Oligonucleotide

1. Specific oligonucleotides used in BER repair assays (see Table 1).
2. T4 polynucleotide kinase.
3. [γ-^{32}P] ATP.
4. 10× T4 polynucleotide kinase buffer.
5. Native loading dye: 0.25 μg/ml bromophenol blue, 2.5 μg/ml xylene cyanol, 30 % glycerol in 1×TBE.

Table 1
Oligonucleotides used in assays for DNA repair activities

Lesion	Enzyme	Top strand sequence
8oxoG	OGG1	5′-ATA TAC CGC G**8**C CGG CCG ATC AAG CTT ATT
dUracil	UNG	5′-ATA TAC CGC G**U**C CGG CCG ATC AAG CTT ATT
5-OHdU	NTH1	5′-ATA TAC CGC G**5**C CGG CCG ATC AAG CTT ATT
5-OHdU bubble	NEIL1	5′-GCT TAG CTT GGA ATC GTA TCA TGT A**5**A CTC GTG TGC CGT GTA GAC CGT GCC
AP	APE1	5′-ATA TAC CGC GG**F** CGG CCG ATC AAG CTT ATT
GAP	Pol γ	5′-CTG CAG CTG ATG CGC **X**GT ACG GAT CCC CGG GTA C
dUracil	BER	5′-ATA TAC CGC G**U**C CGG CCG ATC AAG CTT ATT

The modified bases are shown in *bold*. *Underlined* bases indicate unpaired area. 8 = 8-oxo-guanine (8oxodG); U = deoxyuracil; 5 = 5-OH-cytosine (5-OHdU); F = tetrahydrofuran; X = gap; BER: whole BER pathway

2.4. BER Assays

2.4.1. DNA Glycosylases and AP Endonuclease

1. 90 fmol ^{32}P-end-labeled duplex oligonucleotides containing specific lesions.
2. Reactions contain 20 mM HEPES pH 7.6, 5 mM EDTA, 75 mM KCl, 1 mM $MgCl_2$, 5 % glycerol, 5 mM DTT, and 0.1 mg/mL BSA.
3. Assay termination solutions: 5 mg/ml proteinase K, 10 % SDS.
4. Formamide loading dye: 80 % formamide, 10 mM EDTA, 1 mg/ml xylene cyanol FF, and 1 mg/ml bromophenol blue.

2.4.2. Polymerase Gap-Filling Assay

1. 1 pmol non-labeled duplex oligonucleotide containing a single nucleotide gap.
2. Reactions contain: 50 mM Tris–HCl, pH 7.4, 50 mM KCl, 1 mM DTT, 5 mM $MgCl_2$, 5 % glycerol, 5 μM dCTP, 4 μCi of α^{32}P-dCTP.
3. Assay termination solution: formamide loading dye.

2.4.3. Repair Synthesis Incorporation Assay

1. 100 ng non-labeled duplex deoxyuracil (dU)-containing oligonucleotide.
2. Reactions contain: 40 mM HEPES, pH 7.8, 0.1 mM EDTA, 5 mM $MgCl_2$, 0.2 mg/ml BSA, 50 mM KCl, 1 mM DTT, 40 mM phosphocreatine, 100 μg/ml phosphocreatine kinase, 2 mM adenosine triphosphate (ATP), 20 μM deoxynucleotide triphosphates (dNTPs), 4 μCi ^{32}P-deoxycytidine triphosphate (dCTP), 3 % glycerol.
3. Assay termination solutions: 5 mg/ml proteinase K, 10 % SDS.

4. DNA precipitation: 2.8 mM glycogen, 110 mM ammonium acetate, 95 % ethanol.
5. Formamide loading dye.

2.5. Electrophoresis and Result Analysis

1. Polyacrylamide solution: 40 % polyacrylamide solution (19:1 acrylamide:bisacrylamide), 10 % ammonium persulfate, Tetramethylethylenediamine (TEMED).
2. 10× TBE solution: 1 M Tris–HCl, pH 8.4, 0.9 M boric acid, 10 mM EDTA.
3. Storage phosphor screens.
4. Phosphorimager scanner (Molecular diagnostics) or equivalent.
5. Quantity one® (Bio-rad) or equivalent software.

3. Methods

3.1. Isolation of Mitochondria from Rodent Tissues

Isolation of mitochondria from different mammalian tissues is generally performed by differential centrifugation. However, different tissues require specific conditions, particularly related to centrifugation speeds, treatment of the samples during the homogenization process, such as the use of proteases when isolating mitochondria from muscular tissue, or the use of density gradients for the purification of particular subpopulation of mitochondria such as synaptosomic mitochondria. Here we describe in detail the protocols for isolation of liver and brain mitochondria (Fig. 1). Specific conditions for other tissues such as heart, skeletal muscle, or kidney can be found elsewhere (15, 16).

An important issue during mitochondrial purification is to maintain the integrity of the mitochondria during the whole isolation process. Moreover, since BER activity is present both in nuclei and mitochondria, it is essential to avoid nuclear contamination when investigating mitochondrial BER activities. In order to avoid the presence of nuclear proteins in the mitochondrial preparations, mild protease treatment can be included during homogenization of the tissues. All the steps must be carried out at 4 °C, and all material should be precooled.

3.1.1. Isolation of Mitochondria from Liver

1. Excise liver. After washing with MSHE buffer, mince with scissors and transfer to a glass-glass homogenizer.
2. Add 10 volumes (w/v) of ice-cold MSHE buffer and homogenize in the presence of 1.4 U of subtilisin A per milliliter of MSHE buffer.
3. Transfer to a centrifuge tube and spin for 10 min at 8,500 × *g*.

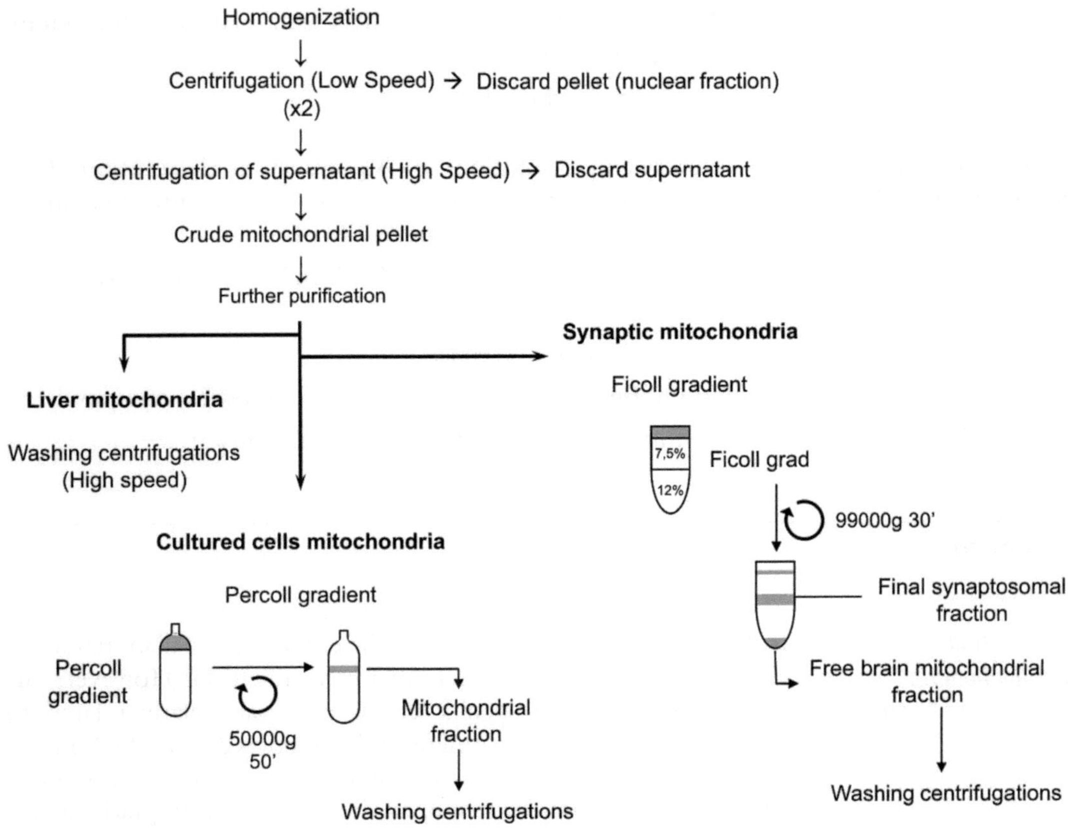

Fig. 1. Scheme of the different steps required for mitochondrial purification from various sources (See text for details).

4. Discard the supernatant and resuspend the pellet in same volume as in step 2 with MSHE buffer containing protease inhibitors (0.15 mM spermine, 0.75 mM spermidine, 1 mM PMSF, 5 mM DTT, 1:1,000 protease inhibitor cocktail set III; Calbiochem) (see Note 1).
5. Transfer to a centrifuge tube and spin 10 min at 500 × *g*.
6. Discard the pellet, transfer the supernatant to a new centrifuge tube and repeat the 500 × *g* centrifugation.
7. Spin the supernatant at 9,500 × *g* for 10 min.
8. Resuspend the pellet in ice-cold MSHE buffer containing protease inhibitors and centrifuge again at 8,500 × *g* for 10 min.
9. Resuspend the final mitochondrial pellet in mitochondrial buffer.

3.1.2. Isolation of Brain Mitochondria

1. Excise the whole brain, or a specific brain region. Wash in ice-cold MSHE buffer, mince tissue with scissors and transfer to a glass-glass homogenizer.
2. Add 10 volumes (w/v) of ice-cold MSHE buffer and homogenize in the presence of 1.4 U of subtilisin A per milliliter of MSHE buffer.

3. Transfer to a centrifuge tube and spin for 10 min at 8,500 × *g*.
4. Discard the supernatant and resuspend the pellet in same volume as in step 2 with MSHE buffer containing protease inhibitors.
5. Transfer to a centrifuge tube and spin 10 min at 500 × *g*.
6. Discard the pellet, transfer the supernatant to a new centrifuge tube, and repeat the 500 × *g* centrifugation.
7. Spin the supernatant at 10,000 × *g* for 10 min.
8. Resuspend the pellet with ice-cold MSHE buffer containing protease inhibitors. Centrifuge again at 9,500 × *g* for 10 min.
9. Resuspend the pellet once again in ice-cold MSHE buffer containing proteases inhibitors. Centrifuge again at 8,500 × *g* for 10 min.
10. Resuspend the final mitochondrial pellet in mitochondrial buffer.

3.1.3. Isolation of Synaptic Mitochondria

The previous protocol describes the isolation of total mitochondria from brain tissue. Moreover, different protocols have been described in order to isolate mitochondria specifically derived from neuronal origin (17–20). We have recently described the synaptic mtDNA repair mechanisms by synaptosomal purification (21). Synaptosomes are subcellular membranous structures corresponding to isolated synapses generated during mild homogenization of brain tissue. They contain the presynaptic side, where synaptic mitochondria are located. The following protocol for purification of synaptosomes allows simultaneous isolation of non-synaptic mitochondria, also called free brain mitochondria.

1. Gently homogenize brain tissue in STE buffer with a glass-glass homogenizer. Perform all steps in the presence of protease inhibitors.
2. After centrifugation at 1,300 × *g* for 10 min, discard the pellet, which represents the nuclear fraction.
3. Spin the supernatant at 17,000 × *g* for 10 min.
4. Resuspend the pellet in STE buffer containing protease inhibitors and layer on a discontinuous ficoll gradient 7.5–12 %.
5. After centrifugation at 99,000 × *g* for 40 min, the synaptosomal fraction is obtained at the 7.5–12 % interface, while the pellet comprises the free brain mitochondria fraction.
6. Obtain the final synaptosomal and free brain mitochondrial fractions by washing with STE buffer containing protease inhibitors and spinning at 18,500 × *g* and 9,800 × *g*, respectively, for 10 min.
7. Resuspend the final pellets in mitochondrial buffer.

3.2. Isolation of Mitochondria from Cell Cultures

Depending on the cell type, 40 or more 15 cm culture plates of near confluent cells must be used for sufficient mitochondrial yield.

1. Wash each plate twice with 10 ml of PBS buffer.
2. Scrape off cells from the plates using MSHE buffer. Keep on ice.
3. Spin all the collected cells at 500 × *g*. Resuspend in total volume of 10 ml MSHE buffer containing protease inhibitors and homogenize with approximately 50 strokes in a glass-glass homogenizer.
4. Spin the homogenate for 12 min at 500 × *g* and discard the pellet.
5. Spin the supernatant at 9,500 × *g* for 10 min.
6. Resuspend the pellet in ice-cold MSHE buffer containing protease inhibitors. Centrifuge again at 8,500 × *g* during 10 min.
7. Resuspend the pellet once again in ice-cold MSHE buffer containing protease inhibitors. Centrifuge again at 9,500 × *g* for 10 min.
8. Resuspend the crude mitochondrial pellet in MSHE buffer containing protease inhibitors.
9. In order to avoid any contamination by nuclear proteins, the crude mitochondrial pellet is further purified by density Percoll gradient centrifugation. Layer the mitochondrial suspension on top of a 10 ml of Percoll gradient (50 % Percoll and 50 % 2×-MSHE buffer). Centrifuge at 50,000 × *g* for 50 min.
10. Remove the mitochondrial band that appears one-third of the distance from the top of the centrifuge tube. Dilute in a new centrifuge tube with 10 volumes of ice-cold MSHE buffer containing protease inhibitors.
11. Spin at 3,000 × *g* for 15 min.
12. Resuspend the final pure mitochondrial pellet in mitochondrial buffer.

Once the mitochondrial fractions are purified, protein concentration is determined. Mitochondrial fractions can be used immediately or aliquoted and stored at −80 °C. Repetitive freeze-thaw cycles should be avoided, because it will lead to loss of enzymatic activity. Aliquots can be kept at −80 °C for several months with no significant loss of DNA repair activity.

3.3. Assessing Mitochondrial and Nuclear Proteins

Because BER activities are present in both nuclei and mitochondria, it is critical to make certain that no nuclear contamination perturb when investigating mitochondrial BER activities. The use of subtilisin A during homogenization, or the purification through percoll or ficoll density gradients of the crude mitochondrial fractions help to avoid the presence of nuclear proteins in the final mitochondrial fractions. Nevertheless, it is important to analyze the mitochondrial fractions by Western blotting in order to ensure the absence of nuclear contamination. The use of antibodies against nuclear proteins like lamin B, PCNA, or CDC47, will confirm the absence of nuclear contamination in the mitochondrial fractions. On the other hand, antibodies against mitochondrial proteins such as cytochrome c oxidase subunit II, mitochondrial transcription factor A (TFAM), or VDAC-1 are excellent candidates for assessing mitochondrial protein content.

3.4. Base Excision Repair Assays

As described in Subheading 1, the BER pathway consists of four different enzymatic steps, which can be investigated either separately or as a whole.

3.4.1. 5′-End-Labeling of Lesion-Containing Oligonucleotides with [γ-^{32}P] ATP and Annealing

Since different DNA glycosylases recognize and cleave different DNA lesions, oligonucleotides containing specific lesions are required for assessing specific DNA glycosylase activities. Similarly, the measurement of APE activity requires oligonucleotides containing abasic sites. The in vitro assays for the first two steps of the BER pathway, by DNA glycosylases and APE, require the use of specific 5′-end-labeled oligonucleotides. Table 1 includes different oligonucleotides containing specific lesions to be used for specific DNA glycosylases and APE activities.

1. Mix 2 μl (100 ng) of the lesion-containing oligonucleotide, 2 μl of 10× polynucleotide kinase buffer, 150 μCi of [γ-^{32}P] ATP, and 2 μl of T4 polynucleotide kinase enzyme in a 20 μl reaction.
2. Incubate for 90 min at 37 °C.
3. Stop the reaction by adding 2 μl of 0.5 M EDTA.
4. Remove free unincorporated [γ-^{32}P] ATP using G25 desalting columns (GE Healthcare). Load the reaction on the G25 column and centrifuge for 2 min at 750 × *g*. Discard the column.
5. Add 1 μl (250 ng) of the corresponding complementary strand and 1 μl of 2 M KCl for annealing.
6. Incubate at 90 °C for 5 min and allow the reaction to slowly reach room temperature. Store the 5′-labeled duplex oligonucleotides at −20 °C
7. Check that the substrates have been annealed completely (see Note 2) resolving single- and double- stranded oligonucleotides in a polyacrylamide gel under native conditions (see Subheading 3.4.1).
8. An oligonucleotide size marker (e.g., 8–32 bases) should also be 5′-end-labeled. Add 1 μl of the oligonucleotide size marker, 2 μl of 10× polynucleotide kinase buffer, 150 μCi of [γ-^{32}P]

ATP, and 2 μl of T4 polynucleotide kinase enzyme in a 20 μl reaction and follow steps 2–4 described above; finally, add formamide dye and store the size marker at −20 °C.

3.4.2. Lysis of Mitochondria

Mitochondrial fractions contain intact mitochondria. Mitochondrial BER proteins are located in the mitochondrial matrix and at the inner membrane, and thus, in order to investigate the mitochondrial BER activities, the mitochondrial extracts need to be processed before the in vitro activity assays.

1. Immediately before running the in vitro assays thaw the mitochondria on ice and lyse them by incubating in the presence of 0.05 % Triton X-100 and 300 mM KCl for 3 min.

3.4.3. DNA Glycosylase Assays

As described above, DNA glycosylases catalyze the first step of the BER pathway, the recognition and cleavage of the damaged base. DNA glycosylases recognize specific base modifications, and they have defined substrate specificity. These activity assays rely on the capacity of the DNA glycosylases to cleave a single lesion-containing oligonucleotide at the site of the lesion (Fig. 2). The activity is calculated as the amount of cleaved substrate relative to the total substrate used in the assay. Among DNA glycosylases, UNG, NTH1, OGG1, and NEIL activities are frequently investigated. Take into account that NEIL and OGG1 activities are usually lower than UNG and NTH1 activities; therefore, higher amounts of mitochondrial lysate and/or longer incubation times are generally required.

1. A 20 μl reaction contains: 20 mM HEPES, 5 mM EDTA, 75 mM KCl, 1 mM $MgCl_2$, 5 % glycerol, 5 mM DTT, 0.1 mg/ml BSA, 90 fmol of the corresponding 5′-end-labeled duplex

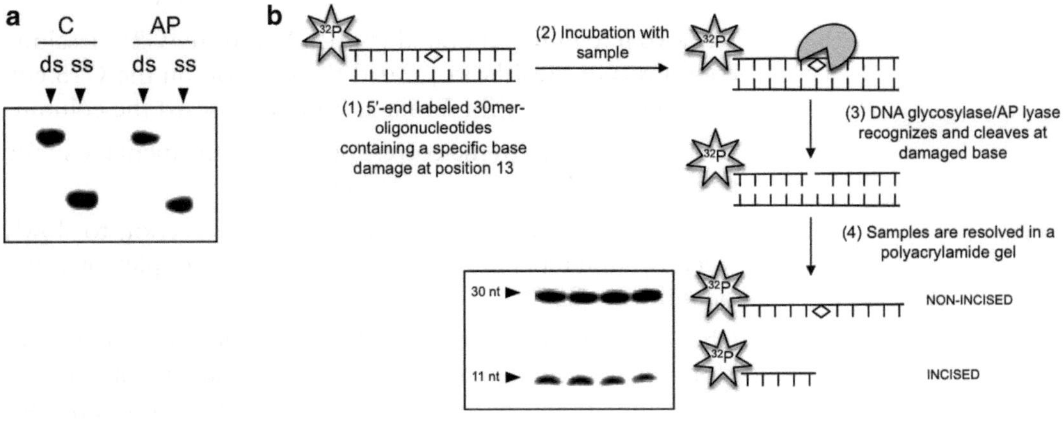

Fig. 2. (**a**) Representative gel shows annealing efficiency after running the 5′-end-labeled single-stranded oligonucleotide (ss) and the corresponding 5′-end-labeled double-stranded oligonucleotide (ds) under native conditions; a control oligonucleotide (C) and a THF-containing oligonucleotide (AP) were loaded. (**b**) Scheme showing the principle behind the incision assay. Representative gel shows incision activity in liver mitochondria after incubating 250 ng of mitochondrial extract with THF-containing oligonucleotide under the conditions described in the text.

oligonucleotide, and the adequate amount of mitochondrial lysate, 1–10 μg (UNG and NTH1) or 10–100 μg (OGG1 and NEIL) (see Notes 3 and 4).

2. Incubate at 37 °C for 1 h (UNG and NTH1) or 3 h (OGG1 and NEIL).
3. Stop the reactions with 4 μg of proteinase K and 0.4 % SDS followed by incubation at 55 °C for 30 min.
4. Add 20 μl of formamide loading dye and incubate for 5 min at 90 °C (see Note 5).
5. Load the samples on a 20 % denaturing polyacrylamide gel and run at 15 W for 1 h.

3.4.4. AP Endonuclease Assay

After recognition and cleavage of the damaged base, the resulting AP site is processed by an AP-endonuclease (APE), which creates a strand break with a 3′-hydroxyl end and a 5′-(dRP) residue. In order to investigate the activity of the AP endonuclease, an oligonucleotide containing tetrahydrofuran (THF) is often used (Table 1). Instead of THF, which is an abasic site analog, oligonucleotides containing a true abasic site can also be used. However, although true abasic sites can be generated by incubation of recombinant monofunctional DNA glycosylases with specific oligonucleotides, we prefer to use THF containing oligonucleotides in APE activity assays, as it is more stable. Similarly to the DNA glyosylase assays, this activity assay relies on the capacity of the AP endonuclease to generate a strand break at the abasic site. The AP endonuclease activity is calculated as the amount of cleaved substrate relative to the total used in the assay

1. A 20 μl reaction contains: 20 mM HEPES, 5 mM EDTA, 75 mM KCl, 1 mM $MgCl_2$, 5 % glycerol, 5 mM DTT, 0.1 mg/ml BSA, 90 fmol of the corresponding 5′-end-labeled duplex oligonucleotide, and the adequate amount of mitochondrial lysate (0.1–1 μg of protein).
2. Incubate at 37 °C for 30 min.
3. Stop the reactions with 4 μg of proteinase K and 0.4 % SDS followed by incubation at 55 °C for 30 min.
4. Add 20 μl of formamide loading dye and incubate for 5 min at 90 °C.
5. Load the samples in a 20 % denaturing polyacrylamide gel and run at 15 W for 1 h.

3.4.5. DNA Polymerase γ Activity

In mitochondria, the next step in the BER pathway after AP processing is catalyzed by the DNA polymerase γ (pol γ). In contrast to DNA glycosylases and APE activity assays, where a 5′-end-labeled duplex oligonucleotide is used, pol γ activity is quantified by using a non-labeled duplex oligonucleotide containing a single-nucleotide gap. The assay measures the incorporation of radioactive nucleotides at the gap during repair synthesis.

1. A 10 μl reaction contains: 50 mM Tris–HCl (pH 7.4), 50 mM KCl, 1 mM DTT, 5 mM $MgCl_2$, 5 % glycerol, 5 μM dCTP, and 1 pmol of non-labeled duplex gap oligonucleotide and 4 μCi of α^{32}P-dCTP. The amount of mitochondrial extract used in these reactions can be as low as 1 μg. However, similarly to all the assays described above, the amount of mitochondrial protein used in the assay should be optimized.
2. Incubate for 1 h at 37 °C.
3. Terminate reactions by adding formamide dye and heating at 90 °C for 10 min.
4. Load the samples on a 20 % denaturing polyacrylamide gel and run at 15 W for 1 h.

3.4.6. Repair Synthesis and Ligation Assay

While all the previous assays described above are specific for individual steps of the BER pathway, the repair synthesis incorporation assay measures the activity of the BER as a whole. Similarly to the DNA polymerase activity assay, a non-labeled duplex oligonucleotide is used. In the repair synthesis and ligation assay a lesion-containing oligonucleotide is used. We normally use a dUracil-containing oligonucleotide (see Note 6). The efficiency of the whole pathway is measured as the final product resulting from the incorporation of radioactive nucleotides and the ligation of the DNA strand (Fig. 3).

1. A 50 μl reaction contains: 40 mM HEPES, pH 7.8, 0.1 mM EDTA, 5 mM $MgCl_2$, 0.2 mg/ml BSA, 50 mM KCl, 1 mM DTT, 40 mM phosphocreatine, 100 μg/ml phosphocreatine kinase, 2 mM ATP, 20 μM dNTPs, 4 μCi ^{32}P-dCTP, 3 % glycerol, and 100 ng non-labeled duplex dU-containing oligonucleotide. The amount of lysate extract is usually in the range between 10 and 50 μg per reaction.
2. Incubate the reaction mixtures at 37 °C for 3 h.
3. Terminate the reaction by adding 12.5 μg of proteinase K and 2.5 μl of 10 % SDS and incubate at 55 °C for 30 min.
4. Precipitate DNA by addition of 1 μg glycogen, 4 μl of 11 M ammonium acetate, 60 μl of 96 % ethanol, and overnight incubation at −20 °C.
5. Pellet DNA by centrifugation (1 h at 4 °C), dry and resuspend in 20 μl of formamide dye, load on a denaturing 20 % polyacrylamide gel, and run at 15 W for 1 h.

3.5. Electrophoresis

3.5.1. Native Conditions

Before using the labeled duplex oligonucleotides, it is important to check that all the substrate has been completely annealed. In order to check the annealing efficiency native gels of the single-stranded

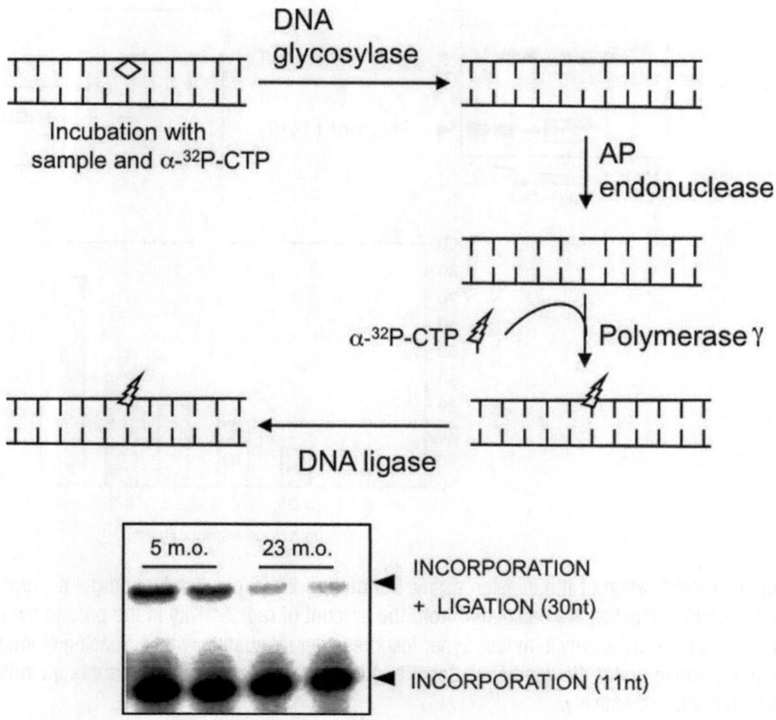

Fig. 3. Scheme showing the principle behind the repair synthesis incorporation assay. Representative gel shows total BER activity (repair synthesis and ligation assay) in synaptosomal fractions (50 μg) from 5-month-old to 23-month-old mice under the conditions described in the text.

5′-end labeled oligonucleotides and the corresponding double-stranded 5′-end labeled oligonucleotides should be resolved in native conditions. It is important to confirm that the substrates are annealed to completion because various DNA glycosylases do not recognize single-stranded substrates.

The single- and double-stranded 5′-labeled oligonucleotides are mixed with native loading dye and loaded on a 20 % polyacrylamide (acrylamide/bis-acrylamide 19:1) gel in 1× TBE and electrophoresed in 1× TBE buffer at 20 mA for approximately 3 h.

3.5.2. Denaturing Conditions

All the activity assays are quantified after resolving the oligonucleotide substrates under denaturing conditions. Samples are loaded on 20 % polyacrylamide (acrylamide/bis-acrylamide 19:1) gels containing 7.5 M urea. Samples are loaded immediately after incubation at 90 °C for 10 min and resolved in 1× TBE buffer at 15 W for approximately 1 h (see Note 7).

3.6. Result Analysis

After electrophoresis, gels are exposed to a storage phosphor screen and after exposure overnight at 4 °C the signals are acquired using a PhosphorImager or equivalent scanner (see Note 8). The radioactivity in the lanes is quantified using Quantity one software (Bio-Rad) or equivalent software.

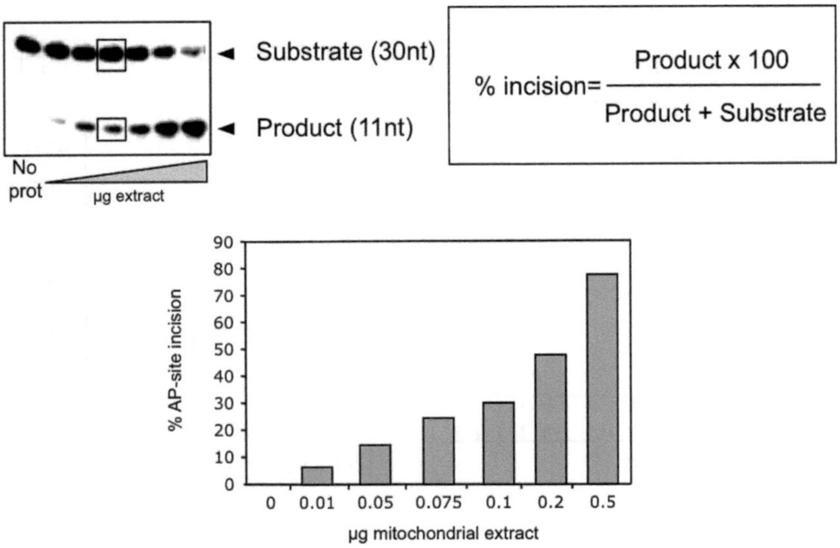

Fig. 4. Quantification of the incision assay. Denaturing 20 % polyacrylamide gel separates oligonucleotides according to size. Incision activities are calculated from the amount of radioactivity in the products relative to the total. Representative gel shows incision activity in mouse synaptosomes after incubation with increasing amounts of extract with THF-containing oligonucleotide under the conditions described in the text. *Boxes* mark the areas quantitated for the amount of substrate and product, respectively.

DNA glycosylase and AP endonuclease activities are quantified as the amount of radioactivity in the product lane relative to the total radioactivity in the lane (Fig. 4). When measuring gap filling and repair synthesis incorporation activities are calculated as the amount of radioactivity due to ^{32}P-α-dCTP incorporated in each lane.

4. Notes

1. Always add protease inhibitors to isolation buffers immediately prior to use.
2. Although some DNA glycosylases, such as UNG, can recognize DNA lesions when present on single-stranded DNA, various DNA repair enzymes do not recognize single-stranded substrates. Choose single- or double-stranded DNA according to the aim of the investigation.
3. Previous to final BER activity assays, optimal conditions of incubation time and extract concentration have to be determined. Titrations with increasing amounts of mitochondrial extracts must be performed to determine the extract concentration in which the activity is linear to protein concentration. Similar preliminary curves should be made for increasing incubation times.

4. For every assay it is necessary to include at least two controls in order to estimate nonspecific activity: (1) Lesion-containing oligonucleotide incubated with no extract and (2) non-damaged oligonucleotide (control oligonucleotide) incubated with the same amount of extract as used for the damaged-oligonucleotide.
5. Bifunctional glycosylases may produce a strand break themselves, whereas monofunctional glycosylases produce an abasic site, which is converted to a strand break when heated in the loading buffer.
6. We have used both a regular double-stranded dU-containing oligonucleotide and a 30 base-pair hairpin loop containing deoxyuracil at position 15.
7. For better resolution we use 1.5 mm gels of at least 16 cm length. Before loading the samples pre-warm the gel 30 min at 15 W. It is recommended to rinse the wells with 1× TBE to eliminate accumulated urea during the pre-warming in order to improve resolution.
8. If phosphor screens and phosphorimager scanner are not available, signals can also be acquired using X-ray films.

Acknowledgements

The authors gratefully acknowledge the financial support from the Velux Foundation (Danish Aging Research Center) and The Danish Research Councils to TS, and from CAM/UCM (CCG10-UCM/SAL-4798) to RG. We also thank the members of the Laboratory for DNA repair and Aging at Dep. of Molecular Biology and Genetics, Aarhus University and the members of the Laboratory of Molecular Gerontology at National Institutes of Health, NIH for longstanding collaborations.

References

1. Grollman AP, Moriya M (1993) Mutagenesis by 8-oxoguanine: an enemy within. Trends Genet 9:246–249
2. Kavli B, Otterlei M, Slupphaug G, Krokan HE (2007) Uracil in DNA—general mutagen, but normal intermediate in acquired immunity. DNA Repair (Amst) 6:505–516
3. Croteau DL, Bohr VA (1997) Repair of oxidative damage to nuclear and mitochondrial DNA in mammalian cells. J Biol Chem 272:25409–25412
4. Hamilton ML, Guo Z, Fuller CD, Van Remmen H, Ward WF, Austad SN, Troyer DA, Thompson I, Richardson A (2001) A reliable assessment of 8-oxo-2-deoxyguanosine levels in nuclear and mitochondrial DNA using the sodium iodide method to isolate DNA. Nucleic Acids Res 29:2117–2126
5. Barja G, Herrero A (2000) Oxidative damage to mitochondrial DNA is inversely related to maximum life span in the heart and brain of mammals. FASEB J 14:312–318
6. Schon EA (2000) Mitochondrial genetics and disease. Trends Biochem Sci 25:555–560
7. Beal MF (2005) Mitochondria take center stage in aging and neurodegeneration. Ann Neurol 58:495–505

8. Melov S (2004) Modeling mitochondrial function in aging neurons. Trends Neurosci 27:601–606
9. Krishnan KJ, Reeve AK, Samuels DC, Chinnery PF, Blackwood JK, Taylor RW, Wanrooij S, Spelbrink JN, Lightowlers RN, Turnbull DM (2008) What causes mitochondrial DNA deletions in human cells? Nat Genet 40:275–279
10. Weissman L, de Souza-Pinto NC, Stevnsner T, Bohr VA (2007) DNA repair, mitochondria, and neurodegeneration. Neuroscience 145: 1318–1329
11. Gredilla R, Bohr VA, Stevnsner T (2010) Mitochondrial DNA repair and association with aging—an update. Exp Gerontol 45: 478–488
12. Seeberg E, Eide L, Bjoras M (1995) The base excision repair pathway. Trends Biochem Sci 20:391–397
13. Bohr VA (2002) DNA damage and its processing. Relation to human disease. J Inherit Metab Dis 25:215–222
14. Bohr VA (2002) Repair of oxidative DNA damage in nuclear and mitochondrial DNA, and some changes with aging in mammalian cells. Free Radic Biol Med 32:804–812
15. Stuart JA, Karahalil B, Hogue BA, Souza-Pinto NC, Bohr VA (2004) Mitochondrial and nuclear DNA base excision repair are affected differently by caloric restriction. FASEB J 18:595–597
16. Karahalil B, Hogue BA, de Souza-Pinto NC, Bohr VA (2002) Base excision repair capacity in mitochondria and nuclei: tissue-specific variations. FASEB J 16:1895–1902
17. Lai JC, Clark JB (1979) Preparation of synaptic and nonsynaptic mitochondria from mammalian brain. Methods Enzymol 55:51–60
18. Kristian T, Hopkins IB, McKenna MC, Fiskum G (2006) Isolation of mitochondria with high respiratory control from primary cultures of neurons and astrocytes using nitrogen cavitation. J Neurosci Methods 152:136–143
19. Brown MR, Sullivan PG, Geddes JW (2006) Synaptic mitochondria are more susceptible to Ca^{2+} overload than nonsynaptic mitochondria. J Biol Chem 281:11658–11668
20. Schrimpf SP, Meskenaite V, Brunner E, Rutishauser D, Walther P, Eng J, Aebersold R, Sonderegger P (2005) Proteomic analysis of synaptosomes using isotope-coded affinity tags and mass spectrometry. Proteomics 5: 2531–2541
21. Gredilla R, Weissman L, Yang JL, Bohr VA, Stevnsner T (2010) Mitochondrial base excision repair in mouse synaptosomes during normal aging and in a model of Alzheimer's disease. Neurobiol Aging 33(4):694–707

Part IV

DNA Strand Breakage and Repair

Chapter 21

O-Hydroxylamine-Coupled Alkaline Gel Electrophoresis Assay for the Detection and Measurement of DNA Single-Strand Breaks

April M. Luke and Jun Nakamura

Abstract

The ability to detect and measure DNA single-strand breaks has been the aim of numerous assays developed to assess genotoxicity. These methods often rely on alkaline conditions to denature DNA. However, alkaline treatment of DNA also introduces artifactual SSBs through the cleavage of alkali-labile sites resulting in confounded data. Here, we describe a modified alkaline gel electrophoresis assay coupled with a neutral *O*-hydroxylamine to obtain the measurement of true SSB formation.

Key words: Agarose, Alkaline, Electrophoresis, *O*-hydroxylamine, Single-strand breaks

1. Introduction

The quantitative measurement of DNA lesions, such as single-strand breaks (SSBs), has played a major role in assessing genotoxicity. Principal methods used over the years to detect SSBs include alkaline sucrose gradient sedimentation, alkaline filter elution, and gel electrophoresis, specifically the single cell gel electrophoresis (comet) assay (1). Alkaline conditions are utilized in these assays to encourage the denaturation of DNA to allow for better detection of SSBs. However, the alkaline conditions can cleave alkali-labile sites (ALSs), such as apurinic/apyrimidinic (AP) sites, present within the DNA resulting in artifactual SSB formation. Such overestimation of SSB formation by these alkaline-based assays may compromise the reliability of data (2). Here, we demonstrate an alkaline gel electrophoresis protocol (3) with modifications to detect and measure true SSB formation. The aldehydic group of AP sites has been demonstrated to react with *O*-hydroxylamines to form a stable complex that is resistant to cleavage through β-elimination by

Lotte Bjergbæk (ed.), *DNA Repair Protocols*, Methods in Molecular Biology, vol. 920,
DOI 10.1007/978-1-61779-998-3_21, © Springer Science+Business Media New York 2012

enzymatic activity (i.e., AP or dRp lyases) or by high pH (4–6). Coupling an *O*-hydroxylamine to the alkaline gel electrophoresis assay incorporates the prevention of artifactual SSB formation while maintaining the alkaline conditions needed for efficient DNA denaturation. Prior to denaturation, DNA samples are reacted with *O*-(tetrahydro-2*H*-pyran-2-yl)hydroxylamine (OTX), a neutral *O*-hydroxylamine, to stabilize AP sites against alkaline cleavage. The OTX-treated DNA is completely denatured under strong alkaline conditions, loaded into an agarose gel, and subjected to alkaline electrophoresis. Following neutralization and staining, an image is obtained and analyzed to determine the percent fragmentation within each sample. Thus, the OTX-coupled alkaline gel electrophoresis (OTX-AGE) assay provides a specific means for quantifying true SSB formation.

2. Materials

Prepare all solutions at room temperature using deionized water and analytical grade reagents. Store solutions at room temperature unless otherwise indicated. Follow safety regulations from manufacturer when handling reagents and disposing of waste.

2.1. DNA Sample

1. *O*-hydroxylamine stock solution: 200 mM OTX. Add about 4 mL water to a graduated cylinder. Weigh 117.15 mg OTX. Transfer the OTX to the cylinder and mix. Adjust the volume to 5 mL with water, mix again, and aliquot into 1 mL units (see Note 1). Store at −20 °C between uses.
2. HEPES buffer: 50 mM HEPES, pH 7.0. Add about 40 mL water to a glass beaker. Weigh 1.1915 g of HEPES and transfer to the beaker. Mix and adjust the pH to 7.0. Add water to bring the volume to 50 mL.
3. NaOH stock solution: 500 mM NaOH. Add about 40 mL of water to a glass beaker. Weigh 1.0 g of NaOH. Transfer the NaOH to the beaker and mix. Adjust the volume to 50 mL with water.

2.2. Agarose Gel

1. Gel buffer: 50 mM NaCl, 1 mM EDTA, pH 7.0. Add about 490 mL water to a glass beaker. Weigh 1.461 g of NaCl and 0.186 g of EDTA and transfer to the beaker. Mix and adjust the pH to 7.0. Adjust volume to 500 mL with water.

2.3. Electrophoresis

1. Running buffer: 30 mM NaOH, 1 mM EDTA, pH 12.4. Add about 990 mL water to a glass beaker. Weigh 1.2 g of NaOH and 372.24 mg of EDTA. Transfer the NaOH and EDTA to the beaker and mix. Adjust the volume to 1 L with water.

Prepare the buffer the day of electrophoresis and store at 4 °C (see Note 2). Depending on the electrophoresis chamber, several liters of running buffer may need to be prepared (see Note 3).

2. Loading buffer: 10 mM NaOH, 95 % Formamide (v/v), 0.05 % bromophenol blue (w/v), 0.05 % xylene cyanol (w/v). Add 95 mL of formamide to a glass beaker. Weigh 40 mg of NaOH, 50 mg of bromophenol blue, and 50 mg of xylene cyanol and transfer to the beaker. Mix well and adjust the volume to 100 mL with water if needed.
3. Neutralization buffer: 400 mM Tris–HCl, pH 7.5. Add about 800 mL of water to a glass beaker. Weigh 48.46 g of Tris and transfer to the beaker. Mix and adjust the pH to 7.5. Add water to bring the volume to 1 L.

2.4. Imaging

1. TE buffer: 10 mM Tris–HCl, 1 mM EDTA, pH 7.5–8.0.
2. SYBR Gold staining solution: 1× concentration. Add 50 μL of SYBR Gold (10,000×) to 500 mL of TE buffer and mix. Cover container with aluminum foil and store at 4 °C (see Note 4).

3. Methods

3.1. 0.7 % Agarose Gel Preparation

1. Mix 934 mg low melting point agarose, 6.66 mL gel buffer, and 126.6 mL dd-H_2O in an Erlenmeyer flask. Heat the mixture in the microwave being sure to stop and gently swirl the beaker every 15–30 s (see Note 5). Heat the mixture until the agarose is melted.
2. Place a thermometer into the agarose mixture and place the beaker into a container filled with room temperature water. Allow the mixture to cool to approximately 60 °C.
3. Slowly pour the mixture into the gel form to avoid introducing air bubbles into the agarose (see Note 6). Insert a 12-well gel comb into the liquid agarose without introducing air bubbles.
4. Loosely cover the top of the gel form and comb with aluminum foil while the agarose solidifies (see Note 7).

3.2. DNA Sample Preparation

1. Slowly draw the DNA sample up and down a pipette tip using a pipettor (see Note 8). Remove 3–5 μL of DNA and place within a 50 μL microcentrifuge tube (see Note 9). Add 1.2 μL of the 200 mM OTX stock solution to the tube and adjust the volume to 8 μL with 50 mM HEPES buffer (see Note 10). Gently mix the sample using a pipettor as described above and then briefly centrifuge the tube to concentrate the sample at the bottom.

2. Place the tube into a heat block set at 37 °C for 1 h. Cover the heat block with aluminum foil to protect the sample from light.
3. After 1 h, remove the tube from the heat block and briefly centrifuge. To the tube, add an additional 1.2 μL of the 200 mM OTX stock solution and then add 3.2 μL of the 500 mM NaOH stock solution. Adjust the volume to 16 μL with 50 mM HEPES buffer (see Note 11). Gently mix the sample using a pipettor and then briefly centrifuge the tube as described in step 1.
4. Place the tube back into the heat block set at 37 °C for 20 min. Replace the aluminum foil over the block to protect the sample from light.
5. After 20 min, remove the tube from the heat block and briefly centrifuge. Add 8 μL of the loading buffer to the tube, bringing the final volume to 24 μL. Gently mix the sample using a pipettor and then briefly centrifuge the tube as described in step 1.

3.3. Alkaline Agarose Gel Electrophoresis

1. Set up the electrophoresis chamber and pump in a 4 °C cold room equipped with a power source. Fill the electrophoresis chamber with chilled running buffer. Remove the gel comb from the formed gel and submerge the gel into the running buffer. Place the gel into the chamber with the wells towards the negative electrode. Allow the gel to soak in the running buffer for 30 min prior to loading the samples (see Note 12).
2. Using a gel loading pipette tip, slowly deliver the prepared sample into a well allowing it to settle to the bottom (see Note 13). Designate a well for the marker DNA (see Note 14). After loading all the samples, turn on the pump to circulate the buffer. Electrophorese at a constant 30 V for 16 h (see Note 15).
3. Following electrophoresis, gently remove the gel from the form and submerge it into neutralizing buffer at room temperature for 20 min (see Note 16). Briefly rinse the neutralized gel twice with deionized water.

3.4. Gel Imaging and Analysis

1. Place the gel into a plastic container with enough 1× SYBR Gold stain solution to completely immerse the gel. Cover the container with aluminum foil to protect the stain from light and place it on a platform rocker to gently agitate the solution. Incubate the gel in the stain solution for 40 min at room temperature.
2. After 40 min, briefly rinse the gel with deionized water before taking an image of the gel. Allow the gel to destain by replacing it into a plastic container filled with deionized water. Cover the container with aluminum foil and agitate the water by placing the container on the platform rocker. After the initial image, take additional images of the gel every hour for up to 6 h (see Note 17).

Fig. 1. SSB detection in DT40 cells exposed continuously to 1 mM MMS for 2 h. Visualization of SSB formation in the cell lines through OTX-coupled AGE analysis. Images obtained using SYBR Gold and the Kodak Image Station 440CF system. *Lane 1*—Marker; *Lanes 2* and *3*—PBS control; *Lanes 4* and *5*—1 mM MMS exposure for 2 h; *Lanes 2* and *4*—OTX-coupled AGE analysis; *Lanes 3* and *5*—AGE analysis without OTX.

3. Use gel analysis software to determine the extent of DNA migration (see Note 18). Generate a profile of intensity values for each lane (see Note 19). DNA in the region ≤23.1 kb is considered fragmented (1) (Fig. 1). Determine the integrated intensity value for the complete lane and the region ≤23.1 kb. Calculate the percent fragmented compared to the total lane integrated intensity value. Percent fragmentation quantitatively represents the extent of SSB formation within each sample.

4. Notes

1. OTX is neutral and does not require titration to a neutral pH. Ensure OTX is thoroughly dissolved before aliquoting into 1 mL units.
2. We find that it is best to prepare this buffer fresh each time. The freshly prepared buffer is stored at 4 °C in preparation for electrophoresis, which is also performed at that temperature.
3. Prepare enough buffer to allow the gel to be submerged with no less than 1 cm of distance between the top of the gel and the buffer level.
4. SYBR Gold is light sensitive. Prepare and handle the stock solution under low light. SYBR Gold diluted in TE buffer may be stored for several weeks if protected from light and kept at 4 °C.
5. Allow the mixture to heat without boiling. Gently swirl the beaker to encourage the agarose to melt and the mixture to become uniform.

6. It is important to remove any air bubbles that may have formed during pouring before the agarose solidifies. We find that it is easiest to remove them using a glass pipette and rubber bulb to pull out the trapped air.
7. Covering the gel form prevents debris in the air from settling into the forming agarose.
8. Gently mixing the DNA in this manner encourages the sample to be uniformly viscous.
9. Use the same amount of DNA with each sample that will be loaded into the agarose gel. Place samples on ice when they are not being handled.
10. Bringing the volume to 8 μL will make the final OTX concentration become 30 mM, the optimal concentration we determined for this method.
11. Bringing the volume to 16 μL will make the final NaOH concentration become 100 mM, the optimal concentration we determined for this method. Adding the additional 1.2 μL of OTX will allow the final OTX concentration within the sample to remain at 30 mM.
12. For time efficiency, we find it best to allow the gel to soak in the running buffer while the DNA samples are being prepared. Be sure to remove any air bubbles that may be trapped within the sample slots of the submerged gel.
13. While loading the gel, be sure to introduce the sample into only one well. Do not allow the sample to overflow into a neighboring well. We find it best to lower the gel loading pipette tip into the well and gently release the sample. The loading buffer within the sample will help it settle to the bottom of the well. Be sure not to introduce any air bubbles with the pipette that will cause the sample to rise out of the well.
14. We use an HINDIII digest of lambda DNA as the marker and prepare it in the same manner as the other DNA samples.
15. The dye front in the loading buffer will separate during the 16 h of electrophoresis. The bromophenol blue will reach the bottom of the gel and the xylene cyanol will migrate near the 4,000 base pairs DNA marker.
16. We place the container on platform rocker to allow the neutralization buffer to gently wash over the gel during incubation.
17. We find that the best images are obtained after 2–3 h of destaining. At that point in the destaining process, enough stain has been removed from the agarose resulting in less background signal while leaving a brighter signal from the DNA within the lanes. Images may still be obtained after 6 h of destaining; however, the signal from the DNA will weaken.

18. We use the AlphaEaseFC gel documentation software from Alpha Innotech, Inc.
19. The profiles are generated with an automatic correction for background intensity. We also subtract the integrated intensity values of a blank lane as background from the intensity values of each DNA lane.

Acknowledgements

This work was supported by National Institute of Environmental Health Science grants T32-ES07126, P42-ES05948, and P30-ES10126 and National Cancer Institute grant T32-CA72319.

References

1. Georgiou C, Papapostolou I, Grizntzalis K (2009) Protocol for the quantitative assessment of DNA concentration and damage (fragmentation and nicks). Nat Protoc 4:125–131
2. Anderson D, Plewa MJ (1998) The international comet assay workshop. Mutagenesis 13:67–73
3. Drouin R, Gao S, Homquist GP (1996) Agarose gel electrophoresis for DNA damage analysis. In: Pfeifer GP (ed) Technologies for detection of DNA damage and mutations. Plenum Press, New York, pp 37–43
4. Nakamura J et al (1998) Highly sensitive apurinic/apyrimidinic site assay can detect spontaneous and chemically induced depurination under physiological conditions. Cancer Res 58:222–225
5. Nakamura J, Swenberg JA (1999) Endogenous apurinic/apyrimidinic sites in genomic DNA of mammalian tissues. Cancer Res 59:2522–2526
6. Horton JK et al (2000) Protection against methylation-induced cytotoxicity by DNA polymerase beta-dependent long patch base excision repair. J Biol Chem 275:2211–2218

Chapter 22

DNA Double-Strand Break Damage and Repair Assessed by Pulsed-Field Gel Electrophoresis

Helen E. Bryant

Abstract

Pulsed-field gel electrophoresis (PFGE) is a technique for resolving large DNA molecules (up to 10 Mb). Using multiple pairs of electrodes DNA is subject to an alternating electric field through a solid agarose matrix. As the current changes direction the reorientation time of DNA is proportional to molecular weight; thus fragments are separated in the gel based on their size. Here we describe the use of PFGE to analyze DNA double-strand break formation and repair in human chromosomal DNA.

Key words: Pulsed-field gel electrophoresis, PFGE, DNA double-strand breaks, Genotoxic, Apoptosis, DNA damage, DNA repair, Method, Protocol

1. Introduction

In conventional gel electrophoresis a constant current is used to sieve small molecules of DNA through a slab of agarose. However as large pieces of DNA exist as randomly coiled molecules they need significant unraveling before they can pass into and weave their way through the gel. Such unraveling and weaving (also known as reptation) means that passage through the gel is no longer proportional to molecular weight. The practical resolution limit for conventional agarose gel electrophoresis is therefore about 50 kb. To overcome this limit Schwartz and Cantor developed pulsed-field gel electrophoresis (PFGE) (1). Cells are embedded in agarose plugs, then subject to proteinase K digestion leaving the DNA intact in the plug. After washing away the protein, plugs are set into an agarose gel prior to PFGE. During electrophoresis the reptation of the DNA molecules through a matrix is exploited; the agarose gel is subject to a homogenous electric field (2) that alternates in direction; as the direction alters the DNA molecules reorientate

Lotte Bjergbæk (ed.), *DNA Repair Protocols*, Methods in Molecular Biology, vol. 920,
DOI 10.1007/978-1-61779-998-3_22, © Springer Science+Business Media New York 2012

to the new direction, maximum mobility only occurring when molecules are fully aligned with the field. The time taken to reorientate is proportional to DNA molecular weight; however once reorientation has occurred, migration through the gel in a size-dependent manner is lost. The frequency of switching between fields is therefore set to optimize the time spent undergoing reorientation considering the resolution required; here an initial switch time of 60 s is used; this increases to 240 s over 24 h. The result is that DNA zig-zags across the gel and that net electrophoretic mobility is dependent on size. To achieve optimal separation an obtuse reorientation angle is required; 120° is usually considered adequate (3) and is used in this protocol. After electrophoresis DNA is visualized using ethidium bromide. Here we describe how double-strand breaks induced by endogenous or exogenous agents can be visualized (4) and in a variation on this how repair of such breaks over time can also be seen (5, 6).

2. Materials

2.1. Buffers

1. Phosphate-buffered saline: Dissolve 8 g of NaCl, 0.2 g KCl, 1.44 g Na_2HPO_4, and 0.24 g KH_2PO_4 in 800 ml distilled H_2O. Adjust pH to 7.4, adjust volume to 1 L with additional distilled H_2O, and sterilize by autoclaving. Store at room temperature.
2. 0.5 M EDTA: Dissolve 186.1 g $Na_2EDTA{\cdot}2H_2O$ in 800 ml distilled water. pH to 8.0 with NaOH (~20 g of NaOH pellets). EDTA will dissolve at pH 8.0. Adjust volume to 1 L with distilled water. Sterilize by autoclaving and store at room temperature.
3. Proteinase K solution: Dissolve 100 mg *N*-laurylsarcosyl and 10 mg Proteinase K into 10 ml 0.5 M EDTA. Scale up as required (1 ml/plug) and store at room temperature.
4. 10× TBE: Dissolve 108 g Tris base and 55 g Boric acid in 800 ml distilled H_2O, add 40 ml 0.5 M EDTA (pH 8.0), adjust volume to 1 L, sterilize by autoclaving, and store at room temperature.
5. 0.5× TBE: Dilute 250 ml 10× TBE into 4.75 L distilled water.
6. TE10: Dissolve 1.21 g Tris base in 900 ml distilled water, add 20 ml 0.5 M EDTA, adjust volume to 1 L, sterilize by autoclaving, and prechill in fridge before use.
7. TE100: Dissolve 1.21 g Tris base in 700 ml distilled water, add 200 ml 0.5 M EDTA, adjust volume to 1 L, sterilize by autoclaving, and prechill in fridge before use.
8. RNAse A solution: Dissolve 1 mg RNAse powder in 10 ml distilled water, aliquot into 1 ml lots, and store at −20 °C.

9. Staining solution: Add 25 μl 10 mg/ml stock ethidium bromide to 500 ml 0.5× TBE.
10. Marker: Commercially available CHEF DNA size marker of *Saccharomyces cerevisiae* chromosomes comes in agarose blocks that can be cut into plugs.
11. Preparation of agarose for plugs: Use InCert® Agarose, a low gelling (≤30°C) and melting (≤70°C) temperature, high-purity agarose. Before starting set a water bath at 50–60 °C. Weigh out 0.15 g agarose powder, add 10 ml room temperature PBS to a Falcon tube, add the agarose powder to the PBS while gently agitating, weigh the tube containing PBS and agarose and note weight, loosen lid on tube, heat in a microwave oven with the power setting on medium using several short (10 s) intervals with gentle swirling in between heating to resuspend the powder, continue till there are no visible particles (see Note 1). Reweigh tube, add warm distilled water to obtain the initial weight, and mix. Place tube in water bath until needed.
12. Preparation of electrophoresis gel: Use Chromosomal grade agarose, dissolve 2.8 g agarose in 350 ml 0.5 %TBE, heat on medium setting in microwave for 5 min or until agarose is completely dissolved. Allow to cool till comfortable to hold in hands, swirling periodically to prevent the gel from setting. Once at an appropriate temp, add a stir bar and transfer to heated stir block set at around 40 °C. For pouring gel see method below.

2.2. Equipment

In addition to standard lab equipment you need,

1. Water bath set at 50–60 °C.
2. Microwave oven.
3. Refrigerated centrifuge to spin 15 ml Falcon tubes at 200 × *g*.
4. The protocol below has been optimized for use with the BioRad CHEFIII system; other commercial systems are available or a system can be homemade (7). Items 5–7 below are accessories to the CHEFIII system.
5. Disposable PFGE plug molds (see Note 2).
6. Standard PFGE casting stand and comb.
7. Screened cap to fit a 50 ml Falcon tube.

3. Methods

3.1. Treatment of Cells

1. Seed cells at 1–4 × 10^6/100 mm dish (see Note 3).
2. Leave 4 h to attach.
3. Treat cells with double-strand break-inducing agent as required (see Notes 4 and 5).

3.2. Preparation of Agarose Plugs

1. Before starting make up 1.5 % InCert agarose and proteinase K buffer (see above), precool the centrifuge, and collect ice. Also label (see Note 6) and prepare the plug molds by ensuring that they are clean and that the bottom is well stuck.
2. Taking a maximum of four plates at a time, trypsinize cells gently off the plate (see Note 7).
3. Resuspend cells in 10 ml media and place on ice.
4. Count harvested cells.
5. Pellet 1×10^6 cells/plug by centrifuging the appropriate volume at 4 °C for 3 min at $200 \times g$ in a 15 ml Falcon tube.
6. Remove supernate from cells.
7. Gently resuspend each pellet in 50 μl PBS.
8. To one tube at a time, add 50 μl warm Incert agarose, gently resuspend, and then immediately pipette 80 μl into the plug mold (see Notes 8 and 9).
9. Once you have made all the plugs, leave at 4 °C for 10 min to set.
10. Meanwhile add 1 ml/plug of proteinase K buffer to each of the appropriate number of 50 ml Falcon tubes (see Note 10).
11. Using the plastic insert provided push plug out of mold into proteinase K buffer and incubate at 50 °C for 48 h.

3.3. Washing Plugs

1. Wash plugs 2× 30 min in cold TE10 (see Note 11).
2. Add 100 μl 1 mg/ml RNAseA to each tube and incubate at 37 °C for 1 h.
3. Wash 1× 30 min TE10.
4. Wash 1× 30 min TE100 (see Note 12).
5. Store in TE100 at 4 °C (see Note 13).

3.4. Electrophoresis

1. Rinse PFGE tank out with distilled water, then add 0.5× TBE buffer, turn on pump, and set chiller to 14 °C (Fig. 1) (see Note 14).
2. Prepare electrophoresis gel (see above).
3. While the agarose is cooling, place each plug on a tooth of the comb (Fig. 2) (see Notes 15–17) and allow to dry a little (see Note 18).
4. Position the comb (with plugs on) in the casting stand (Fig. 3), pour gel around to ¾ depth of the plugs, and allow gel to set (see Note 19). Remove comb and top up gel so that plugs are covered; the plugs should stay in place in the gel. Allow the rest of gel to set.

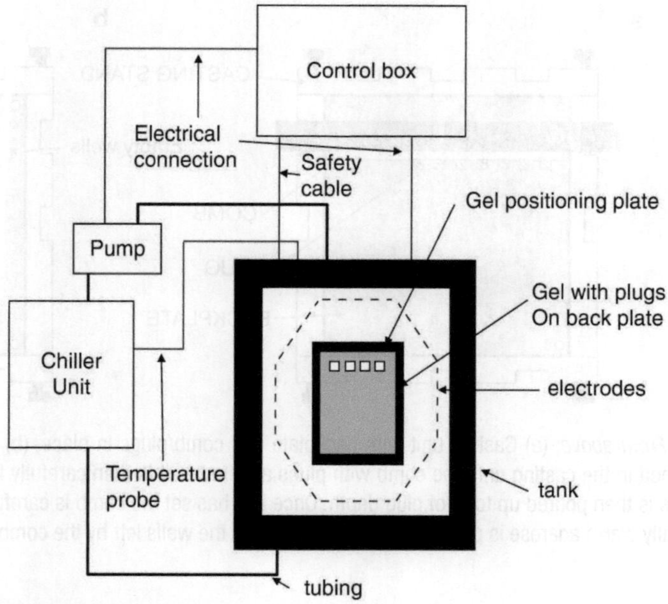

Fig. 1. BioRAD CHEFIII system setup.

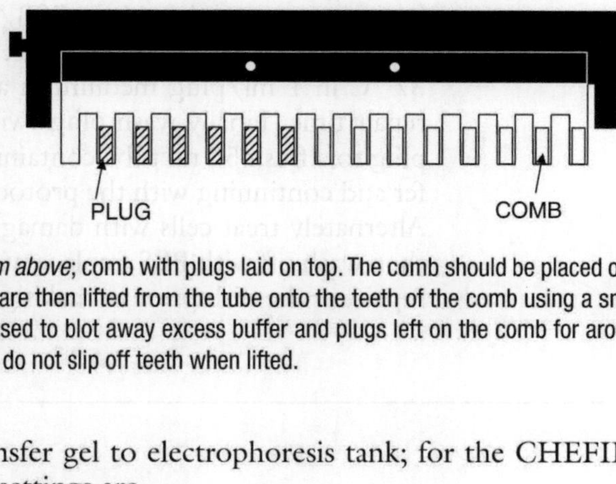

Fig. 2. *From above*; comb with plugs laid on top. The comb should be placed on the bench top; plugs are then lifted from the tube onto the teeth of the comb using a small spatula. Tissue is used to blot away excess buffer and plugs left on the comb for around 30 min, until plugs do not slip off teeth when lifted.

5. Transfer gel to electrophoresis tank; for the CHEFIII system the settings are

PBS protocol	Block 1
Initial switch time	60 s
Final switch time	240 s
Run time	24 h
Volts/cm	4 V
Included angle	120

6. After electrophoresis transfer the gel to staining solution overnight. Visualize on UV light box (see Note 20).

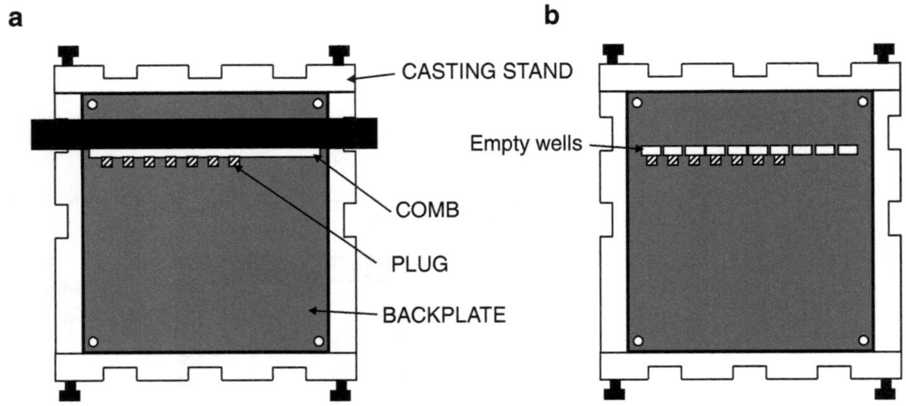

Fig. 3. *From above*; (**a**) Casting unit with backplate and comb/plugs in place, (**b**) with comb removed. The backplate is positioned in the casting unit; the comb with plugs adhered to it is then carefully lowered into the casting unit (**a**). Warm agarose is then poured up to ¾ of plug depth. Once this has set the comb is carefully removed leaving the plugs in place (**b**); finally warm agarose is poured over the plugs, filling the wells left by the comb and covering the plugs.

3.5. Repair Assay

As above BUT treat enough cells to make multiple plugs, then at step 7 in Subheading 3.2 resuspend pellet in warm medium and not PBS, and at step 8 in Subheading 3.2 resuspend in medium/incert agarose rather than PSB/incert agarose using the same ratios. Then after step 9 in Subheading 3.2 incubate each plug at 37 °C in 1 ml/plug medium in a 50 ml Falcon tube for desired repair time. Finally wash plugs with PBS before transferring each plug to a fresh 50 ml tube containing 1 ml/plug proteinase K buffer and continuing with the protocol at step 11 in Subheading 3.2. Alternately treat cells with damaging agent on the 100 mm dish, then wash off with PBS, replace with normal medium, and allow to repair on the dish prior to making plugs in the standard way.

4. Notes

1. When the agarose comes to a boil, handle solution very carefully. Allow to stand briefly at room temperature before releasing the air by gentle swirling.
2. Although PFGE molds are sold as disposable molds, you can wash them out and reuse them using a roll of thin electrical tape.
3. Seed cells so that after treatment at least 1×10^6 cells remain.
4. Remember to include a no treatment control.
5. The limit for detection is approximately the number of DSBs produced by 5 Gy ionizing radiation.
6. Use a non-permanent marker if you want to reuse the molds.
7. Do not bang the plates; this can induce DSBs.

8. Pipette gently without introducing any bubbles.
9. To fill the mold touch pipette to the inside edge of the mold and gently push the agarose in so that it slides down the edge.
10. Duplicate or triplicate plugs can go in the same tube with 2 ml or 3 ml of proteinase k buffer, respectively.
11. To wash remove the normal Falcon tube lid and replace with the screened cap. The screened cap screws onto a 50 ml Falcon tube; it contains holes that allow you to pour away liquid without losing the plug; you can then pour the next wash solution into the tube through the screened cap; finally remove the screened cap and replace with normal lid for the duration of the wash.
12. After washing the plugs should become clear.
13. Samples are stable for at least 1 month at 4 °C.
14. If the buffer is not sufficiently chilled the gel will not run properly and a series of smudged bands will be seen. A common error is to turn on the chiller without the pump; when you remember to turn on the pump buffer will not circulate because buffer has frozen in the chiller. Turn off chiller, wait for 30 min, and then try turning on pump again.
15. Include marker plug, and cut a section from the slab supplied.
16. A small spatula is useful to transfer plugs to the comb.
17. Blot away excess liquid using tissue.
18. To check if dry enough gently tilt the comb; the plugs should stay on the comb; if they begin to slide return comb to horizontal and wait longer. This takes about 20 min.
19. Return the remaining agarose to the stir block.
20. This can be improved by Southern blotting and hybridization to radiolabeled fragmented total genomic DNA probe.

References

1. Schwartz DC, Cantor CR (1984) Separation of yeast chromosome-sized DNAs by pulsed field gradient gel electrophoresis. Cell 37:67–75
2. Gardiner K, Laas W, Patterson D (1986) Fractionation of large mammalian DNA restriction fragments using vertical pulsed-field gradient gel electrophoresis. Somat Cell Mol Genet 12:185–195
3. Chu G, Vollrath D, Davis RW (1986) Separation of large DNA molecules by contour-clamped homogeneous electric fields. Science 234:1582–1585
4. Saleh-Gohari N, Bryant HE, Schultz N, Parker KM, Cassel TN, Helleday T (2005) Spontaneous homologous recombination is induced by collapsed replication forks that are caused by endogenous DNA single-strand breaks. Mol Cell Biol 25:7158–7169
5. Bryant HE, Ying S, Helleday T (2006) Homologous recombination is involved in repair of chromium-induced DNA damage in mammalian cells. Mutat Res 599:116–123
6. Ying S, Myers K, Bottomley S, Helleday T, Bryant HE (2009) BRCA2-dependent homologous recombination is required for repair of Arsenite-induced replication lesions in mammalian cells. Nucleic Acids Res 37:5105–5113
7. Herschleb J, Ananiev G, Schwartz DC (2007) Pulsed-field gel electrophoresis. Nat Protoc 2:677–684

Chapter 23

Quantitative Amplification of Single-Stranded DNA

Eva-Maria Holstein and David Lydall

Abstract

Single-stranded DNA (ssDNA) intermediates play an important role in processes such as DNA replication and homologous recombination, DNA damage responses, and DNA repair. Using quantitative amplification of ssDNA (QAOS), ssDNA arising during various cellular processes in complex genomes can be quantified at numerous single-copy and repetitive loci. QAOS is a useful tool to gain insights into the cellular processes that involve ssDNA and the roles of proteins in regulating ssDNA production and responses to ssDNA.

Key words: QAOS, Single-stranded DNA, Telomere, CST, Nuclease, PCR, Resection

1. Introduction

Single-stranded DNA (ssDNA) intermediates occur during various normal cellular processes such as DNA replication and recombination. In addition, ssDNA is crucial for stimulating checkpoint pathways to activate cell cycle arrest in response to DNA damage and ssDNA plays an important role in DNA repair pathways (1–3).

Telomeres, located at the end of linear eukaryotic chromosomes, terminate in a 3′ single-stranded overhang that resembles a DNA double-strand break. However, telomeres do not usually activate checkpoint or DNA repair pathways as the chromosome ends are masked by various telomere-capping proteins that bind to the telomeric ends. Examples for telomere-capping proteins are the Cdc13-Stn1-Ten1 or the Yku complex in budding yeast (4, 5). Loss of any of these proteins can lead to extensive ssDNA accumulation near telomere ends which in turn induces cell cycle arrest and DNA repair pathways (1, 5–8). Similarly, loss of the Pot1 telomere-capping protein in *Schizosaccharomyces pombe* leads to telomeric 5′ resection and DNA damage checkpoint activation (9).

Lotte Bjergbæk (ed.), *DNA Repair Protocols*, Methods in Molecular Biology, vol. 920,
DOI 10.1007/978-1-61779-998-3_23, © Springer Science+Business Media New York 2012

Combining telomere-capping mutations (e.g., temperature-sensitive allele *cdc13-1*, null-mutant *yku70Δ*) with mutations in DNA damage response genes allows identification of the roles of the corresponding checkpoint or DNA damage repair proteins in stimulating or inhibiting production of ssDNA at uncapped telomeres and in activating cell cycle arrest (1, 8, 10–16).

The Quantitative Amplification of Single-Stranded DNA (QAOS) method described here uses qPCR to measure ssDNA accumulation at various single-copy and repetitive loci near the telomeric ends in budding yeast telomere-capping mutants. QAOS is a fast and reproducible approach to identify and quantify ssDNA intermediates arising during cellular processes (8). QAOS should in principle work in other yeasts, bacteria, and mammalian cells.

Using high-quality DNA purified from yeast, QAOS is initiated by annealing of a tagging primer to ssDNA at low temperature. The tagging primer consists of a locus-specific sequence at the 3′ end and an artificial sequence (tag) at the 5′ end (Fig. 1a). Primer extension then creates a novel tagged sequence whose amount is proportional to the initial amount of ssDNA. The hybrid sequence is amplified by qPCR at temperatures that allow binding of specific primers (Reverse and Tag), but not the initial tagging primer. A fluorescent qPCR probe that hybridizes to the target DNA is used to help quantify the ssDNA at the measured locus. The quantity of ssDNA is calculated relative to the amount of ssDNA in a reference sample (ssDNA standards), using a standard curve.

Here we describe in detail the protocol and primers to detect ssDNA on both strands of chromosome V-R and VI-R in budding yeast.

2. Materials

1. Isopropanol.
2. 70 % (V/V) ethanol in water.
3. TE buffer: 10 mM Tris–HCl, pH 8.0, 1 mM EDTA.
4. Nuclei isolation buffer (NIB): 17 % (V/V) glycerol, 50 mM MOPS, 150 mM potassium acetate, 2 mM $MgCl_2$, 500 μM spermidine, 150 μM spermine. For 1 L: 170 mL glycerol, 10.463 g MOPS, 14.72 g potassium acetate, 2 mL 1 M $MgCl_2$, 0.55 mL 0.9 M spermidine (stored at −70°C), 52 mg spermine (added as dry powder); adjust to 1 L by water, dissolve all components, adjust pH to 7.2, filter-sterilize, and store in 50 mL aliquots in the dark at 4°C.
5. Sodium azide: 10 % (V/V) in water, store at RT.
6. RNAse A: Dissolve pancreatic RNAse at 10 mg/mL in 0.01 M sodium acetate, pH 5.2. Place in boiling water bath for 15 min

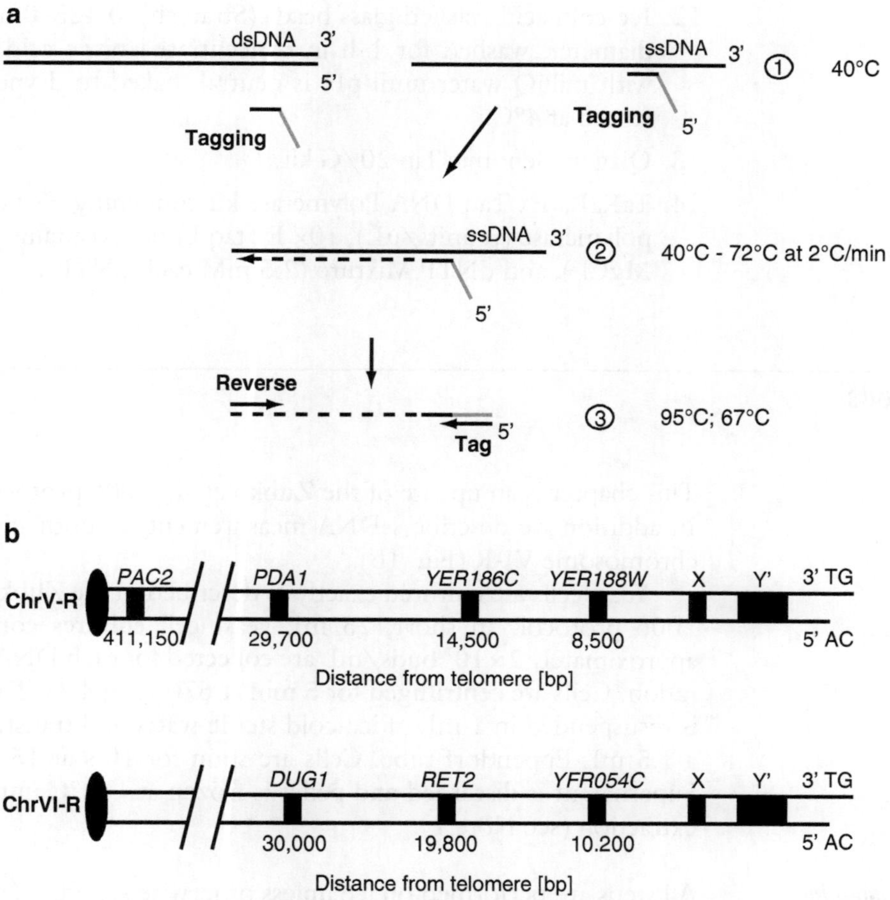

Fig. 1. Schematic diagram of QAOS. (**a**) (1) Tagging primer, containing an artificial tag, binds to ssDNA with a locus-specific sequence during the first step at 40°C. The primer does not bind to dsDNA. (2) Ex Taq Polymerase extends the tagging primer during a slow ramp from 40 to 72°C, creating a hybrid sequence containing the tag. Primer extension increases the hybrid product's length and therefore affinity for ssDNA allowing it to stay annealed to the ssDNA sequence as temperature increases. (3) The hybrid sequence is amplified using tag and reverse primer and stringent PCR temperatures. A fluorescent probe is used to quantify the amount of amplified DNA. (**b**) Location of loci used in QAOS on chromosome V-R and VI-R.

and then allow to cool to room temperature. Adjust pH by adding 0.1 volume of 1 M Tris–HCl, pH 7.4. Aliquot and store at −20°C. Do not freeze/thaw.

7. G2 with RNAse: 800 mM guanidine hydrochloride, 30 mM EDTA, 30 mM Tris–HCl, 5 % Tween 20, 0.5 % Triton X-100, pH 8.0, 200 μg/mL RNAse A (1 mL of 10 mg/mL RNAse per 50 mL tube of G2, add RNAse before use; buffer should then be stable for 6 months when stored at 4°C).
8. Proteinase K: Dissolve lyophilized powder in sterile water to a 20 mg/mL concentration and store in aliquots at −20 or −80°C.
9. QBT: 750 mM NaCl, 50 mM MOPS, 15 % isopropanol, 0.15 % Triton X-100, pH 7.0.
10. QC: 1 M NaCl, 50 mM MOPS, 15 % isopropanol, pH 7.0.
11. QF: 1.25 M NaCl, 50 mM Tris–HCl, 15 % isopropanol, pH 8.5.

12. Ice-cold acid-washed glass beads (Stratech). 0.425–0.6 mm in diameter, washed for 1 h in concentrated nitric acid, rinsed with milliQ water until pH is neutral, baked to dryness, and stored at 4°C.
13. Qiagen Genomic Tip 20/G kit.
14. TaKaRa Ex Taq DNA Polymerase kit containing Ex taq DNA polymerase (5 units/μL), 10× Ex taq buffer (contains 20 mM $MgCl_2$), and dNTP Mixture (2.5 mM each dNTP).

3. Methods

This chapter is an update of the Zubko et al., 2006 protocol (17). In addition we describe ssDNA measurement on both strands of chromosome VI-R (Fig. 1b).

Yeast cells are cultured exactly as described in the Zubko et al., 2006 protocol. In short, 25 mL yeast cell cultures containing approximately 2×10^7 buds/mL are collected for each DNA preparation. Cells are centrifuged for 5 min at $670 \times g$ at 4°C. The pellet is resuspended in 1 mL of ice-cold sterile water and transferred to a 1.5 mL Eppendorf tube. Cells are spun for 10 s at $15{,}000 \times g$. Supernatant is discarded and pellet is frozen at −80°C until DNA extraction (see Note 1).

3.1. DNA Isolation

All steps are performed on ice unless otherwise stated.

1. Thaw cell pellet (previously kept at −80°C) on ice and resuspend the pellet in 1.0 mL NIB.
2. Transfer the mixture to a 2 mL screw cap tube with skirt (e.g., Sarstedt) and spin for 8 s in a micro-centrifuge at 13,200 rpm ($15{,}000 \times g$).
3. Discard supernatant and resuspend in 600 μL of NIB by vortexing.
4. Add cold acid-washed glass beads to about 1.5 mL mark on the tube.
5. Lyse cells through repeated breakage cycles in a tissue homogenizer (Precellys24 lysis and homogenization ribolyser, Bertin technologies) for 5 s at 5,500 power setting per cycle. Remove from homogenizer and place in ice water in between cycles (at least 1 min). 90 % cell breakage is required and breakage efficiency is examined by phase-contrast microscopy (see Note 2).
6. For each lysate prepare a 1.5 mL Eppendorf tube. Remove caps and bottoms from tubes using a dog nail clipper or razor blade (Fig. 2).
7. Use a 21 gauge syringe needle to puncture the bottom of Sarstedt tubes containing the cell lysate and glass beads and

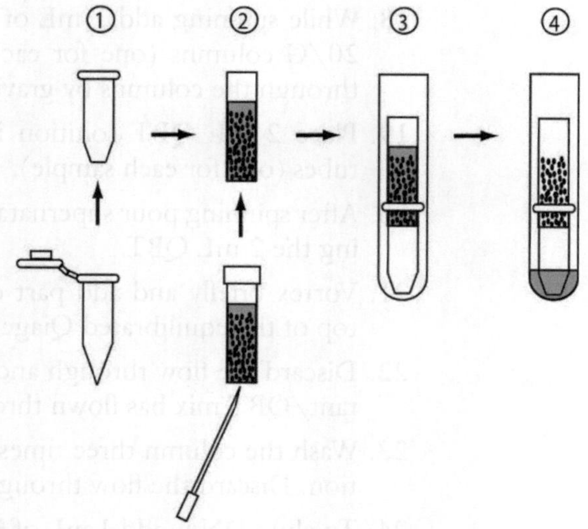

Fig. 2. Collection of lysate after cell breakage. (1) Caps and bottoms from 1.5 mL Eppendorf tube are removed. (2) Bottom of 2 mL screw cap Sarstedt tubes containing the lysates is punctured using a 21 gauge syringe needle. (3) The Sarstedt tube is fitted into the Eppendorf tube and both tubes are put into a 15 mL round-bottomed Falcon tube. (4) Lysate accumulates at the bottom of the 15 mL Falcon tube during centrifugation.

place the tubes securely on the top of the clipped Eppendorf tubes (Fig. 2).

8. Put pairs of tubes in a round-bottomed 15 mL Falcon tube (e.g., BD Biosciences) (Fig. 2).
9. Spin in precooled swing out centrifuge (CS 6R centrifuge, $670\times g$, 2 min) (Fig. 2).
10. Wash the beads twice by adding 1 mL NIB to the top of the 2 mL screw cap Sarstedt tube and spinning at $670\times g$ for 2 min.
11. Remove and discard the Sarstedt/Eppendorf pair of tubes using tweezers.
12. Vortex the 15 mL Falcon tube containing the cell debris for 2 s.
13. Remove lids from the Falcon tubes (they do not fit in the rotor; keep them in a clean place) and spin tubes in a fixed angle F0850 rotor (containing appropriate sleeves) in a Beckman Allegra 64R centrifuge for 20 min at $6,500\times g$ and 4°C.
14. Remove the supernatant by aspiration and resuspend nuclear/cell pellet in 2 mL G2 buffer containing RNAse by vortexing.
15. Place the lids back on the tubes. Incubate at 37°C for 30 min in a water bath and mix every 10 min by gently shaking the tubes.
16. Add 50 μL of 20 mg/mL proteinase K to each tube. Incubate for 1 h at 37°C and every 20 min by gently shaking the tubes.
17. Remove lids from the tubes and spin in F0850 rotor for 10 min at $6,500\times g$ at 4°C.

18. While spinning add 1 mL of QBT solution to top of Qiagen 20/G columns (one for each sample) and let solution flow through the columns by gravity.
19. Place 2 mL QBT solution in new Falcon round-bottomed tubes (one for each sample).
20. After spinning pour supernatant into the Falcon tubes containing the 2 mL QBT.
21. Vortex briefly and add part of the supernatant/QBT mix to top of the equilibrated Qiagen 20/G column.
22. Discard the flow through and repeat step 21 until all supernatant/QBT mix has flown through the column.
23. Wash the column three times by adding 3 × 1 mL of QC solution. Discard the flow through.
24. To elute DNA, add 1 mL of 50°C pre-warmed QF solution to column and collect flow through in a fresh 15 mL Falcon.
25. Add another 1 mL of 50°C pre-warmed QF solution to column.
26. Precipitate DNA at room temperature by adding 1.4 mL of isopropanol to the Falcon tube containing the DNA eluted in QF.
27. Vortex gently and spin in F0850 rotor for 20 min at 7,700 × *g* and 4°C. A small pellet is visible in most cases (see Note 3).
28. Carefully discard supernatant and remove as much liquid as possible by inverting the tube onto a paper towel for 1 min.
29. Wash the pellet by adding 1 mL of 70 % ethanol, and vortex gently for 1 s.
30. Spin in F0850 rotor for 20 min at 7,700 × *g* and 4°C.
31. Carefully discard supernatant and re-spin tubes for 1 min. Remove the rest of the liquid by aspiration.
32. Dissolve DNA by adding 400–600 μL TE to DNA pellet, tightly cap the tube, and incubate on New Brunswick TC7 roller at 23°C overnight. Then store samples at 4°C.
33. Dilute 10 μL of each DNA prep to 90 μL of TE, vortex, and leave overnight at 4°C to equilibrate. The diluted DNA will be used in Subheading 3.2 to measure yield of DNA by qPCR.

3.2. qPCR

3.2.1. Measuring Yield of DNA by qPCR

1. Measure all samples in triplicate in a 96-well PCR plate. Pipette 10 μL of each sample to the bottom of a 96-well PCR plate according to Table 1. Use TE for the no template control (NTC) and use concentrations of 0.4 ng/μL, 2 ng/μL, and 4 ng/μL of genomic DNA standards (STD) (Subheading 3.3.1) (see Note 4).
2. Make PCR master mix as described in Table 2. Use *PAC2* forward and reverse primers and the fluorescent probe shown in Table 3 for quantification of DNA samples.

Table 1
96-well plate set up for PCR to measure total DNA or ssDNA from four yeast strains (A, B, C, and D) sampled every 40 min. Each PCR is performed in triplicate. STD wells contain standards. NTC contain no template control

	1	2	3	4	5	6	7	8	9	10	11	12
A	A/0	A/0	A/0	A/40	A/40	A/40	A/80	A/80	A/80	A/120	A/120	A/120
B	A/160	A/160	A/160	A/200	A/200	A/200	A/240	A/240	A/240	B/0	B/0	B/0
C	B/40	B/40	B/40	B/80	B/80	B/80	B/120	B/120	B/120	B/160	B/160	B/160
D	B/200	B/200	B/200	B/240	B/240	B/240	C/0	C/0	C/0	C/40	C/40	C/40
E	C/80	C/80	C/80	C/120	C/120	C/120	C/160	C/160	C/160	C/200	C/200	C/200
F	C/240	C/240	C/240	D/0	D/0	D/0	D/40	D/40	D/40	D/80	D/80	D/80
G	D/120	D/120	D/120	D/160	D/160	D/160	D/200	D/200	D/200	D/240	D/240	D/240
H	STD1	STD1	STD1	STD2	STD2	STD2	STD3	STD3	STD3	NTC	NTC	NTC

Table 2
PCR master mix for *PAC2* DNA quantification and QAOS. The above mixture is sufficient to add 15 μL master mix to 96 individual 10 μL DNA samples (and contains 5 % extra, to cover pipetting errors). This recipe permits each 250 unit batch of Takara Ex Taq polymerase to be used for four individual 96-well plates

	Final concentration	Stock concentration	Volume (μL) [QPCR]	Volume (μL) [QAOS]
Water			985	960
Tagging primer	30 nM	3 μM (100×)	–	25
Tag	300 nM	30 μM (100×)	25	25
Reverse primer	300 nM	30 μM (100×)	25	25
Probe	200 nM	20 μM (100×)	25	25
Ex Taq buffer	0.97×	10×	245	245
dNTPs	194 nM	2.5 mM	195	195
Ex Taq polymerase	0.024 u/μL	5 u/μL	12	12
Total			1,512	1,512

Table 3
Oligonucleotides for measuring total DNA and ssDNA on the TG-strand near telomeres. Oligonucleotides are classified as tags, tagging primers, probes, and reverse primers (see Fig. 1a). For tagging primers the lower case sequence corresponds to the region that hybridizes to the genome. The *bold regions* overlap between the tag and the tagging primer

Primer	Loci	Sequence	Type of primer
M 442	*PAC2*	AATAACGAATTGAGCTATGACACCAA	Forward
M 573	*PAC2*	AGCTTACTCATATCGATTTCATACGACTT	Reverse
M 326	*PAC2*	CTGCCGCGTTGGTCAAGCCTCAT	Probe
M 239		AACCAGCGCAGCGGCATG**TGT**	Tag
M 272	*PDA1*	AACCAGCGCAGCGGCATG**tgt**gatgatggaa	Tagging
M 320	*PDA1*	TGGAGATGGCTTGTGACGCCTTG	Reverse
M 698	*PDA1*	TGGCATTCTCGATACCGACAGCAATGGCCTC	Probe
M 313		GATCTCGAGCTCGATATCGGATCC**ATT**	Tag
M 314	*YER186C*	GATCTCGAGCTCGATATCGGATCC**att**tacggcgga	Tagging
M 96	*YER186C*	AATCTCGCCTAACAAAAAAGGCTTCTTAGTG	Reverse

(continued)

Table 3 (continued)

Primer	Loci	Sequence	Type of primer
M 325	*YER186C*	AGGCAAATAACGGCAAGCCCTCTCC	Probe
M 315		AAGGAGCGCAGCGCCTGTA**CCA**	Tag
M 444	*YER188W*	AAGGAGCGCAGCGCCTGTA**cca**tagcgtgat	Tagging
M 319	*YER188W*	AACGTACAGGTTACGATCGCGTCATTTTA	Reverse
M 427	*YER188W*	TAGCCGTTATCATCGGGCCCAAAACCGTA TTCATTG	Probe
M 315		AAGGAGCGCAGCGCCTGTA**CCA**	Tag
M 513	*X-repeat*	AAGGAGCGCAGCGCCTGTA**cca**cattttaatatct	Tagging
M 512	*X-repeat*	ATTGAGTGGATAGTAGATGGTGAAAAAGTG GTATAACG	Reverse
M 510	*X-repeat*	TCATTCGGCGGCCCCAAATATTGTATAACTGCCC	Probe
M 520		TGCCCTCGCATCGCTCTC**GAA**	Tag
M 521	*Y'-5000*	TGCCCTCGCATCGCTCTC**gaa**acaaagtcagtga	Tagging
M 517	*Y'-5000*	GTCCTGGAACGTTGTCACGAAAAAGC	Reverse
M 516	*Y'-5000*	TGCTAGGCCGAACGACAGCTCTACGATGCGTACTT	Probe
M 316		TGCCCTCGCATCGCTCTC**ACA**	Tag
M 243	*Y'-600*	TGCCCTCGCATCGCTCTC**aca**gccctatcag	Tagging
M 237	*Y'-600*	GAGATCAGCTTGCGCTGGGAGTTACC	Reverse
M 526	*Y'-600*	ACAGGAATGCCGTCCAATGCGGCACTTTAGA	Probe
M827		AAGGAGCGCAGCGCCTGTA**CCA**	Tag
M1517	*DUG1*	AAGGAGCGCAGCGCCTGTA**cca**agatgcctt	Tagging
M1513	*DUG1*	AAGCAGCCATCGTCTTCATACCACCA	Reverse
M1515	*DUG1*	AATGAGCACCATCATCGCCTCTAC	Probe
M1252		TGTTGCGTGCAGGGTGGA**ACC**	Tag
M1512	*RET2*	TGTTGCGTGCAGGGTGGA**acc**tccattaagat	Tagging
M1508	*RET2*	AATGCATCTTCATATGGAGCTTCGATAGTGAAT	Reverse
M1510	*RET2*	AGCGCACCTGCATCGTTGGCAGCAA	Probe
M313		GATCTCGAGCTCGATATCGGATCC**ATT**	Tag
M1701	*YFR054C*	GATCTCGAGCTCGATATCGGATTC**att**tcattttcagtaa	Tagging
M1503	*YFR054C*	AATTGATCTGATTCAGCTGCAATAATATCGGTATTT	Reverse
M1505	*YFR054C*	TGGTACCTTCCATGCTCTTTGCCGAACTGCA	Probe

3. Add sterile water and all primers and probes to a 2 mL screw cap Eppendorf tube and incubate at 95°C for 5 min. Place on ice water for 3 min.
4. Add the Ex Taq buffer, dNTPs, and the Ex Taq Polymerase according to Table 2. Close the tube and mix the components by inverting the tube for about 20 times. Spin briefly and keep the tube on ice.
5. Add 15 μL of the PCR master mix to each 10 μL DNA sample using an automatic single-channel pipette. Use a single filter-protected tip to add the mastermix to all 96 samples. Hold the pipette at the same position for loading every well and direct the master mix onto internal wall of wells to minimize cross-contamination.
6. Cover the 96-well plate with adhesive film and centrifuge for a couple of seconds to settle the reaction mixtures.
7. PCR conditions for amplification of the *PAC2* locus are as follows:

   Step 1: 94°C for 5 min (1 cycle). Step 2: 95°C for 15 s; 63°C for 1 min (40 cycles).

3.2.2. Analysis of Real-Time PCR Data and Concentration Adjustment

1. After PCR is finished omit any individual samples from analysis that differ by more than 1 C_t from other replicates (<5 omissions per experiment).
2. Export the data into an excel spreadsheet and determine the concentration of DNA samples using the DNA standards. Use the mean value of the triplicates to estimate the DNA concentration.
3. Adjust concentration of the non-diluted samples (Subheading 3.1, step 32) to 2 ng/μL using TE. Vortex and leave overnight at 4°C to equilibrate.
4. Confirm the DNA concentration by repeating the *PAC2* real-time PCR measurement (Subheading 3.2.1). For measurement of ssDNA, the concentration of all samples should be close to 2 ng/μL and definitely between 1.5 and 3 ng/μL.
5. Generate the mean value of the measured DNA concentrations for each triplicate. Divide 2 by that number to generate a correction factor for that triplicate (e.g., if the mean value of the DNA is 2.5 ng/μL, the correction factor would be 2/2.5 = 0.8). The correction factors will be used for analysis of the ssDNA measurement in Subheading 3.2.3.
6. Prepare 96-well plates by pipetting 10 μL of the adjusted DNA samples in rows A to G as described in Table 1 using a multichannel pipettor. Add ssDNA standards to row H.
7. Cover the plates with adhesive film and store at −20°C or −80°C until used for QAOS described below (Subheading 3.2.3).

Table 4
Oligonucleotides for ssDNA on the AC-strand near telomeres. Oligonucleotides are classified as tags, tagging primers, probes, and reverse primers (see Fig. 1a). For tagging primers the lower case sequence corresponds to the region that hybridizes to the genome. The *bold* regions overlap between the tag and the tagging primer

Primer	Loci	Sequence	Type of primer
M 315		AAGGAGCGCAGCGCCTGTA**CCA**	Tag
M 271	*PDA1*	AAGGAGCGCAGCGCCTGTA**cca**tctatctgttg	Tagging
M 288	*PDA1*	ACGGCTTTCACTGAGGCACCTCTC	Reverse
M 698	*PDA1*	TGGCATTCTCGATACCGACAGCAATGGCCTC	Probe
M 313		GATCTCGAGCTCGATATCGGATCC**ATT**	Tag
M 311	*YER186C*	GATCTCGAGCTCGATATCGGATCC**att**ctctcctta	Tagging
M 321	*YER186C*	GCAAGTAGGAAGCATCCCTTCAAGTCATT	Reverse
M 325	*YER186C*	AGGCAAATAACGGCAAGCCCTCTCC	Probe
M 233		ATGCCCGCACCGCCTCA**TTG**	Tag
M 242	*Y'-600*	ATGCCCGCACCGCCTCA**ttg**cgctggga	Tagging
M 236	*Y'-600*	CCGAAATGTTTTATTGCAGAACAGCCCTAT	Reverse
M 526	*Y'-600*	ACAGGAATGCCGTCCAATGCGGCACTTTAGA	Probe
M418		ATGCTCGCAGAGCCCCTGGA**TCT**	Tag
M1516	*DUG1*	ATGCTCGCAGAGCCCCTGGA**tct**tcataccacca	Tagging
M1514	*DUG1*	TGACTTTCCAAGATGCCTTGAACACTAGTGTC	Reverse
M1515	*DUG1*	AATGAGCACCATCATCGCCTCTAC	Probe
M417		ATGGGCGCAGGGCCTGGA**TTC**	Tag
M1511	*RET2*	ATGGGCGCAGGGCCTGGA**ttc**atatggagctt	Tagging
M1509	*RET2*	CAAGAAATGGGAACCTCCATTAAGATTAGCAA	Reverse
M1510	*RET2*	AGCGCACCTGCATCGTTGGCAGCAA	Probe
M417		ATGGGCGCAGGGCCTGGA**TTC**	Tag
M1506	*YFR054C*	ATGGGCGCAGGGCCTGGA**ttc**agctgcaata	Tagging
M1504	*YFR054C*	TCCTGCAATTTGCAGACGCAGGTT	Reverse
M1505	*YFR054C*	TGGTACCTTCCATGCTCTTTGCCGAACTGCA	Probe

3.2.3. Measuring ssDNA by QAOS

ssDNA qPCR measurements are similar to quantification of DNA content in Subheading 3.2.1, except that the standards are all at 2 ng/μL but contain different amounts of ssDNA. In addition, a tagging primer is used that binds specifically to ssDNA at the measured genomic locus and a different temperature profile is used for PCR (Tables 3 and 4). A schematic outline of QAOS and the locations of the loci used are shown in Fig. 1.

1. Thaw 96-well plates containing the DNA samples and ssDNA standards on ice.
2. For the PCR master mix add primers and probes to sterile water as described in Table 2 and incubate at 95°C for 5 min followed by 3 min on ice. Primers to detect ssDNA on the TG (3′) strand near telomeres are shown in Table 3 and to detect ssDNA on the AC (5′) in Table 4.
3. Add the Ex Taq buffer, dNTPs, and the Ex Taq Polymerase according to Table 2. Close the tube and mix the components by inverting the tube for about 20 times. Spin briefly and keep the tube on ice.
4. Add 15 μL of the PCR master mix to each 10 μL DNA sample.
5. Cover the 96-well plate with adhesive film and centrifuge for a couple of seconds to settle the reaction mixtures.
6. PCR conditions for measuring ssDNA are as follows:

   Step 1: 40°C for 5 min (1 cycle). Step 2: ramp to 72°C at 2°C/min (1 cycle). Step 3: 94°C for 4 min (1 cycle). Step 4: 95°C for 15 s; 67°C for 1 min (40 cycles).
7. After PCR has finished omit any individual samples from analysis that differ by more than 1 C_t from other duplicates (see Note 5).
8. Export the data into an excel spreadsheet and calculate the mean value for each triplicate.
9. Multiply the mean value of each triplicate times the corresponding correction factor (generated in Subheading 3.2.2, step 5) to get the true level of ssDNA.
10. Calculate the 95% confidence interval as a measure of error:

    95% confidence interval of the standard error of the mean $= 1.96 \times (\mathrm{STD}/\sqrt{n})$

    with STD = Standard deviation

    n = number of samples (as samples were measured in triplicate: $n = 3$).

3.3. Preparation of Standards and Primers

3.3.1. Prepare Yeast Genomic DNA Standards

1. Arrest Wild-type cells (DLY62: *MATa bar1Δ*) containing *bar1Δ* mutation using α factor to 20 nM.
2. Prepare DNA as described in Subheading 3.1 and adjust to 4 ng/μL, 2 ng/μL, and 0.4 ng/μL in TE buffer to use as genomic DNA standards.
3. Determine concentration of the initial gold standard by measuring the 260 nm/280 nm ratio using a spectrophotometer.
4. Prepare at least 1,600 μL of 2 ng/μL DNA to make ssDNA standards as described in Subheading 3.3.2.

Table 5
Preparation of ssDNA standards

ssDNA (%)	Total	Boiled DNA (μL)	Non-boiled DNA (μL)
51.2	500	256	244
3.2	500	16	484
0.8	500	4	496

Table 6
qPCR primer preparation

	100× Stock (μM)	200 μM stock (μL)	TE (μL)
Tagging primers	3	6	394
Tags	30	60	340
Forward/reverse primers	30	60	340
Probes	20	40	360

3.3.2. Prepare ssDNA Standards

1. Boil 300 μL of the 2 ng/μL genomic DNA in a screw cap Eppendorf tube for 7 min at 98°C.
2. Place tube on ice water and every 15 s vortex for 2 s. Return to ice water in between vortexing and repeat this step three times. Then leave on ice for at least 5 min.
3. Mix boiled and non-boiled DNA according to Table 5 to make 51.2 %, 3.2 %, and 0.8 % ssDNA standards.
4. Vortex standards for 1 min and centrifuge briefly. Vortex briefly, centrifuge again, and then leave standards on ice for 30 min.
5. Vortex standards for 1 min, centrifuge briefly, and use for RT-PCR measurements. Store standards at 80°C, thaw on ice, and vortex for 20 s prior to use.

3.3.3. Preparation of Oligonucleotides

1. Make 200 μM stocks of lyophilized oligonucleotides. Resuspended stock is stored at −20°C.
2. Make working (100×) stocks by diluting oligonucleotides further with TE according to Table 6.

3.3.4. Probe Design

All probes are labeled with FAM at the 5′end and are quenched with TAMRA at the 3′ end.

1. Ideal $T_M = 76–78°C$.
2. GC content: 20–80 %.
3. Length: 9–40 bases. Generally pick the shortest probe.

Table 7
List of tested tags that are used to design tagging primers

Primer	Sequence
M97	GATCTCGAGCTCGATATCGGATCCATT
M137	TCTTGCGTGCAGGGTCCGCACC
M139	TGTTGCGTGCAGGGTGGAACC
M233	ATGCCCGCACCGCCTCATTG
M234	AAGGAGCGCAGCGCCTGTACCA
M235	TGCCCTCGCATCGCTCTCACA
M239	AACCAGCGCAGCGGCATGTGT
M417	ATGGGCGCAGGGCCTGGATTC
M418	ATGCTCGCAGAGCCCCTGGATCT
M514	ATCCAGCGCAGCGGCATGTG
M520	TGCCCTCGCATCGCTCTCGAA

4. No G on the 5′ end.
5. <4 contiguous G's.
6. Must not have more G's than C's.

3.3.5. Primer Design

Primers to detect ssDNA on the TG (3′) strand near telomeres are shown in Table 3 and to detect ssDNA on the AC (5′) in Table 4 (see Note 6). A list of tested tags is shown in Table 7.

1. Ideal $T_M = 66–69°C$ for tag and reverse primer.
2. Ideally there is <2°C difference in T_M between the two primers.
3. GC content: 20–80 %.
4. Length: 9–40 bases.
5. Maximum of 2 G/C in the last five base pairs at the 3′ end.
6. Tagging primers contain the tag and a 10–15 bp long tail with the T_M of the tail being between 20 and 43°C, but preferably about 30°C. Tail and tag contain a 3 bp overlapping sequence.

3.3.6. Amplicon Design

1. Length: 50–150 bp.
2. 3′ end of primer as close to the probe as possible without overlapping.

Primer design for the *PDA1* locus to amplify ssDNA on the 3′ TG strand is shown in Fig. 3. Test primers by using boiled (ssDNA) and non-boiled (dsDNA) as templates in QAOS. Aim to have a more than 10 C_T difference between ssDNA and dsDNA samples.

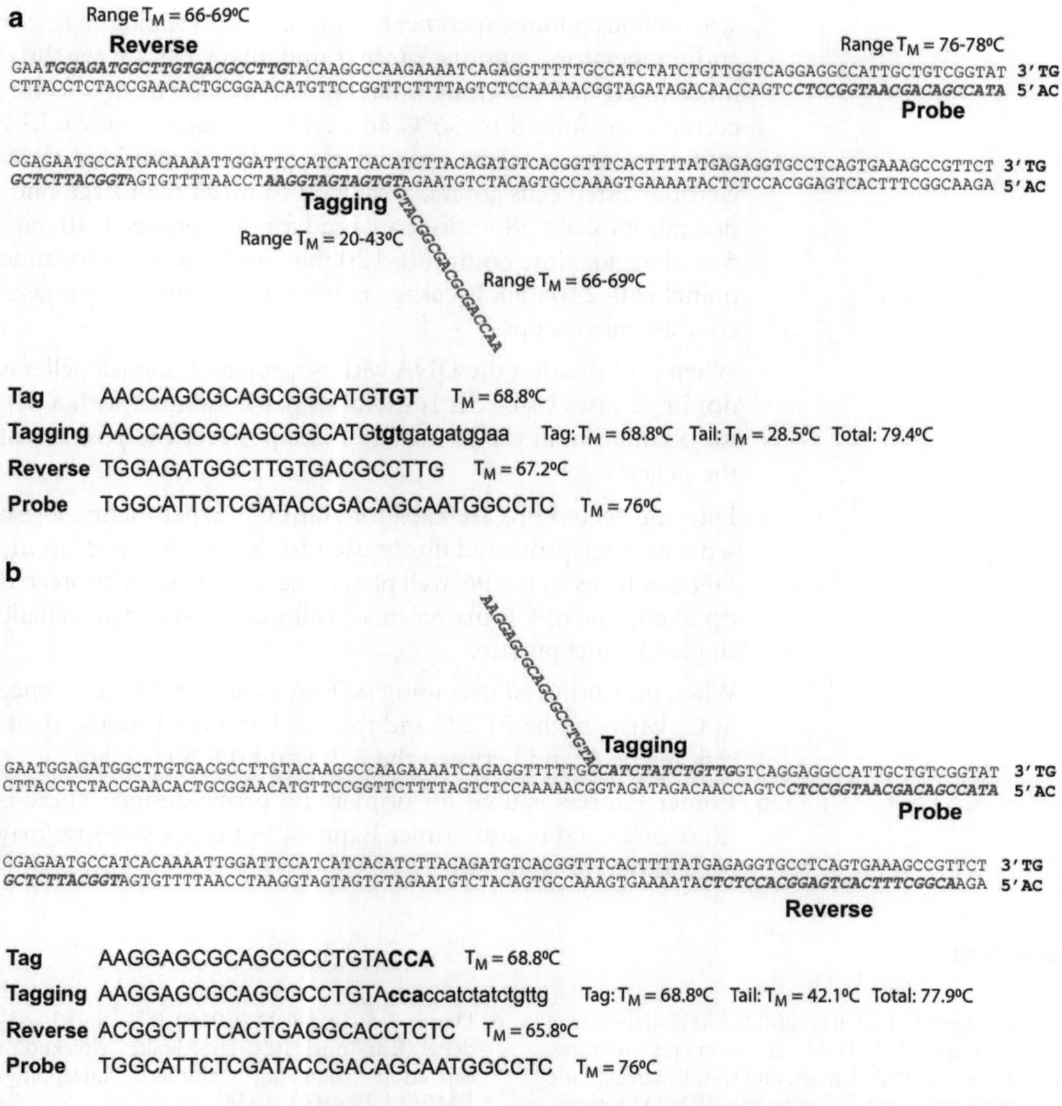

Fig. 3. Primer design for QAOS at the *PDA1* locus on chromosome V-R. (**a**) Primers used to measure ssDNA on the TG-strand. (**b**) Primers used to measure ssDNA on the AC-strand. The same probe is used for measurement of both DNA strands. For tagging primers the lower case sequence corresponds to the region that hybridizes to the genome. The *bold* regions overlap between the tag and the tagging primer.

4. Notes

1. The number of DNA preparations that can be done at a time mainly depends on the space in the fixed angle centrifuge rotor. Using a Beckman Allegra 64R centrifuge with a fixed angle F0850 rotor we are readily able to perform two batches of eight DNA preparations at a time, over an 8-h working day.
2. The number of required sessions to obtain more than 90 % cell breakage is dependent on the experiment. We typically perform

synchronous culture experiments using telomere capping-deficient and temperature-sensitive *cdc13-1* mutants. Cells are synchronized in G1 at 36°C using α factor. After release from α factor, cultures are shifted to 36°C and cells containing the *cdc13-1* mutations arrest over a time course of 4 h. Unbudded alpha factor-arrested cells are more difficult to break than large budded mitotic cells. ≥8 sessions are used for time points 0–40 min, 6 sessions for time points 80–120 min, and 5 sessions for time points 160–240 min. Breakage efficiency is examined by phase-contrast microscopy.

3. When precipitating the DNA with isopropanol, a small pellet is not in all cases visible. It is useful to mark the tube with a pen for orientation in the centrifuge to help detect the position of the pellet.
4. Filter-protected tips are used for all qPCR experiments. Use a separate filter-protected tip for each triplicate when adding the DNA samples to the 96-well plate. Use a single filter-protected tip to add the qPCR master mix to all wells using an automatic single-channel pipette.
5. When measuring ssDNA using ssDNA standards, the difference in C_T between the 51.2 % and the 3.2 % standard should theoretically be 4, and between the 3.2 % and 0.8 % standard 2.
6. Primer Express is used for primer and probe design. There is no strong need to use Primer Express, but other software may use a different algorithm.

References

1. Maringele L, Lydall D (2002) EXO1-dependent single-stranded DNA at telomeres activates subsets of DNA damage and spindle checkpoint pathways in budding yeast yku70Delta mutants. Genes Dev 16:1919–1933
2. Niu H, Chung WH, Zhu Z, Kwon Y, Zhao W, Chi P, Prakash R, Seong C, Liu D, Lu L, Ira G, Sung P (2010) Mechanism of the ATP-dependent DNA end-resection machinery from Saccharomyces cerevisiae. Nature 467: 108–111
3. Zou L, Elledge SJ (2003) Sensing DNA damage through ATRIP recognition of RPA-ssDNA complexes. Science 300:1542–1548
4. Bertuch AA, Lundblad V (2003) The Ku heterodimer performs separable activities at double-strand breaks and chromosome termini. Mol Cell Biol 23:8202–8215
5. Grandin N, Damon C, Charbonneau M (2001) Ten1 functions in telomere end protection and length regulation in association with Stn1 and Cdc13. EMBO J 20:1173–1183
6. Dewar JM, Lydall D (2010) Pif1- and Exo1-dependent nucleases coordinate checkpoint activation following telomere uncapping. EMBO J 29:4020–4034
7. Foster SS, Zubko MK, Guillard S, Lydall D (2006) MRX protects telomeric DNA at uncapped telomeres of budding yeast cdc13-1 mutants. DNA Repair (Amst) 5:840–851
8. Booth C, Griffith E, Brady G, Lydall D (2001) Quantitative amplification of single-stranded DNA (QAOS) demonstrates that cdc13-1 mutants generate ssDNA in a telomere to centromere direction. Nucleic Acids Res 29: 4414–4422
9. Pitt CW, Cooper JP (2010) Pot1 inactivation leads to rampant telomere resection and loss in one cell cycle. Nucleic Acids Res 38:6968–6975
10. Addinall SG, Holstein E-M, Lawless C, Yu M, Chapman K, Banks P, Ngo H-P, Maringele L, Taschuk M, Young A, Ciesiolka A, Lister AL, Wipat A, Wilkinson DJ, Lydall D (2011) Quantitative fitness analysis shows that NMD

proteins and many other protein complexes suppress or enhance distinct telomere cap defects. PLoS Genet 4:e1001362

11. Lee K, Zhang Y, Lee SE (2008) Saccharomyces cerevisiae ATM orthologue suppresses break-induced chromosome translocations. Nature 454:543–546
12. Negrini S, Ribaud V, Bianchi A, Shore D (2007) DNA breaks are masked by multiple Rap1 binding in yeast: implications for telomere capping and telomerase regulation. Genes Dev 21:292–302
13. Ngo HP, Lydall D (2010) Survival and growth of yeast without telomere capping by Cdc13 in the absence of Sgs1, Exo1, and Rad9. PLoS Genet 6:e1001072
14. Jia X, Weinert T, Lydall D (2004) Mec1 and Rad53 inhibit formation of single-stranded DNA at telomeres of Saccharomyces cerevisiae cdc13-1 mutants. Genetics 166:753–764
15. Lazzaro F, Sapountzi V, Granata M, Pellicioli A, Vaze M, Haber JE, Plevani P, Lydall D, Muzi-Falconi M (2008) Histone methyltransferase Dot1 and Rad9 inhibit single-stranded DNA accumulation at DSBs and uncapped telomeres. EMBO J 27:1502–1512
16. Zubko MK, Guillard S, Lydall D (2004) Exo1 and Rad24 differentially regulate generation of ssDNA at telomeres of Saccharomyces cerevisiae cdc13-1 mutants. Genetics 168:103–115
17. Zubko MK, Maringele L, Foster SS, Lydall D (2006) Detecting repair intermediates in vivo: effects of DNA damage response genes on single-stranded DNA accumulation at uncapped telomeres in budding yeast. Methods Enzymol 409:285–300

Chapter 24

Simple, Non-radioactive Measurement of Single-Stranded DNA at Telomeric, Sub-telomeric, and Genomic Loci in Budding Yeast

James M. Dewar and David Lydall

Abstract

Single-stranded DNA (ssDNA) is a DNA repair, replication, and recombination intermediate and a stimulus for checkpoint kinase-dependent cell cycle arrest. Current assays to detect ssDNA generated in vivo are indirect, laborious, and generally require the use of radioactivity. Here, we describe simple, quantitative approaches to measure ssDNA generated in yeast, at single- and multi-copy chromosomal loci and in highly repetitive telomeric sequences. We describe a fluorescence in-gel assay to measure ssDNA in the telomeric TG repeats of telomere cap-defective budding yeast *yku70Δ* and *cdc13-1* mutants. We also describe a rapid method to prepare DNA for Quantitative Amplification of ssDNA, used to measure ssDNA in single-copy and repetitive sub-telomeric loci. These complementary methods are useful to understand the important roles of ssDNA in yeast cells and could be readily extended to other cell types.

Key words: In-gel assay, QAOS, FIGA, Single-stranded DNA, Telomere, Resection

1. Introduction

Much insight into telomere function, DSB repair, and meiosis has been gained by a measuring single-stranded DNA (ssDNA) generated in vivo, particularly in the model organism budding yeast (1–6). The most commonly used methods to measure ssDNA are variants on the Southern blot, relying largely on disappearance of dsDNA restriction fragments, as a proxy for ssDNA generation (7–12). Such methods can make use of strand-specific probes and/or alkaline gel electrophoresis to resolve ssDNA for detection by Southern Blot (2, 7, 13). Southern blot-based techniques are time-consuming, require comparatively large amounts of DNA, and each locus examined requires a particular combination of restriction enzyme and probe, hindering the simple examination of ssDNA at multiple loci.

Lotte Bjergbæk (ed.), *DNA Repair Protocols*, Methods in Molecular Biology, vol. 920,
DOI 10.1007/978-1-61779-998-3_24,

Quantitative Amplification of ssDNA (QAOS), a qPCR-based method to accurately quantify ssDNA at genomic loci (14–19) (see accompanying Chapter 23 by Holstein and Lydall), is an alternative to Southern blots. The major difficulty with QAOS has been making the high-quality DNA preparations that are necessary for its success (14, 20). Finally, in-gel assays measure ssDNA in telomeric DNA after hybridizing a small, radioactive oligonucleotide probe to the TG-rich ssDNA overhangs at telomeres (13, 21–28).

Here, we describe a rapid, simple yeast genomic DNA extraction procedure (Subheading 3.2) that can be used in a simple fluorescent in-gel assay (FIGA). FIGA permits quantitative, nonradioactive measurement of ssDNA in telomeric TG repeats (Subheading 3.3). The same DNA preparations are of sufficient quality and yield to perform hundreds of QAOS assays to examine ssDNA throughout the yeast genome (Subheading 3.4). Our methods will be useful to those trying to understand the role of ssDNA in yeast and could be applied to other prokaryotic or eukaryotic cell types.

2. Materials

1. Lysis Buffer: 2 % Triton X-100, 1 % SDS, 100 mM NaCl, 10 mM Tris–HCl, pH 8.0, 100 mM EDTA, pH 8.0. For 50 mL: 1 mL Triton X-100, 5 mL SDS (10 %W/V in MQ H_2O), 5 mL NaCl (1 M), 1 mL Tris–HCl pH 8.0 (0.5 M), 5 mL EDTA pH 8.0 (0.5 M), sterile H_2O to 50 mL.
2. TE: 10 mM Tris–HCl, pH 8.0, 1 mM EDTA pH 8.0. By dilution of 1 M Tris–HCl, pH 8.0 and 0.5 M EDTA, pH 8.0.
3. Phenol:chloroform:isoamyl alcohol (25:24:1), equilibrated with TE.
4. RNAse A: Dissolve pancreatic RNAse at 10 mg/mL in 0.01 M sodium acetate, pH 5.2. Place in boiling water bath for 15 min and then allow to cool to room temperature. Adjust pH by adding 0.1 volume of 1 M Tris–HCl, pH 7.4. Aliquot and store at −20 °C. Do not freeze/thaw.
5. RNAse/TE: Dilute RNAse to 7.5 μg/mL in TE (75 μL 10 mg/mL RNAse to 10 mL TE).
6. Ice-cold acid-washed glass beads (e.g., Stratech): 0.425–0.6 mm in diameter, washed for 1 h in concentrated nitric acid, rinsed with milliQ water until pH is neutral, baked to dryness, and stored at 4 °C.
7. 3 M NaAc, pH 5.2.
8. Phase lock Gel Light 2 mL tubes, 5 Prime.
9. Cooled microfuge.
10. Precellys 24 shaker, Bertin technologies.

11. 2 mL screw cap tube with skirt (e.g., Sarstedt).
12. CY5 AC Oligo (500 nM): 5′ (CY5)CCCACCACACACA CCCACACCC 3′, purified by HPLC (e.g., Sigma) dissolved in water at 200 μM, and diluted 1 in 400 in water to give a 500 nM working stock (see Note 1).
13. GE Healthcare Typhoon Imager.
14. SYBR safe stain (Invitrogen).
15. TBE: Tris–Borate–EDTA buffer (e.g., Biorad).
16. Orange G loading buffer (6×): 0.125 g Orange G, 0.75 g Ficoll Type 400, 12 mL 0.5 M EDTA, pH 8.0 water to 50 mL.

3. Methods

These methods are complementary to those described by Holstein and Lydall in this issue. The DNA preparation procedure we describe is significantly faster than that described in the accompanying chapter and is of sufficient quality to allow numerous individual FIGA or QAOS assays.

3.1. Sample Collection

Yeast strains are usually in the W303 genetic background and are *RAD5*$^{+}$.

1. Collect culture equivalent up to approximately 1×10^9 nuclei and chill on ice water for 5 min (e.g., 50 mL at 2×10^7 cells/mL). Approximately 2 mL of a stationary-phase culture also works well and the method is tolerant to wide ranges in cell pellet sizes from 7×10^7 to 1.5×10^9 cells.
2. Pellet cells in a centrifuge/microcentrifuge prechilled to 4 °C. Spin either at 670 × *g* for 5 min (cultures 10–50 mL) or 15,000 × *g* for 1 min (cultures ≤2 mL).
3. While spinning, chill 1.5 mL Eppendorf tubes on ice.
4. Resuspend cells in 1 mL of ice-cold H_2O by vortexing and transfer the suspension to prechilled 1.5 mL Eppendorf tubes by decanting.
5. Pellet cells at 4 °C in a microcentrifuge for 15 s, 15,000 × *g*. Decant the supernatant, spin again briefly (approximately 10 s, 15,000 × *g*), and remove the residual liquid by aspiration.
6. Keep cell pellets on ice and transfer to −80 °C for storage.

3.2. DNA Isolation

1. Chill lysis buffer on ice water.
2. Thaw cell pellets and keep on ice.
3. While pellets thaw, add 0.6 g of glass beads (acid-washed, 425–600 μm, Stratech) to a 2 mL Skirted screw tube (Sarstedt) for each sample and chill tubes containing glass beads on ice.

4. Resuspend thawed cell pellets in 400 μL of lysis buffer by gently pipetting up and down with a 1,000 μL pipette tip. Transfer the resuspended cell pellet to the screw-cap tube containing 0.6 g glass beads.
5. Add 400 μL of phenol:chloroform:isoamyl alcohol (25:24:1) to each tube by pipetting from the lower (phenol) layer using positive displacement pipette (e.g., Eppendorf Multipette Stream). Cap the lids tightly, and invert gently to mix, until the glass beads can be seen to flow freely from one end of the tube to the other.
6. Lyse cells at 5.5 power setting for 20 s using a Precellys 24 automated homogeniser and then chill on ice water while the machine cools.
7. Repeat the lysis step at 5.5 power for 20 s, twice more (3× in total), and then chill samples on ice water in between breakage cycles.
8. Spin tubes briefly (approximately 10 s) at 4 °C in a microcentrifuge to remove phenol from the lids of the tubes.
9. Add 400 μL of TE to each tube, cap the lids tightly, and then vortex briefly to mix.
10. Spin the tubes in a microcentrifuge chilled to 4 °C at 15,000 × *g* for 5 min to separate the aqueous (upper) and organic (lower) phases (see Note 2).
11. Prepare 2 mL tubes of phase lock gel, light (2.0 mL phase lock gel, light, e.g., VWR International), by centrifuging in a microcentrifuge at 15,000 × *g* for 30 s and label one tube for each sample.
12. Use a 1,000 μL pipette tip to transfer ≤750 μL of each aqueous phase to a phase lock tube.
13. Add 750 μL of phenol:chloroform:isoamyl alcohol (25:24:1) to each phase lock tube and mix gently, but thoroughly, by inverting ten times to mix the aqueous phase and phenol.
14. Centrifuge the phase lock tubes in a microcentrifuge at 15,000 × *g* for 5 min.
15. Transfer the aqueous phase to a fresh 2 mL Eppendorf tube (e.g., Fisher). Add 100 % Ethanol to fill the tube, gently invert to mix, and leave on the bench at room temperature for 5 min to precipitate nucleic acids.
16. Pellet nucleic acids in a microcentrifuge for 3 min at 15,000 × *g*. Decant the ethanol off, spin briefly (approximately 15 s) at 15,000 × *g*, and remove the residual ethanol by aspiration.
17. Air-dry the nucleic acid pellet for 5 min and make up TE/RNAse.

18. Add 806 μL of TE/RNAse to each pellet and incubate tubes in a 37 °C water bath for 30 min, flicking tubes to mix after 5 min and vortexing briefly after 15 min.
19. To each tube, add 26 μL of 3 M NaAc, pH 5.2, vortex, and then add 100 % ethanol to fill the tube. Invert the tube gently to mix and allow DNA to precipitate at room temperature for 15 min.
20. Pellet DNA in a microcentrifuge for 3 min at 15,000×*g*. Decant the supernatant, spin briefly (approximately 15 s) at 15,000×*g*, and remove the residual liquid by aspiration.
21. Allow DNA pellets to air dry in the fume hood for 30 min.
22. Add 40 μL of TE and incubate tubes in a 37 °C water bath, flicking tubes to mix after 5 min and vortexing briefly after 15 min before incubating on ice for 1 h.
23. Determine yield by spectroscopy (nanodrop) and store at −80 °C.

3.3. FIGA

1. Set up 20 μL restriction digests using master mix:

10 μL	DNA at 0.5–1.0 mg/mL in TE (5–10 μg total)
10 μL	XhoI master mix: 0.1 μL Xho 1 (20 units/μL), 2 μL NEB Buffer 4 (10×), 0.2 μL BSA (100×), 7.7 μL water

2. Incubate tubes in a 37 °C water bath overnight (see Note 3).
3. Incubate the tubes at 65 °C for 20 min to heat-inactivate XhoI.
4. While heat inactivation occurs, make up a 1 % agarose gel in 0.5× TBE. We use the Wide Mini-Sub cell GT system (Bio Rad) and pour 90 mL gels in 15×10 cm trays using 30-well combs. This gives 30 1.5 mm wide lanes per gel, each of which can accommodate 20 μL of loaded sample.
5. After heat inactivation has occurred, place on ice. To each tube, add 1 μL of CY5 AC Oligo (500 nM) and cap the lid, centrifuge briefly, flick to mix, and incubate in a 37 °C water bath for 10 min.
6. Incubate tubes on ice for 30 min to further allow the oligo to anneal to TG ssDNA.
7. While the probe anneals prepare blank samples DNA ladder as follows:

10 μL	TE or 4 ng 1 kb DNA ladder (Invitrogen) in TE
1 μL	CY5 AC Oligo (500 nM)
10 μL	NEB Buffer 4 (2×), BSA (2×), water

8. After annealing add 4 μL of Orange G Loading Buffer (6×) to all tubes (samples, blanks, and ladder), and spin briefly to bring the samples to the bottom of each tube.
9. Load all samples, blanks, and ladder onto the agarose gel in 0.5× TBE with the samples in the middle lanes, and one DNA ladder lane either side of the gel.
10. Run the gel at 5 V/cm for 90–120 min.
11. Once the gel has finished running, pour off the 0.5× TBE running buffer and save for step 16, rinse the gel in water, and place on Typhoon imager.
12. Place the gel facing upwards on a GE Healthcare Typhoon Trio imager. Scan the gel using appropriate settings for CY5, at 50 μm resolution, with the highest sensitivity setting and a focal plane 3 mm above the glass platten. Save the xxx.gel file for analysis (see Note 4).
13. Once the gel has been imaged for CY5, place in a fresh sandwich box and add enough SYBR Safe/0.5× TBE to cover the gel (typically 200 mL of used 0.5× TBE from step 14, with 20 μL of 10,000× SyBr safe stock). Incubate with gentle agitation for 30 min at room temperature.
14. Rinse the gel briefly in water and detect the SYBR Safe using a gel documentation system (we use a FUJI LAS-4000) or Typhoon Imager. Ensure that a long enough exposure is performed that the wells of the gel are all visible.
15. At this point the gel can either be discarded or transferred to a nylon membrane for further analysis by Southern Blot.
16. Image analysis and quantification: The gel file of the CY5 image is analyzed in ImageJ (v1.41o) (29). First, background correction is applied using the "Subtract Background" command for a light background, with a rolling ball radius of 20–50 pixels, using a Sliding Paraboloid with Smoothing disabled. The Window/Level command is then used to reduce the visual window to approximately 1,000 and the level is reduced until background signal from the gel became visible. To quantify the intensity in each lane, the gel is rotated so that all lanes are vertical, then a rectangular selection of one lane width, extending from the highest to the lowest regions of signal on the gel, is moved horizontally across the gel, and the mean pixel intensity of each lane is taken using the "measure" command. The rectangular selection is then narrowed and used to measure the mean pixel intensity between each lane. The signal of each lane is determined as the mean pixel intensity of the lane, minus the mean of the mean pixel intensity either side, and expressed relative to the signal of a *yku70Δ* on the same gel. Signal is then normalized using *PAC2* qPCR measurements; see below.

3.4. QAOS

QAOS is performed on samples diluted to 2 ng/μL as described in Chapter 23. This is approximately a 1,000-fold dilution of the concentration obtained in phenol preps. To quantify DNA concentration in diluted samples by qPCR, TaqMan primers/probe to detect *PAC2* are used as previously described (20).

4. Notes

1. A 0.05 μmol synthesis scale of CY5-labeled oligo currently costs approximately £50.00 and yields approximately 150 μL of 200 μM stock, sufficient for detection of approximately 60,000 samples by in-gel assay.
2. When inserting the tubes into the microcentrifuge, make sure that the label faces to one side so that the view of the interface will not be obstructed after spinning.
3. Alternatively, it is fine to perform the digest for 8 h, store at −20 °C, and thaw on ice before heat inactivation of the enzyme.
4. A high-resolution scan of the gel should take approximately 30 min. It can also be useful to do an initial scan with a resolution of 1,000 μm, which takes approximately 2 min and provides rapid confirmation that the In-Gel Assay has worked.

References

1. Garvik B, Carson M, Hartwell L (1995) Single-stranded DNA arising at telomeres in cdc13 mutants may constitute a specific signal for the RAD9 checkpoint. Mol Cell Biol 15:6128–6138
2. Lydall D, Nikolsky Y, Bishop DK, Weinert T (1996) A meiotic recombination checkpoint controlled by mitotic checkpoint genes. Nature 383:840–843
3. Lydall D, Weinert T (1995) Yeast checkpoint genes in DNA damage processing: implications for repair and arrest. Science 270:1488–1491
4. Buhler C, Borde V, Lichten M (2007) Mapping meiotic single-strand DNA reveals a new landscape of DNA double-strand breaks in Saccharomyces cerevisiae. PLoS Biol 5:e324
5. White CI, Haber JE (1990) Intermediates of recombination during mating type switching in Saccharomyces cerevisiae. EMBO J 9:663–673
6. Lee SE, Moore JK, Holmes A, Umezu K, Kolodner RD, Haber JE (1998) Saccharomyces Ku70, mre11/rad50 and RPA proteins regulate adaptation to G2/M arrest after DNA damage. Cell 94:399–409
7. Mimitou EP, Symington LS (2008) Sae2, Exo1 and Sgs1 collaborate in DNA double-strand break processing. Nature 455:770–774
8. Mimitou EP, Symington LS (2010) Ku prevents Exo1 and Sgs1-dependent resection of DNA ends in the absence of a functional MRX complex or Sae2. EMBO J 29:3358–3369
9. Nicolette ML, Lee K, Guo Z, Rani M, Chow JM, Lee SE, Paull TT (2010) Mre11-Rad50-Xrs2 and Sae2 promote 5′ strand resection of DNA double-strand breaks. Nat Struct Mol Biol 17:1478–1485
10. Zhu Z, Chung WH, Shim EY, Lee SE, Ira G (2008) Sgs1 helicase and two nucleases Dna2 and Exo1 resect DNA double-strand break ends. Cell 134:981–994
11. Chung WH, Zhu Z, Papusha A, Malkova A, Ira G (2010) Defective resection at DNA double-strand breaks leads to de novo telomere formation and enhances gene targeting. PLoS Genet 6:e1000948
12. Terentyev N, Johnson M, Neale A, Bishop-Bailey M, Goldman AS (2010) Evidence that MEK1 could positively promotes interhomologue

double-strand break repair. Nucleic Acids Res 38:4349–4360

13. Bonetti D, Clerici M, Anbalagan S, Martina M, Lucchini G, Longhese MP (2010) Shelterin-like proteins and Yku inhibit nucleolytic processing of Saccharomyces cerevisiae telomeres. PLoS Genet 6:e1000966
14. Booth C, Griffith E, Brady G, Lydall D (2001) Quantitative amplification of single-stranded DNA (QAOS) demonstrates that cdc13-1 mutants generate ssDNA in a telomere to centromere direction. Nucleic Acids Res 29:4414–4422
15. Maringele L, Lydall D (2002) EXO1-dependent single-stranded DNA at telomeres activates subsets of DNA damage and spindle checkpoint pathways in budding yeast *yku70Δ* mutants. Genes Dev 16:1919–1933
16. Zubko MK, Guillard S, Lydall D (2004) Exo1 and Rad24 differentially regulate generation of ssDNA at telomeres of Saccharomyces cerevisiae cdc13-1 mutants. Genetics 168: 103–115
17. Lazzaro F, Sapountzi V, Granata M, Pellicioli A, Vaze M, Haber JE, Plevani P, Lydall D, Muzi-Falconi M (2008) Histone methyltransferase Dot1 and Rad9 inhibit single-stranded DNA accumulation at DSBs and uncapped telomeres. EMBO J 27:1502–1512
18. Dewar JM, Lydall D (2010) Pif1- and Exo1-dependent nucleases coordinate checkpoint activation following telomere uncapping. EMBO J 29:4020–4034
19. Ngo HP, Lydall D (2010) Survival and growth of yeast without telomere capping by Cdc13 in the absence of Sgs1, Exo1, and Rad9. PLoS Genet 6:e1001072
20. Zubko MK, Maringele L, Foster SS, Lydall D (2006) Detecting repair intermediates in vivo: effects of DNA damage response genes on single-stranded DNA accumulation at uncapped telomeres in budding yeast. Methods Enzymol 409:285–300
21. Chandra A, Hughes TR, Nugent CI, Lundblad V (2001) Cdc13 both positively and negatively regulates telomere replication. Genes Dev 15:404–414
22. Bertuch AA, Lundblad V (2004) EXO1 contributes to telomere maintenance in both telomerase-proficient and telomerase-deficient Saccharomyces cerevisiae. Genetics 166:1651–1659
23. Dionne I, Wellinger RJ (1996) Cell cycle-regulated generation of single-stranded G-rich DNA in the absence of telomerase. Proc Natl Acad Sci USA 93:13902–13907
24. Vodenicharov MD, Wellinger RJ (2006) DNA degradation at unprotected telomeres in yeast is regulated by the CDK1 (Cdc28/Clb) cell-cycle kinase. Mol Cell 24:127–137
25. Vodenicharov MD, Laterreur N, Wellinger RJ (2010) Telomere capping in non-dividing yeast cells requires Yku and Rap1. EMBO J 17:3007–3019
26. Miyake Y, Nakamura M, Nabetani A, Shimamura S, Tamura M, Yonehara S, Saito M, Ishikawa F (2009) RPA-like mammalian Ctc1-Stn1-Ten1 complex binds to single-stranded DNA and protects telomeres independently of the Pot1 pathway. Mol Cell 36:193–206
27. Surovtseva YV, Churikov D, Boltz KA, Song X, Lamb JC, Warrington R, Leehy K, Heacock M, Price CM, Shippen DE (2009) Conserved telomere maintenance component 1 interacts with STN1 and maintains chromosome ends in higher eukaryotes. Mol Cell 36:207–218
28. Celli GB, de Lange T (2005) DNA processing is not required for ATM-mediated telomere damage response after TRF2 deletion. Nat Cell Biol 7:712–718
29. Abramoff MD, Magelhaes PJ, Ram SJ (2004) Image processing with ImageJ. Biophotonics Int 11:36–42

Chapter 25

Monitoring DNA Recombination Initiated by HO Endonuclease

Neal Sugawara and James E. Haber

Abstract

DNA double-strand breaks (DSBs) have proven to be very potent initiators of recombination in yeast and other organisms. A single, site-specific DSB initiates homologous DNA repair events such as gene conversion, break-induced replication, and single-strand annealing, as well as nonhomologous end joining, microhomology-mediated end joining, and new telomere addition. When repair is either delayed or prevented, a single DSB can trigger checkpoint-mediated cell cycle arrest. In budding yeast, expressing the HO endonuclease under the control of a galactose-inducible promoter has been instrumental in the study of these processes by providing us a way to synchronously induce a DSB at a unique site in vivo. We describe how the HO endonuclease has been used to study the recombination events in mating-type (*MAT*) switching. Southern blots provide an overview of the process by allowing one to examine the formation of the DSB, DNA degradation at the break, and formation of the product. Denaturing gels and slot blots as well as PCR have provided important tools to follow the progression of resection in wild-type and mutant cells. PCR has also been important in allowing us to follow the kinetics of certain recombination intermediates such as the initiation of repair DNA synthesis or the removal of nonhomologous Y sequences during *MAT* switching. Finally chromatin immunoprecipitation has been used to follow the recruitment of key proteins to the DSB and in subsequent steps in DSB repair.

Key words: HO, Meganuclease, Recombination, Double-strand break, Gene conversion, Mating-type switch, Chromatin immunoprecipitation

1. Introduction

Cells can repair double-strand breaks (DSBs) by several pathways such as nonhomologous end joining or by several homologous recombination pathways such as gene conversion, single-strand annealing, and break-induced replication (1). The study of these processes has been facilitated by the use of a site-specific meganuclease, HO endonuclease, that recognizes two unique sequences in the yeast *Saccharomyces cerevisiae*, the alternative *MAT*a and *MAT*α loci.

Lotte Bjergbæk (ed.), *DNA Repair Protocols*, Methods in Molecular Biology, vol. 920,
DOI 10.1007/978-1-61779-998-3_25, © Springer Science+Business Media New York 2012

HO can be efficiently induced using the UAS from the *GAL1–GAL10* genes by adding galactose to a yeast culture grown under non-fermenting conditions (using YP-lactate or YP-raffinose media). When controlled by the *GAL1–GAL10* UAS, *HO* can be synchronously expressed and used to rapidly produce a single well-defined DSB in almost all the cells in a yeast culture (2, 3). Because of this synchrony it is possible to follow the subsequent recombination steps by isolating DNA from the culture and analyzing it by Southern blots, PCR, and chromatin immunoprecipitation (ChIP). Furthermore, expression can be down-regulated by addition of 2 % glucose to turn off HO cutting. Because HO endonuclease contains a PEST sequence (not found in the I-SceI meganuclease) it is rapidly degraded (4). As a consequence when cells are shifted from a galactose-containing medium to glucose, HO is absent in <30 min.

While we have used the HO endonuclease to study a number of recombination pathways, we focus here on the mating-type (*MAT*) switching mechanism in yeast to illustrate how HO can be used to study this process (5). During *MAT* switching HO endonuclease creates a DSB at *MAT* that induces gene conversion using donor sequences located at either *HML* or *HMR*. Both *HML* and *HMR* are maintained in transcriptionally silent heterochromatin and are not cleaved by HO; however, they can serve as donors for repair of the HO-cleaved DNA from *MAT*. As shown in Fig. 1, *MAT* is in part homologous with *HML* (W, X, and Z) and *HMR* (W and Z). The *MAT*-specific sequences are composed of either Y**a** or Yα, which determine the *MAT* of the cell.

Figure 25.1 illustrates the recombination steps involved in switching a cell from *MAT***a** to *MAT*α using the donor cassette *HML*α. HO endonuclease creates a DSB within the *MAT* locus at the junction of the Y and Z sequences. This cut is processed by 5′ to 3′ resection of the DSB end, creating 3′ single-stranded tail that recruits first the RPA single-strand DNA-binding protein and then the Rad51 recombinase. The DSB also promotes the recruitment of a number of other chromatin-remodeling and repair proteins and also the ATR/ATM (Mec1 and Tel1)-dependent phosphorylation of the histone H2A (γ-H2AX). The Rad51-coated 3′ tail at *MAT***a** engages in a search for homology and thus strand invades *HML-Z* region and forms a D-loop. DNA synthesis can be initiated from the 3′ end of the invading strand copying the Yα sequence from *HML*α. According to the synthesis-dependent strand annealing model (SDSA) of DSB repair, the elongating, newly copied strand dissociates from the donor and anneals with the *MAT-X* sequence and promotes second strand synthesis to complete repair of the DSB after clipping off the nonhomologous Y sequence. Alternatively a double Holliday junction intermediate can be created and then resolved into a gene conversion event occasionally accompanied by crossing-over. Ultimately a final product is formed in

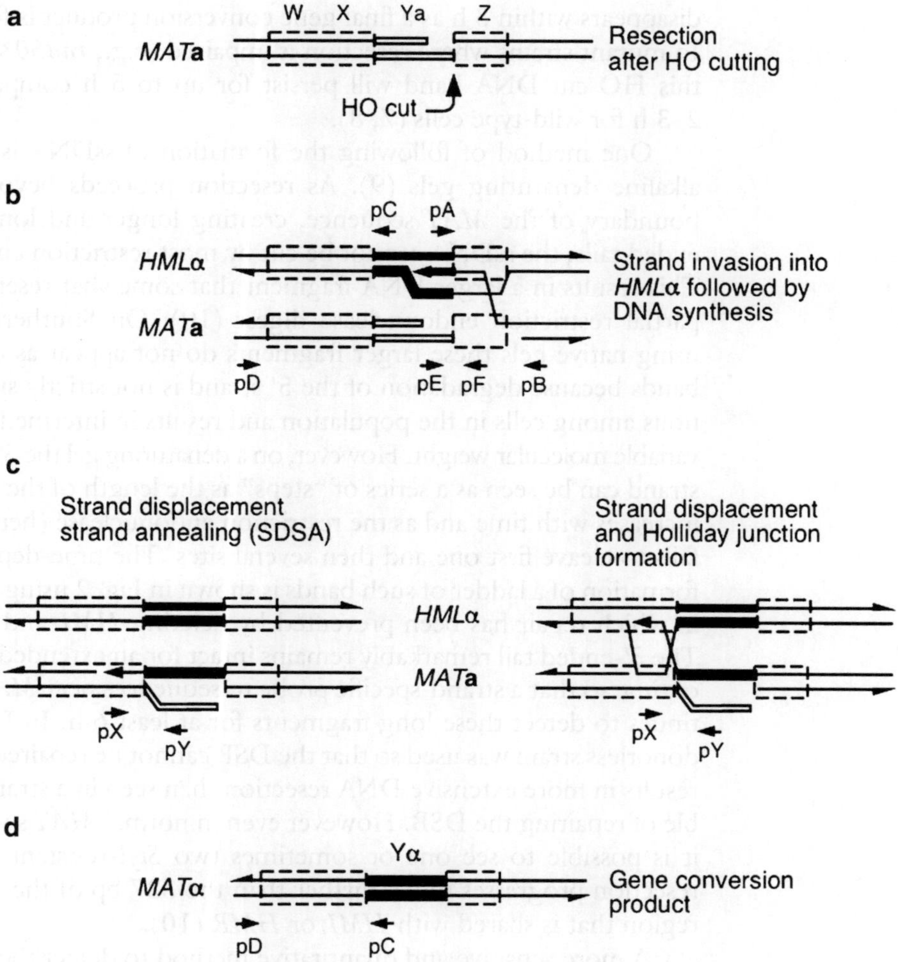

Fig. 1. Models of gene conversion during mating-type switching. (**a**) HO cuts DNA at the junction between Y and Z regions of *MAT* which is followed by 3′ to 5′ resection. (**b**) The 3′ tail strand invades *HML*α using the homologous sequences in *MAT*-Z. The 3′ tail primes new DNA synthesis and extends the strand into the *MAT*-Yα region (*thick line*) of *HML*. This event can be detected by PCR using the primers pA and pB. (**c**) In the strand displacement and strand annealing model (SDSA), the newly synthesized DNA is displaced from *HML* and anneals with the homologous sequences from *MAT*-X. In the double Holliday junction model strand extension proceeds across *MAT*-Y and creates a Holliday junction that is subsequently resolved by endonucleases. In either model the tail composed of Y**a** sequences (*white box*) is clipped off and can be monitored by PCR using the pX and pY primers. (**d**) Completion of gene conversion resulting in an *MAT*α product as seen on a Southern blot occurs coincidently with the product of the primers pC and pD.

which the *MAT*-Ya sequence has been replaced with Yα. Consistent with SDSA being the predominant mechanism, nearly all the newly copied DNA is found in the recipient, *MAT* locus (6).

The overall process can be followed on a Southern blot as shown in Fig. 2. In this blot the DNA was hybridized with a probe from a unique sequence distal to the *MAT*-Z sequence. Initially there are two bands representing DNA substrate for HO endonuclease and the adjacent restriction fragment. The intact *MAT*a DNA disappears rapidly after HO cleavage and is accompanied by the appearance of the transient HO-cut fragment. This fragment disappears within 3 h as a final gene conversion product is formed. In mutant strains where resection is impaired (e.g., *rad50* strains), this HO-cut DNA band will persist for up to 5 h compared to 2–3 h for wild-type cells (7, 8).

One method of following the formation of ssDNA is to use alkaline denaturing gels (9). As resection proceeds beyond the boundary of the *MAT* sequence, creating longer and longer 3′-ended tails, the ssDNA cannot be cut by most restriction enzymes. This results in a larger DNA fragment that somewhat resembles a partial restriction endonuclease digest (10). On Southern blots using native gels these larger fragments do not appear as discrete bands because degradation of the 5′ strand is not strictly synchronous among cells in the population and results in intermediates of variable molecular weight. However, on a denaturing gel the 3′-ended strand can be seen as a series of "steps" as the length of the ssDNA increases with time and as the restriction endonuclease (here, *Sty*I) fails to cleave first one and then several sites. The time-dependent formation of a ladder of such bands is shown in Fig. 2 using a strain in which repair has been prevented by deleting *HML* and *HMR*. The 3′-ended tail remarkably remains intact for an extended period of time so that a strand-specific probe to sequences near *MAT* continues to detect these long fragments for at least 6 h. In Fig. 2, a donorless strain was used so that the DSB cannot be repaired which results in more extensive DNA resection than seen in a strain capable of repairing the DSB. However even in normal *MAT* switching it is possible to see one or sometimes two *Sty*I-resistant sites as resection progresses much further than the 327 bp of the *MAT-Z* region that is shared with *HML* or *HMR* (10).

A more sensitive and quantitative method to detect the rate of formation of ssDNA is to use slot blots, since the DNA is concentrated into a small area on the blot. DNA can be bound to positively charged nylon membranes and hybridized with single-stranded probes. If native DNA is loaded onto the membrane, then the ssDNA selectively binds to the membrane and an increase in the signal can be detected as ssDNA is formed during resection. If denatured DNA is loaded, then a loss of signal will be observed

Fig. 2. (continued) when the DNA is run on an alkaline gel. DNA from a donorless *rad52* strain was used. This strain contains *GAL::HO* on a *CEN* plasmid and consequently does not cut *MAT* in the entire cell population. (**d**) Time course DNA from a strain switching from *MAT***a** to *MAT*α was analyzed by real-time PCR. Primers spanning the HO-cut site show that cutting is efficient since almost all of the PCR signal is lost by 1 h. The locations of the pX and pY primers that assay the removal of the *MAT*-Y**a** tail are shown in Fig. 25.1. High-resolution PCR assays in Hicks et al. (11) showed that the initial loss of the Y-tail occurred coincidently with strand extension assayed by using primers pA and pB. Signals were normalized to an independent site (*CEN8*). The HO-cut, *MAT*-Y tail and *MAT*-adjacent signals were normalized to the 0-h time point. DNA was extracted from a time course independent from the one shown in *panel B*.

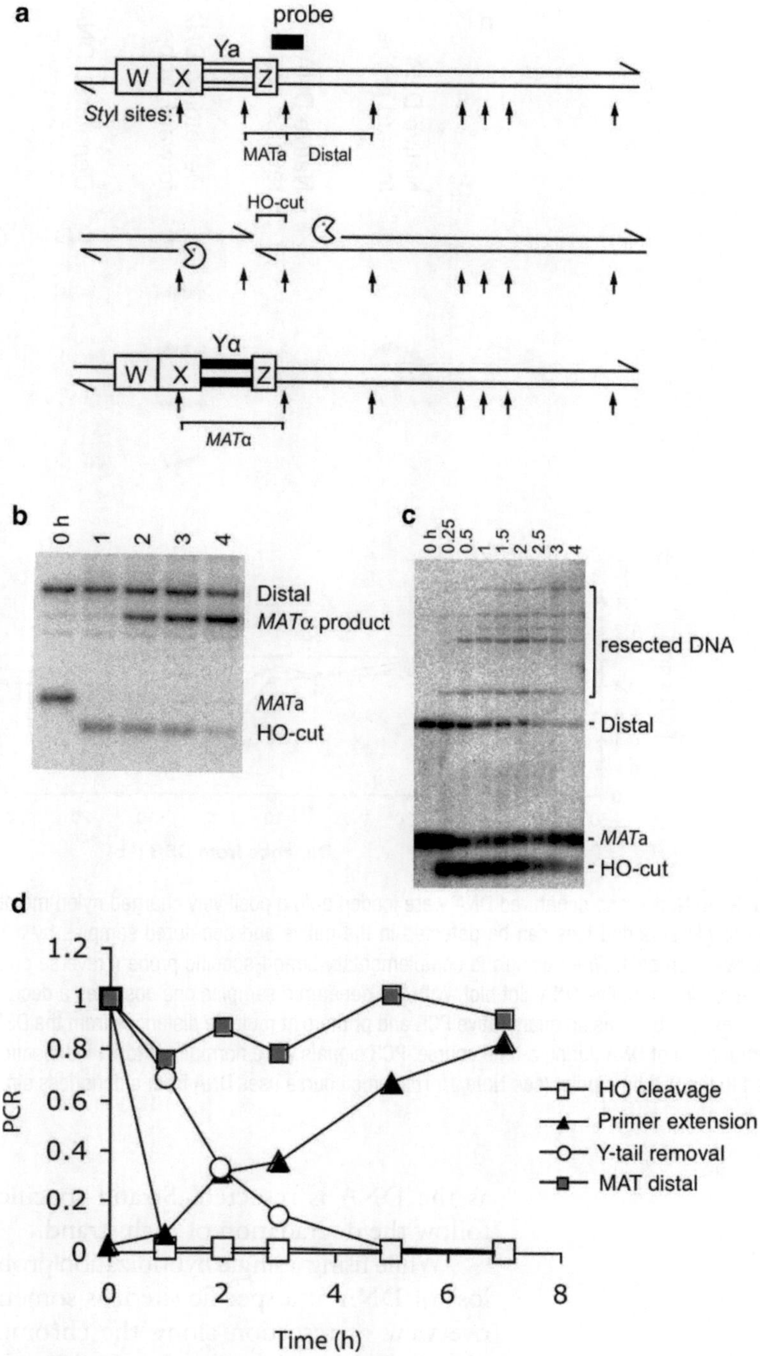

Fig. 2. Examination of *MAT* switching by Southern blots and PCR. (**a**) During a time course DNA is extracted and cut with *Sty*I (*arrows*). Southern blots are probed with an *MAT*-distal probe (*black box*) that overlaps two *Sty*I restriction fragments. *MAT*-Y**a** sequences (*white*) are replaced with *MAT*-Yα sequences (*black*). (**b**) In a native gel *MAT***a** (0.93 kb) is cut by HO resulting in a 0.72 kb band. The *MAT*-Yα sequence lacks an *Sty*I site and results in a product band of larger size (1.88 kb). (**c**) Restriction enzymes such as *Sty*I fail to cut at sites that have become single stranded. This results in a ladder of bands that increase in size (3, 5, 5.7, 6.8, and 10 kb) as the time course progresses and which can be detected

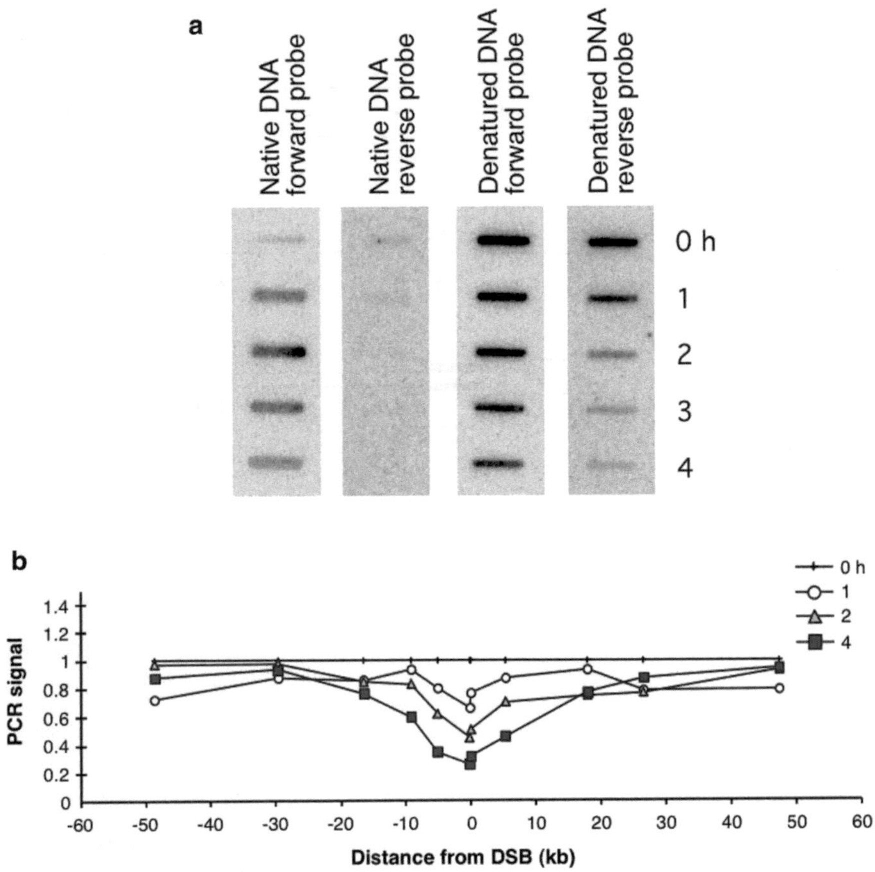

Fig. 3. (**a**) Native and denatured DNA were loaded onto a positively charged nylon membrane using a slot blot apparatus. The single-stranded tails can be detected in the native and denatured samples by using an RNA strand-specific probe ("forward probe"). When using the complementary strand-specific probe ("reverse probe") the resected DNA cannot be detected in the native DNA slot blot. With the denatured samples one observes a decrease in signal as 3′ to 5′ resection progresses. (**b**) By using quantitative PCR and primers at multiple distances from the DSB one can gain an overview of the degradation of DNA during a time course. PCR signals were normalized to an independent locus (*CEN8*) and then normalized to the 0-h time point (see Note 1). This time course uses DNA from a donorless strain cut at the *MAT* locus.

as the DNA is resected. Strand-specific probes can be used to follow the degradation of each strand.

While using a single hybridization probe is suitable for examining loss of DNA at a specific site it is sometimes desirable to gain an overview of resection along the chromosome at multiple points after DSB induction. One method is to use PCR to measure the loss of DNA at multiple points along a chromosome. PCR primers can be selected at various distances from a DSB in a donorless strain and the qPCR signal can be quantitated, normalized, and plotted as a function of distance and time (Fig. 3). Here, the reduction in the DNA signal will only decrease to about 0.5, as the 3′-ended strand remains intact (see Note 1).

PCR has also been extensively used to follow the kinetics of intermediates of *MAT* gene conversion as well as other processes.

Primers spanning the HO-cut site (pE and pF in Fig. 1b) can be used to monitor the formation and efficiency of HO cutting. Loss of the PCR signal can be seen in Fig. 2d.

The initiation of new DNA synthesis after strand invasion can be observed by using primers pA and pB as diagrammed in Fig. 1b. Primers pA and pB generate a unique product when the *MAT* ssDNA tail invades the donor and synthesizes a short region of the *MAT*-Y sequence. Appearance of this product can be observed as early as 30 min after the appearance of the DSB (10, 11).

Another key step in *MAT* switching is the nucleolytic removal of the original *MAT*-Y sequence (tail clipping) so that the second end of the DSB can act as a primer to complete the replacement of Y sequences. This step can also be monitored by PCR, by observing the loss of an amplification product that spans the junction between homologous and nonhomologous sequences (primers pX and pY in Fig. 2c). This approach has made it possible to demonstrate that tail clipping only occurs after new DNA synthesis has been initiated, as would be expected if the excision of the tail only occurred after annealing of the resected second end, containing the nonhomologous Y tail, with the displaced, extended first end (Fig. 1) (11).

Finally the completion of gene conversion can be followed by using primers pC and pD (Fig. 1b, d). This amplification product appears coincidently with the completion of switching as monitored by Southern blots (10).

ChIP has become a powerful tool to examine the recruitment of recombination proteins, end-joining proteins, chromatin remodelers, and many other proteins to the end of a DSB (12). In this method, proteins that are recruited to the DSB can be cross-linked to the DNA with formaldehyde and immunoprecipitated with antibodies specific to proteins of interest. DNA can be released from the chromatin by reversal of the cross-links and detected by using PCR primers specific for *MAT*. The recruitment of RPA to the exposed ssDNA end can be detected by anti-RPA antibodies and serves as an independent way to follow the extent of resection of the DSB end (13). The recruitment of Rad51 recombinase to form a presumed filament on ssDNA can be measured in the same way, using an anti-Rad51 antibody. The roles of other proteins in Rad51 filament assembly, such as Rad52, Rad54, Rad55, and Rad57, can also be assayed in this fashion, by using knockout mutations of these mediators (14, 15).

In addition, ChIP also allows one to monitor the time it takes for an Rad51 filament to encounter and bind to homologous donor sequences (Fig. 4) (14). Here, PCR primers specific for *MAT* can detect recruitment of Rad51 to *MAT* after DSB formation and an *HML*-specific primer can demonstrate the association of the Rad51::*MAT*-Z filament with *HML*α as strand invasion takes place. For *MAT***a** switching to *MAT*α, using *HML*α, the time from the formation of Rad51 at the cut *MAT* locus to synapsis with *HML* takes about 15 min (11).

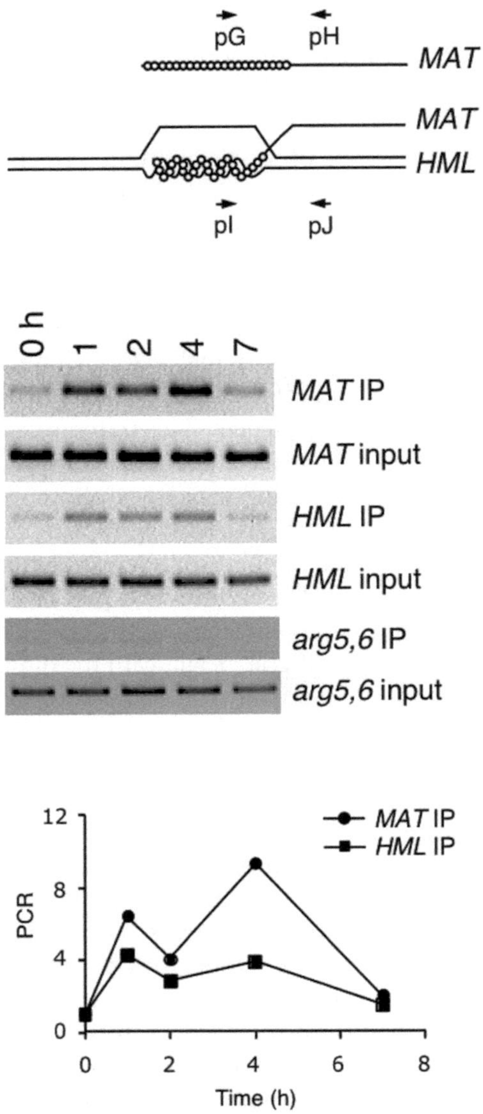

Fig. 4. Chromatin from an HO-induced strain undergoing *MAT***a** to *MAT*α gene conversion was cross-linked with formaldehyde. The chromatin was isolated and precipitated with anti-Rad51 antibodies. The cross-links were reversed and the released DNA was used in PCR reactions using primers specific for *MAT* (pG and pH) or for *HML* (pI and pJ). PCR products were run on 2 % agarose gels stained with ethidium bromide (reverse images are shown). The PCR signals were quantitated densitometrically and normalized to the input PCR signal using primers from an independent locus (*arg5,6*) (see Note 6).

2. Materials

2.1. Introduction of GAL::HO into Yeast Strains

The *HO* gene placed under the control of a galactose-inducible *GAL1–GAL10* promoter was described by Jensen and Herskowitz (2, 3). This *GAL::HO* gene can be introduced into yeast strains on centromeric plasmids such as pGAL-HO (*URA3*) (3, 16), pFH800 (*TRP1*) (17), or pJH727 (*LEU2*) (A. Plessis), but it can be more stably introduced into a chromosomal site such as *ade3* by homologous recombination (18). Here, a *URA3*-marked integrating plasmid carrying the *GAL::HO* gene is targeted into the *ade3* gene; the subsequent excision of the plasmid sequences, *URA3* and *ADE3*, can be accomplished by using 5-fluoro-orotic acid (19), leaving *GAL::HO* in chromosome 7. The plasmid contains an autonomous replication sequence and transformants need to be screened for stable integration events. Alternatively, the *GAL::HO* gene can be integrated at *ade3* using the *NATMX* cassette (pEAI201) (E. Alani).

For the substrate of HO endonuclease, the native *MAT* sequence can be used; alternatively one can insert cloned segments containing the *MAT***a** or *MAT*α cleavage site at other locations. The most commonly used site is a 117 bp subset composed of the *Bgl*II to *Hinc*II restriction fragment derived from *MAT***a** (20). However, it should be noted that the *HMR***a** locus of the genomic sequence reference strain, S288c, and its commonly used BY4743 derivatives, carry a very poorly cleaving "stuck" mutation in the Z11 position (21). Cleavage sites as small as 46 bp are efficiently cleaved (22), but a 21-bp recognition site is only poorly cut and most often results in the nicking of one strand (23).

For some experiments it is useful to delete one or both of the donor loci. Note that the silencer regions surrounding the *MAT* sequences at *HML* and *HMR* contain cryptic origins of replication (ARS sequences) so that transformants designed to replace *HML* and *HMR* coding sequences with a selectable marker often yield circularized and autonomously replicating fragments.

Strains should be tested to avoid *gal3* mutations (identified by their failure to grow on YP-galactose medium under anaerobic conditions) and respiratory-deficient (petite) strains which do not induce *GAL::HO* (24, 25).

2.2. Induction of HO Endonuclease and Extraction of DNA

1. YP-lactate medium: 3 % Lactic acid neutralized to pH 5.5 using NaOH, 1 % yeast extract, 2 % peptone. Add 8 mL of adenine solution (0.5 % adenine in 0.05 M HCl) per liter of medium if desired. YP-raffinose can be used interchangeably with YP-lactate in the following procedures: 1 % yeast extract, 2 % peptone, 40 μg/mL adenine, 2 % filter-sterilized raffinose added after autoclaving.
2. YPD: 2 % Glucose, 1 % yeast extract, 2 % peptone, 40 μg/mL adenine.

3. YP-galactose plates: 1 % Yeast extract, 2 % peptone, 40 μg/mL adenine, 2 % filter-sterilized galactose added after autoclaving, 2.5 % agar. Plates can be stored at 4 °C.
4. Selective medium (per liter): Dissolve into water 6.7 g yeast nitrogen base without amino acids, 20 g dextrose, 20 g agar, 0.87 g amino acid/base mixture, omitting one of the following supplements: 0.8 g adenine, 0.8 g methionine, 2 g phenylalanine, 8 g threonine, 0.9 g tryptophan, 1.2 g tyrosine, and 0.8 g uracil; add all but one of the ingredients to make the desired selective mixture. Autoclave.
5. 20 % Galactose (containing <0.01 % glucose from Sigma): Dissolve galactose at room temperature in water and filter sterilize. Galactose isomerizes at elevated temperatures.
6. 20 % Dextrose (autoclaved).
7. Acid-washed glass beads, 0.5 μm diameter.
8. DNA extraction buffer: 100 mM Tris–HCl, pH 8.0, 50 mM EDTA, 2 % sodium dodecyl sulfate (SDS).
9. Equilibrated phenol, pH 8.0.
10. 3 M sodium acetate, pH 5.5.
11. Isopropanol.
12. Ribonuclease A (RNase A).
13. TE: 10 mM Tris–HCl, pH 7.5, 1 mM EDTA.
14. 70 % Ethanol.

2.3. Alkaline Gel Electrophoresis

1. Alkaline gel: 30 mM NaCl, 2 mM EDTA, agarose, water.
2. 1 % Alkaline buffer: 30 mM NaOH, 2 mM EDTA.
3. 6× Alkaline-loading buffer: 0.3 M NaOH, 6 mM EDTA, 18 % Ficoll (type 400), 0.15 % bromocresol green, 0.25 % xylene cyanole FF (store at 4 °C or freeze at −20 °C).
4. 3 M sodium acetate, pH 5.5.
5. 0.5 M EDTA, pH 8.0.
6. 70 and 100 % ethanol.
7. TAE or TBE electrophoresis buffer with ethidium bromide (0.5 μg/mL):
   TAE: 0.04 M Tris–acetate, 1 mM EDTA.
   TBE: 0.09 M Tris–borate, 2 mM EDTA.

2.4. Single-Stranded DNA Slot Blot Assay

1. Gel-loading dye: 0.25 % Bromophenol blue, 0.25 % xylene cyanol, 15 % Ficoll type 400.
2. 20× SSC buffer: 3 M NaCl (175 g/L), 0.3 M Na_3 citrate (88 g/L), adjusted to pH 7.0 with 1 M HCl.
3. 0.4 N NaOH.

4. 3 M sodium acetate, pH 5.5.
5. 96 Well microtiter dish.
6. Positively charged nylon membrane used for Southern hybridizations.
7. Slot blot apparatus.

2.5. PCR of Recombination Intermediates

1. PCR master mix: 1× Taq buffer, 200 μM each dNTP, 20 nM of each primer.
2. Real-time PCR master mix: 1× SYBR Green buffer plus hot start Taq, 20 nM of each primer.
3. Primers (Figs. 1 and 3):

To detect strand invasion and extension use (shown as pA and pB in Fig. 1) GCAGCACGGAATATGGGACT (*MAT*-Y**a**) and ATGTGAACCGCATGGGCAGT (*MAT* distal primer).

To monitor completion of *MAT* switching use pC and pD in Fig. 1: AGATGAGTTTAAATCCAGCA (*MAT*-Yα) and TGTTG TCTCACTATCTTGCC (*MAT* proximal).

For real-time PCR alternatives to pA and pB that result in a shorter PCR product use TCGTCAACCACTCTACAAAACCA (*MAT*-Yα) and GGTTAAGATAAGAACAAAGAATGATGCT (*MAT* distal).

To monitor HO cutting use primers pE and pF: GATCTAAATAAATTCGTTTTCAATGATTAAAATAG (*MAT*-Y**a**) and TCGTCAACCACTCTACAAAACCA (*MAT*-Z). The latter primer must be used in a strain deleted for *HMR*-Y**a** sequences. Alternatively, the *MAT* distal primers listed above may be used. To monitor tail clipping at the *MAT* X/Y junction use TGTTACACTCTCTGGTAACTTAGG (*MAT*-X) and CCGCCATACTACAAATATCATCC (*MAT*-Y).

To monitor possible resection distal to *MAT* and also for ChIP at *MAT* (pG and pH in Fig. 4) use AGTTACCTTCGGCTTCACAA and CACCCAAGAAGGCGAATAAG.

As a control use primer sequences located near Cen8: TGACAAAACTCTCCCTTAGTGC and CTCCAACAATTACACATCCACA.

For ChIP at *HML* (pI and pJ in Fig. 4) use AGTAGCT TTCGGATGGCACA and CGTCTTGCTCTTGTTCCCAA.

For assaying degradation of DNA at various distances from the DSB at *MAT* on the centromere-proximal side: 0.186 kb CCAAATTCACAGGATAGCGTCTGG, GAAGTAGTCCCATA TTCCGTGCTG; 5.079 kb ATTGCGACAAGGCTTCACCC, CACATCACAGGTTTATTGGTTCCC; 9.144 kb TCAGGG TCTGGTGGAAGGAATG,CAAAGGTGGCAGTTGTTGAACC; 16.446 kb CGTCTTCTCAGCGAACAACAGC, GCAATAA CCCACGGAAACACTG; 29.611 kb TCGTCGTCGCCAT CATTTTC, GCCCAAGTTTGAGAGAGGTTGC.

For assaying degradation of DNA at various distances from the DSB at *MAT* on the centromere-distal side: 0.205 kb CCTGGTTTTGGTTTTGTAGAGTGG, GAGCAAGACGATGG GGAGTTTC; 5.744 kb CCAAGGAACTAATGATCTAAGCACA, ACCAGCAGTAATAAGTCGTCCTGA; 9.544 kb TGGATC ATGGACAAGGTCCTAC, GGCGAAAACAATGGCACTCT; 18.229 kb CGGTCCTCGATTTTGTTACCTTC, GCAAGG ATATTCCTGCCTTTTTC; 26.891 kb GGAAAGACTGGCT CATCAAAAC, ACATTCTCAGAGAGAACCTCCA; 47.711 kb CGGTCCTCGATTTTGTTACCTTC, GCAAGGATATTCCT GCCTTTTTC.

2.6. Chromatin Immunoprecipitation

1. 2.5 M glycine.
2. 5 mg/mL glycogen.
3. 4 M lithium chloride.
4. 0.1 M phenylmethylsulfonyl fluoride (PMSF).
5. 37 % Formaldehyde.
6. 20 mg/mL proteinase K in 10 mM Tris–HCl, pH 7.5.
7. Tris-buffered saline (TBS): 20 mM Tris–HCl, pH 7.6, 150 mM NaCl.
8. Lysis buffer: 50 mM HEPES pH 7.5, 1 mM EDTA, 140 mM NaCl, 1 % Triton X-100, 0.1 % sodium deoxycholate, 1 mg/mL bacitracin, 1 mM benzamidine, 1 mM PMSF.
9. High-salt lysis buffer: Same as lysis buffer except 0.5 M NaCl.
10. Wash buffer: 10 mM Tris–HCl, pH 7.5, 1 mM EDTA, 0.25 M LiCl, 0.5 % IGEPAL CA-630, 0.5 % sodium deoxycholate.
11. 10 mM Tris–HCl, pH 7.5, 1 mM EDTA.
12. Elution buffer: 50 mM Tris–HCl, pH 7.5, 1 mM EDTA, 1 % SDS.
13. TE/0.67 % SDS: 10 mM Tris–HCl, pH 7.5, 1 mM EDTA, 0.67 % SDS.
14. TE/1 % SDS: 10 mM Tris–HCl, pH 7.5, 1 mM EDTA, 1 % SDS.

3. Methods

3.1. Induction of Mating-Type Switching and Extraction of Time Course DNA

1. Grow cells from a single colony overnight in 2 mls at 30 °C in YPD. Selective medium should be used if the strain contains a plasmid that needs to be maintained and propagated. For some strains that grow slowly it may be necessary to a use a larger inoculum.
2. Spin down cells and discard the supernatant. Resuspend the pellet in an equal or larger volume of YP-lactate medium. Spin

the cells down again and resuspend in 4 mls of YP-lactate medium to remove residual glucose. Shake the cells at 30 °C for 3–10 h or until the cells begin to grow again. Strains undergo a lag phase where they do not grow upon the shift to YP-lactate medium. The pre-growth in lactate medium in this step is intended to overcome the lag phase and makes it easier to obtain the proper cell density of the final culture particularly when multiple strains are being grown in parallel.

3. Inoculate YP-lactate medium with enough cells to give 5×10^6 cells/mL the next day. Shake the culture vigorously overnight at 30 °C using a flask large enough to provide sufficient aeration (for example, 0.5 L in a 2 L flask with a foam plug). Most strains have a doubling time of about 3 h in YP-lactate.
4. When cells have grown to a density between 5×10^6 cells/mL and 10^7 cells/mL take a 50 mL sample for the uninduced control and then add 20 % galactose to a final concentration of 2 % to the remaining culture. Continue shaking culture at 30 °C and take 50 mL samples at the necessary time points. Centrifuge the cells down and resuspend the cells in 1 mL of distilled water, transfer to a microcentrifuge tube, spin, and discard the supernatant. The pellet may be stored at −70 °C at this point.
5. For experiments studying *MAT* switching, add 1/10th volume of 20 % glucose to down-regulate the HO induction after 30 min (or 60 min) to prevent recutting of the HO-cut site after gene conversion has occurred. For strains where *MAT* gene converts to a *mat*-inc mutation or in recombination assays where the HO-cut site is removed (such as in single-strand annealing), addition of glucose is unnecessary. Usually *MAT* gene conversion assays are carried out for 5 h. For longer time courses the glucose will get used up and allow recutting by HO endonuclease. In this case, an additional aliquot of glucose (1/10th volume of 20 % glucose) can be added at 6 h to maintain HO repression.
6. At the initial time point, before addition of galactose, and at the desired time points, dilute a sample of the culture by 10^{-4} and plate on YPD. Usually 50 μL per plate for three plates is adequate, although more should be plated if viability is expected to be low due to inability to repair the break. After the colonies grow they can be analyzed to determine (1) the fraction of colonies with recombined substrates, (2) the viability of the strain after HO induction, or (3) the stability of the *GAL::HO* plasmid or other plasmids if they were used. Cells can also be plated on YP-galactose plates for strains where *MAT* gene converts to a *mat*-inc mutation or in recombination assays where the HO-cut site is removed by recombination.
7. Resuspend the cells in 400 μL of extraction buffer and mix with 500 μL of glass beads and 400 μL of phenol. The glass

bead/phenol mixture may be prepared beforehand and stored at 4 °C or on ice (spin briefly in a microcentrifuge to get the phenol to the bottom of the tube).

8. Vortex tube for 2 min on high. The extent of vortexing needs to be determined empirically since longer times lead to DNA shearing and shorter times will lead to lower yields of DNA.
9. Centrifuge the samples in a microcentrifuge for 10 min. The samples can be allowed to sit on ice for 1 min before centrifugation to improve the phase separation.
10. Carefully remove the top aqueous layer to a new tube while avoiding the interface. Most variations on carrying out phenol extractions will work.
11. Add 40 μL of 3 M sodium acetate pH 5.5 and 600 mL isopropanol. Mix. Low-temperature (–20 °C) precipitations are not recommended.
12. Spin in a microcentrifuge for 10 min. Discard the supernatant.
13. Add 200 μL TE containing 10 mg of RNase A to each tube. Incubate at 37 °C for 30–60 min with occasional, minimal agitation to resuspend the pellet. RNA is mostly digested when the pellet dissolves.
14. Add 30 μL of 5 M NaCl and 200 μL of isopropanol and centrifuge for 10 min. The DNA pellet may be small. Discard the supernatant. Carefully rinse the pellet with 70 % ethanol and dry the pellet. Fully dissolve the pellet in TE (200 μL for each 50 mls of cells).
15. Digest 30–50 % of the sample with restriction enzymes for use on Southern blots (see Note 2).

3.2. Detection of Single-Stranded DNA Using Alkaline Denaturing Gels

1. Melt agarose in 50 mM NaCl and 1 mM EDTA, pour into a gel tray, and allow to solidify. Submerge the gel in a gel box in 50 mM NaOH and 1 mM EDTA and allow to equilibrate for 30 min or longer. This procedure is used because agarose will decompose if boiled under alkaline conditions. Alternatively melt the agarose using water and cool to 60 °C. At that temperature the agarose can be adjusted to 50 mM NaOH and 1 mM EDTA by addition of concentrated NaOH and EDTA. Ethidium bromide is omitted because it does not efficiently bind to DNA under these conditions.
2. Prepare the DNA samples by adjusting the solution to 0.3 M sodium acetate and 5 mM EDTA followed by addition of 2 volumes of ethanol to precipitate the DNA. EDTA is included to prevent the magnesium from precipitating out under alkaline conditions. After chilling the samples on dry ice and centrifuging in a microcentrifuge (10 min), the supernatant is discarded and the pellet is rinsed in 70 % ethanol. As much of

the supernatant as possible should be removed before drying the pellet. Dissolve the pellet in alkaline gel loading buffer. Size markers can be denatured by directly adding alkaline gel loading buffer to a small volume of plasmid or phage DNA digested by restriction enzymes.

3. After loading the DNA, a glass plate can be (optionally) placed on the gel to prevent the dye from diffusing from the agarose during the course of the run. This is generally not necessary for short electrophoresis runs. Because of the large currents that can be generated with denaturing gels, gels are usually run slowly at lower voltages (see Note 3).
4. After the DNA has migrated far enough, remove the gel and if desired, stain the gel with ethidium bromide (0.5 μg/mL) in 1× TAE electrophoresis buffer using two 15-min washes. The DNA will be faint because the DNA is single stranded.
5. Soak the gel in 0.25 N HCl for 6–7 min with gentle agitation.
6. Rinse the gel with water and soak the gel in 0.5 N NaOH and 1.5 M NaCl for 30 min with gentle agitation.
7. Rinse briefly with water and apply the gel to the DNA transfer apparatus. The DNA can be blotted onto either neutral or positively charged nylon membranes using 2× or 10× SSC by various methods such as a capillary transfer table or a vacuum blotter.
8. Southern hybridizations can be carried out according to different protocols. We hybridize according to the protocol of Church and Gilbert (26). We prepare RNA probes according to Melton et al. (27) and random-primed DNA probes according to Feinberg and Vogelstein (28).

3.3. Detection of Single-Stranded Tails by Using Slot Blots

Three sets of samples are loaded in a typical slot blot experiment described below. When using native samples, only the single-stranded tails bind to the membrane. The denatured samples can be used for assessing the degradation of the 5′ to 3′ strand. Finally various controls can be included to assure accuracy.

1. Native samples: Dilute 1 μg of yeast genomic DNA from the time course samples to 200 μL and adjust the final concentration to 10× SSC. This will be enough for two slot blots, one for each strand-specific probe.
2. Denatured samples: Denature 0.1 μg of genomic DNA by adding NaOH to a final concentration of 0.4 N NaOH. Neutralize with 3 M sodium acetate pH 5.5.
3. Quantitation samples and other controls: Digest a plasmid containing the target sequence with a restriction enzyme and prepare samples ranging in amount from 0.05 to 1.0 pg of hybridizable sequence. Denature the samples with 0.4 N NaOH and then neutralize the samples using sodium acetate

as described in the previous step. These samples will be useful in deriving a calibration curve if the signals need to be quantitated using light-based densitometry.

For other controls include a sample with non-denatured plasmid DNA (0.01 ng of hybridizable sequence), samples of nonradioactive RNA transcripts complementary to the RNA probes that will be used (0.001 ng of hybridizable sequence), and, if possible, denatured genomic DNA lacking hybridizable sequences as a negative control.

4. Load the samples into the wells of the slot blot or dot blot apparatus and transfer the DNA to a nylon membrane. Repeat this with a second nylon membrane so that there is one blot for each complementary strand-specific probe.
5. After the samples have transferred to the membrane, rinse the sides of each well with 500 μL of 10× SSC.
6. The DNA is then cross-linked to the membrane with UV light and hybridized according to the protocol of Church and Gilbert (26) or other suitable hybridization protocol.

3.4. Chromatin Immunoprecipitation

3.4.1. Chromatin Cross-Linking with Formaldehyde

1. When the culture is between 5×10^6 and 1×10^7 cells/mL, collect a 45 mL sample in a 50 mL conical tube (see Note 4). Cross-link the chromatin by adding 1.2 mls of 37 % formaldehyde. Incubate the sample on a platform-rocker for 10 min at room temperature (see Note 5).
2. Dilute an aliquot of cells by 10^{-4} and plate 100 μL on several plates (before induction and at 4 or 5 h after induction of *HO*).
3. Quench the formaldehyde by adding 2.5 mL of 2.5 M glycine and incubate for 5 min at room temperature with occasional rocking.
4. Centrifuge the tubes for 5 min, discard the supernatant, and resuspend the cells in 1 mL of ice-cold TBS in a microcentrifuge tube. Repeat the wash twice with TBS while removing as much supernatant as possible. After the final spin, remove the supernatant. At this point the pellet may be frozen and stored at −70 °C.

3.4.2. Cell Lysis and Sonication

1. Resuspend the cell pellet in 0.55 mL of cold lysis buffer. Transfer the cell lysate to a microcentrifuge tube containing 0.5 mL of glass beads on ice. Lyse the cells on a multi-tube vortexer for 30–60 min at 4 °C until 90 % lysis is achieved (see Note 5).
2. Make a very small hole in the bottom of the tubes with a red hot syringe needle, heated with a Bunsen burner. Place a tube inside of a 15 mL conical centrifuge tube such that the microcentrifuge tube rests at the lip of the conical tube. Repeat for each sample and centrifuge the tubes to allow the whole cell lysate to escape through the small hole.

3. Resuspend the pellet in the supernatant at the bottom of the conical tube and transfer the lysate to 2 mL microcentrifuge tubes taking care not to transfer any glass beads. Sonicate the samples on ice until the DNA has an average size of 500 bp (see Note 5).
4. Centrifuge the lysate for 10 min at 4 °C, transfer the supernatant to a new tube, and repeat the spin. Transfer the supernatant to a new tube on ice and freeze at −70 °C. The supernatants can be analyzed by western blots if desired.

3.4.3. Chromatin Immunoprecipitation

1. For the non-immunoprecipitated (input) samples, aliquot 50 μL of lysate to a new tube. Store these samples on ice until the immunoprecipitated (IP) sample is ready for reversal of the cross-links.
2. For the IP sample, add 350 μL of lysate to a chilled tube containing the antibody (see Note 5). Freeze the remaining lysate and save for use as the pre-IP control on a western blot. Incubate the IP tubes at 4 °C for 1 h.
3. Gently spin protein G or protein A agarose beads down at 800 × *g* for 1 min in a 2 mL tube and remove the supernatant. Wash the beads twice with 1 mL of lysis buffer. Resuspend the beads in the original volume using lysis buffer. Add 30 μL of the protein G or protein A agarose beads to each IP sample. Incubate the samples at 4 °C with slow agitation for 1 h.
4. Pellet the beads for 1 min at 800 × *g*. Remove the supernatant and save at −70 °C. This can be analyzed with a western blot to check the efficiency of the antibody binding to the protein. Wash the beads with the following buffers for 5 min at room temperature, centrifuging at 800 × *g* for 1 min:
   Two times with lysis buffer.
   One time with high-salt lysis buffer.
   One time with wash buffer.
   One time with TE.
5. Add 100 μL of elution buffer and incubate at 65 °C for 15 min to elute the precipitate.
6. Pellet the beads for 1 min at 15,800 × *g* and save the eluate in a fresh tube. Add 150 μL of TE/0.67 % SDS solution to the tubes containing the beads and incubate at 65 °C for 10 min. Pellet the beads for 1 min and combine with the first eluates. Add 200 μL of TE/1 % SDS solution to the Non-IP samples and incubate both the IP and Non-IP samples at 65 °C for 6 h to reverse the cross-links.
7. Add 250 μL TE containing proteinase K (0.4 mg/mL) and glycogen (0.1 mg/mL). Incubate the samples for at least 2 h at 37 °C.

3.4.4. DNA Precipitation and PCR

1. Add 55 μL of 4 M LiCl to the samples and transfer to 500 μL of phenol in new tubes on ice. Vortex briefly, centrifuge, and transfer the aqueous phase to new tubes. Precipitate the DNA with 1 mL of ethanol and centrifugation. Wash the pellet with 70 % ethanol. Centrifuge for 10 min and discard the supernatant. Dissolve the DNA in TE after the pellets have dried.
2. The DNA can be analyzed using conventional or real-time PCR. For conventional PCR, a typical program is 1 min 94 °C, 26 cycles of 15 s 94 °C, 20 s 55 °C, and 20 s 72 °C, and 5 min at 72 °C. PCR products are chosen to be about 300 bp long and can be visualized on a 2 % agarose gel. Adjust the conditions to obtain signals in the linear range of a calibration curve prepared with dilutions of the Non-IP 0-h sample. If using real-time PCR, the products should be 100–300 bp in length and the 0-h Non-IP sample can be used for the standardization curve. A hot-start Taq polymerase is recommended to minimize competing PCR products. For controls, Non-IP and IP samples from an independent locus should be amplified and quantitated. A sample not treated with formaldehyde provides a control to determine if protein binding may have occurred after lysis. PCR signals should be normalized to the PCR signal from an independent locus (see Notes 6 and 7).

3.5. Monitoring Recombination Intermediates by PCR

DNA extracted from the time course or from ChIP can be analyzed by PCR. Real-time PCR can be used when the PCR products can be designed to be short (<300 bp). When the products are large (>300 bp), such as when using the pA and pB primers in Fig. 25.1, then conventional end-point PCR is used where the products are run on agarose gels and then quantitated by densitometry.

1. Dilute the genomic DNA from a time course 1:20 with TE and aliquot the samples in PCR tubes.
2. For a standardization curve, prepare four serial dilutions (1:5) of DNA from the preinduction sample (0 h). The dilutions should be adjusted as needed to encompass the range of PCR signals acquired.
3. Prepare a master mix of all the reaction components except the genomic DNA utilizing a hot start Taq for conventional PCR or an SYBR Green hot start Taq for real-time PCR. Add the master mix to the sample DNA.
4. For conventional PCR use 94 °C for 1 min and 25 cycles of 94 °C for 10 s, 55 °C for 20 s, and 72 °C for 0:20 to 1:30 min. The number of cycles should be adjusted so that the PCR signals remain in the linear range based on the signals from the serial dilutions. For real-time PCR use 94 °C for 10 min and 40 cycles of 94 °C for 10 s, 55 °C for 20 s, and 72 °C for 20 s.

The length of the initial cycle at 94 °C should be adjusted to accommodate the manufacturer's recommendation for the activation of the hot start Taq.

4. Notes

1. In principle qPCR measurements of resected DNA should give a value of 0.5 relative to double-stranded DNA. In practice lower values (0.2–0.5) are observed in samples taken at 6–12 h after the DSB is made. Analysis of strains that start to repair a DSB at 6 h by single-strand annealing (29) shows that viability is approximately 85 % which indicates that the 3′ tail containing the homologous repeats is relatively stable. This leads us to conclude that a small percentage of the ssDNA is not recovered during the DNA extraction in this protocol and that there is a loss of stability of the tail near the DSB at later time points (≥8 h).
2. The DNA extraction protocol based on glass beads described above has the advantage of breaking the cells open and rapidly stopping any recombination in progress. A disadvantage, however, is that traces of some of the components of the extraction buffer (EDTA, SDS, phenol) may inhibit downstream applications such as restriction digestion or PCR. This problem can sometimes be overcome by diluting out the inhibitors (e.g., using 300–600 μL for restriction digests or by diluting PCR reactions 1:50 instead of 1:20). An additional DNA precipitation may also alleviate this problem.
3. One problem encountered with alkaline denaturing gels is that moderate voltages will generate large currents that in turn generate excessive heat. Several steps can be taken to alleviate this. The amount of current drawn can be reduced by pouring the gel as thin as possible and by using a minimal amount of running buffer. Run the gel at a low voltage with circulating buffer. Gels can be run in a cold room at 4 °C. Also part way through the run, the gel-running buffer can be replaced with fresh buffer.
4. When collecting cells for ChIP over a long time course (>4 h), the cell pellets will become much larger, either because the cells have repaired the DSB and have resumed growth or because the cells have arrested in G2 of the cell cycle and are acquiring more mass. Since this may have an effect on cross-linking, the volume of cells harvested can be reduced and diluted with YP-lactate by taking OD_{600} measurements and diluting the cells accordingly.

5. Optimizing ChIP: (1) The extent of cross-linking can be increased by increasing the incubation time with formaldehyde. This may also decrease the specificity of cross-linking and decrease the ability to use ChIP to follow protein recruitment during a time course. (2) This protocol aims for lysis of about 90 % of the cells. The duration of vortexing can be optimized by using a microscope to assess the amount of lysis. (3) The sonication dosage should be adjusted to yield an average DNA size of 0.5 kb or less with the size ranging up to approximately 1 kb or less. The size range of the sonicated DNA can be checked by reversing the cross-links of the input samples, treating with RNase A and running the DNA on a 0.8 % agarose gel. (4) Multiple antibody dosages can be tested to determine the dosage where a sufficient amount of antibody is present in all of the time point immunoprecipitations.
6. Normalization of ChIP data: There are multiple methods of normalizing ChIP data of which we describe two. The first method normalizes to the IP samples of the control locus and the second method normalizes to the input samples of the control locus. The first method normalizes the IP time point signals to the input signal at 0 h for the locus of interest (e.g., the *MAT* site) and normalizes the independent control locus in the same manner. This is conveniently accomplished by using dilutions of the 0-h input sample for the standardization curve. The ratios of the *MAT* IP to control IP signal for each time point are calculated and then normalized to the 0-h time point. The second method normalizes the IP signals to the input signal at 0 h for both the *MAT* and control loci as described above. The ratio of the *MAT* IP signal to the control input signal can then be calculated for each time point. This method is useful if the signals from the 0-h IP and/or control IPs are very low or if the *MAT* and control IP signals are not completely independent. One can imagine, for example, that recruitment of a low-abundance protein to a DSB may reduce its recruitment to the supposedly independent locus.
7. The NATMX cassette is sometimes used in yeast molecular biology as a selectable marker conferring drug resistance to nourseothricin (30). This is a GC-rich sequence and we have noticed that it sometimes gives rise to anomalously high qPCR readings when using primer sequences near the cassette (within 5 kb). The addition of 1 % DMSO to the PCR reaction aids in obtaining proper readings when using this sequence.

References

1. Pâques F, Haber JE (1999) Multiple pathways of recombination induced by double-strand breaks in *Saccharomyces cerevisiae*. Microbiol Mol Biol Rev 63:349–404
2. Jensen R, Sprague GF, Herskowitz I (1983) Regulation of yeast mating-type interconversion: feedback control of HO gene expression by the mating-type locus. Proc Natl Acad Sci U S A 80:3035–3039
3. Jensen RE, Herskowitz I (1984) Directionality and regulation of cassette substitution in yeast. Cold Spring Harb Symp Quant Biol 49:97–104
4. Kaplun L, Ivantsiv Y, Bakhrat A, Raveh D (2003) DNA damage response-mediated degradation of HO endonuclease via the ubiquitin system involves its nuclear export. J Biol Chem 278:48727–48734
5. Haber JE (1998) Mating-type gene switching in *Saccharomyces cerevisiae*. Annu Rev Genet 32:561–599
6. Ira G, Satory D, Haber JE (2006) Conservative inheritance of newly synthesized DNA in double-strand break-induced gene conversion. Mol Cell Biol 26:9424–9429
7. Ivanov EL, Sugawara N, White CI, Fabre F, Haber JE (1994) Mutations in *XRS2* and *RAD50* delay but do not prevent mating-type switching in *Saccharomyces cerevisiae*. Mol Cell Biol 14:3414–3425
8. Sugawara N, Haber JE (1992) Characterization of double-strand break-induced recombination: homology requirements and single-stranded DNA formation. Mol Cell Biol 12:563–575
9. Sambrook J, Fritsch EF, Maniatis T (1989) Molecular cloning: a laboratory manual. Cold Spring Harbor Laboratory Press, New York, NY
10. White CI, Haber JE (1990) Intermediates of recombination during mating type switching in *Saccharomyces cerevisiae*. EMBO J 9:663–673
11. Hicks WM, Yamaguchi M, Haber JE (2011) Real-time analysis of double-strand DNA break repair by homologous recombination. Proc Natl Acad Sci U S A 108:3108–3115
12. Collas P (2009) The state-of-the-art of chromatin immunoprecipitation. Methods Mol Biol 567:1–25
13. Wang X, Haber JE (2004) Role of *Saccharomyces* single-stranded DNA-binding protein RPA in the strand invasion step of double-strand break repair. PLoS Biol 2:E21
14. Sugawara N, Wang X, Haber JE (2003) In vivo roles of Rad52, Rad54, and Rad55 proteins in Rad51-mediated recombination. Mol Cell 12:209–219
15. Wolner B, van Komen S, Sung P, Peterson CL (2003) Recruitment of the recombinational repair machinery to a DNA double-strand break in yeast. Mol Cell 12:221–232
16. Herskowitz I, Jensen RE (1991) Putting the HO gene to work: practical uses for mating-type switching. Methods Enzymol 194:132–146
17. Nickoloff JA, Singer JD, Hoekstra MF, Heffron F (1989) Double-strand breaks stimulate alternative mechanisms of recombination repair. J Mol Biol 207:527–541
18. Sandell LL, Zakian VA (1993) Loss of a yeast telomere: arrest, recovery, and chromosome loss. Cell 75:729–739
19. Boeke JD, Trueheart J, Natsoulis G, Fink GR (1987) 5-Fluoroorotic acid as a selective agent in yeast molecular genetics. Methods Enzymol 154:164–175
20. Kostriken R, Strathern JN, Klar AJ, Hicks JB, Heffron F (1983) A site-specific endonuclease essential for mating-type switching in *Saccharomyces cerevisiae*. Cell 35:167–174
21. Ray BL, White CI, Haber JE (1991) Heteroduplex formation and mismatch repair of the "stuck" mutation during mating-type switching in *Saccharomyces cerevisiae*. Mol Cell Biol 11:5372–5380
22. Pâques F, Haber JE (1997) Two pathways for removal of nonhomologous DNA ends during double-strand break repair in *Saccharomyces cerevisiae*. Mol Cell Biol 17: 6765–6771
23. Cortes-Ledesma F, Aguilera A (2006) Double-strand breaks arising by replication through a nick are repaired by cohesin-dependent sister-chromatid exchange. EMBO Rep 7:919–926
24. Evans IH, Wilkie D (1976) Mitochondrial factors in the utilization of sugars in *Saccharomyces cerevisiae*. Genet Res 27:89–93
25. Mahler HR, Wilkie D (1978) Mitochondrial control of sugar utilization in *Saccharomyces cerevisiae*. Plasmid 1:125–133
26. Church GM, Gilbert W (1984) Genomic sequencing. Proc Natl Acad Sci U S A 81:1991–1995

27. Melton DA, Krieg PA, Rebagliati MR, Maniatis T, Zinn K, Green MR (1984) Efficient in vitro synthesis of biologically active RNA and RNA hybridization probes from plasmids containing a bacteriophage SP6 promoter. Nucleic Acids Res 12:7035–7056

28. Feinberg AP, Vogelstein B (1984) A technique for radiolabeling DNA restriction endonuclease fragments to high specific activity. Addendum. Anal Biochem 137:266–267

29. Vaze MB, Pellicioli A, Lee SE, Ira G, Liberi G, Arbel-Eden A, Foiani M, Haber JE (2002) Recovery from checkpoint-mediated arrest after repair of a double-strand break requires Srs2 helicase. Mol Cell 10:373–385

30. Wach A, Brachat A, Pohlmann R, Philippsen P (1994) New heterologous modules for classical or PCR-based gene disruptions in *Saccharomyces cerevisiae*. Yeast 10:1793–1808

Chapter 26

Monitoring I-SceI-Induced Double-Strand Break Repair in DT40 Cells

David du Plessis and Julian E. Sale

Abstract

The chicken B lymphocyte cell line DT40 is now a well-established system for studying DNA repair in vertebrates. Central to its utility is its ready genetic tractability allowing for the creation of isogenic mutants in any chosen gene. The combination of this genetic tractability with the I-SceI-induced double-strand break (DSB) repair assay, pioneered by the laboratory of Maria Jasin, has proven particularly powerful for dissecting the pathways of DSB repair in a vertebrate setting.

Key words: DT40, DNA double-strand break, I-SceI, Homologous recombination, Nonhomologous end-joining

1. Introduction

The use of double-strand break (DSB) repair assays based on the rare-cutting endonuclease I-SceI from *Saccharomyces cerevisiae* was pioneered by the laboratory of Maria Jasin. Initially applied to mammalian cells (1, 2), the assay was introduced into DT40 in 2001 (3). Although there now exist a number of variants of the assay (reviewed in ref. 4), the basic principle is simple. A reporter gene, initially neomycin phosphotransferase (2) and later green fluorescent protein (GFP) (5), is inactivated by the incorporation of the 18 base-pair recognition site for the I-SceI endonuclease. Downstream, a truncated, and therefore also inactive, copy of the reporter is provided to act as a recombination donor for the repair of an I-SceI-induced break. Successful homology-directed repair results in reconstitution of an active reporter gene resulting in either the acquisition of resistance to neomycin or expression of GFP (Fig. 1). Recently, the use of a GFP-based construct, such as DR-GFP (5) has been preferred because of the ease with which subsequent analysis is performed.

Lotte Bjergbæk (ed.), *DNA Repair Protocols*, Methods in Molecular Biology, vol. 920,
DOI 10.1007/978-1-61779-998-3_26, © Springer Science+Business Media New York 2012

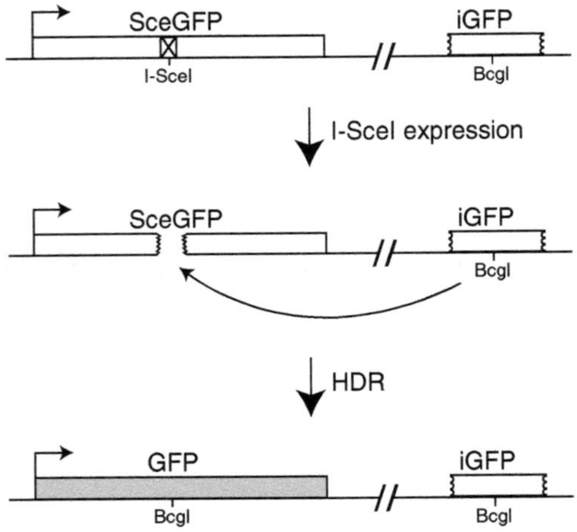

Fig. 1. Summary of the basic I-SceI break repair assay (adapted from ref. 5). SceGFP is GFP disrupted by the insertion of the 18 base pair I-SceI restriction site. 3′ of this a truncated GFP with sequence spanning the I-SceI site is provided as a donor sequence for homology directed repair. Successful repair results in restoration of the correct open reading frame of GFP and introduction of the BcgI site from iGFP into SceGFP and the expression of functional GFP.

Critical to the reproducibility and success of the assay is a matched integration of the reporter construct in the DT40 lines to be compared. This can be achieved either by the random integration of a single copy of the reporter into DT40 with subsequent disruption of the genes under study in this line (6, 7) or the targeted integration of a single copy of the reporter into a neutral genomic locus, such as ovalbumin, that is not expressed in DT40 (3).

This method therefore primarily deals with I-SceI DSB reporters based on GFP, in particular the DR-GFP reporter (5), and focuses on the practical approach to measuring I-SceI-induced homologous recombination (HR) in established DT40 reporter cell lines.

2. Materials

1. I-SceI reporter construct vector or a DT40 cell line with the reporter construct stably integrated as a single copy (see Note 1).
2. pCβASce I-SceI-NLS expression vector (8) (see Note 2).
3. ptdTomato-C1 (Clontech) or other suitable RFP expression plasmid (see Note 3).
4. Amaxa Nucleofector II® (Lonza AG).
5. Cell line Nucleofector® kit T (Lonza AG).
6. Supplemented RPMI (prewarm to 37 °C): 500 mL RPMI 1640 + Glutamax (Invitrogen), 35 mL (7 %) fetal bovine serum, 15 mL (3%) chicken serum, 50 μM final concentration beta-mercaptoethanol, penicillin 10 U/mL and streptomycin 0.1 mg/mL.

7. Cell counter/bright line hemocytometer.
8. Flow cytometer with a minimum of: 488 nm excitation laser, forward- and side-scatter detectors and filters, two fluorescence PMT detectors with 530 and 585 nm bandpass filters.
9. Laminar flow hood.
10. 37 °C, 10 % CO_2, humidified incubator (see Note 4).
11. 15 mL centrifuge tubes.
12. 6-well tissue culture plates.
13. Sterile 1× TE buffer: 10 mM Tris–HCl pH 8.0, 1 mM EDTA.
14. Sterile filtered pipette tips for tissue culture work.
15. Sterile unplugged pipettes and a vacuum trap and pump for aspiration.
16. Sterile 1× PBS: 140 mM NaCl, 10 mM sodium phosphate pH 7.4.
17. 5 mL polystyrene round-bottomed FACS tubes (e.g., Falcon).

3. Methods

This method will focus on the use of the most commonly used I-SceI DSB reporter construct in DT40, DR-GFP. It will assume that reporter lines carrying this construct are available (see Note 1).

3.1. Amaxa Nucleofection

1. The propagation of DT40 is covered in detail elsewhere (9). We have routinely cultured DT40 at 37 °C although they will grow happily at temperatures up to 41 °C (see Note 4). All handling steps of the protocol are performed in a laminar flow hood. This protocol assumes you are familiar with the operation of the Amaxa Nucleofector II® and your flow cytometer.
2. Reporter DT40 lines should be maintained at between 0.2 and 1×10^6 cells/mL and ideally should exhibit >90 % viability as assessed by trypan blue exclusion.
3. Add Supplement 1 to the Amaxa nucleofector solution according to the kit instructions, seal and mix well by inversion, then prewarm to room temperature.
4. Mix 27 μg pCβASce and 3 μg ptdTomato (9:1 ratio, 30 μg DNA total) per transfection (see Note 3), then make up to a final volume per transfection of 10 μL with TE. Prepare enough of this assay master mix to perform the entire experiment in triplicate.
5. Prepare a ptdTomato-only ("Red-only") control by diluting 3 μg ptdTomato in 10 μL final volume TE buffer. pMax-GFP, supplied in the Amaxa kit, serves a GFP positive control and is prepared in the same way. These controls serve to allow accurate gating and compensation of the FACS plots.

6. Label the wells of a 6-well plate with the names of the samples to be transfected, ensuring you prepare wells for the RFP only, GFP only and untransfected controls (see Note 5). Add two million wild type cells in 5 mL supplemented RPMI to your untransfected control well. Fill the other wells with 5 mL supplemented RPMI. Store the sealed plate in the incubator until ready to transfect.
7. Aliquot three million cells per transfection into prelabeled 15 mL centrifuge tubes. Spin down at 500×*g* for 5 min (see Note 6). During the spin, remove the Amaxa electroporation vials from their packets and prepare your work area. (see Note 7)
8. Set the Nucleofector II to program B-23.
9. Aspirate the supernatant from one transfection sample, removing as much medium as possible (see Note 8).
10. Resuspend the pellet in 100 μL supplemented Amaxa nucleofector solution by gentle pipetting.
11. Add 10 μL of 9:1 pCβASce/ptdTomato master mix and mix thoroughly by gentle pipetting with a 200 μL pipette tip, then immediately transfer to an Amaxa cuvette, seal and transfect on program B-23.
12. Using the sterile Pasteur pipette supplied in the Amaxa kit, transfer the full volume of the transfected sample to the medium in its respective well of the labeled 6-well plate.
13. Repeat steps 9–12 for each other sample in turn. (see Note 9)
14. Incubate at 37 °C with 10 % CO_2 (see Note 4) and 95 % relative humidity in the dark for 24 h.

3.2. FACS Analysis to Determine the Efficiency of Homology Directed Repair

1. Resuspend your untransfected control sample and transfer 2.5 mL to a polystyrene FACS tube. Resuspend the other samples and transfer the entire volume of each to 5 mL polystyrene FACS tubes. Centrifuge for 7 min at 500×*g*, aspirate supernatant, resuspend in 250–500 μL PBS, and place tubes on ice. They are now ready to analyze by flow cytometry.
2. Gating is performed as shown in Fig. 2. 100,000 live cells should be counted to give the most accurate possible result.
3. Calculate the transfection efficiency-corrected percentage of homologous repair using the following equation:

$$\text{HDR}(\%) = \left[(Q2 + Q4)/(Q1 + Q2)\right] \times 100$$

where Q1–4 correspond to the number of cells in the four quadrants indicated in Fig. 2.

4. It is possible to carry out further analysis of the results of repair of SceGFP by PCR and Southern blotting. Full details can be found in ref. 4.

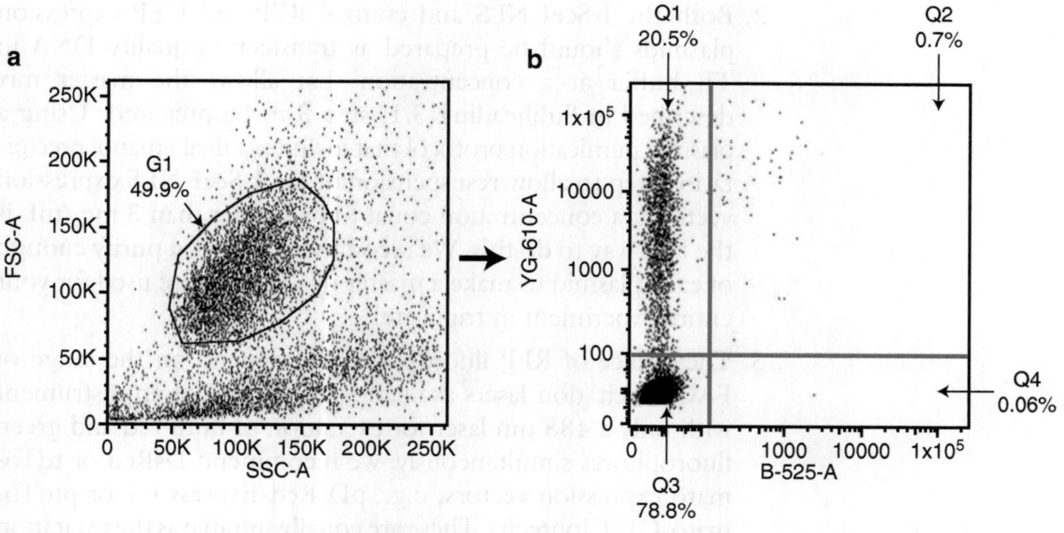

Fig. 2. FACS gating for the analysis of GFP-positive cells. (**a**) First gate on forward (FSC-A) and side scatter (SSC-A). Plot the cells in G1 with tdTomato expression (YG-610-A) on the *Y*-axis and GFP (B-525-A) on the *X*-axis. Draw quadrant gates Q1–4 as indicated.

4. Notes

1. A number of different I-SceI reporter constructs have been described that are able to monitor different aspects of double-strand break repair. This method focuses on the most commonly used of these in DT40, pDR-GFP (5). The details of the variants of this assay are beyond the scope of this method but have been thoroughly explored elsewhere (4). However, the practical method of transiently expressing the I-SceI endonuclease is the same in all cases. In all of these assay systems, it is important to note that the efficiency of HR in response to an I-SceI DSB is highly dependent on the site of integration of the reporter construct. Therefore, it is essential that reporter cell lines that are isogenic for reporter construct integration be used in all experiments. Generation of a parental reporter DT40 line with which to study the effects of gene disruptions on HR is beyond the scope of this protocol. Briefly, it requires either targeting of the reporter construct into a neutral locus such as ovalbumin (3) or the random and stable integration of the reporter construct (6, 7). In the latter case, it is necessary to assay copy number by Southern blot and ensure a selection of a single-copy integrant. It is also prudent to ensure that the "wild type" reporter line exhibits a suitable I-SceI-induced HR efficiency before proceeding to create mutants in this background. The use of non-isogenic random integrants is not safe and will not yield interpretable results.

2. Both the I-SceI-NLS and control RFP and GFP expression plasmids should be prepared as transfection-quality DNA in TE buffer at a concentration that allows the master mix described in Subheading 3.1, step 3 to be prepared. Using a plasmid purification protocol that includes a final ethanol precipitation step to allow resuspension of the I-SceI-NLS expression vector at a concentration equal to or greater than 3 mg/mL is the best way to do this. We recommend that you purify enough of each plasmid to make a master mix that can be used for your entire experiment in triplicate.
3. The choice of RFP fluorophore will depend on the range of FACS excitation lasers available. In the case of an instrument with only a 488 nm laser for excitation of both red and green fluorophores simultaneously, we recommend DsRed or tdTomato expression vectors, e.g., pDsRed-Express-C1 or ptdTomato-C1 (Clontech). These are equally suitable as the excitation maximum of DsRed closer to 488 nm but tdTomato has a higher quantum yield, so the voltage and gain on your detectors will need to be roughly the same. mCherry cannot be used. If a laser that can specifically excite red fluorescent proteins (e.g., a 561 nm yellow-green diode laser) is available, we recommend the use of tdTomato. mCherry can be used with a 561 nm laser, but is not recommended.
4. DT40 can be grown at temperatures between 37 and 41 °C, reflecting the higher core temperature of chickens. We have grown our cells routinely at 37 °C and in 10 % CO_2. Doubling time at 37 °C is about 11 h compared with about 9 h at 41 °C. We are not aware of any systematic studies of the effect of temperature on HR frequency in DT40, but the temperature at which experiments are carried out should be born in mind when comparing different studies.
5. In the first replicate and final replicates of the assay, it is wise to assess the percentage of GFP-positive cells that have arisen spontaneously, i.e., without transfection of I-SceI. Spontaneous recombination does occur, so a few GFP positive clones will most likely be found in untransfected clones, but typically they will make up less than 0.01 % of cells, which usually equates to less than 1 % of the experimentally measured GFP-positive cells following I-SceI expression. Generally therefore, these cells will not significantly affect the results and can safely be disregarded.
6. Transfection efficiency appears to be reduced the longer cells stay in a pellet. Attempting to transfect more than six samples at a time is not advisable. Transfect your samples in two or more groups if you have more than six samples.
7. Amaxa nucleofection solution is toxic to DT40 cells after around 5 min. Work quickly and perform each transfection in

its entirety from Subheading 3.1, step 9–12 before starting the next one.

8. It helps to partially invert the tube in the late stages of aspiration to allow you to aspirate away from the cell pellet without disturbing it.
9. Perform the RFP only and GFP only control transfections using 10 μL of the respective master mixes. Control transfections should be performed as the final two transfections in a set of six, as transfection efficiency is less important for these samples, and is almost always very high in any case.

Acknowledgements

Work in the laboratory is supported by the Medical Research Council, Association for International Cancer Research and The Fanconi Anemia Research Fund.

References

1. Rouet P, Smih F, Jasin M (1994) Expression of a site-specific endonuclease stimulates homologous recombination in mammalian cells. Proc Natl Acad Sci U S A 91:6064–6068
2. Rouet P, Smih F, Jasin M (1994) Introduction of double-strand breaks into the genome of mouse cells by expression of a rare-cutting endonuclease. Mol Cell Biol 14:8096–8106
3. Fukushima T, Takata M, Morrison C, Araki R, Fujimori A, Abe M, Tatsumi K, Jasin M, Dhar PK, Sonoda E, Chiba T, Takeda S (2001) Genetic analysis of the DNA-dependent protein kinase reveals an inhibitory role of Ku in late S-G2 phase DNA double-strand break repair. J Biol Chem 276:44413–44418
4. Weinstock DM, Nakanishi K, Helgadottir HR, Jasin M (2006) Assaying double-strand break repair pathway choice in mammalian cells using a targeted endonuclease or the RAG recombinase. Methods Enzymol 409:524–540
5. Pierce AJ, Johnson RD, Thompson LH, Jasin M (1999) XRCC3 promotes homology-directed repair of DNA damage in mammalian cells. Genes Dev 13:2633–2638
6. Bridge WL, Vandenberg CJ, Franklin RJ, Hiom K (2005) The BRIP1 helicase functions independently of BRCA1 in the Fanconi anemia pathway for DNA crosslink repair. Nat Genet 37:953–957
7. Szüts D, Simpson LJ, Kabani S, Yamazoe M, Sale JE (2006) Role for RAD18 in homologous recombination in DT40 cells. Mol Cell Biol 26:8032–8041
8. Richardson C, Moynahan ME, Jasin M (1998) Double-strand break repair by interchromosomal recombination: suppression of chromosomal translocations. Genes Dev 12:3831–3842
9. Buerstedde JM, Takeda S (2006) Reviews and protocols in DT40 research, vol 40. Springer, Berlin, Heidelberg, New York

Chapter 27

I-SceI-Based Assays to Examine Distinct Repair Outcomes of Mammalian Chromosomal Double Strand Breaks

Amanda Gunn and Jeremy M. Stark

Abstract

Chromosomal double strand breaks (DSBs) can be repaired by a number of mechanisms that result in diverse genetic outcomes. To examine distinct outcomes of chromosomal DSB repair, a panel of human cell lines has been developed that contain *GFP*-based reporters with recognition sites for the rare-cutting endonuclease I-SceI. One set of reporters is used to measure DSB repair events that require access to homology: homology-directed repair, homology-directed repair that requires the removal of a nonhomologous insertion, single strand annealing, and alternative end joining. An additional reporter (EJ5-GFP) is used to measure end joining (EJ) between distal DSB ends of two tandem I-SceI sites. These Distal-EJ events do not require access to homology, and thus are distinct from the repair events described above. Indeed, this assay provides a measure of DSB end protection during EJ, via physical analysis of Distal-EJ products to determine the frequency of I-SceI-restoration. The EJ5-GFP reporter can also be adapted to examine EJ of non-cohesive DSB ends, using co-expression of I-SceI with a non-processive 3′ exonuclease (Trex2), which can cause partial degradation of the 4 nucleotide 3′ cohesive overhangs generated by I-SceI. Such co-expression of I-SceI and Trex2 leads to measurable I-SceI-resistant EJ products that use proximal DSB ends (Proximal-EJ), as well as distal DSB ends (Distal-EJ). Therefore, this co-expression approach can be used to examine the relative frequency of Proximal-EJ versus Distal-EJ, and hence provide a measure of the fidelity of end utilization during repair of multiple DSBs. In this report, the repair outcomes examined by each reporter are described, along with methods for cell culture, transient expression of I-SceI and Trex2, and repair product analysis.

Key words: I-SceI, End joining, NHEJ, alt-EJ, Homologous recombination, Double strand break

1. Introduction

Efficient repair of chromosomal double strand breaks (DSBs) is critical for genome stability and promotes cellular resistance to clastogenic cancer therapeutics (1). Understanding the factors and pathways that influence the efficiency and fidelity of DSB repair will provide insight into cancer etiology and therapeutics. Here, methods are described for a series of reporter cell lines to quantify

Lotte Bjergbæk (ed.), *DNA Repair Protocols*, Methods in Molecular Biology, vol. 920,
DOI 10.1007/978-1-61779-998-3_27, © Springer Science+Business Media New York 2012

the relative frequency of diverse repair outcomes with distinct mutagenic consequences.

A panel of human cell lines (HEK293 and U2OS) has been generated, each of which harbors a chromosomally integrated copy of an individual reporter. Each reporter contains an inactive expression cassette for green fluorescent protein (GFP) that is interrupted by one or more recognition sites for the rare-cutting endonuclease I-SceI (2, 3). I-SceI generates a defined DSB within its 18 base pair (bp) recognition sequence, resulting in DSB ends with 4 nucleotide (nt) 3′ cohesive overhangs (4). The individual reporters are designed such that a defined DSB repair outcome leads to restoration of a *GFP* expression cassette. Accordingly, the *GFP*-marked repair outcome is measured by transiently expressing I-SceI, culturing the cells to allow completion of repair, then determining the percentage of *GFP+* cells by fluorescence-activated cell sorting (FACS) analysis. The duration of cell culture is determined by the number of days required to achieve maximal repair frequencies, which for these cell lines is 3 days post-transfection (2, 3). Some of the reporter assays also involve physical analysis of repair products using PCR amplification with primers that flank the I-SceI recognition site(s) (2, 3).

Beginning with DR-GFP, DRins-GFP, SA-GFP, and EJ2-GFP, these reporters are designed to examine a series of repair outcomes that utilize homology (Fig. 1a–e). DR-GFP and DRins-GFP are used to quantify two distinct homology-directed repair (HDR) events, whereas SA-GFP and EJ2-GFP are used to measure single strand annealing (SSA), and alternative end joining (alt-EJ), respectively (2, 3, 5, 6). With DR-GFP, an I-SceI-induced DSB in the upstream *SceGFP* cassette, followed by HDR that uses the downstream homologous template (*iGFP*) to prime nascent DNA synthesis, restores the *GFP+* cassette (5) (Fig. 1a). The derivative reporter, DRins-GFP, includes a nonhomologous 464 bp insertion in the *SceGFP* cassette that is removed during HDR (3) (Fig. 1b). The SA-GFP reporter contains a 5′ *GFP* fragment separated 2.7 kilobases (kb) from a 3′ *GFP* fragment that contains an I-SceI recognition site (6). The two *GFP* fragments share 266 nt of homology that can bridge the I-SceI-induced DSB during SSA, thereby restoring a functional *GFP+* cassette (Fig. 1c). While this repair outcome could also be caused by HDR with crossing-over (HDR/CO), the low frequency of HDR/CO (7, 8) indicates that these rare events do not significantly contribute to the formation of the *GFP+* product in SA-GFP. Finally, EJ2-GFP contains an I-SceI site flanked by 8 nt of homology that can bridge the I-SceI-induced DSB during alt-EJ, thereby restoring a functional *GFP+* cassette, causing a short 35 nt deletion, and generating an XcmI recognition site (2) (Fig. 1d).

The terms HDR, SSA, and alt-EJ are used in this context to describe a defined repair outcome, while also invoking the mechanism

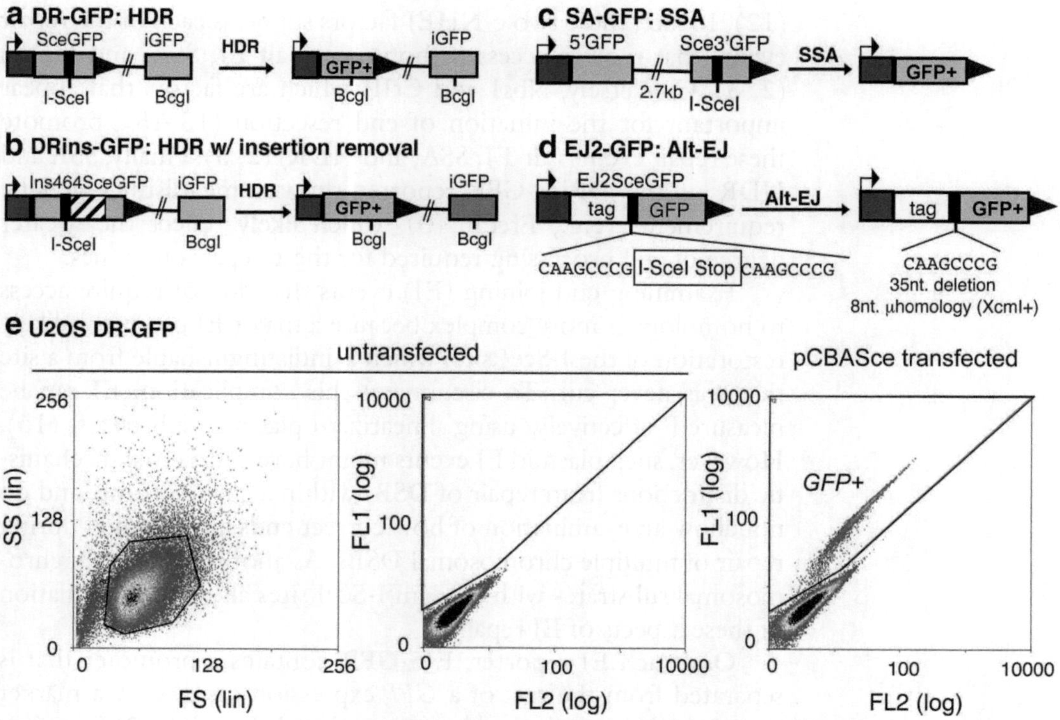

Fig. 1. I-SceI-based reporters for repair outcomes that involve access to homology. (**a**) DR-GFP contains the *SceGFP* cassette that is interrupted by a single I-SceI site, along with a 5′ and 3′ truncated fragment of GFP (*iGFP*). Homology-directed repair (HDR) of the I-SceI-induced DSB, using *iGFP* as the template, can lead to a GFP+/BcgI+ product. (**b**) DRins-GFP is similar to DR-GFP, with the exception of a 464 nt insertion (hatched box within *Ins464SceGFP*) that must be removed during HDR to generate the GFP+/BcgI+ product. (**c**) SA-GFP contains a 5′ fragment of GFP (*5′GFP*), and a 3′ fragment of GFP (*Sce3′GFP*) that contains an I-SceI site. The *GFP* fragments are separated by 2.7 kb and share 266 nt of homology. Repair of the DSB in *Sce3′GFP* by single-strand annealing (SSA) will lead to a *GFP+* product. (**d**) EJ2-GFP contains an expression cassette for a tagged version of *GFP* that is interrupted by an I-SceI site and a series of stop codons, which is flanked by 8 nt of homology. Repair by alternative end joining (alt-EJ) that deletes the stop codons, restores the *GFP* coding frame, and bridges the 8 nt of flanking homology, leads to a *GFP+* cassette that is marked with an XcmI recognition site. (**e**) Representative FACS profiles of the U2OS DR-GFP cell line. Shown (*left*) is a plot of untransfected cells using forward scatter (FS, *x*-axis) and side scatter (SS, *y*-axis), which is used to generate a gate of cells (within polygon). A logarithmic plot of these gated cells is shown (*middle*) using FL2 (orange fluorescence, *x*-axis) and FL1 (green fluorescence, *y*-axis). Also shown (*right*) is an FL2/FL1 plot for pCBASce transfected cells that were gated using a FS/SS plot, as for the untransfected sample. Autofluorescence is detected on the diagonal, where cells showing increased green fluorescence above autofluorescence are gated to determine the percentage of *GFP+* cells. Items in this figure are reproduced from Bennardo et al. (2009) with permission from *PLoS Genetics*.

of repair. Repair outcome alone is not sufficient to assign mechanism; however, these repair outcomes show distinct genetic requirements that allow some degree of mechanistic classification. For instance, the strand exchange factor Rad51 specifically promotes HDR, which is consistent with the unique requirement for HDR of nascent DNA synthesis using a homologous template (9). In contrast, Rad51 suppresses SSA and is dispensable for alt-EJ (2, 6). Also, alt-EJ is independent of Xrcc4 and Ku70 (2, 3, 10, 11), which are two factors integral to the classical nonhomologous end joining (c-NHEJ) pathway that mediates V(D)J recombination

(12). In fact, these two c-NHEJ factors suppress each of the repair events that require access to homology (alt-EJ, SSA, and HDR) (2, 3). Conversely, Nbs1 and CtIP, which are factors that appear important for the initiation of end resection (13–15), promote these repair events (alt-EJ, SSA, and HDR) (2, 3). Finally, SSA and HDR of the DRins-GFP reporter show some distinct genetic requirements (e.g., Ercc1) (3), which likely reflect the greater degree of end processing required for these repair outcomes.

Examining end joining (EJ) events that do not require access to homology is more complex because a major EJ product is likely restoration of the I-SceI site, which is indistinguishable from a site that was never cut. To circumvent this complication, EJ can be measured effectively using linearized plasmid substrates (16). However, such plasmid EJ events might have substantial mechanistic distinctions from repair of DSBs within a chromosome, and do not allow an examination of how correct ends are matched during repair of multiple chromosomal DSBs. As another approach, chromosomal substrates with tandem I-SceI sites allow an examination of these aspects of EJ repair.

One such EJ reporter, EJ5-GFP, contains a promoter that is separated from the rest of a *GFP* expression cassette by a marker gene (*puro*) that is flanked by two tandem I-SceI sites (2) (Fig. 2a). EJ that uses the distal DSB ends (Distal-EJ) restores the *GFP* expression cassette. Because the two I-SceI sites in EJ5-GFP are in the same orientation, one possible Distal-EJ product is the restoration of an I-SceI site using the 3′ cohesive DSB ends. Such I-SceI restoration can be quantified by PCR amplification of the Distal-EJ repair junction of *GFP*+ sorted cells, followed by I-SceI digestion analysis (Fig. 2a, b, primers p1 and p2). Furthermore, the pattern of deletions/insertions during EJ can be examined by subcloning the I-SceI-resistant amplification products for sequence analysis.

Factors that protect DSB ends during EJ are likely important for I-SceI-restoration. Accordingly, the c-NHEJ factors Ku70 and Xrcc4, as well as Brca1, are each important for I-SceI-restoration; whereas Nbs1 is important for loss of the I-SceI site (Fig. 2b) (2, 3, 10, 11). Thus, examining the frequency of I-SceI restoration, and the repair junctions via sequencing analysis, can provide a complementary method to EJ2-GFP for identifying factors or growth conditions that influence the degree of end protection during EJ. Notably, the overall frequency of Distal-EJ using the EJ5-GFP reporter is not substantially affected by genetic loss of Ku70 or Xrcc4 (2, 3, 17). These findings indicate that alt-EJ pathways are sufficient to generate Distal-EJ products with an intact *GFP* expression cassette, albeit with larger deletions. In summary, the EJ5-GFP reporter can be used to examine the frequency of EJ repair that is independent of access to homology, which provides a useful contrast to the other reporters. Furthermore, examining the Distal-EJ repair junctions by PCR analysis can provide insight into the degree of end protection during EJ.

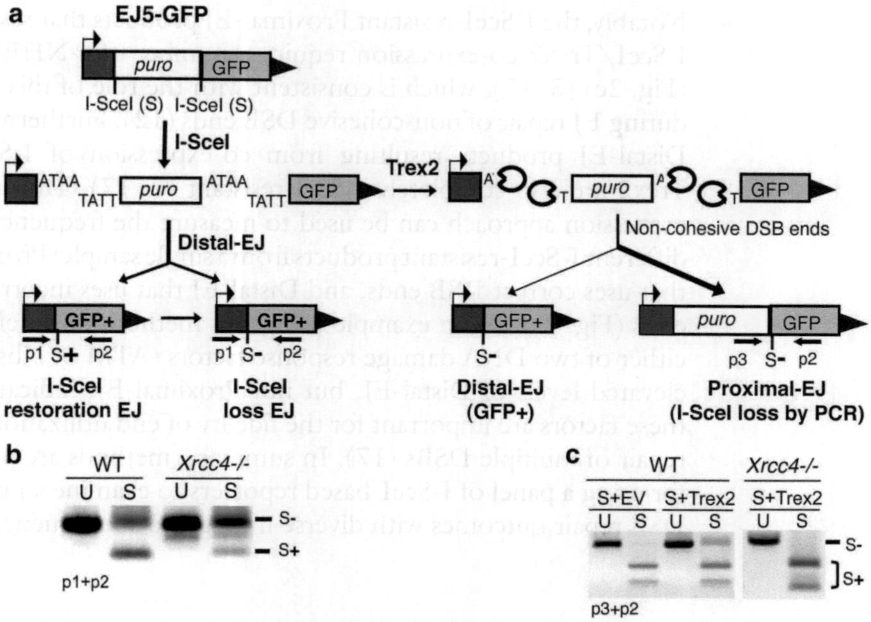

Fig. 2. A reporter for end joining between tandem I-SceI-induced DSBs. (**a**) The EJ5-GFP reporter system for measuring multiple aspects of end joining (EJ). EJ5-GFP contains the pCAGGS promoter separated from the rest of a *GFP* expression cassette by a puromycin selection cassette (*puro*) flanked by I-SceI sites. Shown (*left*) are Distal-EJ events following expression of I-SceI alone, which lead to restoration of the *GFP* cassette. Such Distal-EJ can also restore an I-SceI site if the DSB ends are protected during EJ, which can be quantified by PCR amplification with primers p1 and p2, followed by I-SceI digestion analysis. Also shown (*right*) are EJ events following co-expression of I-SceI and Trex2. Again, Distal-EJ is marked with restoration of the *GFP* cassette, but in this case, all Distal-EJ products show loss of the I-SceI site (S-), due to Trex2 activity. Also, Trex2 expression leads to siginificant loss of the I-SceI site (S-) during Proximal-EJ, which can be quantified by PCR amplification using p3 and p2, and subsequent I-SceI digestion analysis. (**b**) Xrcc4 is important for I-SceI restoration during Distal-EJ. Wild type (WT) and Xrcc4−/− mouse ES cells with the EJ5-GFP reporter were transfected with the I-SceI expression vector, and subsequently *GFP*+ cells (Distal-EJ events) were enriched using FACS sorting. Shown are amplification products from these sorted samples (primers p1/p2) that were uncut (U) or I-SceI digested (S). (**c**) Co-expression of I-SceI and Trex2 leads to significant levels of I-SceI-resistant Proximal-EJ products, which are promoted by Xrcc4. The cell lines described in (**b**) were transfected with I-SceI in conjunction with either a Trex2 expression vector, or empty vector (EV). Shown are Proximal-EJ products that were amplified (p3/p2) and left uncut (U) or digested with I-SceI (S). Items in this figure are reproduced from Bennardo et al. (2009), Bennardo and Stark (2010), with permission from *PLoS Genetics*.

Finally, the EJ5-GFP reporter can also be used to examine EJ outcomes that use correct DSB ends versus incorrect DSB ends (Proximal-EJ and Distal-EJ, respectively) (17). Proximal-EJ is difficult to measure with I-SceI expression alone because EJ that restores the I-SceI site cannot be differentiated from the uncut reporter. To address this limitation, the EJ5-GFP reporter system has been adapted by co-expressing I-SceI with a non-processive 3′ exonuclease (Trex2) (18, 19), which can cause partial degradation of the 4 nt 3′ cohesive overhangs generated by I-SceI (3, 17). Co-expression of I-SceI with Trex2 leads to a high level of I-SceI-resistant Proximal-EJ products that show short deletions of the I-SceI overhang region (3). Thus, Proximal-EJ that leads to loss of the I-SceI site can be quantified by PCR amplification and subsequent I-SceI digestion analysis (Fig. 2a, c, primers p2 and p3).

Notably, the I-SceI-resistant Proximal-EJ products that result from I-SceI/Trex2 co-expression require a number of c-NHEJ factors (Fig. 2c) (3, 17), which is consistent with the role of this pathway during EJ repair of non-cohesive DSB ends (12). Furthermore, the Distal-EJ products resulting from co-expression of I-SceI and Trex2 are also completely I-SceI-resistant (3, 17). Thus, this co-expression approach can be used to measure the frequency of two different I-SceI-resistant products from a single sample: Proximal-EJ that uses correct DSB ends, and Distal-EJ that uses incorrect DSB ends (Fig. 2a). As an example using this method, cells deficient in either of two DNA damage response factors (ATM or Nbs1) show elevated levels of Distal-EJ, but not Proximal-EJ, indicating that these factors are important for the fidelity of end utilization during repair of multiple DSBs (17). In summary, methods are presented for using a panel of I-SceI-based reporters to examine a number of DSB repair outcomes with diverse mutagenic consequences.

2. Materials

2.1. Plasmids and Primer Sequences

1. Expression vectors: I-SceI (pCAGGS-I-SceI, called pCBASce), Trex2 (pCAGGS-Trex2), GFP (pCAGGS-NZEGFP), and empty vector (pCAGGS-BSKX) (3, 5). Each vector contains an ampicilin-resistant gene, and can be amplified in *E. coli* using Circlegrow with ampicilin, purified using a Qiagen QiaFilter plasmid preparation, and dissolved in sterile TE (10 mM Tris pH 8.0, 1 mM EDTA pH 8.0).
2. Primer sequences for examining EJ5-GFP repair products. Distal-EJ product analysis: p1 (KNDRF) 5′-CTGCTAACCAT GTTCATGCC, p2 (KNDRR) 5′-AAGTCGTGCTGCTTCAT GTG (2, 20). Proximal-EJ product analysis (3′ I-SceI site): p3 (EJ5PurF) 5′-AGCGGATCGAAATTGATGAT and p2 (3).

2.2. Cell Lines

1. A panel of HEK293 cell lines (clone HEK293A7 from New England Biolabs), each with an integrated reporter (DR-GFP, SA-GFP, EJ2-GFP, or EJ5-GFP), were described previously (2).
2. A panel of U2OS cell lines (HTB-96, ATCC, Manassas), each with an integrated reporter (DR-GFP, DRins-GFP, SA-GFP, EJ2-GFP, or EJ5-GFP), were generated using the same integration protocol as for the HEK293 cells (2).

2.3. Growth Media, Transfection, and Fixation

1. Growth medium: 500 ml DMEM high glucose with L-glutamine, 55 ml fetal bovine serum, and 6 ml Pen/Strep solution (10,000 U/ml penicillin, 10,000 μg/ml streptomycin). Including 8 μg/ml plasmocin and 2 μg/ml puromycin can promote cell line stability during long-term culturing. For

antibiotic-free medium, omit Pen/Strep and plasmocin. For the entire transfection experiment, omit puromycin.

2. Freezing medium: 50 % Growth media, 40 % fetal bovine serum, 10 % dimethylsulfoxide.
3. Trypsin: 1× Trypsin-EDTA in Hanks Salts.
4. Poly-lysine: 0.01 % Solution.
5. Transfection reagents: Lipofectamine 2000, RNAiMAX, and Optimem without phenol red (Invitrogen).
6. Formaldehyde solution: 10 % Formaldehyde (w/v in aqueous phosphate buffer).
7. Sterile tissue culture plates: 10 cm plate, 2 cm^2 vessel (typical well of a 24-well plate), 4 cm^2 vessel (typical well of a 12-well plate).

2.4. Physical Analysis of I-SceI-Resistant Repair Products

1. P-lysis buffer: 20 mM Tris pH 8.0, 85 mM NaCl, 15 mM EDTA, 0.5 % SDS.
2. RNase: DNase-free RNase from bovine pancrease, 500 μg/ml.
3. Proteinase K: Recombinant proteinase K, 14–22 mg/ml, PCR grade.
4. Phenol: Saturated phenol pH 6.6 ± 0.2.
5. Chloroform.
6. 70 % Ethanol in water.
7. Isopropanol.
8. 0.1× TE: 1 mM Tris pH 8, 0.1 mM EDTA pH 8.
9. PCR SuperMix High Fidelity (Invitrogen).
10. Column for PCR purification: Illustra GFX PCR DNA and Gel Band Purification Kit (GE).
11. I-SceI enzyme and incubation buffers (New England Biolabs).
12. Agarose gel: Low EEO agarose, 1× TBE gel running buffer (90 mM Tris-borate, 2 mM EDTA).
13. Loading buffer: 15 % Ficoll in gel running buffer, 30 % glycerol, 0.01 % bromophenol blue, 50 mM EDTA.

3. Methods

3.1. Culturing HEK293 and U2OS Cells, and siRNA Transfection

1. U2OS cells are grown as adherent cells directly on tissue culture plastic, using growth medium, in a tissue culture incubator (37 °C and 5 % CO_2). HEK293 cells adhere more efficiently if plates are first coated with Poly-lysine: add 1–3 ml per 10 cm plate to coat the growth surface and remove prior to adding growth medium. Poly-lysine solution can be reused to coat several plates.

2. Keep stocks of cells between 20 and 80 % confluent. To passage, remove growth medium, add 1 ml of trypsin per 10 cm plate to cover the growth surface, wait 3–5 min, dilute with growth medium, and add to a fresh plate with growth medium.
3. For shRNA-mediated disruption of individual factors in HEK293 cell lines, infect with lentiviral shRNA expression cassettes prior to I-SceI transfection. Such infection is beyond the scope of this review.
4. For use of siRNA in U2OS cells, treat with siRNA 48 h prior to I-SceI transfection. Reverse transfect the U2OS cells by plating 1×10^5 cells in antibiotic-free media per 2 cm^2 culture vessel that already contains preformed siRNA transfection complexes (e.g., 10 pmol siRNA duplex mixed with 1.8 μl RNAiMAX in 100 μl Optimem for 25 min). Let the transfection proceed for 20 h and passage to 4 cm^2 cutlure vessel (see Note 1). 48 h after the siRNA transfection, transfect with I-SceI (or I-SceI and Trex2) with 10 pmol of siRNA duplex (see below).

3.2. Transient I-SceI Expression with or Without Trex2

1. 24 h pre-transfection, plate 1×10^5 cells per 4 cm^2 culture vessel using 2 ml growth medium. Disperse the cells thoroguly by pipetting, add the cells to the growth medium in the culture vessel, then spread the cells by shaking the plate gently and repeatedly (see Note 1).
2. The day of transfection, aspirate the medium, wash with 1 ml antibiotic-free medium or PBS, and add 1 ml fresh antibiotic-free medium. Incubate for 2 h before adding transfection complexes.
3. Prepare transfection complexes. In one tube, dilute nucleic acid into 100 μl Optimem. Suggested: 0.8 μg of I-SceI expression vector along with 10 pmol siRNA and/or 0.4 μg secondary expression vector (e.g., Trex2 expression vector, or internal transfection control plasmid, see Note 2). In a second tube, dilute 3.6 μl Lipofectamine 2000 into 100 μl Optimem, and mix by pipetting. Invert Lipofectamine 2000 several times before use, and do not vortex any of the above solutions. Let diluted Lipofectamine sit for 5 min at room temperature. Next, add diluted Lipofectamine 2000 to the diluted nucleic acid, mix by pipetting, and incubate for 25 min at room temperature. Add 200 μl of transfection complexes to the cells in the 4 cm^2 culture vessel, without changing the media. Shake the plate back and forth several times to mix, and return to the tissue culture incubator for 3 h. Aspirate off transfection complexes, add 1 ml of antibiotic-free medium, shake gently and aspirate to wash off residual transfection reagent. Add 4 ml growth medium (containing antibiotics) and culture for 3 days (approximately 72 h) prior to end-point analysis (see Note 2).

4. Chemical treatment. To address the effect of a chemical on these assays, add the compound right after transfection, at the 3 h media change. Subsequently, leave the cells alone until preparation for FACS (3 days). Removal of the transfection reagents prior to adding the chemical is likely important to avoid transfection of the chemical, which could cause high intracellular levels of the compound.

3.3. FACS Analysis

1. Three days (approximately 72 h) after transfection, analyze the cells by FACS.
2. Aspirate medium from each of the transfected wells, and add 200 μl trypsin per well. Incubate at room temperature 3–5 min. After incubation, add 200 μl growth medium to each transfected well and disperse cells by pipetting.
3. Immediatedly after dispersing the trypsinized sample, mix 2:1 with 10 % formaldehyde (e.g., 400 μl sample added to 200 μl 10 % formaldehyde). Vortex immediately for 2–3 s at medium speed. Samples are ready for FACS analysis, and are stable up to 4 h at room temperature or on ice (see Note 3).
4. FACS analysis. Use a plot of forward scatter versus side scatter to gate for events that are consistent with individual cells. Analyze these gated cells in a plot for green fluorescence (FL1) on the *y*-axis, and orange or red fluorescence (FL2 or FL3) on the *x*-axis. This plot allows a distinction between cells with high autofluorescence versus green fluorescent cells. Namely, cells expressing GFP show an increase in the FL1 signal above autofluorescence (Fig. 1e).

3.4. PCR Analysis to Examine I-SceI-Resistant Products

1. Cell lysate for genomic DNA (gDNA). For Proximal-EJ analysis, on the day of FACS, trypsinize the cells as described above, use half the sample for fixation and FACS, and use the other half for gDNA extraction. For Distal-EJ analysis, sort live *GFP*+ cells using FACS, culture to expand the population of sorted cells, and collect approximately 10^6 cells using trypsinization. Spin cells at 150 × *g* for 5 min to pellet the cells, pour off the aqueous, and add 500 μl PBS. Repeat spin and pour off aqueous, keeping approximately 50 μl of the PBS. Vortex to disperse pellet, then add 500 μl P-lysis buffer. Samples are stable in this buffer for at least 1 month at room temperature.
2. Preparing gDNA via phenol/chloroform extraction. Add 1 μl RNase to the sample in 500 μl P-lysate buffer, invert 25×, and incubate 5 min at room temperature. Next, add 5 μl proteinase K, invert 25×, and incubate at room temperature for 1 h. Add 30 μl 5 M NaCl. Add 500 μl saturated phenol, shake tubes gently, but thoroughly, for 10 min at room temperature, then centrifuge at 10,000 × *g* for 5 min. Transfer the aqueous (top) layer to a fresh microcentrifuge tube and repeat this step twice

more, for a total of three phenol extractions. Extract once using 500 μl chloroform, using the same procedure as the phenol extraction. Transfer the aqueous layer to a fresh microcentrifuge tube, and add 500 μl isopropanol to precipitate gDNA. Shake for 10 min and centrifuge at 10,000 × *g* for 5 min. After this spin, there should be a visible pellet at the bottom of the tube. Carefully aspirate without disturbing the pellet, and add 1 ml 70 % ethanol. Spin again at 10,000 × *g* for 5 min, and aspirate the ethanol carefully before air-drying the pellet (≤10 min). Add 50 μl 0.1× TE to the pellet, and heat samples at 55 °C for 30–60 min with regular vortexing to resuspend the pellet. Store at 4 °C, but reheat at 55 °C for 30 min prior to PCR.

3. PCR conditions. Analysis of Distal-EJ junctions from *GFP*+ sorted cells can be performed with p1 and p2. Analysis of 3′ Proximal-EJ can be performed with p3 and p2. Create a Primer Mix with each primer at 10 μM in sterile water. Next, add 1.5 μl Primer Mix and 1 μl gDNA (approximately 200 ng) to 22.5 μl PCR HiFi Supermix. Thermocycler conditions: 94 °C 3 min, cycle 32 times (94 °C 45 s, 63 °C 45 s, 68 °C 1 min 45 s), 68 °C 7 min, store at 4 °C.
4. Product purification and quantitative I-SceI digestion. Column purify PCR product prior to analysis according to manufacturer's instructions. Elute purified PCR product in 22 μl 10 mM Tris pH 8. Add 3 μl 10× I-SceI buffer, 3 μl 10× BSA, and 1 μl I-SceI enzyme (5 U/μl), then incubate at 37 °C for 1 h. Add an additional 1 μl I-SceI and incubate for a second hour, and repeat for a total of 3 μl I-SceI and 3 h. Add 5 μl loading buffer to digested samples, and run samples on a 2 % agarose gel for 120 min at 100 V. Stain the gel evenly with 0.5 μg/ml ethidium bromide in gel running buffer for 15 min prior to capturing image.
5. Determine the percentage of I-SceI-resistant products by quantifying the relative intensity of the I-SceI-resistant and I-SceI-sensitive products within the same sample (see Note 4). The Distal-EJ product that restores the I-SceI site (EJ5-GFP, p1 and p2, Fig. 2a, b) is 380 bp, and is digested by I-SceI into two products 281 and 99 bp (calculated from the middle of the 3′ overhang generated by I-SceI). Determine the relative staining intensity values for the 380 and 281 bp fragments using gel imaging software. To account for the smaller size of the 281 bp fragment, multiply the staining value by 1.35 (normalized 281 bp fragment). Divide the value of the 380 bp fragment by the sum of the values for the 380 bp and normalized 281 bp fragment, and multiply by 100 to determine the percentage of I-SceI site loss. Similarly, the Proximal-EJ product that restores the I-SceI site (p1 and p3, Fig. 2a, c) is 748 bp,

and is digested by I-SceI into two products, 467 and 281 bp (calculated as above). In this case, determine the relative staining intensity values for the 748 and 467 bp fragments, and multiply the value of the 467 bp fragment by 1.6 to generate the normalized 467 bp fragment. Divide the value of the 748 bp fragment by the sum of the values for the 748 bp and normalized 467 bp fragment, and multiply by 100 to determine the percentage of I-SceI-resistant Proximal-EJ products.

3.5. Analysis of Repair Junctions

1. The I-SceI-resistant PCR products can be isolated from the gel for sequence analysis using the same kit as for purification of the PCR product. These isolated fragments can be used directly for ligation into a cloning/sequencing vector (e.g., TA cloning vector pCR2.1, Invitrogen), according to the manufacturer's instructions. Incubation of the purified fragments with polymerase immediately before ligation (e.g., 5 μl purified fragment and 20 μl PCR HiFi Supermix, 68 °C 20 min) can substantially improve ligation efficiency.

4. Notes

1. Even dispersal of the cells. Suggested cell densities are based on even dispersal of the cells to allow sufficient surface area for growth during the experiment. At a minimum, gently shake each plate back and forth, and then side to side, a dozen times each. Optimally, return the plate to the incubator and allow the cells to settle for 10 min, and subsequently remove for another set of shaking. Repeat the incubation and shaking an additional three times.
2. Controlling for transfection efficiency. Since the frequency of any repair event is directly affected by the efficiency of I-SceI-induced DSBs, as well as the survival of transfected cells, it is important to control for variations in these parameters. For example, it is important to determine whether a change in the frequency of *GFP+* cells using EJ2-GFP reporter, following treatment with a specific RNAi reagent, is due to an effect on alt-EJ repair, rather than I-SceI activity and/or survival after transfection. Parallel transfections using a GFP expression vector (e.g., pCAGGS-NZEGFP), or co-expressing another fluorescent marker with I-SceI (e.g., co-transfection of expression cassettes for mCHERRY or dsRED), can control for such alterations in transfection efficiency or survival. Since transfection efficiency with these cell lines show very little variation between replicates, a parallel transfection with the GFP

expression vector is usually sufficient. Alternatively, the I-SceI protein expressed from pCBASce includes a hemagglutinin (HA) tag, which can be used to quantify expression levels by immunoblotting with commercially available anti-HA antibodies. In addition to the above transfection controls, focusing on RNAi reagents, chemicals, or cellular growth conditions that differentially affect a subset of repair outcomes is suggested. For example, a reagent that causes a decrease in the frequency of *GFP*+cells with the DR-GFP reporter, but not with the EJ5-GFP reporter, cannot be readily explained by changes in transfection or I-SceI activity.

3. Formaldehyde fixation before FACS. Analysis can also be performed on live cells resuspended in media without fixation, though cells can eventually start to aggregate. Accoringly, fixation is suggested immediately after trypsinization and resuspension to avoid aggregation. Should *GFP*+cells be absent in the analysis, ensure that the correct concentration of formaldehyde is being used (10 %, w/v, in PBS stock solution, 3 % final). To analyze fixed cells more than 4 h after fixation and up to 2 days later, wash out the fixative. Specifically, incubate the cells in fixative for 30 min, spin at 150 × *g* for 5 min to pellet the cells, remove the fixative, and resuspend in 500 μl PBS. Repeat spin and pour off aqueous, resuspend in 500 μl PBS, and store at 4 °C in the dark.
4. Quantitative I-SceI digestion. Be sure to include a control for complete cutting by I-SceI, usually gDNA from untransfected cells. For the analysis of Distal-EJ products (p2 and p3), use untransfected cells and the Proximal-EJ primers (p1 and p2). Ensure that the amount of PCR amplification product from the control digestion is equivalent or greater than the transfected samples. If the control sample is not completely digested, digest a smaller amount of the PCR amplification products, and ensure that each addition of I-SceI enzyme includes pipetting to mix thoroughly. These assays involve quantifying the relative ratio of two products within the same sample, which remains consistent during amplification. In any case, defined mixtures of control and transfected cells can be used to ensure the assay is quantitative within twofold (5, 17).

Acknowledgements

This work was supported by NIH grant RO1CA120954 to J.M.S. Thanks to Nicole Bennardo, Meilen Muñoz, and Anita Cheng for technical contributions.

References

1. Wyman C, Kanaar R (2006) DNA double-strand break repair: all's well that ends well. Annu Rev Genet 40:363–383
2. Bennardo N, Cheng A, Huang N, Stark JM (2008) Alternative-NHEJ is a mechanistically distinct pathway of mammalian chromosome break repair. PLoS Genet 4:e1000110
3. Bennardo N, Gunn A, Cheng A, Hasty P, Stark JM (2009) Limiting the persistence of a chromosome break diminishes its mutagenic potential. PLoS Genet 5:e1000683
4. Moure CM, Gimble FS, Quiocho FA (2003) The crystal structure of the gene targeting homing endonuclease I-SceI reveals the origins of its target site specificity. J Mol Biol 334:685–695
5. Pierce AJ, Johnson RD, Thompson LH, Jasin M (1999) XRCC3 promotes homology-directed repair of DNA damage in mammalian cells. Genes Dev 13:2633–2638
6. Stark JM, Pierce AJ, Oh J, Pastink A, Jasin M (2004) Genetic steps of mammalian homologous repair with distinct mutagenic consequences. Mol Cell Biol 24:9305–9316
7. Stark JM, Jasin M (2003) Extensive loss of heterozygosity is suppressed during homologous repair of chromosomal breaks. Mol Cell Biol 23:733–743
8. Johnson RD, Jasin M (2000) Sister chromatid gene conversion is a prominent double-strand break repair pathway in mammalian cells. EMBO J 19:3398–3407
9. Stark JM, Hu P, Pierce AJ, Moynahan ME, Ellis N, Jasin M (2002) ATP hydrolysis by mammalian RAD51 has a key role during homology-directed DNA repair. J Biol Chem 277:20185–20194
10. Guirouilh-Barbat J, Huck S, Bertrand P, Pirzio L, Desmaze C, Sabatier L, Lopez BS (2004) Impact of the KU80 pathway on NHEJ-induced genome rearrangements in mammalian cells. Mol Cell 14:611–623
11. Guirouilh-Barbat J, Rass E, Plo I, Bertrand P, Lopez BS (2007) Defects in XRCC4 and KU80 differentially affect the joining of distal nonhomologous ends. Proc Natl Acad Sci U S A 104:20902–20907
12. Lieber MR (2010) The mechanism of double-strand DNA break repair by the nonhomologous DNA end-joining pathway. Annu Rev Biochem 79:181–211
13. Sartori AA, Lukas C, Coates J, Mistrik M, Fu S, Bartek J, Baer R, Lukas J, Jackson SP (2007) Human CtIP promotes DNA end resection. Nature 450:509–514
14. Stracker TH, Petrini JH (2011) The MRE11 complex: starting from the ends. Nat Rev Mol Cell Biol 12:90–103
15. Yang Y-G, Saidi A, Frappart P-O, Min W, Barrucand C, Dumon-Jones V, Michelon J, Herceg Z, Wang Z-Q (2006) Conditional deletion of Nbs1 in murine cells reveals its role in branching repair pathways of DNA double-strand breaks. EMBO J 25: 5527–5538
16. Iliakis G, Rosidi B, Wang M, Wang H (2006) Plasmid-based assays for DNA end-joining in vitro. Methods Mol Biol 314:123–131
17. Bennardo N, Stark JM (2010) ATM limits incorrect end utilization during non-homologous end joining of multiple chromosome breaks. PLoS Genet 6:e1001194
18. Chen M-J, Ma S-M, Dumitrache LC, Hasty P (2007) Biochemical and cellular characteristics of the 3′→5′ exonuclease TREX2. Nucleic Acids Res 35:2682–2694
19. Perrino FW, Harvey S, McMillin S, Hollis T (2005) The human TREX2 3′→5′-exonuclease structure suggests a mechanism for efficient nonprocessive DNA catalysis. J Biol Chem 280:15212–15218
20. Nakanishi K, Yang YG, Pierce AJ, Taniguchi T, Digweed M, D'Andrea AD, Wang ZQ, Jasin M (2005) Human Fanconi anemia monoubiquitination pathway promotes homologous DNA repair. Proc Natl Acad Sci U S A 102: 1110–1115

Chapter 28

Studying Repair of a Single Protein-Bound Nick In Vivo Using the Flp-Nick System

Ida Nielsen, Anni H. Andersen, and Lotte Bjergbæk

Abstract

The Flp-nick system is a simple *in vivo* system developed for studying the cellular responses to a protein-bound nick at a single genomic site in the budding yeast *Saccharomyces cerevisiae*. The Flp-nick system takes advantage of a mutant Flp recombinase that can introduce a nick at a specific Flp recombinase recognition target (FRT) site, which has been integrated into the yeast genome. Upon cleavage at the FRT site, the Flp mutant becomes covalently linked to the 3′ DNA end at the nick in an irreversible manner, as the mutant fails to accomplish the required religation process. Thus, the established damage mimics a stabilized topoisomerase I-DNA cleavage complex. DNA topoisomerases are ubiquitous enzymes that relieve topological stress in the DNA arising during DNA replication or transcription. During this process, they make transient enzyme-DNA cleavage complexes, which normally are reversed by a rapid ligation step. However, aberrant long-lived enzyme-DNA complexes may occur frequently due to either endogenous or exogenous damage, and the cellular repair machinery therefore needs to be able to eliminate this type of damage. This chapter describes how to establish the Flp-nick system in *S. cerevisiae*, how to detect FlpH305L at the FRT site using a modified chromatin immunoprecipitation assay where formaldehyde fixation is omitted, and how to monitor nicking at the FRT site by alkaline denaturing gel analysis.

Key words: Flp recombinase, Topoisomerase I-DNA cleavage complexes, Double-strand break repair, Protein-DNA adduct, Single-strand break

1. Introduction

To study repair of a protein-DNA adduct associated with a nick in the DNA, we developed the Flp-nick system taking advantage of a mutant form of the site-specific Flp recombinase (1). In wild type yeast cells, the *FLP* gene is encoded by the 2 μm circle and is thought to be responsible for maintaining multiple copies of this extrachromosomal element *in vivo* (2). Along with the Cre recombinase of bacteriophage P1, the Flp recombinase has been used as a powerful tool for genetic manipulations of higher organisms (3). The wild type Flp recombination site or FRT (Flp recognition

Lotte Bjergbæk (ed.), *DNA Repair Protocols*, Methods in Molecular Biology, vol. 920,
DOI 10.1007/978-1-61779-998-3_28, © Springer Science+Business Media New York 2012

target) consists of three 13 bp symmetry elements, two of which are present in inverse orientation and flank an 8 bp spacer. The third element present in tandem with one of the other repeats can be deleted without significant effects on target-site function (4) (see Fig. 1a). The normal recombination reaction catalyzed by Flp requires two DNA substrates and four recombinase monomers. A monomer of Flp recombinase harbors a fractional active site that is made functionally complete by association of at least two monomers which share catalytic residues between them (5, 6). Initiation of recombination requires strand cutting by the active site tyrosine in Flp (Tyr343) at one end of the spacer in each DNA substrate, which generates a nick with a free 5′-hydroxyl group and a 3′ phosphotyrosine linkage between Flp recombinase and the 3′-end of the DNA. During the strand-joining step, the 5′-hydroxyl group attacks the phosphotyrosine bonds across substrates, whereby a Holliday junction substrate is formed. By repeating the strand cleavage and exchange steps at the other end of the spacer, the

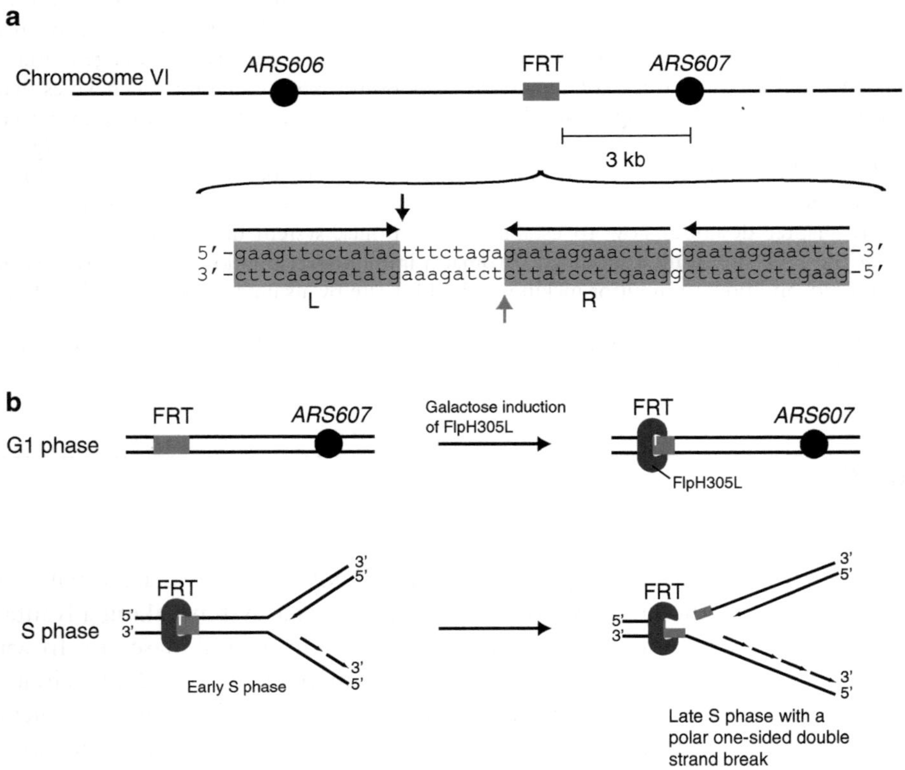

Fig. 1. (**a**) Schematic representation of the Flp-nick system. The FRT site is placed close to *ARS607* in a nontranscribed area. The FRT site consists of three DNA elements that are specifically recognized by the Flp recombinase. *L* and *R* indicate the *left* and *right* sides of the FRT site, respectively. *Vertical black arrow* indicates the position of the major scissile phosphodiester bond, whereas the *vertical gray arrow* indicates the less frequent cleavage site. (**b**) Illustration of events at the Flp-nick site during DNA replication. Upon induction of the flpH305L protein, cleavage will occur at the FRT site generating a protein-DNA adduct. During DNA replication (S phase) the nick is converted to a polar one-sided double-strand break when the replication fork collides with the Flp-nick damage. For convenience only the Flp monomer that makes the nick is shown, although two Flp monomers will bind to the FRT site.

junction is resolved into two reciprocal recombinant DNA products. Due to the *trans*-horizontal manner of cleavage by Flp recombinase, strand cutting can be initiated when one FRT site is occupied by two Flp monomers (7). Although such a complex can organize two shared active sites, in reality only one of the two sites is organized at any given time (8, 9). Hence, only single-strand breaks are made within a substrate.

The strand breakage and religation step of Flp recombinase follows the enzymatic mechanism utilized by type 1B topoisomerases. Topoisomerase I (Top1) is a type 1B enzyme, which is highly conserved and known to relieve topological stress in the DNA arising during DNA replication and transcription. Top1 establishes a transient covalent 3′-phosphotyrosyl-enzyme-DNA cleavage complex, which is reversed by a rapid ligation step during relaxation of topological stress. However, aberrant long-lived stabilized protein-DNA complexes are generated when the 5′-hydroxyl DNA end required for the religation step is misaligned. This occurs under conditions of endogenous and exogenous DNA lesions or when cells are treated with agents that block the religation step, e.g., the drug camptothecin (10).

To obtain a long-lived covalent protein-DNA complex, we use a "step-arrest" mutant of the Flp recombinase (FlpH305L), which cleaves DNA but does not accomplish the religation reaction (11). An integration plasmid carrying the *flp(H305L)* gene under control of the *GAL10* promoter is recombined into the yeast genome (12). This allows for efficient and controlled expression of the Flp mutant upon addition of galactose to the yeast culture under non-fermenting conditions (YP + raffinose media). We have generated a series of FRT integration plasmids with different selection markers based on the pFA6a plasmids (13), which enables us to integrate the FRT site at any given location in the yeast genome by homologous recombination. So far our studies with the Flp-nick system have involved integration of the FRT site close to an early firing origin *ARS607* in a nontranscribed region (see Fig. 1b). The Flp-nick damage, which is generated in our system upon induction of *flp(H305L)*, will be converted into a polar one-sided double-strand break upon replication fork collision (see Fig. 1b). The system has the advantage that it can easily be modified to study Flp-nick damage in any chromosomal context.

2. Materials

2.1. Medium and Plates

1. YPD: 2 % Bacto peptone, 1 % yeast extract, 2 % D-(+)-glucose monohydrate, 0.0025 % adenine hemisulfate salt. Autoclave.
2. YP + raffinose: 2 % Bacto peptone, 1 % yeast extract, 2 % D-(+)-raffinose pentahydrate, 0.0025 % adenine hemisulfate salt. Autoclave.

3. YP + raffinose pH 3.5: As for YP + raffinose but adjust pH to 3.5 with 6 M HCl before autoclaving (see Note 1).
4. YPD plates: 2 % Bacto peptone, 1 % yeast extract, 2 % D-(+)-glucose monohydrate, 0.0025 % adenine hemisulfate salt, 2 % agar, 1 NaOH pellet/L. Autoclave, cool down and pour the plates.
5. YPD + G418 plates: As for YPD plates. When the medium has cooled down G418 solution is added to a concentration of 200–800 μg/mL and the plates are poured. The concentration varies with the batch of G418 powder and strain.
6. YP + agar: 2 % Bacto peptone, 1 % yeast extract, 0.0025 % adenine hemisulfate salt, 2 % agar, 1 NaOH pellet/L. Autoclave.
7. YP + 3 % galactose plates: Prepare YP + agar. When cooled down, add galactose from 20 % solution to 3 % and pour the plates.
8. 20 % Galactose: Dissolve galactose in water at room temperature. Filter sterilize. Do not autoclave. Store at 4°C.
9. SC selection plates: Prepare SC-5 + agar: 0.17 % yeast nitrogen base w/o ammonium sulfate and amino acids, 0.5 % ammonium sulfate, 2 % D-(+)-glucose monohydrate, 0.12 % dropout mix (see Subheading 2.1, step 10), 2.17 % agar, 1 NaOH pellet/L. Autoclave. After autoclaving, cool down and add the required amino acid solutions (see Subheading 2.1, step 11), leaving one to make the desired selective plates. Pour the plates.
10. Dropout mix: 1.2 g arginine, 6 g aspartic acid, 6 g glutamic acid, 1.8 g lysine, 1.2 g methionine, 3 g phenylalanine, 22.5 g serine, 12 g threonine, 1.8 g tyrosine, 9 g valine.
11. Amino acid solutions: 1 % Leucine, autoclave, add 0.01 % to medium. 1 % Histidine, autoclave, add 0.002 % to medium. 0.2 % tryptophane, filter sterilize, keep in dark at 4°C, add 0.002 % to medium. 0.2 % adenine, autoclave, add 0.002 % to medium. 0.1 % Uracil, autoclave, add 0.0035 % to medium.

2.2. Regular Agarose Gels

1. 10× DNA loading buffer: 20 % Ficoll (type 400), 0.1 M Na_2EDTA pH 8.0, 1 % sodium dodecyl sulfate (SDS), 0.25 % bromophenol blue, 0.25 % xylene cyanol.
2. 10× TBE-buffer: 890 mM Trizma base, 890 mM boric acid, 20 mM Na_2EDTA pH 8.0.
3. Agarose.
4. Ethidium bromide: Make stock of 10 mg/mL and store dark at room temperature. Use 0.5 μg/mL.
5. UV lamp with camera.

2.3. General Methods

2.3.1. PCR Product for Transformation

1. Primers: Dissolve to 100 μM in H_2O.
2. Taq PCR master mix kit from QIAGEN or equivalent.
3. 50 mM $MgCl_2$.
4. Plasmid DNA for template (see Note 2).
5. 3 M sodium acetate pH 5.2. Autoclave.
6. Ethanol: 96 and 70 %.
7. 1× TE-buffer: 10 mM Tris–HCl pH 7.5, 1 mM Na_2EDTA pH 8.0. Autoclave.

2.3.2. Yeast Transformation

1. 10× TEL: 0.1 M Tris–HCl pH 7.5, 10 mM Na_2EDTA pH 8.0, and 10.2 g lithium acetate per 100 mL buffer. Autoclave. Use in fume hood.
2. 40 % PEG4000 in TEL: 40 % polyethylene glycol 4000 dissolved in 1× TEL. Autoclave.
3. Carrier DNA from salmon sperm (10 mg/mL).
4. 42°C heating block.
5. Appropriate plates.

2.3.3. Isolation of Genomic DNA by Phenol/Chloroform Extraction

1. 2 mL tubes with screw cap.
2. Glass beads, acid-washed.
3. Extraction buffer: 2 % Triton-X-100, 1 % SDS, and 100 mM NaCl in 1× TE-buffer.
4. Phenol/chloroform/isoamyl alcohol (25:24:1): Mix by inverting 20 mL phenol and 20 mL chloroform/isoamyl alcohol. Keep in a light protected disposable tube at 4°C. Make the day before use at the latest as it takes a night to properly separate the two phases. Use the lower phase, and use in fume hood wearing gloves.
5. Hybaid RiboLyser or equivalent to crack cells.
6. 1× TE-buffer: 10 mM Tris–HCl pH 7.5, 1 mM Na_2EDTA pH 8.0. Autoclave.
7. 3 M sodium acetate pH 5.2. Autoclave.
8. Ethanol: 96 and 70 %.

2.3.4. PCR on Genomic DNA or Colonies

1. Primers: Dissolve to 100 μM in H_2O.
2. Taq PCR master mix kit from QIAGEN or equivalent.
3. 50 mM $MgCl_2$.

2.4. Engineering a Flp-Nick Strain

1. pFV17Δ-H305L (see Note 3).
2. Enzymes: BglII (10,000 U/mL), BstEII (10,000 U/mL), SacII (20,000 U/mL), SpeI (10,000 U/mL), XbaI (20,000 U/mL). Reaction buffers and 10× BSA solution are supplied by distributor.

3. GFX-kit from GE Healthcare or equivalent.
4. 1× TE-buffer: 10 mM Tris–HCl pH 7.5, 1 mM Na_2EDTA pH 8.0. Autoclave.
5. W303 yeast strain (see Note 4).
6. Primers: Dissolve to 100 μM in H_2O.
7. Oligoes corresponding to the FRT site: Dissolve to 1 μg/μL.
8. 10 mM ATP.
9. T4 polynucleotide kinase (10,000 U/mL, buffer supplied from distributor).
10. 2 M KCl.
11. Polystyrene rack to float on water.
12. pFA6a-KanMX4 plasmid (see Note 2).
13. Calf intestinal alkaline phosphatase (CIP, 10,000 U/mL).
14. T4 DNA ligase (400,000 U/mL, buffer supplied from distributor).
15. Competent *E. coli* strain.
16. Appropriate plates.
17. DYEnamic ET terminator reagent kit or equivalent.

2.5. Verification and Induction of the Flp-Nick System

1. Alpha-factor: Dissolve to 6 mg/mL in 10 mM HCl, 1 mM β-mercaptoethanol and 0.2 mM Na_2EDTA pH 8.0. Store aliquots at −20°C.
2. 20 % Galactose (see Subheading 2.1, step 8).

2.6. Chromatin Immunoprecipitation to Monitor FlpH305L at the FRT Site

1. pFA6a-3HA-HIS3MX6 plasmid (see Note 2).
2. Primers: Dissolve to 100 μM in H_2O.
3. 10× PBS: 8 % NaCl, 0.2 % KCl, 1.42 % $Na_2HPO_4{\cdot}2H_2O$, 0.2 % KH_2PO_4. Adjust pH to 7.0. Autoclave and store at 4°C.
4. Lysis buffer: For 500 mL: 25 mL 1 M HEPES pH 7.0, 14 mL 5 M NaCl, 1 mL 0.5 M Na_2EDTA pH 8.0, 50 mL 10 % Triton-X-100, and add water to 500 mL. Use autoclaved HEPES, NaCl and Na_2EDTA. Add 0.025 g sodium deoxycholate and protease inhibitors per 25 mL lysis buffer before the ChIP procedure is started. Store the lysis buffer at 4°C.
5. Wash buffer: For 500 mL: 5 mL 1 M Tris–HCl pH 8.0, 50 mL 5 M NaCl, 25 mL 10 % Igepal, 1 mL 0.5 M Na_2EDTA pH 8.0, and add water to 500 mL. Use autoclaved Tris–HCl, NaCl, and Na_2EDTA. Add 0.12 g sodium deoxycholate and protease inhibitors per 25 mL wash buffer before the ChIP procedure is started. Store the wash buffer at 4°C.
6. Protease inhibitors: Make stocks of the protease inhibitors and keep them at −20°C in small aliquots. PMSF is made fresh every time.

(a) Phenylmethanesulfonyl fluoride (PMSF): Make stock of 100 mM in isopropanol. Weigh off 8.7 mg powder for 500 μL. Working concentration is 0.2 mM.

(b) Benzamidine: Make stock of 1 M in H_2O. Working concentration is 1 mM.

(c) Leupeptin: Make stock of 0.5 mg/mL in H_2O. Working concentration is 0.5 μg/mL.

(d) Antipain: Make stock of 10 mM in H_2O. Working concentration is 20 μM.

(e) Pepstatin A: Make stock of 2 mg/mL in 100 % DMSO. Working concentration is 1 μg/mL.

(f) Nα-Tosyl-L-Lysine Chloromethyl Ketone Hydrochloride (TLCK): Make stock of 20 mg/mL in H_2O. Working concentration is 100 μg/mL.

(g) N-p-Tosyl-L-Phenylalanine Chloromethyl Ketone (TPCK): Make stock of 20 mg/mL in 96 % ethanol. Working concentration is 100 μg/mL.

7. Albumin from bovine serum (BSA) powder.
8. Sodium deoxycholate monohydrate powder.
9. Magnetic Dynabeads: Sheep anti-mouse IgG with a concentration of 4×10^8 beads/mL.
10. Primary antibody: HA-probe mouse monoclonal (200 μg/mL).
11. Magnetic rack.
12. 2 mL tubes with screw cap.
13. Glass beads, acid-washed.
14. Hybaid RiboLyser or equivalent.
15. Sonication tubes.
16. Needle.
17. Bunsen burner.
18. Sonicator.
19. 1× TE-buffer: 10 mM Tris–HCl pH 7.5, 1 mM Na_2EDTA pH 8.0.
20. 1× TE + 1 % SDS.
21. Thermoshaker.
22. Safelock tubes.
23. Proteinase K from Finnzymes (20 mg/mL) or equivalent.
24. 5 M lithium acetate.
25. Phenol/chloroform/isoamyl alcohol (25:24:1) (see Subheading 2.3.3, step 4).
26. Ethanol: 96 and 70 %.

27. Glycogen solution from Roche or equivalent.
28. 3 M sodium acetate pH 5.2.
29. SYBR Green qPCR kit from Finnzymes or equivalent.
30. Primers for qPCR (see Subheading 3.5.5, step 1).

2.7. Alkaline Denaturing Gel Analysis to Detect Nicking at the FRT Site

1. QIAGEN Genomic tip 20/G kit.
2. Required buffers and reagents used in the QIAGEN Genomic tip 20/G protocol: G2, Y1, QBT, QC, and QF buffers. See recipes in QIAGEN handbook. Proteinase K solution (make stock of 20 mg/mL), Zymolyase (20,000 U/g) solution (make stock of 1,000 U/mL), RNase A (make stock of 100 mg/mL), isopropanol, 70 % ethanol, 1× TE-buffer.
3. EcoNI (15,000 U/mL) and buffer supplied from distributor.
4. 3 M sodium acetate pH 5.2.
5. 0.5 M Na_2EDTA pH 8.0.
6. Ethanol: 96 and 70 %.
7. 1× Alkaline loading buffer: 50 mM NaOH, 1 mM Na_2EDTA pH 8.0, 2.5 % ficoll (type 400) in water, bromocresol green, xylene cyanol FF.
8. Agarose.
9. 10 M NaOH.
10. 1× Alkaline electrophoresis buffer: 50 mM NaOH, 1 mM Na_2EDTA pH 8.0.
11. Glass plate that fits the gel.
12. 0.25 M HCl.
13. 0.4 M NaOH.
14. Whatman 3 MM paper.
15. Hybond-XL membrane from GE Healthcare or equivalent.
16. Saran wrap.
17. Paper towels.
18. 95°C heating block.
19. Primers and DNA for making PCR product for a probe (see Subheading 3.1.1).
20. GFX-kit from GE Healthcare or equivalent.
21. dCTP $\alpha^{32}P$.
22. Rediprime II Random prime labeling system from GE Healthcare or equivalent.
23. 1× TE-buffer: 10 mM Tris–HCl pH 7.5, 1 mM Na_2EDTA pH 8.0.
24. 0.2 M Na_2EDTA pH 8.0. Dilute 0.5 M Na_2EDTA pH 8.0 in water.

25. 20× SSC: 3 M NaCl, 0.3 M tri-sodium citrate dihydrate.
26. 50× Denhardt's solution: 1 % BSA, 1 % ficoll (type 400), 1 % polyvinylpyrrolidone (PVP). Store as 1 mL aliquots at −20°C.
27. Hybridization buffer: 5× SSC, 5× Denhardt's solution, 0.5 % SDS, 500 μg carrier DNA. Use 10 mL.
28. 2× SSC + 0.1 % SDS.
29. 1× SSC + 0.1 % SDS.
30. 0.1× SSC + 0.1 % SDS.
31. Typhoon Trio or equivalent.

3. Methods

3.1. General Methods

3.1.1. PCR Product for Transformation

1. Design primers that will allow homologous recombination at a specific genomic site in the yeast genome with 3′ and 5′ ends annealing to the plasmid template.
2. Mix 200 μL master mix, 183 μL H_2O, 4 μL of each primer, 1 μL template DNA, which should correspond to 0.5–1 μg DNA, and 8 μL 50 mM $MgCl_2$ and divide into four tubes of 100 μL each. Run the PCR according to the following program:
   (a) Segment 1: 95°C for 3 min.
   (b) Segment 2: 95°C for 1 min, 52°C for 1 min, 72°C for 2 min. Run 35 cycles.
   (c) Segment 3: 72°C for 7 min.
3. Mix 5 μL from each tube with 1 μL 10× DNA loading buffer and apply on a 1 % agarose gel with ethidium bromide and run in 1× TBE to verify size and amount (see Subheading 2.2).
4. If all reactions are correct, pool and precipitate using 1/10 3 M sodium acetate pH 5.2 and 2.5 volumes of 96 % ethanol. Precipitate at −80°C for 1 h.
5. Spin the DNA down at 4°C and wash the pellet in 70 % ethanol.
6. Air dry the pellet and dissolve in 20 μL 1× TE-buffer. We usually use 2 and 8 μL from this in a transformation, together with a control yeast sample with no addition of DNA.

3.1.2. Yeast Transformation

1. Pellet 1×10^9 cells of an exponentially growing culture in YPD (see Subheading 2.1 and Note 4) by centrifugation at 3,800 × *g* for 4 min at 4°C. Wash the cells twice in 20 mL H_2O and once in 20 mL 1× TEL, before resuspending in 500 μL 1× TEL. Incubate at 30°C for 30 min.

2. Mix 100 μL cells with 50 μg carrier DNA and various amounts of PCR fragments or digested plasmids. The carrier DNA should be heated for 5 min at a 95°C heating block and placed on ice until use. Incubate at 30°C for 30 min.
3. Add 330 μL 40 % PEG4000 in TEL and mix by pipetting. Incubate at 30°C for 30 min to 1 h.
4. Heat shock the cells for 10 min at 42°C and then place the tubes on ice for 10 min.
5. Wash the cells twice with 1 mL sterile water and resuspend in 100 μL sterile water before plating on appropriate selection plates. Incubate at 30°C for 2–3 days.

3.1.3. Isolation of Genomic DNA by Phenol/Chloroform Extraction

Because of the phenol/chloroform everything is carried out in the fume hood wearing gloves.

1. Pellet 1 mL cells from an o/n culture growing in YPD (see Subheading 2.1) in 2 mL tubes with screw cap.
2. Remove the supernatant and add glass beads (2× volume of pellet), 200 μL extraction buffer and 200 μL phenol/chloroform.
3. Bead beat for 2× 45 s at settings 6.5 in a ribolyser.
4. Add 200 μL 1× TE-buffer, vortex, and spin at 15,800 ×*g* in a microcentrifuge for 5 min.
5. Transfer the water phase (usually 400–500 μL) to another tube and add 400 μL phenol/chloroform, vortex, and spin at 15,800 ×*g* for 5 min in a microcentrifuge.
6. Transfer the water phase (200 μL) to a new tube and add 20 μL 3 M sodium acetate pH 5.2 and 500 μL 96 % ethanol. Precipitate at −80°C for 1 h.
7. Spin the DNA down at 15,800 ×*g* in a microcentrifuge for 30 min at 4°C, remove the supernatant, and wash the pellet in 70 % ethanol.
8. Let the pellet air dry completely before dissolving the DNA in 50 μL H_2O or 1× TE-buffer.
9. 1 μL or 1 μL of a 10× dilution is enough to use in a PCR reaction.

3.1.4. PCR on Genomic DNA or Colonies

1. For every sample to be tested prepare a mix consisting of 12.5 μL master mix, 11.5 μL H_2O, 0.5 μL 50 mM $MgCl_2$, and 0.25 μL of each primer. We usually use a pair of primers that anneal to the inserted fragment and 500–1,000 bp upstream of the inserted fragment. Add 1 μL DNA (undiluted or 10× diluted) or skip the purification of genomic DNA and add a very small amount of cells to the PCR-mix (see Note 5).

2. Run the PCR according to the following program:
   (a) Segment 1: 94°C for 4 min.
   (b) Segment 2: 94°C for 1 min, 52°C for 1 min, 72°C for 2 min. Run 35 cycles.
   (c) Segment 3: 72°C for 7 min.
3. Add 3 μL 10× DNA loading buffer to each tube and apply 10–15 μL on a 1 % agarose gel with ethidium bromide and run in 1× TBE (see Subheading 2.2).
4. A positive sample should show an intense band at the expected size.

3.2. Engineering a Flp-Nick Strain

3.2.1. Making a Yeast Strain Containing FlpH305L

1. Digest the FlpH305L bearing plasmid pFV17Δ-H305L (see Note 3), which contains FlpH305L under a *GAL10* promoter and a functional *LEU2* gene, with 2 μL BstEII in the presence of 1× BSA and 1× reaction buffer in a total volume of 20 μL at 60°C for 2 h. BstEII will linearize the plasmid by cutting in the *LEU2* gene. We use around 5 μg pFV17Δ-H305L.
2. Purify the digested linearized plasmid using the GFX-kit and elute with 30 μL 1× TE-buffer.
3. Transform into the *leu2-3,-112* locus of a W303 *S. cerevisiae* yeast strain as described in Subheading 3.1.2, using different amounts of digested pFV17Δ-H305L. Plate on SC-leucine plates (see Subheading 2.1). This will result in a functional *LEU2* gene and an unfunctional *leu2* gene. Let the plates grow for 2–3 days at 30°C.
4. Restreak the colonies to a new SC-leucine plate and test for insertion of pFV17Δ-H305L by colony PCR (see Subheading 3.1.4) using primers CGTTTCTATTATGAA TTTCA and GGTGGTTAGCAATCGTCTTA that anneal in the *flp(H305L)* gene and in the genome upstream of the wild type *LEU2* gene, resulting in a product of 1,250 bp. Run the products on a 1 % agarose gel with ethidium bromide in 1× TBE (see Subheading 2.2).

3.2.2. Cure the FlpH305L Yeast Strain of the 2 μm Plasmid

The 2 μm circle is found naturally in most *S. cerevisiae* yeast strains. Among others it contains the wild type *FLP* gene and two FRT sites (Flp recognition target sites). To avoid that the wild type Flp and FRT sites interfere with the inserted FRT site, the FlpH305L transformed strain is cured for the 2 μm circle.

1. Restreak single colonies on YP + 3 % galactose plates three times (see Subheading 2.1).
2. Test the colonies for presence (cir^+) or absence (cir^0) of the 2 μm circle by colony PCR (see Subheading 3.1.4) using primers CCATCCACCTTTCGCTCCTT and CGAGAGA CTGCTTGCTTGTA located in the *REP1* gene of the 2 μm

circle. This will result in a PCR product of around 1,000 bp only if the cells are still Cir^+. Alternatively, perform a Southern blot to confirm lack of 2 μm.

3.2.3. Construction of Plasmid Containing the FRT Site

pFA6a or pAG plasmids containing the FRT site can be constructed as described below or alternatively they can be obtained from our laboratory upon request.

In order to generate FRT containing pFA6a plasmids, two oligoes corresponding to the upper and lower strand of a full length FRT sequence are annealed and cloned into the pFA6a plasmid. The oligoes are designed with two different cloning sites in each end to make the FRT site fit into the pFA6a or pAG plasmids.

1. Two oligoes corresponding to the upper and lower strand of the FRT site are synthesized. We used oligoes containing a full FRT site (see Fig. 1a) and SacII and SpeI sites in the ends.

   Oligo A: CTAGTGAAGTTCCTATACTTTCTAGAGAAT AGGAACTTCCGAATAGGAACTTCCCGC.

   Oligo B: GGGAAGTTCCTATTCGGAAGTTCCTATTC TCTAGAAAGTATAGGAACTTCA.
2. Dissolve each oligo to a concentration of 1 μg/μL in H_2O.
3. For each oligo, make a kinase reaction consisting of: 1 μg oligo, 1× kinase buffer, 1 mM ATP, and 1 μL T4 polynucleotide kinase. Mix in a volume of 20 μL and leave at 37°C for 30–45 min.
4. Annealing of oligo A and B: Heat water in a beaker to 80°C. Mix the two kinase reactions and add 2 μL 2 M KCl. Put the Eppendorf tube into a polystyrene rack so that the reaction is in the water. Cool down to 30°C.
5. Digest of vector: We use the pFA6a-KanMX4 vector for cloning (see Note 2). Around 3–5 μg of the vector is digested with 1 μL SacII and 1 μL SpeI in a reaction containing 1× reaction buffer and 1× BSA in a volume of 20 μL. Leave at 37°C for 2 h, add 1 μL CIP and leave for 1 more hour at 37°C. CIP will remove the 5′ phosphate group of the digested plasmid to prevent religation of the plasmid. Run the digested plasmid on a 1 % agarose gel with ethidium bromide in 1× TBE (see Subheading 2.2) and purify using the GFX-kit, eluting with 10 μL 1× TE-buffer.
6. Ligation of vector and FRT site: Make five reactions containing different amounts of FRT site (no FRT site, undiluted, 10× diluted, 50× diluted, and 100× diluted). The reactions consist of 1/10 digested vector, FRT site, 1× ligase buffer, 1/10 T4 DNA ligase, and 1 mM ATP. Mix in a volume of 10 μL and leave at 16°C o/n. Next day, the reactions are transformed into a competent *E. coli* strain by standard procedures.

7. Plasmid is isolated by standard miniprep procedure and the product is digested to confirm insertion of the FRT site. The FRT site contains a XbaI restriction site and this is used together with a BglII site in the vector, resulting in products of 1,500 and 2,480 bp. Digest 1–2 ng plasmid with 1× BSA and 1× reaction buffer in a volume of 10 μL for 2 h at 37°C. Run the digested plasmids on a 1 % agarose gel with ethidium bromide in 1× TBE (see Subheading 2.2).
8. To verify correct insertion and sequence of the FRT site, the plasmid is sequenced over the inserted FRT site using primer CCTCGACATCATCTGCCC with the DYEnamic ET terminator reagent kit following that protocol.

3.2.4. Transformation of the FRT Site Into FlpH305L Cir^0 Yeast Strain

When designing primers for PCR-mediated insertion remember to take into account that Flp prefers to cleave on the left site of the FRT site. Therefore, depending on direction of insertion, either leading or lagging strand may be targeted for nicking.

1. Make a PCR product (see Subheading 3.1.1) using primers with homologous ends to the region of interest. We use primers: CCTGATTCATCTATTAGCCGTATAGAGAATATACCGCATCACGACTCACTATAGGGAGAC and GCTTTCATGGATCGAGTATTTGTTTGTCGAAAAGGATCAGAACGCGGCCGCCAGCTGAA with homologous regions approximately 3 kb from *ARS607*, for transforming the FRT site into the FlpH305L Cir^0 strain. The underline is the part of the primer that corresponds to the genomic DNA.
2. Transform the FRT site into the FlpH305L Cir^0 strain (see Subheading 3.1.2). Plate on YPD plates and replica plate to YPD + G418 plates the next day (see Subheading 2.1).
3. Make genomic DNA and run PCR to verify correct insertion of the FRT site (see Subheadings 3.1.3 and 3.1.4), using primers GGTGATGTGATCGTATTCAA and CCGTGCGGCCATCAAAATG locating in the genome and in the plasmid, respectively.

3.3. Verification of the Flp-Nick System

To verify that the Flp-nick system works correctly, we normally generate an *RAD52* deletion in the Flp-nick strain. This deletion will result in a near-dead phenotype on galactose plates. Suppressors are likely to arise and will result in fast-growing cells.

1. Generate an *RAD52* deletion strain as described in the protocols in Subheadings 3.1.1 and 3.1.2 for PCR-mediated gene disruption and transformation. Verify the disruption using the protocols for genomic DNA and PCR described in Subheadings 3.1.3 and 3.1.4. Primers for making the deletion and PCR should be designed.

2. The positive colonies are tested by spot assay in the following way. Grow colonies o/n in 2 mL YP + raffinose at 30°C (see Subheading 2.1), measure OD_{600} and dilute all cultures to $OD_{600} = 1$ in 1 mL YP + raffinose. Make a 10× dilution row of each culture (1:1; 1:10; 1:100; 1:1,000; 1:10,000; 1:100,000). Spot 4 μL from each dilution on YPD and YP + 3 % galactose plates (see Subheading 2.1). Incubate at 30°C for 1–3 days.

3.4. Growth of Yeast Culture and FlpH305L Induction

The Flp-nick system can be controlled by adding or removing galactose. Together with an arrest of the cells this gives the advantage that the Flp-nick damage can be studied in a specific part of the cell cycle, e.g., the S phase. Here follows the general arrest and induction protocol for the Flp-nick system. Modifications are made to this protocol depending on which experiment is done and how many cells are needed.

1. Grow cells o/n in YP + raffinose at 30°C (see Subheading 2.1).
2. Dilute the cells and let them grow one generation in YP + raffinose pH 3.5 (see Subheading 2.1 and Note 6).
3. Add alpha-factor (1:3,000) and treat for 60 min at 30°C. Alpha-factor is a mating pheromone that affects MATa cells to inhibit DNA synthesis thereby arresting in G1.
4. Add galactose to 3 % (see Note 7) and additional alpha-factor (1:6,000) and treat for 2 h at 30°C.
5. Wash the cells once with water.
6. Release the cells into YPD (see Subheading 2.1) at 25°C to slow down the replication process, and take out samples throughout the S/G2 phase. We also recommend taking out samples for flow cytometry (see Note 8).

3.5. Chromatin Immunoprecipitation to Monitor FlpH305L at the FRT Site

In order to verify that FlpH305L actually binds and get covalently attached to the FRT site, chromatin immunoprecipitation (ChIP) can be performed. In the general ChIP procedure, formaldehyde is used to cross-link proteins to DNA, but as FlpH305L forms long-lived covalent protein-DNA complexes upon DNA cleavage, formaldehyde treatment is omitted. For the ChIP assay, a strain with a tag on FlpH305L is used (see Note 9).

3.5.1. Growth of Yeast Cells

Grow the cells as described in Subheading 3.4, with these modifications:

1. Grow cells until 1.0×10^7 cells/mL.
2. Take out 25×10^7 cells from the random growing culture, to be used as a negative control for detection of FlpH305L at the FRT site.

3. Take out 25×10^7 cells from the G1 arrested, galactose-induced culture to be used as a control for the presence of FlpH305L at the FRT site before replication is allowed.
4. After induction take out 1.0×10^7 cells for a trichloroacetic acid preparation to be used as a control of the presence of FlpH305L-3HA and perform a western blot.
5. Take out 25×10^7 cells at different time points during the replication process, e.g., every 20 min for up to 160 min.
6. Spin down the samples, wash in 1 mL 1× PBS, and store at −20°C in 2 mL tubes with screw cap until the ChIP procedure is performed.

3.5.2. Preparation of Buffers for ChIP

Before the ChIP procedure is started prepare the buffers.

1. Weigh off PMSF and dissolve in isopropanol.
2. Add sodium deoxycholate monohydrate and protease inhibitors only to the volume of lysis- and wash buffer that is needed during the experiment (see Notes 10 and 11).
3. Weigh off BSA and dissolve in 1× PBS. Usually, it is enough to make 10–15 mL.

3.5.3. Coupling of Antibody to Beads

First the antibody is coupled to magnetic beads. For each sample, we use 100 μL magnetic Sheep anti-mouse IgG beads (e.g., 8 samples = 800 μL).

1. Wash the beads in 1 mL 1× PBS + 5 mg/mL BSA by pipetting up and down and then placing the tube in a magnetic rack and allowing the beads to move to the magnet before removing the buffer. Repeat this wash three times.
2. After wash dissolve the beads in 1 mL 1× PBS + 5 mg/mL BSA and divide into two tubes of 500 μL each.
3. Add antibody to one of the tubes. We use 2 μL of HA-probe mouse monoclonal/50 μL original volume beads (e.g., 8 samples = 16 μL antibody).
4. Leave the two tubes (beads + antibody and beads-antibody) to rotate on a wheel for 2 h at 4°C.

3.5.4. Preparation of Extract

While the antibody is coupled to the beads the cell extracts are prepared. Keep the samples cold so prepare everything on ice.

1. To each sample add glass beads (approximately twice as many as the volume of the cells) and 700 μL lysis buffer.
2. Crack the cells in a ribolyser using setting 6.5 for 3–4 times 45 s. Leave 1 min breaks in between every round of ribolysing and place the samples on ice. The breakage can be checked under microscope. Cracked cells will look transparent. Continue with the ribolysing until ~90 % of the cells are cracked.

3. Warm a needle in a Bunsen burner flame and quickly make a hole at the bottom of the tube to spin out the extract. Place the tube in a sonication tube and spin at 3,800 × *g* for 5 min at 4°C. If necessary, place the sonication tube into another tube that fits in the centrifuge.
4. Transfer the extract, including the pellet, to an Eppendorf tube and spin at 15,800 × *g* in a microcentrifuge for 15 min at 4°C. Wash the sonication tube with water and reuse at step 6.
5. Remove the supernatant. Keep 40 μL of the supernatant in another tube for western blotting. This sample should not contain any protein (see Note 12).
6. Dissolve the pellet in 700 μL lysis buffer. It can be very hard to dissolve. When dissolved, transfer the extract back to the sonication tube and add 700 μL lysis buffer (the volume is now 1.4 mL).
7. Sonicate the samples to obtain DNA fragments of 500–1,000 bp (see Note 13).
8. Transfer the extract to a new Eppendorf tube and spin at 4,500 × *g* in a microcentrifuge for 2 min at 4°C.
9. From each supernatant take out 40 μL for the input sample. Choose one sample where FlpH305L-3HA will be expected and take out 40 μL for western blotting.
10. Divide the extract from each sample into two Eppendorf tubes (samples +ab and −ab) where beads +/− antibody will be added.
11. Wash the beads +/− antibody once in wash buffer and twice in lysis buffer using a volume of 800 μL. Resuspend the beads in lysis buffer to the original volume (e.g., 8 samples = 400 μL for beads + antibody and 400 μL for beads-antibody).
12. Divide the beads + antibody and the beads-antibody into the +ab and −ab samples, respectively. Add 45–50 μL to each sample and make sure to add an equal volume into each sample.
13. Incubate on a rotating wheel for 2 h at 4°C.
14. After the incubation, place the samples into the magnetic rack and take out 40 μL for western blotting from selected +ab and −ab samples. Wash the samples twice in lysis buffer, once in wash buffer and once in 1× TE-buffer. Use a volume of 750 μL. Between each washing step incubate the samples for 5 min with shaking of at 4°C.
15. Disrupt the coupling between beads and antibody by adding 60 μL 1× TE buffer + 1 % SDS and shake at 100 × *g* at 65°C for 10 min.
16. Transfer the supernatant into a safe-lock tube (see Note 14) and repeat step 15. Take 5 μL out of selected +ab and −ab samples for western blotting.

17. Combine the supernatants from the two rounds of uncoupling (the volume is now ~120 μL).
18. To reverse the cross link, add 210 μL 1× TE-buffer + 1 % SDS to the input samples and 130 μL 1× TE-buffer + 1 % SDS to the +ab and −ab samples and incubate o/n at 65°C, e.g., in a hybridization oven or waterbath.
19. Next day treat the samples with proteinase K for 3 h at 37°C by adding 240 μL 1× TE-buffer and 20 μL proteinase K to each sample.
20. Isolate the DNA by adding 50 μL 5 M lithium acetate and 450 μL phenol/chloroform.
21. Vortex the samples and spin for 5 min at 15,800 × *g* in a microcentrifuge.
22. Transfer the water phase (~480 μL) to a new Eppendorf tube and add 1 mL of an ethanol, glygogen, and sodium acetate solution. To make a stock for 25 samples, mix 25 mL 96 % ethanol, 50 μL glycogen, and 1.1 mL 3 M sodium acetate pH 5.2.
23. Precipitate at −80°C for at least 2 h or o/n. Precipitation takes longer for small DNA fragments.
24. Spin the samples for 45 min at 15,800 × *g* at 4°C in a microcentrifuge, and carefully remove the supernatant. The pellet is small and may be very loose.
25. Add 200 μL 70 % ethanol to the pellet and spin for 2 min at 15,800 × *g* in a microcentrifuge.
26. Dry the samples and dissolve the pellets in 60 μL H_2O.

3.5.5. Quantitative PCR

To investigate the enrichment of the FRT site, the recovered DNA is analyzed by quantitative PCR (qPCR). Two different primer sets locating close to the FRT site and in a telomeric region are used.

1. The primer concentrations are optimized to have an efficiency as close to 100 % as possible, meaning that the amount of DNA is doubled in every qPCR cycle. We test the primers with different or the same concentrations of the forward and reverse primer together with a series of eight DNA dilutions where the DNA is diluted four times in each step (1:1–1:16,384). We use primers AAGTTCGACATGGCTTCAG and TCGATAAAACGATGCAACCA which localize 3 kb from *ARS607* around *ROG3* start site and TGGCAAGGGTAAAAACCAGT and ACCCACACCACCTCACTAGC which localize in the telomeric region of the right arm of chromosome VI and serves as a negative control. The primers are designed to give a product of around 160 bp. It might be necessary to dilute the input and +ab samples 100 times to make the amount of DNA fit in the area of the standard curve where the primers have 100 % efficiency. Remember to take the dilution into account when calculating fold increases.

2. We use the following qPCR program using Stratagenes MX3000P with SYBR Green master mix:
   (a) Segment 1: 95°C for 10 min.
   (b) Segment 2: 95°C for 30 s, 55°C for 30 s, 72°C for 30 s. Repeat this for 40 cycles.
   (c) Segment 3: 95°C for 1 min, 55°C for 30 s, 95°C for 30 s.
3. The fold increase is calculated by this formula: $2^{(\text{ct value -ab} - \text{ct value +ab})}$, assuming that the primer efficiency is 100 %.

3.6. Alkaline Denaturing Gel Analysis to Detect Nicking at the FRT Site

Purified DNA from Flp-nick cells are analyzed on alkaline denaturing gels to visualize nicking at the FRT site.

3.6.1. Growth of Yeast Cells

Grow the cells as described in Subheading 3.4, with these modifications:

1. Grow cells until 2.0×10^7 cells/mL.
2. Take out 2.0×10^9 cells from the G1 arrested culture before galactose induction. This sample is used as a negative control, where no break should be detected.
3. Take out 2.0×10^9 cells from the G1 arrested, galactose-induced culture. This sample is used as a positive control to detect the single-strand break.
4. 2.0×10^9 cells are taken out, e.g., 30, 60, and 90 min after release into S phase.
5. The cells are pelleted and stored at −80°C until DNA extraction.

3.6.2. Preparation of Genomic DNA

1. The genomic DNA is isolated using the QIAGEN Genomic tip 20/G protocol with the following modifications. The samples are treated with proteinase K for 2 h at 37°C. After 1 h the same amount of proteinase K is added as before. The DNA can be left to dissolve slowly shaking at 30°C for 1–3 days. Apply 1 μL on a 1 % agarose gel with ethidium bromide and run in 1× TBE to see if the DNA has been dissolved (see Subheading 2.2).
2. Digest 10–15 μg DNA with EcoNI for 4 h at 37°C using 4 μL enzyme with 1× enzyme buffer in a total volume of 200 μL. This digest will result in a DNA fragment of approximately 2,750 bp around the nick, when the FRT site is placed in the *ARS606-ARS607* region.
3. Apply 10 μL of the digest reaction on a 1 % agarose gel with ethidium bromide and run in 1× TBE to check that all samples are digested (see Subheading 2.2).

4. Precipitate the digested DNA with 0.3 M sodium acetate pH 5.2, 5 mM Na_2EDTA pH 8.0 and two volumes of 96 % ethanol at –80°C for at least 1 h. Wash the pellet with 70 % ethanol and dissolve it in 20 μL 1× alkaline loading buffer.

3.6.3. Preparation of and Running the Gel

1. Prepare a 1 % agarose solution and cool to 60°C, add NaOH to 50 mM and Na_2EDTA pH 8.0 to 1 mM and pour the gel. We use a gel of 19 cm.
2. Run the gel in 1× alkaline electrophoresis buffer (see Note 15) at 4°C for approximately 1 h at 1 V/cm. Place a glass plate on the gel, to prevent loss of the color, and electrophoresis is carried out at 4°C for 18–22 h at 0.5 V/cm (see Note 16).

3.6.4. Southern Blot

The gel is not stained with ethidium bromide because it does not bind very well to single-stranded DNA.

1. Incubate the gel in 0.25 M HCl for 30 min with gentle agitation, wash with water and incubate in 0.4 M NaOH for 30 min with gentle agitation.
2. Assemble the Southern blot in a tray with 0.4 M NaOH. Place a platform in the tray and soak two Whatman 3 MM papers in 0.4 M NaOH before placing them on the platform. They should have the width of the gel and be long enough, so the ends can hang into the NaOH. Remove air bubbles.
3. Place the gel on the Whatman papers with the opening of the slots facing down. Remove air bubbles.
4. Equilibrate a Hybond-XL membrane that fits with the size of the gel in 0.4 M NaOH for a couple of minutes and place it on top of the gel. Cut off a corner to keep track of orientation and remove air bubbles.
5. Soak two gel-sized Whatman 3 MM papers in 0.4 M NaOH and place them on the membrane. Remove air bubbles.
6. Place saran wrap around the edges of the gel to avoid that the buffer go through any paper towels that might touch the long pieces of Whatman 3 MM paper.
7. Build up a big stack of paper towels, cut to the size of the gel, and add some weight. Make sure that the transfer tower is leveled.
8. Transfer o/n.
9. The membrane is hybridized with a probe recognizing the right side of the nick, which equals a DNA fragment of around 2 kb. Design primers that will generate a PCR product hybridizing to the region of interest. We use primers ATTGAGACTTTATGGGAAGA and CCACAAATGG TGAAATAAAA which locates in the *ROG3* gene and gives a PCR product of around 1,000 bp. We use genomic DNA

purified from wild type cells as template in the PCR reaction (see Subheading 3.1.1). The PCR product is gel purified and labeled using the Rediprime II Random prime labeling system from GE Healthcare using dCTP α^{32}P. The membrane can also be hybridized with a probe that recognizes the left side of the nick, which equals a DNA fragment of around 750 bp (see Note 17).

10. After transfer wash the membrane with 2× SSC.
11. Prepare 10 mL hybridization buffer. Heat the carrier DNA on a 95°C heating block for 5 min and cool on ice, before adding it to the hybridization buffer. If the SDS in the hybridization buffer precipitates, the buffer can be warmed at 65°C.
12. Place the membrane in a hybridization tube with the DNA facing in and add the hybridization buffer. Pre-hybridize for 1 h at 65°C. We use a Hybaid oven.
13. Denature the probe for 5 min at a 95°C heating block and cool on ice. Add it to the hybridization buffer and hybridize o/n.
14. Next day wash the membrane for 10 min in 150 mL 2× SSC + 0.1 % SDS at room temperature, for 15 min in 125 mL 1× SSC + 0.1 % SDS at 65°C, and for 10 min in 125 mL 0.1× SSC + 0.1 % SDS at 65°C.
15. Dry the filter on paper towels and expose.
16. The film is developed in a Typhoon Trio+.

4. Notes

1. For YP + raffinose pH 3.5 medium, we use meat peptone powder from Lab M, because the medium will otherwise be unclear after autoclaving. However, there is precipitate in the bottle after autoclaving. We avoid using the part of the medium with the precipitate. For medium that does not need to be pH adjusted, we use a granulated meat peptone.
2. We use the pFA6a and the pAG vectors for cloning, gene knockouts, and gene tagging. Most of the plasmids can be obtained from EUROSCARF.
3. Upon request plasmid pFV17Δ-H305L can be obtained from our laboratory. This plasmid was generated by Marc Gartenberg and first published in (12).
4. We use strains that are derivatives of W303-1A from R. Rothstein.
5. We normally first perform colony PCR, and if this does not give positive hits, we make genomic DNA and perform PCR

on this. We often see that although colony PCR does not work we get positive results with PCR on genomic DNA.

6. If the strain has a slow growth phenotype, the culture is diluted less.
7. We use galactose solutions that are no older than 14 days.
8. We take out a sample for flow cytometry from the random growing culture, the G1 arrested culture and from every time point when the cells are released to determine cell cycle phase. We take out approximately 0.5×10^7 cells and use a FACSCalibur for the flow cytometry and WinMDI for analyzing the results.
9. We used pFA6a-3HA-HIS3MX6 to make a PCR product (see Subheading 3.1.1) with primers: GAGGTACTAGACTACCTTTCATCCTACATAAATAGACGCATACGGATCCCCGGGTTAATTAA, and TGAGAAGAACGGCATAGTGCGTGTTTATGCTTAAATGCGTACGAATTCGAGCTCGTTTAAAC which is transformed (see Subheading 3.1.2) into the Flp-nick strain using SC-histidine plates for selection (see Subheading 2.1). The underline is the part of the primer that corresponds to the genomic DNA. The colonies are tested by PCR using genomic DNA (see Subheadings 3.1.3 and 3.1.4) and primers: AATTTTTGAGGAATTCTGAA and CCGTGCGGCCATCAAAATG locating in the *flp(H305L)* gene and in the plasmid which will result in a product of around 1,050 bp. The HA-tagged Flp-nick strain can be obtained from our laboratory upon request.
10. One sample uses 5.1 mL lysis buffer and 1.5 mL wash buffer excluding buffers for the magnetic beads.
11. Add the protease inhibitors to the wash buffer right before it is required to wash the incubated extract, beads, and antibody.
12. Only take out samples for western blotting from one or two samples where the presence of FlpH305L-3HA will be expected. The molecular weight of FlpH305L with the 3HA tag will be of the same size as the long IgG chain from the primary antibody, and it can therefore be hard to see FlpH305L on the western blot in the samples with antibody.
13. We sonicate using a Branson Sonifier 250, where we sonicate three times for 20 s at setting 1 and twice for 20 s at setting 2. Cool the samples on ice between each sonication time. If the sample foam, stop the sonication as this can denature the proteins. Remember to use protective goggles and earmuffs when sonicating.
14. Use safe-lock tubes as phenol/chloroform extraction will be carried out in these tubes the next day.

15. Alkaline running buffer should be mixed fresh each time.
16. It is a good idea to load a marker composed of different sizes of DNA recognized by the probe on the gel, e.g.:
    (a) 2,400 bp: ATTGAGACTTTATGGGAAGA and AAATA TAAACAAGACAAAGC.
    (b) 1,600 bp: ATTGAGACTTTATGGGAAGA and TTTAAG CCCGAGGGCACTGG.
    (c) 1,000 bp: ATTGAGACTTTATGGGAAGA and CCAC AAATGGTGAAATAAAA (the same as the probe).

    We use genomic DNA purified from wild type cells as template in the PCR reaction (see Subheading 3.1.1).
17. Primers for probe recognizing the left side of the nick: TCCAACATGGATGCTGATTT and GGATGGCGGCGTT AGTATCG. Make PCR product using genomic DNA purified from the Flp-nick strain as this probe locates to the KanMX4 selection marker.

Acknowledgement

We thank Marc Gartenberg, Robert Wood Johnson Medical School, for plasmid pFV17Δ-H305L. This research was supported by grants from the Danish Research Counsil (FNU 272-07-0366), Aase and Einar Danielsens Foundation, Augustinus Foundation and Dagmar Marshall Foundation to L.B., and the Danish Cancer Society to A.H.A. and L.B.

References

1. Nielsen I, Bentsen IB, Lisby M, Hansen S, Mundbjerg K, Andersen AH, Bjergbaek L (2009) A Flp-nick system to study repair of a single protein-bound nick in vivo. Nat Methods 6:753–757
2. Volkert FC, Broach JR (1986) Site-specific recombination promotes plasmid amplification in yeast. Cell 46:541–550
3. Jones JR, Shelton KD, Magnuson MA (2005) Strategies for the use of site-specific recombinases in genome engineering. Methods Mol Med 103:245–257
4. Andrews BJ, Proteau GA, Beatty LG, Sadowski PD (1985) The FLP recombinase of the 2 micron circle DNA of yeast: interaction with its target sequences. Cell 40:795–803
5. Chen JW, Lee J, Jayaram M (1992) DNA cleavage in trans by the active site tyrosine during Flp recombination: switching protein partners before exchanging strands. Cell 69:647–658
6. Chen JW, Yang SH, Jayaram M (1993) Tests for the fractional active-site model in Flp site-specific recombination. Assembly of a functional recombination complex in half-site and full-site strand transfer. J Biol Chem 268: 14417–14425
7. Lee J, Jayaram M, Grainge I (1999) Wild-type Flp recombinase cleaves DNA in trans. EMBO J 18:784–791
8. Qian XH, Inman RB, Cox MM (1990) Protein-based asymmetry and protein-protein interactions in FLP recombinase-mediated site- specific recombination. J Biol Chem 265: 21779–21788
9. Voziyanov Y, Lee J, Whang I, Lee J, Jayaram M (1996) Analyses of the first chemical step in Flp

site-specific recombination: synapsis may not be a pre-requisite for strand cleavage. J Mol Biol 256:720–735
10. Pommier Y, Redon C, Rao VA, Seiler JA, Sordet O, Takemura H, Antony S, Meng L, Liao Z, Kohlhagen G et al (2003) Repair of and checkpoint response to topoisomerase I-mediated DNA damage. Mutat Res 532:173–203
11. Parsons RL, Prasad PV, Harshey RM, Jayaram M (1988) Step-arrest mutants of FLP recombinase: implications for the catalytic mechanism of DNA recombination. Mol Cell Biol 8:3303–3310
12. Tsalik EL, Gartenberg MR (1998) Curing *Saccharomyces cerevisiae* of the 2 micron plasmid by targeted DNA damage. Yeast 14:847–852
13. Longtine MS, McKenzie A 3rd, Demarini DJ, Shah NG, Wach A, Brachat A, Philippsen P, Pringle JR (1998) Additional modules for versatile and economical PCR-based gene deletion and modification in *Saccharomyces cerevisiae*. Yeast 14:953–961

Chapter 29

Measurement of Diversification in the Immunoglobulin Light Chain Gene of DT40 Cells

Julian E. Sale

Abstract

The immunoglobulin loci of the genetically tractable chicken B cell line DT40 provide a unique opportunity to study the cellular response to endogenously generated DNA damage in a chromosomal context. Abasic sites generated by the concerted action of Activation-Induced Deaminase (AID) and Uracil DNA Glycosylase result in both homologous recombination-dependent gene conversion and translesion synthesis-dependent point mutations. The system has provided important insights into both the early stages of AID-dependent immunoglobulin gene diversification and into the relationship between pathways of DNA damage bypass. Here we describe the assays that can be employed to monitor the rate and pattern of immunoglobulin gene diversification at the light chain locus of DT40.

Key words: DT40, Immunoglobulin genes, Homologous recombination, Gene conversion, Somatic hypermutation, Translesion synthesis, Abasic sites, Mutation rate

1. Introduction

Chicken DT40 cells have been used extensively over the past 20 years for studying DNA repair and many other cellular processes (reviewed in ref. 1). In addition to their genetic tractability (2), they exhibit continuous diversification of their immunoglobulin (Ig) loci by gene conversion and non-templated point mutation (3, 4). Studies in DT40 have allowed detailed insight not only into the mechanism of Ig gene repertoire development but also into pathways of DNA damage bypass, notably the relationship between translesion synthesis and homologous recombination (5).

Sequence diversification of the Ig loci of the chicken is triggered by the action of Activation-Induced Deaminase (AID) (6, 7). Unlike mice and humans which generate a significant level of Ig gene diversity by V(D)J recombination, the chicken uses

Lotte Bjergbæk (ed.), *DNA Repair Protocols*, Methods in Molecular Biology, vol. 920,
DOI 10.1007/978-1-61779-998-3_29, © Springer Science+Business Media New York 2012

V[D]J recombination in a restricted way simply to generate a functional heavy and light chain from a very limited number of segments (8). Following productive rearrangement, AID acts by deaminating deoxycytosine bases within the rearranged heavy and light chain variable genes to produce uracil. Uracil is recognized in the V gene DNA by Uracil DNA Glycosylase (UNG) and is excised to produce an abasic site. In the chicken, these abasic sites trigger two mutagenic outcomes, homeologous gene conversion and mutagenic bypass by translesion synthesis, which together drive AID-dependent Ig diversification.

Gene conversion is likely to be triggered by single-stranded gaps generated at the V gene abasic sites during replication (9). In the light chain, the donor sequences used during gene conversion are derived from an array of 25 pseudogenes located upstream of the rearranged V gene, which is called VL1 (10). Ig gene conversion is a somewhat unusual form of homologous recombination in that it is readily able to utilize the homeologous pseudogenes resulting in the incorporation of tracts of sequence change into VL1. The process described in chickens appears to be faithfully recapitulated in DT40 cells (3, 4).

Effective Ig gene conversion is dependent on canonical recombination proteins such as RAD54, BRCA2, and the RAD51 paralogues (11–13). However, as noted above, not all mutations in the chicken Ig genes can be accounted for by gene conversion (10, 14). Non-templated point mutations are formed by direct bypass of the abasic sites by translesion synthesis. These mutations are reduced by disruption of the specialized translesion polymerase REV1 and by prevention of PCNA monoubiquitination, a key step in the recruitment of translesion polymerases to the replication fork (15, 16). Thus, examination of the rate and pattern of diversification of the DT40 light chain gene can be used to study the relative use of translesion synthesis and recombination during the replicative bypass of a naturally occurring chromosomal DNA lesion (for a more extensive review see ref. 5).

While the pattern of Ig locus diversification is best assessed by direct sequencing, it is also possible to monitor the rate of diversification by following the gain or loss of surface IgM (sIgM). The rate of Ig gene conversion can be assessed by following the gain of sIgM expression in a subclone carrying a defined frameshift in VL1, for example CL18 (3). Overall diversification can also be monitored by following the generation of sIg-loss variants in sIg-positive clones (13). However, it is worth bearing in mind that the relationship between the surface expression of Ig and mutation of the V gene is more complex than simple inactivation of expression. This is particularly evident in mutants that exhibit very high levels of non-templated point mutation. Here loss of surface expression frequently results not from nonsense mutation of the variable genes but from mutations that prevent pairing of the heavy and light chains. This results in intracellular accumulation of unpaired heavy and light chains, which can result in decreased viability in turn impacting on the accuracy with which sIg loss variants are detected. Thus, the accumulation of sIg-loss variants in point mutator mutants (e.g., mutants of the RAD51 paralogues) may not provide an accurate assessment of mutation "rate."

2. Materials

2.1. DT40 Culture, Clone Propagation, and Freezing

1. DT40 wild-type cells (or an sIg-negative clone such as CL18 if gene conversion rates are to be assessed) plus mutants to be tested.
2. DT40 culture medium pre-warmed: RPMI 1640 medium with Glutamax® (Invitrogen) plus 7 % fetal calf serum (FCS) (e.g., Hyclone or PAA), 3 % chicken serum (Sigma), 50 μM 2-mercaptoethanol, penicillin 10 U/mL, and streptomycin (0.1 mg/mL). Full details of DT40 culture techniques can be found elsewhere (1).
3. Laminar flow cabinet for tissue culture.
4. Humidified incubator with CO_2.
5. Benchtop centrifuge for 24-well plates and 5 mL tubes.
6. Cell freezing mix (FCS:chicken serum:dimethylsulfoxide in the ratio 63:27:10).
7. Dry ice.

2.2. sIg Staining and Cell Sorting

1. Ice-cold phosphate-buffered saline (PBS):140 mM NaCl, 10 mM Sodium phosphate, pH 7.4.
2. Ice-cold PBS plus 1 % FCS.
3. Anti-chicken IgM-FITC (e.g., Bethyl Laboratories).
4. 25 μg/mL propidium iodide in PBS.
5. 5 mL polystyrene round-bottom tubes (e.g., Falcon).
6. Round-bottomed 96-well tissue culture plate.
7. 24-well tissue culture plate.
8. Flow cytometer capable of two-color (green/red) detection.
9. Cell sorter, e.g., DAKO-Cytomation MoFlo (see Note 1).
10. Benchtop microfuge for 1.5 mL tubes.

2.3. DNA Preparation, Cloning, and Sequencing

1. Reagents for gDNA extraction, e.g., Gentra Puregene® kit.
2. Ultrapure water (purified to 18 MΩ/cm at 25 °C).
3. High-fidelity thermostable DNA polymerase, e.g., *Pfu* or KOD.

4. General purpose cloning vector (e.g., pBluescript) and suitable competent cells (e.g., DH5α).
5. Sequencing primers for cloning vector (e.g., M13F and M13R).

3. Methods

The propagation of DT40, using the medium described in Subheading 2.1, is covered in detail elsewhere (1). We have routinely cultured DT40 at 37 °C although it will grow happily at temperatures up to 41 °C (see Note 2). Lines and subclones should be maintained at between 0.2 and 1×10^6 cells/mL.

3.1. Isolating and Expanding sIg+ve or sIg−ve Subclones

The first step is to isolate single-cell clones. This is most conveniently and accurately achieved by staining for surface IgM and using a cell sorter to isolate desired sIg-positive or -negative cells. If a cell sorter is not available, then isolate clones by limiting dilution (see Note 1).

1. Pellet approximately 5×10^6 cells in a 5 mL round-bottomed polystyrene tube at 250 g for 5 min in a cooled benchtop centrifuge. Remove the supernatant and wash once with ice-cold PBS. Remove PBS, resuspend cells by gentle vortexing in the residual PBS, and add 100 μL PBS/1 % FCS with anti-chicken IgM-FITC conjugate at an appropriate dilution determined by titration (see Note 3). Mix thoroughly by gentle vortexing or by flicking the tube. Incubate on ice for 30–60 min.
2. Wash the cells in 4 mL ice-cold PBS/1 % FCS and then resuspend in 0.5 mL cold PBS/1 % FCS. Keep on ice until ready to sort.
3. Sort sIg-positive or sIg-negative cells into 96-well round-bottom plates containing 200 μL fresh DT40 medium at 1 cell per well (see Note 4).
4. Transfer to incubator. Clones will normally be visible in 7–10 days (see Note 5). When the clones are about 2–3 mm across at the bottom of the well, transfer using a 200 μL pipette to a 24-well plate containing 1 mL DT40 medium. One 24-well plate is usually sufficient for each line to be assayed. The clones are then propagated for a fixed time in the 24-well plate, usually 4 weeks unless AID is overexpressed (see Note 6). It can be helpful to use a large-volume multichannel pipette for feeding the cells while in the 24-well plates. Try to keep all wells at the same density of between 0.2×10^6 and 1×10^6 per mL and avoid overgrowth.

5. It is worth considering making a duplicate 24-well plate that can be frozen in case of problems during the propagation of the clones and for future reference. To freeze cells in the plate, spin the plate at 250 g in a benchtop centrifuge for 5 min. Remove as much medium as possible, leaving the cells undisturbed. Add 500 μL freezing mix and swirl to resuspend the cells. Seal the plate with sellotape and place on top of a layer of dry ice covered with three layers of tissue paper. Once frozen, transfer to a –80 °C freezer. Cells that have been frozen carefully and stored in a stable freezer will keep for at least 2 months. To defrost, thaw as rapidly as possible in a 37 °C incubator. Spin at 250 g for 5 min and remove the freezing mix. Resuspend in 1 mL fresh medium at 37 °C.

3.2. FACS Analysis of sIg Loss or Gain

1. Once the cells are ready to analyze, transfer 0.5 mL (which ideally will be 0.5×10^6 cells) to a round-bottomed polystyrene tube. Spin down, wash in 4 mL ice-cold PBS/1 % FCS, spin down again, aspirate almost all of the supernatant, and resuspend in the residual liquid. Add 50 μL PBS/1 % FCS with anti-chicken IgM–FITC conjugate (see Note 7). Incubate on ice for 30–60 min.
2. Wash once with ice-cold PBS/1 % FCS and then resuspend in PBS/1 % FCS with propidium iodide at 25 μg/mL.
3. For flow cytometric analysis, gate first on forward and side scatter. Then gate on cells that exclude PI, and then plot sIg-FITC signal on a histogram plot (Fig. 1). The position of the sIg–ve population can be deduced from appropriate controls, e.g., unstained cells. To determine the percentage of sIg-negative cells (assuming that the starting clones were sIg positive) use a caliper set for a fixed fold-intensity change (8–12× is usually appropriate) (Fig. 1c). The same approach can be applied to clones that started out sIg negative and have generated sIg-positive cells. Exclude any clone that is predominantly opposite in terms of sIg staining to the starting population as it is most likely to represent a sorting error.

3.3. Sequence Analysis of the Immunoglobulin Light Chain

It is rarely profitable to randomly sequence the VL from unselected DT40 cells unless there is good reason to believe that the gene conversion or point mutation rate will be significantly increased. In the case of reversion of an initially sIg-negative subclone carrying a VL frameshift sequencing sIg-positive cells can help reveal the spectrum of pseudogene usage. Examination of sIg loss variants from an initially sIg-positive clone can reveal deviations from the normal balance of gene conversion and hypermutation.

1. If sequence data is required, pick six or more clones (see Note 8) from the 24-well plate and expand them into tissue culture flasks in order to have sufficient cells from which to isolate gain or loss variants by cell sorting. The number of cells to be grown will depend on the size of the gain or loss variant population

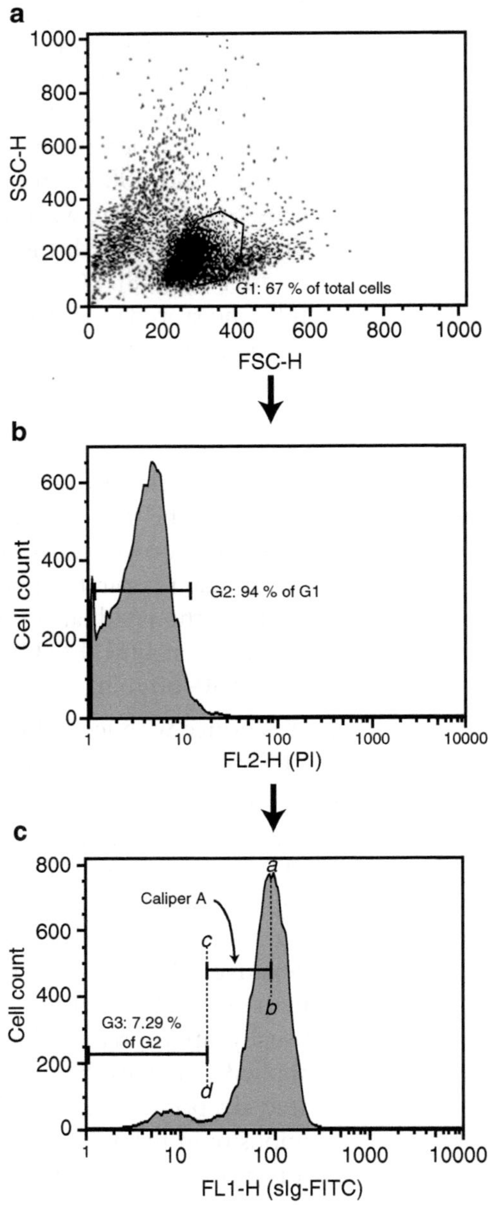

Fig. 1. FACS gating for analysis of sIg-loss variants. (**a**) First gate on forward (FSC) and side scatter (SSC) to exclude the majority of dead cells. This is Gate 1 (G1). (**b**) Then gate for exclusion of propidium iodide. This ensures the exclusion of the few remaining dead cells falling within the FSC/SSC gate. This is Gate 2 (G2). Exclusion of dead cells is particularly important in cases where there are very few sIg-loss variants and when the DT40 line under study exhibits reduced viability. (**c**) Determination of the proportion of sIg-loss variants. Set a caliper for a fixed fold-reduction in staining intensity (Caliper A). In this case Caliper A is set at eightfold. Place the right-hand end at the center of the sIg-positive peak (*dashed line ab*). Determine the percentage of G1-gated sIg-loss variants falling below the *left-hand extent* of Caliper A (*dashed line cd*) as a proportion of Gate 2.

and the accuracy of the cell sorter being used. Stain a sample of cells as described above and sort about 10,000 of the sIg variants into a clean 1.5 mL Eppendorf tube containing 500 μL PBS. If possible, it is worth putting an aliquot of the sorted cells back though a flow cytometer to assess the purity of the sort.

2. Seal the tube and gently vortex it to ensure that all the cells are suspended in the PBS and not left on the side of the tube. Pellet the cells in a benchtop microcentrifuge at 800×*g* for 1 min. The pellet will be just visible to the naked eye.
3. Prepare genomic DNA from the cells. We use the Gentra Puregene® (Qiagen) method. Resuspend the genomic DNA pellet in ultrapure water at the equivalent of 500 cells/μL.
4. Amplify VL1 using the primers CVLF5 (5′-CAGGAGCT CGGCTCTGTCCCATTGCTGCGCGG) and CVLR3 (5′-GCGCAAGCTTCCCCAGCCTGCCGCCAAGTCCAA) in the following reaction using a high-fidelity thermostable polymerase, e.g., *Pfu* Turbo (Stratagene).

Genomic DNA (equivalent to 5,000 cells)	10 μL
Primer F5 @ 10 pmol/μl	2 μL
Primer R3 @10 pmol/μL	2 μL
dNTP @ 10 mM each	2 μL
10× Pfu buffer	10 μL
Pfu Turbo polymerase	1 μL
Ultrapure water	73 μL

and thermal cycling conditions

90 °C	Hot start (see Note 9)	×8 cycles
95 °C	1 min	
68–60 °C	1 min; decrease by 1 °C/cycle over the first 8 cycles	
72 °C	2 min 30 s	

94 °C	30 s	×22 cycles
60 °C	1 min	
72 °C	2 min 30 s	
72 °C	5 min	
4 °C	Pause	

5. Check the PCR reaction by running out 10 μL of the reaction on a 1 % agarose gel and if successful proceed to step 6 (see Note 10).
6. The resultant PCR products are gel purified and can be cloned into pBluescript using the SacI and HindIII sites located in the PCR primers (see Note 11).

7. Transform competent *Escherichia coli* (e.g., DH5α) with the ligation product and employ Blue/White selection by plating out on plates containing IPTG and X-Gal (see Note 12). Before sequencing check the efficiency of ligation by selecting ten colonies, performing a standard plasmid miniprep, and ensuring an insert in the plasmid of the correct size in at least eight or nine of the ten plasmid preps.
8. When satisfied that the majority of bacterial clones contain the amplified V gene sequence, up to 50 colonies from each of the selected DT40 subclones using the M13R -20 primer (see Note 8).

3.4. Analysis of the Pattern of Immunoglobulin Diversification

1. Align the sIg variant sequences obtained from each cellular subclone using an algorithm such as Clustal. In the case of sIg-positive revertants from a starting frameshifted sIg-negative cell such as CL18 it should now be possible to identify the gene conversion tracts resulting in restoration of the light chain reading frame. To confirm which donor is being used search for the new sequence within the database of 25 light chain pseudogenes (see Note 13). Note that there may also be associated non-templated point mutations.
2. In the case of sIg loss variants generated from sIg-positive clones it is necessary to identify which mutations within VL are due to gene conversion and which are due to non-templated point mutation. Use the algorithm shown in Fig. 2 to assign as many sequence changes as possible to one of the three categories, gene conversion, non-templated mutation, and ambiguous (13). The ambiguous category comprises point mutations that nonetheless have a potential donor sequence ≥9 bp long in one of the pseudogenes (see Note 14).
3. To derive a pattern of diversification exclude any sequence change that occurs more than once in any one cellular subclone to avoid any issues with overrepresentation of events that occurred early in the expansion of the clone (see Note 8). As an additional refinement it is useful in certain circumstances to determine whether sIg loss is due to larger structural rearrangements of VL1 by Southern blotting of the locus (see Note 15).

4. Notes

1. If a fluorescence activated cell sorter is not available, it is possible to simply derive clones by limiting dilution (17). However it is wise to exclude clones whose Ig staining has apparently become almost entirely opposite to that of the expected starting population. These are likely to have arisen by selection of a starting cell with the incorrect sIg phenotype.

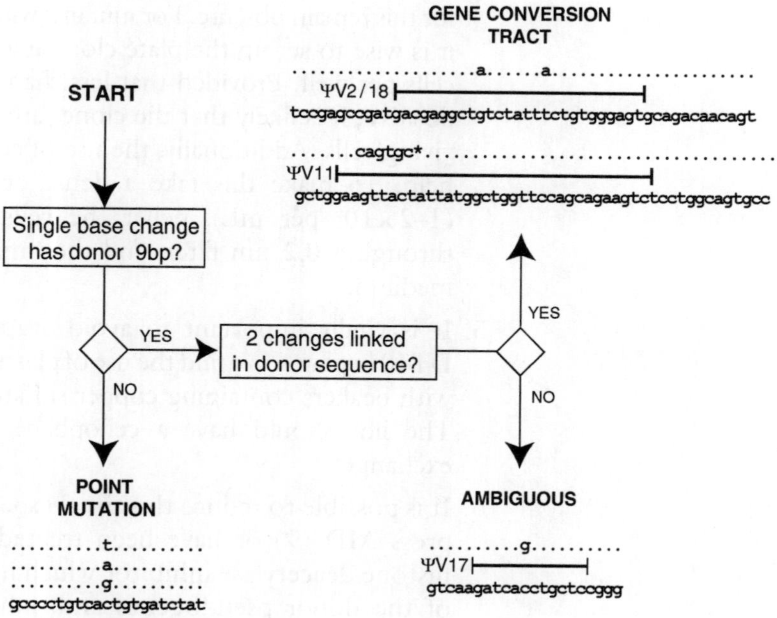

Fig. 2. Algorithm for classification of sequence alterations in VL1. For each mutation identified, the database (Table 1) of pseudogenes is searched for a donor sequence ≥ 9 bp long. 9 bp was chosen arbitrarily as a cutoff based on the smallest probable gene conversion donor tract reported (21). If no donor is found the change is classified as a non-templated point mutation. If a donor is found then the adjacent sequence is examined for linked changes that can be explained by the same donor. If such changes are found the linked changes are classified as a gene conversion tract. If no linked sequence changes are present the mutation is classified as ambiguous since it is not possible to tell whether it was generated by gene conversion or not.

2. DT40 can be grown at temperatures between 37 °C and 41 °C, reflecting the higher core temperature of chickens. We have grown our cells routinely at 37 °C, while growing the cells at 37 °C results in a doubling time of c. 11 h, as opposed to c. 9 h at 41 °C, we have not found any qualitative differences in the pattern of Ig diversification. However, the number of cell cycles completed in a given time will be higher at the elevated temperature and this may need to be taken into account when assessing Ig locus diversification.
3. Usually a 1:25 to 1:50 dilution of the Bethyl antibody recommended above is adequate, although it is important to titrate the antibody concentration to ensure saturation of the surface IgM and clear discrimination of the sIg-positive and -negative cells and lower dilutions may be necessary. We have also found the anti-chicken IgM antibody (clone M-1) from Southern Biotech to work well.
4. We have found that round-bottomed plates work much better than flat-bottomed ones for this sort of cloning. The reasons

for this remain obscure. For mutants with poor cloning efficiency, it is wise to set up the plate cloning at three, five, or even ten cells per well. Provided that less than about 20 % of wells are positive, it is likely that the clones are derived from single surviving cells. Additionally, the use of conditioned medium may help. To make this take a dense culture of wild-type cells ($1–2 \times 10^6$ per mL), pellet the cells, filter the supernatant through a 0.2 μm filter, and then mix this 50:50 with fresh medium.

5. It is vitally important to avoid evaporation when culturing DT40 in plates. We find the use of plastic pizza boxes humidified with beakers containing copper sulfate solution to be helpful. The lids should have a cellophane window to allow CO_2 exchange.

6. It is possible to reduce this time, especially if the cells overexpress AID (9) or have been treated with trichostatin A, a histone deacetylase inhibitor which increases the accessibility of the donor pseudogenes and enhances the rate of gene conversion (18).

7. Again the ratio of antibody concentration to number of cells needs to be titrated and it is important to ensure that this ratio is the same for each sample. This is most easily achieved by keeping the cell density and volume constant between samples.

8. In an expanding population of mutating B cells a sequence change that happens early will be propagated through the population and will therefore be overrepresented in prevalence when the mutation load is assessed. However, a mutation that happens early may also be a mutation that happens more frequently and therefore reconstructing the mutation spectrum from a single "snapshot" of mutation can be misleading. A detailed mathematical consideration of this problem is set out in Luria and Delbrück's classic paper on the kinetics of acquisition of bacteriophage resistance in *E. coli* (19). While it is sometimes possible to create a cladogram from mutations collected from a single cellular clone, in practical terms it is safest to expand the database of mutations by sequencing from multiple subclones and counting any individual change only once from a particular subclone. Clearly a balance needs to be struck between acquiring enough sequences from any one subclone to obtain an accurate reflection of diversity and obtaining information from independent subclones to allow detection of true mutation "hot spots" and bias in pseudogene usage. We have found that for wild-type DT40 cultured for a month from cloning it is usually possible to obtain good information from 50 sequences each from six subclones. Obviously this ratio will need to be adjusted depending on the rate of diversification, but in general little is usually to be gained by

obtaining more than 50 V gene sequences from a single subclone.

9. In the case of Pfu polymerase, adding the polymerase to the PCR mix at 90 °C improves the specificity of the reaction. Many more recent enzyme mixes come with systems (e.g., anti-polymerase antibodies) to prevent the polymerase working until the reaction has been heated up. In these instances, the hot start step is unnecessary. If performing a classical hot start reaction it is important to avoid cross contamination between reactions and the use of a tube opening tool is recommended.
10. This PCR reaction will often pick up both the rearranged and unrearranged alleles. These are readily distinguishable by size, 530 bp in the case of the rearranged VL1 and 2,324 bp in the case of the unrearranged allele.
11. It is worth bearing in mind that gel purification of the restricted VL1 PCR products may limit the detection of large deletions. If detecting these rare events is important it is worth considering removing the restriction enzymes using a Sephacryl S200 column.
12. Blue/white selection helps identify *E. coli* colonies harboring pBluescript containing an insert. We have noted that the correct colonies, containing the 530 bp VL1 fragment, are frequently pale blue rather than the expected white.
13. The sequences of the chicken light chain pseudogenes can be found in NCBI with accession number AH002536.1. These sequences come from the inbred CB strain of White Leghorn chicken (10). The DT40 cell line is, however, derived from the F_1 hybrid Hy-line SC strain whose parental strains are designated G4 and S3. The one rearranged VL1 allele is derived from the G4 strain, while the allele from the S3 strain is unrearranged. The known differences between the G4 pseudogenes and CB pseudogenes are shown in bold in Table 1.
14. Automated analysis is possible and an example of an implementation of the algorithm by Svend Petersen-Mahrt and the Bioinformatics and Biostatistics unit at Cancer Research UK can be found at https://bioinformatics.cancerresearchuk.org/~mitter01/IgV_Analyser_TEST/query.html
15. Larger structural rearrangements of the VL loci can be detected by Southern blotting (20). *KpnI* digestion and blotting with a VL1 probe allows the rearranged and unrearranged alleles to be distinguished, the monitoring of gene conversion by introduction and loss of a *KpnI* site in VL1 (4), and the detection of larger deletions or rearrangements (20).

Table 1
Sequences of VL1 and the 25 VL pseudogenes from the CB strain of chicken (10) with the differences identified in the G4 strain (22) (from which the rearranged allele of DT40 is derived) highlighted in bold

VL1 GGGGCTGGGCTAGGAGTGGGCAGGGAGGTGTGGGGAGAGGTGTAGGTCGTGGCTGG GAGCGCAGGGAGTTATTTGCATAGGGGGGCGTGCTGCGGAAGGACGCGGGTATAAAA GGGCATCGAGGTCCCCGGCACAGCCCCATCGGCGTGGGGACACACAGCTGCTGGGA TTCCGCCATGGCCTGGGCTCCTCTCCTCCTGGCGGTGCTCGCCCACACCTCAGGTA CTCGTTGCGCCCGGTCGGGGACTGTGGGCACGGGGCTCTGTCCCACTGCTGCGCGGG CAGGGCTGTGCGTGCGGGGCCGTCACTGATTGCCGTTTTCTCCCCTCTCTCCTCTCCC TCTCCAGGTTCCCTGGTGCAGGCAGCGCTGACTCAGCCGTCCTCGGTGTCAGCGAA CCCGGGAGGAACCGTCAAGATCACCTGCTCCGGGGATAGCAGCTACTATGGCTGGTACC AGCAGAAGGCACCTGGCAGTGCCCCTGTCACTGTGATCTATGACAACACCAACAGACCC TCGAACATCCCTTCACGATTCTCCGGTTCCAAATCCGGCTCCACAGCCACATTAACCAT CACTGGGGTCCGAGCCGACGACAATGCTGTCTATTACTGTGCGAGTACAGACAGCAGC AGTACTGCAGCACGGTGACACAAAGCAATGGGGAAATGATACAAAAACCTCCTGCCAG TGCAACGAGCAGCTG
VL1 AATCTGATTATCAACCACTATGGCTGGTACCAGCAGAAGGCACCTAGCAGTGTCCCTGT CACTCTGATCTACTATGATGATGAGAGACCCTCGAACATCCCTTCACGATTCTCCGGTT CCAAATCCAGCTCCACACACACATTAACCATCACTGGGGTCCAAGCCGATGACGAGGCT GTCTATTACTGTGGGAATGAAGACAGCAGCGGTACTGGTGCCCCAGTGACACAGAGCAA TGGAGATATGATACACAGACCTGACATCTC
VL2 AATTGCTCAGATCCATTTTCTCCCCTCTCTCCTCTCCCTCTCCAGGTTCCCTGGTGCAGG CAGCGCTGACTCAGCCGGCCTCGGTGTCAGCAAACCTGGGAGGAACCGTTGAGATCA CCTGCTCCGGGGGTTACAGCGGCTATGCCTGGCAGTGCCCCTGTCACTGTGATCTAT AATGGCAATAACAGACCCTCGGACATCCCTTCACGATTCTCCGGTTCTGGATCCGGC TCCACAAACACATTAACCATCACTGGGGTCCAAGCCGAGGACGAGGCTATCTATTACT GTGGGAGCATAGACAGCAGCATCACAGCAATAACACCACTGCCAGGTTCATGCTGCTG TAAAGCATAAACATAATGTCTG
VL3 TGGCTGGTTCCTGGCAGTGCCCCTGTCTCTGTGATCTATAGCAGCACCAACAGACCCTC AAACATCCCTTCACGATTCTCCGGTTCCAAATCTGGCTCCACACACACATTAACCATCA CTGGGGTCCAAGCCGATGACGAGGCTGTCTATTACTGTGGGAGCACAGACAGCAGCT ATGTTGGTATCACAGTGACATGGAGCAATTAAAAATGATGCATAACCTTTTGCACA
VL4 ATAAGCCCCCCTCTCTCCTGTCCCTCTCCAGGTTCCCTGGTGCAGGCAGCGCTGACT CAGCCGTCCTCGGTGTCAGCAAACCTGGGAGGAACCGTCAAGATCACCTGCTCCGGG AGTGACAGCAGCTATGGTTACAGCTGGCACCAGCAGAAGTCTCCTGGCAGTGCCCCT GTCACTCTGATCTATAGCAACGACAAGAGACCCTCGGACATCCCTTCACGATTCTCC GGTTCCGCATCCGGCTCCACAGCCACATTAACCATCACTGGGGTCCAAGTCGAGGA CGAGGCTGTCTATTACTGTGCAAACTTCGACAGCAGCAGTACTGCTGTTAGTGACA GGGGGGA

(continued)

Table 1 (continued)

VL5 GAGCCCCATTTTCTCTCCTCTCCCTCTCCAGGTTCCCTGGTGCAGGCAGCAGTGACTCA GCCGGCCTCGGTGTCAGCAAACCCAGGAGAAACCGTCAAGATCACCTGCTCCGGGGG TGGTAGCAGCAGCTACTATGGCTGGTACCAGCAGAAGTCTCCTGGCAGTGCCCTTGTC ACTGTGATCTATGAAAGCACCAAGAGACCCTCGAACATCCCTTCACGATTCTCCGGTT CCACATCTGGCTCCACAAGCACATTAACCATCACTGGGGTCCAAGCCGAGGACGAGGC TGTCTATTTCTGTGGGAGCTACGACAGCAGTGCTGGTTATGTTGGTGGCACTGTTGGA
VL6 CACATCTGCTCCAGGATTGTTAACAACTACGGCTGGTTCCAGCAGAAGTCACCTGGCAGT GCTCCTGTCACTGTGATTTATGGTAGCACCAGTAGACCCTCAAACATCCCTTCACGATT CTCCGGTTCCAAGTCCGGCTCCACAGCCACATTAACCATCACTGGGGTCCGAGCCGA GGACGAGGCTGTCTATTACTGTGGGAGTGCAGACAGCAGTGGTGCTGGTCCCATGGT GACACACAGCAGTGAGGAAGTGATGCACACAACCTCCTGTCAG
VL7 GAGGCCCATTTTCTCTTCTCTCCCTCTCCAGGTTCCCTGGTGCAGGCAGCAGTGACTCAG CCGGCCTCGGTGTCAGCAAACCTGGGAGGAACCGTCGAGATCACCTGCTCCGGGAGTA GTGGCAGCTAGGCTGGTATCAGCAGAAGTCACCTGGCAGTGCCCCTGTCACTGTGATC TATTACAACGACAAGAGACCCTCGGACATCCCTTCACGATTCTCCGGTTCCAAATCCGG CTCCACGGGCACATTAACCATCACTGGGGTCCAAGCCGAGGACGAGGCTGTCTATTACT GTGGGAGCAGGGACAGCAGCTATGTTGGTATGGTGCTACAGAGAGG
VL8 ATCCTTGTCCCTCTCCAGGTTCCCTGGTGCAGGCAGCGCTGACTCAGCCGGCCTCGGTG TCAGCAAACCCAGGAGAAACCGTCAAGATCACCTGCTCCGGGGGTGGCAGCTATGCT GGAAGTTACTATTATGGCTGGTACCAGCAGAAGGCACCTGGCAGTGCCCCTGTCACTG TGATCTATGACAACACCAACAGACCCTCGAACATCCCTTCACGATTCTCCGGTTCCCTAT CCGGCTCCACAAACACATTAACCATCACTGGGGTCCAAGTCGAGGACGAGGCTATCTA TTTCTGTGGGAGCTACGAAGGCAGCACTAGTGCTGTATGGTGACAATGAT
VL9 GGGATCAGTCAGTCCGGGAGTAGCAATGAGTATGGCTGGTTCCAGGAGAAACCTGGTAG TGGTGCTGTGACTGTGATCCACAGTGACACCAACAGACCCTCGAACATCCCTTCACAA TTCTCCTGTTCCCCATCTGGCTCCACGCACAAGCTAACCATCACTGAGGTCCAAGCTG AGGACAAGGCTGTATATTACCGTGGAGACTGAAAGGCAGCAGTGATCTATTTTGACAAT GTGTAGTAGGAACA
VL10 TTTTCTCTCCCTCTCCAGGTTCCCTGGTGCAGGCAGCGCTGACTCAGCCGGCCTCGGTG TCAGCAAACCTGGGAGGAACCGTCAAGATCACCTGCTCCGGGGGTGGTGGCAGCTAT GGCTGGTTCCAGCAGAAGTCTCCTGGCAGTGCCCCTGTCACTCTGATCTATAACAAC AACAACAGACCCTCGGACATCCCTTCACGATTCTCCGGTTCCAAATCCGGCTCCACA GCCACATTAACCATCACTGGGGTCCAAGCCGAGGACGAGGCTGTCTATTTCTGTGGT GGCTACGACAGCAGCACTTATGCTGGTATTGCCTCAGCGGCAGGTGAGGAACTGATG GATTGATCAGGCTT

(continued)

Table 1
(continued)

VL11 CTCTACTGAGCCCCATTTTCTCCCCTCTCCTCTCCAGGTTCCCTGGTGCAGGCAGCACTG ACTCAGCCAGCCTCAGTGTCAGCAAACCTAGGAGAAACTCAGGATCACCTGCTCCGGGA GTGGCAGTAACAGTGCTTATGGCTGGTTCCAGCAGAAGTCACCGGGCAGTGCCCTTGT CACTGTGATCTACTGGAATGACAAGAGACCCTCAAACATCCCTTCACGATTCTCCGGTT CCAAGTCCGGCTCCACAGCCACATTAACCATCACTGGGGTCCGAGCCGAGGACGAGGC TGTCTATTACTGTGGGAGTGCAGACAGCAGTGGTGCTGGTCCCATGGTGACACACAGC AGTGAGGGAAGCGATGCACACACCTCCTGTCAG
VL12 AAACACTGAGCCCCATTTTCTCCCCTCTCCCTCTCCAGGTTCCCTGGTGCAGGCAGCGC TGACTCAGCCGGCCTCAGTGTCAGCAAACCTGGGAGGAACCGTCAAGCTCACCTGCT CCGGGGGTAGCAGTGGCTGTGGTTATGGCTGGTATCAGCAGAAGTCACCTGGCAGTG CCCCTGTCACTCTGATCTATTACAACAACAAGAGACCCTCGGACATCCCTTCACGATT CTCCGGTTCCAAATCCGGCTCCACGGGCACATTAACCATCACTGGGGTCCAAGCCGA GGACGAGGCTGTCTATTACTGTGGGAATGCAGACAGCAGCAGTACTGATACCACAGTG ACTCAGATGTAATACACAAACCTCCCGGCAG
VL13 TCCCCTCTCCCTATGCAGGTTCCCTGGTGCAGGCAGCGCTGACTCTGCCAGACTCATTAT CAGCAAATCTGGGAGAAACTGTCAGGATAACCTGGTCTGGAGGTAGCTATGATTATGGTT GGCACCAGCAGAAATCACCTGGCAGTGCCCCTGTCACTGTGATCTATAGCAACAACAAG AGACCCTCGGACATCCCTTCACGATTCTCCGGTGCCCTATCCGGCTCCACAGCCACATT AACCATCACTGGGGTCCGAGCCGAGGACGAGGCTGTCTATTTCTGTGGTGCCTGGG AAAGCAGCAGTAATCCTGGTATGGTGATAGGGGGCGAATTGGGGAGAG
VL14 CCACACTGATGCCGTTTTCTCCCCTCTCTCCTCTCCCTCTCCAGGTTCCCTGGTGCAGGC AGCACTGACTCAGCAGCCGGCCTCGGTGTCAGCAAACCCAGGAGAAACCGTCAAGATC ACCTGCTCTGGGGGCAGCTATAGCTATGGCTGGTTCCAGCAGAAGTCTCCTGGCAGTG CCCTTGTCACTGTGATCTACTGGGATGATGAGAGACCCTCGGGCATCCCTTCACGATT CTCCGGTTCCACATCCGGCTCCACGGGCACATTAACCATCACTGGGGTCCGAGCTGA GGACGAGGCTGTCTATTACTGTGGGAGCTTCGACAGCAGCACTGATAGTGGATATGCT GGTGGTGGCACTGCGG
VL15 GGGATCAGTCACTCTGGGAGTAGCCATGAGTATGGCTGGTTCCAGGAGAAACCTGGTAGT GGTGCTGTGACTGTGATCCACAGTGACACCAACAGACCCTCGAACATCCCTTCACAATT CTCCAGTTCCACC
VL16 GAGCTTCATTTTCTCTCCTCTCCCCCTCCAGGTTCCCTGGTGCAGGCAGCACCGACTCA GGCAGCACCGACTCAGCTGGCCTTCATGTCAGCAAACCCAGGAGAAACCGTCAAGATCA CCTGCTCAGGCAGTAGCGATGCTTATGGCTGGTACCAGCAGAAGTCTCCTGGCAGTGCCCC TGTCACTGTGATCTATGGTAACACCGGTAGACTCTT**AA**GCATCCCTTCACTAATGGGG
VL17 CCATTTTCTCCAGGTTCCCTGGTGCAGGCAGTGCTGACTCAGCTGGCCTTGATGTCAGT GCATCTGCGTGTTACTGATTACAGGATCAGCTGCTCTGGGAGTAGTGGTGACTATGGC AGTAACTGTCACGGGCAGCACCAGC**AA**AGGCA**T**CTGGCAGTGC**T**CCT**T**TC**TA**TGTGC TCTACC**TGAA**CAGCAACAGACCC**TCA**AACATCCCTTCACGATTCTCCGGTTCTGGATCC GGCTCCACAAACACATTAACCATCACTGGGGTCCGAGCCGAGGACGAGGCTGTCTATTA CTGTGGTGGCTACGACGGCAGCACTGATGCTGGTAGTGATGGGGAGATGTG

(continued)

Table 1 (continued)

VL18 ATTCACTGAGCTCCATTTTCTCCCCTCTCTCCTCTCCCTCTCCAGGTTCCCTGGTGCAGG CAGCAGTGACTCAGCCGGCCTCGGTGTCAGCAAACCTGGGAGGAACCGTCGAGATCA CCTGCTCCGGGGGTAGCAGCGATGATGGAAGTTATTACTATAGCTGGCACCAGCAGAA GTCTCCTGGCAGTGCCCCTGTCACTGTGATCTATAGCAACAACCAGAGACCCTCGAAC ATCCCTTCACGATTCTCCGGTTCCACATCCGGCTCCACAAGCACATTAACCATCACTGG GGTCCAAGCCGACGACGAGGCTATCTATTACTGTGGGAGCTGGGATAGCAGCACTGAA GCTGTAGGGTGATGATGATTAA
VL19 TTTGTTGATTTTCTCTCCTCTCCCTCTCCAGGTTCCCTGGTGGAGGCAGAACTCATTCAG CCGACCT CGGTGTCAGCAAACCCTGGGGAAACTGTCAAGATCACCTGCAGTGGAGGTA TCGGCCAGTGGTAT GGCTGGTACCAGCAGAAGTCACCTGGCAGTGTGCCTGTCACTGT GATCAATGCTAACACCAACAG ACCCTCGAACATCCCTTCCCGATTCTCCGGTTCCCTCT CCGTATTCACACACACATTAACCATCAC TGGGGTCCAAGCCGAGGATGAGGCTGTCTA TTACTGTGGTGCCTACGACAGCAGCTATGTTGGTATAGGTCACGGGGGAGA
VL20 AACCTGACTCAGCCAGCCTCAGTGTCAGCAAACCCAGGAGAAACTGTCAAGATCACCTGCT CTGGTAGCAGCAGCAGCTACTATGGCTGGATCCAGCAGAAATCACCTGGCAGTGCCCCT GTCACTGTGATTTATGATAGCAGCAGCAGACCCTCGGACATCCCTTCACGATTCTCCGG TTCCAAATTCGGCTCCACGGCCACATTAACCATCACTGGGGTCCAAGTCGAGGACGAGG CTGTCTATTACTGTGGTGACTGGGACAGGAGCAATAGTGCTGGTATAGGTGTGTGGG AAGATGCGGGGGTGATACAGAATGCGATTCAA
VL21 CCTGTTTTCTCTCCTCTCCCTCTCCAGGTTCCCTGGTGCAGGCAGCGCTGACTCAGCC GGCCTCGGTGTCAGCGAACCTGGGAGGAACAGTCAAGATCACCTGCACCAGGGATGA CAGCTGGTATGGCTGGCACCTG
VL22 CCACAGACCATCTTTTACTACAGCGGGTACCAGCAGAAGACGCCTGGCAGTGCATCTGT CACTGTGATCTATGCTGACACCAACAGACCCACCAACACCACGGGG
VL23 GAGCTCCATTTTGTCTCCTGTCCCTCTCCAGGTTCCCTGGTGCAGGCAGCGCTGACTCA GCCATCCTCGGTGTCAGCAAACCTGGGAGGAACCGTCGAGATCACCTGCTCCGGGGG TAGCAACAACTATGGCTGGTATCAGCAGAAGTCTCCTGGCAGTGCCCCTGTCACTGTG ATCTATCAAAACACCAAGAGGCCCTCGGGTCTCTCTTCATAACTTTCTTGTTCCTCAT CCGGCTCTGCAAACACATTAACCATCACTGGGGTCCAAGCCAAGGATGAGGCTGTCT ATTACTGTGGTGGCTGGGATAGCTGCTGCCTGAGTCAGTGCTGCATGCACCAGG
VL24 ACTCACTGAGCTCCATTTTCTCCCCTCTCTCCTCTTGACTCAGCCGGCCTCGGTGTCAG CGAACCCGGGAGAAACCGTTGAGATCACCTGCTCCGGGGATAGCAGCTACTATGGCT GGTACTATGGCTGGTACCAGCAGAAGGCACCTGGCAGTGCCCCTGTCACTCTGATCT ATGACAACACCAACAGACCCTCGAACATCCCTTCACGATTCTCCGGCTCCCCAGGCA GAACCACAAGCACATTAACCATCACTGGGGTCCAAGTCGAGGACGAGGCTGTCTATT ACTGGGGGAACTGGGATGACAGCACTGATGCTGTATGGTGACGATGATTCA
VL25 GACAAATAGATGTGTACAGCTGCTCTGGGAGTAACAACAACAACTATGGCTGGTACCAG CAGAAGACACCTGGCAGTGCCCCTCTCACTGTGATCTATCAAAACACAAAGAGACCTT CAGGTATATCTTTATACTTCAGCTGG

Acknowledgements

I would like to thank David du Plessis for comments on the manuscript. Work in the laboratory is supported by the Medical Research Council, Association for International Cancer Research, and The Fanconi Anemia Research Fund.

References

1. Buerstedde JM, Takeda S (2006) Reviews and protocols in DT40 research, vol 40. Springer, Berlin
2. Buerstedde JM, Takeda S (1991) Increased ratio of targeted to random integration after transfection of chicken B cell lines. Cell 67:179–188
3. Buerstedde JM, Reynaud CA, Humphries EH, Olson W, Ewert DL, Weill JC (1990) Light chain gene conversion continues at high rate in an ALV-induced cell line. EMBO J 9:921–927
4. Kim S, Humphries EH, Tjoelker L, Carlson L, Thompson CB (1990) Ongoing diversification of the rearranged immunoglobulin light-chain gene in a bursal lymphoma cell line. Mol Cell Biol 10:3224–3231
5. Sale JE (2004) Immunoglobulin diversification in DT40: a model for vertebrate DNA damage tolerance. DNA Repair (Amst) 3:693–702
6. Arakawa H, Hauschild J, Buerstedde JM (2002) Requirement of the activation-induced deaminase (AID) gene for immunoglobulin gene conversion. Science 295:1301–1306
7. Harris RS, Sale JE, Petersen-Mahrt SK, Neuberger MS (2002) AID is essential for immunoglobulin V gene conversion in a cultured B cell line. Curr Biol 12:435–438
8. Reynaud CA, Anquez V, Dahan A, Weill JC (1985) A single rearrangement event generates most of the chicken immunoglobulin light chain diversity. Cell 40:283–291
9. Nakahara M, Sonoda E, Nojima K, Sale JE, Takenaka K, Kikuchi K, Taniguchi Y, Nakamura K, Sumitomo Y, Bree RT, Lowndes NF, Takeda S (2009) Genetic evidence for single-strand lesions initiating Nbs1-dependent homologous recombination in diversification of Ig v in chicken B lymphocytes. PLoS Genet 5:e1000356
10. Reynaud CA, Anquez V, Grimal H, Weill JC (1987) A hyperconversion mechanism generates the chicken light chain preimmune repertoire. Cell 48:379–388
11. Bezzubova O, Silbergleit A, Yamaguchi-Iwai Y, Takeda S, Buerstedde JM (1997) Reduced X-ray resistance and homologous recombination frequencies in a RAD54-/- mutant of the chicken DT40 cell line. Cell 89:185–193
12. Hatanaka A, Yamazoe M, Sale JE, Takata M, Yamamoto K, Kitao H, Sonoda E, Kikuchi K, Yonetani Y, Takeda S (2005) Similar effects of Brca2 truncation and Rad51 paralog deficiency on immunoglobulin V gene diversification in DT40 cells support an early role for Rad51 paralogs in homologous recombination. Mol Cell Biol 25:1124–1134
13. Sale JE, Calandrini DM, Takata M, Takeda S, Neuberger MS (2001) Ablation of XRCC2/3 transforms immunoglobulin V gene conversion into somatic hypermutation. Nature 412:921–926
14. Parvari R, Ziv E, Lantner F, Heller D, Schechter I (1990) Somatic diversification of chicken immunoglobulin light chains by point mutations. Proc Natl Acad Sci U S A 87:3072–3076
15. Arakawa H, Moldovan GL, Saribasak H, Saribasak NN, Jentsch S, Buerstedde JM (2006) A role for PCNA ubiquitination in immunoglobulin hypermutation. PLoS Biol 4:e366
16. Simpson LJ, Sale JE (2003) Rev1 is essential for DNA damage tolerance and non-templated immunoglobulin gene mutation in a vertebrate cell line. EMBO J 22:1654–1664
17. Buerstedde JM (2006) Subcloning DT40 by limiting dilution. Subcell Biochem 40:393–394
18. Seo H, Masuoka M, Murofushi H, Takeda S, Shibata T, Ohta K (2005) Rapid generation of specific antibodies by enhanced homologous recombination. Nat Biotechnol 23:731–735
19. Luria SE, Delbruck M (1943) Mutations of bacteria from virus sensitivity to virus resistance. Genetics 28:491–511
20. Szüts D, Simpson LJ, Kabani S, Yamazoe M, Sale JE (2006) Role for RAD18 in homologous recombination in DT40 cells. Mol Cell Biol 26:8032–8041
21. McCormack WT, Thompson CB (1990) Chicken IgL variable region gene conversions display pseudogene donor preference and 5′ to 3′ polarity. Genes Dev 4:548–558
22. McCormack WT, Hurley EA, Thompson CB (1993) Germ line maintenance of the pseudogene donor pool for somatic immunoglobulin gene conversion in chickens. Mol Cell Biol 13:821–830

Chapter 30

Live Cell Microscopy of DNA Damage Response in *Saccharomyces cerevisiae*

Sonia Silva, Irene Gallina, Nadine Eckert-Boulet, and Michael Lisby

Abstract

Fluorescence microscopy of the DNA damage response in living cells stands out from many other DNA repair assays by its ability to monitor the response to individual DNA lesions in single cells. This is particularly true in yeast, where the frequency of spontaneous DNA lesions is relatively low compared to organisms with much larger genomes such as mammalian cells. Single cell analysis of individual DNA lesions allows specific events in the DNA damage response to be correlated with cell morphology, cell cycle phase, and other specific characteristics of a particular cell. Moreover, fluorescence live cell imaging allows for multiple cellular markers to be monitored over several hours. This chapter reviews useful fluorescent markers and genotoxic agents for studying the DNA damage response in living cells and provides protocols for live cell imaging, time-lapse microscopy, and for induction of site-specific DNA lesions.

Key words: Homologous recombination, Checkpoint, Fluorescence microscopy

1. Introduction

A broad palette of genetically encoded fluorophores have been developed for fluorescence microscopy with the most commonly used being derivatives of the green and red fluorescent proteins (GFP and RFP) from *Aequoria victoria* and *Discosoma* sp., respectively (1). The available variants of these molecules vary with respect to excitation and emission spectra, brightness, photostability, pKa, and oligomeric state in vitro. However, in vivo additional factors may affect the fluorescence capabilities of fluorophores. For example, both GFP and RFP require molecular oxygen for the maturation of the nascent polypeptide into a chromophore by cyclization and oxidation of central amino acids before the protein becomes fluorescent (2, 3). Codon usage is also of great importance for the proper expression and folding of the fluorophore, and notably the optimal codon usage in *Saccharomyces cerevisiae* is different from

Lotte Bjergbæk (ed.), *DNA Repair Protocols*, Methods in Molecular Biology, vol. 920,
DOI 10.1007/978-1-61779-998-3_30, © Springer Science+Business Media New York 2012

that of human cells, for which many GFP and RFP variants have been optimized. Finally, it is advisable to avoid fluorophores that oligomerize because these can compromise biological function of the fusion protein. We have good experience with three specific variants of CFP (clone W7) (4), YFP (clone 1OC) (3) and RFP (clone yEmRFP) (5). To facilitate PCR-based cloning-free tagging of proteins at their endogenous chromosomal loci with these fluorophores, we have engineered PCR template vectors that use the *Kluyveromyces lactis URA3* for selection of transformants. The *K. lactis URA3* marker is subsequently popped out to leave a clean integration of the fluorophore encoding open reading frame. The relevant plasmids (pWJ1162, pWJ1163, pWJ1164, pWJ1165, pNEB30, and pNEB31) are listed in Table 1 and the tagging procedure described elsewhere (13).

The detection of fluorescent fusion proteins expressed from their endogenous promoter depends on multiple factors besides the expression level including protein turnover, incubation temperature, and localization of the protein. Specifically, the number of fluorescent molecules available for detection is defined by the balance between maturation and protein turnover, which may vary with experimental conditions. For the CFP, YFP, and RFP variants used in this study, the maturation half-time was previously estimated to 49, 39, and 15 min, respectively (14, 15). Moreover, high temperature negatively affects the folding of some GFP and RFP variants (1, 16). In this chapter, we present a selection of DNA repair and checkpoint markers that are easily detected with most wide-field microscopes equipped with a CCD camera (Subheading 2, Table 2 and Fig. 1). The selected proteins are primarily involved in recombinational DNA repair and checkpoint activation.

2. Materials

2.1. Media and Sugars

1. SC (synthetic complete) powder: 100 g yeast nitrogen base without amino acids and without NH_4SO_4, 293.11 g NH_4SO_4, 1.2 g adenine sulfate, 1.2 g L-arginine sulfate, 1.2 g L-histidine-HCl, 1.8 g L-isoleucine, 3.6 g L-leucine, 1.8 g L-lysine-HCl, 1.2 g L-methionine, 3 g L-phenylalanine, 1.2 g L-tryptophan, 1.8 g L-tyrosine, 1.2 g uracil, 9 g L-valine (18). Grind in ball mill overnight.
2. SC+Ade liquid medium: 7.25 g SC powder, 0.1 g adenine sulfate, and 20 g glucose in 1,000 mL. Adjust pH to 5.8 with NaOH/HCl.
3. Galactose: Filter-sterilize 30 % (w/v) stock solution in H_2O.

Table 1
Yeast strains and plasmids

Marker	Strain/plasmid name	Genotype[a]	References
Rad52-G-CFP	ML187-2D	*MATα TRP1 lys2Δ*	(6)
Rad52-G-CFP	pWJ1218	pRS413-based *HIS3/CEN*	This study
Rad52-G-YFP	W5094-1C	*MATa trp1-1 LYS2*	(6)
Rad52-G-YFP	pWJ1213	pRS413-based *HIS3/CEN*	(7)
Rad52-G-YFP	pWJ1344	pRS415-based *LEU2/CEN*	(7)
Rad52-G-yEmRFP	NEB322	*MATa trp1-1 LYS2*	This study
$Mec1_{1-64}$-YFP-AA-$Mec1_{65-2,369}$	ML41-4B	*MATa trp1-1 LYS2*	This study
YFP-GDPGG-Tel1	ML129-6B	*MATa trp1-1 LYS2*	This study
Ddc2-AAAA-YFP	W5088-4A	*MATa trp1-1 LYS2*	(8)
Ddc2-AAAA-CFP	ML311-18C	*MATα TRP1 lys2Δ*	This study
Mre11-LLAKKRKG-YFP	ML104-6B	*MATa trp1-1 LYS2*	(8)
Rfa1-AAAAAAAAG-YFP	ML147-2B	*MATa trp1-1 LYS2*	(8)
Rfa1-AAAAAAAAG-CFP	ML67-4B	*MATα TRP1 lys2Δ*	This study
YFP-AAAAAAAA-Sml1	W4622-14D	*MATa trp1-1 LYS2 bar1::LEU2*	(9)
Ddc1-AAAA-YFP	W5059-8A	*MATa trp1-1 LYS2*	This study
YFP-GGPGG-Rad9	SMG242	*MATa trp1-1 LYS2*	(10)
$Mec1_{1-64}$-YFP-AA-$Mec1_{65-2,369}$ Ddc2-AAAA-CFP	ML292-6C	*MATa trp1-1 LYS2*	This study
YFP-AAAAAAAA-Sml1 Rad52-G-yEmRFP	IG101-12D	*MATa trp1-1 LYS2*	This study
Mre11-LLAKKRKG-YFP Rfa1-AAAAAAAAG-CFP	SS54-9B	*MATa trp1-1 LYS2 RAD52-RFP*	This study
CFP-3′-K.l.URA3	pWJ1162	pRS423-based *HIS3/2μ*	(11)
CFP-5′-K.l.URA3	pWJ1163	pRS423-based *HIS3/2μ*	(11)
YFP-3′-K.l.URA3	pWJ1164	pRS423-based *HIS3/2μ*	(11)
YFP-5′-K.l.URA3	pWJ1165	pRS423-based *HIS3/2μ*	(11)
yEmRFP-3′-K.l.URA3	pNEB31	pRS423-based *HIS3/2μ*	This study
yEmRFP-5′-K.l.URA3	pNEB30	pRS423-based *HIS3/2μ*	This study

[a]All strains used in this study are *ADE2 RAD5* derivatives of W303-1A (*BAR1 ade2-1 can1-100 ura3-1 his3-11,15 leu2-3,112 trp1-1 LYS2 rad5-535*) (12)

Table 2
DNA damage response markers, exposure time, and biological function

Protein	Construct[a]	Biological function	Localization	Exposure time[b]	References
Rad52	Rad52-G-CFP, Rad52-G-YFP, Rad52-G-yEmRFP	Early marker of HR	Diffuse nuclear, partially excluded from the nucleolus. Relocalizes to foci during HR	1 s	(5, 6)
Mec1	$Mec1_{1-64}$-YFP-AA-$Mec1_{65-2,369}$	Single-stranded DNA damage checkpoint	Localizes from diffuse nuclear to foci in response to DNA damage	3 s	(8)
Tel1	YFP-GDPGG-Tel1	Double-strand break checkpoint	Localizes from diffuse nuclear to foci in response to DSBs	7 s	(17)
Ddc2	Ddc2-AAAA-YFP	Single-stranded DNA damage checkpoint	Localizes from diffuse nuclear to foci in response to DNA damage	3 s	(17)
Ddc2	Ddc2-AAAA-CFP	Single-stranded DNA damage checkpoint	Localizes from diffuse nuclear to foci in response to DNA damage	1 s	(17)
Mre11	Mre11-LLAKKRKG-YFP	Double-strand break recognition and processing	Localizes from diffuse nuclear to foci in response to DSBs	3 s	(17)
Rfa1	Rfa1-AAAAAAAG-CFP	Single-stranded DNA recognition	Multiple faint replication foci during S phase. Localizes to bright foci in response to DSBs	0.5 s	(17)
Sml1	YFP-AAAAAAAA-Sml1	Downstream target of Rad53-dependent DNA damage checkpoint	Cytoplasmic degraded upon checkpoint activation and during S phase	3 s	(8, 9)
Ddc1	Ddc1-AAAA-YFP	Single-stranded DNA damage checkpoint	Localizes from diffuse nuclear to faint foci in response to DNA damage	3 s	(17)
Rad9	YFP-GGPGG-Rad9	DNA damage checkpoint	Localizes from diffuse nuclear to transient foci in response to DNA damage	3 s	(10)

[a]For microscopy of a single protein, YFP is the preferred fluorophore for tagging, because it gives the best signal-to-noise ratio. For imaging three proteins in the same cell, it is advisable to tag the protein of the lowest abundance with YFP to maximize sensitivity, the most abundant protein with CFP to reduce phototoxicity, and the third protein with RFP
[b]The optimal exposure time depends on the fluorophore, the abundance of the protein, the sensitivity of the camera and the intensity of the excitation light, which should be adjusted so that no significant photobleaching occurs during acquisition

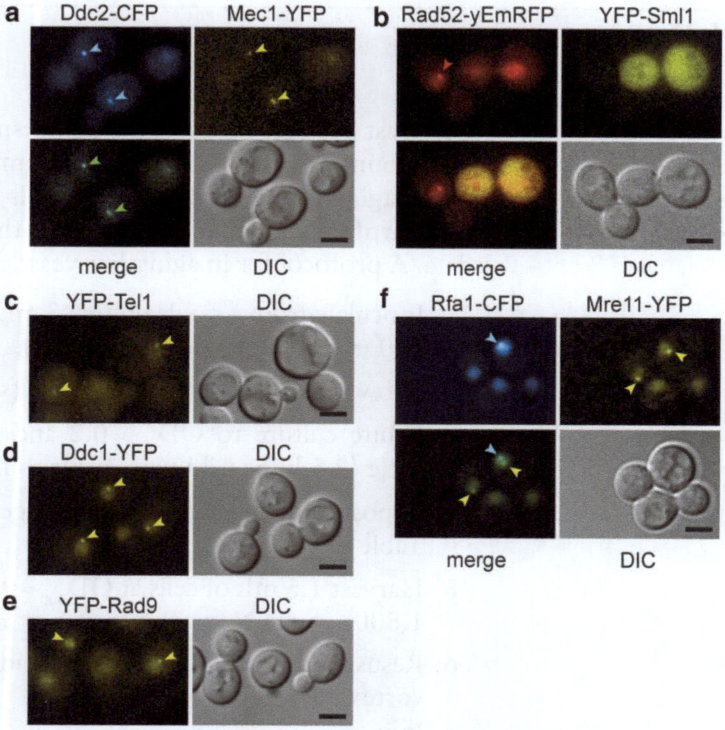

Fig. 1. Representative images of selected DNA repair and checkpoint proteins. Strains were imaged as described in Subheading 3 after treatment with 100 μg/mL zeocin for 1 h. Strains are (**a**) ML292-6C, (**b**) IG101-12D, (**c**) ML129-6B, (**d**) W5059-8A, (**e**) SMG242, and (**f**) SS54-9B. Scale bar, 3 μm.

2.2. Wax for Sealing Microscope Slides

1. Melt and mix 1 volume of petroleum jelly (vaseline), 1 volume of beeswax, and 1 volume of lanolin at 80 °C.
2. For sealing the edges of the microscope cover glass for time-lapse microscopy, the wax mixture can be kept at 65 °C. The molten wax is applied along the edges of the cover glass with a flat metal spatula pre-heated over a Bunsen burner or similar.

2.3. Reagents and Chemicals

The following chemicals were used: zeocin, camptothecin (CPT), methyl methanesulfonate (MMS), raffinose, galactose, petroleum jelly, beeswax, lanolin, 4-nitroquinoline-1-oxide (4-NQO), and hydroxyurea (HU).

2.4. Microscope Hardware

The essential elements of the microscope hardware are a high sensitivity cooled CCD or EM-CCD camera and a high magnification objective (100×) with high numerical aperture (≥1.4).

3. Methods

3.1. Basic Imaging Protocol for Fluorescence Microscopy

For yeast live-cell imaging, the best result is obtained by culturing and mounting cells in filter-sterilized minimal (SC) medium. Live cell imaging is preferred over fixed cells, because fixation may generate artifacts or otherwise decrease the quality of the obtained data. A protocol for imaging live yeast cells is described here:

1. Inoculate from a single colony 3 mL of SC medium containing 100 μg/mL adenine (see Note 1).
2. Grow shaking overnight at 25 °C (see Note 2).
3. Dilute culture to $OD_{600} = 0.2$ and grow for at least one cell cycle (2.5 h for wild type) prior to microscopy.
4. Expose the culture to DNA damage using one of the agents in Table 3.
5. Harvest 1.5 mL of cells at $OD_{600} = 0.4–0.6$ by centrifugation at 1,500 × *g* (see Note 3).
6. Resuspend the pellet of cells in 50 μL of medium by vortexing.
7. Cells are mounted on standard glass slides and covered by cover glass appropriate to the optics of the microscope (see Note 4). For most strains, the cells are immobilized on the slide simply by adjusting the volume applied (usually 2–3 μL), so that the cells settle in a monolayer with the cells touching both the slide and the cover glass.
8. Capture 11 optical sections separated by 0.4 μm around the focal plane of the cell (see Note 5).

3.2. Time-Lapse Microscopy

Imaging of cells over time requires that phototoxicity is minimized and that favorable growth conditions can be maintained (see Note 6). Phototoxicity can be reduced by decreasing the fluorescence exposure time or intensity of the excitation light, and by reducing the number of optical sections acquired. The easiest way to maintain favorable growth conditions is to reduce the cell density on the slide, so that only a single cell is present in the field of view, thereby prolonging the time that the cell can grow before nutrients are locally exhausted. For this reason it is not recommended to concentrate the cells by centrifugation before mounting. For long-term imaging (>30 min), the edges of the cover glass are sealed with wax to prevent evaporation (see Subheading 2.2 and Note 7). The protocol is:

1. Turn on the microscope 1–2 h before the time-lapse is launched (see Note 8).
2. Cells are cultured and mounted essentially as described above (see Subheading 3.2).

Table 3
Compounds for DNA damage induction

Genotoxic agent[a]	Dosage	Time of treatment	Targets/lesions	Primary repair pathway(s)	References
HU	100–200 mM	30 min	Inhibits ribonucleotide reductase leading to depletion of dNTP pools and replication fork stalling	Replication restart by homologous recombination	(19, 20)
MMS	0.03 %	1 h	Methylates DNA on N^7-deoxyguanine and N^3-deoxyadenine, which leads to replication fork stalling and collapse	Replication restart by homologous recombination, DSB repair	(21, 22)
IR	1 DSB per haploid genome per 2 krad (20 Gy)	Seconds–minutes	In addition to DSBs, approximately 17 SSBs are generated per DSB	Homologous recombination	(23)
4-NQO	10 ng/mL	1 h	Guanine adducts at positions C^8 and N^2	Nucleotide excision repair, translesion synthesis, HR	(24, 25)
CPT	5 μg/mL	1 h	Stabilizes covalent topoisomerase I-DNA intermediates, which are converted to DSBs during DNA replication	Excision repair and homologous recombination	(26, 27)
Zeocin	100 μg/mL	1 h	DSBs	Homologous recombination	(28)

[a] *HU* hydroxyurea, *MMS* methyl methanesulfonate, *IR* ionizing radiation, *4-NQO* 4-nitroquinoline-1-oxide, *CPT* camptothecin, *SSB* single-strand break, *DSB* double-strand break, *HR* homologous recombination

3. The appropriate acquisition parameters for time-lapse microscopy should be determined empirically. With our microscopy setup, we are able to image Rad52-YFP with minimal phototoxicity for 20 time points over 4 h by reducing the number of optical sections from 11 to 9 (0.5 μm between sections), by reducing the intensity of the excitation light by inserting a neutral density filter in the fluorescence path that allows only 10 % transmission, and by reducing the exposure time from 1 to 0.5 s.

3.3. Site-Specific DNA Lesions

The advantage of site-specific, inducible DNA lesions is that the location of the lesion can be marked fluorescently and therefore followed during repair. Fluorescence marking of the DNA lesion is achieved by inserting a tandem array of DNA binding sites for a fluorescently tagged protein, e.g., the Lac or Tet repressor, adjacent to the DNA lesion as described (13). Different types of site-specific DNA lesions can be introduced. For example, the HO and I-SceI meganucleases are used to induce double-strand breaks (DSBs) (21); a mutant of the Flp recombinase (Flp-H305L) irreversibly cleaves one DNA strand forming a covalent Flp-DNA intermediate (29), which causes replication fork stalling and collapse; and a mutant of the I-AniI homing endonuclease (I-AniI-K227M) exhibits site-specific nicking (30). Strains in which a single site-specific endonuclease restriction site in the genome has been marked fluorescently, such as the HO or the I-SceI sites, are transformed with a plasmid expressing the appropriate endonuclease from an inducible promoter, e.g., pJH132 (pGAL-HO) or pWJ1108 (pGAL-I-SceI), respectively (8, 31).

1. Transform the appropriate plasmid (expressing HO or I-SceI) into the assay strain by the LiAc method (32). Select transformants on the appropriate glucose-containing minimal medium.
2. Inoculate fresh transformants in 2 mL of the appropriate minimal, raffinose-based medium (2 %, w/v, raffinose, autoclaved) (see Note 9).
3. Grow shaking at 25 °C for 24 h.
4. Dilute culture to $OD_{600} = 0.2$ and grow for one cell cycle (3.5 h for wild type) prior to endonuclease induction.
5. Add galactose (sterile-filtered, stock solution 30 %, w/v) to a final concentration of 2–3 % (w/v). This induces expression of the endonuclease.
6. Grow shaking at 25 °C for another 90 min prior to microscopy.

4. Notes

1. The supplementation of SC medium with 100 μg/mL adenine is necessary only for *ade2* mutant strains, which will otherwise accumulate a red pigment that is strongly autofluorescent. However, this medium can also be used for *ADE2* wild-type cells.
2. It is important that the culture is well aerated, because maturation of GFP and RFP requires molecular oxygen (2, 3).

3. At this point it is not necessary to use autoclaved microcentrifuge tubes. In fact, yeast cells tend to adhere to the side of some microcentrifuge tubes, if they have been autoclaved, whereas cells will pellet to the bottom of untreated tubes.
4. A high cell density in the field of view is desired to maximize data acquisition. However, at high cell density and growth rate (glucose), the medium quickly becomes saturated with carbon dioxide, which precipitates as gas bubbles that may displace cells. Therefore, the optimal cell density must often be determined empirically.
5. The yeast nucleus has a diameter of approximately 2 μm and the cell 5 μm. The number of optical sections required to sample the entire volume of the cell is determined by the Nyquist theorem. Ideally, the sampling distance should be half the depth of focus, but in our experience a sampling distance of 0.3–0.4 μm is sufficient for most applications. However, undersampling may cause some structures to be overlooked.
6. Phototoxicity is of particular importance when studying DNA damage response, because photolesions can be introduced into the DNA especially when imaging CFP, which has the excitation wavelength with the highest energy compared to YFP and RFP.
7. Nail polish or other solutions containing organic solvents should be avoided when sealing slides for live cell imaging, because the solvents may harm the cells.
8. Most microscopes will heat up and expand during use, which will cause the focal plane to change. The temperature of the microscope will usually reach equilibrium within 1–2 h of use.
9. It is important to use fresh transformants in order to avoid selecting cells in which mutations have accumulated in the endonuclease recognition site. Such mutations tend to appear in old transformants due to leakage of the *GAL1-10* promoter.

Acknowledgment

We thank Dr. Neta Dean, Stony Brook University, for sharing the yEmRFP construct. This work was supported by Fundação para a Ciência e a Tecnologia (SS), The Danish Agency for Science, Technology and Innovation (ML, NEB), the Villum Kann Rasmussen Foundation (ML), and the European Research Council (ML).

References

1. Shaner NC, Steinbach PA, Tsien RY (2005) A guide to choosing fluorescent proteins. Nat Methods 2:905–909
2. Remington SJ (2006) Fluorescent proteins: maturation, photochemistry and photophysics. Curr Opin Struct Biol 16:714–721
3. Ormo M, Cubitt AB, Kallio K, Gross LA, Tsien RY, Remington SJ (1996) Crystal structure of the *Aequorea victoria* green fluorescent protein. Science 273:1392–1395
4. Heim R, Tsien RY (1996) Engineering green fluorescent protein for improved brightness, longer wavelengths and fluorescence resonance energy transfer. Curr Biol 6:178–182
5. Keppler-Ross S, Noffz C, Dean N (2008) A new purple fluorescent color marker for genetic studies in *Saccharomyces cerevisiae* and *Candida albicans*. Genetics 179:705–710
6. Lisby M, Rothstein R, Mortensen UH (2001) Rad52 forms DNA repair and recombination centers during S phase. Proc Natl Acad Sci U S A 98:8276–8282
7. Alvaro D, Sunjevaric I, Reid RJ, Lisby M, Stillman DJ, Rothstein R (2006) Systematic hybrid LOH: a new method to reduce false positives and negatives during screening of yeast gene deletion libraries. Yeast 23:1097–1106
8. Torres-Rosell J, Sunjevaric I, De Piccoli G, Sacher M, Eckert-Boulet N, Reid R, Jentsch S, Rothstein R, Aragon L, Lisby M (2007) The Smc5-Smc6 complex and SUMO modification of Rad52 regulates recombinational repair at the ribosomal gene locus. Nat Cell Biol 9:923–931
9. Andreson BL, Gupta A, Georgieva BP, Rothstein R (2010) The ribonucleotide reductase inhibitor, Sml1, is sequentially phosphorylated, ubiquitylated and degraded in response to DNA damage. Nucleic Acids Res 38:6490–6501
10. Germann SM, Oestergaard VH, Haas C, Salis P, Motegi A, Lisby M (2011) Dpb11/TopBP1 plays distinct roles in DNA replication, checkpoint response and homologous recombination. DNA Repair (Amst) 10:210–224
11. Reid R, Lisby M, Rothstein R (2002) Cloning-free genome alterations in *Saccharomyce cerevisiae* using adaptamer-mediated PCR. Methods Enzymol 350:258–277
12. Zhao X, Muller EG, Rothstein R (1998) A suppressor of two essential checkpoint genes identifies a novel protein that negatively affects dNTP pools. Mol Cell 2:329–340
13. Eckert-Boulet N, Rothstein R, Lisby M (2011) Cell biology of homologous recombination in yeast. Methods Mol Biol 745:523–536
14. Gordon A, Colman-Lerner A, Chin TE, Benjamin KR, Yu RC, Brent R (2007) Single-cell quantification of molecules and rates using open-source microscope-based cytometry. Nat Methods 4:175–181
15. Shaner NC, Campbell RE, Steinbach PA, Giepmans BN, Palmer AE, Tsien RY (2004) Improved monomeric red, orange and yellow fluorescent proteins derived from *Discosoma* sp. red fluorescent protein. Nat Biotechnol 22: 1567–1572
16. Lim CR, Kimata Y, Oka M, Nomaguchi K, Kohno K (1995) Thermosensitivity of green fluorescent protein fluorescence utilized to reveal novel nuclear-like compartments in a mutant nucleoporin NSP1. J Biochem (Tokyo) 118:13–17
17. Lisby M, Barlow JH, Burgess RC, Rothstein R (2004) Choreography of the DNA damage response; spatiotemporal relationships among checkpoint and repair proteins. Cell 118: 699–713
18. Sherman F, Fink GR, Hicks JB (1986) Methods in yeast genetics. Cold Spring Harbor Laboratory, Cold Spring Harbor, NY
19. Reichard P (1988) Interactions between deoxyribonucleotide and DNA synthesis. Annu Rev Biochem 57:349–374
20. Lopes M, Cotta-Ramusino C, Pellicioli A, Liberi G, Plevani P, Muzi-Falconi M, Newlon CS, Foiani M (2001) The DNA replication checkpoint response stabilizes stalled replication forks. Nature 412:557–561
21. Lisby M, Mortensen UH, Rothstein R (2003) Colocalization of multiple DNA double-strand breaks at a single Rad52 repair centre. Nat Cell Biol 5:572–577
22. Beranek DT, Weis CC, Swenson DH (1980) A comprehensive quantitative analysis of methylated and ethylated DNA using high pressure liquid chromatography. Carcinogenesis 1: 595–606
23. Friedland W, Jacob P, Paretzke HG, Merzagora M, Ottolenghi A (1999) Simulation of DNA fragment distributions after irradiation with photons. Radiat Environ Biophys 38:39–47
24. Fronza G, Campomenosi P, Iannone R, Abbondandolo A (1992) The 4-nitroquinoline 1-oxide mutational spectrum in single stranded DNA is characterized by guanine to pyrimidine transversions. Nucleic Acids Res 20:1283–1287
25. Wiltrout ME, Walker GC (2011) The DNA polymerase activity of *Saccharomyces cerevisiae* Rev1 is biologically significant. Genetics 187: 21–35

26. Deng C, Brown JA, You D, Brown JM (2005) Multiple endonucleases function to repair covalent topoisomerase I complexes in *Saccharomyces cerevisiae*. Genetics 170:591–600
27. Eng WK, Faucette L, Johnson RK, Sternglanz R (1988) Evidence that DNA topoisomerase I is necessary for the cytotoxic effects of camptothecin. Mol Pharmacol 34:755–760
28. Moore CW, McKoy J, Dardalhon M, Davermann D, Martinez M, Averbeck D (2000) DNA damage-inducible and *RAD52*-independent repair of DNA double-strand breaks in *Saccharomyces cerevisiae*. Genetics 154:1085–1099
29. Nielsen I, Bentsen IB, Lisby M, Hansen S, Mundbjerg K, Andersen AH, Bjergbaek L (2009) A Flp-nick system to study repair of a single protein-bound nick in vivo. Nat Methods 6:753–757
30. McConnell Smith A, Takeuchi R, Pellenz S, Davis L, Maizels N, Monnat RJ Jr, Stoddard BL (2009) Generation of a nicking enzyme that stimulates site-specific gene conversion from the I-AniI LAGLIDADG homing endonuclease. Proc Natl Acad Sci U S A 106:5099–5104
31. Jensen RE, Herskowitz I (1984) Directionality and regulation of cassette substitution in yeast. Cold Spring Harbor Symp Quant Biol 49:97–104
32. Gietz D, St Jean A, Woods RA, Schiestl RH (1992) Improved method for high efficiency transformation of intact yeast cells. Nucleic Acids Res 20:1425

Chapter 31

Use of Gene Targeting to Study Recombination in Mammalian Cell DNA Repair Mutants

Jennifer J. Rahn, Gerald M. Adair, and Rodney S. Nairn

Abstract

The study of gene function has been greatly facilitated by the development of strategies to modify genomic DNA. Gene targeting is one of the most successfully applied techniques used to examine the roles of specific genes in a wide variety of model systems from yeast to mammals. Our laboratory has pioneered the use of the Chinese hamster ovary (CHO) cell culture model system to study pathways of DNA repair and recombination at the hemizygous CHO *APRT* locus. By using a simple and effective gene targeting method, we have generated a number of DNA repair-deficient cell lines that have been used in targeted recombination experiments to investigate pathways of recombinational repair in somatic mammalian cells. These methods can be readily customized to generate a variety of cell lines deficient in specific genes of interest and can be applied to study the roles of other DNA repair proteins in pathways of mammalian recombinational repair.

Key words: CHO cells, Gene targeting, Recombinational repair, *APRT*, *ERCC1*

1. Introduction

Genetic manipulation by modification of genomic DNA is a powerful tool used widely to interrogate cellular pathways in a variety of systems. Gene targeting, one of the most successful methods by which such modification can be undertaken, is defined as homologous recombination between a transfected DNA sequence and an endogenous chromosomal gene locus (the "target"). Gene targeting strategies and procedures have been developed both for yeast and higher eukaryotic cells, and enable target gene correction, disruption, deletion, replacement, or site-directed modification of virtually any gene or chromosomal locus for which a cloned DNA sequence is available (1–4). Many of the first experiments using gene targeting were undertaken by investigators wishing to obtain mutant mice defective at a selected target gene locus

Lotte Bjergbæk (ed.), *DNA Repair Protocols*, Methods in Molecular Biology, vol. 920,
DOI 10.1007/978-1-61779-998-3_31, © Springer Science+Business Media New York 2012

("knock-out") and were performed utilizing mouse embryonic stem (ES) cells (2, 5). Subsequently, approaches to introduce more subtle genetic alterations were developed and applied to the generation of mutant mouse disease models (5–7). Chicken DT40 cells, which exhibit a very high capacity for efficient gene targeting, have also been successfully developed and used as gene targeting recipients to study the roles of genes involved in DNA repair (8–10). Comprehensive reviews have been published describing gene targeting in yeast, DT40 and ES cells (1–5, 9, 10). It is not within the scope of this chapter to review the general area of gene targeting; instead, here we describe some specific methods and the use of gene targeting as an approach to study mechanisms of recombination in cultured mammalian cells. Our particular focus is on the development and use of gene targeting to generate and analyze DNA repair-deficient Chinese hamster ovary (CHO) cell lines to reveal interactions between DNA repair and recombination pathways in mammalian cells.

In budding yeast, *Saccharomyces cerevisiae*, both the nucleotide excision repair (NER) and mismatch repair (MMR) pathways have components that have been shown to have additional roles in specialized recombination pathways. The heterodimeric Rad10-Rad1 complex forms a structure-specific endonuclease responsible for making the 5′-side incision during NER (11). Additionally, this enzyme complex has been shown to act in the single-strand annealing (SSA) recombination pathway to remove single-stranded 3′ DNA "tails" of nonhomologous sequence, and has a similar activity in precisely removing lengthy nonhomologous DNA sequences that block extension of 3′-OH homologous ends during plasmid-chromosome gene targeting (12–15). MutS-β, a heterodimer composed of Msh2/Msh3, is another example of an enzyme complex with dual functionality as it has a critical role in MMR and is also required in SSA in yeast (16). We have investigated the roles of mammalian counterparts to these proteins (i.e., Ercc1, Xpf, and Msh3) in specialized mammalian DNA repair pathways, such as SSA, in CHO cell lines by developing and applying gene targeting approaches. We and our colleagues developed the first gene targeting methods used to modify the endogenous hamster adenine phosphoribosyltransferase (*APRT*) locus, and then used this system to investigate mechanisms of homologous gene targeting at this locus in somatic cells (17–23). We have also used targeted gene disruption to construct *ERCC1* knock-out mutants in DNA repair-proficient parental CHO cell lines (24–27) as well as in other CHO cell DNA repair mutants (our unpublished results). Other investigators have used derivative approaches to knock out genes such as *FANCG* and *RAD51D* in CHO cells (28, 29) for studies of recombination and repair. Additionally, we have successfully used two-step gene replacement targeting strategies to modify endogenous target loci in a number of CHO DNA repair mutants, creating isogenic pairs of cell lines for subsequent studies of targeted and mitotic recombination at the endogenous *APRT* gene locus (26, 27, 30). Recently, we used an RNAi approach to generate isogenic CHO cell lines depleted for the Werner RecQ DNA helicase, and applied gene targeting assays to examine the role of *WRN* in both targeted and mitotic recombination pathways (31).

In the following subsections, we describe some of the underlying principles of gene targeting, including gene disruption, correction and modification, applied specifically to CHO cells, and the utility of the hemizygous CHO *APRT* gene as a model target locus for gene targeting studies designed to investigate mechanisms of homologous recombination in somatic mammalian cells, citing some specific examples. The advantages of using CHO cells for such studies include:

1. CHO cells retain a relatively stable, pseudo-diploid karyotype, are easily grown and maintained, and have a high cloning efficiency.
2. CHO cells are widely used in somatic cell genetics, and a large data base exists in the literature describing mutagenic and cytotoxic responses of CHO cells to a variety of physical and chemical agents (32, 33).
3. A significant number of gene loci are hemizygous in CHO cell lines, including several DNA repair genes (34, 35); hemizygous gene loci greatly simplify both the generation of targeted gene knock-outs and molecular analysis of gene alterations generated from gene targeting.

1.1. Targeted Knock-Out of a DNA Repair Gene Locus in Cultured Mammalian Cells

Target gene disruption of the *ERCC1* gene in CHO cells is used in the discussion below as an example of the application of gene targeting to create DNA repair-deficient mutants in somatic cells; however, as noted above, other DNA repair genes have also been successfully targeted in CHO cells (28, 29). Isogenic sequence (i.e., DNA isolated from the particular cell line to be used for gene targeting) is necessary for efficient targeted recombination in mammalian cells (36, 37). Therefore, targeted gene disruption of the CHO *ERCC1* gene necessitated cloning *ERCC1* from CHO cells. A human cDNA *ERCC1* probe (38) was used to screen a CHO cell genomic library constructed in Lambda FIXII (Stratagene, La Jolla, CA), resulting in recovery of the CHO *ERCC1* gene. A fragment containing exons 4 and 5 was subcloned and used to construct a targeting vector (see Fig. 1a); this vector has been used for targeted *ERCC1* knock-out in several wild-type CHO cell lines (e.g., CHO-K1, CHO-ATS49tg, RMP41; see refs. 24–27), as well as two DNA repair-deficient CHO cell mutants (cg*XPD* and cg*MSH3*), generating compound mutants defective in more than one DNA repair gene (our unpublished results).

Several parameters of gene targeting should be taken into consideration when designing a knock-out vector. A study of gene targeting

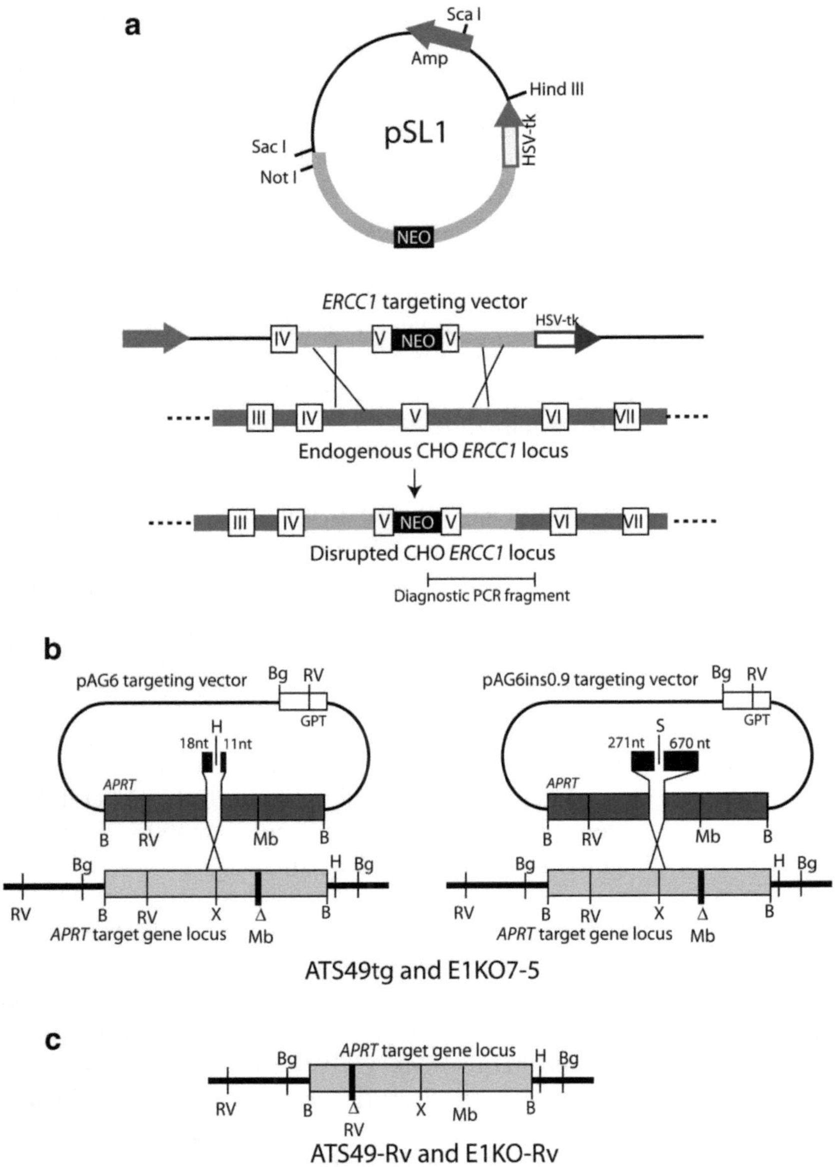

Fig. 1. Targeting vectors and gene targets in CHO cell lines. (**a**) Targeted gene knock-out strategy for gene disruption of the *ERCC1* locus in CHO cells. (**b**) *APRT* targeting vectors, also showing the *APRT* exon 5 *Mbo*II deletion locus. (**c**) *APRT* exon 2 *EcoRV* deletion target locus.

at the *HPRT* locus in mouse ES cells (36) showed that the use of isogenic targeting vector sequence/target gene homology is more important for efficient targeted gene disruption than targeting vector configuration (i.e., insertion vector configuration, having a double-strand break (DSB) within target gene sequence homology in plasmid DNA, vs. replacement vector configuration, with a DSB in the targeting vector outside target gene sequence homology in plasmid DNA). In CHO cells, we have found that although the overall efficiency of gene targeting is similar for both integration-type and replacement-type vectors (20, 39), the distribution of types of gene targeting events can be dramatically affected by targeting vector configuration. Use of a replacement-type targeting vector greatly reduces targeted integration recombinants compared to gene replacement/conversion recombinants (see Fig. 2 for generalized structures of integration and conversion recombinants at the model *APRT* locus). Therefore, integration-type targeting vectors (insertion vectors) are most efficient for two-step gene replacement or modification strategies (39–41), which require targeted integration in the first step, whereas replacement-type vectors may be preferable for gene disruption (i.e., knock-out) gene targeting approaches.

Another factor critical for efficient gene targeting is the length of target gene sequence homology in the targeting vector. A minimum length of sequence homology flanking each side of the positive selection marker (e.g., *NEO*) is required in a replacement-type targeting vector (see Fig. 1a). Targeting efficiency decreases rapidly with decreasing length of sequence homology below about 500 bp (23), suggesting that a minimum of 500–600-bp of homologous sequence flanking each side of *NEO* (or whatever selectable marker is used) should be designed into the knock-out vector; typically, several kilobases of homologous sequence are used to flank each side of the selection marker (42). Negative selection markers, such as *HSV-tk*, are used in positive–negative selection protocols to enrich the population of transformed cells for targeted homologous recombinants (43). Selection against *HSV-tk* using ganciclovir or FIAU applied to the transfected cell population kills nonhomologous recombinants that have randomly integrated the targeting vector into their genomic DNA; gene replacement events that have lost the *HSV-tk* marker during recombination are thus enriched in a positive–negative selection protocol using *NEO* and ganciclovir or FIAU (see Fig. 1a). Although very large enrichment ratios using positive–negative selection (up to three orders of magnitude) have been reported (43), the degree of enrichment attainable depends upon the target locus, the cell type, and the targeting vector; in practice, we have consistently achieved ~6 to 12-fold enrichment for targeting the CHO *ERCC1* locus with the positive–negative gene disruption vector shown in Fig. 1a.

1.2. Use of APRT Targeting and Modification to Study Recombination Deficiencies in Mammalian Cells

We developed practical mammalian gene targeting assay systems utilizing hemizygous, CHO *APRT* deletion mutants, such as CHO-ATS49tg, for studying homologous recombination in cultured mammalian cells (17–24, 26, 27, 31, 39). Our work, and work from other laboratories, has demonstrated the utility of the hemizygous CHO *APRT* gene as a model target locus for studies designed to probe the mechanisms of mammalian recombination (17–24, 26, 27, 44–47). Targeted recombination/*APRT* gene correction assays, employing either "ends-in" integration-type or

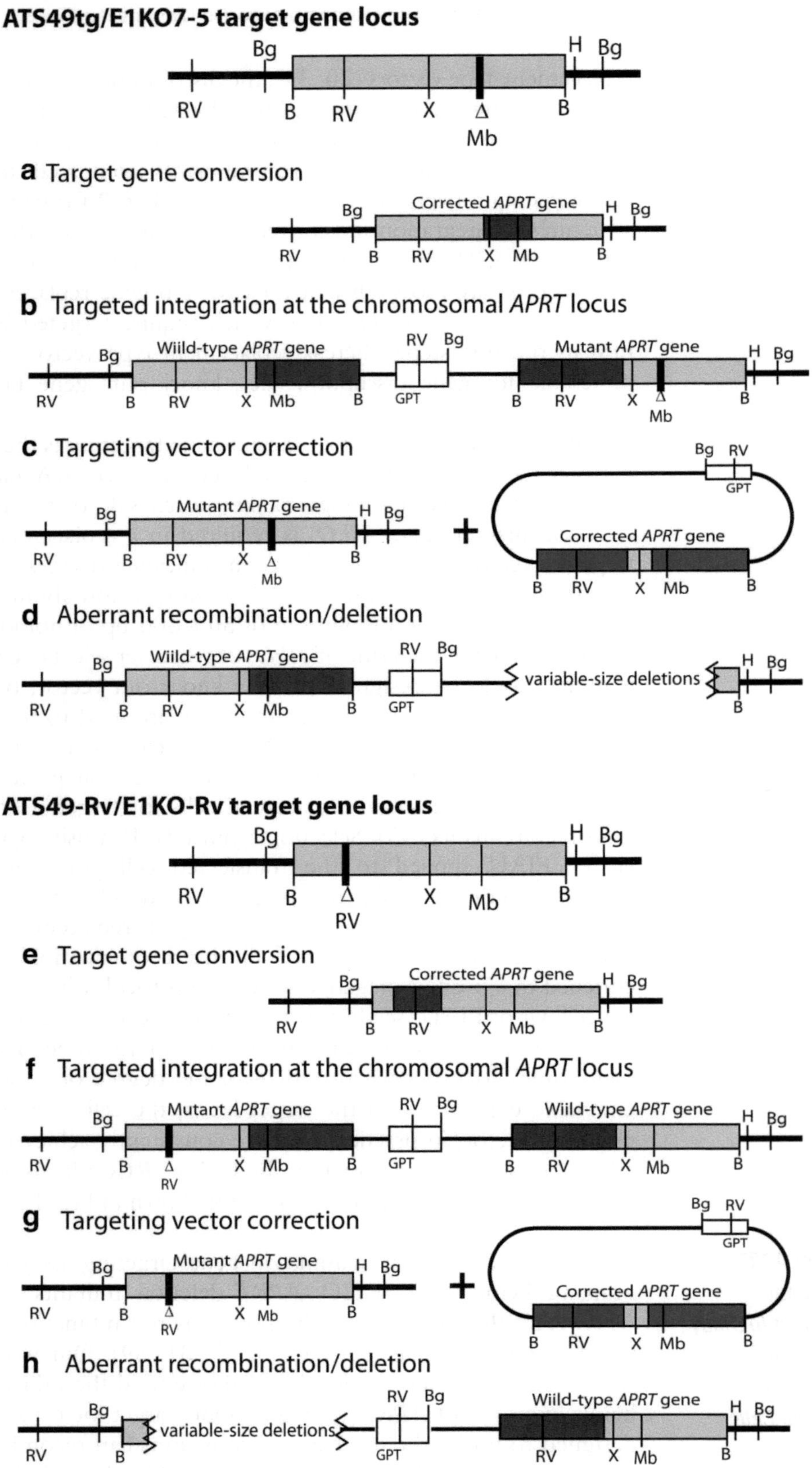

Fig. 2. Structures of targeted recombinants at CHO cell *APRT* loci. (**a**–**d**) Targeted recombinants generated at the exon 5 CHO-ATS49tg/E1KO-7 *APRT* locus. (**e**–**h**) Targeted recombinants generated at the exon 2 CHO-ATS49-Rv/E1KO-Rv *APRT* locus.

"ends-out" replacement-type targeting vector configurations, allow direct selection of APRT+ recombinants. Use of a hemizygous, nonrevertible, deletion-inactivated *APRT* gene as a target locus ensures that APRT+ cells can arise only by targeted homologous recombination at the endogenous *APRT* gene locus. The small size of the CHO *APRT* gene (which spans only ~2.2 kb from promoter to polyadenylation signal), the absence of *APRT* pseudogenes, and hemizygosity for the target gene locus all greatly facilitate molecular analysis of CHO cell APRT+ recombinants. Additionally, modification of the targeting vector to contain regions of nonhomologous sequence of varying lengths (e.g., pAG6ins0.9 vector shown in Fig. 1b) can facilitate experiments interrogating the role of blocking nonhomologous sequences on gene targeting outcomes (24, 30).

In CHO *APRT* gene correction targeting experiments employing conventional "ends-in" integration vectors, with two arms of *APRT* targeting homology flanking a DSB, and a hemizygous *APRT* target locus with a 3-bp exon 5 *Mbo*II site deletion or a 2-bp exon 2 *Eco*RV site deletion (17–20, 23, 24, 30, 31), we have consistently recovered three distinct classes of APRT+ recombinants, reflecting three different types of targeted homologous recombination events:

1. Target gene conversions.
2. Targeted integrations at the chromosomal *APRT* locus.
3. Targeting vector correction recombinants (which still retain an uncorrected *APRT* gene at the targeted chromosomal locus).

These three recombinant classes are readily distinguishable on the basis of their diagnostic restriction fragment patterns, differing somewhat depending on the specific inactivated *APRT* deletion mutant allele used as a target. Target gene convertants, in which the inactivating deletion has been corrected by unidirectional transfer of sequence information from *APRT* sequence in the targeting vector, result from precise replacement of the deleted bases and restoration of the exon 5 *Mbo*II restriction site in the cases of cell lines ATS49tg and E1KO-7 (Fig. 2a) or the exon 2 *Eco*RV restriction site in the case of cell lines ATS49-Rv and E1KO-Rv (Fig. 2e) yielding a functional *APRT* gene. Targeted integration of the targeting vector at the chromosomal gene locus by a single-reciprocal exchange/crossover event results in a duplication of the *APRT* gene sequences (including at least one wild-type copy of the *APRT* gene) (Fig. 2b, f). Targeting vector correction recombinants (Fig. 2c, g) retain an uncorrected *APRT* gene (with the 3-bp *Mbo*II site deletion or the 2-bp *Eco*RV site deletion) at the targeted chromosomal locus, but have also acquired an ectopically integrated wild-type *APRT* gene. In this recombinant class, the targeting vector *APRT* sequence has been corrected by unidirectional transfer of sequence information from the target gene locus, likely by one-sided strand invasion and target gene-templated extension of the vector *APRT* sequence, either with or without gap-repair/ vector recircularization. The corrected targeting vector, now carrying a wild-type *APRT* gene, is then randomly integrated at an ectopic site.

We also recovered a fourth class of targeted recombinant (Fig. 2d, h) that appears to be related to a targeted integration event, but results in large deletions in one of the *APRT* duplications. These recombinants were recovered only in *ERCC1*-deficient cell lines (i.e., E1KO-7 and E1KO-Rv), and only from the pAG6ins0.9 targeting vector. Identification of these aberrant recombinants provided the first experimental evidence that Ercc1-Xpf, like Rad10-Rad1 in yeast, was required to correctly process recombination intermediates arising from gene targeting in which nonhomologous DNA sequence blocked homologous DNA ends and prevented templated extension (24). The structures also suggested that the mechanisms for generating targeted integrations and these aberrant deletion recombinants were related.

The structures shown in Fig. 2 illustrate how differences among specific *APRT* mutant cell lines can be exploited in order to examine the effect of varying gene targeting parameters on the targeting outcomes. ATS49tg and its isogenic *ERCC1*-deficient derivative (E1KO7-5) bear a single copy of *APRT* inactivated by a 3-bp deletion in exon 5 abolishing an *Mbo*II restriction site (Fig. 1b) (48). In order to examine what effect altering the position of this inactivating mutation might have, APRT$^+$ convertants were selected from these two cell lines, each was electroporated with the restriction enzyme *Eco*RV, and clones were then selected for loss of *APRT* function. Two clones (one from ATS49 and one from E1KO-7) were selected that contained identical 2-bp inactivating deletions in exon 2 and resulted in the loss of a unique *Eco*RV site (clones ATS49-Rv and E1KO-Rv) (Fig. 1c). These two pairs of nearly isogenic cell lines permitted examination of the effects on recombination of altering the physical distance between the DSB in the targeting vector and the inactivating mutation in the *APRT* target gene, and also of changing the position of the target gene mutation from 3′ to 5′ relative to the DSB in the targeting vector. Results of this study (30) confirmed expectations for similar effects of Ercc1-Xpf on gene conversion tracts in mammalian cells as for Rad10-Rad1 in yeast. In addition, the effects of "targeting polarity," defined as the location of the target gene mutation being either upstream (3′) or downstream (5′) to the DSB in the targeting vector, allowed our testing of the hypothesis that aberrant deletion recombinants arose as the result of Ercc1-Xpf being defective during targeted integration. The recombinant structures shown in Fig. 2 clearly show that the polarity of the inactive *APRT* heteroallele in targeted integrants is the same as the variable-sized deletion in aberrant recombinants, and depends on the orientation of the target gene mutation, since only the recombinants shown in Fig. 2b, d were recovered from cells with the *Mbo*II deletion in *APRT* exon 5 and only the recombinants shown in Fig. 2f, h were recovered from cells with the *Eco*RV deletion in exon 2 (30).

Two-step targeted modification of the CHO *APRT* locus has also been extremely useful in generating cell lines harboring intrachromosomal recombination substrates for investigation of mitotic recombination. In this approach, conventional gene targeting produces a targeted integration correcting the endogenous *APRT* target, as shown in Fig. 2b, f. In a second step, selection of "pop-out" recombinants leads to recovery of a cell line in which the original mutant allele has been replaced with wild-type *APRT* (18). Merrihew et al. (45) designed a targeting vector in which an Flp recombination target (FRT) sequence in intronic *APRT* sequence was similarly introduced in two steps to generate a CHO cell line with the FRT-containing *APRT* allele, suitable for further modification by site-specific Flp recombination. This general approach has been used to construct CHO cell lines used in studies of the effects of telomere and triplet repeats (44, 46) and site-specific strand breaks (47) on recombination between intrachromosomal *APRT* heteroalleles, as well as for a series of studies investigating the effects of *ERCC1* and *XPF* deficiencies on mitotic recombination (26, 27, 49). We recently used two-step Flp modification of the *APRT* locus to construct a CHO cell line in which the *WRN* RecQ helicase is stably depleted, and characterized a mitotic hyperrecombination phenotype resulting from *WRN* deficiency (31). In this study, we also used the *APRT*-targeted correction assay in *WRN* knock-down CHO cells to show that *WRN* function was required for efficient nonhomologous recombination. Combining these different *APRT* recombination assays in isogenic CHO cell lines enabled our demonstration of the complexity of pleiotrophic *WRN* recombination phenotypes using a rigorous experimental design. Thus these approaches, based on gene targeting to correct and/or modify a small, endogenous, nonessential, and easily selectable gene locus, offer a wide variety of experimental systems to investigate DNA repair and recombination in somatic mammalian cells.

2. Materials

2.1. Knock-Out and APRT Gene Targeting

1. αMEM-10% FCS medium: α-Modified minimal essential medium (50) with penicillin (50 U/mL) (Invitrogen), and streptomycin (50 μg/mL) (Invitrogen). Media was prepared according to the suppliers' instructions and was supplemented with 10% fetal calf serum (FCS) for culturing conditions

(10%-MEM). Some cell manipulations and dilutions used αMEM without serum (SF-MEM).

2. Puck's solution A (Solution A): Physiological saline (0.8% NaCl) in 5 mM KCl, 4.2 mM $NaHCO_3$, containing 0.1% dextrose. A 10× stock solution is prepared by dissolving 80 g NaCl, 4 g KCl, 10 g dextrose, and 3.5 g $NaHCO_3$ in 1 L of H_2O; a 1× working solution is used after 1:10 dilution with H_2O and filter sterilization. Store both 10× and 1× solutions at 4°C.
3. Trypsin-EDTA: 0.25% Trypsin, 25 mM Tris–HCl, 0.5 mM EDTA in Solution A, pH. 7.4. Dissolve 3 g Trizma base in 460 mL of Solution A and add 20 mL of 1 N HCl; dissolve 0.2 g $Na_2EDTA{\cdot}2H_2O$ in 20 mL of water; dissolve 2.5 g trypsin in 450 mL of Solution A, while stirring. Stir all three solutions individually until homogeneous, mix the Trizma and EDTA solutions together to dissolve, then add trypsin suspension, and adjust to pH 7.4 if necessary. Adjust the final volume to 1 L with H_2O, filter-sterilize, and dispense in 10-mL aliquots. Store frozen (do not refreeze).
4. Tris-EDTA buffer (TE): 10 mM Tris–HCl, 1 mM EDTA, pH 8.0. For a 10× stock solution, dissolve 12.1 g Trizma base and 3.7 g $Na_2EDTA{\cdot}2H_2O$ in 950 mL of H_2O; adjust to pH 8.0 with HCl and bring the final volume to 1 L; store at 4°C. For a 1× working solution, dilute 1:10 with H_2O and filter-sterilize.
5. 10% Sodium dodecyl sulfate (10% SDS): Dissolve 100 g of SDS in 1 L (final volume) of hot water with gentle stirring; autoclave and store at room temperature.
6. 5 M lithium acetate: Dissolve 51 g $LiCl_2H_3O_2{\cdot}2H_2O$ in 100 mL H_2O (final volume); autoclave and store at 4°C.
7. Crystal violet staining solution: A 10× stock solution is prepared by dissolving 20 g crystal violet in 400 mL of 90% ethanol. A 1× working solution for staining is prepared by 1:10 dilution with 90% ethanol. Store at room temperature.
8. Restriction endonucleases: Restriction endonucleases were obtained from various sources, and used according to the suppliers' directions.
9. Acid wash: 42 mL 12 N HCl in 2 L H_2O.
10. Alkaline wash: Dissolve 40 g NaOH and 175.2 g NaCl in 2 L H_2O.
11. Neutral wash: Dissolve 121.2 g Trizma base and 350.4 g NaCl in ~1.5 L H_2O. Adjust pH to 7.5 with ~70 mL 12 N HCl. Bring volume to 2 L with H_2O.
12. 20× SSC (0.3 M sodium citrate, 3 M sodium chloride, 2 mM citric acid): Dissolve 87.7 g sodium citrate, 175.3 g NaCl, and 0.36 g anhydrous citric acid in 1 L H_2O.

13. 20× SSPE: Dissolve 175.3 g NaCl, 27.6 g NaH_2PO_4, and 9.4 g EDTA in 800 mL H_2O. Adjust pH to 7.4 with ~27 mL 10 N NaOH. Sterilize by autoclaving.
14. 100× Denhardt's solution: Dissolve 2 g Ficoll 400, 2 g polyvinylpyrrolidone (PVP) and 2 g bovine serum albumin (BSA) in 100 mL H_2O. Aliquot and store at −20°C.
15. Prehybridization solution: Mix together 50 mL formamide, 30 mL 20× SSPE, 5 mL 100× Denhardt's solution, 2.5 mL 20% SDS, 2.5 mL 10 mg/mL denatured salmon sperm DNA, and 10 mL H_2O. Mix well and aliquot into 40 mL portions. Store at −20°C.
16. High salt wash: 2× SSC, 0.1% SDS. Combine 50 mL 20× SSC and 2.5 mL 20% SDS in a final volume of 500 mL.
17. Low salt wash: 0.1× SSC, 0.1% SDS. Combine 2.5 mL 20× SSC and 2.5 mL 20% SDS in a final volume of 500 mL.

2.2. Knock-Out Gene Targeting in CHO Cells

1. Cell line: CHO-ATS49tg or derivative cell lines; other CHO cell lines are also suitable (see Note 1).
2. *ERCC1* targeting vector: pSL1 (see Note 2).
3. High ionic strength electroporation buffer (HEP buffer): 20 mM *N*-2-hydroxyethylpiperizine-*N*′-(2-ethanesulfonic acid) (HEPES), 137 mM NaCl, 5 mM KCl, 0.7 mM Na_2HPO_4, 6 mM dextrose, pH 7.05. Just prior to electroporation, a fresh solution is prepared by dissolving 238.3 mg HEPES, 400 mg NaCl, 54 mg dextrose, 18.5 mg KCl, and 9.5 mg $Na_2HPO_4{\cdot}7H_2O$ in 40 mL H_2O; adjust the pH to 7.05 with 1 N NaOH. Adjust the final volume to 50 mL prior to filter sterilization.
4. G418/FIAU selection medium: 400 μg/mL G418 (Geneticin), 0.2 μM FIAU (1-(2-deoxy-2-fluoro-b-D-arabinofuranosyl-5-iodo)-uracil) in αMEM-10% DFCS. Adjust the G418 (Gibco-BRL) to 100% activity by weight, and dissolve directly in SF-MEM to 20 (active) mg/mL (50×), then neutralize (to color) with 1 N NaOH, filter-sterilize, and store at 4°C. Prepare FIAU as a 200 mM (1,000×) stock by dissolving 4.6 mg of FIAU in 0.5 mL of fresh 1 N NaOH, and then bringing the volume to 60 mL with PBS; after filter-sterilizing, aliquot, and store frozen. Prepare selection medium containing G418 alone, or both G418 and FIAU, by appropriate dilutions directly into αMEM-10% DFCS (see Note 3). FIAU was the generous gift of Bristol-Myers Squibb Pharmaceutical Research Institute (Wallingford, CT).
5. DNA lysis solution: 1% Proteinase K. Dissolve proteinase K lyophilized powder at 10 mg/mL in H_2O; stir or shake gently to avoid foaming. Prepare just before use.

6. PCR primers: Forward Neo 1: 5′-CCGCTTTTCTGGATTC ATCGAC-3′, and forward Neo 2: 5′-GCTTCCTCGTGCTT TACGGTATC-3′ are nested in the pMC1neopA cassette (53); reverse *ERCC1*: 5′-TGCCACCCCTGACCACAT ATACAC-3′ was designed in a region of *ERCC1* not represented in the targeting vector, pSL1 (see Note 4). We use commercially prepared primers from Operon (Houston, TX) or Genosys (Alameda, CA).
7. PCR reagents: Reagents for PCR, including Taq DNA polymerase, were commercially purchased as a kit from Perkin-Elmer (GeneAmp PCR Reagent Kit). Taq Extender and TaqStart antibody were purchased from Stratagene and Clontech (Palo Alto, CA), respectively.

2.3. APRT Gene Targeting in CHO Cells

1. Cell lines: CHO-ATS49tg and E1KO7-5 or CHO-ATS49-Rv and E1KO-Rv (see Note 5).
2. *APRT* targeting vectors: pAG6 and/or pAG6ins0.9 (see Note 6).
3. Low ionic strength electroporation buffer (KLEP buffer): Just prior to electroporation, prepare fresh 10 mM stock phosphate buffer by combining 95 μL 1 M monobasic potassium phosphate (KH_2PO_4), 405 μL 1 M dibasic potassium phosphate (K_2HPO_4), and 49.5 mL sterile water. Confirm a pH of 7.4 (adjust if necessary) and filter-sterilize. Prepare fresh KLEP by dissolving 2.33 g sucrose in 17.5 mL 10 mM stock phosphate buffer. Add 25 μL 1 M $MgCl_2$ and bring final volume to 25 mL with sterile water. Adjust the pH to 7.4 and filter-sterilize. Keep at 4°C.
4. ALASA selection medium (ALASA): 25 μM alanosine, 50 μM azaserine, 100 μM adenine in 10%-MEM (see Note 7). Prepare alanosine as a 12.5 mM (500×) stock by dissolving 93 mg alanosine in 1 mL of 1 N fresh NaOH; bring to 50 mL with Solution A, filter-sterilize and store frozen in small aliquots. Alanosine (NSC-529469) can be obtained from the Drug Synthesis and Chemistry Branch of the National Cancer Institute. Prepare azaserine as a 25 mM (500×) stock by dissolving 216.5 mg azaserine in 1 mL of 1 N fresh NaOH; bring to 50 mL with Solution A, filter-sterilize, and store frozen in small aliquots. Prepare adenine as a 5 mM (50×) stock by dissolving 429 mg of adenine·HCl in 5 mL 1 N NaOH and bring volume to 500 mL with Solution A. Filter-sterilize and store at room temperature.
5. HAT selection medium (HAT): 100 μM hypoxanthine, 2 μM methotrexate, 50 μM thymidine in 10%-MEM (see Note 8). Prepare hypoxanthine as a 5 mM (50×) stock by dissolving 341 mg hypoxanthine in 5 mL 1 N NaOH; bring volume to

450 mL with Solution A, add 5 mL 1 N HCl and adjust final volume to 500 mL with Solution A, filter-sterilize, and store at room temperature. Prepare thymidine as a 25 mM (500×) stock by dissolving 302.5 mg thymidine in 50 mL Solution A; filter-sterilize, and store at 4°C. Prepare methotrexate as a 1 mM (500×) stock by dissolving 22.8 mg methotrexate in 1 mL 1 N NaOH; adjust volume to 50 mL with Solution A. Filter-sterilize and freeze in small aliquots (light-sensitive).

3. Methods

3.1. Preparation of Targeting Vectors for Electroporation

Targeting vector DNAs are prepared for electroporation in an identical manner whether they are gene knock-out vectors or *APRT* targeting vectors. For gene knock-out experiments, the targeting vector is linearized at a unique restriction site such as *Sca*I in the plasmid backbone (i.e., opposite plasmid homologous *ERCC1* DNA sequence containing *NEO* and flanked by HSV-*tk* sequences; see Fig. 1a). For *APRT*-targeted recombination assays utilizing the pAG6 or pAG6ins0.9 integration vector, the vector is linearized by cutting at a unique *Hind*III (pAG6) or *Sal*I (pAG6ins0.9) restriction site within the heterologous insert that disrupts the *APRT* targeting sequence homology (see Fig. 1b; Note 6).

1. Linearize the vector by overnight restriction endonuclease digestion of plasmid DNA, using 3–4 U of restriction enzyme/μg of DNA, and employing the specific reaction buffers and conditions recommended by suppliers.
2. Take a small sample of the digested vector and check for complete linearization by gel electrophoresis.
3. After complete digestion has been confirmed, treat the linearized targeting vector DNAs with a solution of fresh proteinase K (200 μg/mL), 1% SDS, 1 M lithium acetate ($LiC_2H_3O_2 \cdot 2H_2O$) for 2 h at 55°C.
4. Allow the reactions to cool to room temperature.
5. Add one volume of Tris-saturated phenol pH 8.0 and mix well; centrifuge in a microfuge to separate the organic phase; collect the upper aqueous phase in a new tube.
6. Add one volume of chloroform:isoamyl alcohol (24:1) and mix well; centrifuge in a microfuge to separate the organic phase; collect the upper aqueous phase in a new tube.
7. Recover the DNA from the separated aqueous phase by addition of two volumes of absolute ethanol, incubate at −20°C for 2 h, and centrifuge in a microcentrifuge.

8. After rinsing with ice-cold 70% ethanol and air-drying for 10–15 min, resuspend the pellet completely in sterile water to achieve a DNA concentration of 500 μg/mL. Store frozen (see Note 9).

3.2. Growth and Preparation of Cell Cultures

CHO cell cultures are grown and prepared for electroporation in an identical manner whether targeted gene knock-out or *APRT*-targeted recombination assays are to be performed.

1. Harvest cells from exponentially growing (~50–60% confluent) monolayer cultures of CHO cells grown in T-150 flasks (Corning, NY) at 37°C in a humidified atmosphere of 5% CO_2/95% air. Cell monolayers are rinsed twice with pre-warmed, sterile Solution A and exposed to 2 mL of trypsin-EDTA, which is aspirated and discarded after 60–90 s. Tap the flasks forcefully several times and resuspend the detached cells in 10 mL of ice-cold 10%-MEM by vigorous and repeated pipetting. After low-speed centrifugation, put the cell pellets on ice and use within 15–20 min (see Note 10).
2. Based on cell counts obtained using a Z2 Coulter Counter (Beckman Coulter, Miami, FL), determine the volume of cell suspension to spin down in a 15-mL sterile conical polypropylene centrifuge tube in order to provide sufficient cells for one electroporation (1×10^7 cells/cuvette for knock-out electroporations) (see Note 11).
3. Using a sterile, plastic 1-mL pipette, aseptically resuspend each cell pellet in 0.9 mL of sterile, ice-cold, HEP electroporation buffer containing restriction enzyme-linearized targeting vector DNA. For gene knock-out experiments, a final concentration of 1×10^7 cells and 20 μg of DNA/mL HEP buffer should be achieved (see Note 12).
4. Using a sterile, plastic 1-mL pipette, immediately transfer 0.85 mL of the CHO cell/targeting vector DNA suspension in EP buffer to a prechilled, sterile electroporation cuvette (BioRad, 4-mm electrode gap) and place on ice for a 10-min, pre-electroporation incubation.
5. Electroporate by delivering a single pulse of 500 V at 25 μF to the cuvette, using a BioRad Gene Pulser apparatus, and immediately place the cuvette back on ice for a 10 min, post-electroporation incubation (see Note 13).
6. Using a sterile plastic pipette, dilute the electroporated cell suspension by pipetting the contents of the cuvette into 8.5 mL cold 10%-MEM (also add one 1 mL rinse of the cuvette with 10%MEM), yielding a final cell concentration of ~$1–2 \times 10^6$ cells/mL.
7. Make a further dilution with ice-cold SF-MEM to yield a final volume of 60 mL. Plate 0.2 mL of this suspension to each of

three hundred 100-mm tissue culture dishes each containing 10 mL of 10%-MEM made with dialyzed FCS (DFCS). Incubate the dishes at 37°C for 16–24 h in a humidified atmosphere of 5% CO_2/95% air (see Note 14).

3.3. Targeted Recombination Assays at the APRT Locus in CHO Cells

The following procedure was carried out regardless of the cell line being targeted or the vector being used. The experiment is constructed in sets of three electroporation cuvettes per cell line/targeting vector combination (i.e., three cuvettes make up one set). However, this can be modified depending on the availability of cells and targeting vector.

1. Label and fill cell culture plates with 10%-MEM: 60 plates (100 mm) per set for ALASA selection, 12 plates (100 mm) per set for HAT selection, 12 plates (60 mm) per set for plating efficiency (PE) estimates, 6 of these for "pre-electroporation" and 6 for "post-electroporation."
2. Prepare KLEP buffer and set up the electroporation apparatus (BioRad Gene Pulser) for 25 μF capacitance and 0.5 kV.
3. Prepare five dilution tubes containing 9 mL 10%-MEM and one additional tube with 7.2 mL 10%-MEM. Keep these on ice.
4. Heat the linearized plasmid DNA at 55°C for 30 min then place on ice.
5. Combine 3 mL of KLEP buffer with 33 μg of DNA (11 μg/mL per cuvette) and keep on ice.
6. Place the appropriate number of electroporation cuvettes (4-mm gap) on ice.
7. Harvest CHO cells from exponentially growing (~50–60% confluent) monolayer cultures grown in T-150 flasks (Corning, NY) at 37°C in a humidified atmosphere of 5% CO_2/95% air (see Note 10). Cell monolayers are rinsed twice with pre-warmed, sterile Solution A and exposed to 2 mL of trypsin-EDTA, which is aspirated and discarded after 60–90 s. Tap the flasks forcefully several times, and resuspend the detached cells in 10 mL of ice-cold 10%-MEM by vigorous and repeated pipetting. After low-speed centrifugation, put the cell pellets on ice and use within 15–20 min.
8. Combine cell pellets for use in the same set in 20 mL cold 10%-MEM and pull through a 20-cm^3 syringe fitted with a 22-gauge needle and push out into a new tube filled with 20 mL cold 10%-MEM.
9. Determine the number of cells in the suspension using a Z3 Coulter Counter (Beckman Coulter, Miami, FL), and calculate the volume of cell suspension required to obtain 2×10^7 cells. Pipette this amount into three 15-mL sterile conical polypropylene centrifuge tubes and pellet the cells (see Note 11).

At this time remove 1×10^6 cells to a separate tube ("pre-electroporation") and keep on ice (this will be used to make a dilution series for plating efficiency calculations).

10. Aspirate all the media from the cell pellets and lightly loosen by flicking the bottom of the tube. Using a sterile, plastic 1-mL pipette, resuspend each cell pellet in 0.9 mL KLEP buffer with plasmid DNA, and transfer 0.85 mL to the prechilled electroporation cuvette. Incubate on ice 10 min (see Note 15).
11. Electroporate by delivering a single pulse of 0.5 kV at 25 μF to the cuvette, using a BioRad Gene Pulser apparatus, and immediately place the cuvette back on ice for 10 min. Repeat for each cuvette in the set (see Note 16).
12. Combine cells from each cuvette into 29 mL cold 10%-MEM, rinsing the cuvettes with a small amount of this media. The final volume will be 31.5 mL (see Note 17).
13. Remove 0.8 mL of this suspension and add to the reserved dilution tube containing 7.2 mL 10%-MEM; this will result in 1.62×10^5 cells/mL.
14. Plate 0.5 mL of the 1.62×10^5 cells/mL suspension into each of the 12 designated HAT plates, resulting in 8.1×10^4 cells per plate.
15. Plate 0.5 mL of the 31.5 mL suspension into each of the 60 designated ALASA plates, resulting in 8.1×10^5 cells per plate.
16. Return to the aliquoted 1×10^6 "pre-electroporation" tube and carry the dilution down to 1×10^3 using the dilution tubes prepared earlier. Plate 0.2 mL into each of the six designated "pre-electroporation" plates resulting in 200 cells per plate.
17. Return to the remaining suspension from step 13 (1.62×10^5 cells/mL) and carry the dilution down to 1.62×10^3 cells/mL using the prepared dilution tubes. Plate 0.15 mL into each of the six designated "post-electroporation" plates resulting in 243 cells per plate.
18. Incubate dishes at 37°C for 48 h in a humidified atmosphere of 5% CO_2/95% air before removing the 10%-MEM and replacing it with the appropriate selection media (either HAT or ALASA). Plating efficiency plates do not receive a media change.

3.4. Selection and Screening of Targeted Gene Knock-Out Clones

1. After 24 h, add G418 to each of the 300 dishes to achieve a final concentration of 400 μg/mL (adjusted to 100% activity). After an additional 24 h of incubation, add FIAU to a final concentration of 0.2 μM to all except 16 plates. Refeed dishes with 10%-MEM DFCS containing G418 and FIAU (except for 16 dishes that are refed with G418 alone) every 3–4 days until visible colonies are evident, approximately 12–14 days post-electroporation.

2. Pick well-isolated G418/FIAU-resistant colonies by gentle scraping with a Pipetman P-200 while pipetting up 200 μL of culture medium using sterile, wide-bore yellow tips. To ensure that cell clones are independent, pick only one colony per dish. For each set of six picked and resuspended colonies, inoculate two 6-well dishes: 50 μL per well to a reserved master dish, and 150 μL to wells of a corresponding dish to be used for DNA isolation for screening by PCR.
3. Stain the plates after marking picked colonies with a felt pen on the bottom of the dishes; enrichment can be calculated by counting the total number of colonies from the 16 plates that were subjected to G418 selection alone, normalizing to a per plate value, and comparing to the number of colonies per plate from the G418/FIAU selected cells.
4. After 3–5 days, cell monolayers growing in all wells of 6-well DNA dishes are rinsed twice with Solution A. Treat with trypsin-EDTA for 1 min. Aspirate the trypsin solution and discard. Resuspend cells by adding 10%-MEM to each well and vigorously pipetting. Pool cell suspensions for each dish into one tube (Falcon 1063, Becton Dickinson, Franklin Lanes, NJ), harvest cells by low speed centrifugation, wash once with PBS, and resuspend in 50 μL of PBS. Lyse the cells by addition of 200 μL of H_2O and heating for 10 min at 93°C. To the cell lysate, add 10 μL of fresh 1% proteinase K, incubate for 30 min at 55°C, heat for 10 additional min at 93°C to inactivate proteinase K, centrifuge, and aspirate the supernatant for PCR analysis.
5. The resulting lysate is used in a two-step nested PCR strategy as shown in Fig. 1a. A 5′ primer (forward Neo 1: 5′-CCGCTT TTCTGGATTCATCGAC-3′) designed to anneal to *NEO* sequence within the targeting vector, and a 3′ primer (reverse *ERCC1*: 5′-TGCCACCCCTGACCACATATACAC-3′) designed using *ERCC1* sequence downstream from exon 5 and not present in the targeting vector are used with 5 μL of cell lysate from the previous step in a 50 μL total reaction volume containing 1 μL Taq Extender (Statagene) and 1 μL TaqStart Antibody (Clontech). Perform 20 cycles of PCR (1 min 30 s at 95°C; 1 min at 56°C; 3 min at 73°C). Then, perform a second round of PCR with 5 μL of first round amplimer using a nested *NEO* primer (forward Neo 2: 5′-GCTTCCTC GTGCTTTACGGTATC-3′). After electrophoresis on 0.6% agarose gels, identify pooled colony samples exhibiting a diagnostic 1.8 kb targeting fragment (see Fig. 1a).
6. To verify successful gene targeting, individually trypsinize and transfer to T-25 flasks the six cultures from each reserved master dish that correspond to the positive pooled sample(s). Repeat the PCR analysis as described in step 5 to identify which

of the six clones exhibited the diagnostic recombinant fragment. Southern analysis (see Subheading 3.6) can be used to confirm disruption of the target gene by the *NEO* sequence present in the knock-out vector.

3.5. Selection of APRT-Targeted Recombinant Clones

1. Colonies arising on PE dishes can be fixed, stained, and counted after ~8 days of incubation.
2. Colonies arising on HAT selection dishes can be fixed, stained, and counted after ~12 days of incubation.
3. Representative ALASA^r APRT$^+$ recombinant clones can be picked for analysis after ~14 days of incubation (see Note 18), by gently scraping off and pipetting up cells from individual colonies using a Pipetman P-200 and sterile, wide-bore yellow tips, as described in step 2 of Subheading 3.4. These colonies are plated into T-25 culture flasks. The rest of the colonies can then be fixed, stained, and counted.
4. The selected colonies are maintained as individual clones and passaged up to T-75 flasks when appropriate. When flasks reach ~90% confluency, cells are harvested and pelleted for DNA extraction; refeeding flasks allows recovery of cells for cryopreservation.
5. *APRT* targeting frequencies are calculated, taking into account the plating efficiencies, and normalized to the frequencies of untargeted events, as determined from the HAT selection plates (see Note 19).

3.6. Southern Analysis of APRT-Targeted Recombinant Clones

1. Isolate genomic DNAs from each independent ALASA^r recombinant clone, as well as from each parental cell line (e.g., ATS49tg, E1KO7-5, ATS49-Rv, E1KO-Rv) used for *APRT* targeting experiments. We used the Promega Genomic Wizard DNA extraction kit but any method yielding high quality genomic DNA should be satisfactory.
2. Subject approximately 10 μg genomic DNA from each recombinant clone (or parental cell line) to overnight restriction endonuclease digestion by *Mbo*II, *Hind*III/*Sac*I, or *Bgl*II/*Xho*I, using 2–3 U of restriction enzyme/μg of DNA, employing the specific reaction buffers/conditions recommended by suppliers.
3. Following restriction endonuclease digestion, separate DNA by electrophoresis (560 Vh) in 0.8% agarose gels. To prepare the gel for transfer, wash the gel for 10 min in acid buffer, rinse with water, wash 30 min in alkaline buffer, rinse with water, wash 30 min in neutralizing buffer, and finally rinse with water. Transfer by capillary blotting to nitrocellulose or nylon membranes.
4. Fix the DNA to the membrane by cross-linking or baking as appropriate for the membrane type used. Prehybridize membrane

for at least 3 h in 10 mL prehybridization solution. Hybridize the blot with nick-translated or random-primed ^{32}P-labeled *APRT* gene-specific probes by adding the denatured probe directly to the prehybridization solution. For *Mbo*II digests, a 1.4-kb *Eco*RI–*Xba*I *APRT* fragment probe is used; a 3.9-kb *BamH*I fragment probe that includes the entire CHO *APRT* gene sequence is used for *Hind*III/*Sac*I and *Bgl*II/*Xho*I digests (see Fig. 2).

5. Allow the hybridization to proceed overnight at 42°C. The next day carry out the following washes: one wash with high salt buffer 30 min at 42°C, one wash with high salt buffer 30 min at 55°C, two washes with low salt buffer 30 min each at 55°C, while monitoring with a hand-held Geiger counter. Expose the membrane to audioradiography film for 2–7 days.
6. The various classes of targeted *APRT* recombinants (Fig. 2) can be readily identified on the basis of their distinctive and diagnostic restriction fragment patterns (see Note 20).

4. Notes

1. CHO cell lines such as CHO-ATS49tg are hemizygous for the *ERCC1* gene (25, 35). We have successfully used CHO-ATS49tg, CHO-K1, and several other CHO cell lines ((24–27, 30), and unpublished) for *ERCC1* gene knock-out experiments as described in this chapter.
2. Construction of the *ERCC1* targeting vector pSL1 is described in (25). In principle, a repair gene targeting vector with the replacement-type structure illustrated in Fig. 1a could be constructed using *NEO* and *HSV-tk* markers and used to knock-out a variety of NER or DSBR genes in CHO or other cultured mammalian cell lines. For example, Thompson and co-workers have developed several related strategies (28, 29).
3. Some positive–negative selection gene targeting protocols use FIAU in selection media containing non-dialyzed serum; however, we find that FIAU counter-selection is more stringent and economical (requiring much less FIAU) if dialyzed serum is used. Alternatively, ganciclovir (Sigma) can be substituted for FIAU at 1 μM final concentration. As for FIAU, dialyzed serum is used in selection media.
4. These forward Neo PCR primers, nested in the *NEO* cassette from plasmid pMC1neopA (51) should be useful for any PCR screening strategy in which *NEO* is used as the gene disruption marker and the reverse PCR primer is designed from a target

gene sequence outside of the homologous sequence in the targeting vector, and downstream (3′) from the exon to be disrupted.

5. CHO-ATS49tg is an APRT$^-$/HPRT$^-$ Chinese hamster cell line derived from CHO-AT3-2 (19). CHO-AT3-2 cells are hemizygous for the *APRT* locus (52). CHO-ATS49tg cells contain a single, mutationally altered copy of the *APRT* gene, with a 3-bp deletion in exon 5 that has eliminated an *Mbo*II restriction site (19, 48), providing a convenient restriction fragment length polymorphism for monitoring whether targeted recombination events have resulted in correction of the *APRT* target gene defect. E1KO7-5 is a CHO *ERCC1* knock-out cell line generated by targeted disruption of the hemizygous *ERCC1* gene locus in CHO-ATS49tg cells (24). ATS49-Rv and E1KO-Rv were generated as described in ref. 30 and differ from ATS49tg and E1KO7-5 in the position of the *APRT* inactivating mutation. Where ATS49tg and E1KO7-5 have a 3-bp deletion in exon 5 abolishing an *Mbo*II restriction site, ATS49-Rv and E1KO-Rv have a 2-bp deletion in exon 2 abolishing an *Eco*RV restriction site. We have also used several other CHO-AT3-2- or CHO-AA8-4-derived cell lines with hemizygous, mutant *APRT* alleles suitable for targeted correction for gene targeting experiments (our unpublished results).
6. The pAG6ins0.9 *APRT* targeting vector (see Fig. 1b) was derived from pSV2gpt (53) in several steps that involved the construction of several intermediate plasmids, using conventional cloning methods (24). The pAG6ins0.9 targeting vector contains 3.9-kb of *APRT* targeting homology, including a full-length Chinese hamster *APRT* gene that has been disrupted by a 949-bp heterologous insertion into the exon 3 *Xho*I site (24). Linearization of this targeting vector by cleavage at a unique *Sal*I site within the large heterologous insertion creates an "ends-in" targeted integration vector configuration in which both arms of *APRT* targeting homology (that would ordinarily flank each side of the DSB) are blocked by long terminal nonhomologies (see Fig. 1b). When this *Sal*I-linearized vector is electroporated into cells, resection of the 5′-ended strands on each side of the DSB leaves long (271- or 670-nt), 3′-ended, single-stranded nonhomologous "tails" on each arm of the targeting vector. These 3′-end-blocking, nonhomologous tails must be removed to permit target gene-templated DNA synthesis and extension of homologous ends following strand invasion, in order for targeted integration (crossover) or vector correction (conversion) recombination products to be generated from homologous recombination intermediates (24). A *GPT* cassette on the pAG6ins0.9 targeting vector allows determination of the frequencies of untargeted vector integration (HAT^r,

GPT+ transfectants), thereby providing a reliable internal control for monitoring electroporation efficiency in *APRT* targeting experiments. Targeting vector pAG6 is similar except that it has short regions of nonhomologous sequence and when cleaved with *Hind*III, has short (18- or 11-nt) 3′ nonhomologous tails (24).

7. Alanosine and azaserine act as inhibitors of different steps in the de novo pathway for purine biosynthesis. APRT is a purine salvage pathway enzyme that allows cells to utilize adenine for purine biosynthesis by catalyzing the conversion of adenine to AMP. ALASA selection medium will effectively kill APRT− cells such as CHO-ATS49tg (which are unable to utilize exogenous adenine as a purine source), but will support the growth of APRT+ recombinants.
8. Amethopterin (methotrexate) inhibits the de novo pathways for both purine and pyrimidine biosynthesis. Both *HPRT* and its bacterial equivalent, *GPT*, are purine salvage pathway enzymes that allow cells to utilize hypoxanthine or guanine for purine biosynthesis. HAT selection medium will effectively kill HPRT− cells such as CHO-ATS49tg (which are unable to utilize exogenous hypoxanthine as a purine source), but will support the growth of GPT+ transfectants.
9. All steps subsequent to ethanol precipitation should be performed aseptically in a tissue culture hood. DNA pellets may be left to air-dry for 20 min in a sterile hood, and then resuspended using sterile technique.
10. To achieve maximal and reproducible transformation efficiency in gene targeting experiments, it is important that the cells are in exponential growth at the time cell cultures are trypsinized and harvested for electroporation.
11. Cells should be harvested such that only one cell pellet of $1–2 \times 10^7$ cells/centrifuge tube is used for one cuvette; we have experienced poorer results when we tried resuspending sufficient cells for several cuvettes in a single centrifuge tube and then distributing the suspension to multiple cuvettes.
12. It is important to aspirate the supernatant fluid as completely as possible after centrifugation. Prior to resuspending cells in electroporation buffer, flick the conical centrifuge tube containing the cell pellet sharply several times to loosen the pellet. Then, resuspend the cells very carefully by pipetting back-and-forth slowly several times, without introducing bubbles.
13. For knock-out electroporation in HEP buffer, time constants under these conditions should be ~0.6. The 10 min post-electroporation incubation on ice is not necessary for knock-out electroporations in HEP buffer.

14. 1–2 Cuvettes represent a standard size experiment for knock-out gene targeting, resulting in 300–600 100-mm dishes. In our experience, one cuvette will produce approximately 180–200 G418/FIAU doubly resistant colonies after positive–negative selection.
15. For *APRT*-targeted correction recombination experiments, it is important that the cell pellet be directly suspended in KLEP buffer (without a rinse step) after carefully draining off the supernatant and aspirating any remaining traces of liquid from the sides of the tube. Inclusion of a KLEP buffer rinse step markedly reduces cell survival after electroporation; this procedure is also followed for knock-out gene targeting experiments using HEP buffer.
16. For *APRT* gene targeting in KLEP buffer, time constants under these conditions should be ~5.5–6.0.
17. It is important to: (a) keep this cell suspension (and any dilutions to be used for plating) on ice to minimize cell clumping or attachment and (b) make sure that the cells are kept well-suspended throughout the plating procedure, and that the cells are uniformly distributed after plating into tissue culture dishes.
18. To ensure the independence of all $APRT^+$ recombinants analyzed, only a single $ALASA^r$ colony from each independent cell population should be picked for expansion, cryogenic preservation, and Southern blot analysis.
19. For *APRT* targeting experiments, a standard size experiment assessing the effects of one targeting vector structure on targeted homologous recombination in one cell line is 3–6 cuvettes, resulting in 18–36 100-mm dishes prior to plating into ALASA selection. Two or more cell lines, and one or two targeting vector configurations are typically assayed at the same time in parallel. Plating efficiency counts are used to correct the targeting efficiency calculations. Some cell death is expected due to manipulation of the cells and the process of electroporation and these losses must be taken into account and applied to the calculated frequencies. Routinely, we observe about 50% killing between the pre- and post-electroporation plating efficiency counts; any substantial deviation from this difference is cause for discarding the experiment. A simple calculation is done by dividing the number of colonies obtained from the post-electroporation PE plates by the number of colonies expected (based on the number of cells plated). This "plating efficiency" correction is then applied to the calculated recombination frequencies to obtain normalized frequencies. We calculate both $ALASA^r$ and HAT^r frequencies, correct for plating efficiency (using the post-electroporation counts), and "normalize" the $APRT^+$ targeting frequencies to the GPT^+

nontargeted frequencies by dividing the ALASA frequencies by the HAT frequencies; this normalization helps to control variation from experiment to experiment.

20. Restriction digest patterns will differ depending on the cell line pair (i.e., ATS49tg/E1KO-7 vs. ATS49-Rv/E1KO-Rv) used in the targeted recombination assays. Examples of Southern blots can be found in refs. 24 and 30, which provide extensive discussion concerning deducing recombinant structures from Southern blot patterns. Briefly, when ATS49tg or E1KO7-5 (bearing the 3-bp exon 5 deletion abolishing the *Mbo*II site) are used, $APRT^+$ recombinants having arisen by target gene conversion (Fig. 2a) show loss of the 2.0-kb endogenous *Mbo*II fragment and reappearance of a 1.5-kb *Mbo*II fragment characteristic of a wild-type Chinese hamster *APRT* locus (24). They retain normal, unchanged (wild-type) *APRT* restriction fragment patterns for all other restriction enzymes, including *Hind*III (which yields a single 8.6-kb restriction fragment), and *Bgl*II/*Xho*I (which yields 2.2- and 2.4-kb fragments). ATS49-Rv- or E1KO-Rv (bearing the 2-bp deletion in exon 2 abolishing an *Eco*RV restriction site)-derived recombinants will differ only when digested with *Hind*III/*Eco*RV. Gene convertants (Fig. 2e) will yield two bands at 2.2- and 3.2-kb instead of one at 5.5-kb in the parental cell lines. All other digests will remain as described for ATS49tg convertants. ATS49tg- and E1KO7-5-derived $APRT^+$-targeted integration recombinants (Fig. 2b) show diagnostic restriction fragment patterns for *Hind*III (13- and 4.0-kb fragments), and *Bgl*II/*Xho*I (2.2-, 3.9-, 4.6-, and 2.4-kb fragments), after hybridization with the 3.9-kb *BamH*I fragment probe. Similarly, targeted integrations obtained from ATS49-Rv or E1KO-Rv (Fig. 2f) reveal the same restriction patterns described for ATS49tg/E1KO7-5 after digestion with *Hind*III and *Bgl*II/*Xho*I. ATS49tg-/E1KO7-5-derived $APRT^+$ vector correction recombinants (Fig. 2c) will contain both the original, endogenous, 2.0-kb (mutant) *Mbo*II fragment and a newly acquired 1.5-kb *Mbo*II fragment characteristic of a wild-type CHO *APRT* gene. After digestion with *Hind*III and hybridization with the 3.9-kb *BamH*I fragment probe, vector correction recombinants show both an 8.6-kb *Hind*III fragment characteristic of the endogenous *APRT* locus, and a novel fragment of indeterminate size, representing the randomly integrated, wild-type (corrected vector) *APRT* sequence. After digestion with *Bgl*II/*Xho*I, vector correction recombinants will show 2.2- and 2.4-kb fragments characteristic of the endogenous *APRT* locus, plus two or more novel fragments, produced by ectopic integration of the corrected, vector-derived *APRT* sequence. ATS49-Rv-/E1KO7-5-derived targeted vector corrections

(Fig. 2g) will have similar restriction patterns except that double digest with *Hind*III/*Eco*RV will yield a wild-type (mutant) band pattern in addition to additional bands of various sizes (30). Aberrant integration/deletion recombinants (Fig. 2d, h), in which a variable portion of the integrated vector sequence appears to have been deleted, are only recovered from pAG6ins0.9 targeting experiments in *ERCC1*$^-$ cell lines. These recombinants can display widely differing restriction fragment patterns, depending on the size of the deleted region (24, 30).

Acknowledgements

We thank Becky Brooks for assistance in manuscript preparation. This work was supported by PHS grant CA097175 from the National Cancer Institute.

References

1. Capecchi MR (1989) Altering the genome by homologous recombination. Science 244: 1288–1292
2. Ramirez-Solis R, Davis AC, Bradley A (1993) Gene targeting in embryonic stem cells. Methods Enzymol 225:855–878
3. Rothstein R (1991) Targeting, disruption, replacement, and allele rescue: integrative DNA transformation in yeast. Methods Enzymol 194:281–301
4. Waldman AS (1992) Targeted homologous recombination in mammalian cells. Crit Rev Oncol Hematol 12:49–64
5. Capecchi MR (1989) The new mouse genetics: altering the genome by gene targeting. Trends Genet 5:70–76
6. Friedberg EC, Meira LB (2006) Database of mouse strains carrying targeted mutations in genes affecting biological responses to DNA damage. Version 7. DNA Repair (Amst) 5:189–209
7. Wijnhoven SW, van Steeg H (2003) Transgenic and knockout mice for DNA repair functions in carcinogenesis and mutagenesis. Toxicology 193:171–187
8. Dhar PK, Sonoda E, Fujimori A, Yamashita YM, Takeda S (2001) DNA repair studies: experimental evidence in support of chicken DT40 cell line as a unique model. J Environ Pathol Toxicol Oncol 20:273–283
9. Takata M, Ishiai M, Kitao H (2009) The Fanconi anemia pathway: insights from somatic cell genetics using DT40 cell line. Mutat Res 668:92–102
10. Winding P, Berchtold MW (2001) The chicken B cell line DT40: a novel tool for gene disruption experiments. J Immunol Methods 249: 1–16
11. Friedberg EC, Walker GC, Siede W, Wood RD, Schultz RA, Ellenberger T (2006) DNA repair and mutagenesis, 2nd edn. ASM Press, Washington, DC
12. Fishman-Lobell J, Haber JE (1992) Removal of nonhomologous DNA ends in double-strand break recombination: the role of the yeast ultraviolet repair gene *RAD1*. Science 258:480–484
13. Ivanov EL, Haber JE (1995) *RAD1* and *RAD10*, but not other excision repair genes, are required for double-strand break-induced recombination in *Saccharomyces cerevisiae*. Mol Cell Biol 15:2245–2251
14. Schiestl RH, Prakash S (1988) *RAD1*, an excision repair gene of *Saccharomyces cerevisiae*, is also involved in recombination. Mol Cell Biol 8:3619–3626
15. Schiestl RH, Prakash S (1990) *RAD10*, an excision repair gene of *Saccharomyces cerevisiae*, is involved in the *RAD1* pathway of mitotic recombination. Mol Cell Biol 10:2485–2491
16. Sugawara N, Paques F, Colaiacovo M, Haber JE (1997) Role of *Saccharomyces cerevisiae* Msh2 and Msh3 repair proteins in double-strand

break-induced recombination. Proc Natl Acad Sci U S A 94:9214–9219
17. Adair GM, Nairn RS (1995) Impact on human diseases and cancer, in gene targeting. In: Vos J-M (ed) DNA repair mechanisms. Biomedical Landes Company, Austin, pp 301–328
18. Adair GM, Nairn RS, Wilson JH, Scheerer JB, Brotherman KA (1990) Targeted gene replacement at the endogenous *APRT* locus in CHO cells. Somat Cell Mol Genet 16:437–441
19. Adair GM, Nairn RS, Wilson JH, Seidman MM, Brotherman KA, MacKinnon C, Scheerer JB (1989) Targeted homologous recombination at the endogenous adenine phosphoribosyltransferase locus in Chinese hamster cells. Proc Natl Acad Sci U S A 86:4574–4578
20. Nairn RS, Adair GM, Porter T, Pennington SL, Smith DG, Wilson JH, Seidman MM (1993) Targeting vector configuration and method of gene transfer influence targeted correction of the *APRT* gene in Chinese hamster ovary cells. Somat Cell Mol Genet 19:363–375
21. Porter T, Pennington SL, Adair GM, Nairn RS, Wilson JH (1990) A novel selection system for recombinational and mutational events within an intron of a eukaryotic gene. Nucleic Acids Res 18:5173–5180
22. Sargent RG, Merrihew RV, Nairn R, Adair G, Meuth M, Wilson JH (1996) The influence of a (GT)29 microsatellite sequence on homologous recombination in the hamster adenine phosphoribosyltransferase gene. Nucleic Acids Res 24:746–753
23. Scheerer JB, Adair GM (1994) Homology dependence of targeted recombination at the Chinese hamster *APRT* locus. Mol Cell Biol 14:6663–6673
24. Adair GM, Rolig RL, Moore-Faver D, Zabelshansky M, Wilson JH, Nairn RS (2000) Role of *ERCC1* in removal of long non-homologous tails during targeted homologous recombination. EMBO J 19:5552–5561
25. Rolig RL, Layher SK, Santi B, Adair GM, Gu F, Rainbow AJ, Nairn RS (1997) Survival, mutagenesis, and host cell reactivation in a Chinese hamster ovary cell *ERCC1* knock-out mutant. Mutagenesis 12:277–283
26. Sargent RG, Meservy JL, Perkins BD, Kilburn AE, Intody Z, Adair GM, Nairn RS, Wilson JH (2000) Role of the nucleotide excision repair gene *ERCC1* in formation of recombination-dependent rearrangements in mammalian cells. Nucleic Acids Res 28:3771–3778
27. Sargent RG, Rolig RL, Kilburn AE, Adair GM, Wilson JH, Nairn RS (1997) Recombination-dependent deletion formation in mammalian cells deficient in the nucleotide excision repair gene *ERCC1*. Proc Natl Acad Sci U S A 94:13122–13127
28. Tebbs RS, Hinz JM, Yamada NA, Wilson JB, Salazar EP, Thomas CB, Jones IM, Jones NJ, Thompson LH (2005) New insights into the Fanconi anemia pathway from an isogenic FancG hamster CHO mutant. DNA Repair (Amst) 4:11–22
29. Hinz JM, Tebbs RS, Wilson PF, Nham PB, Salazar EP, Nagasawa H, Urbin SS, Bedford JS, Thompson LH (2006) Repression of mutagenesis by Rad51D-mediated homologous recombination. Nucleic Acids Res 34:1358–1368
30. Rahn JJ, Rowley B, Lowery MP, Coletta LD, Limanni T, Nairn RS, Adair GM (2011) Effects of varying gene targeting parameters on processing of recombination intermediates by ERCC1-XPF. DNA Repair (Amst) 10:188–198
31. Rahn JJ, Lowery MP, Della-Coletta L, Adair GM, Nairn RS (2010) Depletion of Werner helicase results in mitotic hyperrecombination and pleiotropic homologous and nonhomologous recombination phenotypes. Mech Ageing Dev 131:562–573
32. Gottesman MM (1985) Molecular cell genetics. Wiley, New York, NY
33. O'Neill JP, Couch DB, Machanoff R, San Sebastian JR, Brimer PA, Hsie AW (1977) A quantitative assay of mutation induction at the hypoxanthine-guanine phosphoribosyl transferase locus in Chinese hamster ovary cells (CHO/HGPRT system): utilization with a variety of mutagenic agents. Mutat Res 45:103–109
34. Siciliano MJ, Stallings RL, Humphrey RM, Adair GM (1986) Mutation in somatic cells as determined by electrophoretic analysis of mutagen-exposed Chinese hamster ovary cells. In: de Serres FJ (ed) Chemical mutagens. Plenum, New York, NY, pp 509–531
35. Thompson LH, Bachinski LL, Stallings RL, Dolf G, Weber CA, Westerveld A, Siciliano MJ (1989) Complementation of repair gene mutations on the hemizygous chromosome 9 in CHO: a third repair gene on human chromosome 19. Genomics 5:670–679
36. Deng C, Capecchi MR (1992) Reexamination of gene targeting frequency as a function of the extent of homology between the targeting vector and the target locus. Mol Cell Biol 12:3365–3371
37. te Riele H, Maandag ER, Berns A (1992) Highly efficient gene targeting in embryonic stem cells through homologous recombination with isogenic DNA constructs. Proc Natl Acad Sci U S A 89:5128–5132
38. van Duin M, de Wit J, Odijk H, Westerveld A, Yasui A, Koken MH, Hoeijmakers JH, Bootsma

D (1986) Molecular characterization of the human excision repair gene ERCC-1: cDNA cloning and amino acid homology with the yeast DNA repair gene RAD10. Cell 44: 913–923

39. Adair GM, Scheerer JB, Brotherman A, McConville S, Wilson JH, Nairn RS (1998) Targeted recombination at the Chinese hamster *APRT* locus using insertion versus replacement vectors. Somat Cell Mol Genet 24:91–105
40. Hasty P, Ramirez-Solis R, Krumlauf R, Bradley A (1991) Introduction of a subtle mutation into the Hox-2.6 locus in embryonic stem cells. Nature 350:243–246
41. Valancius V, Smithies O (1991) Testing an "in-out" targeting procedure for making subtle genomic modifications in mouse embryonic stem cells. Mol Cell Biol 11:1402–1408
42. Hasty P, Rivera-Perez J, Bradley A (1991) The length of homology required for gene targeting in embryonic stem cells. Mol Cell Biol 11:5586–5591
43. Mansour SL, Thomas KR, Capecchi MR (1988) Disruption of the proto-oncogene int-2 in mouse embryo-derived stem cells: a general strategy for targeting mutations to non-selectable genes. Nature 336:348–352
44. Kilburn AE, Shea MJ, Sargent RG, Wilson JH (2001) Insertion of a telomere repeat sequence into a mammalian gene causes chromosome instability. Mol Cell Biol 21:126–135
45. Merrihew RV, Sargent RG, Wilson JH (1995) Efficient modification of the *APRT* gene by FLP/FRT site-specific targeting. Somat Cell Mol Genet 21:299–307
46. Meservy JL, Sargent RG, Iyer RR, Chan F, McKenzie GJ, Wells RD, Wilson JH (2003) Long CTG tracts from the myotonic dystrophy gene induce deletions and rearrangements during recombination at the *APRT* locus in CHO cells. Mol Cell Biol 23:3152–3162
47. Sargent RG, Brenneman MA, Wilson JH (1997) Repair of site-specific double-strand breaks in a mammalian chromosome by homologous and illegitimate recombination. Mol Cell Biol 17:267–277
48. Smith DG, Adair GM (1996) Characterization of an apparent hotspot for spontaneous mutation in exon 5 of the Chinese hamster *APRT* gene. Mutat Res 352:87–96
49. Talbert LL, Coletta LD, Lowery MG, Bolt A, Trono D, Adair GM, Nairn RS (2008) Characterization of CHO *XPF* mutant UV41: influence of *XPF* heterozygosity on double-strand break-induced intrachromosomal recombination. DNA Repair (Amst) 7: 1319–1329
50. Stanners CP, Eliceiri GL, Green H (1971) Two types of ribosome in mouse-hampster hybrid cells. Nat New Biol 230:52–54
51. Thomas KR, Capecchi MR (1987) Site-directed mutagenesis by gene targeting in mouse embryo-derived stem cells. Cell 51:503–512
52. Adair GM, Stallings RL, Nairn RS, Siciliano MJ (1983) High-frequency structural gene deletion as the basis for functional hemizygosity of the adenine phosphoribosyltransferase locus in Chinese hamster ovary cells. Proc Natl Acad Sci U S A 80:5961–5964
53. Mulligan RC, Berg P (1981) Selection for animal cells that express the *Escherichia coli* gene coding for xanthine-guanine phosphoribosyltransferase. Proc Natl Acad Sci U S A 78:2072–2076

Chapter 32

In Vitro Rejoining of Double Strand Breaks in Genomic DNA

George Iliakis, Emil Mladenov, and Nge Cheong

Abstract

Recent genetic and biochemical studies have provided important insights into the mechanism of nonhomologous end joining (NHEJ) pathways in higher eukaryotes, and have facilitated the functional characterization of several of its components including DNA-PKcs, Ku, DNA ligase IV, XRCC4, XLF/Cernunnos, and Artemis. Nevertheless, there is evidence that as of yet uncharacterized repair factors may contribute to the efficiency of NHEJ, for example by modulating the activity of known factors. Also, the discovery of alternative pathways of NHEJ that function as backup to the classical DNA-PK-dependent pathway of NHEJ has added yet another dimension in the set of activities involved. The biochemical characterization of NHEJ in higher eukaryotes has benefited significantly from in vitro plasmid-based end joining assays. However, because of differences in the organization and sequence of genomic and plasmid DNA, and because multiple pathways of NHEJ are operational, it is possible that different factors are preferred for the rejoining of DSBs induced in plasmid versus genomic DNA organized in chromatin. Here, we describe an in vitro assay that allows the study of DSB rejoining in genomic DNA. The assay utilizes as a substrate DSBs induced by various means in genomic DNA prepared from agarose-embedded cells after appropriate lysis. Two extremes in terms of state of DNA organization are described: "naked" DNA and DNA organized in chromatin. We describe the protocols developed to carry out and analyze these in vitro reactions, including procedures for the preparation of cell extract and the preparation of the substrate DNA ("naked" DNA or nuclei).

Key words: DNA repair, Double strand breaks, Chromatin structure, In vitro assay, Nonhomologous end joining

1. Introduction

A large number of studies suggests that double strand breaks (DSBs) induced in the DNA by ionizing radiation or chemical agents are critical lesions which if unrepaired, or if misrepaired, may kill a cell, or cause its transformation to a cancer cell. DSBs can also be induced by endogenous oxidative stress, by interference of the replication fork with single strand breaks, as well as during meiosis and V(D)J recombination. Cells have therefore developed

Lotte Bjergbæk (ed.), *DNA Repair Protocols*, Methods in Molecular Biology, vol. 920,
DOI 10.1007/978-1-61779-998-3_32, © Springer Science+Business Media New York 2012

efficient repair mechanisms for dealing with this type of lesion and restoring DNA integrity.

At least three enzymatically distinct processes are utilized to repair DNA DSBs in eukaryotic cells: homology directed repair (HDR), nonhomologous end joining (NHEJ), and a recently identified process of alternative end joining with a quasi-backup functions (B-NHEJ) (1–4). HDR removes DSBs by utilizing homologous DNA segments, on the same or different DNA molecules, and restores faithfully the original DNA sequence around the break. Both pathways of NHEJ remove DSBs from the genome by simply joining DNA ends without any homology requirements and without ensuring sequence restoration around the break. Even the joining of the original ends and thus the restoration of the original DNA molecule is not ensured through any of these repair pathways, allowing thus the formation of chromosome exchanges. B-NHEJ is more prone to errors than the classical NHEJ both in terms of sequence restoration, as well as in terms of rejoining the original ends. While homologous recombination is the preferred mechanism of DSB repair in prokaryotes and in lower eukaryotes such as yeast, NHEJ appears to be extensively used in higher eukaryotes (5), with classical NHEJ normally dominating and B-NHEJ taking over whenever classical NHEJ fails for whatever reason. Aspects of regulation in the selection among these DSB repair pathways are at the moment the subject of intensive research.

Recent genetic and biochemical studies have provided important insights into the mechanisms of NHEJ in higher eukaryotes, and have helped to characterize several important components of the NHEJ apparatus including DNA-PKcs, Ku, DNA ligase IV, XRCC4, and XLF/Cernunnos (1, 3). Evidence exists however that additional factors are required for the successful removal in vivo of DSBs by the DNA-PK-dependent pathway of end joining and several aspects of regulation in the function of these proteins remain uncharacterized (6–8). In addition, there is evidence that cells deficient in the classical, DNA-PK-dependent pathway of NHEJ do not compensate their repair defect by increasing utilization of HDR, but rely instead on backup pathways of NHEJ to remove ionizing radiation-induced DNA DSBs (5, 9–11). It is likely, therefore, that the characterization of components of the classical pathway of NHEJ, as well as the biochemical characterization of B-NHEJ will benefit from in vitro assays of DNA end joining.

The biochemical characterization of NHEJ in higher eukaryotes has significantly benefited from in vitro plasmid-based end joining assays (12). These assays are very useful and have generated information essential for our understanding of the mechanisms utilized by the cell to rejoin DNA ends. However, because of differences in the organization and sequence of genomic and plasmid DNA, and because multiple pathways are available for end joining in mammalian cells, it is possible that different factors are preferred for the rejoining of DSBs induced in plasmid versus genomic DNA. For this reason, we developed and describe here an in vitro assay that allows the study of DSB rejoining in genomic DNA. The assay utilizes as a substrate DSBs induced by various means (e.g., ionizing radiation, restriction endonucleases, bleomycin, etc.) in genomic DNA prepared from agarose-embedded cells after appropriate lysis. At present, two extremes in terms of state of DNA organization have been tested: "naked" DNA and DNA organized in chromatin (13–15). The former stage is generated by complete lysis of cells embedded in agarose using detergents and proteases, while the latter stage is generated by gentle lysis that leaves nuclei intact. Although not yet fully tested, intermediate stages of chromatin organization can also be generated, by gradually extracting nuclear proteins and histones with increasing salt concentrations.

The unique feature of the assay lies in the fact that the agarose fiber network that encapsulates the intact cell generates a "cage" that protects and restricts the expansion and mobility of the cellular DNA, or chromatin, particularly when freed upon lysis from the nucleus. This "cage" is expected to preserve essential features of DNA organization, especially after treatment to induce DSBs, and also to prevent extensive entangling. In this way, essential elements of the in vivo DNA repair process are maintained, including the large molecular weight DNA, the low number of DSBs (ends) present per Mbp, the high local DNA concentration, the sequence context and, for nuclei-based or derivative assays, the possibility of retaining elements of chromatin structure.

Here, we describe the protocols developed to carry out these in vitro reactions for DSB rejoining. The required procedures include: (1) preparation of HeLa-cell extract; (2) preparation of the DNA ("naked" DNA or nuclei); (3) generation of DSBs in the DNA; (4) assembly of in vitro repair reactions; (5) assay for DNA damage using pulsed-field gel electrophoresis. Figure 1 shows a graphic outline of these steps.

2. Materials

2.1. Preparation of HeLa-Cell Extract

1. Minimum essential medium (MEM) modified for suspension cultures (S-MEM), supplemented with 5% fetal or iron supplemented bovine serum and antibiotics (penicillin 100 U/mL, streptomycin 100 μg/mL).
2. Hypotonic buffer solution: 10 mM HEPES-KOH, pH 7.5 at 4°C (stock 0.6 M, pH 7.5 at room temperature (RT)), 1.5 mM $MgCl_2$ (stock 0.5 M), 5 mM KCl (stock 3 M). Immediately before use add 0.2 mM PMSF (stock 100 mM in ethanol; can be stored at −20°C for more than a month), and 0.5 mM dithiothreitol (DTT) (stock 1 M in H_2O; store at −20°C).

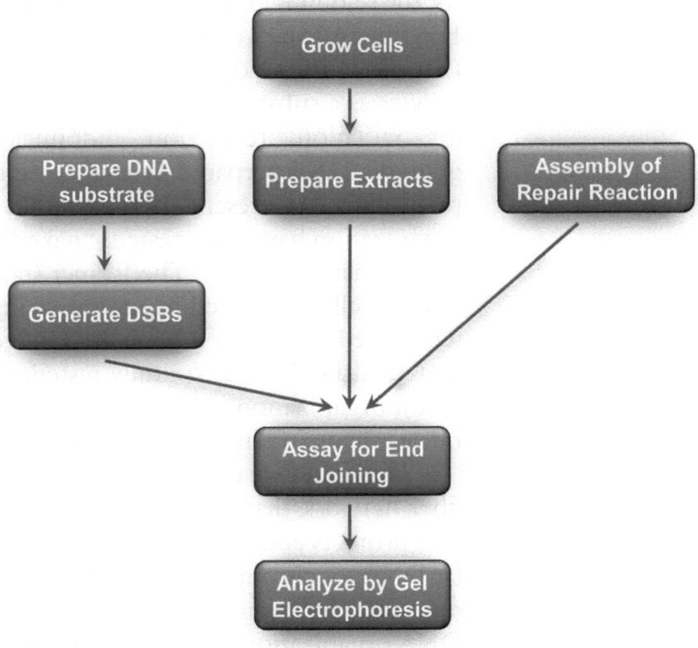

Fig. 1. Outline of the individual steps required to analyze the repair of genomic DSBs in vitro.

3. High salt buffer: 100 mM HEPES-KOH, pH 7.5, 1.4 M KCl, and 15 mM $MgCl_2$.
4. Dialysis buffer: 25 mM Tris–HCl, pH 7.5 at 4°C (stock 1 M, pH 7.5 at 4°C), 10% glycerol, 50 mM NaCl (stock 5 M), 1 mM EDTA (stock 0.5 M, pH 8.0). Immediately before use add 0.2 mM PMSF, and 0.5 mM DTT.
5. Microcarrier spinner flasks of 30 L nominal volume (e.g., Bellco Glass Inc.).
6. Microcarrier magnetic stirrers (e.g., Bellco Glass Inc.).
7. Tissue culture dishes (100 mm).
8. Dounce homogenizer with B pestle, 50 mL.

2.2. Preparation of "Naked" DNA from Agarose-Embedded Cells

1. McCoys 5A growth medium (Gibco BRL) supplemented with 5% bovine serum, 100 U/mL penicillin, and 100 μg/mL streptomycin.
2. HEPES-buffered McCoy's 5A growth medium. Prepared from McCoy's 5A medium by reducing the $NaHCO_3$ concentration to 5 mM and by adding HEPES to a final concentration of 20 mM.
3. ^{14}C-thymidine, specific activity ~50 mCi/mmol.
4. 500 μM thymidine in water.
5. 1% Solution of InCert agarose (FMC) in serum-free HEPES-buffered McCoy's 5A growth medium (no serum added). Incubate at 42°C in a water bath to prevent solidification.

6. 3.5 mm diameter glass tubes cut in 15–20 cm lengths. Seal one end with tape. Precool by immersing into ice.
7. Lysis solution: 10 mM Tris–HCl, pH 7.5 at 50°C, 100 mM EDTA, pH 8.0, 50 mM NaCl, 1% *N*-lauryl sarcosyl (NLS—stock 10% in H_2O). Just before use add 0.2 mg/mL proteinase.
8. Washing buffer: 10 mM Tris–HCl, pH 7.5 at 4°C, 100 mM EDTA, pH 8.0, and 50 mM NaCl.
9. RNase A solution: 0.1 mg/mL RNase A in 10 mM Tris–HCl, pH 7.5 at 37°C, 100 mM EDTA, pH 8.0, 50 mM NaCl.

2.3. Preparation of Nuclei from Agarose-Embedded Cells

1. Saponin lysis solution: 50 mM Tris–HCl, pH 8.0 at RT, 100 mM EDTA, pH 8.0, 100 mM KCl, 0.05% saponin, 10% glycerol. Add 200 mM β-mercaptoethanol immediately before use.
2. Nuclei washing solution: 10 mM Tris–HCl, pH 8.0 at RT, 100 mM EDTA, pH 8.0. Add 1 mM PMSF immediately before use.
3. Nuclei pre-reaction washing solution: 10 mM Tris–HCl, pH 8.0 at RT. Add 1 mM PMSF just before use.

2.4. Generation of DSBs in the DNA

The requirements will vary depending on the experimental protocol. See Subheading 3.4 for more details.

2.5. Assembly of In Vitro Repair Reactions: "Naked" DNA

1. Pre-reaction washing buffer: 40 mM HEPES-KOH, pH 7.5 at 37°C, 1 mM EDTA, pH 8.0, 20 mM KCl, 3 mM $MgCl_2$.
2. Reaction buffer: 40 mM HEPES-KOH, pH 7.5 at 37°C, 20 mM KCl, 3 mM $MgCl_2$, 10 μM (each) dNTPs (stock 2 mM each), 1.5 mM ATP (stock 100 mM), 1 mM β-mercaptoethanol. To assemble reactions, prepare a 10× solution. Dilute with water to the final volume. Account for the volume of the extract (and of other components that may be added in the reaction) by reducing proportionally the amount of water added.
3. Lysis solution: 10 mM Tris–HCl, pH 7.5 at 50°C, 100 mM EDTA, 50 mM NaCl, 2% NLS, 0.1 mg/mL proteinase.
4. Washing solution: 10 mM Tris–HCl, pH 7.5 at 37°C, 100 mM EDTA, pH 8.0, 50 mM NaCl.
5. RNase A treatment solution: 0.1 mg/mL RNase A in 10 mM Tris–HCl, pH 7.5 at 37°C, 100 mM EDTA, 50 mM NaCl.

2.6. Assembly of In Vitro Repair Reactions: Nuclei

1. Nuclei pre-reaction buffer: 20 mM Tris–HCl, pH 7.5 at RT, 2 mM EDTA, pH 8.0, 50 mM KCl, 5 mM $MgCl_2$.
2. Nuclei reaction buffer: 20 mM Tris–HCl, pH 7.5 at RT, 2 mM EDTA, pH 8.0, 50 mM KCl, 5 mM $MgCl_2$, 10 μM dNTPs, 1.5 mM ATP, 1 mM β-mercaptoethanol. See Subheading 2.5 for details in the use of this solution to assemble reactions.

3. Lysis solution: 10 mM Tris–HCl, pH 8.0 at RT, 100 mM EDTA, pH 8.0, 2% NLS, 0.1 mg/mL proteinase.
4. Washing solution: 10 mM Tris–HCl, pH 8.0 at RT, 100 mM EDTA, pH 8.0, 50 mM NaCl.
5. RNase A treatment solution: 0.1 mg/mL RNase A in 10 mM Tris–HCl, pH 7.5 at 37°C, 100 mM EDTA, pH 8.0, 50 mM NaCl.

2.7. Assay for DNA Damage Using Pulsed-Field Gel Electrophoresis

1. 0.5× TBE: 45 mM Tris–HCl, 45 mM boric acid, 1 mM EDTA, pH 8.2.
2. Agarose, Seakem, FMC.

3. Methods

3.1. Preparation of Cell Extract

The method described here allows the preparation of cell extract from 10 L of cell suspension. Higher or lower amounts of extract can be prepared by appropriate scaling.

1. Grow HeLa cells at 37°C for 3 days in twenty-five 100 mm tissue culture dishes prepared at an initial density of 3×10^6 cells/dish in 20 mL S-MEM supplemented with serum and antibiotics. The final density after 3 days of growth should be ~20×10^6 cells/dish, giving a total of 5×10^8 cells in 25 dishes (see Note 1).
2. Trypsinize the cells from all dishes and resuspend in 10 L of pre-warmed complete growth medium in a 30 L nominal volume microcarrier flask, thoroughly pre-gassed with 5% CO_2 in air. The initial cell concentration should be ~5×10^4 cells/mL. Place in a warm room at 37°C; provide adequate stirring (~60 rpm).
3. Allow cells to grow for 4 days, to a final concentration of $4–6 \times 10^5$ cells/mL. Do not allow cells to exceed this concentration.
4. Collect cells by centrifugation (8 min at $2{,}500 \times g$). Collection should be fast and is best done using a refrigerated centrifuge that can accept 1 L bottles (e.g., Beckman J6-MI). All further processing should be carried out at 0–4°C.
5. Rinse twice in PBS and centrifuge (5 min at $500 \times g$). Determine the packed cell volume (pcv) (~12–15 mL total).
6. Resuspend cell pellet in five pcv volumes of hypotonic buffer solution and centrifuge quickly (10 min, $1{,}200 \times g$). Cells swell and the pcv approximately doubles.
7. Determine the new pcv. Resuspend cell pellet in an equal volume of hypotonic buffer and disrupt in a Dounce homogenizer

(20 strokes, pestle B). It is advisable to test cell disruption using a phase contrast microscope.

8. Add 0.11 volumes of high salt buffer and centrifuge 3,000×*g* for 20 min.
9. Carefully remove supernatant and centrifuge at 100,000×*g* for 1 h.
10. Place the resulting extract (S100) in dialysis tubing with an MW cutoff of 10–14 kDa and dialyze overnight against 50–100 volumes of dialysis buffer.
11. Collect extract. Centrifuge at 10,000×*g* for 20 min to remove precipitated protein. Aliquot and snap freeze. Store at −70°C. Keep a small aliquot for determining protein concentration using the Bradford assay (protein assay, Bio-Rad).

3.2. Preparation of "Naked" DNA from Agarose-Embedded Cells

The procedure described here produces "naked" DNA from agarose-embedded cells with no detectable impurities of protein and RNA (16). The outline is for the preparation of DNA from the human lung carcinoma cell line A549, but it can be modified to prepare DNA from other cell lines (see Note 2).

1. Plate 2×10^6 A549 cells per 100 mm dish for 6 days in 20 mL McCoy's 5A supplemented with serum, antibiotics, as well as with 0.01 μCi/mL ^{14}C-thymidine (methyl-^{14}C-thymidine) and 2.5 μM cold thymidine. Cells reach a plateau after this period of growth, with more than 95% accumulated in G1/G0 phase (if possible verify by flow cytometry). There will be ~2×10^7 cells/dish.
2. Trypsinize cells, collect by centrifugation (5 min 500×*g*), and wash once in serum-free HEPES-buffered McCoy's 5A growth medium.
3. Resuspend cells in serum-free medium at a concentration of 6×10^6 cells/mL. Ensure no clumps are present at this stage. If clumps are visible under the microscope, they should be carefully disrupted using a narrow bore Pasteur pipette, or a syringe with a 19 G needle.
4. Mix with an equal volume of 1% InCert agarose. The final cell concentration is 3×10^6 cells/mL, and the final agarose concentration is 0.5%.
5. Quickly pipette suspension into precooled 3.5 mm diameter glass tubes and incubate on ice until solidification (2–5 min).
6. Gently remove solidified agarose from the tubes and cut 5 mm long blocks (~50 μL per block; ~1.4×10^5 cells per block; ~1.0 μg DNA per block, assuming 5 pg DNA per diploid human G1 cell—A549 cells have 1.4 fold more DNA than human diploid cells; local DNA concentration in the "cage" generated by a cell with a diameter of ~15 μm: ~4 mg/mL). This operation

should be carried out in the cold room. Immerse agarose blocks into an excess of serum-free HEPES-buffered McCoy's 5A growth medium, enough to completely cover them.

7. Transfer agarose blocks to lysis solution (5 blocks per mL) and incubate at 4°C for 45 min.
8. Transfer to 50°C and incubate for 12 h in a moderately shaking water bath (longer incubation is also possible if desired, or if more convenient).
9. Carefully remove lysis solution and rinse once in washing solution (5 blocks per mL).
10. Remove washing solution, add fresh washing solution, and repeat washing for further 6 h with three changes in a cold room with moderate shaking using an orbital shaker.
11. Carefully remove washing solution and treat with RNase A for 1 h at 37°C in a moderately shaking water bath (5 blocks per mL).
12. Rinse agarose blocks in washing solution and transfer to fresh washing solution. Incubate in washing solution at 4°C for 4 more hours under continuous shaking. Exchange washing solution every 2 h.
13. Transfer agarose blocks into fresh lysis solution and incubate at 50°C for 12 h in a moderately shaking water bath (proteinase concentration can be reduced to 0.1 mg/mL at this stage).
14. Wash as described above in steps 9–11. After completion of these steps, RNA and protein are not detectable in the agarose blocks. Agarose blocks are now ready to be used in in vitro repair reactions; they can be stored in washing buffer at 4°C in a dark place for at least 12 months (see Note 3).

3.3. Preparation of Nuclei from Agarose-Embedded Cells

1. Grow A549 cells to a plateau phase as described in step 1 of the protocol described under Subheading 3.2.
2. Trypsinize, wash, and embed cells in agarose as described in steps 2–6 of the protocol described under Subheading 3.2.
3. Carefully transfer agarose blocks to saponin lysis solution and incubate for 45 min at 4°C; then transfer to 50°C for 2 h in a moderately shaking water bath.
4. Remove lysis solution and rinse agarose blocks in nuclei washing solution.
5. Add fresh nuclei washing solution and incubate for 1 h at 37°C in a shaking water bath under moderate shaking. Repeat washing procedure, if necessary. At this stage, agarose blocks are nearly free of cytoplasm, and are ready to be used in repair reactions. Agarose blocks so treated should be used immediately. If necessary, they can be stored overnight at 4°C. Longer storage is not advisable and, if adopted, it should be tested (see Note 3).

3.4. Generation of DSBs in the DNA

DBSs can be induced, either in "naked" DNA or in nuclei embedded in agarose, by different agents. The choice of agent will be determined by the studies to be performed. Ionizing radiation can be used to randomly induce DSBs with non-ligatable ends. If ligatable ends are desired, treatment with rare-cutting restriction endonucleases can be utilized. Aspects of these treatments are described next.

Exposure to X-rays or γ-rays can be carried out *after* embedding the cells in agarose as described above in steps 1–6 of Subheading 3.2, and *before* lysis to generate "naked" DNA or nuclei. Doses in the range of 20–40 Gy should be used with the detection assays outlined here. Alternative detection assays should be utilized if experiments requiring lower doses must be performed. When intact, G1-phase, A549 cells are exposed to radiation, approximately 25 DSBs are produced per cell per Gy of 250 kV X-rays (1.65 mm Al filter). Radiation exposure is carried out with the agarose blocks immersed in serum-free HEPES-buffered McCoy's 5A medium, on ice. After irradiation the agarose blocks are lysed for the preparation of either nuclei or "naked" DNA, as outlined in the corresponding protocols.

When experiments are performed with "naked" DNA, it is more convenient to expose the agarose blocks to radiation *after* completion of the lysis procedure, shortly before the actual experiment. Because "naked" DNA is more sensitive to breakage by ionizing radiation than DNA organized in chromatin, approximately 3–5 fold lower doses should be used (5–10 Gy) in order to generate a number of DSBs equivalent to that produced in intact cells exposed to doses between 10 and 50 Gy (16).

When rare-cutting restriction endonucleases are used to generate DSBs, digestions should be carried out in the buffer recommended by the supplier of the enzyme (frequently cutting restriction endonucleases generate extensive DNA fragmentation that is only partly amenable to ligation under the conditions of the assay). It is advisable to wash the plugs in the enzyme buffer for 2–4 h with at least one change before proceeding with the actual digestion. The agarose barrier makes it necessary to increase the units of enzyme, as well as the treatment time employed. Increase of the treatment time to 4 h and of the enzyme activity units 5–10-fold above what would have been used to digest DNA in solution, provide in general satisfactory results.

3.5. Assembly of In Vitro Repair Reactions: "Naked" DNA

Once cellular extracts and agarose blocks with DSBs are prepared, in vitro reactions to assay rejoining can be assembled.

1. Wash the required number of plugs in pre-reaction buffer (5 agarose blocks per mL) for 2 h at 4°C under continuous, moderate shaking using an orbital shaker. Repeat washing procedure 2–3 times.

2. Transfer agarose blocks (one per 50 μL reaction) in 1.5 mL Eppendorf tubes containing 50 μL (total volume) reaction

buffer supplemented with cell extract (50–100 μg/reaction) (see Notes 4 and 5). Maintain reactions at 4°C during these preparation steps.

3. Incubate at 37°C in a shaking water bath for various periods of time, as required by the experimental protocol.
4. After the required incubation period for repair, transfer agarose blocks into cold lysis solution (0.5 mL per agarose block, in 10 mL round bottom tubes) for 30 min and then to 50°C for 2 h in a moderately shaking water bath.
5. Wash agarose blocks for 1 h at 37°C in washing buffer (2.5 mL per agarose block). Remove washing buffer and treat with RNase A solution (0.5 mL per agarose block) for 1 h at 37°C in a moderately shaking water bath.
6. Remove RNase solution and wash agarose blocks in washing buffer for 2 h at 4°C in a moderately shaking water bath (optional). Repeat wash step with fresh washing buffer (optional). At this point, agarose blocks are ready to be loaded for pulsed-field gel electrophoresis.

3.6. Assembly of In Vitro Repair Reactions: Nuclei

1. Wash the required number of plugs in nuclei pre-reaction buffer for 2 h at 4°C.
2. Transfer agarose blocks in 1.5 mL Eppendorf tubes containing 50 μL nuclei reaction buffer supplemented with cell extract (50–100 μg/reaction). Keep reactions at 4°C during these preparation steps.
3. Incubate at 37°C for various periods of time, as required by the experimental protocol.
4. After the incubation period for repair, transfer agarose blocks into 1 mL cold lysis solution (in 10 mL round bottom tubes) for 30 min and then to 50°C for 12 h in a moderately shaking water bath.
5. Wash agarose blocks for 1 h at 37°C in washing buffer. Repeat wash step with fresh washing buffer. Remove washing buffer and treat with 1 mL RNase A solution for 1 h at 37°C in a moderately shaking water bath.
6. Remove RNase solution and wash agarose blocks in washing buffer for 2 h at 4°C in a moderately shaking water bath (optional). Repeat washing step with fresh washing buffer (optional). At this point, agarose blocks are ready to be loaded for pulsed-field gel electrophoresis.

3.7. Assay for DNA Damage Using Pulsed-Field Gel Electrophoresis

A variety of pulsed-field gel electrophoresis techniques have been developed to quantitate DSBs in cellular DNA. Any one of these methods can, in principle, be used to evaluate in vitro repair in the reactions described above. The choice will depend upon equipment availability and the requirements of the individual experiment. We routinely use Asymmetric Field Inversion Gel Electrophoresis (AFIGE) when a large number of samples need to be evaluated and precise sizing of DNA is not required. When sizing of DNA fragments is considered important, we use Clamped Homogeneous Electric Field (CHEF) gel electrophoresis (16). A detailed description of these methods is beyond the scope of this protocol, and only the parameters routinely used in our laboratory for the quantification of DSBs using the above assays are presented next.

3.7.1. AFIGE

1. Cast a 0.5% agarose (SeaKem or Bio-Rad) gel (250 mL, 20×25×0.5 cm) with 0.5 μg/mL ethidium bromide in 0.5× TBE using the appropriate combs (20 wells, 3.5×6 mm). Allow gel to solidify at RT, transfer to 4°C for 1 h for further solidification. Remove combs just before loading.
2. Load gel by inserting agarose blocks into the wells. Avoid trapping of air in the gel during this step. Close wells with 0.5% agarose.
3. Place gel into the electrophoresis box (horizontal gel electrophoresis system, model H4, Gibco/BRL), containing 2.5 L of precooled (10°C) 0.5× TBE. Cooling is achieved by a cooling unit and a circulating pump.
4. Run at 10°C for 40 h by applying cycles of 1.25 V/cm for 900 s in the direction of DNA migration, and 5.0 V/cm for 75 s in the reverse direction.
5. After completion of electrophoresis, place gel under a UV table and take a photograph for documentation.
6. Dry gel by carefully placing it on filter paper (model 583 gel dryer filter paper, 34×45 cm, Bio-Rad) in a gel dryer (1 h at 80°C), under vacuum (20–25 mmHg).
7. Expose a PhosphorImager screen to the dried gel for 24–48 h (see Note 6).
8. Quantitate DSBs present in the samples by evaluating the amount of activity present in the entire sample over that present in the lane (fraction of activity released; FAR). This parameter is directly related to the number of DSBs present in the DNA.
9. Plot results as a function of time. A typical gel and its quantification are shown in Fig. 2.

3.7.2. CHEF Gel Electrophoresis

CHEF gel electrophoresis is performed in a CHEF DRII apparatus (Bio-Rad) using a 0.8% agarose gel (SeaKem or Bio-Rad) in 0.5× TBE at 10°C for 68 h at 45 V with 60 min pulses. After completion of electrophoresis, the gel is stained with 0.5 μg/mL ethidium bromide. Other details are as described for AFIGE.

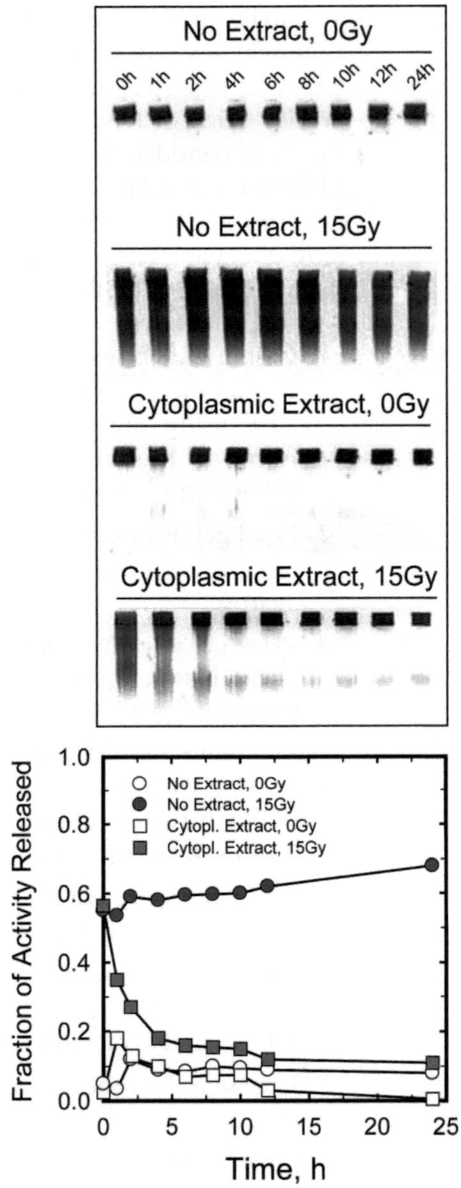

Fig. 2. Analysis of DSBs rejoining in genomic DNA. The *upper panel* shows the series of four gels stained with ethidium bromide. The lower panel shows the results of their quantitation using a phosphor Imager. *Open symbols* represent non irradiated samples, and *closed symbols* represent samples exposed to 15 Gy of X-rays just before assembling the repair reactions. A reduction of FAR indicates an increase in genomic DNA size as expected after DSB repair. Notice that the FAR is very low for samples not exposed to radiation, suggesting low nuclease activity. Notice also that in the irradiated samples the FAR decreases only when extract is present in the reaction.

4. Notes

1. The preparation of a good extract depends strongly on the quality of the cells used. When cell growth is not optimal, or when cells overgrow, high nuclease activity may severely compromise the use of the extract in repair reactions. To ensure optimal growth we routinely take the following measures: (a) carefully test different batches of serum to find one with good growth characteristics. HeLa cells have a generation time of <20 h when grown as a monolayer, and <24 h when grown in suspension, under optimal growth conditions. (b) Use cells grown in dishes to start the suspension cultures for extract preparation. This helps reduce clumping occurring after extensive growth in suspension. (c) We follow cell growth daily, and collect cells for extract preparation when they reach a concentration of $4–6\times10^5$ cells/mL. (d) We measure cell cycle distribution by flow cytometry. A high percentage of S-phase cells (~25% for HeLa cells), suggests that the cell culture is still in an active state of growth.
2. The choice of A549 cells for substrate preparation is arbitrary and is guided by their property to produce cultures with high percentage of G0/G1 cells when reaching the plateau phase of growth. Other cell lines can also be used if desired, but the protocol may require some modifications to accommodate the specific properties of these cells. When optimizing for the preparation of "naked" genomic DNA, the parameter to watch is degradation of non-damaged DNA during incubation under standard reaction conditions, but in the absence of extract; the FAR should be <10%. Relatively high FAR values under these conditions indicate the action of residual nucleases which should be removed by further lysis.
3. A large number of agarose blocks with "naked" DNA can be prepared as a set and used, as needed, over a period of more than 1 year. This gives a great degree of flexibility in the planning of this type of repair experiments. Extensive storage of nuclei embedded in agarose is not possible.
4. Since usually a large number of reactions need to be assembled, we prepare a 10× stock of the reaction buffer by mixing stocks of the individual components. Subsequently, we set up reactions by adding the appropriate amount of 10× reaction buffer (5 μL for a 50 μL reaction), the desired amount of extract, and enough H_2O to reach a final volume of 50 μL.
5. Accurate pH control is essential for the preparation of extracts with high repair and low nuclease activity. Unless otherwise stated, the pH values given in the protocol are for the indicated temperatures. It is important to keep in mind that the pH of Tris-buffered solutions changes significantly with temperature

(it increases as the temperature decreases). Thus, a solution adjusted to a pH of 7.0 at RT will reach a pH of nearly 7.5 at 4°C. We found it very helpful to adjust the pH at the temperature under which each solution will eventually be used.

6. In the method described here, cells are radioactively labeled with ^{14}C-thymidine for quantification. Quantification of radioactively labeled DNA can be carried out either using a scintillation counter or a PhosphorImager. We prefer the latter instrument as it is less labor intensive. It is also possible to use fluorescence detection methods for quantification (e.g., a FluorImager). If extreme sensitivity of detection is required, Southern blotting can also be considered. These alternative detection methods can easily be fitted in the described protocol. However, care should be exercised to ensure linear response for the parameter measured within the range of interest. Otherwise, appropriate calibration curves should be generated.

References

1. Valerie K, Povirk LF (2003) Regulation and mechanisms of mammalian double-strand break repair. Oncogene 22:5792–5812
2. Mladenov E, Iliakis G (2011) Induction and repair of DNA double strand breaks: the increasing spectrum of non-homologous end joining pathways. Mutat Res 711:61–72
3. Lieber MR (2010) The mechanism of double-strand DNA break repair by the nonhomologous DNA end-joining pathway. Annu Rev Biochem 79:1.1–1.31
4. Lieber MR (2010) NHEJ and its backup pathways in chromosomal translocations. Nat Struct Mol Biol 17:393–395
5. Iliakis G, Wang H, Perrault AR, Boecker W, Rosidi B, Windhofer F, Wu W, Guan J, Terzoudi G, Pantelias G (2004) Mechanisms of DNA double strand break repair and chromosome aberration formation. Cytogenet Genome Res 104:14–20
6. Baumann P, West SC (1998) DNA end-joining catalyzed by human cell-free extracts. Proc Natl Acad Sci U S A 95:14066–14070
7. Udayakumar D, Bladen CL, Hudson FZ, Dynan WS (2003) Distinct pathways of nonhomologous end joining that are differentially regulated by DNA-dependent protein kinase-mediated phosphorylation. J Biol Chem 278:41631–41635
8. Huang J, Dynan WS (2002) Reconstruction of the mammalian DNA double-strand break end-joining reaction reveals a requirement for an Mre11/Rad50/NBS1-containing fraction. Nucleic Acids Res 30:1–8
9. Wang H, Zeng Z-C, Bui T-A, Sonoda E, Takata M, Takeda S, Iliakis G (2001) Efficient rejoining of radiation-induced DNA double-strand breaks in vertebrate cells deficient in genes of the RAD52 epistasis group. Oncogene 20:2212–2224
10. Wang H, Perrault AR, Takeda Y, Qin W, Wang H, Iliakis G (2003) Biochemical evidence for Ku-independent backup pathways of NHEJ. Nucleic Acids Res 31:5377–5388
11. Perrault R, Wang H, Wang M, Rosidi B, Iliakis G (2004) Backup pathways of NHEJ are suppressed by DNA-PK. J Cell Biochem 92:781–794
12. Labhart P (1999) Nonhomologous DNA end joining in cell-free systems. Eur J Biochem 265:849–861
13. Ganguly T, Iliakis G (1995) A cell-free assay using cytoplasmic cell extracts to study rejoining of radiation-induced DNA double-strand breaks in human cell nuclei. Int J Radiat Biol 68:447–457
14. Cheong N, Iliakis G (1997) In vitro rejoining of double strand breaks induced in cellular DNA by bleomycin and restriction endonucleases. Int J Radiat Biol 71:365–375
15. Cheong N, Perrault R, Iliakis G (1998) In vitro rejoining of DNA double strand breaks: a comparison of genomic-DNA with plasmid-DNA-based assays. Int J Radiat Biol 73:481–493
16. Cheong N, Okayasu R, Shah S, Ganguly T, Mammen P, Iliakis G (1996) In vitro rejoining of double-strand breaks in cellular DNA by factors present in extracts of HeLa cells. Int J Radiat Biol 69:665–677

Chapter 33

An In Vitro DNA Double-Strand Break Repair Assay Based on End-Joining of Defined Duplex Oligonucleotides

Kamal Datta, Shubhadeep Purkayastha, Ronald D. Neumann, and Thomas A. Winters

Abstract

DNA double-strand breaks (DSBs) are caused by endogenous cellular processes such as oxidative metabolism, or by exogenous events like exposure to ionizing radiation or other genotoxic agents. Repair of these DSBs is essential for the maintenance of cellular genomic integrity. In human cells, and cells of other higher eukaryotes, DSBs are primarily repaired by the nonhomologous end-joining (NHEJ) DSB repair pathway. Most in vitro assays that have been designed to measure NHEJ activity employ linear plasmid DNA as end-joining substrates, and such assays have made significant contributions to our understanding of the biochemical mechanisms of NHEJ. Here we describe an in vitro end-joining assay employing linear oligonucleotides that has distinct advantages over plasmid-based assays for the study of structure–function relationships between the proteins of the NHEJ pathway and synthetic DNA end-joining substrates possessing predetermined DSB configurations and chemistries.

Key words: Nonhomologous end-joining, DNA double-strand break, Multiply damaged site, DNA repair, Radiation, In vitro DNA double-strand break repair assay, Duplex oligonucleotides, DSB end-joining substrates, DNA repair assay methods

1. Introduction

In vitro assays have made important contributions to our current understanding of essentially all DNA repair pathways. Homologous recombination (HR) and NHEJ are the two major pathways involved in the repair of DNA double-strand breaks (DSBs), the most detrimental type of DNA damage (1, 2). In human cells, both HR and NHEJ are active in dividing cells, however, HR appears to have a limited DSB repair role and acts mainly during the S and G_2/M cell cycle phases. In contrast, NHEJ is active throughout the cell cycle and is the mechanism by which most DSBs are repaired. In nondividing cells, NHEJ is the primary

Lotte Bjergbæk (ed.), *DNA Repair Protocols*, Methods in Molecular Biology, vol. 920,
DOI 10.1007/978-1-61779-998-3_33, © Springer Science+Business Media New York 2012

mechanism of DSB repair. This dominant role of the NHEJ pathway in the repair of DNA DSBs in both dividing and nondividing cells, highlights the importance of in vitro assays that have been designed and developed to permit the control and analysis of all aspects of this repair reaction, from defined molecular structures for the DSB substrates that are to be joined, to the choice of proteins to be employed as the source of end-joining activity.

In the majority of in vitro NHEJ assays developed to date, the substrate DNAs that have been used are obtained by restriction enzyme cleavage of plasmid DNA, while end-joining proteins have been derived from cell or tissue extracts.

During the last 10–15 years, significant efforts have been devoted to identifying and understanding the proteins that are involved in NHEJ. As a result of these efforts, the current model for NHEJ recognizes a core complex of five proteins for the primary pathway. In this model, DSB ends are recognized and bound very quickly by a Ku70/Ku80 heterodimer, which serves to protect the damaged ends from degradation and acts to recruit the other proteins of the pathway. Typically, DNA-PK_{cs} (the catalytic subunit of the DNA-dependent protein kinase) is recruited next, and upon binding to the Ku heterodimer and DNA end, forms the active heterotrimeric DNA-dependent protein kinase, DNA-PK. Following alignment and subsequent recruitment of DNA ligase IV/XRCC4 to each DSB end, such that a tetrameric complex is formed, ligation of the ends proceeds restoring DNA integrity (2–4). This pathway is generally referred to as classical NHEJ, and is considered to be the standard core NHEJ reaction pathway (Fig. 1).

In recent years, additional factors have been identified that may also have important roles in the NHEJ process depending on the context under which the end-joining reaction occurs (e.g., VDJ recombination vs. repair of DSBs formed by DNA damaging agents). These additional potential protein cofactors include DNA polymerases μ and λ, the artemis endonuclease, and the DNA ligase IV stimulatory factor XLF/Cernunnos (5–7).

In addition, factors such as PARP-1, XRCC1, DNA ligase III, and histone H1 have been shown to participate in an alternative, or backup NHEJ pathway (7–9). Furthermore, it has also been indicated that MRE11, Rad51, and NBS1 may have some context dependent roles in classical NHEJ, as well as possible roles in an alternative backup NHEJ pathway (10). Although many of these factors were discovered by genetic studies, it has been the use of in vitro assays that have validated many of their specific roles in NHEJ.

Although much progress has been made in understanding the functions of the core pathway proteins in NHEJ, the full range of proteins that participate in the reaction have yet to be determined. This is especially true when considering the repair of structurally complex DSBs such as those formed by genotoxic agents like

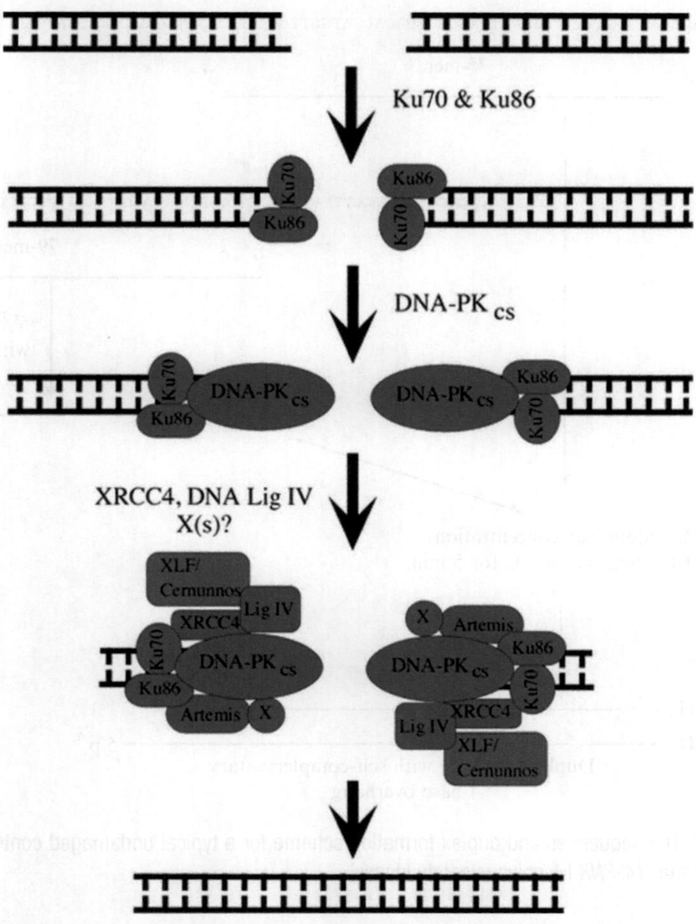

Fig. 1. Schematic representation of the steps thought to constitute the "classical" pathway of NHEJ, as well as the major factors known to participate in the process. The 'X' in the scheme represents proteins involved in the process that are yet to be determined.

radiation, the repair of which is likely to require proteins capable of removing nucleotide fragments from DSB ends as well as permitting repair of DSB ends with nearby damaged nucleotides. We and others have shown that DSBs induced by radiation or other genotoxic agents not only exhibit scission of the two DNA strands, but may also be accompanied by associated damage vicinal to the broken ends (11–13). Although restriction enzyme cut plasmids typically possess defined DSB ends, these ends are chemically simple and lack the structural complexity and damage diversity observed for radiation-induced DSBs (11, 14).

Here we describe an in vitro DSB repair assay using duplex linear oligonucleotides (Fig. 2). We have previously established that the substrate DNA used in our assay binds to the core proteins of the NHEJ pathway, and when exposed to human cell extracts undergoes an end-joining process that is dependent on these proteins (15). Being a synthetic oligonucleotide, the substrate

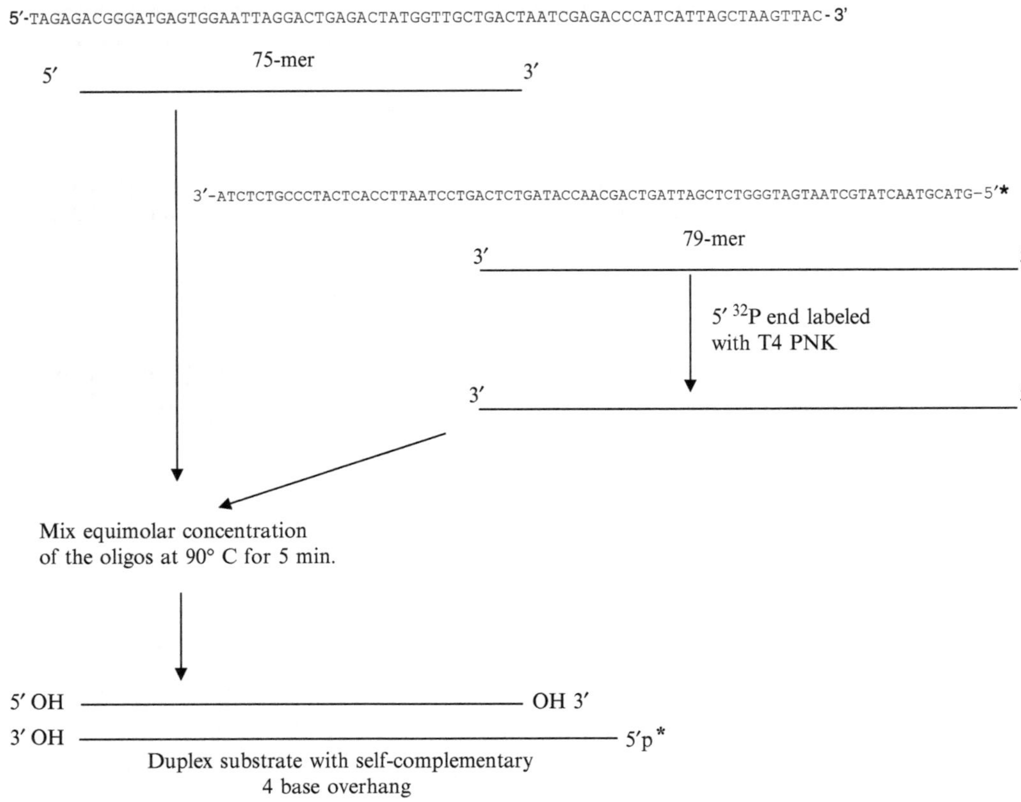

Fig. 2. The sequences and duplex formation scheme for a typical undamaged control duplex oligonucleotide end-joining substrate. *T4-PNK* T4-polynucleotide kinase.

DNA used in this assay cannot only be assembled quickly and easily, it can be chemically modified to reflect a full range of genotoxic lesions and structurally complex combinations observed in DNA that has been damaged by authentic DNA damaging agents such as ionizing radiation (Fig. 3). Furthermore, these substrates are designed such that end-joining preferentially takes place at only one end of the structure, thus simplifying interpretation and analysis of results obtained for the effects of structural modifications of the substrates on NHEJ, as well as making them useful in identifying new proteins that may be important in the processing of these modifications (16). The assay described here employs a 75 bp (Fig. 2) oligonucleotide substrate standardized by us for NHEJ (15). Although nuclear extracts may be used as a source of end-joining proteins for the assay, we have frequently used HeLa whole cell extracts as the source of these proteins.

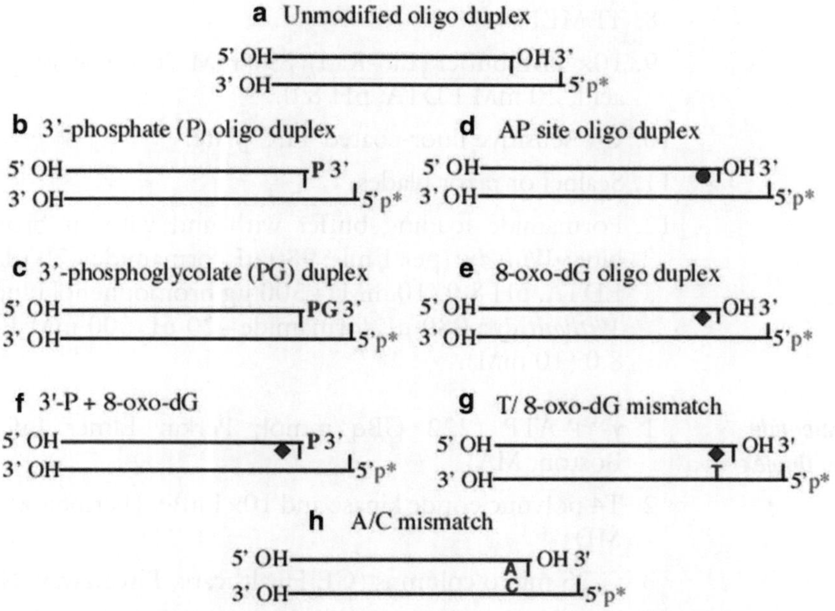

Fig. 3. Examples of duplex oligonucleotide constructs that we have created and employed in the assay (16) to simulate features of genotoxic lesions and structurally complex lesion combinations observed in DNA DSBs that have been caused by authentic DNA damaging agents such as ionizing radiation.

2. Materials

Sterile MilliQ quality water (18 MΩ resistance at 25°C (sddH$_2$O)) should be used for all procedures, as should analytical grade reagents.

2.1. Oligonucleotide Synthesis

All oligonucleotide synthesis was carried out using an ABI-394 DNA/RNA synthesizer (Applied Biosystems, Foster City, CA, USA). All solid phase deoxynucleotide columns and all other reagents for oligonucleotide synthesis were purchased from Glen Research (Sterling, VA).

2.2. Oligonucleotide Purification Electrophoresis

1. Life Technologies slab sequencing gel electrophoresis apparatus (Model No. S2).
2. Glass plates.
3. 1 mm spacers.
4. 5-Well 1 mm comb.
5. BioRad PowerPac™ 3000 power supply with temperature probe.
6. SequaGel sequencing system obtained from National Diagnostics (Atlanta, GA).
7. Ammonium persulfate.

8. TEMED.
9. 10× TBE buffer (Bio-Rad): 890 mM Tris–base, 890 mM boric acid, 20 mM EDTA, pH 8.0.
10. UV sensitive fluor-coated TLC plate.
11. Scalpel or razor blades.
12. Formamide loading buffer with and without bromophenol blue: *With dye* (per 1 ml): 980 μL formamide + 20 μL 500 mM EDTA, pH 8.0 (10 mM) + 500 μg bromophenol blue (0.05%). *Without dye*: 980 μL formamide + 20 μL 500 mM EDTA, pH 8.0 (10 mM).

2.3. Oligonucleotide Labeling and Duplex Formation

1. γ-^{32}P-ATP (222 GBq/mmol; Perkin Elmer Life Science; Boston, MA).
2. T4 polynucleotide kinase and 10× buffer (Fermentas; Hanover, MD).
3. G-25 micro columns (GE Healthcare; Piscataway, NJ).
4. Novex 6% TBE polyacrylamide gels (Invitrogen; Carlsbad, CA).

2.4. Cell Culture

1. Dulbecco's modified Eagle's medium (DMEM).
2. Fetal bovine serum.
3. Nonessential amino acids (10 mM).
4. Glutamine (100 mM).
5. Penicillin/streptomycin ((10,000 U/mL) (Invitrogen; Carlsbad, CA)).
6. HeLa cells were obtained from American Type Tissue Culture (ATCC, Manassas, VA).
7. Laminar flow cell culture hood.
8. Cell culture incubator maintained at 37°C in air plus 5% CO_2.
9. Cell extraction buffer: 10 mM HEPES, pH 7.9, 60 mM KCl, 1 mM DTT, 1 mM EDTA, 1 mM Pefabloc, 1 μg/mL aprotinin, 1 μg/mL leupeptin, 10 μg/mL bestatin.

2.5. Repair Assay Components

1. Bio-Rad Protein assay system (Hercules, CA).
2. T4 DNA ligase (Fermentas; Hanover, MD).
3. 5× End-joining reaction mix buffer: 500 mM Tris–HCl, pH 7.6, 25 mM $MgCl_2$, 5 mM ATP, 5 mM dithiothreitol (DTT), 25% polyethyleneglycol (PEG) 8000.
4. 1 M Tris–HCl, pH 8.0.
5. Thermal cycler, accurate heater blocks, or circulating water baths capable of maintaining steady temperatures of between 17 and 65°C.

6. 10× Stock protease inhibitor cocktail containing 10 μg/mL aprotinin, 10 μg/mL leupeptin, 100 μg/mL bestatin. Pefablock should be added to a final concentration of 1 mM (from a 10 mM stock) in all final 10× protease mixtures and to all 1× reaction mixes/extraction buffers, where needed, just prior to using these components.

2.6. Repair Reaction Electrophoresis

1. Novex mini electrophoresis system.
2. 8% TBE gels (Invitrogen; Carlsbad, CA), and/or self-prepared 20% acrylamide denaturing 0.25 mm sequencing gels (Life Tech Model S2) prepped using the SequaGel sequencing system.
3. Bio-Rad PowerPac™ 3000.
4. 1× TBE buffer, pH 8.4.
5. Formamide loading buffer with and without bromophenol blue.
6. Fuji PhosphorImager 2500 and FujiFilm Image Gauge v3.45 software.

3. Methods

3.1. Oligonucleotides Synthesis and Purification

1. Without going into the lengthy process of synthesis and purification, the unmodified oligonucleotides used in this assay can be purchased directly from commercial sources (see Note 1). As we synthesized our own oligonucleotides, the process is described briefly here.
2. Oligo sequences were entered into ABI-394 DNA/RNA synthesizer, appropriate solid phase deoxynucleotide columns were selected, and all chemicals needed by the synthesizer to complete the syntheses were loaded onto the machine as per the standardized ABI synthesizer protocols.
3. Run synthesizer overnight and recover synthesis products in standard 3 mL borosilicate glass vials and stopper with Teflon lined screw caps.
4. Deprotect the newly synthesized oligonucleotides for 8 h at 55°C in a heater block, then recover and allow to cool to room temperature (RT).
5. Transfer to 1.5 mL plastic Eppendorf style tube and pierce cap multiple times with a 26 gauge hypodermic needle, freeze in liquid nitrogen, then lyophilize with centrifugation.
6. Dissolve lyophilize oligonucleotides in 50 μL of $sddH_2O$ and pass through a G-50 spin column for desalting according to the manufacturer's instructions.

7. Run a 12% acrylamide preparatory (1 mm thickness) PAGE gel in 1× TBE. Pre-run the gel for 30 min, or until the gel temperature reaches 45–50°C. Mix the crude desalted oligonucleotide synthesis products with an equal volume of sequencing gel loading buffer without bromophenol blue dye, and incubate at 90°C for 3 min. Load samples onto the gel and run at 70 W constant power to achieve a steady running temperature of 55°C (this can also be accomplished by running with a constant set temperature of 55°C when using the gel plate temperature probe option). As a run length reference marker, load one lane with loading buffer (containing bromophenol blue) alone.
8. Run gel for 1–2 h, or until the marker dye reference has migrated 65–75% of the gel length. Carefully separate the gel plates using a plate separator or similar tool, and transfer the gel onto a piece of saran wrap. Place a UV-light sensitive fluor-coated TLC plate below the gel and view oligonucleotide bands by UV shadowing using a handheld short-wave UV light (254 nm).
9. To recover band-purified oligonucleotides, cut the shadowed area of the isolated full-length oligonucleotides from the preparative gel using a sterile scalpel or an ethanol flamed razor blade. Transfer the gel pieces to 15 mL round-bottom polypropylene tubes and crush thoroughly using a glass pestle. Add either 1 mL of sddH$_2$O, or gel extraction buffer (typically TE pH 8.0) and allow to stand at RT overnight.
10. Oligonucleotides extracted from the gel in the previous step are recovered by spin-filtration using a Costar Spin-X centrifuge tube filter (0.22 μm pore). Load extraction mixture (supernatant and gel debris) into an appropriate number of 500 μL Spin-X devices and centrifuge at 1,400×*g* for 5 min. Recover the filtrates containing the band-purified oligonucleotides and transfer to sterile 1.5 mL Eppendorf tubes.
11. After piercing the caps of the tubes containing the recovered filtrates multiple times with a 26-gauge hypodermic needle, the samples should be frozen on dry ice and lyophilized to dryness under centrifugation. Dissolve the recovered powder in 100 μl of sddH$_2$O and pass through a G-50 spin column to desalt. Determine the concentration of the recovered band-purified oligonucleotides by UV absorbance (see Note 2), and store at −80°C until used.

3.2. Oligonucleotide Labeling and Duplex Formation

1. To 5′-^{32}P end label the 79-mer oligonucleotide that will eventually serve as the lower strand for the duplex substrate, mix 10 pmoles of 79-mer oligonucleotide (Fig. 2), 1 μL 10× T4 PNK buffer (500 mM Tris–HCl (pH 7.6 at 25°C), 100 mM MgCl$_2$, 50 mM DTT, 1 mM spermidine), 50 pmoles of

γ-^{32}P-ATP, 1 μL (10 U/μL) T4 PNK, and sddH_2O to a final volume of 10 μL.

2. Incubate the mixture at 37°C for 45 min in a thermocycler.
3. Incubate at 70°C for 10 min to heat inactivate T4 PNK.
4. Prepare a G-25 spin column as per the manufacturer's instructions and load the column drop-wise to the very center of the bed, being careful to not allow the sample to touch the column walls or flow around the edges of the top surface of the matrix (all spin columns should always be loaded in this manner). Centrifuge the column as indicated below to remove the unincorporated γ-^{32}P-ATP.
5. Place the column in a DNase-free 1.5 mL tube and spin it for 2 min at 735 × *g* in a fixed angle microfuge.
6. The filtrate containing the purified oligonucleotide sample is collected in the 1.5 mL tube and is ready to be used for duplex formation (see Note 3).
7. To form duplexes, mix the labeled 79-mer (10 pmoles) with an equimolar (10 pmoles) amount of band-purified upper strand 75-mer oligonucleotide (Fig. 2) in 10 mM Tris–HCl, pH 8.0 and add sddH_2O to make the volume 20 μL.
8. Incubate at 90°C for 5 min in a thermocycler.
9. Allow the reaction mix to gradually cool to RT. This can best be achieved by creating a cool-down program for your thermocycler such that the duplex annealing reaction cooling rate is reproducible and controlled, and proceeds at a rate of no more than 1°C every 2 min until reaching 20°C (or RT). The product of this annealing reaction is the duplex oligonucleotide end-joining substrate that will be used in the DSB end-joining reactions described below. The annealed duplexes can be stored at −80°C until used (see Note 4).
10. To confirm duplex formation, load 5 μL of the sample mixed with formamide loading buffer onto a 6% nondenaturing polyacrylamide gel and run in 1× TBE buffer at 40 W constant until the dye front reaches the lower third of the gel. In a parallel lane, run the single-stranded 5′-end-labeled 79-mer as a negative control.
11. Duplex formation should be documented by Phosphorimager. Assuming that the 5′-end-labeling efficiency of the lower-strand 79-mer was determined following the initial end-labeling reaction, the images obtained by phosphorimaging can be analyzed by suitable gel analysis software (e.g., Image Gauge software) and the duplex yields can be quantified as a function of the amount of 79-mer expected in the imaged duplex. The duplex will move slower than the 79-mer oligo and the duplex band will appear above the monomer (Fig. 4).

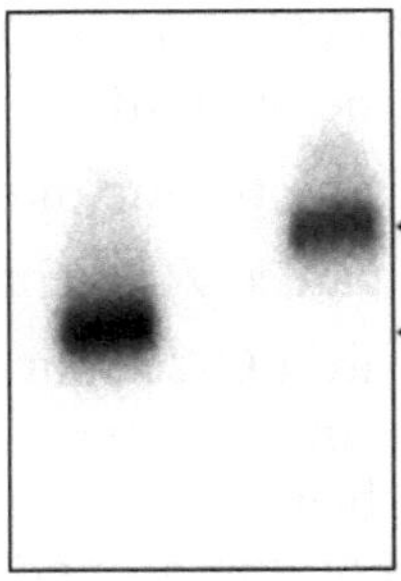

Fig. 4. Duplex formation. The faster migrating band in the left lane is the single-stranded 5′-^{32}P end-labeled 79-mer monomer and moves faster than the annealed 75-mer/79-mer duplex shown in the lane on the *right*.

3.3. Cell Culture and Preparation of Whole Cell Extracts

1. To prepare complete medium, mix 50 mL heat-inactivated fetal bovine serum (10%, v/v), 5 mL nonessential amino acids (1%, v/v), and 5 mL penicillin/streptomycin (1%, v/v) with 500 mL DMEM.
2. Filter the complete medium through a 0.22 μm filter using a sterile prepackaged vacuum filtration device such as a Corning filter system.
3. Quickly thaw a frozen working-stock vial of HeLa cells (0.5 mL cryopreserved cell suspension, typically at ~1–2 × 10^7 cells/mL) at 37°C. Place into a 15 mL conical bottom Falcon tube and slowly add 10 mL of complete medium in a drop-wise fashion with periodic gentle vortex mixing.

   Centrifuge at 800 × *g* for 5 min.
4. Discard the medium, resuspend the cell pellet in 10 mL of fresh complete medium, and then transfer the cell suspension to a T-75 tissue culture flask.
5. Incubate cells at 37°C in air plus 5% CO_2 environment and grow as monolayers up to 80–90% confluence.
6. Harvest cells by aspirating off the medium and washing the monolayers with sterile PBS. Following aspiration of the PBS, 1.0 mL of 0.25% (w/v) trypsin solution warmed to 37°C is added and the flasks are returned to the incubator for several minutes (3–4) to facilitate trypsinization and release of the cells from the flask.
7. Trypsinized monolayers should be observed under the microscope to monitor cell detachment.
8. Once detached, add 10 mL of complete medium into the trypsinized cells. Make a single cell suspension by pipetting up and down.
9. Count cells using an automated or manual cell counter and record.

10. Centrifuge cell suspension at 800 × *g* for 5 min and discard the supernatant media.
11. Wash the cell pellet by resuspending and dispersing in ice-cold PBS. Pellet cells by centrifugation at 800 × *g* for 5 min and remove supernatant by aspiration. Repeat this step a second time.
12. Resuspend the cell pellet in cell extraction buffer at a density of 8×10^7 cells/mL.
13. Lyse cells by the freeze/squeeze method: place the cells in dry ice ethanol bath until frozen solid, then immediately transfer the cells to a 37°C water bath for 1–2 min with gentle agitation/mixing (or until fully thawed) and repeat the process for a total of three cycles. After the third thaw cycle, centrifuge at 16,000 × *g* for 30 min at 4°C.
14. Carefully collect the supernatant without disturbing the pelleted cell debris and discard the cell debris pellet. The supernatant constitutes the HeLa whole cell extract (WCE) and should be flash frozen in liquid nitrogen (LN_2) in small aliquots (20–100 μL). The aliquots can then be stored at −80°C until needed (see Note 5).
15. Hold back a small (100 μL) aliquot on ice for protein concentration determination.
16. Determine protein concentration by a reliable method such as the Bradford method using a Bio-Rad protein assay kit according to the manufacturer's instructions.

3.4. Repair Assay Procedure

1. Prepare on ice (see Note 6), a volume of repair reaction premix sufficient for the number of reactions to be run by multiplying the following single reaction composition by the amount of reactions you are planning to run (see Note 7). Combine the following for each 10 μL end-joining reaction to be run. To a 0.5 mL thin-walled PCR tube add: 2 μL of 5× end-joining reaction mix buffer, 1 μL of 10× protease inhibitor cocktail mixture, 10 ng substrate oligo duplex DNA and sddH_2O such that the final premix volume is between 10 and 20% of the total complete reaction volume of 10 μL (i.e., 8–9 μL). This mixture constitutes the end-joining reaction buffer premix. Once the premix has been prepared, add of up to 5 μg HeLa WCE protein and/or purified proteins as appropriate per reaction (see Note 8). Bring the final volume up to 10 μL with sddH_2O (see Note 9), and initiate the end-joining reactions by transferring the reaction tube from ice to a thermocycler set for the reaction temperature of choice ranging between 17 and 37°C (see Note 10).
2. Incubate the reaction mix at 17°C for 30 min (see Note 11) in a thermocycler (see Note 12).

3. Stop the reaction by incubating the mix at 65°C for 15 min.
4. Add 8 μL of formamide loading buffer (with dye) to each sample and incubate at 90°C for 5 min, then immediately transfer to ice (see Note 13).
5. Resolve the end-joined repair products in an 8% (0.4 mm) denaturing polyacrylamide gel, or in a 20% (0.25 mm) denaturing polyacrylamide gel (depending on resolution needs). Run the gel at 40 W constant power in 1× TBE buffer (pH 8.4), or at a constant temperature of 70°C if a temperature sensing electrophoresis apparatus is being used.
6. Visualize gels (Fig. 5) by phosphorimaging (in our case we use a Fuji PhosphorImager 2500 and analyze images using Fuji Image Gauge (v3.45) software).
7. To quantify the relative product yields (amounts of end-joined 154-mer dimers produced) of the reactions, computer-aided densitometric quantification of the ^{32}P-radiolabeled oligonucleotide substrate and product bands observed in each lane should be done as follows:

   % Relative end-joining = ((dimer band density (product))/(dimer band density + monomer band density (substrate))) * 100.

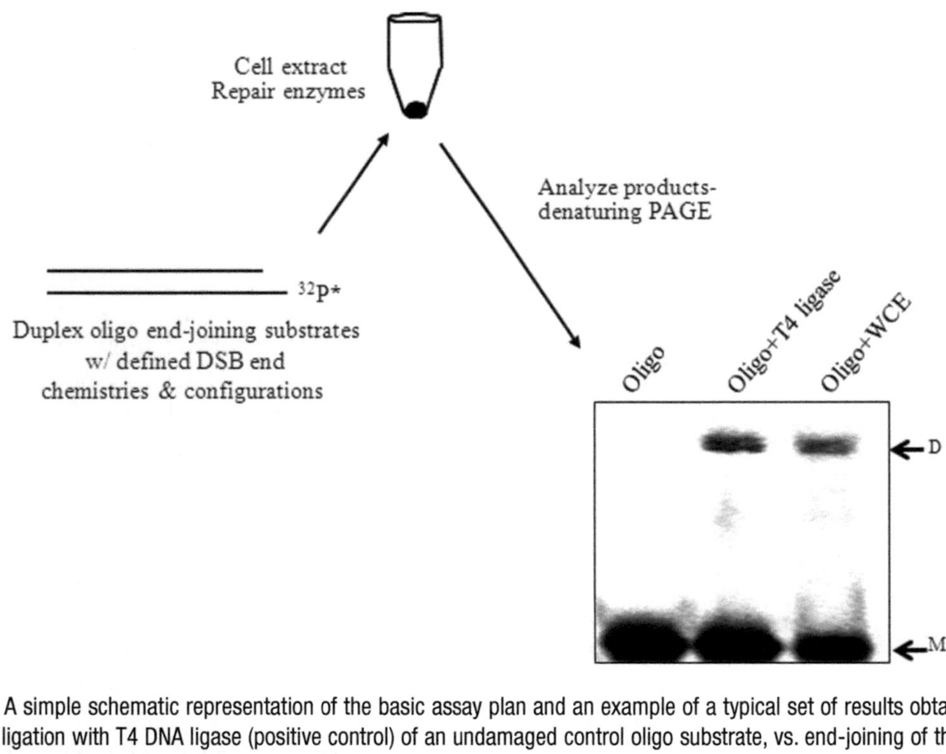

Fig. 5. A simple schematic representation of the basic assay plan and an example of a typical set of results obtained for direct ligation with T4 DNA ligase (positive control) of an undamaged control oligo substrate, vs. end-joining of the same substrate by HeLa WCE proteins. The 154 bp dimer products of the reaction are indicated (D), the unreacted duplex substrate monomer oligo of the negative control reaction is also indicated (M).

8. Repair fidelity may also be measured in this assay as a function of the sensitivity of the end-joined dimer products of the reaction to re-cleavage by the restriction enzyme CviQI (see Note 14) at the site of the self-complementary 4-base overhang at the joining site of the original substrate monomers (see Note 15).

4. Notes

1. Purified oligos can also be purchased directly from commercial sources such as IDT, Invitrogen, genescript, Blue Heron, etc. Specialized syntheses and/or postproduction chemistry may not be available from all vendors and may require special orders/contracts.
2. Determine oligonucleotide concentrations spectrophotometrically by UV absorbance at 260 nm (in a 1 cm cuvette) according to the following formula:

$$[\text{concentration, in mM}] = (A_{260} \times \text{dilution factor} \times 33{,}000) / \text{MW of oligo},$$

where "33,000" is a conversion factor from mass to concentration, and "MW" is the molecular weight of the oligonucleotide, which can be calculated using the formula $\text{MW} = (A \times 312.2) + (C \times 288.2) + (G \times 328.2) + (T \times 303.2) - 61$.
3. At this stage, end-labeling efficiency can be assessed by scintillation counting in order to obtain an objective measure of the amount of duplex formed in subsequent steps of this protocol, as well as to quantitatively establish the amount of duplex substrate being used in end-joining DSB repair reactions.
4. However, to minimize potential radiolysis and other damage that might result from prolonged storage of the radiolabeled oligonucleotides, end-labeled duplexes should be employed as substrates in DSB end-joining repair reactions as soon as practical.
5. Avoid repeated freezing and thawing of WCEs, as this will typically seriously degrade the end-joining activity of the samples after as little as 2–3 freeze thaw cycles. However, if samples must be refrozen and reused, always quick-thaw and immediately place on ice, keep on ice while in use, then flash-freeze in LN_2 before returning to −80°C for storage.
6. All reactions mixes, buffers, and protein solutions are always prepared on ice and kept on ice while in use. When not in use, reaction mixes and buffers are typically stored cold or frozen at −20°C, while most protein solutions containing enzymatically active or DNA functional proteins are generally flash frozen in LN_2 and stored at −80°C.

7. Prep the premix, such that all components except a predetermined volume of sddH$_2$O equivalent to the expected maximum volume of cell extract or repair proteins to be incorporated per reaction are combined in a single pot and mix.
8. This assay was developed to permit the easy incorporation of substrates designed to possess various chemical modifications of our choice (genotoxic damage simulants, e.g., Fig. 3), and to permit monitoring of the effects these modifications on DSB end-joining reactions, particularly those of the human NHEJ pathway. Although we have standardized the assay with HeLa whole cell extracts, repair proteins used in the assay may be from any source. We have used extracts of other cell lines (16), nuclear extracts, and purified proteins, as well as tissue extracts (unpublished observations) as enzyme sources in our experiments. However, when a previously untested source of end-joining proteins is being evaluated, the protein concentration required for the assay has to be optimized for each extract type. Furthermore, different extraction protocols may be required for different sources (e.g., tissue samples).
9. The volume held back from this single pot premix should not exceed 10–20% of the total complete reaction mix final volume. For example, if a total of 10, 10 μL reactions are to be run, no more than 10–20 μL of sddH$_2$O volume should be held back from the final premix volume. Therefore, in the example given here, if 20 μL of sddH$_2$O volume is withheld from the reaction premix (i.e., 20% of the total final reaction volume), then the final volume of the premix will be 80 and 8 μL of this premix should be aliquoted to each end-joining reaction tube to be run. Cell extract and/or other repair proteins to be analyzed should be added to each reaction tube on ice as needed, such that their volumes do not exceed 2 μL, any volume <2 μL should be made up with sddH$_2$O as needed on a case-by-case basis. Furthermore, to accommodate any variability in pipetting accuracy, the total number of reactions planned for should always exceed the actual number to be run by at least 1–2 reaction volumes. This extra volume will prevent the experimenter from running out of premix due to small fractional losses during pipetting prior to completing their experimental set-up. Finally, use of this single-pot reaction premix strategy will maximize the uniformity of each individual end-joining reaction to the next being performed within each experiment/experimental set-up.
10. In our experience, the optimal reaction temperature for WCEs permitting the best efficiency with the highest product yields and least substrate degradation is 17°C. However, depending on need, we have demonstrated effective results when running reactions at temperatures ranging between 17 and 37°C. Also, when dealing with large numbers of reactions, this initiation

method provides more uniform reaction start times than those obtained by sequentially adding enzymes to pre-warmed reaction mixes.

11. Reaction times may vary as needed, up to incubation times of 16 h and beyond with little or no loss of signal for most reactions run at 17°C.
12. For negative controls, run mock repair assays with duplex alone without added repair proteins. Make up missing protein volume with sddH_2O. Positive controls for dimer formation are generated by direct ligation of the duplex oligonucleotide monomer substrates with T4 DNA ligase (1 U/reaction) instead of cell extract.
13. Load the whole repair assay reaction mix onto the gel for better product detection.
14. The duplex is designed in such a way that the 5′ overhang, when end-joined, generates a unique CviQI restriction site.
15. The substrate DNA duplex is designed in such a way that one end possesses an unligateable blunt 5′-OH end, while the other consists of a ligateable (3′-OH/5′-PO_4) self-complementary 4 base 5′-over hang on which the overhanging 5′-end is phosphorylated with ^{32}P for phosphorimaging detection. This configuration allows end-joining at the self-complementary 5′-overhang end, but no end-joining at the opposite blunt end. Consequently, the only end-joining products formed in this assay are 154 bp dimers of the duplex substrate monomers. The simplicity of this arrangement makes analysis and quantification of the end-joining assay results easy and unambiguous.

Acknowledgements

This work was supported in part by the Intramural Research Program of the NIH through the Warren Grant Magnuson Clinical Center.

References

1. Chu G (1997) Double strand break repair. J Biol Chem 272:24097–24100
2. Jackson SP (2002) Sensing and repairing DNA double-strand breaks. Carcinogenesis 23:687–696
3. Lieber MR, Ma Y, Pannicke U, Schwarz K (2003) Mechanism and regulation of human non-homologous DNA end-joining. Nat Rev Mol Cell Biol 4:712–720
4. Lee KJ, Huang J, Takeda Y, Dynan WS (2000) DNA ligase IV and XRCC4 form a stable mixed tetramer that functions synergistically with other repair factors in a cell-free end-joining system. J Biol Chem 275:34787–34796
5. Ahnesorg P, Smith P, Jackson SP (2006) XLF interacts with the XRCC4-DNA ligase IV complex to promote DNA nonhomologous end-joining. Cell 124:301–313

6. Darroudi F, Wiegant W, Meijers M, Friedl AA, van der Burg M, Fomina J, van Dongen JJ, van Gent DC, Zdzienicka MZ (2007) Role of Artemis in DSB repair and guarding chromosomal stability following exposure to ionizing radiation at different stages of cell cycle. Mutat Res 615:111–124
7. Akopiants K, Zhou RZ, Mohapatra S, Valerie K, Lees-Miller SP, Lee KJ, Chen DJ, Revy P, de Villartay JP, Povirk LF (2009) Requirement for XLF/Cernunnos in alignment-based gap filling by DNA polymerases lambda and mu for nonhomologous end joining in human whole-cell extracts. Nucleic Acids Res 37:4055–4062
8. Rosidi B, Wang M, Wu W, Sharma A, Wang H, Iliakis G (2008) Histone H1 functions as a stimulatory factor in backup pathways of NHEJ. Nucleic Acids Res 36:1610–1623
9. Wang H, Rosidi B, Perrault R, Wang M, Zhang L, Windhofer F, Iliakis G (2005) DNA ligase III as a candidate component of backup pathways of nonhomologous end joining. Cancer Res 65:4020–4030
10. Zha S, Boboila C, Alt FW (2009) Mre11: roles in DNA repair beyond homologous recombination. Nat Struct Mol Biol 16:798–800
11. Datta K, Neumann RD, Winters TA (2005) Characterization of a complex 125I-induced DNA double-strand break: implications for repair. Int J Radiat Biol 81:13–21
12. Datta K, Neumann RD, Winters TA (2005) Characterization of complex apurinic/apyrimidinic-site clustering associated with an authentic site-specific radiation-induced DNA double-strand break. Proc Natl Acad Sci U S A 102:10569–10574
13. Ward JF (1994) The complexity of DNA damage: relevance to biological consequences. Int J Radiat Biol 66:427–432
14. Ward JF (1985) Biochemistry of DNA lesions. Radiat Res Suppl 8:S103–S111
15. Datta K, Neumann RD, Winters TA (2006) An in vitro nonhomologous end-joining assay using linear duplex oligonucleotides. Anal Biochem 358:155–157
16. Datta K, Purkayastha S, Neumann RD, Pastwa E, Winters TA (2011) Base damage immediately upstream from double-strand break ends is a more severe impediment to nonhomologous end joining than blocked 3′-termini. Radiat Res 175:97–112

Part V

DNA Damage Tolerance and Checkpoints

Chapter 34

Assays of Bypass Replication of Genotoxic Lesions in Cell-Free Extracts

Nana Nikolaishvili-Feinberg and Marila Cordeiro-Stone

Abstract

The in vitro replication assay described here measures bidirectional replication of a circular double- stranded DNA template upon initiation at the SV40 origin. It models a single eukaryotic replication unit (replicon) and recapitulates the biochemical steps involved in the catalysis of both leading and lagging strand synthesis during semiconservative DNA replication. Except for the SV40 large T antigen, all other proteins necessary for initiation and assembly of functional replication forks are provided by the cell-free extract. This assay can be used to demonstrate bypass replication of genotoxic lesions. It supports replication across a specific damaged site on the template DNA (i.e., translesion synthesis) by specialized DNA polymerases. This chapter illustrates the efficient translesion synthesis of UV-induced thymine dimers by DNA polymerase eta.

Key words: Translesion synthesis, Pyrimidine dimers, Ultraviolet light, DNA polymerase eta, SV40 large T antigen, DNA replication, Human cells

1. Introduction

Bypass replication of genotoxic lesions denotes the capability of organisms to complete genome replication in spite of the presence of template lesions. This term does not specify the mechanism by which the template lesion is tolerated. Some pathways depend either on fork reversal and template switching to bypass the lesion, or homologous recombination to patch daughter strand gaps left opposite the lesion (reviewed in refs. 1, 2). However, one of the most efficient mechanisms of DNA damage tolerance is translesion synthesis (TLS) through the recruitment of specialized DNA polymerases to stalled replication forks. Depending on the structure of the template lesion, one or two of these polymerases are used to overcome the replication block (reviewed in refs. 3, 4). For instance,

Lotte Bjergbæk (ed.), *DNA Repair Protocols*, Methods in Molecular Biology, vol. 920,
DOI 10.1007/978-1-61779-998-3_34, © Springer Science+Business Media New York 2012

the most common DNA lesion induced by ultraviolet light (UV), the cyclobutane pyrimidine dimer, can be bypassed by DNA replication forks following polymerase switching from the replicative DNA polymerase (delta or epsilon) to DNA polymerase eta. This enzyme, however, is not capable of replicating past the UV-induced [6–4] pyrimidine–pyrimidone adduct. A current model for TLS of these adducts calls for the action of two polymerases, one to add a potentially mispaired nucleotide opposite the 3′ base of the adduct, which could be done by DNA polymerase eta (or a different Y-family polymerase), and another to extend the 3′ end beyond the adduct, a step most likely catalyzed by DNA polymerase zeta (5). DNA polymerase kappa, on the other hand, is required for TLS of benzo[a]pyrene-diol-epoxide adducts (6).

The SV40 origin/large T antigen-dependent in vitro replication assay has been a powerful tool in the investigation of essential DNA replication factors. Historically, it supported the characterization of the minimum number of proteins needed to carry out the complete duplication of double-stranded DNA molecules (reviewed in ref. 7). The circular DNA substrates are quite stable under storage conditions and during the in vitro incubation time used to detect fully replicated daughter DNA molecules. These substrates carry a single SV40 origin of replication to which the SV40 large T antigen binds and directs the assembly of human proteins into functional replication complexes. The dependence of the assay on the minimal SV40 origin sequence and large T antigen increases significantly the efficiency of in vitro replication. It should be recognized, however, that the SV40 large T antigen is also a potent helicase; therefore, the initiation step and the rate of displacement of the replication fork on the undamaged DNA are not the same as if only proteins present in the human cell extract were involved in all in vitro replication steps. Even though histones do bind to the double-stranded circular DNA substrates, the resulting structure in vitro does not fully reproduce nuclear chromatin.

The opportunity to insert a defined DNA lesion at a specific site on the double-stranded DNA substrate created the unique opportunity to investigate mechanisms of TLS and the specialized proteins involved in this process (8–12). The assay supports efficient TLS of the cyclobutane thymine dimer ([c,s]TT) by DNA polymerase eta (13). It was instrumental in documenting the TLS defect in xeroderma pigmentosum variant (XP-V) fibroblasts (12) and in supporting the discovery of DNA polymerase eta (14). The protocol described here was used to compare the replication in vitro of double-stranded circular DNA substrates carrying a site-specific [c,s]TT or a [6–4] pyrimidine–pyrimidone adduct between two adjacent thymines ([6–4]TT). Semiconservative replication of double-stranded DNA molecule containing a single lesion leads to the generation of two daughter molecules. Even after successful TLS, only one of these would represent the product of the damaged strand. Therefore, another modification of the substrate was needed to facilitate discrimination of the two daughter molecules. This was accomplished by inserting a single *Pst*I site nearby the position selected for the site-specific lesion (15). When the DNA substrates carrying a site-specific photolesion were constructed in vitro, the same strand also included a mismatch in the *Pst*I recognition site. Upon replication of the heteroduplex, the daughter molecule containing the original undamaged strand inherits a perfect *Pst*I recognition site, hence is sensitive to linearization by this restriction enzyme. If TLS takes places and the damaged strand is incorporated into a closed circular duplex molecule, this product is resistant to digestion by *Pst*I and can be separated from the linearized daughter and other replication intermediates by electrophoresis in an ethidium bromide-containing agarose gel.

2. Materials

2.1. Cell-Free Extract

1. A pellet containing 10^8–10^9 highly proliferative, mammalian cells (see Notes 1 and 2).
2. Hypotonic buffer: 20 mM Hepes, pH 7.5, 5 mM KCl, 1.5 mM $MgCl_2$, 1 mM dithiothreitol (DTT).
3. Prechilled Dounce homogenizer.
4. Phase microscope and glass slides.
5. Trypan blue.
6. Refrigerated microcentrifuge reaching 10,000 × *g*.

2.2. DNA Substrates

1. A synthetic deoxyoligonucleotide in which the base(s) to be targeted for modification is present only once (see Note 3). In the example illustrated here, the 8-mer includes in its center two adjacent thymines: 5′GTATTATG (16).
2. Dilution buffer: 10 mM Tris–HCl, 1 mM EDTA, pH 8.0 (TE).
3. Germicidal UV lamp emitting predominantly 254 nm radiation (UVC).
4. CH_3CN (acetonitrile; see Note 4).
5. 1 M stock solution of K_2HPO_4/KH_2PO_4, pH 6.8.
6. High performance liquid chromatography (HPLC) equipment, preparative C-18 column, and reversed-phase Sep-Pak C_{18} classic cartridge (Waters, Inc.).
7. SpeedVac concentrator.
8. UV spectrophotometer.
9. Synthetic deoxyoligonucleotides (a 15-mer and a 17-mer) for assembly of the unmodified and lesion-containing 8-mers into larger molecules (40-mers) (see Fig. 1).

Synthetic Deoxyoligonucleotides

1) 8-mer with a photolesion ([c,s]TT or [6-4]TT) at the central thymines (TT)

5′ GTATTATG

2) 17-mer to be ligated at the 5′ end of the lesion-containing oligomer (the T at this position results in a point mutation disrupting a *Pst* I recognition site)

5′ GTTATCCGCTCACTGCT

3) 15-mer to be ligated at the 3′ end of the lesion-containing oligomer

5′ TCTAGTCACAATTCC

4) 29-mer scaffold

5′ TGTGACTAGACATAATACAGCAGTGAGCG

Assembly and ligation

Mutant *Pst*I

5′ GTTATCCGCTCACTGCTGTATTATGTCTAGTCACAATTCC
3′ GCGAGTGACGACATAATACAGATCAGTGT

The ligated 40-mer is used to build the double-stranded circular DNA substrates for in vitro replication reactions.

Fig. 1. Sequence of the synthetic deoxyoligonucleotides used to assemble the larger 40-mers containing a site-specific photolesion and a mismatch in the *Pst*I recognition sequence. The 40-mers including a [c,s]TT, a [6–4]TT, or undamaged TT were used for in vitro construction of closed circular heteroduplexes as substrates for in vitro replication and determination of translesion synthesis activity.

10. T4 polynucleotide kinase and reaction buffer: 70 mM Tris–HCl, pH 7.6, 10 mM $MgCl_2$, 5 mM DTT.
11. T4 DNA ligase and reaction buffer: 50 mM Tris–HCl, pH 7.5, 10 mM $MgCl_2$, 10 mM DTT, 1 mM ATP.
12. Synthetic 40-mer of the same sequence of the assembled 40-mers containing the DNA lesion (see Figs. 1 and 2).
13. Polyacrylamide gel electrophoresis (PAGE) apparatus and 15 % polyacrylamide-7 M urea gels (see Note 4).
14. Ethidium bromide (10 mg/mL) (see Note 4).
15. Diffusion buffer: 100 mM Tris–HCl, pH 8.0, 1 mM EDTA, 500 mM NaCl.
16. Synthetic deoxyoligonucleotide primer, 5′GGAATTGTGA CTA.
17. Klenow fragment of DNA polymerase I and its buffer: 10 mM Tris–HCl, pH 7.9, 50 mM NaCl, 10 mM $MgCl_2$, 1 mM DTT.
18. Deoxynucleoside triphosphates (dNTPs).
19. Formamide (see Note 4).
20. T4 endonuclease V and reaction buffer: 25 mM NaH_2PO_4/Na_2HPO_4, pH 6.8, 100 mM NaCl, 1 mM EDTA.

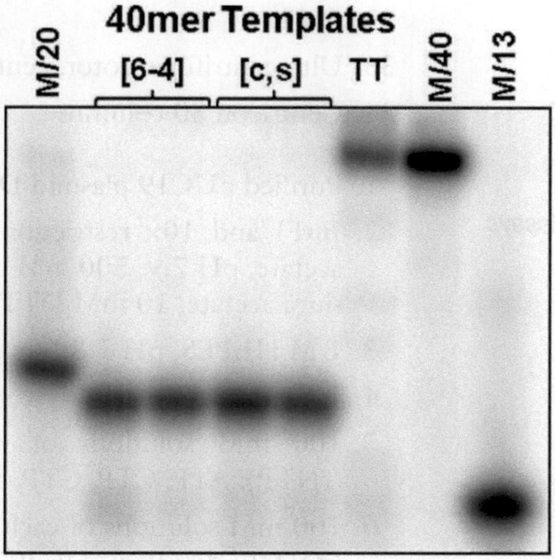

Fig. 2. Primer extension by Klenow on 40-mer templates. The primer was a ^{32}P-5′-labeled 13-mer and the templates were 40-mers containing a [6–4]TT, a [c,s]TT, or undamaged TT at positions 19 and 20 from the 3′ end. These 40-mers were assembled for incorporation into the double-stranded circular DNA heteroduplexes shown in Fig. 5. M/13, M/20, and M/40 correspond to end-labeled oligonucleotides used as size markers. Only the primers annealed to the undamaged 40-mer were fully extended. When the template contained a photoproduct, extension was blocked at the lesion. (Reproduced from ref. 15 with permission from the American Chemical Society).

21. 1 mg/mL bovine serum albumin (BSA).
22. 1 M piperidine (see Note 4).
23. ^{32}P-radiation detection system: X-ray film or a digital imaging system, such as the STORM 840 PhosphoImager™ (Molecular Dynamics, Inc., Sunnyvale, CA).
24. M13mp2SV bacteriophages with the SV40 origin of replication to the left (oriL) or to the right (oriR) of the lacZα gene (17–19).
25. *Escherichia coli* dut$^-$ ung$^-$ strain.
26. Synthetic 59-mer that includes the designed sequence to be inserted by site-directed mutagenesis flanked by regions of homology to the lacZα gene in M13mp2SV bacteriophage (see Fig. 4 and Note 14).
27. Annealing buffer for in vitro mutagenesis and construction of double-stranded closed circular molecules: 20 mM Tris–HCl, pH 7.8, 2 mM $MgCl_2$, 50 mM NaCl.
28. Reaction buffer for in vitro second strand synthesis; 1×: 10 mM Tris–HCl, pH 7.8, 25 mM $MgCl_2$, 2 mM DTT, 10 mM ATP, 0.5 mM of each dNTPs (prepare fresh as a 10× solution).
29. 50 % glycerol.
30. T4 DNA polymerase.
31. *E. coli* ung$^+$ strain.

32. CsCl.
33. Ultracentrifuge, rotor, centrifuge tubes.
34. Centricon 30 columns.

2.3. In Vitro Replication Assays

1. Purified pUC19 plasmid DNA.
2. *Bsr*FI and 10× restriction digestion buffer: 200 mM Tris–acetate, pH 7.9, 500 mM potassium acetate, 100 mM magnesium acetate, 10 mM DTT.
3. 1 M HEPES, pH 7.8.
4. 1 M $MgCl_2$.
5. 100 mM solutions of each ribonucleoside triphosphates (rNTP): ATP, CTP, GTP, and UTP.
6. 100 mM solutions of each deoxyribonucleoside triphosphates (dNTP): dATP, dCTP, dGTP, and dTTP.
7. 1 M creatine phosphate in 20 mM HEPES, pH 7.8.
8. 2.5 mg/mL creatine phosphokinase in 20 mM HEPES, pH 7.8.
9. 1 M stock solutions each of NaH_2PO_4 and Na_2HPO_4.
10. Purified SV40 large T antigen (Molecular Biology Resources, Inc., Milwaukee, WI).
11. [α-^{32}P]dCTP (>3,000 Ci/mmol).
12. 10 % SDS.
13. 0.5 M EDTA.
14. Proteinase K at 10 mg/mL in 10 mM Tris–HCl, pH 8.0. Store 0.5–1 mL aliquots frozen at −20 °C.
15. QIAEX II Gel Extraction System (Qiagen).
16. *Pst*I and 10× restriction digestion buffer: 500 mM Tris–HCl, pH 7.9, 1 M NaCl, 100 mM $MgCl_2$, 10 mM DTT.
17. Digital imaging system and software for quantification of signal intensities (e.g., STORM 840 PhosphoImager™ and ImageQuaNT™ software—Molecular Dynamics, Inc.).

3. Methods

3.1. Preparation of Cell-Free Extracts

The procedure described here was adapted from methodology developed by other laboratories (20–22) (see Note 1).

1. Collect the cell pellet by following routine procedures for the selected cell line (see Note 2).
2. Resuspend the pellet in hypotonic buffer using at most 1× the volume of the cell pellet and transfer to a prechilled Dounce homogenizer.

3. Incubate the cell suspension on ice for 30–35 min. Dilute a small aliquot (5 μL) of swollen cells in Trypan blue solution on a glass slide (control for lysis efficiency in step 4).
4. Dounce the swollen cells with 8–12 strokes until 90–95 % of the cells are lysed, as judged by inspection under a phase microscope of a small aliquot of cells stained with Trypan blue.
5. Centrifuge at 10,000 × *g* for 10 min to clarify the cell lysate.
6. Remove a small aliquot for determination of protein concentration (see Note 5)
7. Divide the lysate into smaller aliquots (~100–200 μL, depending on protein concentration, to avoid freezing and thawing more than two times) and store them at −80 °C.

3.2. DNA Substrates

3.2.1. Preparation of Deoxyoligonucleotides Containing a UV-Induced Lesion Between Two Adjacent Thymines [16]

1. Dissolve 2 μmol of the 8-mer (see Fig. 1) in 10 mL of TE, add the solution to a 10-cm Petri dish on ice and expose it to 254 nm light (~2 mW/cm^2) for 2 h, mixing the solution every 15 min.
2. Fractionate the irradiated 8-mer by HPLC, using a preparative C-18 column and a 60-min 6–10 % gradient of CH_3CN (see Note 4) in 75 mM K_2HPO_4/KH_2PO_4, pH 6.8, at a flow rate of 0.6 mL/min.
3. The fractions expected by their chromatographic behavior to contain oligonucleotides with a [c,s]TT or a [6–4]TT at the di-thymine site (16) are pooled, concentrated by SpeedVac, and desalted by reversed-phase chromatography using a Sep-Pak C_{18} classic cartridge. Desalted fractions are evaporated, resuspended in water, and processed through one or two additional HPLC purification cycles.
4. Absence of cross-contamination of lesion-containing 8-mers with each other or the undamaged oligonucleotide is confirmed by HPLC analysis of small aliquots and the concentration of the purified samples is determined in a UV spectrophotometer from the absorbance at 260 nm (the aliquot used in the spectrophotometer is discarded).

3.2.2. Assembly of Lesion-Containing 8-mers into Larger Oligonucleotides (40-mers)

Ligate the unmodified 8-mer, and the 8-mers with a [c,s]TT or a [6–4]TT (see Note 3), to a 17-mer and a 15-mer (see Note 6) at its 5′ and 3′ ends, respectively, using a synthetic 29-mer scaffold (see Fig. 1). The resulting 40-mer is used to generate the DNA substrate for in vitro replication assays (see Subheading 3.2.5).

1. Prepare phosphorylation reactions for each of the three 8-mers and the 15-mer; mix 5 nmol of the oligo with 100 units of T4 polynucleotide kinase in 50 μL of T4 DNA ligase buffer (see Notes 7 and 8); incubate at 37 °C for 60 min, then heat inactivate the enzyme at 65 °C for 20 min.

2. Perform the annealing and ligation reactions by preparing solutions in T4 DNA ligase buffer containing approximately 1 nmol of one of the phosphorylated 8-mers (already in T4 DNA ligase buffer—see previous step) with 1.5 nmol of each of the other three oligonucleotides: 17-mer, phosphorylated 15-mer (already in T4 DNA ligase buffer—see previous step), and 29-mer scaffold. Heat the solutions to 70 °C in a water bath then let the water bath cool slowly to room temperature. Add 120 units of T4 DNA ligase and incubate overnight at 16 °C.
3. Concentrate the ligation reaction mixture using SpeedVac and separate the ligated 40-mer from the scaffold and unligated oligomers on a 15 % PAGE-7 M urea gel. Load onto separate lanes of the same gel the scaffold oligomer and a synthetic 40-mer (same size and sequence of the expected product of ligation) as size markers. Run the gel at 350 V for 2.5–3 h. Stain the marker lanes with ethidium bromide (see Note 4) and use them to guide the recovery of the ligated 40-mer product from the unstained portion of the same gel (which was not exposed to a UV box).
4. Elute the 40-mer from the gel slices (cut in small pieces) by overnight incubation in a small volume (~0.6 mL for ~300 mg gel slice) of diffusion buffer at 4 °C. Next day, separate the buffer with the eluted 40-mer from the gel; add 0.25–0.30 mL of buffer to the gel and incubate at 50 °C for 1 h. Add this second aliquot of buffer to the first and concentrate the sample using SpeedVac; desalt using reversed-phase Sep-Pak C_{18} classic cartridge and dry desalted oligo in SpeedVac. Resuspend the 40-mers in 50–100 μL of TE.
5. Run a small aliquot on a 15 % PAGE-7 M urea gel to check for integrity and purity. Determine the approximate concentration of the 40-mer solution by comparison to known amounts of the synthetic 40-mer loaded onto parallel lanes of the same gel.

3.2.3. Characterization of 40-mers Containing a Single [c,s]TT or [6–4]TT

1. Label the primer sequence 5′GGAATTGTGACTA with ^{32}P by incubating 8 pmol in 50 μL of T4 polynucleotide kinase buffer with 10 units of the enzyme and 3 μCi [γ-^{32}P]ATP at 37 °C for 60 min. Inactivate the enzyme by incubation at 65 °C for 20 min.
2. Prepare primer-extension reactions for each of the 40-mers assembled in Subheading 3.2.2; anneal 1.5 pmol of the ^{32}P-labeled primer with 10 pmol of a 40-mer in 10 μL reaction for the Klenow fragment of DNA polymerase I. Add 1 unit of the Klenow enzyme and 100 μM of each of the four dNTPs; incubate at 37 °C for 20 min. Stop the primer-extension reaction by adding 20 μL of formamide (see Note 4). Load ~40,000 cpm

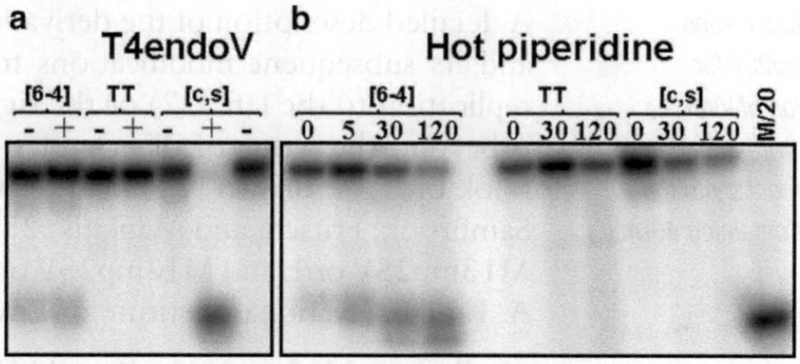

Fig. 3. Identity and purity of photolesion-containing 40-mers. (**a**) T4 endonuclease V recognizes cyclobutane pyrimidine dimers in double-stranded DNA, but not [6–4] adducts. Therefore, cleavage of the phosphodiester bond by this enzyme demonstrates the presence of the [c,s]TT in the digested 40-mer. Complete digestion of the [c,s]TT 40-mer and resistance to digestion of the [6–4]TT 40-mer and the TT 40-mer indicate the absence of cross-contamination. (**b**) Hot piperidine cleaved only the [6–4]TT 40-mer but not the [c,s] TT 40-mer or the TT 40-mer. Together with the results shown in Fig. 2, these data demonstrate that the 40-mers used to assemble the heteroduplexes added to the in vitro replication reactions included either the named photolesion or undamaged dipyrimidines at the specified site.

of each sample onto a 15 % PAGE-7 M urea gel. Carry out electrophoresis at 400 V for 2 h (see Fig. 2 and Note 9).

3. In separate reactions, phosphorylate 8 pmol of each of the photoproduct-containing 40-mers and the undamaged 40-mer (see step 1 in Subheading 3.2.3).
4. Mix 2 pmol of each 5′ ^{32}P-labeled 40-mer with 3 pmol of 29-mer scaffold (see Fig. 1) in 9 μL of deionized H_2O and 1 μL of 10× T4 endonuclease V buffer. Heat the solution to 70 °C and let it cool slowly to room temperature. Divide the reaction in two 5 μL samples; complete each sample to 10 μL by adding 10× T4 endonuclease V buffer (0.5 μL), deionized water (1.5 μL), 1 mg/mL BSA (1 μL), and 2 μL of T4 endonuclease V (970 ng/μL) or 1× T4 endonuclease V buffer. Incubate at 37 °C for 40 min and analyze digestion products in 15 % PAGE-7 M urea (see Fig. 3a and Notes 10 and 12).
5. Dissolve 2 pmol of each 5′ ^{32}P-labeled 40-mer in 400 μL of 1 M piperidine (see Note 4), separate into four aliquots of 100 μL, and incubate each at 95 °C for 0, 5, 30 (see Fig. 3b), or 120 min. At the end of the incubations, samples are frozen in dry ice and lyophilized. Dried samples are resuspended in 50 μL of water and lyophilized three more times to remove the piperidine (23). Finally, the samples are resuspended in 95 % formamide and electrophoresed in 15 % PAGE-7 M urea (see Fig. 3b and Notes 11 and 12).
6. Gels are exposed to an X-ray film or phosphor screen and scanned by a PhosphorImager™ (Molecular Dynamics, Sunnyvale, CA) (see Figs. 2 and 3).

3.2.4. Modification of M13mp2SV for Production of Viral (+) Strands with Sequences Complementary to the Lesion-Containing 40-mers

A detailed description of the derivation of bacteriophage M13mp2 and its subsequent modifications to include the SV40 origin of replication to the left (17) or the right (18, 19) of the lacZα gene is beyond the scope of this chapter. The reader is directed to the book by T. A. Brown (24) or the laboratory manual published by Sambrook, Fritsch, and Maniatis (25) for background information. M13mp2SV oriL and M13mp2SV oriR were gifts from Dr. Thomas A. Kunkel (National Institute of Environmental Health Sciences).

1. Prepare M13mp2SV oriL and M13mp2SV oriR single-stranded DNA [(+) strand] from phage grown in *E. coli* dut⁻ ung⁻ strain (e.g. CJ236 or KT852; see Note 13) (26, 27).
2. Assemble the designed 59-mer (see Fig. 4 and Note 14) for site-directed mutagenesis of M13mp2SV bacteriophages.

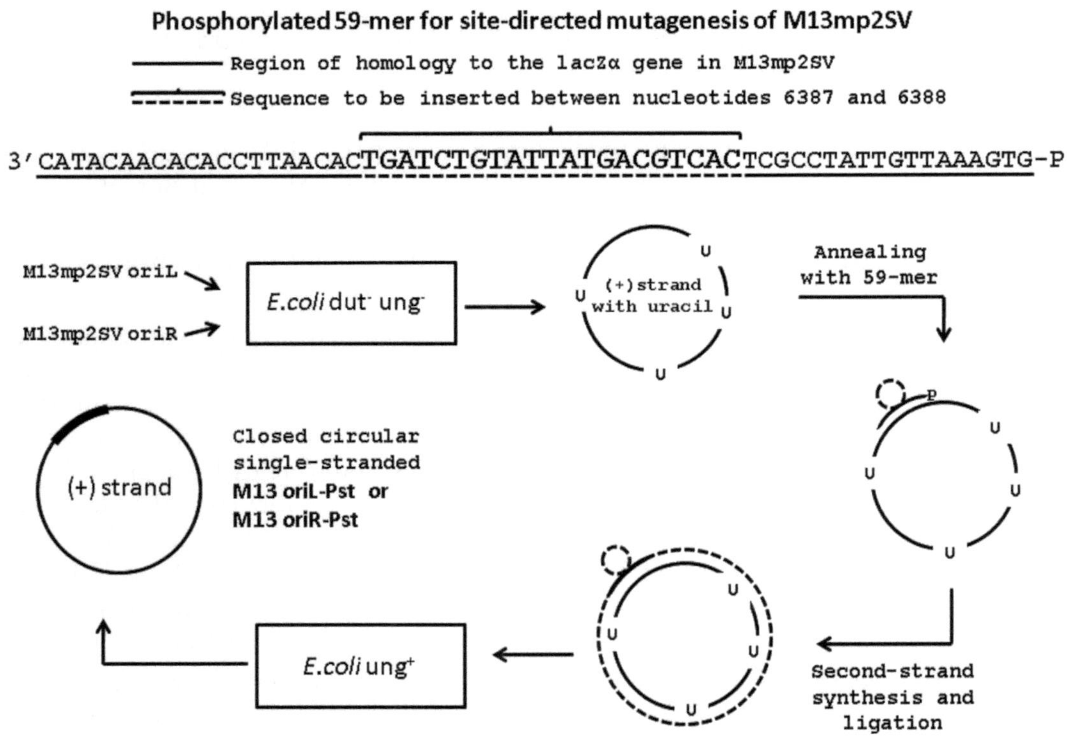

Fig. 4. Strategy used for site-directed mutagenesis of M13mp2SV oriL or oriR to generate closed circular single-stranded derivatives with 21 nucleotides inserted into the LacZα gene. The (+) strand of bacteriophage replicated in *E. coli* dut⁻ ung⁻ has uracil replacing thymine at several sites. The 5′-phosphorylated 59-mer, containing the designed insertion flanked by sequences complementary to the lacZα gene, is annealed to the uracil-containing single-stranded circular DNA. The 21-mer forms a small loop that is held in place by the lacZα complementary sequences, which also prime the synthesis of the second strand of DNA by the T4 DNA polymerase. This polymerase dissociates from the DNA upon reaching the 5′-phosphorylated end of the annealed primer, which is then ligated to the newly synthesized strand by DNA ligase. Transfection of the resulting double-stranded DNA into a competent *E. coli* ung⁺ host results in nicking of the uracil-containing strand and replication of the in vitro synthesized strand. The phage particles produced by the bacteria now contains the closed circular single-stranded DNA [(+) strand] of the mutagenized M13 oriL-Pst or M13oriR-Pst, which includes the sequences complementary to the undamaged or the lesion-containing 40-mers (see Figs. 1 and 5).

3. Mix 2 μg of the viral (+) strand containing uracil substitutions to 120 pmol of 5′ phosphorylated 59-mer (see step 1 in Subheading 3.2.2) in 50 μL of annealing buffer (see step 27 in Subheading 2.2, Fig. 4, and Notes 14 and 15). Heat the solution to 61 °C and let it cool very slowly to room temperature.
4. Add 6 μL of 10× reaction buffer (see step 28 in Subheading 2.2), 1.2 μL 50 % glycerol, 3 units of T4 DNA polymerase (see Note 16), and 2–3 units of T4 DNA ligase; first incubate for 60 min at 37 °C, then for 14 h at 14 °C. Terminate the polymerization reaction at 65 °C for 15 min (see Note 17). An aliquot of the reaction is fractionated in 0.8 % agarose-ethidium bromide gel to check for the presence of covalently closed circular DNA as the major product; control lanes include single-stranded and double-stranded closed circular molecules.
5. Transfect or electroporate 5–10 μL of the products of the in vitro DNA synthesis described above into competent *E. coli* ung^+ host.
6. Analyze the phage particles in a plaque-forming assay (25). Mutant phage (containing the insertion of the designed 21-mer into the lacZα gene—see Fig. 4) give rise to white plaques, while phage lacking the insertion forms blue plaques.
7. Purify phage DNA from several white plaques (25) and sequence the mutagenized lacZα gene using a primer upstream of nucleotide 6,368 (see Fig. 4).
8. Purify the mutagenized M13mp2SV-oriL-Pst or M13mp2SV-oriR-Pst closed single-stranded DNA from phage particles (25) obtained from sequence-verified genomes.

3.2.5. Construction of Double-Stranded Closed Circular Heteroduplexes Containing a Site-Specific Lesion in One Strand and Nearby, in the Opposite Strand, a PstI Restriction Enzyme Recognition Sequence

1. The 40-mers containing a TT-, [c,s]TT-, or [6–4]TT- and a point mutation (T replacing A) in the *Pst*I recognition site (see Subheading 3.2.2) are phosphorylated by T4 polynucleotide kinase in T4 DNA ligase buffer (see step 1 in Subheading 3.2.2).
2. Each phosphorylated 40-mer (20–40 pmol) is annealed to 6 pmol (~30 μg) of closed circular, single-stranded oriL-Pst or oriR-Pst (see step 3 in Subheading 3.2.4) in 400 μL of annealing buffer (see step 27 in Subheading 2.2) at 61 °C (see end of Note 15), followed by cooling at room temperature.
3. Set up second-strand synthesis: add 50 μL of reaction buffer (see step 28 in Subheading 2.2) to the above annealing solution; add 30–40 units of T4 DNA polymerase and 40 units of T4 DNA ligase. Do not vortex; mix slowly using pipet tips; keep tubes 10 min at room temperature; then incubate for 60 min at 37 °C, and for 14 h at 14 °C. Terminate the polymerization reaction at 65 °C for 15 min (see Note 17). Check reaction efficiency in 0.8 % agarose gel.
4. Purify the closed circular DNA duplex by CsCl density gradient centrifugation in the presence of ethidium bromide.

3.2.6. Purification of In Vitro Synthesized Plasmid DNA in CsCl Gradients

1. Add 4.1 g CsCl to a 15-mL sterile capped tube.
2. Add DNA (products of reactions described above) diluted into 4.1 mL of 10 mM Tris–HCl, 1 mM EDTA, pH 8.0, and 360 μL of ethidium bromide (stock is 10 mg/mL).
3. Mix well and transfer to a nitrocellulose tube fitting an ultracentrifuge rotor (Beckman VTi 80 vertical rotor or equivalent).
4. Centrifuge at 510,000 × g at r_{max} (80,000 rpm in the VTi 80 rotor) at 25 °C for 3 h; allow the rotor to decelerate without the use of the break (see Note 18).
5. Collect the covalently closed circular DNA band by puncturing the side of the tube just below the closed circular DNA band (higher mobility than other reaction products; visible under UV light) (25).
6. Dilute the sample to 2 mL with 10 mM Tris–HCl, 1 mM EDTA, and pH 8.0 (see Note 19).
7. Use a Centricon 30 column to desalt and concentrate the DNA.
8. Repeat steps 6 and 7.
9. Check the concentration and integrity of the purified covalently closed circular DNA by electrophoresis in 0.8–1 % agarose gel containing 0.2 μg/mL ethidium bromide. Load in one lane a known volume of the purified sample and in adjacent lanes increasing amounts of purified double-stranded oriL-Pst or oriR-Pst plasmid DNA from a solution of known concentration, as determined from its absorbance at 260 nm ($A260 = 1$ corresponds to 50 μg/mL of double-stranded DNA) (see Note 20).

3.3. In Vitro Replication

The closed circular heteroduplexes produced by the methods described in Subheading 3.2 contain a site-specific [c,s]TT or [6–4]TT and a mismatch at the single *Pst*I recognition site at a defined position relative to the SV40 origin of replication (see Fig. 5 and Note 21).

3.3.1. Preparation of an Internal Standard

Quantification of replication products in the presence or absence of a site-specific template lesion is aided by the use of an internal standard, which provides the means to control for experimental variations in DNA recovery among different samples within the same experiment, as well as the extent of digestion with diagnostic restriction enzymes.

1. Prepare pUC19 double-stranded plasmid DNA from *E. coli* cultures (25).
2. Linearize pUC19 by restriction digestion with *Bsr*FI (see Note 22).
3. End-label the linear molecules with Klenow, dNTPs, and [α-^{32}P]dCTP (see Note 23).

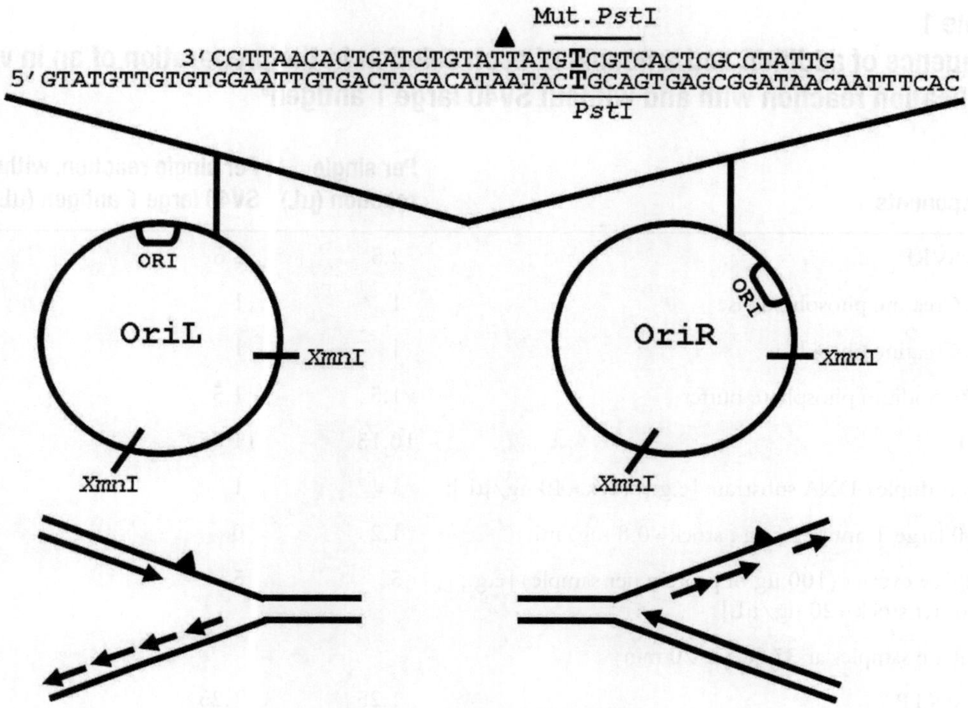

Fig. 5. Double-stranded circular heteroduplexes used as substrates for bypass replication assays. A site-specific photoproduct is placed on the template for the leading (oriL) or the lagging (oriR) strand when the correspondent 40-mer is annealed to M13 oriL-Pst or M13oriR-Pst, followed by second-strand synthesis and ligation in vitro. Three oriL and three oriR constructs were prepared, each containing a T:T mismatch in the recognition site for *Pst*I and a [6–4]TT, a [c,s]TT, or undamaged di-thymines at the position denoted by the *filled triangle*. The location of the SV40 origin of replication to the left (oriL) or to the right (oriR) of the photoproduct is indicated as ORI. The recognition sites for *Xmn*I are shown for orientation purpose only. The distance between the photolesion and the center of SV40 ORI (*Bgl*I site) is 409 bp in the oriL and 392 bp in the oriR constructs. Their total size is 7,419 bp. (Reproduced with modifications from ref. 15 with permission from the American Chemical Society).

3.3.2. In Vitro Replication Reactions

Each in vitro replication reaction must contain the final concentration of the following components: 30 mM HEPES, pH 7.8, 7 mM $MgCl_2$, 4 mM ATP, 200 μM each of the other three rNTPs, 100 μM each of dATP, dGTP and dTTP, 50 μM dCTP, 100–150 μCi/mL [α-^{32}P]dCTP, 40 mM creatine phosphate, 100 μg/mL creatine phosphokinase, 15 mM sodium phosphate, pH 7.5, 1.6 μg/mL DNA substrate, 3–4 mg/mL of protein from human cell-free extract, with or without 40 μg/mL SV40 large T antigen. The steps described below streamline the preparation of identical reactions (usually 25 μL each); the volume of each item added (Table 1) will depend on the number of reactions planned for each experiment.

1. Prepare a 10× standard in vitro reaction buffer (SVRB) containing 300 mM HEPES, pH 7.8, 70 mM $MgCl_2$, 40 mM ATP, 2 mM CTP, 2 mM GTP, and 2 mM UTP. Store in 50 μL aliquots frozen at −20 °C.

Table 1
Sequence of addition and amounts of each solution in the preparation of an in vitro replication reaction with and without SV40 large T antigen[a]

Components	Per single reaction (μL)	Per single reaction, without SV40 large T antigen (μL)
10× SVRB	2.5	2.5
25× Creatine phosphokinase	1	1
25× Creatine phosphate	1	1
16.7× Sodium phosphate buffer	1.5	1.5
H_2O	10.15	11.35
Heteroduplex DNA substrate [e.g., stock = 40 ng/μL]:	1	1
SV40 large T antigen [e.g., stock = 0.8 mg/mL]	1.2	0
Cell-free extract (100 μg of protein per sample) [e.g., extract stock = 20 μg/μL]	5	5
Incubate samples at 37 °C for 20 min		
20× dNTP	1.25	1.25
[α^{32}P]dCTP	0.4	0.4

[a] Scale up the volumes for each component in accordance with the objectives of the experiment and the number of reactions needed

2. Mix the dNTPs for a 20× solution: 2 mM dATP, 2 mM dGTP, 2 mM TTP, and 1 mM dCTP (see Note 24). Store in 50–100 μL aliquots frozen at −20 °C.
3. Prepare 1 M creatine phosphate (25× solution) in 20 mM HEPES, pH 7.8. Store in 50–100 μL aliquots frozen at −20 °C.
4. Prepare 2.5 mg/mL creatine phosphokinase (25× solution) in 20 mM HEPES, pH 7.8. Store in 1 mL aliquots refrigerated at 4 °C.
5. Prepare 100 mL of 250 mM sodium phosphate, pH 7.5 by mixing 20.35 mL of 1 M Na_2HPO_4, 4.65 mL of 1 M NaH_2PO_4, and 75 mL of deionized water.
6. Make a solution of 2 mg/mL proteinase K, 50 mM EDTA and 2 % SDS to stop the in vitro replication reaction (by addition of an equal volume). This stop solution should be made fresh and in amounts sufficient for each experiment (usually 25 μL per reaction).
7. Mix the components of the in vitro replication reactions in the sequence shown in Table 1. Adjust the volumes of each component in accordance with the number of reactions planned and the objectives of a given experiment.

8. Include in the experimental plan an in vitro replication reaction lacking SV40 large T antigen (see Note 25).
9. Incubate the in vitro replication reactions at 37 °C for 20 min before the addition of dNTPs and [α^{32}P]dCTP (see Note 26).
10. Add dNTPs and [α^{32}P]dCTP to each reaction and continue the 37 °C incubation for 30–120 min.
11. At the end of the chosen incubation period, the reaction is stopped by the addition of an equal volume (25 μL) of 2 % SDS, 2 mg/mL proteinase K, and 50 mM EDTA.
12. Add *Bsr*FI-linearized, ^{32}P-end-labeled pUC19 DNA to each reaction (use a minimum of 15,000 to 20,000 cpm per reaction).
13. Purify the DNA from each in vitro reaction using the QIAEX II Gel Extraction System (Qiagen), or an equivalent protocol.
14. Divide the purified DNA (eluted in TE) into two equal volume samples; add 1/10 volume of 10× buffer for *Pst*I.
15. Add 60 units of *Pst*I to one of the paired samples and the equivalent volume of 1× *Pst*I buffer to the other.
16. Incubate at 37 °C for 3 h to overnight.
17. Load the samples on 1 % agarose gel containing 0.2 μg/mL ethidium bromide to fractionate the replication products and internal standard.
18. Expose gel to a phosphor screen and scan in a PhosphorImager™ (see Figs. 8, 9, and 10).

3.3.3. Identification of Bypass Replication Products

The diagrams in Fig. 6 illustrate the expected products of in vitro replication of the designed synthetic substrates; their relative positions on an agarose gel after electrophoresis in the presence of ethidium bromide are indicated in Fig. 7.

1. Determine the signal intensity for specific DNA bands from pixel volume (corrected for background), as quantified by the ImageQuaNT™ software (Molecular Dynamics, Inc.).
2. Divide the signal intensity in the band representing a DNA replication product of interest by the signal intensity in the pUC19 band present in the same lane of the gel.
3. Calculate the contribution of each parental strand of the heteroduplex substrate in directing the synthesis of closed circular daughter molecules (Form IV) from the normalized amount of Form IV DNA, before and after *Pst*I digestion (see Note 27). The amount of *Pst*I-resistant Form IV DNA (product of the strand carrying the mismatch and the photolesion) represents the true product of TLS. Form IV molecules sensitive to *Pst*I digestion are converted to linear molecules (Form III) and represent the product of replication of the undamaged strand (see Note 28).

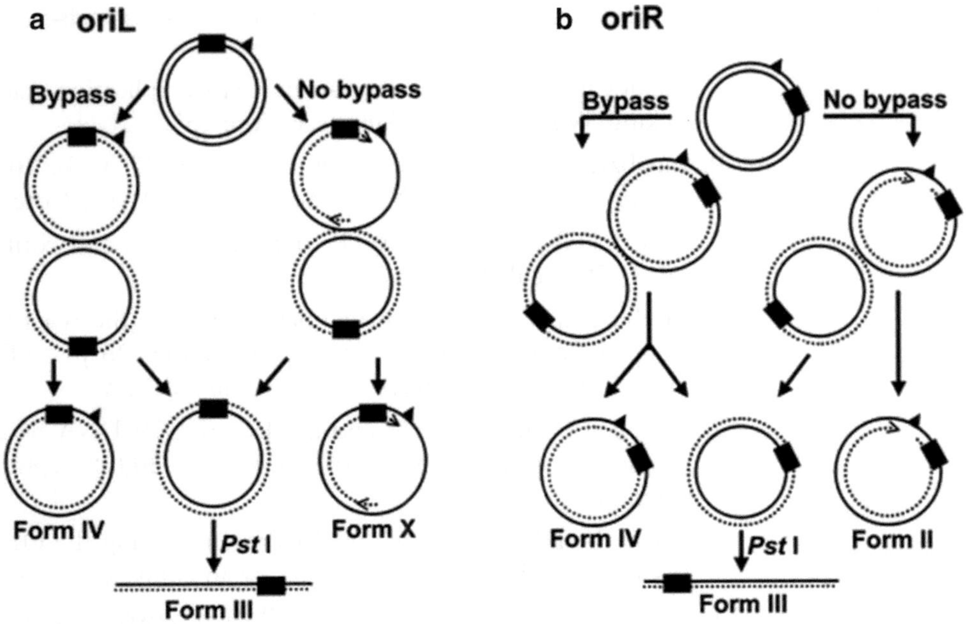

Fig. 6. Models depicting the effects on replication products of a photolesion on the template to the (**a**) leading strand (oriL) or the (**b**) lagging strand (oriR) of nascent DNA. The SV40 origin of replication in the closed circular molecules is represented by a *filled rectangle* and the position of the photoproduct is indicated by the *filled triangle*. *Solid lines* depict circular parental strands and *dotted lines* the DNA strand synthesized in vitro. *Arrowheads* indicate the direction of DNA polymerization. Only the product of replication of the undamaged daughter strand in the oriL or oriR construct is linearized by *PstI* (Form III). The closed circular double-stranded daughter molecule generated from the photolesion-containing strand is resistant to digestion and represents the true product of translesion synthesis (Form IV). In the absence of bypass replication, the partially replicated oriL molecule (Form X) shows low mobility in the agarose-ethidium bromide gel; in the oriR construct, the block of lagging strand synthesis generates a gapped molecule that behaves in the agarose-ethidium bromide gels as a nicked double-stranded circular molecule (Form II). (Reproduced from ref. 15 with permission from the American Chemical Society).

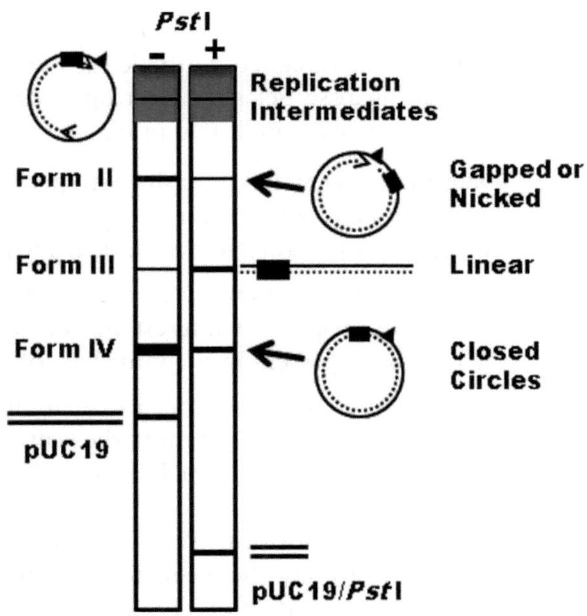

Fig. 7. Diagram illustrating the expected positions of different products of in vitro replication in 1 % agarose gel in the presence of ethidium bromide.

4. Notes

1. Extracts containing essential DNA replication factors and high DNA synthesis activity are obtained from cells growing logarithmically (large fraction of cells in S phase) either in suspension or as attached monolayers. Swelling the cells in hypotonic buffer allows for sufficient amounts of replication factors to leak out of the nucleus, but limits the extraction of DNA replication inhibitors (28, 29).

2. Neoplastic or transformed cell lines are best suited for this purpose, but active extracts can be obtained from normal fibroblasts immortalized by ectopic expression of the catalytic subunit of human telomerase and retaining a stable and diploid karyotype. Stock cultures should be shown to be free of mycoplasma contamination; this can be done regularly with commercial test kits, such as PlasmoTest™ from InvivoGen. Immortalized or SV40-transformed human fibroblasts grow well in Eagle's or Dubecco's minimal essential medium (MEM), supplemented with 10 % fetal bovine serum, and 2 mM L-glutamine. Antibiotics (50 μg/mL gentamicin, or 100 μg/mL streptomycin and 100 units/mL of penicillin) are never added to stock cultures, only when expanding them to collect the large number of cells needed for extract preparation. Sub-confluent cultures of adherent cells are incubated with 0.1–0.5 % trypsin for 2–5 min; individual cells are resuspended in culture medium containing serum (to inactivate the trypsin) and centrifuged at 800–1,000 × *g* for 5–10 min. Pellets of HeLa S3 cells grown in suspension are usually available from tissue culture core facilities at research universities or from the National Cell Culture Center (Minneapolis, MN—http://www.nccc.com/intro.html). At the UNC Lineberger Comprehensive Cancer Center, HeLa S3 cells are grown in S-MEM supplemented with 2 mM L-glutamine, 5 % horse serum, and 5 % fetal bovine serum. Cultures are maintained between 1×10^5 and 1×10^6 cells per mL with replacement and/or dilution of medium two to three times per week.

3. The deoxyoligonucleotide sequence is selected in accordance with the structure of DNA lesion of interest and the most efficient protocol is used to induce the site-specific DNA lesion. The main limitation is that the DNA lesion must be chemically stable during the subsequent purification steps, substrate assembly and conditions for in vitro replication. In the case illustrated here, the 8-mer sequence was selected on the basis of published analytical methods for preparation and purification of oligos containing a single UV-induced lesion between two adjacent thymines (16).

4. Several reagents must be handled with care (consult Material Safety Data Sheets) and according to environmental health and safety requirements: (a) Acetonitrile—use adequate ventilation. (b) Acrylamide hazard can be minimized by using commercial solutions, such as Biorad 40 % acrylamide/Bis solution, instead of powder products. (c) Ethidium bromide is mutagenic and moderately toxic. Ethidium bromide solutions for gel staining and ethidium bromide-containing agarose gel electrophoresis buffer should be prepared fresh. Ethidium bromide solutions can be decontaminated by passing through filters containing a bed of activated carbon. (d) Formamide should be used under a chemical fume hood. (e) Piperidine; keep solutions in glass containers inside a fume hood. (f) Best practices in the handling of ^{32}P-labeled reagents should be followed, including the use of plexiglass shields for protection from radiation.
5. Total protein concentration in cell-free lysates can be determined with the Bradford assay (Biorad), using BSA as a standard.
6. The deoxyoligonucleotide used in this protocol must be chosen in accordance with the sequence and engineered characteristics of the double-stranded circular DNA substrate to be added to the in vitro replication reactions.
7. The 17-mer oligo does not need to be phosphorylated at this time.
8. For nonradioactive phosphorylation of oligos, T4 ligase buffer [50 mM Tris–HCl, pH 7.5, 10 mM $MgCl_2$, 10 mM DTT, 1 mM ATP] is used because (a) T4 polynucleotide kinase exhibits 100 % activity in this buffer and (b) it facilitates the preparation of the subsequent ligase reaction (no buffer exchange). The amount of T4 polynucleotide kinase per reaction depends on oligo concentration; one supplier of this enzyme (New England BioLabs, Inc.) recommends the use of 10 units of T4 polynucleotide kinase for 300 pmol of 5′ ends per 50 μL reaction. However, we achieved sufficient phosphorylation by using 100 units of the enzyme and 5 nmol of the substrate.
9. The presence of a DNA lesion on the 40-mer template strand will stop the primer extension reaction by the Klenow fragment of DNA polymerase I. In the case of the 40-mers illustrated in Fig. 1, the [c,s]TT or [6–4]TT are at nucleotides 19 and 20 from the 3′ end. Therefore, the only product of the primer-extension reaction should be 18 nucleotides long. Only the control 40-mer (undamaged) should support the synthesis of a ^{32}P-labeled product that is 40 nucleotides long (Fig. 2). Therefore, the absence of this longer product when the reaction

included one of the lesion-containing 40-mers demonstrates that the assembled 8-mer was devoid of contamination with undamaged oligonucleotide. The nature of the lesion, however, is not disclosed by this assay. Determining the absence of cross-contamination between the [c,s]TT- and the [6–4]TT-containing 8-mers requires other analytical methods (Fig. 3).

10. Cleavage by T4 endonuclease V demonstrates the presence of a [c,s]TT in the 40-mer (Fig. 3a). This enzyme binds to cyclobutane pyrimidine dimers in double-stranded DNA; it cleaves the N-glycosylic bond of the 5′ pyrimidine of the dimer and then breaks the phosphodiester bond 3′ to the abasic site; this enzyme does not recognize the [6–4] photoproduct (30). T4 endonuclease cleavage of the [c,s]TT-containing 40-mer labeled with ^{32}P at its 5′ end resulted in its complete digestion into a radioactive fragment migrating the same distance in a denaturing polyacrylamide gel as a synthetic 20-mer (Fig. 3a). As expected, the control (undamaged) 40-mer or the one containing the [6–4]TT were resistant to digestion by T4 endonuclease V (the light radioactivity signal detected at approximately the same migrating distance as the 20-mer size marker in the [6–4]TT sample appears to be the product of chemical cleavage and not the presence of contaminating [c,s]TT-40-mer) (see Note 11).
11. Only the [6–4]TT-containing 40-mer is sensitive to cleavage by hot piperidine (23) (Fig. 3b). This oligomer has a small contaminant that could have been generated during the preparative steps described in Subheading 3.2.2. This small contaminant did not interfere with the assembly of the circular heteroduplex substrate used in the in vitro replication reactions.
12. Together the results illustrated in Fig. 3 demonstrate that the purified lesion-containing 8-mer used to assemble the 40-mers with or without a specific photoproduct are not contaminated with each other.
13. The use of an *E. coli* dut$^-$ (deficient in dUTP pyrophosphatase) ung$^-$ (deficient in uracil N-glycosylase) strain in this step results in the incorporation of uracil in place of thymine in the bacteriophage DNA [(+) single strand] used for the in vitro synthesis of its complementary strand [(−) strand]. This characteristic will result in cleavage of the (+) strand in an *E. coli* ung$^+$ strain and favor the in vivo replication of the (−) strand for production of the phage containing the desired site-directed insertion (27) (see Fig. 4).
14. The 59-mer shown in Fig. 4 was designed to insert the sequence 5′ACTAGACATAATA<u>CTGCAG</u>TG between nucleotides 6,387 and 6,388 within the lacZα gene of M13mp2SV oriL or oriR (the underlined sequence corresponds to the *Pst*I recognition

sequence). By flanking the 21-mer with nucleotides complementary to phage sequences 6,368–6,387 on the left and 6,388–6,426 on the right, the noncomplementary loop is stabilized and second strand synthesis can be accomplished in vitro by extension with T4 DNA polymerase.

15. The high molar ratio of 59-mer to circular single-stranded DNA template was used to drive the hybridization of the complex 59-mer to each (+) strand and maximize yield. Control reactions without addition of the 59-mer should be done to determine that the template DNA is not contaminated with pieces of single-stranded DNA that could prime the synthesis of the second strand and yield double-stranded circular DNA product without the desired insertion. Only a single *Pst*I recognition site was inserted into the final product of site-directed mutagenesis, indicating that only one 59-mer hybridized to each template molecule, despite the high molar ratio used. A 100:1 molar ratio was used initially; however, it was determined in later experiments that the yield of double-stranded, closed circular DNA molecules was the same when 8:1 or even 3:1 molar ratios were used. The lower ratios maximize the use (minimize the waste) of the modified oligos.
16. T4 DNA polymerase or native T7 DNA polymerase (the T7 gene 5 protein/*E. coli* thioredoxin complex) are used for the second-strand synthesis during the derivation of modified new vectors by oligonucleotide-directed mutagenesis because these enzymes do not perform strand displacement that could dislodge the designed primer (27).
17. This sequence of incubations can be programmed in and carried out within a machine routinely used for polymerase chain reactions (PCR).
18. Conditions for the CsCl gradient ultracentrifugation can be adapted to other types of rotors. The initial CsCl concentration and speed of centrifugation should allow the generation of a linear gradient providing the best separation of closed circular duplex DNA from open circles and linear molecules.
19. The use of pH 8.0 in this dilution buffer and subsequent steps is very important to maintain closed circular forms and minimize nicked ones. When purified closed circular DNA was placed in pH 7.7–7.8 buffer, the fraction of nicked circular forms increased during storage.
20. The closed circular, double-stranded plasmid DNA purified from a bacterial culture is negatively supercoiled, while the one prepared by in vitro synthesis is relaxed. Therefore, the plasmid DNA migrates faster in the gel. Note also that when purifying double-stranded plasmid DNA from bacterial cultures in CsCl gradients, it is recommended that after mixing well the DNA

with CsCl (see step 3 in Subheading 3.2.6) the solution be incubated at 4 °C for 20–40 min and any insoluble material removed by a low spin centrifugation (8,000 × *g*; 6,000 rpm in a Marathon 22 KBR centrifuge) for 25 min at 4 °C before the ultracentrifugation step (this is not necessary when purifying closed circular DNA synthesized in vitro).

21. The closed circular heteroduplexes derived from M13mp2SV-oriL-Pst have the SV40 origin of replication to the left of the site-specific lesion. During bidirectional replication of this substrate, the lesion will be on the template for the leading strand. In the M13mp2SV-oriR-Pst derivatives, the SV40 origin of replication is to the right of the DNA lesion; hence, the damaged site will be on the template for the lagging strand (see Fig. 5). In both cases, the closed circular DNA duplex derived from the undamaged strand during semiconservative DNA replication will contain a perfect *Pst*I recognition sequence, while in the product of the damage strand this sequence will be mutated (see Fig. 6).
22. There is a single recognition site for *Bsr*FI in pUC19, which is located approximately 180 ° from the single recognition site for *Pst*I. Therefore, when pUC19 is linearized by digestion with *Brsf*I, the *Pst*I site will be centrally located.
23. Upon *Bsr*FI digestion, a 3′-GGCC-5′ overhang is generated, which Klenow can use to incorporate ^{32}P-dCMP opposite the two template guanines.
24. Cold dCTP is added to the in vitro DNA replication reactions at one half of the concentration of the other dNTPs because [α^{32}P]dCTP is used to radiolabel the reaction products. The lower dCTP concentration increases the incorporation of the radiolabeled precursor without compromising the rate of in vitro polymerization.
25. Reactions lacking the SV40 large T antigen serve as a negative control (see Figs. 8, 9, and 10). This protein is essential for the recruitment of replication factors to the SV40 origin of replication (20). Therefore, any incorporation of [^{32}P]CMP in the closed circular DNA substrate in the absence of SV40 large T antigen would not be due to semiconservative DNA replication.
26. The pre-incubation of the closed circular DNA substrate with cell-free extracts and SV40 large T antigen in the absence of dNTPs allows for assembly of the initiation complex at the SV40 origin of replication. Delay in elongation (because of the absence of dNTPs) provides for partial synchronization (elimination of the initial lag period) and higher yields of replication products after short incubation periods (31).

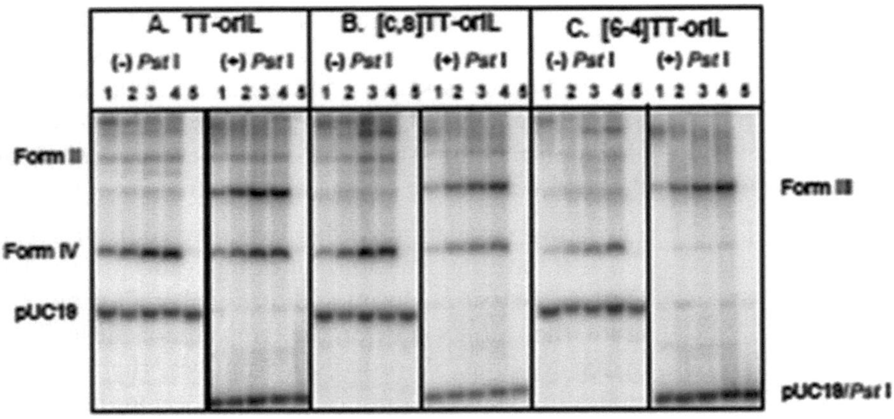

Fig. 8. Time course of in vitro replication of oriL constructs by a HeLa cell extract. Reactions contained either the control molecule TT-oriL lacking a photoproduct (A), or the heteroduplexes carrying a site-specific [c,s]TT (B) or a [6–4]TT (C) on the template to the leading strand. Reaction mixtures containing all components, except dNTPs, were pre-incubated for 20 min at 37 °C. Following the addition of [^{32}P]dCTP and dNTPs, the reactions were incubated for 0.5, 1, 1.5, and 2 h (*lanes 1–4*, respectively). Reactions containing all components, except SV40 Tag, were incubated for 2 h and 20 min (*lanes 5*). Equal amounts of radiolabeled linear pUC19 were added to the reactions at the end of the incubation period and prior to DNA purification. One half of each sample was treated with *PstI*. Undigested [(−) *PstI*] and digested [(+) *PstI*] samples were fractionated by electrophoresis in 1 % agarose-ethidium bromide gels. The positions of replication Form II (nicked or gapped circular molecules), Form III (linear molecules), Form IV (closed circular molecules), full-length pUC19 and *PstI*-digested pUC19 are indicated. In the samples not treated with *PstI*, results show a time-dependent increase in Form IV DNA for each substrate; as expected, the rate of synthesis was the highest for the undamaged substrate, followed closely by the [c,s]TT-oriL construct; however, the rate of synthesis of radiolabeled form IV from the [6–4]TT-oriL was much lower [15]. Digestion with *PstI* revealed that efficiency of translesion synthesis of the [6–4]TT under these in vitro conditions was very low (at most 10 % of control) while [c,s]TT TLS was 70 % as efficient as replicating the undamaged molecule [15]. (Reproduced from ref. 15 with permission from the American Chemical Society).

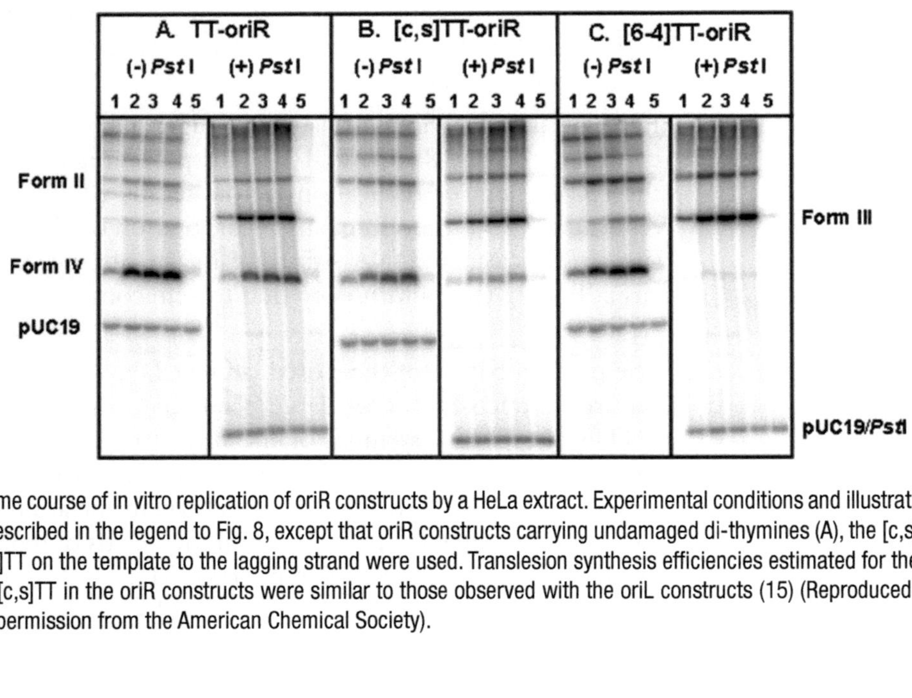

Fig. 9. Time course of in vitro replication of oriR constructs by a HeLa extract. Experimental conditions and illustration labels are as described in the legend to Fig. 8, except that oriR constructs carrying undamaged di-thymines (A), the [c,s]TT (B) or the [6–4]TT on the template to the lagging strand were used. Translesion synthesis efficiencies estimated for the [6–4]TT and the [c,s]TT in the oriR constructs were similar to those observed with the oriL constructs (15) (Reproduced from ref. 15 with permission from the American Chemical Society).

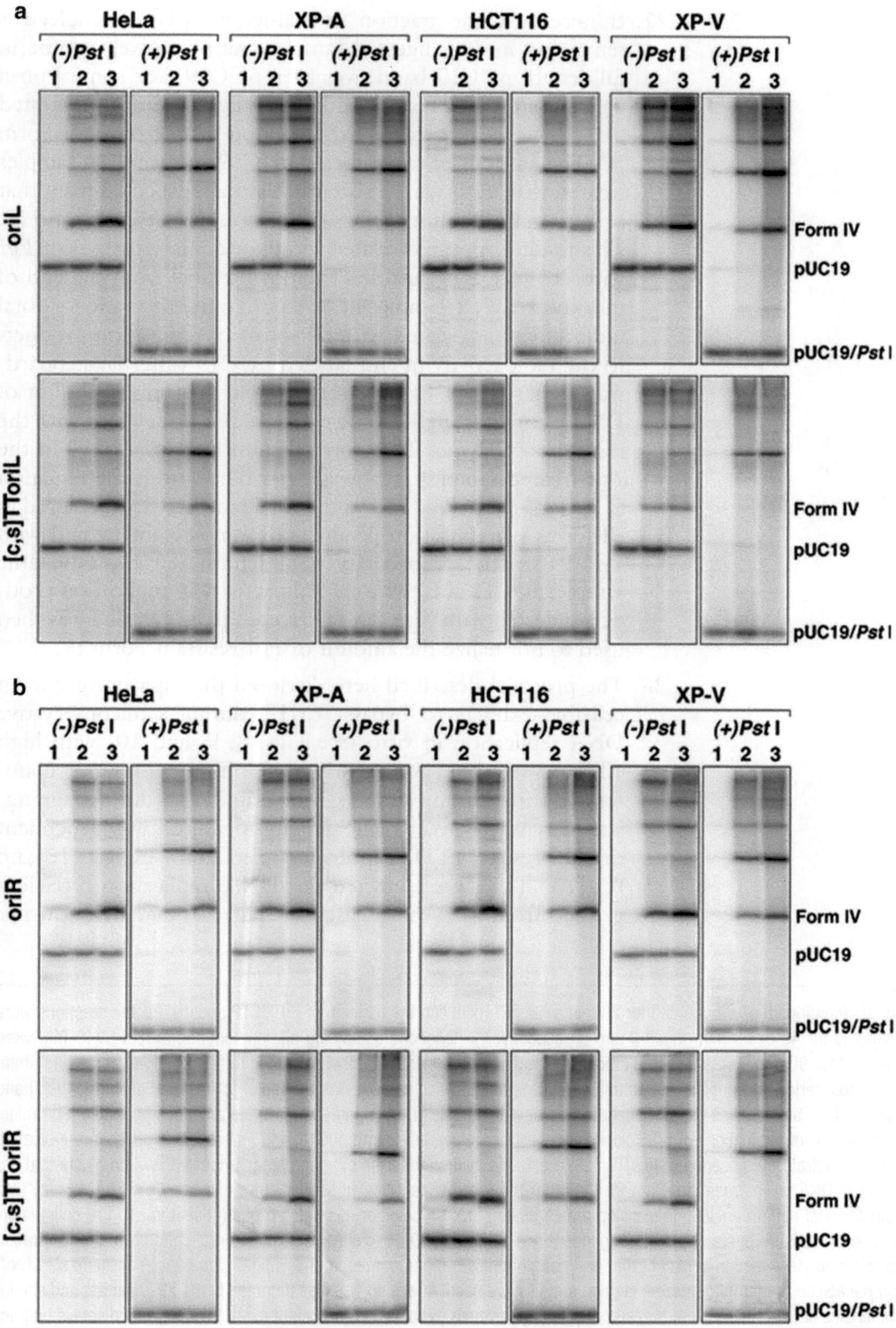

Fig. 10. In vitro replication of circular heteroduplexes by cell-free extracts from different human cells. (**a**) OriL constructs: Reaction mixtures contained either the control molecule with the T:T mismatch, but lacking the dimer (oriL, *top panels*), or the heteroduplex carrying both ([c,s]TToriL). Reaction mixtures containing all components, except dNTPs, were

27. If necessary, the fraction of full-length pUC19 molecules remaining in *Pst*I-digested samples (ratio of pixel volume in full-length pUC19 band over total pUC19, i.e., sum of pixel volumes in bands corresponding to digested and undigested pUC19) is used to correct the fraction of *Pst*I-resistant Form IV. The amount of radiolabeled pUC19 recovered in samples treated with *Pst*I was consistently less than expected from that present in the aliquot of the same sample not treated with *Pst*I. This finding appeared related to minor contaminations of *Pst*I with exonucleases capable of removing label from the end of the linear pUC19 molecules. Accordingly, the ratio of total radioactivity associated with all forms of replication products to the radioactivity in end-labeled pUC19 (internal standard) was higher in the *Pst*I-digested sample than in the aliquot of DNA purified from the same reaction, but not treated with the restriction enzyme. Therefore, the amount of pUC19 in the *Pst*I-treated samples (P2) was determined from the equation $T1 / P1 = T2 / P2$. T1 was the total radioactivity measured in all forms of replication products in samples not treated with *Pst*I, P1 was the radioactivity in full-length pUC19 in the same sample, and T2 was the total radioactivity in replication products digested with *Pst*I. This corrected pUC19 value was then used to normalize the amount of *Pst*I-resistant Form IV.
28. The protocol described here disclosed the capacity of human cell-free extracts to bypass [*c,s*]TT during semiconservative DNA replication in vitro (see Figs. 8, 9, and 10) with high efficiency (84 % on average when this photolesion was encountered during leading strand synthesis and 64 % during lagging-strand synthesis (32)). The activity measured was dependent on the presence of DNA polymerase eta in the extract (13); no [c,s]TT TLS was detected in DNA replication-proficient extracts from XP-V fibroblasts (see Fig. 10) that are known to

Fig. 10. (continued) pre-incubated for 20 min at 37 °C. Following the addition of [^{32}P]dCTP and dNTPs, the reactions were incubated for 45 or 90 min (*lanes 2* and *3*, respectively). Reactions containing all components, except SV40 Tag, were extended for 90 min (*lanes 1*). Equal amounts of radiolabeled linear pUC19 were added to the reactions at the end of the incubation period and prior to DNA purification. One half of each sample was treated with *PstI*. Undigested [(−)*PstI*] and digested [(+)*PstI*] samples were fractionated by electrophoresis in 1 % agarose gels in the presence of ethidium bromide. The positions of replication Form IV (closed circular molecules), full-length pUC19, and *PstI*-digested pUC19 are indicated. Results illustrate the absence of [c,s]TT TLS in extracts from xeroderma pigmentosum variant (XP-V) fibroblasts that are devoid of DNA polymerase eta. Data for the other three cell lines revealed similar efficiencies for [c,s]TT TLS; they also indicated that neither nucleotide excision repair (absent in XP-A cells) or mismatch repair (absent in HCT116 cells) interfered with measurements of pol eta-dependent [c,s]TT TLS. (**b**) OriR constructs. Experimental conditions and illustration labels are as described in **a**, except that oriR constructs carrying undamaged di-thymines (*top panels*) or the [c,s]TT (*bottom panels*) on the template for the lagging strand were used. Again, no TLS was detected in the XP-V extract and similar efficiencies of [c,s]TT TLS were measured for the other three extracts. TLS proficient extracts as a group displayed higher efficiency for bypass replication of a [c,s]TT on the leading-strand template than in the lagging-strand template (32). (Reproduced from ref. 32 with permission from the Elsevier).

lack this polymerase (13, 32). It is not yet clear if the low efficiency of [6–4]TT TLS observed in vitro (see Figs. 8 and 9) reflects only an intrinsic lower rate for bypass replication of this photolesion in human cells or a low abundance (or absence) in the extract of the specialized polymerase(s) that catalyze this reaction in vivo (5).

Acknowledgement

This research was supported by the US Public Health Service Award CA55065 from the National Cancer Institute, National Institutes of Health. We thank Dr. Tadayoshi Bessho for assistance in preparing lesion-containing oligonucleotides and Dr. Stephen Chaney for access to HPLC equipment (Department of Biochemistry and Biophysics, UNC Chapel Hill). We are grateful to Dr. Thomas Kunkel (NIEHS) for the gift of M13mp2SV oriL and oriR. We thank Dr. John J. McNulty for reading the manuscript.

References

1. Unk I, Hajdu I, Blastyak A, Haracska L (2010) Role of yeast Rad5 and its human orthologs, HLTF and SHPRH in DNA damage tolerance. DNA Repair (Amst) 9:257–267
2. Branzei D (2011) Ubiquitin family modifications and template switching. FEBS Lett. doi:10.1016/j.febslet.2011.04.053
3. Livneh Z, Ziv O, Shachar S (2010) Multiple two-polymerase mechanisms in mammalian translesion DNA synthesis. Cell Cycle 9: 729–735
4. Lange SS, Takata K, Wood RD (2011) DNA polymerases and cancer. Nat Rev Cancer 11:96–110
5. Shachar S, Ziv O, Avkin S, Adar S, Wittschieben J, Reissner T, Chaney S, Friedberg EC, Wang Z, Carell T, Geacintov N, Livneh Z (2009) Two-polymerase mechanisms dictate error-free and error-prone translesion DNA synthesis in mammals. EMBO J 28:383–393
6. Avkin S, Goldsmith M, Velasco-Miguel S, Geacintov N, Friedberg EC, Livneh Z (2004) Quantitative analysis of translesion DNA synthesis across a benzo[a]pyrene-guanine adduct in mammalian cells: the role of DNA polymerase kappa. J Biol Chem 279:53298–53305
7. Waga S, Stillman B (1998) The DNA replication fork in eukaryotic cells. Annu Rev Biochem 67:721–751
8. Thomas DC, Veaute X, Kunkel TA, Fuchs RP (1994) Mutagenic replication in human cell extracts of DNA containing site-specific N-2-acetylaminofluorene adducts. Proc Natl Acad Sci U S A 91:7752–7756
9. Thomas DC, Veaute X, Fuchs RP, Kunkel TA (1995) Frequency and fidelity of translesion synthesis of site-specific N-2-acetylamino fluorene adducts during DNA replication in a human cell extract. J Biol Chem 270:21226–21233
10. Carty MP, Lawrence CW, Dixon K (1996) Complete replication of plasmid DNA containing a single UV-induced lesion in human cell extracts. J Biol Chem 271:9637–9647
11. Svoboda DL, Vos JM (1995) Differential replication of a single, UV-induced lesion in the leading or lagging strand by a human cell extract: fork uncoupling or gap formation. Proc Natl Acad Sci U S A 92:11975–11979
12. Cordeiro-Stone M, Zaritskaya LS, Price LK, Kaufmann WK (1997) Replication fork bypass of a pyrimidine dimer blocking leading strand DNA synthesis. J Biol Chem 272: 13945–13954
13. Nikolaishvili-Feinberg N, Jenkins GS, Nevis KR, Staus DP, Scarlett CO, Unsal-Kacmaz K, Kaufmann WK, Cordeiro-Stone M (2008) Ubiquitylation of proliferating cell nuclear antigen

and recruitment of human DNA polymerase eta. Biochemistry 47:4141–4150
14. Masutani C, Araki M, Yamada A, Kusumoto R, Nogimori T, Maekawa T, Iwai S, Hanaoka F (1999) Xeroderma pigmentosum variant (XP-V) correcting protein from HeLa cells has a thymine dimer bypass DNA polymerase activity. EMBO J 18:3491–3501
15. Nikolaishvili-Feinberg N, Cordeiro-Stone M (2001) Bypass replication in vitro of UV-induced photoproducts blocking leading or lagging strand synthesis. Biochemistry 40:15215–15223
16. Zhao X, Kao JL, Taylor JS (1995) Preparation and characterization of a deoxyoligonucleotide 49-mer containing a site-specific thymidylyl-(3′,5′)-deoxyadenosine photoproduct. Biochemistry 34:1386–1392
17. Roberts JD, Kunkel TA (1988) Fidelity of a human cell DNA replication complex. Proc Natl Acad Sci U S A 85:7064–7068
18. Roberts JD, Thomas DC, Kunkel TA (1991) Exonucleolytic proofreading of leading and lagging strand DNA replication errors. Proc Natl Acad Sci U S A 88:3465–3469
19. Roberts JD, Izuta S, Thomas DC, Kunkel TA (1994) Mispair-, site-, and strand-specific error rates during simian virus 40 origin-dependent replication in vitro with excess deoxythymidine triphosphate. J Biol Chem 269:1711–1717
20. Li JJ, Kelly TJ (1984) Simian virus 40 DNA replication in vitro. Proc Natl Acad Sci U S A 81:6973–6977
21. Li JJ, Kelly TJ (1985) Simian virus 40 DNA replication in vitro: specificity of initiation and evidence for bidirectional replication. Mol Cell Biol 5:1238–1246
22. Roberts JD, Kunkel TA (1993) Fidelity of DNA replication in human cells. Methods Mol Genet 2:295–313
23. Smith CA, Taylor JS (1993) Preparation and characterization of a set of deoxyoligonucleotide 49-mers containing site-specific cis-syn, trans-syn-I, (6–4), and Dewar photoproducts of thymidylyl(3′->5′)-thymidine. J Biol Chem 268:11143–11151
24. Brown TA (2010) Gene cloning and DNA analysis: an introduction. Wiley-Blackwell, Chichester
25. Sambrook J, Fritsch EF, Maniatis T (1989) Molecular cloning: a laboratory manual cold spring harbor laboratory press. Cold Spring Harbor, New York
26. Kunkel TA (1985) Rapid and efficient site-specific mutagenesis without phenotypic selection. Proc Natl Acad Sci U S A 82:488–492
27. Kunkel TA, Bebenek K, McClary J (1991) Efficient site-directed mutagenesis using uracil-containing DNA. Methods Enzymol 204:125–139
28. Stillman BW, Gluzman Y (1985) Replication and supercoiling of simian virus 40 DNA in cell extracts from human cells. Mol Cell Biol 5:2051–2060
29. Stillman B (1986) Chromatin assembly during SV40 DNA replication in vitro. Cell 45:555–565
30. Friedberg EC, Walker GC, Siede W (1995) DNA repair and mutagenesis. ASM Press, Washington, D.C, pp 1–58
31. Wobbe CR, Dean FB, Murakami Y, Weissbach L, Hurwitz J (1986) Simian virus 40 DNA replication in vitro: study of events preceding elongation of chains. Proc Natl Acad Sci U S A 83:4612–4616
32. Cordeiro-Stone M, Nikolaishvili-Feinberg N (2002) Asymmetry of DNA replication and translesion synthesis of UV-induced thymine dimers. Mutat Res 510:91–106

Chapter 35

Quantitative Measurement of Translesion DNA Synthesis in Mammalian Cells

Omer Ziv, Noam Diamant, Sigal Shachar, Ayal Hendel, and Zvi Livneh

Abstract

Translesion DNA synthesis (TLS) is a DNA damage tolerance mechanism, in which specialized low-fidelity DNA polymerases bypass lesions that interfere with replication. This process is inherently mutagenic due to the miscoding nature of DNA lesions, but it prevents double strand breaks, genome instability, and cancer. We describe here a quantitative method for measuring TLS in mammalian cells, based on non-replicating plasmids that carry a defined and site-specific DNA lesion in a single-stranded DNA region opposite a gap. The assay is responsive to the cellular composition of TLS DNA polymerases, and TLS regulators. It can be used with a broad variety of cultured mammalian cells, and is amenable to RNAi gene silencing, making it a useful tool in the study of TLS in mammalian cells.

Key words: Translesion DNA Synthesis, TLS, DNA repair, Mutagenesis, DNA damage tolerance, Site-specific DNA damage, Gapped plasmid, Carcinogenesis

1. Introduction

Translesion DNA synthesis (TLS) is a mechanism by which mammalian cells overcome replication obstacles in the DNA. It is carried out by specialized DNA polymerases, which are characterized by low fidelity, and the ability to bypass a broad variety of DNA lesions. TLS is inherently mutagenic due to the miscoding of most DNA lesions, and therefore leads to the formation of point mutations. Yet, its activity prevents collapse of replication forks and the generation of double strand breaks, which may lead to genome instability and a variety of diseases most notably cancer. Thus, TLS prevents DNA breakage at the cost of point mutations, which is a less deleterious outcome of DNA damage (1–5).

Lotte Bjergbæk (ed.), *DNA Repair Protocols*, Methods in Molecular Biology, vol. 920,
DOI 10.1007/978-1-61779-998-3_35, © Springer Science+Business Media New York 2012

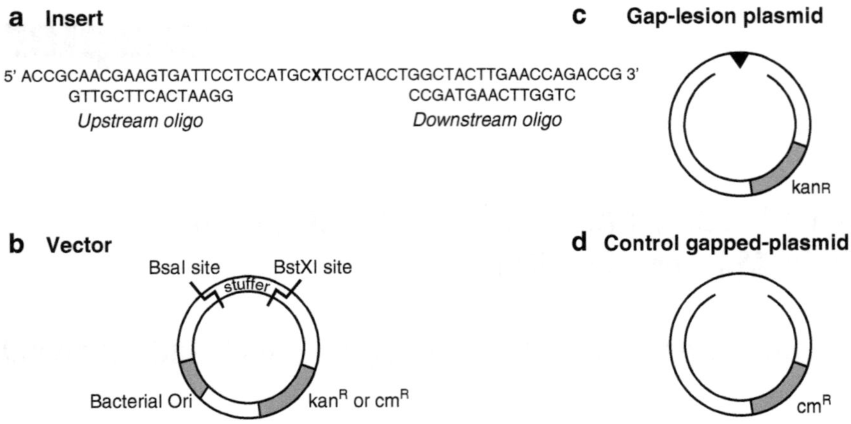

Fig. 1. Outline of the gap-lesion plasmid and the components needed for its construction. (**a**) An oligonucleotide insert carrying a defined and site-specific lesion (marked by X). (**b**) The vector into which the insert is ligated after removing the stuffer fragment by cleavage with restriction enzymes BsaI and BstXI. (**c**) The gap-lesion plasmid. The *black triangle* represents the lesion. (**d**) The control gapped plasmid without a lesion.

Several methods based on the progression of DNA replication can be used to evaluate TLS in mammalian cells (e.g., ref. 6); however, those methods are of low resolution, and do not specifically assay TLS. The measurement of induced mutagenesis in a chromosomal gene (e.g., *HPRT*; e.g., (1, 7)) can provide high-resolution data about mutagenic TLS, but is blind to accurate TLS. Several plasmid-based assay systems (e.g., ref. 8), and in particular those carrying defined and site-specific lesions (9–11) were very useful to elucidate the mutagenic specificity of certain DNA lesions in mammalian cells (12, 13); however, they are limited by their inability to provide a quantitative measurement of the extent of TLS.

Here we describe a quantitative method to measure specifically TLS, uncoupled from replication and the chromosomal context. It is based on a non-replicating plasmid, which carries a defined and site-specific DNA lesion in a single-stranded DNA (ssDNA) region opposite a gap (gap-lesion plasmid) (14, 15). Briefly, mammalian cells are transfected with a gap-lesion plasmid, along with a control, gapped plasmid without a lesion (Fig. 1c–d). The gap-lesion plasmid contains a kanamycin-resistance gene, while the control gapped-plasmid contains a chloramphenicol-resistance gene. After allowing for TLS to occur in the cells, covalently closed plasmids are extracted, introduced into an indicator TLS-defective *Escherichia coli* strain, and plated in parallel on LB-kanamycin and LB-chloramphenicol plates. The ratio of kan^R/cm^R *E. coli* colonies is a measure of the efficiency of TLS in the mammalian cells, and sequence analysis of plasmids extracted from individual kan^R colonies provides data on DNA sequence changes during TLS (Fig. 2). It should be noticed that the assay is designed for DNA lesions that block replication, and therefore need TLS polymerases for

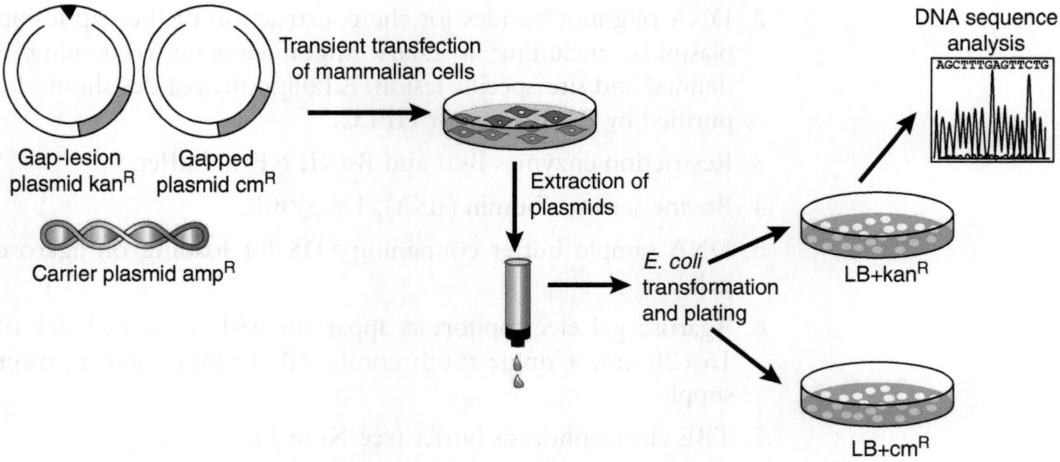

Fig. 2. Outline of the quantitative TLS assay. Mammalian cells are co-transfected with an equi-molar mixture of a gap-lesion plasmid (kan^R), and a control gapped-plasmid without a lesion (cm^R). A third plasmid (amp^R) serves as a carrier to improve transfection efficiency. Following an incubation period of 1–24 h to allow TLS, the plasmids are extracted under alkaline conditions followed by renaturation, after which only covalently closed plasmids end up non-denatured. The plasmids are then introduced into an indicator *E. coli recA*-deficient strain, and plated in parallel on LB-kanamycin and LB-chloramphenicol plates for analysis. The ratio of kan^R/cm^R transformants represents the extent of gap repair, which is mostly via TLS. Individual kan^R colonies are picked, and their plasmid contents analyzed for mutations in the DNA region corresponding to the original site of the gap. To obtain the extent of TLS from the extent of gap repair, the latter are multiplied by the percentage of TLS events out of the total events, as determined by the DNA sequence analysis.

bypass. These include many common lesions such as abasic (apurinic/apyrimidinic) sites, cyclobutane pyrimidine dimers, 6–4 photoproducts, cisplatin-GG intrastrand adducts, benzo[*a*]pyrene-G adducts, and many more. Weakly blocking lesions, such as 8-oxoguanine, require special adaptation, and the use of specialized indicator *E. coli* strains (16).

The assay was found to be a very useful model system to study TLS in mammalian cells, perhaps because it is a good model system of postreplication gaps. It is responsive to the TLS DNA polymerases content of cells (15, 17–19), and is affected by the activity of TLS regulators such as PCNA ubiquitination (20), p53, and p21 (21). Having both TLS extent and mutability readouts enables to study in details the mechanism by which TLS operates in vivo, leading, for example, to the demonstration that TLS operates via two-polymerases reactions (18, 19).

2. Materials

2.1. Preparation of a Gap-Lesion Plasmid

1. Plasmids pSKSL-kan^R and pSKSL-cm^R (15), or other plasmids suitable for the preparation of the gap-lesion plasmid and the control gapped-plasmid.

2. DNA oligonucleotides for the construction of the gap-lesion plasmids, including a DNA oligonucleotide containing a defined and site-specific lesion. All oligonucleotides should be purified by PAGE-urea or HPLC.
3. Restriction enzymes BsaI and BstXI, NEB3 buffer.
4. Bovine serum albumin (BSA), 1 mg/mL.
5. DNA sample buffer containing SDS for loading on agarose gels.
6. Agarose gel electrophoresis apparatus with a gel bed size of 15 × 20 cm, a single-tooth comb, 13-cm long, and a power supply.
7. TBE electrophoresis buffer (see Note 1).
8. Electro-elution system (Elutrap), Schleicher & Schuell Bio-Science.
9. 3 M NaOAc, pH 5.2.
10. Ethanol absolute.
11. TE buffer: 10 mM Tris–HCl, pH 8.0, 1 mM EDTA.
12. T4 Polynucleotide Kinase (PNK), PNK buffer.
13. 10–25 mM ATP.
14. 5 M NaCl, dissolved in double-distilled water (DDW).
15. T4 DNA ligase and ligase buffer.
16. SpeedVac apparatus.
17. Formamide.
18. Polyacrylamide gel 12 %, 8 M urea, glass plates, spacers, and a comb to build a preparative gel (see Note 2).
19. γ^{32P}-ATP and appropriate equipment to carry out radioactive reactions (see Note 3).
20. UVC hand-lamp, and thin layer chromatography (TLC) plastic plates (20 × 20 cm; Merck).
21. Spin-X column (Costar).
22. Micro-biospin-6 column (Bio-Rad).

2.2. Preforming the TLS Assay

1. Tissue culture facility.
2. Cultured mammalian cells, growth medium, trypsin-EDTA solution.
3. Carrier plasmid DNA (expressing antibiotics resistance other than kan^R or cm^R).
4. DNA Transfection reagent.
5. Phosphate-buffered saline (PBS).
6. Plasmid Miniprep kit (see Note 4).

7. *E. coli recA*-deficient cells competent for electroporation.
8. Electroporator, 0.2 cm cuvette.
9. SOC medium or other rich bacterial growth medium.
10. LB-agar plates containing 50 μg/mL kanamycin and LB-agar plates containing 30 μg/mL chloramphenicol.
11. A light table and a bacterial colony counter.

3. Methods

3.1. Preparation of the DNA Vector

The vector for constructing the gap-lesion plasmid contains a kanamycin-resistance gene, while for the control gaped-plasmid the vector contains a chloramphenicol-resistance gene, each under a constitutive bacterial promoter (Fig. 1b, see Note 5). Both vectors contain a bacterial origin of replication, but not a eukaryotic one. They should also contain single cleavage sites for restriction nucleases BsaI and BstXI (see Note 6) separated by a stuffer fragment 400–1,000 nucleotides long (see Note 7). The vector should be no bigger than 3.5 kb in length (excluding the stuffer; see Note 7).

1. Double digest 1 mg of each plasmid with both BsaI and BstXI restriction enzymes. Each reaction mixture contains 600 units of each enzyme, 1× NEB3 buffer, and 100 μg/mL BSA in a final volume of 2 mL. Divide each reaction into eight 250 μL aliquots, incubate at 37 °C for 4 h, and heat-inactivate at 65 °C for 20 min. Analyze the extent of digestion by electrophoresis on a 0.8 % agarose gel (see Note 8).
2. Add 0.25 volumes of fivefold concentrated sample buffer containing SDS, and load on four preparative 0.8 % agarose gels with a single-tooth comb each (a single 13 cm well). Run in TBE buffer at 60 V for about 12 h (see Note 1). The DNA should run a distance of approximately 10 cm.
3. Excise the desired band from the gel, and elute the DNA using an electro-elution system, at 200 V for 3 h. Collect the DNA into a 1.5 mL tube.
4. Ethanol-precipitate the DNA (see Note 9) and dissolve in TE buffer to a final concentration of 0.5–2 mg/mL. This usually produces 300–400 μg DNA vector.

3.2. Elongation of DNA Oligonucleotide Containing a Lesion

The oligonucleotide carrying the defined and site-specific lesion has to be 54–65 bases long and flanked by sequences complementary to the Bsa1 and BstX1 restriction sites (Fig. 1a). For some types of DNA lesions, such as an basic site, such an oligonucleotide can be ordered from a commercial source, and then purified by PAGE-urea, and used for building the gap-lesion plasmid, as described in

```
                         ligation        ligation
                            ↓               ↓
        left oligo          CATGCXTCCTAC          right oligo
5' ACCGCAACGAAGTGATTCCTC                CTGGCTACTTGAACCAGACCG 3'
              CACTAAGGAGGTACGAAGGATGGACCGATGAA
                          scaffold
```

Fig. 3. The oligonucleotides used for elongating a short oligonucleotide carrying the site-specific lesion. A specific example of a short oligonucleotide carrying the site-specific lesion (marked by X) is presented (*bold type*), along with the two flanking oligonucleotides ("left" and "right") and the scaffold oligonucleotide. Whereas the "left" and "right" oligonucleotides can be generic, the scaffold oligonucleotide must be altered to base pair with the specific lesion-oligonucleotide that is being used. *Arrows mark* the sites were the ligation reaction take place.

the next step. For other types of lesions, collaborations must be established with chemists who study specific DNA lesions. The oligonucleotides that contain certain DNA lesions are often very short (11–13 nucleotides in length) to ease the task of getting them with high purity. In such cases they must be extended before they can be used to construct the gap-lesion plasmid. All oligonucleotides used in this step should be purified by PAGE-urea or HPLC.

1. In three separate reactions, phosphorylate 1.5 nmol of the lesion-oligonucleotide, 1.8 nmol of left-oligonucleotide, and 1.8 nmol of right-oligonucleotide (Fig. 3, see Note 10). Each reaction mixture contains 50 units of T4 PNK, and 1× T4 ligase buffer in a final volume of 90 μL (see Note 11). Incubate at 37 °C for 1 h, and heat-inactivate at 65 °C for 20 min.
2. Unite the three phosphorylation reactions, and add 1.8 nmol of scaffold-oligonucleotide. Add NaCl to a final concentration of 150 mM (total volume of approximately 300 μL). Anneal the four oligonucleotides by heating to 70 °C, and then cooling down slowly to room temperature at a rate of approximately 1 °C per minute (see Note 12).
3. Add ATP to a final concentration of 1 mM, and 4,000 units of T4 DNA ligase. Ligate the oligonucleotides at 16 °C overnight, and then heat-inactivate at 65 °C for 10 min (see Note 3).
4. Reduce the volume of the reaction in a SpeedVac apparatus to approximately 40 μL, or just before sediments appear. Add an equal volume of formamide and load on a 12 % polyacrylamide gel-8 M urea. Run in TBE buffer at 35 mA for 3–4 h (see Note 2). Cut the band from the gel by UV-shadowing with a UVC lamp and a TLC plate (Fig. 4, see Note 13).
5. Crush the gel slice in a 2 mL test tube and add 1 mL DDW. Shake the tube at room temperature overnight. Gradually load the liquid and gel slices into a spin-X column and centrifuge at

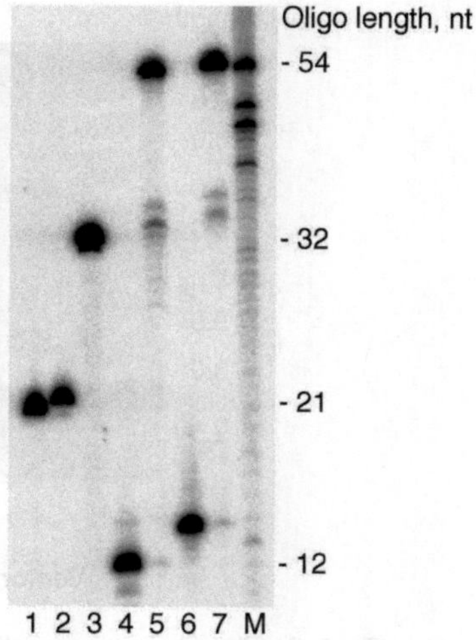

Fig. 4. Results of a small-scale elongation reaction with ^{32}P-labeled oligonucleotides, for the preparation of a 54 nucleotides-long oligonucleotide carrying a site-specific lesion. The oligonucleotides and reaction products were fractionated by PAGE-8 M urea, after which the gel was dried, and the oligonucleotides visualized by phosphorimaging. *Lane 1*, the "left" oligonucleotide. *Lane 2,* the "right" oligonucleotide. *Lane 3*, the scaffold oligonucleotide. *Lane 4*, a 12-nucleotides long oligonucleotide without a lesion, serving as a control for the elongation reaction. *Lane 5*, products of the elongation reaction of the ^{32}P-labeled 12 nucleotides long oligonucleotide without a lesion. *Lane 6*, a 12 nucleotides long DNA oligonucleotide with a site-specific cisplatin-GTG adduct. *Lane 7*, products of the elongation reaction of the ^{32}P-labeled 12 nucleotides long DNA oligonucleotide with a site-specific cisplatin-GTG-adduct. A non-purified 54 nucleotides long DNA-oligonucleotide served as a size marker (*lane* marked M).

full speed for 1–3 min. Transfer the flow-through to a new 1.5 mL tube.

6. Reduce the volume to 100–150 μL using a SpeedVac apparatus. Exchange the buffer using a micro-biospin-6 column according to the manufacturer's instructions, and then determine the concentration of the oligonucleotide using a NanoDrop instrument (see Note 14). Keep the oligonucleotide at −20 °C.

3.3. Construction of the Gap-Lesion Plasmid

1. In three separate reactions, phosphorylate 150 pmol of the elongated lesion-oligonucleotide, 240 pmol of the upstream-oligonucleotide and 240 pmol of the downstream-oligonucleotide (Fig. 1a, see Note 15). Each reaction mixture should contain 10 units of T4 PNK, 1× PNK buffer, and 1 mM ATP in a final volume of 20 μL. Incubate at 37 °C for 1 h, and then heat-inactivate at 65 °C for 20 min.

2. Combine the three phosphorylation reactions and add NaCl to a final concentration of 150 mM. Anneal by heating to 70 °C, and then cooling down slowly to room temperature at the rate of approximately 1 °C per minute (see Note 11)

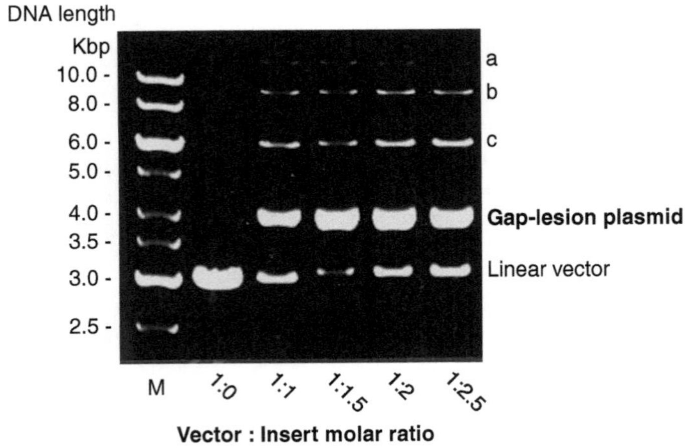

Fig. 5. Small-scale ligation reactions for the construction of a gap-lesion plasmid, carried out at several molar ratios of vector:insert. The ligation products were fractionated by agarose gel electrophoresis in the presence of ethidium bromide, and visualized under long wavelength UV light. The DNA bands marked a–c are ligation byproducts, most likely linear vector-insert-vector duplets, triplets, etc.

3. Ligate the 150 pmol insert to 100 pmol linear vector (see Note 16) in a total volume of 12 mL containing 1× ligation buffer and 60,000 units T4 DNA ligase. Incubate at 16 °C overnight. Do not heat-inactivate the ligation product. Run 20 μL ligation-products on 0.8 % agarose gel to evaluate the extent of ligation. The gapped plasmid runs just above the linear vector (Fig. 5).
4. Ethanol-precipitate the DNA (see Note 9), and then dissolve it in 500 μL TE buffer. Heat to 30–40 °C, and run on 0.8 % agarose gel containing ethidium bromide (EtBr) with a single-tooth comb (see Subheading 3.1).
5. Elute the appropriate band from the gel by electro-elution (see Subheading 3.1) and filter the liquid with a spin-X column.
6. Ethanol-precipitate the DNA, and then dissolve it in TE buffer to a final concentration of 50–100 ng/μL. Heat to 40 °C for 5 min, centrifuge at full speed for 10 min and transfer the liquid to a new tube, making sure to leave out the salt sediments. Divide the new substrate into several aliquots, and keep at −20 °C.

3.4. Quality Control of the Gap-Lesion Plasmid

1. Run 200 ng of the newly synthesized gap-lesion plasmid or control gapped-plasmid on a 0.8 % analytic agarose gel and inspect for purity. Make sure that there is little, if any, broken (linear) DNA (runs faster than the gapped plasmid, similarly to the vector used for its preparation).
2. Transform the *E. coli recA* indicator strain with 50 ng of the newly synthesized gap-lesion plasmid or control gapped-

plasmid, and plate on LB-agar plates containing kanamycin or chloramphenicol. The gap-lesion plasmid should give rise to *E. coli* colonies on LB-kanamycin plate (and not LB-chloram-phenicol plates), whereas the control gapped plasmid on LB-chloram-phenicol plates (but not LB-kanamycin plates).

3. Transform 100 μL of *E. coli recA*-deficient competent cells with an equi-molar mixture of the gap-lesion plasmid, and the control gapped-plasmid (about 5 ng each), using a cold 0.2 cm electroporation cuvette, and a single pulse of 2.5 or 3 kV. Recover the cells in 1 mL SOC medium at 37 °C for 1 h. Prepare 1:10, 1:100, and 1:1,000 serial dilutions in LB medium, and plate 100 μL cells of each dilution in duplicates in parallel on LB-agar plates containing kanamycin and chloramphenicol.
4. Choose the dilution with a countable number of colonies, and calculate the extent of plasmid repair in the bacterial cells by dividing the number of kan^R colonies (containing descendents of the gap-lesion plasmid), by the number of cm^R colonies (containing descendents of the control gapped plasmid). Since *recA*-deficient *E. coli* cells are incapable of performing TLS, there should be few kan^R colonies and many cm^R colonies. The ratio represents the background fraction of gapped plasmid without the lesion, reflecting mainly the purity of the lesion-containing oligonucleotide used for the construction. Good quality preparations have a background of about 1 %; however, if the extent of TLS is high for a particular lesion in mammalian cells, backgrounds up to 5 % can be acceptable.

3.5. Preforming a TLS Assay

5. Plate mammalian cells on 6 cm or 10 cm dishes. Incubate overnight under regular growth conditions. Cells should reach 50–70 % confluence on the day of transfection.
6. Transfect cells with an equi-molar mixture of the gap-lesion plasmid GP-kan^R, the control gapped plasmid GP-cm^R (approximately 50 ng each) and a carrier plasmid. (Fig. 2, see Note 17). As a control for contaminations, a mock transfection is performed with the carrier plasmid alone. Incubate the cells under normal growth conditions for up to 24 h (see Note 18)
7. Wash the cells three times with growth medium or PBS to remove the remaining DNA. Harvest the cells with trypsin-EDTA solution and collect into a 1.5 mL tube. Centrifuge the cells 3 min at 1,000 × *g* to remove the trypsin-EDTA solution. Wash with 1 mL PBS, and centrifuge again at 1000 × *g* to remove the PBS (see Note 19).
8. Extract the plasmids from the mammalian cells under alkaline conditions followed by renaturation. Under this procedure only covalently closed plasmids end up non-denatured (see Note 4). Elute the DNA in DDW and reduce the volume to 20 μL by a SpeedVac apparatus.

9. Transform 100 μL of *E. coli recA*-deficient competent cells with 1–20 μL of the DNA using a cold 0.2 cm electroporation cuvette, and a single pulse of 2.5 kV or 3 kV (see Note 20). Recover the cells in 1 mL SOC medium at 37 °C for 1 h.
10. Plate 50–200 μL cells (see Note 20) in duplicates or preferably triplicates, in parallel on LB-agar plates containing kanamycin or chloramphenicol (see Note 21).
11. Calculate the extent of plasmids repair (see Note 22) by dividing the number of gap-lesion plasmid, GP-*kan*^{R} transformants (average number of colonies obtained from 100 μL bacteria liquid on kanamycin plates) by the number of corresponding transformants of the normalizing gapped plasmid GP-*cm*^{R} (average number of colonies obtained from 100 μL bacteria on chloramphenicol plates).
12. To determine the DNA sequence changes that have occurred during plasmids repair, extract the DNA content from the kan^R colonies, and subject it to DNA sequence analysis at the region in the vicinity of the site of the lesion.
13. To obtain precise values of TLS extent, multiply the plasmid repair extent values by the fraction of TLS events out of all plasmid repair events, based on the DNA sequence analysis (see Note 22).

4. Notes

1. Do not use TAE running buffer instead of TBE since it has a lower buffer capacity, and is less suitable for prolonged procedures.
2. The dimensions of the gel are 39 × 20 cm, and the spacers used are 1.5 mm thick. We use a comb that contains 4 × 2 cm wells. One of the wells serves to load bromophenol-blue dissolved in formamide. Run until the bromophenol-blue approaches 2–3 cm from the edge of the gel.
3. It is recommended to perform the annealing and ligation reactions first on a small scale, with 1.5 pmol of ^{32}P-labeled lesion-oligonucleotide, and 1.8 pmol of each of the other oligonucleotides. Ligation efficiency is usually above 90 % (Fig. 4)
4. Many of the commercial mini-prep kits are based on alkaline conditions and are likely to be suitable for this task.
5. Antibiotic-resistance genes other than kanamycin and chloramphenicol can also be tried, as long as they are bactericidal and

do not allow the growth of satellite colonies (as in the case of ampicillin).

6. The use of non-palindromic restriction sites greatly improves ligation efficiency by reducing the formation of vector–vector duplets, triplets, etc. Other non-palindromic restriction sites apart from BsaI and BstXI can also be used, as long as their restriction reaction and subsequent ligation are highly efficient.
7. The stuffer should be at least 400 bp long to enable clear separation between single and double cleavage products. It is not recommended to design gapped-plasmid larger than 3.5 kb (excluding the stuffer), since it makes its purification more complicated, and reduces yields.
8. Highly efficient digestion reaction is critical to eliminate contamination of single-restricted products that will self-ligate in subsequent stages. In case of incomplete digestion, add additional 2.5 μL of each restriction enzyme into each aliquot and continue digestion until complete. Repeat heat-inactivation.
9. In order to precipitate DNA, add 0.1 volume of 3 M NaOAc pH 5.2, and 2 volumes of 100 % ethanol. Incubate at –20 °C overnight or in dry-ice for 1 h. Centrifuge at maximal speed for 1 h at 4 °C and remove the supernatant. Wash with cold 70 % ethanol, and centrifuge once again at maximal speed for 30 min at 4 °C. Remove the supernatant and dry the DNA pellet.
10. The left and right oligonucleotides should be flanked by sequences complementary to the BsaI and BstXI restriction sites, to allow ligation to the cleaved vector at a later stage (see Fig. 3). The scaffold oligonucleotide must be complementary to the annealed oligonucleotides and should be approximately 30–35 bases in length.
11. The phosphorylation reaction is carried out using T4-ligase buffer since the annealing stage is followed by an addition of T4 DNA ligase, and PNK performs well in ligase buffer. If the DNA lesion is sensitive to dithiothreitol (DTT), as in the case of platinum lesions, use a buffer that does not contain DTT.
12. Gradual cooling down to allow annealing can be preformed either by simply turning off the water bath, or using a PCR machine.
13. The band of the elongated oligonucleotide is the upper one. Below an additional band, equal in strength, of the scaffold-oligonucleotide can be seen (Not seen in Fig. 4 since only products containing ^{32}P are visible in this image). Ensure minimal exposure of the DNA to UV light, to prevent the formation of undesired UV damage. This is extremely important in

the case of cyclobutane pyrimidine dimers (CPD) since UV light can also reverse those DNA lesions.

14. Oligonucleotide molarity is calculated according to Beer–Lambert equation: $C = \text{OD} / \varepsilon$. Determine ε according to the sequence of the oligonucleotide in the following way:

$$\varepsilon_{(\text{liter/mole})} = 15.4 \times (\text{number of } A) + 7.3 \times (\text{number of } C) + 11.7 \times (\text{number of } G) + 8.8 \times (\text{number of } T).$$

15. Upstream and downstream oligonucleotides are 15–20 bases long each, purified by PAGE-urea. They are designed to leave 4 bases overhangs at each side for subsequent ligation to the vector, and 15–20 bases long gap opposite the DNA lesion (Fig. 1a).
16. It is recommended to perform the ligation reaction first on a small scale using 1.5 pmol vector and 1.5, 2.25, or 3 pmol insert (corresponding to vector:insert ratios of 1:1, 1:1.5, and 1:2, respectively) at a total volume of 120 μL containing 1× ligation buffer and 0.5 μL T4 DNA ligase. Run ligation products on 0.8 % agarose gel to evaluate the best vector:insert ligation ratio to be used in the large-scale ligation reaction. The gapped plasmid runs just above the linear vector (Fig. 5).
17. The transfection should be highly efficient and with minimal toxicity for the cells used. We have good experience in transfecting the gapped plasmids using cationic polymer reagents, liposomes, calcium-phosphate, and electroporation.
18. The gap-filling reaction in the control plasmid is very fast and usually happens within the first 1–2 h. The kinetics of the TLS reaction in the gap-lesion plasmid is slower and reaches plateau within 10–20 h, depending on the type of lesion (18).
19. Cell pellets can be frozen in liquid N_2 and stored at –80 °C, or directly used for DNA extraction.
20. It is recommended to transform and plate volumes that yield 100–1000 colonies/plate.
21. A separate plating-control electroporation is conducted with a DNA mixture containing 50 pg of non-damaged gapped plasmid bearing kan^R, and 50 pg of non-damaged gapped plasmid bearing cm^R. 100 μL is then plated in triplicates on both LB-kanamycin and LB-chloramphenicol plates. This control serves to normalize for possible different plating efficiencies on the two antibiotics.
22. The extent of plasmid repair usually gives a reliable estimation of the efficiency of TLS in the mammalian cells. However it may include also non-TLS events, most likely due to failure of TLS, followed by the formation of double-stranded breaks,

and subsequently repair such as nonhomologous end joining or recombination (18). These are observed as plasmid isolates with large deletions or insertions. They usually account for a small fraction of the events scored; however, their fraction may be high under certain conditions, e.g., when a particular TLS protein is missing.

Acknowledgement

ZL is the incumbent of the Maxwell Ellis Professorial Chair in Biomedical Research. This work was supported by grants to ZL from the Flight Attendant Medical Research Institute, Florida, USA, the Leona M. and Harry B. Helmsley Charitable Trust, NY, USA, and the Israel Science Foundation (no. 564/04 and 1136/08).

References

1. Friedberg EC, Walker GC, Siede W, Wood RD, Schultz RA, Ellenberger T (2006) DNA repair and mutagenesis (2nd edn.). ASM Press, Washington, DC
2. Prakash S, Johnson RE, Prakash L (2005) Eukaryotic translesion synthesis DNA polymerases: specificity of structure and function. Annu Rev Biochem 74:317–353
3. Yang W, Woodgate R (2007) What a difference a decade makes: insights into translesion synthesis. Proc Natl Acad Sci U S A 104: 15591–15598
4. Livneh Z (2006) Keeping mammalian mutation load in check. Regulation of the activity of error-prone DNA polymerases by p53 and p21. Cell Cycle 5:1918–1922
5. Livneh Z, Ziv O, Shachar S (2010) Multiple two-polymerase mechanisms in mammalian translesion DNA synthesis. Cell Cycle 9: 729–735
6. Jansen JG, Tsaalbi-Shtylik A, Hendriks G, Gali H, Hendel A, Johansson F, Erixon K, Livneh Z, Mullenders LH, Haracska L et al (2009) Separate domains of Rev1 mediate two modes of DNA damage bypass in mammalian cells. Mol Cell Biol 29:3113–3123
7. Gibbs PE, McGregor WG, Maher VM, Nisson P, Lawrence CW (1998) A human homolog of the Saccharomyces cerevisiae REV3 gene, which encodes the catalytic subunit of DNA polymerase zeta. Proc Natl Acad Sci U S A 95:6876–6880
8. Seetharam S, Kraemer KH, Waters HL, Seidman MM (1990) Mutational hotspot variability in an ultraviolet-treated shuttle vector plasmid propagated in xeroderma pigmentosum and normal human lymphoblasts and fibroblasts. J Mol Biol 212:433–436
9. Loechler EL, Green CL, Essigmann JM (1984) In vivo mutagenesis by O6-methylguanine built into a unique site in a viral genome. Proc Natl Acad Sci U S A 81:6271–6275
10. Napolitano R, Janel-Bintz R, Wagner J, Fuchs RPP (2000) All three SOS-inducible DNA polymerases (Pol II, Pol IV, and Pol V) are involved in induced mutagenesis. EMBO J 19:6259–6265
11. Szüts D, Marcus AP, Himoto M, Iwai S, Sale JE (2008) REV1 restrains DNA polymerase zeta to ensure frame fidelity during translesion synthesis of UV photoproducts in vivo. Nucleic Acids Res 36:6767–6780
12. Moriya M, Takeshita M, Johnson F, Peden K, Will S, Grollman AP (1988) Targeted mutations induced by a single acetylaminofluorene DNA adduct in mammalian cells and bacteria. Proc Natl Acad Sci U S A 85:1586–1589
13. Yoon JH, Prakash L, Prakash S (2009) Highly error-free role of DNA polymerase eta in the replicative bypass of UV-induced pyrimidine dimers in mouse and human cells. Proc Natl Acad Sci U S A 106:18219–18224
14. Avkin S, Adar S, Blander G, Livneh Z (2002) Quantitative measurement of translesion replication in human cells: evidence for bypass of abasic sites by a replicative DNA polymerase. Proc Natl Acad Sci U S A 99:3764–3769
15. Avkin S, Goldsmith M, Velasco-Miguel S, Geacintov N, Friedberg EC, Livneh Z (2004) Quantitative analysis of translesion DNA synthesis

across a benzo[*a*]pyrene-guanine adduct in mammalian cells. The role of DNA polymerase k. J Biol Chem 279:53298–53305

16. Avkin S, Livneh Z (2002) Efficiency, specificity and DNA polymerase-dependence of translesion replication across the oxidative lesion 8-oxoguanine in human cells. Mutat Res 510:81–90
17. Hendel A, Ziv O, Gueranger Q, Geacintov N, Livneh Z (2008) Reduced fidelity and increased efficiency of translesion DNA synthesis across a TT cyclobutane pyrimidine dimer, but not a TT 6-4 photoproduct, in human cells lacking DNA polymerase eta. DNA Repair 7:1636–1646
18. Shachar S, Ziv O, Avkin S, Adar S, Wittschieben J, Reissner T, Chaney S, Friedberg EC, Wang Z, Carell T et al (2009) Two-polymerase mechanisms dictate error-free and error-prone translesion DNA synthesis in mammals. EMBO J 28:383–393
19. Ziv O, Geacintov N, Nakajima S, Yasui A, Livneh Z (2009) DNA polymerase zeta cooperates with polymerases kappa and iota in translesion DNA synthesis across pyrimidine photodimers in cells from XPV patients. Proc Natl Acad Sci U S A 106:11552–11557
20. Hendel A, Krijger PH, Diamant, N, Goren, Z, Langerak P, Kim, Reißner, T, Lee, K, Geacintov N, Carell T, Myung, K, Tateishi S, D'Andrea, A, Jacobs H, and Livneh Z (2011) PCNA ubiquitination is important, but not essential for DNA damage bypass in mammalian cells. PLoS Genetics. Sep;7(9):e1002262
21. Avkin S, Sevilya Z, Toube L, Geacintov NE, Chaney SG, Oren M, Livneh Z (2006) p53 and p21 regulate error-prone DNA repair to yield a lower mutation load. Mol Cell 22:407–413

Chapter 36

Detection of PCNA Modifications in *Saccharomyces cerevisiae*

Adelina A. Davies and Helle D. Ulrich

Abtract

PCNA modifications by members of the ubiquitin family are associated with a range of different transactions during replication of damaged and undamaged DNA. This chapter describes detailed protocols for the detection and isolation of ubiquitin and SUMO conjugates of PCNA from total budding yeast cell lysates, using Ni-NTA affinity chromatography under denaturing conditions. We describe approaches based on the purification of PCNA itself and on the isolation of total ubiquitin or SUMO conjugates. The chapter covers the construction of the appropriate strains, methods for the detection of modified PCNA, and the use of various DNA-damaging agents as well as mutants of PCNA and relevant conjugation enzymes to examine the cellular response to replication stress.

Key words: PCNA, Ubiquitin, SUMO, DNA damage bypass, Postreplication repair, Ni-NTA affinity chromatography, Budding yeast

1. Introduction

1.1. PCNA Modifications in DNA Replication and Damage Bypass

The ability of cells to cope with replication stress is an important factor protecting us from genome instability and cancer (1). In addition to those pathways involved in sensing and repairing DNA lesions, cells have developed strategies of damage tolerance, allowing the bypass of lesions and hence the completion of replication in the presence of damage (2–4). DNA damage tolerance, also called postreplication repair (PRR), involves either the use of specialized DNA polymerases in a mutagenic process called translesion synthesis (5, 6) or an error-free processing of lesions by means of homologous recombination or template switching (4, 7). While the pathway contributes to genome maintenance, its mutagenic

Lotte Bjergbæk (ed.), *DNA Repair Protocols*, Methods in Molecular Biology, vol. 920,
DOI 10.1007/978-1-61779-998-3_36, © Springer Science+Business Media New York 2012

aspects can be a source of genomic instability. In eukaryotes, a major part of PRR is controlled by ubiquitylation of the sliding clamp, PCNA (8). Monoubiquitylation at a unique, highly conserved lysine residue, K164, by the ubiquitin-conjugating enzyme (E2) Rad6 in complex with the ubiquitin protein ligase (E3) Rad18 activates translesion synthesis via recruitment of damage-tolerant polymerases, most of which contain ubiquitin-binding domains that afford a preferential interaction with the modified clamp (9–14). Polyubiquitylation at the same site is achieved via extension of the monoubiquitin moiety by another E2–E3 pair, the heterodimeric E2, Ubc13-Mms2, and Rad5 (8, 15). The modification is required for the error-free template-switching pathway, although the molecular mechanism and the relevance of polyubiquitylated PCNA in this process are not well understood. Both mono- and polyubiquitylation of PCNA are induced by single-stranded (ss)DNA that accumulates at stalled replication intermediates, such as daughter-strand gaps, where the association of Replication Protein A, a ssDNA-binding complex, mediates recruitment of the ubiquitin ligase Rad18 to its site of action (16, 17). The pathway of PRR and the enzymes involved in PCNA ubiquitylation are highly conserved, although in higher eukaryotes the balance between mono- and polyubiquitylation—and as a consequence that between translesion synthesis and template switching—appears to be shifted much more towards the former (18). Moreover, budding yeast PCNA (encoded by *POL30*) undergoes an additional modification that is less conserved than its ubiquitylation. In a damage-independent manner, PCNA is modified by the small ubiquitin-like modifier SUMO (Smt3 in budding yeast) during undisturbed DNA replication (8). The modification is targeted to K164 and to a lesser extent at K127. Whereas the former site does not correspond to the well-known sumoylation consensus motif, ΨKXE, and requires the SUMO-conjugating enzyme Ubc9 and the SUMO ligase Siz1 for modification, K127 falls into a ΨKXE motif and is sumoylated by Ubc9 and another SUMO-specific E3, Siz2 (8, 19). Sumoylation of budding yeast PCNA causes the recruitment of an antirecombinogenic helicase, Srs2, by means of a C-terminal SUMO-interaction motif, and the subsequent inhibition of unscheduled recombination events during replication (20, 21). In addition, the modification causes an enhanced interaction with the genome maintenance factor Elg1 and prevents the association of PCNA with Eco1, a factor involved in cohesion establishment during S phase (22, 23). In contrast to ubiquitylation of PCNA, which requires the clamp to be associated with ssDNA in vivo and is therefore damage-dependent, PCNA sumoylation is constitutive during S phase and merely requires PCNA to reside on DNA for efficient modification both in vivo and in vitro (8, 19).

Reliable detection of ubiquitylated and sumoylated PCNA in vivo is essential for gaining information about how the modifications react to various environmental stimuli or investigating the cross talk of damage bypass and DNA replication with other cellular pathways. In addition, the differential requirements for PCNA modifications mean that the modified forms can serve as markers for DNA-related processes in vivo. Whereas sumoylated PCNA accurately traces the association of PCNA with chromatin and is therefore indicative of ongoing replication, ubiquitylated PCNA is associated with those types of DNA damage that specifically lead to replication problems. Importantly, the modification occurs independently of checkpoint activation and is thus detectable even in a checkpoint-deficient background (16, 17). Finally, excessive amounts of DNA damage cause a large fraction of cellular PCNA to become chromatin-associated and therefore sumoylated in a cell cycle-independent manner (8, 19).

1.2. Detection of Ubiquitin and SUMO Conjugates In Vivo

Key to the successful analysis of any posttranslational modification in vivo is a detection method that faithfully represents the proportion of the modified target after work-up of the samples. As ubiquitin and SUMO conjugates are often inherently unstable due to the presence of deconjugating enzymes in the cell, isolation of conjugates under fully denaturing conditions has proven most suitable for detection (24–33). In addition, PCNA ubiquitylation usually affects only those molecules associated with a stalled replication intermediate and is therefore difficult to detect in crude extracts, simply due to the low abundance of modified forms and possible cross-reactivity of relevant antibodies (16). Ni-NTA chromatography of hexahistidine (His)-tagged proteins, which can be performed in strongly denaturing conditions such as 8 M urea or 6 M guanidine, is therefore considered the most powerful approach for the identification and characterization of ubiquitin and SUMO targets in general.

This chapter describes two complementary approaches, differing in the position of the His-tag. The first approach is based on the use of N-terminally His-tagged PCNA (^{His}PCNA). Pull-down assays with this construct allow a representative isolation of PCNA and all its modifications, whose identities can be verified by immunoblotting. In this approach, the purification step mainly serves to enrich the target protein, thus overcoming the problem of detecting low levels of modified forms against a background of total cellular protein. The method is suitable for analysis of PCNA modification in response to the cell cycle, various environmental conditions or mutant backgrounds. An alternative strategy involves placing the His-tag onto the modifier itself for pull-down of either total ubiquitin (^{His}Ubi) or total SUMO (^{His}Smt3) conjugates and detection of modified PCNA among these by means of an anti-PCNA antibody. While this method does not permit the analysis of ubiquitin and SUMO conjugates simultaneously, it is particularly useful in experiments that involve detecting the modification of a number of different target proteins in parallel or for proteomic approaches. This chapter provides detailed instructions for the preparation of extracts under denaturing conditions and subsequent Ni-NTA affinity chromatography. The protocol is optimized for yeast PCNA, but can easily be adapted to other organisms or conjugates when combined with an appropriate lysis procedure or alternative detection methods. In addition, we discuss relevant controls for the successful isolation of modified PCNA and provide information about relevant constructs and yeast strains for the analysis (see Note 1).

2. Materials

2.1. Cultivation of Yeast Strains

1. YPD medium (liquid): 1 % (w/v) Bacto-yeast extract, 2 % (w/v) Bacto-peptone, 2 % (w/v) glucose in water. Sterilize by autoclaving and store at room temperature.
2. YPD plates: YPD medium as above, containing 2 % (w/v) Bacto-agar.
3. SC medium (liquid): synthetic complete medium lacking individual amino acids. Available from various suppliers (e.g., Sigma-Aldrich), but can also be prepared in the lab (34).
4. SC plates: SC medium as above, containing 2 % (w/v) Bacto-agar.
5. 5FOA plates: SC-Ura medium supplemented with uracil and adenine to a total of 0.04 mg/mL each, containing 1 mg/mL 5-fluoro-orotic acid (5FOA) and 2 % (w/v) Bacto-agar. To make the plates, prepare equal volumes of 4 % (w/v) agar in water (sterilize by autoclaving) and a twofold concentrated stock of all the other ingredients (filter-sterilize). Combine the two solutions, and quickly pour the plates.
6. $CuSO_4$ stock solution: 100 mM in H_2O. Filter-sterilize.
7. DNA-damaging agents: Methyl methanesulfonate (MMS, neat), 4-nitroquinoline-1-oxide (NQO, 3 mg/mL stock solution in DMSO), hydroxyurea (powder), hydrogen peroxide (H_2O_2, 30 % (v/v) solution in H_2O), bleomycin (Bleo, 10 mg/mL stock solution in H_2O), camptothecin (CPT, 2.5 mg/mL stock solution in DMSO).
8. Disposable Petri dishes (90 mm).
9. Culture tubes and flasks in sizes of 15 and 125 mL.
10. Incubators for plates, shake flasks and tubes at 30 °C.
11. Source of ionizing radiation (e.g., ^{137}Cs irradiator).
12. Handheld ultraviolet (UV) lamp (254 nm).

2.2. Strain Constructions

1. Isogenic yeast strains: wild type (*WT*), *rad18Δ*, *siz1Δ*, *siz1Δ siz2Δ* (see Note 2).
2. Plasmids: YIp128-P_{30}-^{His}POL30 (*WT*, *K164R* and *K127R*), YIp128-P_{30}-POL30 (*WT*, *K164R*, *K127/164R*), pBL243 (35), YEplac195 (36), YEp195-CUP-^{His}Ub, YEp195-CUP-^{His}Smt3 (see Note 3).
3. Restriction enzymes and buffers (e.g., New England Biolabs).
4. Sodium acetate solution: 3 M, pH 5.2.
5. Ethanol: absolute and 70 % (v/v) in H_2O.
6. LiT: 100 mM lithium acetate, 10 mM Tris–HCl, pH 7.4; sterilize by autoclaving.
7. PEG/LiT: dissolve 100 g polyethylene glycol 3350 (Sigma-Aldrich) in 100 mL LiT; sterilize by autoclaving (see Note 4).
8. TE buffer: 10 mM Tris–HCl, pH 8.0, 1 mM EDTA, pH 8.0.
9. HT DNA: 10 mg/mL sheared herring testes DNA (e.g., Promega).
10. Rotating wheel for microfuge tubes, Vortex mixer.
11. Heat blocks or water baths at 37 and 42 °C.

2.3. SDS-PAGE and Western Blot Analysis

1. Electrophoresis system for small-scale vertical SDS-polyacrylamide gels with power supply and semidry blotting apparatus.
2. NuPAGE 4–12 % Bis-Tris gradient gels (1.0 mm) and MOPS SDS running buffer, 20× (Invitrogen) or self-prepared gels and Tris–Glycine running buffer according to standard protocols (37).
3. Molecular weight marker: prestained, e.g., PeqGold Protein Marker IV (Peqlab).
4. Blotting buffer I: 300 mM Tris–HCl, pH 10.4, 15 % (v/v) methanol.
5. Blotting buffer II: 30 mM Tris–HCl, pH 10.4, 15 % (v/v) methanol.
6. Blotting buffer III: 25 mM Tris–HCl, pH 9.4, 40 mM 6-aminocaproic acid, 15 % (v/v) methanol.
7. PVDF membrane (Millipore or equivalent).
8. Gel blotting paper: extra thick (1.2 mm), e.g., Whatman GB005.
9. TBS: 50 mM Tris–HCl, pH 7.5, 150 mM NaCl.
10. $TBST_{0.1}$: TBS + 0.1 % (v/v) Tween 20.
11. $TBST_{0.1}$/milk: $TBST_{0.1}$ + 3 % (w/v) skimmed milk powder. This solution can be stored at 4 °C for a few days.

12. Primary antibodies: rabbit polyclonal antibodies against PCNA and Smt3 (both available from our lab); mouse monoclonal antibodies against ubiquitin (e.g., P4D1 Cell Signaling Technologies), the His_6-tag (e.g., HIS-1, Sigma), and 3-phosphoglycerate kinase (PGK, e.g., 22C5, Invitrogen) (see Note 5).
13. Secondary antibodies: goat anti-mouse and goat anti-rabbit horseradish peroxidase conjugates.
14. Chemiluminescence detection kit: ECL (GE Healthcare) or equivalent and autoradiography film (e.g., Amersham Hyperfilm ECL).
15. Heat block or water bath at 60 °C.

2.4. Preparation of Total Cell Extracts by TCA Lysis

1. NaOH/BME: 1.85 M NaOH, 7.5 % (v/v) 2-mercaptoethanol. Prepare immediately before use from a 2 M NaOH stock solution by adding 2-mercaptoethanol to a final concentration of 7.5 %.
2. TCA solution: 55 % (w/v) trichloroacetic acid in H_2O.
3. HU buffer: 8 M urea, 200 mM Tris–HCl, pH 6.8, 1 mM EDTA, 5 % (w/v) SDS, 0.1 % (w/v) bromophenol blue, 1.5 % (w/v) dithiothreitol. This buffer should be stored in 1 mL aliquots at −20 °C.
4. Neutralizing buffer: 1.5 M Tris–HCl, pH 8.8.

2.5. Denaturing Ni-NTA Affinity Chromatography

1. Buffer A: 6 M guanidine HCl, 100 mM sodium phosphate, pH 8.0, 10 mM Tris–HCl, pH 8.0. Store at 4 °C.
2. Buffer C: 8 M urea, 100 mM sodium phosphate, pH 6.3, 10 mM Tris–HCl, pH 6.3. Store at 4 °C.
3. Tween-20: 10 % (w/v) stock solution in H_2O. Store at room temperature.
4. Ni-NTA agarose: Qiagen or equivalent.
5. Imidazole: 1 M stock solution in H_2O. Store at 4 °C.

3. Methods

3.1. General Methods

3.1.1. Yeast Transformation

1. An exponentially growing yeast culture (see Note 6) in YPD medium is pelleted by centrifugation at 3,000 × *g* for 5 min. Aliquots of 10–15 mL per transformation are sufficient. Alternatively, 700 μL of a fresh overnight culture in YPD medium can be used (harvest in a microcentrifuge tube at 16,000 × *g* for 15 s). Transformation efficiencies will be somewhat higher with exponential cultures.

2. Completely remove the medium and resuspend the cells in 100 μL LiT solution per transformation. Mix the plasmid DNA with 10 μL of HT DNA and add the mixture to the cell suspension. Vortex briefly.
3. Add 500 μL of PEG/LiT, vortex for several seconds, and incubate on a rotating wheel for 15–30 min at room temperature.
4. Add 50 μL of DMSO and incubate 15 min in a 42 °C heat block. Shorter incubation times can be used for temperature-sensitive strains.
5. Pellet the cells by centrifugation at 500 × *g* for 30 s, resuspend in 75 μL water and spread onto a selective SC plate. Incubate the plate for 2–3 days at 30 °C.

3.1.2. SDS-Polyacrylamide Gel Electrophoresis and Western Blotting

1. Assemble a precast NuPAGE 4–12 % Bis-Tris gradient gel with MOPS SDS running buffer in a matching electrophoresis chamber. Alternatively, prepare a standard SDS-polyacrylamide gel according to the protocol and assemble in an electrophoresis chamber with Tris–Glycine running buffer (37).
2. Load the denatured samples along side an appropriate molecular weight marker (see Note 7) and run the gel at 150 V for 1 h or until the blue loading dye has reached the bottom of the gel.
3. Prepare a sheet of PVDF membrane in the size of your gel by soaking briefly in methanol, and then rinsing in water and incubating for several minutes in blotting buffer II (see Note 8).
4. Prepare six sheets of blotting paper per gel to fit the size of the gel. Soak two sheets in blotting buffer I and place onto the anode (+) of a semidry blotting apparatus. Soak one sheet in blotting buffer II and add to the stack. Add the PVDF membrane on top of this and then carefully place the acrylamide gel onto the membrane, making sure not to shift its position once it has touched the membrane (see Note 9). Soak three sheets of blotting paper in blotting buffer III and stack onto the gel. Roll over the surface of the stack with a glass pipette or test tube to remove any bubbles from between the layers.
5. Cover the gel stack with the cathode (−) of the blotting apparatus, set the maximum voltage to 25 V and blot at 40 mA for 75 min.
6. Disassemble the blotting apparatus and block unspecific binding sites on the membrane by incubation on a rocking platform in $TBST_{0.1}$/milk for 30 min at room temperature.
7. Incubate with primary antibody in $TBST_{0.1}$/milk for a minimum of 1 h at room temperature or overnight at 4 °C.
8. Wash the membrane three times for 10 min in TBST.

9. Incubate with secondary antibody in $TBST_{0.1}$/milk for 1 h at room temperature.
10. Wash the membrane as before and then incubate for 5 min in TBS.
11. Detect the signals by chemiluminescence using the detection kit according to the manufacturer's instructions. Obtain several different exposures of the membrane (see Note 10).

3.1.3. Small-Scale TCA Lysis

1. Harvest ca. 4×10^7 yeast cells (see Note 6) by centrifugation at 16,000 × *g* for 15 s.
2. Aspirate the supernatant and resuspend the pellet in 500 μL water. Place the tube on ice.
3. Add 75 μL of NaOH/BME, vortex, and incubate for 15 min on ice.
4. Add 75 μL of TCA solution, vortex, and incubate for 10 min on ice.
5. Centrifuge at 16,000 × *g* for 10 min at 4 °C.
6. Aspirate the supernatant and centrifuge again briefly (30 s) in order to collect residual buffer. Remove the last drops of liquid by aspiration.
7. Resuspend the pellet in 30 μL of HU buffer. If the pellets are not sufficiently dry, the suspension will turn green or yellow. In this case, add 1 μL of Neutralizing buffer and make sure that the resulting suspension turns blue.
8. Denature the sample at 60 °C for 10 min.
9. Centrifuge briefly before analyzing by SDS-PAGE. Samples can be stored at −20 °C, but it is important after thawing the samples to completely solubilize the SDS before loading on a gel.

3.1.4. Denaturing Ni-NTA Pull-Down

1. Harvest approximately 10^9 cells of the desired yeast culture (see Note 6) by centrifugation at 3,000 × *g* for 5 min (see Note 11). Remove the supernatant and place the tube on ice. Add 5 mL of ice-cold water, then 0.8 mL of NaOH/BME, resuspend the pellet by vortexing and incubate on ice for 20 min.
2. Add 0.8 mL of TCA solution, vortex, and incubate on ice for 20 min.
3. Centrifuge at 8,000 × *g* for 20 min at 4 °C.
4. Remove the supernatant and centrifuge again for 2 min in order to collect residual buffer. Remove the last drops of liquid by aspiration.
5. Resuspend the pellet in 1 mL of buffer A, transfer to a microcentrifuge tube and incubate in a shaking heat block at room temperature for 1 h to solubilize the precipitate completely.

6. Centrifuge at 16,000 × *g* for 10 min at 4 °C and transfer the supernatant to a fresh tube. Discard the pellet (see Note 12).
7. Prepare a 20 μL aliquot of Ni-NTA agarose by washing three times with buffer A containing 0.05 % Tween-20. When preparing several aliquots, these can be washed together and distributed into individual tubes at the last wash step (see Note 13).
8. Add the prepared cell extract to the tube containing the agarose. Then add 5 μL of 10 % Tween-20 and 15 μL of imidazole solution (see Note 14). Incubate overnight at room temperature on a rotating wheel.
9. Centrifuge the agarose beads at 200 × *g* for 15 s and remove the supernatant. Wash the pellet twice with 1 mL of buffer A containing 0.05 % Tween-20 and four times with 1 mL of buffer C containing 0.05 % Tween-20. Remove the supernatant completely after the last wash.
10. Add 30 μL of HU buffer and denature the sample at 60 °C for 10 min. Centrifuge briefly before analyzing by SDS-PAGE. Samples can be stored at −20 °C, but need to be reheated before loading in this case.

3.1.5. MMS Sensitivity Assays

1. Freshly prepare YPD agar plates containing the desired concentration of MMS. It is advisable to always use a range of at least two different concentrations, determined empirically.
2. Prepare exponential cultures of the relevant yeast strains in YPD medium.
3. Dilute the cultures to $OD_{600} = 0.2$ and prepare three tenfold serial dilutions from these.
4. Spot aliquots of 3.5 μL each (starting with $OD_{600} = 0.2$) onto the plates containing MMS and a control YPD plate.
5. Incubate the plates at 30 °C for 2–3 days before recording the growth pattern by photography or scanning.

3.1.6. UV Sensitivity Assays

This is a protocol for a quick and simple, but rather crude determination of relative UV sensitivities. It merely serves to compare the sensitivities of different strains with each other and should not be used where accurate measurements of UV doses are required.

1. Prepare exponential cultures of the relevant yeast strains in YPD medium.
2. Onto the back of a standard YPD plate, draw the outlines of a square of 6 × 6 cm and label two opposite faces with small tick marks, spaced at 0.5 cm intervals.
3. Dilute the cultures to an $OD_{600} = 0.5$. Place a drop of 12 μL of the first culture onto the plate at one of the unmarked borders of the square and let the drop run to the opposite end by carefully

tilting the plate. Wait until the liquid has been drawn into the plate and the surface appears dry. Then repeat this procedure for the remaining strains until a series of parallel streaks covers the plate.

4. Place the plate, with its lid removed, but covered by a square piece of cardboard or other UV-intransparent material, under a handheld UV lamp mounted at 40–50 cm distance from the table top. The exact distance depends on the strength of the lamp and should be determined empirically. Using gloves and face protection, gradually uncover the plate in a step-wise fashion over a total of 3 min, using the 0.5 cm grid marks on the plate as a guide (i.e., uncovering one 0.5 cm segment every 30 s). This results in the yeast strains being exposed to a gradient of UV radiation (0–3 min). The dose range can be varied by changing this timing. The procedure should be performed in the dark or at red or yellow light to prevent photoreactivation.
5. Incubate the plate at 30 °C in the dark for 2–3 days before recording the growth pattern by photography or scanning.

3.2. Modifications of ^{His}PCNA: Basic Experimental Setup

PCNA modifications will be detected using His-tagged *POL30* alleles and endogenous yeast ubiquitin and Smt3. In this setup, the Ni-NTA pull-downs serve to enrich the modified forms of PCNA against a background of nonspecific bands cross-reacting with the anti-PCNA antibody. This procedure significantly enhances the sensitivity of detecting low-abundance conjugates and allows the visualization of polymeric ubiquitin and SUMO chains on PCNA using relevant antibodies (8, 38). We detect the modification patterns on *WT* PCNA as well as mutants in the two major ubiquitin and/or SUMO acceptor sites, K127 and K164, in the absence and presence of DNA damage inflicted by the alkylating agent methyl methanesulfonate (MMS). A control strain, harboring untagged *POL30*, is processed in parallel to account for nonspecific signals from the antibodies.

3.2.1. Construction of Yeast Strains

His-tagged alleles of *POL30* are introduced into the host strain on an integrative plasmid inserted at the *LEU2* locus, followed by replacement of the endogenous allele by the *URA3* marker (Fig. 1).

1. Digest 0.5 μg of plasmid YIp128-P_{30}-^{His}*POL30* and mutant versions (*K127R* and *K164R*) (see Note 2) with *Afl*II in a total volume of 20 μL each.
2. Precipitate the DNA by adding 2 μL of sodium acetate solution and 50 μL of absolute ethanol and incubation at −20 °C for 30 min. Recover the DNA by centrifugation at 16,000 × *g* for 10 min at 4 °C and removal of the supernatant, rinse the

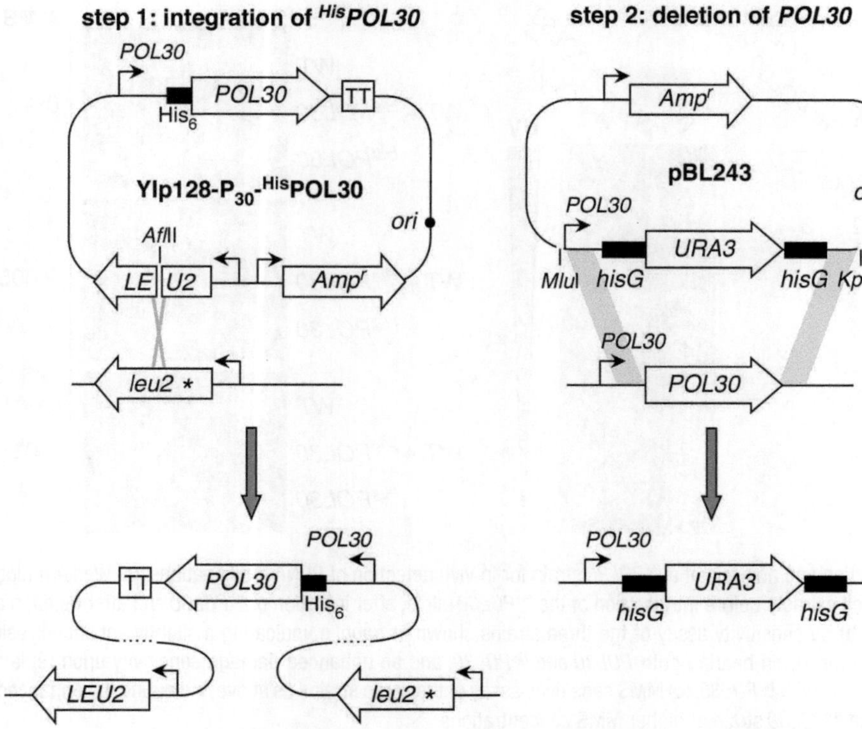

Fig. 1. Strategy for the construction of ^{His}*POL30* strains for in vivo analysis of PCNA modifications. In step 1, transformation of *WT* yeast cells with *Afl*II-digested YIp128-P$_{30}$-^{His}POL30 leads to integration at the *LEU2* locus. In step 2, endogenous *POL30* is deleted by replacement with the *hisG-URA3-hisG* marker, derived from an *Mlu*I-*Kpn*I fragment of the plasmid pBL243. Recombination events are indicated by *gray shading*. *Amp^r*: Ampicillin resistance marker (encoding β-lactamase) for selection in *Escherichia coli*; *ori*: *E. coli* replication origin; TT: transcriptional terminator.

pellets with 70 μL of 70 % ethanol, and centrifuge as before. Resuspend in 5 μL of TE buffer.

3. Use all of the digested DNA for transformation of *WT* yeast cells as described in Subheading 3.1.1. Spread the cells onto SC-Leu plates and incubate at 30 °C for 2–3 days.
4. Inoculate several 2 mL cultures in liquid SC-Leu medium with isolated colonies, shake overnight at 30 °C, and prepare TCA lysates for analysis by anti-PCNA Western blot (see Subheadings 3.1.2–3.1.3). For each allele of ^{His}*POL30*, select a colony that shows an additional band of equal intensity above endogenous PCNA, corresponding to the His-tagged protein (Fig. 2a, lane 2). Restreak these colonies onto a fresh SC-Leu plate and incubate at 30 °C for 2–3 days to obtain single clones (see Note 15).
5. Digest 1 μg of plasmid pBL243 (35) for each strain with *Mlu*I and *Kpn*I and precipitate the DNA as described in step 2.
6. For each of the ^{His}*POL30* alleles, inoculate an overnight culture of the strain from step 4, harboring a correct construct, in

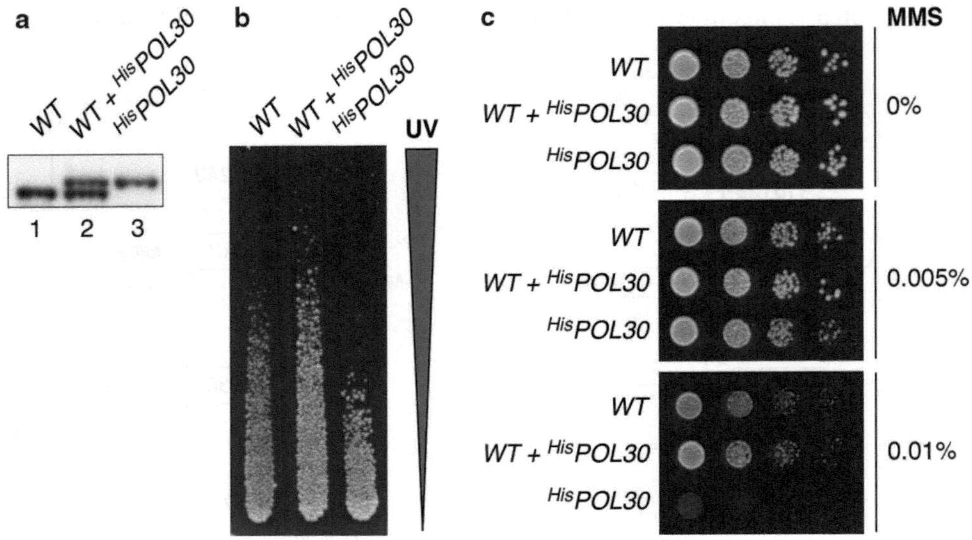

Fig. 2. Construction and analysis of a ^{His}*POL30* strain for in vivo detection of PCNA modifications. (**a**) Western blot showing PCNA in total cell extracts before introduction of the ^{His}*POL30* allele, after insertion of ^{His}*POL30* and after deletion of endogenous *POL30*. (**b**) UV sensitivity assay of the three strains shown in *panel* **a**, indicating a slightly enhanced resistance to DNA damage in the strain bearing both *POL30* and ^{His}*POL30*, and an enhanced damage sensitivity upon replacement of endogenous *POL30* with ^{His}*POL30*. (**c**) MMS sensitivity assay of the same strains as above, indicating the enhanced damage sensitivity of the ^{His}*POL30* strain at higher MMS concentrations.

liquid YPD medium and use this for transformation with 1 μg each of *Mlu*I/*Kpn*I-digested pBL243. Spread onto SC-Ura plates and incubate at 30 °C for 2–3 days.

7. Inoculate several 2 mL cultures in liquid SC-Ura medium with isolated colonies, shake overnight at 30 °C, and prepare TCA lysates for analysis by anti-PCNA Western blot as above. For each ^{His}*POL30* allele, select a clone that has lost the endogenous, untagged PCNA (Fig. 2a, lane 3). Restreak this clone onto a fresh SC-Ura plate and incubate at 30 °C for 2–3 days to obtain single clones (see Note 15).

3.2.2. Pull-Down of ^{His}PCNA Under Denaturing Conditions

1. Use the clones described above, together with an unmodified *WT* strain, for inoculation of 6 mL cultures in YPD and shake at 30 °C overnight.
2. Use the overnight cultures to inoculate three 50 mL cultures per strain in the same medium at a starting cell density of $OD_{600} \approx 0.3$. Shake at 30 °C until the OD_{600} is around 1.0.
3. Add methyl methanesulfonate (MMS) to 0.02 % (v/v) and 0.3 % (v/v), respectively, to two of the cultures of each strain and continue the incubation for an additional 90 min.
4. Harvest 10^9 cells and perform denaturing Ni-NTA pull-downs as described in Subheading 3.1.4.

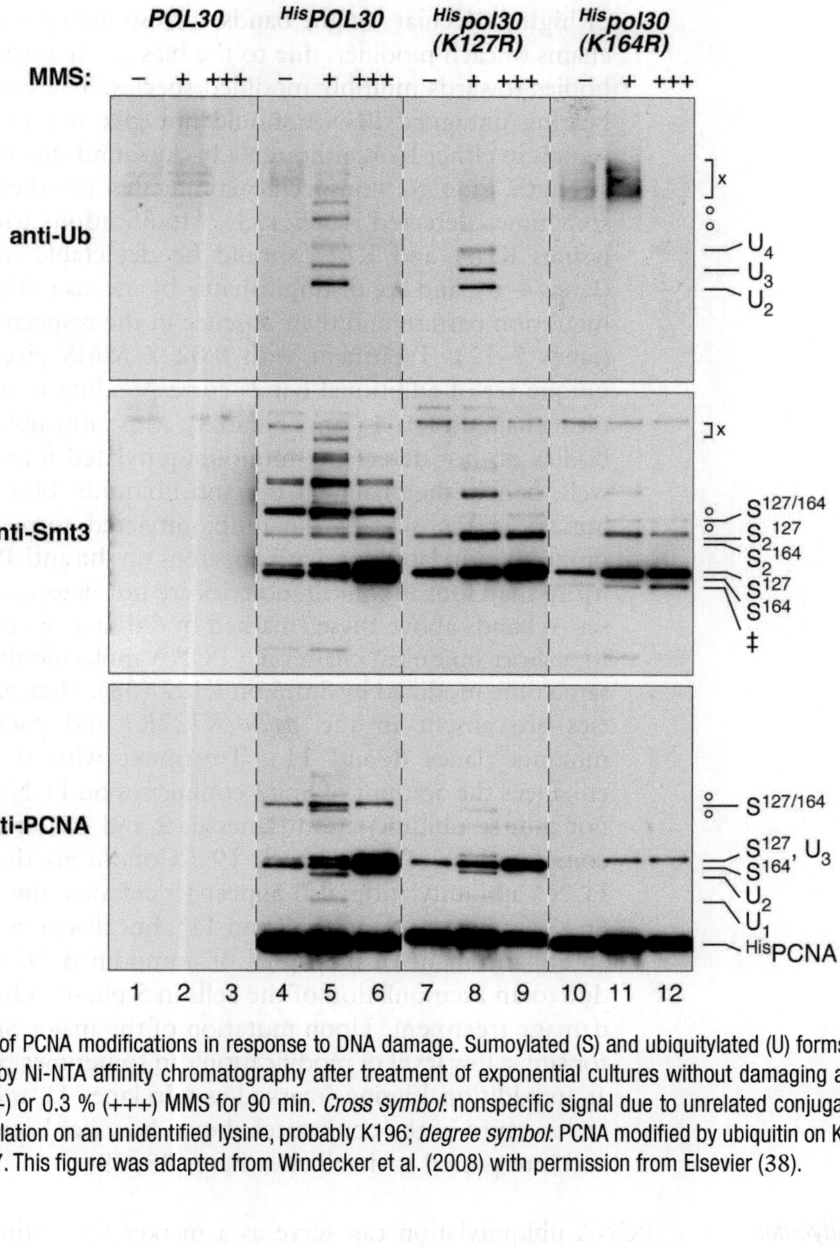

Fig. 3. Pattern of PCNA modifications in response to DNA damage. Sumoylated (S) and ubiquitylated (U) forms of HisPCNA were isolated by Ni-NTA affinity chromatography after treatment of exponential cultures without damaging agent (–) or with 0.02 % (+) or 0.3 % (+++) MMS for 90 min. *Cross symbol*: nonspecific signal due to unrelated conjugates; *double dagger*: sumoylation on an unidentified lysine, probably K196; *degree symbol*: PCNA modified by ubiquitin on K164 and by SUMO on K127. This figure was adapted from Windecker et al. (2008) with permission from Elsevier (38).

3.2.3. Detection of Ubiquitin and SUMO Conjugates

1. Assemble three NuPAGE 4–12 % Bis-Tris gradient gels (see Note 16) and load all pull-down samples onto each of them. Run the gels and blot onto PVDF membranes.
2. Develop one of the membranes with anti-PCNA, one with anti-ubiquitin and one with anti-Smt3 antibodies.
3. Analysis of the pull-down samples by SDS-PAGE and Western blot exemplifies the sensitivity of the detection method (Fig. 3). The anti-PCNA blot shows the physiological ratio of unmodified to modified PCNA under each condition, while the anti-Smt3 and anti-ubiquitin blots allow for the detection

of high-molecular weight bands corresponding to polymeric chains of each modifier, due to the bias of the respective antibodies towards multiply modified species. The control strain bearing untagged PCNA should not give rise to significant signals in either blot, although a background signal of varying strength, due to nonspecific conjugates (marked as ×), is sometimes detected (lanes 1–3). Modifications with Smt3 at lysines K164 and K127 should be detectable in *^His^POL30* (lanes 4–6), and are distinguishable by means of their distinct migration pattern and their absence in the respective mutants (lanes 7–12). Treatment with 0.02 % MMS gives rise to a unique set of additional bands corresponding to a polyubiquitin chain on K164 (lanes 5 and 8). Most anti-ubiquitin antibodies do not detect the monoubiquitylated form of PCNA well; hence, the chain on the anti-ubiquitin blot appears to initiate at the size of the diubiquitylated form, while the monoubiquitylated species is apparent on the anti-PCNA blot. More than four ubiquitin moieties are not detectable, and the set of bands above these (marked by ° in lane 5) corresponds to a short ubiquitin chain on a PCNA molecule that is at the same time modified by Smt3 on K127 (38). Hence, these species are absent in the *pol30(K127R)* and *pol30(K164R)* mutants (lanes 8 and 11). Treatment with 0.3 % MMS enhances the amount of Smt3 conjugates on PCNA, but does not induce ubiquitylation (lanes 6, 9 and 12) and results in complete loss of viability (8, 19). Conditions that result in PCNA ubiquitylation also appear to enhance the amount of Smt3 conjugates (lanes 5, 8 and 11), but this goes along with an enhancement of the signal of unmodified PCNA as well, due to an accumulation of the cells in S phase induced by the damage treatment. Upon mutation of the major Smt3 acceptor site, a low level of modification can sometimes be observed at an additional lysine (marked as ‡ in lanes 11 and 12). The architecture of the conjugates shown here has been described and interpreted in detail elsewhere (38, 39).

3.3. Ubiquitylation of ^{His}PCNA: Response to Different Types of DNA Damage

PCNA ubiquitylation can serve as a marker for distinct types of damage, based on the notion that single-stranded DNA, covered by the RPA complex, is a prerequisite for ubiquitylation at sites of stalled replication intermediates (16). As a consequence, various agents that cause adducts to the bases or replication fork stalling by nucleotide depletion induce the modification, whereas DNA double-strand breaks that are not also associated with base damage do not. This is experimentally verified with the *^{His}POL30 WT* strain constructed as described in Subheading 3.2.1, although it is possible to omit steps 5–7 and work with a strain bearing the His-tagged allele in addition to the endogenous one (see Note 15).

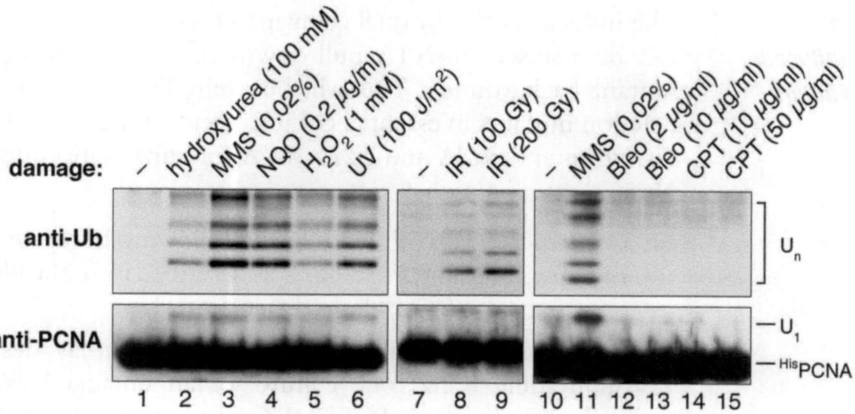

Fig. 4. Pattern of PCNA ubiquitylation in response to different DNA-damaging agents. Ubiquitylated (U) forms of ^{His}PCNA were isolated by Ni-NTA affinity chromatography after treatment of exponential cultures with the indicated damaging agents. *MMS* methyl methanesulfonate, *NQO* 4-nitroquinoline-1-oxide, H_2O_2 hydrogen peroxide, *UV* ultraviolet (254 nm) radiation, *IR* ionizing radiation, *Bleo* bleomycin, *CPT* camptothecin. This figure was reproduced from Davies et al., (2008) with permission from Elsevier (16).

1. Construct a ^{His}*POL30* strain as described in Subheading 3.2.1 (omitting steps 5–7 if desired).
2. Prepare several 50 mL cultures as described in Subheading 3.2.2 and subject them to the desired damage treatments. Hydroxyurea is added to the cultures as a solid, MMS as a neat liquid, and all other damaging agents used here are added as stock solutions (see Subheading 2.1) for 90 min. UV irradiation is performed either by spreading cells onto agar plates and resuspending them in medium after irradiation or by irradiation in a small volume of water in open petri dishes. Incubation is then continued for another 40 min before lysis. γ-irradiation is performed in liquid cultures. Leave one control culture untreated.
3. Harvest 10^9 cells for each culture and perform Ni-NTA pull-down assays as described in Subheading 3.1.4.
4. Analyze the samples by SDS-PAGE and Western blotting using antibodies against PCNA and ubiquitin as described in Subheading 3.1.2.
5. The ubiquitylation pattern reflects the type of damage (Fig. 4). The alkylating agent MMS usually gives a strong and reliable signal (lane 3), as do other adduct-forming agents such as 4-Nitroquinoline-1-oxide (NQO), as well as ultraviolet (UV) irradiation or the oxidizing agent hydrogen peroxide (H_2O_2) (lanes 4–6). Replication fork stalling by nucleotide depletion, achieved by hydroxyurea treatment, generally results in a weaker, but still reproducible response (lane 2). Ionizing radiation results in PCNA ubiquitylation due to the associated base damage (lanes 8 and 9), whereas other double-strand break-inducing agents, such as bleomycin (Bleo) and camptothecin (CPT), do not trigger the modification (lanes 12–15) (16).

3.4. Modifications of ^{His}PCNA: Influence of Cellular Factors

The influence of individual cellular factors on PCNA modification can be assessed Ni-NTA pull-downs of ^{His}PCNA using relevant mutant backgrounds. This will be exemplified here by means of deletion mutants in essential conjugation factors: *rad18Δ* for ubiquitylation and *siz1Δ* and *siz1Δ siz2Δ* for sumoylation defects (see Note 2).

1. Construct ^{His}*POL30* alleles in the following backgrounds: *WT*, *rad18Δ*, *siz1Δ*, and *siz1Δ siz2Δ*, as described in Subheading 3.2.1, omitting steps 5–7 if desired (see Note 15).
2. Prepare 50 mL cultures in YPD medium as described in Subheading 3.2.2: one culture for an untagged *POL30 WT* strain, two for the ^{His}*POL30 WT* strain, and one for each mutant. Add MMS to 0.02 % (v/v), but leave one ^{His}*POL30 WT* culture untreated. Continue incubation at 30 °C for 90 min.
3. Harvest 10^9 cells for each culture and perform Ni-NTA pull-down assays as described in Subheading 3.1.4.
4. Analyze the samples by SDS-PAGE and Western blotting using antibodies against PCNA and ubiquitin as described in Subheading 3.1.2.
5. Analysis of the blots reveals the contributions of each conjugation enzyme to PCNA modification (Fig. 5). SUMO conjugates and monoubiquitylated PCNA are apparent on the anti-PCNA

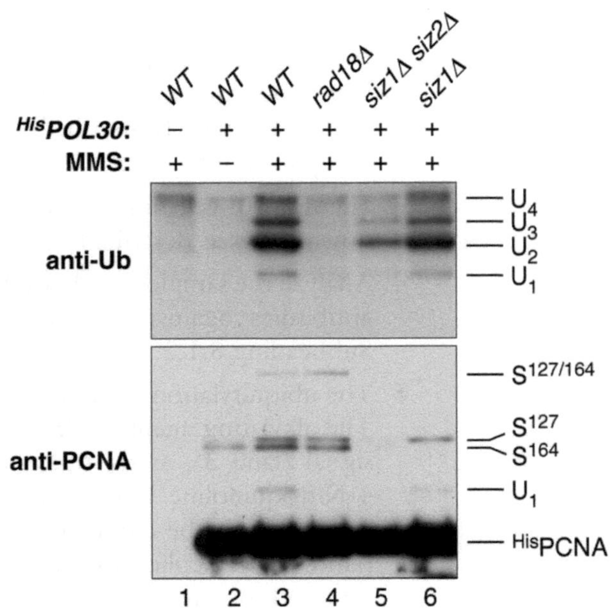

Fig. 5. Pattern of PCNA modifications in mutants of individual conjugation factors. Sumoylated (S) and ubiquitylated (U) forms of ^{His}PCNA were isolated by Ni-NTA affinity chromatography after treatment of exponential cultures of the indicated strain backgrounds with 0.02 % MMS for 90 min where indicated.

blots, whereas ubiquitin chains are detected on the anti-ubiquitin blot. Note that the anti-ubiquitin antibody poorly recognizes the monoubiquitylated forms of PCNA, which results in a significant reduction of the signal compared to the diubiquitylated form (lanes 3 and 6). Lack of the ubiquitin signal in *rad18Δ* (lane 4) indicates that Rad18 is responsible for all ubiquitin conjugation, but does not affect PCNA sumoylation. In contrast, the *siz1Δ* mutant (lane 6) ubiquitylates PCNA normally, but fails to sumoylate PCNA at K164. All sumoylation is abolished in *siz1Δ siz2Δ*, indicating that Siz2 is responsible for modification at K127 (lane 5) (19). Note that in this strain background the ubiquitin signal is reduced as well, to such an extent that the monoubiquitylated form is hardly detectable on either blot. This is due to the lack of a stimulatory effect that the Smt3 moiety on PCNA normally exerts on Rad18, which contains a SUMO interaction motif (J.L. Parker and H.D. Ulrich, unpublished results).

3.5. Modifications of Untagged PCNA: Use of ^{His}Ub and ^{His}Smt3

Modifications of endogenous PCNA will be detected by expression of His-tagged ubiquitin or Smt3 and performing Ni-NTA affinity chromatography to isolate the total pool of conjugates. Among these, modified PCNA (or other target proteins) can then be detected with a specific antibody. This strategy has the advantage of using native, untagged PCNA, but instead involves overexpression of the modifier itself in a tagged form, which might introduce other types of artifacts. In the case of PCNA, both strategies yield very similar modification patterns, indicating that they are unlikely to be artifactual.

3.5.1. Construction of Yeast Strains

POL30 will be replaced by relevant lysine mutants (*K164R* and *K127/164R*) by means of introducing the mutant alleles on an integrative plasmid into the *LEU2* locus and deleting the endogenous *POL30* gene by replacement with the *URA3* marker (9). His-tagged ubiquitin or Smt3 is then introduced on an episomal plasmid. If the desired strain backgrounds are already available, skip steps 1–9 and proceed directly with the transformation of the plasmids encoding His-tagged ubiquitin and Smt3 (step 10).

1. Digest 0.5 μg of plasmid YIp128-P_{30}-POL30 and mutant versions (*K127R* and *K127/164R*) (see Note 3) with *Afl*II in a total volume of 20 μL each.
2. Precipitate the DNA by adding 2 μL of sodium acetate solution and 50 μL of absolute ethanol and incubation at −20 °C for 30 min. Recover the DNA by centrifugation at 16,000 × *g* for 10 min at 4 °C and removal of the supernatant, rinse the pellets with 70 μL of 70 % ethanol, and centrifuge as before. Resuspend in 5 μL of TE buffer.
3. Use all of the digested DNA for transformation of *WT* yeast cells as described in Subheading 3.1.1. Spread the cells onto SC-Leu plates and incubate at 30 °C for 2–3 days.

4. Restreak a couple of colonies onto a fresh SC-Leu plate and incubate at 30 °C for 2–3 days to obtain single clones.
5. Verification of correct integration can be achieved by comparing the amounts of PCNA via Western blot of TCA lysates of individual clones as described in Subheadings 3.1.2 and 3.1.3. In addition, UV streak assays (see Subheading 3.1.6) will show a slightly increased resistance in the strain bearing a second copy of *WT POL30* (Fig. 2b).
6. Digest 1 μg of plasmid pBL243 (35) for each *POL30* allele with *Mlu*I and *Kpn*I and precipitate the DNA as described in step 2.
7. For each of the *POL30* alleles, inoculate an overnight culture of the strain harboring the integrative plasmid in liquid YPD medium and use this for transformation with 1 μg each of *Mlu*I/*Kpn*I-digested pBL243 as described in Subheading 3.1.1. Spread onto SC-Ura plates and incubate at 30 °C for 2–3 days.
8. Inoculate several 2 mL cultures in liquid SC-Ura medium with isolated colonies and shake overnight at 30 °C. Inoculate a 2 mL YPD culture for each of these and grow to exponential phase ($OD_{600} \approx 1$) at 30 °C. Perform damage sensitivity assays as described in Subheadings 3.1.5 and 3.1.6, using an unmodified *WT* strain as control. By means of their damage sensitivities (8, 9), identify clones in which the exchange of endogenous *POL30* for the damage-sensitive mutant alleles was successful (Fig. 6). A correct isogenic *WT POL30* strain

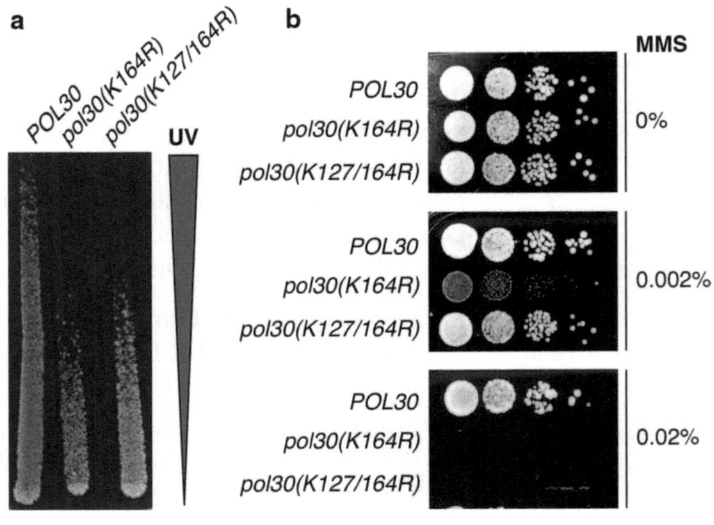

Fig. 6. Analysis of DNA damage sensitivity of PCNA mutants in individual acceptor sites. (**a**) UV sensitivity assays of *POL30*, *pol30(K164R)* and *pol30(K127/164R)* showing increased sensitivity of the acceptor site mutants. (**b**) MMS sensitivity assays of the same strains.

can be identified by verifying native levels of PCNA in TCA lysates as described in step 6 and a UV sensitivity identical to the unmodified *WT* strain. Restreak one correct clone of each genetic background onto a fresh SC-Ura plate and incubate at 30 °C for 2–3 days to obtain single colonies.

9. Select for loss of the *URA3* marker (see Note 3) by growing an overnight culture of each strain in YPD medium and streaking 10–20 μL of each culture onto a 5FOA plate. Incubate at 30 °C for 3 days and restreak individual colonies onto a fresh 5FOA plate to obtain single colonies. Examine damage sensitivities to ensure that they have not changed compared to the parental strains.
10. Transform one correct colony each of *WT* and mutant strains with 0.5 μg of the following plasmids (Subheading 3.1.1): YEplac195 (36), YEp195-CUP-^{His}Ub, and YEp195-CUP-^{His}Smt3. Spread the cells onto SC-Ura medium and incubate for 2–3 days.
11. Inoculate several 2 mL cultures per strain with isolated colonies in liquid SC-Ura medium supplemented with 0.1 mM $CuSO_4$, shake overnight at 30 °C and prepare TCA lysates for analysis by anti-His Western blot (see Subheadings 3.1.2–3.1.3) to assess the expression levels of the His-tagged modifiers. Alternatively, use anti-ubiquitin and anti-Smt3 antibodies on separate blots. Use a control antibody (e.g., anti-PGK) for loading control. Select clones with comparable expression levels.

3.5.2. Pull-Down of ^{His}PCNA Under Denaturing Conditions

1. Use the clones described above for inoculation of 6 mL cultures in SC-Ura + 0.1 mM $CuSO_4$ and shake at 30 °C overnight (see Note 17).
2. Use the overnight cultures to inoculate two 50 mL cultures of each *WT* strain and one 50 mL culture per mutant strain in YPD + 0.1 mM $CuSO_4$ at a starting cell density of $OD_{600} \approx 0.3$. Shake at 30 °C until the OD_{600} is around 1.0.
3. Add 0.02 % (v/v) MMS to one culture per strain and continue the incubation for an additional 90 min. Leave one of each *WT* culture untreated.
4. Harvest 10^9 cells and perform denaturing Ni-NTA pull-downs as described in Subheading 3.1.4.

3.5.3. Detection of Ubiquitin and SUMO Conjugates

1. Assemble a NuPAGE 4–12 % Bis-Tris gradient gel (see Note 16) and load all pull-down samples onto each of them. A useful control sample can be obtained by preparing a small-scale TCA lysate of a *WT* strain and loading that alongside the pull-down samples. Run the gel and blot onto PVDF membrane.
2. Develop the membrane with anti-PCNA antibody.

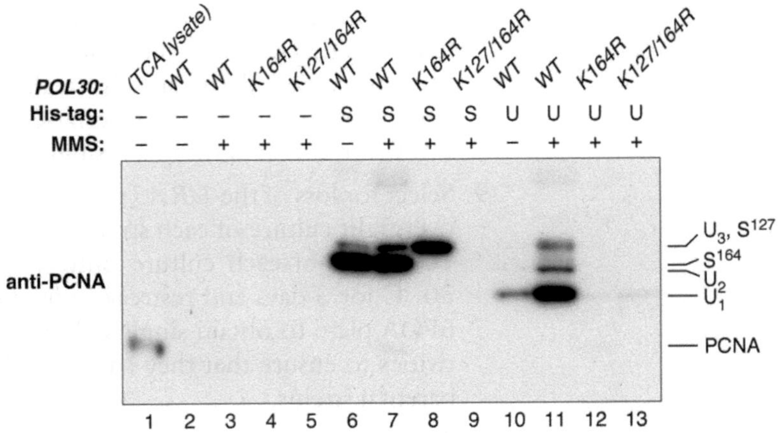

Fig. 7. Pattern of PCNA modifications in mutants of individual acceptor sites, detected by means of His-tagged modifiers. Sumoylated (S) and ubiquitylated (U) forms of endogenous PCNA were isolated by Ni-NTA affinity chromatography of total ^{His}Ub or ^{His}Smt3 conjugates and detected by anti-PCNA Western blot. Exponential cultures were treated with 0.02 % MMS for 90 min prior to lysis where indicated.

3. Analysis of the Western blot reveals the modified forms of PCNA (Fig. 7). The control sample (lane 1), showing unmodified PCNA, serves as a convenient size marker to which the modified forms can be compared. Samples derived from the strains harboring the empty control vector, YEplac195, which are devoid of a His-tagged modifier, should not give rise to any signal (lanes 2–5). The signals derived from the clones bearing ^{His}Smt3 reveal PCNA sumoylated on K164 and K127 (lanes 6–9). Finally, damage-induced polyubiquitylation of PCNA is observed in the strains harboring His-tagged ubiquitin (lanes 10–13). This method reliably detects monoubiquitylated PCNA and reveals its relative abundance compared to the polyubiquitylated species (lane 11). It also shows that even in the absence of exogenous DNA damage a small amount of monoubiquitylated PCNA is detectable in *WT* cells (lane 10). Note that overexpression of His-tagged modifier by induction with $CuSO_4$ does not necessarily enhance the abundance of modified PCNA (see Note 17).

4. Notes

1. Note that there is significant overlap between this chapter and a previously published protocol for the in vivo detection and characterization of SUMO conjugates in *S. cerevisiae*, which uses PCNA as a model SUMO target and describes the

methodology for general Ni-NTA pull-down assays as well as the construction of relevant yeast strains (39). That chapter also describes alternative methods of constructing yeast strains bearing ^{His}*POL30* alleles, which are not discussed here.

2. All strains used in this chapter are available from our lab upon request. They are isogenic to DF5, which has the following genotype: *Mat***a**, *his3-Δ200, leu2-3,2-112, lys2-801, trp1-1(am), ura3-52* (40). Other strain backgrounds will serve as well, but note that some strain backgrounds, such as BY4741 or related ones, do not allow integration of plasmids into the *LEU2* marker, as they harbor complete deletions rather than point mutations of relevant markers. Alternative strategies for replacement of *POL30* need to be applied in that case.

3. The YIp128-P_{30}-POL30 and YIp128-P_{30}-^{His}POL30 series are integrative vectors, derived from the yeast shuttle vector YIplac128 (36). They bear the *LEU2* marker and *POL30* or ^{His}*POL30* alleles under control of the *POL30* promoter (9, 20). The open reading frame is followed by a transcription terminator derived from the two-hybrid vector pGBT9 (Clontech). pBL243 is useful for replacement of the *POL30* gene by a *hisG-URA3-hisG* marker that can be removed by counterselection on 5-fluoroorotic acid (35). YEp195-CUP-^{His}Ub and YEp195-CUP-^{His}Smt3 are episomal (2 μ) vectors based on the yeast shuttle vector YEplac195 (36), carrying the *URA3* marker and His_7-tagged ubiquitin or *SMT3* (in their processed form without C-terminal extensions) under control of the copper-inducible *CUP1* promoter and followed by the pGBT9 transcriptional terminator. All plasmids are available from our lab.

4. It is important to use polyethylene glycol of the correct molecular weight range (PEG 3350). The solution is quite viscous and has to be pipetted slowly. It is sterilized by autoclaving. During this process, a white precipitate may form that will eventually settle at the bottom of the flask. This does not affect transformation efficiency.

5. Several anti-ubiquitin antibodies are commercially available. None of those that we have tested efficiently recognizes monoubiquitylated PCNA. Commercial monoclonal antibodies against PCNA are available (e.g., PC10), but these usually do not recognize budding yeast PCNA well. For detection of small amounts of modified PCNA, we therefore use affinity-purified polyclonal antibody raised against *S. cerevisiae* PCNA (20). Anti-Smt3 antibodies are also commercially available (e.g., Abcam), although we have not tested their usefulness for detection of sumoylated PCNA and generally use a polyclonal antibody developed in our lab (20).

6. The number of yeast cells in a culture can be roughly determined by the optical density at 600 nm (OD_{600}). Although the

ratio between OD_{600} and cell number depends on the size of the cells, an approximate calculation can be performed using the ratio: 1 $OD_{600} \approx 2–10^7$ cells/mL. For accurate measurements of optical densities, yeast cultures should be diluted to a reading not exceeding $OD_{600} = 0.3$. Cultures grown in YPD remain exponential up to $OD_{600} \approx 2$. In SC medium they should not exceed an OD_{600} of 1.

7. When loading samples in HU buffer, make sure to load equal volumes of the buffer into empty wells in order to obtain a straight running pattern. If this is not done, the high concentration of urea in the HU buffer will promote a spreading of the corresponding lanes into adjacent urea-free regions of the gel. Samples in other loading buffers, such as the molecular weight markers, need to be mixed with HU buffer for loading in order to avoid being "squeezed" by adjacent urea-containing lanes during the run.
8. PVDF membranes are preferable to Nitrocellulose membranes because of their better physical stability and the lower propensity of proteins to run through the membrane during blotting. These membranes need to be wetted with methanol due to their hydrophobicity. Once wet, they should never be allowed to dry during the blotting and detection process.
9. If the gel is moved after having touched the membrane, spurious signals are likely to appear on the film. This is particularly relevant for gels bearing pull-down samples, because some proteins, such as ^{His}PCNA, are highly enriched by this procedure and may leave an imprint on the membrane on first contact.
10. The chemiluminescence signal can be quantified on an imager (e.g., Fuji LAS-3000), but it has to be borne in mind that the signal strength is usually not proportional to the amount of protein on the blot. Accurate quantification will require comparison to samples of known concentrations.
11. Depending on the abundance of target protein and the extent of its modification, smaller amounts of cells can often be used for the pull-down. Isolation of tagged proteins from the lysates under the conditions described here is not quantitative, but the amount recovered in the eluate is roughly proportional to the number of cells used for preparing the lysate.
12. Incomplete removal of the supernatant of this step will result in lower pull-down efficiencies because of an inappropriately low pH due to TCA carry-over. To avoid this, the pH of the solution can be checked at this point and—if needed—adjusted. However, if the supernatant after TCA precipitation is removed completely by two steps of centrifugation and aspiration, adjustment should not be necessary.

13. Agarose beads should be centrifuged at low speed to avoid crushing them. All wash solutions should contain some detergent, which prevents the beads from sticking to the sides of the tube. For distributing them with a pipette, cut the end off the plastic pipette tip. Magnetic Ni-NTA beads are also available and are easier to handle, but we have found in direct comparisons that their binding capacity is significantly lower.
14. Addition of low concentrations of imidazole significantly reduces the background signal arising from nonspecific binding to the Ni-NTA beads. However, concentrations higher than 15 mM are not recommended, as they will reduce the pull-down efficiency.
15. Appending tags to the N or C terminus of the *POL30* open reading frame often interferes with the function of the protein. An N-terminal His_6-tag is tolerated and does not cause a growth defect when it replaces endogenous PCNA, but it conveys a measurable DNA damage sensitivity (Fig. 2b, c). Tagged alleles of PCNA should therefore not be used for genetic analysis, but only for the purpose of detecting modifications. If the gene replacement leads to a synthetic interaction with another mutant, ^{His}*POL30* can be co-expressed with the endogenous, untagged allele. Note that this causes a slightly enhanced resistance to damaging agents (Fig. 2b), presumably because of the increased dose of PCNA and the formation of (functional) mixed heterotrimers of tagged and native subunits.
16. Standard SDS-polyacrylamide gels according to Lämmli (37) are equally suitable for analysis of PCNA modifications, but note that the distribution of conjugates will appear different, depending on the type of gel, due to a variability in their running behavior. This is discussed in more detail elsewhere (39).
17. Expression of His-tagged ubiquitin and SUMO from YEp195-CUP-^{His}Ub and YEp195-CUP-^{His}Smt3 is under control of the *CUP1* promoter, inducible by addition of 0.1 mM $CuSO_4$. This causes a significant overexpression of the modifiers and may in consequence lead to an enhancement of conjugation and changes the overall pattern of conjugates (39). However, the *CUP1* promoter is partially active in YPD medium even without addition of $CuSO_4$. Under these conditions, the total pool of ubiquitin or SUMO is not significantly increased compared to the native situation; yet, the amount of tagged modifier is sufficient for pull-down assays. Hence, the isolation of specific ubiquitin or SUMO conjugates can be examined in the presence or absence of $CuSO_4$.

Acknowledgments

The authors would like to thank H. Windecker for contributing images of PCNA conjugates and P. Burgers for providing plasmid pBL243. Work in this lab is supported by Cancer Research UK.

References

1. Hoeijmakers JH (2009) DNA damage, aging, and cancer. N Engl J Med 361:1475–1485
2. Lawrence C (1994) The *RAD6* DNA repair pathway in *Saccharomyces cerevisiae*: what does it do, and how does it do it? Bioessays 16:253–258
3. Friedberg EC (2005) Suffering in silence: the tolerance of DNA damage. Nat Rev Mol Cell Biol 6:943–953
4. Branzei D, Foiani M (2010) Maintaining genome stability at the replication fork. Nat Rev Mol Cell Biol 11:208–219
5. Lehmann AR, Niimi A, Ogi T, Brown S, Sabbioneda S, Wing JF, Kannouche PL, Green CM (2007) Translesion synthesis: Y-family polymerases and the polymerase switch. DNA Repair 6:891–899
6. Waters LS, Minesinger BK, Wiltrout ME, D'Souza S, Woodruff RV, Walker GC (2009) Eukaryotic translesion polymerases and their roles and regulation in DNA damage tolerance. Microbiol Mol Biol Rev 73:134–154
7. Ulrich HD (2011) Timing and spacing of ubiquitin-dependent DNA damage bypass. FEBS Lett 585(18):2861–2867
8. Hoege C, Pfander B, Moldovan GL, Pyrowolakis G, Jentsch S (2002) *RAD6*-dependent DNA repair is linked to modification of PCNA by ubiquitin and SUMO. Nature 419:135–141
9. Stelter P, Ulrich HD (2003) Control of spontaneous and damage-induced mutagenesis by SUMO and ubiquitin conjugation. Nature 425:188–191
10. Watanabe K, Tateishi S, Kawasuji M, Tsurimoto T, Inoue H, Yamaizumi M (2004) Rad18 guides polη to replication stalling sites through physical interaction and PCNA monoubiquitination. EMBO J 23:3886–3896
11. Bienko M, Green CM, Crosetto N, Rudolf F, Zapart G, Coull B, Kannouche P, Wider G, Peter M, Lehmann AR, Hofmann K, Dikic I (2005) Ubiquitin-binding domains in Y-family polymerases regulate translesion synthesis. Science 310:1821–1824
12. Guo C, Tang TS, Bienko M, Parker JL, Bielen AB, Sonoda E, Takeda S, Ulrich HD, Dikic I, Friedberg EC (2006) Ubiquitin-binding motifs in REV1 protein are required for its role in the tolerance of DNA damage. Mol Cell Biol 26:8892–8900
13. Plosky BS, Vidal AE, de Henestrosa AR, McLenigan MP, McDonald JP, Mead S, Woodgate R (2006) Controlling the subcellular localization of DNA polymerases ι and η via interactions with ubiquitin. EMBO J 25:2847–2855
14. Parker JL, Bielen AB, Dikic I, Ulrich HD (2007) Contributions of ubiquitin- and PCNA-binding domains to the activity of polymerase η in *Saccharomyces cerevisiae*. Nucleic Acids Res 35:881–889
15. Parker JL, Ulrich HD (2009) Mechanistic analysis of PCNA poly-ubiquitylation by the ubiquitin protein ligases Rad18 and Rad5. EMBO J 28:3657–3666
16. Davies AA, Huttner D, Daigaku Y, Chen S, Ulrich HD (2008) Activation of ubiquitin-dependent DNA damage bypass is mediated by Replication Protein A. Mol Cell 29:625–636
17. Niimi A, Brown S, Sabbioneda S, Kannouche PL, Scott A, Yasui A, Green CM, Lehmann AR (2008) Regulation of proliferating cell nuclear antigen ubiquitination in mammalian cells. Proc Natl Acad Sci U S A 105:16125–16130
18. Ulrich HD (2009) Regulating post-translational modifications of the eukaryotic replication clamp PCNA. DNA Repair 8:461–469
19. Parker JL, Bucceri A, Davies AA, Heidrich K, Windecker H, Ulrich HD (2008) SUMO modification of PCNA is controlled by DNA. EMBO J 27:2422–2431
20. Papouli E, Chen S, Davies AA, Huttner D, Krejci L, Sung P, Ulrich HD (2005) Crosstalk between SUMO and ubiquitin on PCNA is mediated by recruitment of the helicase Srs2p. Mol Cell 19:123–133
21. Pfander B, Moldovan GL, Sacher M, Hoege C, Jentsch S (2005) SUMO-modified PCNA recruits Srs2 to prevent recombination during S phase. Nature 436:428–433

22. Moldovan GL, Pfander B, Jentsch S (2006) PCNA controls establishment of sister chromatid cohesion during S phase. Mol Cell 23:723–732
23. Parnas O, Zipin-Roitman A, Pfander B, Liefshitz B, Mazor Y, Ben-Aroya S, Jentsch S, Kupiec M (2010) Elg1, an alternative subunit of the RFC clamp loader, preferentially interacts with SUMOylated PCNA. EMBO J 29:2611–2622
24. Panse VG, Hardeland U, Werner T, Kuster B, Hurt E (2004) A proteome-wide approach identifies sumoylated substrate proteins in yeast. J Biol Chem 279:41346–41351
25. Vertegaal AC, Ogg SC, Jaffray E, Rodriguez MS, Hay RT, Andersen JS, Mann M, Lamond AI (2004) A proteomic study of SUMO-2 target proteins. J Biol Chem 279:33791–33798
26. Wohlschlegel JA, Johnson ES, Reed SI, Yates JR 3rd (2004) Global analysis of protein sumoylation in *Saccharomyces cerevisiae*. J Biol Chem 279:45662–45668
27. Zhao Y, Kwon SW, Anselmo A, Kaur K, White MA (2004) Broad spectrum identification of cellular small ubiquitin-related modifier (SUMO) substrate proteins. J Biol Chem 279:20999–21002
28. Zhou W, Ryan JJ, Zhou H (2004) Global analyses of sumoylated proteins in *Saccharomyces cerevisiae*. Induction of protein sumoylation by cellular stresses. J Biol Chem 279:32262–32268
29. Denison C, Rudner AD, Gerber SA, Bakalarski CE, Moazed D, Gygi SP (2005) A proteomic strategy for gaining insights into protein sumoylation in yeast. Mol Cell Proteomics 4:246–254
30. Hannich JT, Lewis A, Kroetz MB, Li SJ, Heide H, Emili A, Hochstrasser M (2005) Defining the SUMO-modified proteome by multiple approaches in *Saccharomyces cerevisiae*. J Biol Chem 280:4102–4110
31. Kirkpatrick DS, Weldon SF, Tsaprailis G, Liebler DC, Gandolfi AJ (2005) Proteomic identification of ubiquitinated proteins from human cells expressing His-tagged ubiquitin. Proteomics 5:2104–2111
32. Peng J, Cheng D (2005) Proteomic analysis of ubiquitin conjugates in yeast. Methods Enzymol 399:367–381
33. Rosas-Acosta G, Russell WK, Deyrieux A, Russell DH, Wilson VG (2005) A universal strategy for proteomic studies of SUMO and other ubiquitin-like modifiers. Mol Cell Proteomics 4:56–72
34. Guthrie C, Fink GR (1991) Guide to yeast genetics and molecular biology, vol 194. Academic, San Diego
35. Ayyagari R, Impellizzeri KJ, Yoder BL, Gary SL, Burgers PM (1995) A mutational analysis of the yeast proliferating cell nuclear antigen indicates distinct roles in DNA replication and DNA repair. Mol Cell Biol 15:4420–4429
36. Gietz RD, Sugino A (1988) New yeast-*Escherichia coli* shuttle vectors constructed with in vitro mutagenized yeast genes lacking six-base pair restriction sites. Gene 74:527–534
37. Laemmli UK (1970) Cleavage of structural proteins during the assembly of the head of bacteriophage T4. Nature 227:680–685
38. Windecker H, Ulrich HD (2007) Architecture and assembly of poly-SUMO chains on PCNA in *Saccharomyces cerevisiae*. J Mol Biol 376:221–231
39. Ulrich HD, Davies AA (2009) In vivo detection and characterization of sumoylation targets in *Saccharomyces cerevisiae*. Methods Mol Biol 497:81–103
40. Finley D, Ozkaynak E, Varshavsky A (1987) The yeast polyubiquitin gene is essential for resistance to high temperatures, starvation, and other stresses. Cell 48:1035–1046

Chapter 37

In Vitro PCNA Modification Assays

Joanne L. Parker and Helle D. Ulrich

Abstract

Modification of the replicative sliding clamp, PCNA, by monoubiquitin, polyubiquitin, and SUMO contributes to the processing of DNA damage during replication. In order to investigate the properties of the relevant conjugation enzymes, their interactions, substrate recognition, and the regulation of their activities, reconstitution of the modification reactions from purified components in vitro is an instructive exercise. Here we describe the purification of the relevant enzymes and accessory proteins from *E. coli* or *S. cerevisiae* as well as protocols for setting up small-scale ubiquitylation and sumoylation reactions with budding yeast PCNA. In addition, we provide a method for the purification of monoubiquitylated PCNA for further biochemical studies.

Key words: PCNA, Ubiquitin, SUMO, Ubiquitin-conjugating enzyme, Ubiquitin protein ligase, DNA damage bypass, Postreplication repair

1. Introduction

Posttranslational modification is an important strategy to modulate a protein's properties and interactions quickly and reversibly without the need for de novo protein synthesis. Over the past decade, members of the ubiquitin family, such as ubiquitin itself and the small ubiquitin-related modifier, SUMO, have been identified as important regulators of genome maintenance (1, 2). Ubiquitin and SUMO are attached to lysine residues of their targets via an isopeptide linkage in a recation catalyzed by an enzymatic cascade of an activating enzyme (E1), a conjugating enzyme (E2) and a protein ligase (E3) (3). Although in principle the two modification systems operate by very similar mechanisms, there are important differences concerning the features that determine substrate selectivity. In the ubiquitin system, a large number of E2s and an even

Lotte Bjergbæk (ed.), *DNA Repair Protocols*, Methods in Molecular Biology, vol. 920,
DOI 10.1007/978-1-61779-998-3_37, © Springer Science+Business Media New York 2012

greater variety of E3s are employed in a combinatorial manner to achieve selectivity (4). In contrast, SUMO is conjugated by a single E2 and a small number of E3s, where the E2 often plays an important part in substrate recognition via a consensus site, ΨKXE (Ψ stands for a bulky aliphatic residue) (5). Hence, in vitro sumoylation reactions sometimes proceed rather nonspecifically, whereas ubiquitylation assays often require an appropriate combination of enzymes to succeed at all.

Posttranslational modifications of the replicative sliding clamp, proliferating cell nuclear antigen (PCNA), control the replication over damaged DNA, a pathway that is also called DNA damage tolerance (1, 2, 6). The system serves as an instructive model case for the study of ubiquitin and SUMO conjugation mechanisms, enzyme specificities, and regulatory schemes. Ubiquitylation is targeted to a single conserved lysine residue, K164 (7). The E2–E3 pair Rad6-Rad18 attaches a single ubiquitin moiety, which can be converted to a K63-linked polyubiquitin chain by a second E2–E3 complex, consisting of the heterodimeric E2 Ubc13-Mms2 and the E3 Rad5 (7, 8). Whereas monoubiquitylated PCNA activates translesion synthesis by damage-tolerant DNA polymerases, polyubiquitylation of PCNA is required for an error-free damage avoidance pathway that likely involves template switching (6). Contributions to translesion synthesis have also been observed for Ubc13-Mms2 and/or Rad5 (9–11). In contrast to ubiquitylation, which is induced by replication stress and DNA damage, SUMO is attached to PCNA in a damage-independent manner when the clamp resides on DNA (7, 12). The E2 Ubc9 cooperates with the E3 Siz1 to modify the same K164 residue that is also targeted by Rad6-Rad18; in addition, a second lysine, K127, which conforms to the ΨKXE consensus motifs, is modified to a smaller extent. In vitro, this is due to E3-independent activity of Ubc9, while in vivo a second E3, Siz2, is responsible for the reaction (12, 13). In contrast to ubiquitylation, which is observable in all eukaryotes examined thus far, sumoylation of PCNA is more limited in scope. In budding yeast, the modification leads to the recruitment of an anti-recombinogenic helicase, Srs2, which prevents unscheduled recombination events during S phase and allows ubiquitin-dependent damage bypass to occur under conditions of damage (14, 15). In addition, it has been shown to promote interaction of PCNA with the alternative clamp loader Elg1 (16) and to interfere with the binding of Eco1, a factor involved in the establishment of cohesion during replication (17).

In vitro reconstitution of PCNA modifications from purified components is useful for studies of the mechanisms and functions of the conjugation enzymes, such as analyses of substrate recognition or interactions and cooperations between the individual factors. In addition, preparation of biochemically useful amounts of modified PCNA is essential for an investigation of the downstream reactions of DNA damage bypass, such as the interactions of modified PCNA with translesion polymerases, its activity in promoting DNA synthesis and its interactions with other downstream effector proteins (18, 19). This chapter describes protocols for purification of the relevant proteins, the purification of monoubiquitylated PCNA, and the set-up of small-scale modification reactions, which can easily be scaled up for the preparation of larger amounts of modified PCNA.

2. Materials

2.1. Cultivation and Transformation of Bacteria

1. *Escherichia coli* host strain: BL21-CodonPlus® (Stratagene).
2. Plasmids: listed in Table 1 (see Note 1).
3. LB medium: 10 g/L Bacto-Tryptone, 5 g/L Bacto-yeast extract, and 10 g/L NaCl in water. Adjust to pH 7.5 and sterilize by autoclaving.
4. TB medium: 12 g Bacto-Tryptone, 24 g Bacto-yeast extract, and 4 mL glycerol. Adjust to 900 mL with water and sterilize by autoclaving, then supplement with 100 mL of a filter-sterilized solution of 0.17 M KH_2PO_4 and 0.72 M K_2HPO_4 in water.
5. Ampicillin stock solution (1,000×): 100 mg/mL in water. Filter-sterilize and store at –20 °C.
6. Kanamycin stock solution (1,000×): 50 mg/mL in water Filter-sterilize and store at –20 °C.
7. IPTG stock solution: 1 M in water. Store at –20 °C.
8. Petri dishes and shake flasks.
9. Shaking incubator.

2.2. Cultivation and Transformation of Yeast

1. *Saccharomyces cerevisiae* host strain: BJ5460 (see Note 2).
2. Plasmids: listed in Table 1 (see Note 1).
3. YPD medium: 1 % (w/v) Bacto-yeast extract, 2 % (w/v) Bacto-peptone, 2 % (w/v) glucose in water. Sterilize by autoclaving and store at room temperature.
4. SC medium: synthetic complete medium lacking individual amino acids. Available from various suppliers (e.g., Sigma-Aldrich), but can also be prepared in the lab (20).
5. Restriction enzymes: *Afl*II or *Eco*RV (e.g., New England Biolabs).
6. Petri dishes and shake flasks.
7. Shaking incubator.

Table 1
Information about the proteins used for in vitro PCNA modification assays, their relevant expression vectors and approximate yields

Protein	Molecular mass (kDa)	Expression vector	Selection marker	Promoter	Approximate yield per liter	Expression host	References
^{His}PCNA	94 (trimer)	pQE30-POL30	*amp^r*	T5	>20 mg	*E. coli*	(13)
PCNA	87 (trimer)	pET11c-POL30	*amp^r*	T7	>20 mg	*E. coli*	(12)
^{His}RFC	66, 42, 40, 39, 37	pBL481-RFC	*amp^r*	T7	5 mg	*E. coli*	(23)
^{His}Ubi	9	pET16b-Ubi	*amp^r*	T7	>20 mg	*E. coli*	Lab stock
^{His}Ubc13	19	pQE30-Ubc13	*amp^r*	T7	>10 mg	*E. coli*	(24)
Mms2	16	pTYB12-MMS2	*amp^r*	T7	3 mg	*E. coli*	(24)
^{His}Smt3	12.6	pET21a-SMT3	*amp^r*	T7	>20 mg	*E. coli*	(30)
Uba2^{His} ^{His}Aos1	72 39	pET11a-UBA2 pET28a-AOS1	*amp^r* *kan^r*	T7 T7	2 mg (complex)	*E. coli*	(31) (32)
Ubc9^{His}	18	pET21b-UBC9	*amp^r*	T7	>10 mg	*E. coli*	(33)
^{His}Siz1(1–508)	61	pQE32-SIZ1(508)	*amp^r*	T5	1 mg	*E. coli*	Lab stock
^{His}Rad18 Rad6	57 20	pYES2-^{His}RAD18 YIp128-ADH-RAD6	*URA3*, *amp^r* *LEU2*, *amp^r*	*GAL1-10* *ADH1*	0.05 mg (complex)	*S. cerevisiae*	(8) (8)
^{His}Rad5	134	pYES2-^{His}RAD5	*URA3*, *amp^r*	*GAL1-10*	0.2 mg	*S. cerevisiae*	(8)

2.3. SDS-Polyacrylamide Gel Electrophoresis and Western Blotting

1. Standard set-up for SDS-polyacrylamide gel electrophoresis (SDS-PAGE), i.e., gel plates, combs, and reagents or ready-made gels, electrophoresis chamber, power supply and running buffer (21).
2. General protein stain: e.g., Coomassie Brilliant Blue.
3. Standard set-up for Western blotting, i.e., semi-dry blotter or tank-blotting apparatus, membrane, and filter papers (21).
4. Anti-PCNA and secondary antibody (see Note 3), detection reagents.

2.4. Protein Purification

1. DTT stock solution: 1 M 1,4-dithiothreitol in water. Store at –20 °C.
2. EDTA stock solution: 0.5 M ethylenediaminetetraacetid acid in water, pH 7.5. Dissolve and adjust the pH by adding solid NaOH pellets while stirring, taking care not to add too much NaOH.
3. PEI: 10 % (w/v) polyethylenimine in water.
4. Ni-NTA buffer: 40 mM Tris–HCl, pH 7.5, 300 mM NaCl, 30 mM imidazole. Optional: 1 mM EDTA, 1 mM DTT (see Note 4).
5. Gel filtration buffer: 40 mM Tris–HCl, pH 7.5, 200 mM NaCl, 1 mM EDTA, 1 mM DTT, 10 % (v/v) glycerol.
6. Intein buffer: 20 mM Tris–HCl, pH 8.0, 500 mM NaCl, 1 mM EDTA.
7. Buffer A: 25 mM Tris–HCl, pH 7.5, 0.5 mM DTT, 1 mM EDTA (see Note 5).
8. Buffer B: 30 mM HEPES, pH 7.4, 1 mM DTT, 0.5 mM EDTA, 10 % (v/v) glycerol.
9. Buffer C: 50 mM Tris–HCl, pH 8.0, 300 mM NaCl, 10 % (v/v) glycerol, supplemented with a protease inhibitor cocktail, e.g., Complete™ (Roche).
10. Buffer D: 40 mM HEPES, pH 7.5, 1 mM EDTA, 1 mM DTT, 10 % (v/v) glycerol.
11. Buffer E: 40 mM HEPES, pH 7.4, 10 % (v/v) glycerol.
12. Ampholytes (e.g., Sigma Ampholyte high resolution pH 3–10).
13. Centrifuges: high-speed centrifuge (e.g., Sorvall RC-5B, rotor SS-34) and ultracentrifuge (e.g., Beckmann Optima L-100 XP, rotors Ti45, Ti70).
14. Sonicator (e.g., Branson).
15. High-pressure cell disruptor (e.g., Constant Systems or French Press).
16. Chromatography system (e.g., Äkta, GE Healthcare).

17. Ni-NTA agarose (e.g., Qiagen or GE Healthcare) (see Note 4).
18. Chitin beads (New England Biolabs).
19. Gel filtration column or resin: Superdex™ 200 (GE Healthcare).
20. Anion exchange columns or resin: HiTrap™ Q, Mono Q (GE Healthcare).
21. Cation exchange columns or resin: HiTrap™ SP, Mono S (GE Healthcare).
22. HiTrap™ Heparin column (GE Healthcare).
23. Centrifugal concentrators: MWCO 3,000–5,000.
24. Disposable polypropylene columns: 1 and 10 mL bed volumes.
25. Dialysis tubing: MWCO 3,000.

2.5. Conjugation Assays

1. 10× Modification assay buffer (MAB): 400 mM HEPES, pH 7.4, 80 mM magnesium acetate, 500 mM NaCl.
2. Plasmid DNA (e.g., pBluescript, supercoiled).
3. Nicking enzyme, e.g., Nb.BsrDI (New England Biolabs).
4. DNA cleanup kit (e.g., QIAquick DNA cleanup system, Qiagen).
5. Ubiquitin-activating enzyme (Uba1): commercially available (e.g., Biomol) or prepared in the lab (22). Store in small aliquots at −80 °C.
6. Ubiquitin: commercially available (e.g., Sigma) (see Note 6).
7. ATP stock solution: 100 mM in water, adjusted to pH 7.0. Store at −20 °C.
8. Benzonase (Merck): 250 U/μL stock solution in water. Store at −20 °C.
9. 5× SDS loading buffer: 250 mM Tris–HCl, pH 6.8, 500 mM DTT, 10 % (w/v) SDS, 0.5 % (w/v) bromophenol blue, 50 % (w/v) glycerol. Store at −20 °C.

3. Methods

3.1. General Methods

3.1.1. Protein Detection and Analysis by SDS-PAGE and Western Blot

No specific instructions are given for the preparation and running of SDS polyacrylamide gels and Western blots, as any standard procedure based on self-made or precast gels and any common transfer technology are appropriate (21). For the purpose of protein purification, samples of the input, the flow-through and relevant protein-containing fractions should always be analyzed by SDS-PAGE, followed by staining with a general protein dye such as Coomassie Brilliant Blue. In order to visualize the conjugation reactions, Western blotting with anti-PCNA antibodies is appropriate, as this allows the detection of unmodified and modified PCNA without interference from conjugation factors or free ubiquitin or SUMO chains, which may also be formed in the course of the reaction.

3.1.2. General Conditions for Protein Production in E. coli

1. Transform the expression host (*E. coli* BL21-CodonPlus) with the relevant plasmid and cultivate on an LB plate, supplemented with the relevant antibiotics, overnight at 37 °C.
2. Inoculate 5 mL liquid LB or TB medium, supplemented with the relevant antibiotics, with a single colony from the selective plate and incubate with shaking overnight at 37 °C (see Note 7).
3. Inoculate a desired volume of LB or TB medium, supplemented with the relevant antibiotics, with a small amount of the overnight culture, resulting in an OD_{600} of less than 0.2, and incubate with vigorous shaking until the OD_{600} has reached 0.6–0.8 (see Note 8).
4. Induce expression by addition of 0.2 mM IPTG and continue the incubation for an additional 4 h at 37 °C.
5. Harvest the cells by centrifugation, using pre-weighed tubes, and resuspend the pellet in the desired buffer, using 3–5 mL per g wet weight. Store the resuspended pellets at −80 °C.

3.1.3. Preparation of an E. coli Lysate

1. Thaw the resuspended pellets and disrupt the cells by sonication on ice.
2. Clarify the lysate by centrifugation at 40,000 × *g* for 20 min at 4 °C.
3. Transfer the supernatant to a fresh tube and add 0.05 % PEI from a 10 % stock solution to precipitate the DNA.
4. Centrifuge at 150,000 × *g* for 45 min at 4 °C and collect the supernatant.

3.1.4. Ni-NTA Affinity Chromatography

1. Incubate the cleared lysate, prepared in Ni-NTA buffer unless otherwise noted, with an appropriate amount of Ni-NTA agarose for 30–60 min on a roller or rotator at 4 °C (see Note 9).
2. Transfer the resin into a chromatography column and wash extensively with Ni-NTA buffer until the absorbance (A_{280}) of the eluate is stable.
3. Elute the protein in Ni-NTA buffer supplemented with 300 mM imidazole, collecting fractions and monitoring the A_{280}.
4. Analyze the fractions by SDS PAGE and Coomassie staining.
5. Pool the relevant fractions and concentrate in a centrifugal concentrator.
6. Purify the protein further by gel filtration chromatography (see Note 9).

3.2. Protein Purifications from E. coli

Information about the individual proteins, their expression vectors, approximate purification scales, and expected yields are given in Table 1.

3.2.1. Purification of PCNA

^{His}PCNA is purified by following the standard conditions for His-tagged proteins as described in Subheadings 3.1.2–3.1.4. Native, untagged PCNA is overexpressed in *E. coli* using plasmid pET11c-POL30 and purified as follows:

1. Transform *E. coli* strain BL21-CodonPlus with plasmid pET11c-POL30 and follow the general protocol described in Subheadings 3.1.2–3.1.3 to induce expression and prepare a lysate, using buffer A supplemented with 100 mM NaCl for resuspension.
2. Apply the cleared lysate to an anion exchange column (e.g., HiTrap™ Q) equilibrated in buffer A + 100 mM NaCl and wash with this buffer until the A_{280} is stable.
3. Elute the protein with a linear gradient of NaCl (100–700 mM). PCNA elutes at ca. 400 mM NaCl.
4. Analyze the relevant fractions by SDS PAGE, combine the PCNA-containing fractions, and dialyze against buffer A + 40 mM NaCl.
5. Apply the dialyzed sample to a cation exchange column (e.g., HiTrap™ SP) equilibrated in buffer A + 40 mM NaCl.
6. Collect the flow-through, which contains PCNA (see Note 10).
7. Load the sample onto a strong anion exchange resin (Mono Q) equilibrated in buffer A + 40 mM NaCl.
8. Elute with a linear gradient of NaCl (40–700 mM). PCNA elutes at ca. 450 mM NaCl.
9. Analyze the relevant fractions by SDS PAGE, combine the PCNA-containing fractions and concentrate in a centrifugal concentrator to a maximum of 5 mg/mL.
10. Apply to a gel filtration column (Superdex 200) equilibrated in gel filtration buffer (see Notes 9 and 10).

3.2.2. Purification of ^{His}RFC

RFC is a protein complex of five subunits, Rfc1, 2, 3, 4, and 5, which are expressed from one single plasmid, pBL481-RFC (23). The complex is purified by means of a His_7-tag on subunit Rfc1 (see Note 11).

1. Transform *E. coli* strain BL21-CodonPlus with plasmid pBL481-RFC.
2. Prepare a starter culture in a small volume of TB medium by incubating it with shaking at 37 °C over the course of a day.
3. Use this starter culture to inoculate a larger volume of TB medium and shake at 22 °C overnight.

4. Induce the following morning (at an OD_{600} of ca. 3–4) with 0.2 mM IPTG and continue incubation for 5 h at 22 °C.
5. Harvest the cells and resuspend in buffer B + 200 mM NaCl.
6. Prepare a clarified lysate as described in Subheading 3.1.3.
7. While stirring the lysate, slowly add solid ammonium sulfate to a final concentration of 0.28 g/mL of lysate. This will precipitate the RFC complex.
8. Collect the precipitate by centrifugation at 9,000 × *g* for 20 min and resuspend in buffer B containing 10 mM imidazole and 0.05 % ampholytes.
9. Perform an Ni-NTA affinity chromatography as described in Subheading 3.1.4, but use buffers containing 0.05 % ampholytes for all steps.
10. Pool the fractions containing RFC and dilute 1:1 with buffer B.
11. Load onto a strong cation exchange column (Mono S) equilibrated in buffer B + 100 mM NaCl and wash with buffer B + 200 mM NaCl.
12. Apply a linear gradient between 200 and 500 mM NaCl in buffer B. RFC elutes at ca. 350 mM NaCl.
13. Dialyze the RFC-containing fractions into buffer B + 200 mM NaCl and snap-freeze.

3.2.3. Purification of ^{His}Ubiquitin

Native bovine ubiquitin is available commercially and serves well for conjugation assays. Recombinant ^{His}Ubiquitin can also be purchased, but it is easy to purify in the lab.

1. Transform *E. coli* strain BL21-CodonPlus with plasmid pET16b-Ubi.
2. Follow the standard procedure described in Subheadings 3.1.2–3.1.3 for cultivation of *E. coli* and preparation of a cleared lysate.
3. Perform a standard Ni-NTA affinity chromatography as described in Subheading 3.1.4 to purify ^{His}Ubiquitin. The protein elutes as a monomer.

3.2.4. Purification of ^{His}Ubc13

1. Transform *E. coli* strain BL21-CodonPlus with plasmid pQE30-UBC13.
2. Follow the standard procedure described in Subheadings 3.1.2–3.1.3 for cultivation of *E. coli* and preparation of a cleared lysate.
3. Perform a standard Ni-NTA affinity chromatography as described in Subheading 3.1.4 to purify ^{His}Ubc13. The protein elutes as a monomer.

3.2.5. Purification of Mms2

Mms2 is produced without an affinity tag because a complex of ^{His}Ubc13 with ^{His}Mms2 or ^{GST}Mms2 exhibits reduced catalytic activity. The protein is produced as a fusion to a self-cleavable intein tag, followed by a chitin-binding domain for affinity purification (24).

1. Transform *E. coli* strain BL21-CodonPlus with plasmid pTYB12-MMS2.
2. Follow the standard procedure described in Subheading 3.1.2 for cultivation of *E. coli*, but induce expression with 1 mM IPTG at 28 °C.
3. Prepare a cleared lysate in intein buffer according to Subheading 3.1.3.
4. Equilibrate chitin beads (ca. 7 mL per liter culture) in intein buffer and incubate with the cleared lysate on a roller or rotator for 90 min.
5. Transfer the slurry to a column and wash extensively with intein buffer (use at least 10 column volumes).
6. Flush quickly with 3 column volumes of intein buffer supplemented with 50 mM DTT and incubate at room temperature for 36 h to afford cleavage.
7. Elute the cleaved Mms2 with gel filtration buffer and concentrate for subsequent gel filtration.

3.2.6. Purification of ^{His}Smt3

1. Transform *E. coli* strain BL21-CodonPlus with plasmid pET21a-SMT3.
2. Follow the standard procedure described in Subheadings 3.1.2–3.1.3 for cultivation of *E. coli* and preparation of a cleared lysate.
3. Perform a standard Ni-NTA affinity chromatography as described in Subheading 3.1.4 to purify ^{His}Smt3.

3.2.7. Purification of ^{His}Aos1-Uba2^{His}

1. Transform *E. coli* strain BL21-CodonPlus with plasmids pET11a-UBA2 and pET28a-AOS1 for co-expression. This can be done simultaneously or in two successive steps.
2. Follow the standard procedure described in Subheadings 3.1.2–3.1.3 for cultivation of *E. coli* and preparation of a cleared lysate.
3. Perform a standard Ni-NTA affinity chromatography as described in Subheading 3.1.4 to purify the heterodimeric complex, eluting from the Ni-NTA column in Ni-NTA buffer containing only 150 mM NaCl and 300 mM imidazole.
4. Apply the eluate to a strong anion exchange column (Mono Q) equilibrated in buffer A + 150 mM NaCl.
5. Develop the column with a linear gradient from 200 to 500 mM NaCl. The complex elutes at ca. 350 mM NaCl.
6. Pool and concentrate the relevant fractions and apply to a gel filtration column (see Note 9). The complex elutes as a heterodimer.

3.2.8. Purification of ^{His}Ubc9

1. Transform *E. coli* strain BL21-CodonPlus with plasmid pET21b-UBC9.
2. Follow the standard procedure described in Subheadings 3.1.2–3.1.3 for cultivation of *E. coli* and preparation of a cleared lysate.
3. Perform a standard Ni-NTA affinity chromatography as described in Subheading 3.1.4 to purify ^{His}Ubc9.

3.2.9. Purification of ^{His}Siz1(1–508)

Full-length Siz1 can be expressed and purified from *E. coli*. However, the truncated protein described here is produced with better yield, and the ligase activity towards PCNA appears equal to that of the full-length protein.

1. Transform *E. coli* strain BL21-CodonPlus with plasmid pQE32-Siz1(508).
2. Prepare a culture in TB medium as described in Subheading 3.1.2 and induce at $OD_{600} \approx 0.6$ with 0.2 mM IPTG, then drop the temperature to 30 °C and harvest after 4 h.
3. Prepare a clarified lysate as described in Subheading 3.1.3.
4. Perform standard Ni-NTA affinity chromatography as described in Subheading 3.1.4, eluting the protein with buffer A containing 200 mM imidazole and 200 mM NaCl.
5. Load the eluate onto a HiTrap™ Q column equilibrated with buffer A + 100 mM NaCl.
6. Wash the column with buffer A + 200 mM NaCl and develop with a linear gradient from 300 to 500 mM NaCl. ^{His}Siz1(1–508) elutes as a broad peak, due to the presence of some degradation products.
7. Pool fractions containing mainly intact protein, concentrate, and perform gel filtration (see Note 9). The protein elutes as a monomer, but some aggregation may be detectable.

3.3. Protein Purifications from S. cerevisiae

As for the proteins produced in *E. coli*, information about the conjugation enzymes, their expression vectors, approximate purification scales, and expected yields are given in Table 1.

3.3.1. Purification of the Rad6-^{His}Rad18 Complex

The procedure described here is used to purify Rad18 in complex with Rad6 from budding yeast overexpressing both genes (8), as production in *E. coli* does not yield soluble, active protein (see Note 12).

1. Use a protease-deficient *S. cerevisiae* strain, e.g., BJ5460, grown in YPD medium, to integrate plasmid YIp128-ADH-RAD6, digested with *Afl*II or *Eco*RV, and select on SC-Leu.
2. Transform the resulting strain with plasmid pYES2-^{His}RAD18 and select on SC-Ura (see Note 13).
3. Grow an overnight culture at 30 °C in SC-Ura medium supplemented with 2 % glucose. Use this to inoculate a large

culture in SC-Ura medium containing 0.1 % glucose to a final OD_{600} of 0.15.

4. When the culture has reached $OD_{600} \approx 0.6$ (ca. 6 h after inoculation), induce production of Rad18 by addition of galactose to a final concentration of 2 % and incubate at 25 °C for an addition 20–22 h.
5. Harvest the cells by centrifugation (15 min at 3,500 × *g*). The pellet can be frozen and stored until required.
6. Resuspend the yeast pellet in 3 mL/g wet weight of buffer C + 20 mM imidazole and lyse the cells by three passes through a high-pressure cell disruptor or French press at 35,000 psi.
7. Prepare a clarified lysate as described for *E. coli* in Subheading 3.1.3.
8. Perform standard Ni-NTA affinity chromatography as described in Subheading 3.1.4, eluting the protein with buffer D supplemented with 150 mM NaCl and 200 mM imidazole. At this point, the preparation still contains many contaminating proteins.
9. Apply the Ni-NTA eluate to a heparin column equilibrated in buffer D + 150 mM NaCl. Wash with buffer D + 200 mM NaCl and elute with a linear gradient up to 500 mM NaCl. The protein elutes over a rather large range.
10. Dialyze the protein into buffer D + 200 mM NaCl or dilute to the appropriate final salt concentration and apply to a strong anion exchange column (Mono Q) equilibrated with the same buffer. Wash with the same buffer and elute with a linear gradient up to 500 mM NaCl. The peak fractions correspond to ca. 350 mM NaCl.
11. Pool and concentrate the relevant fractions and apply to a Superdex 200 gel filtration column equilibrated in buffer D + 150 mM NaCl (see Note 9). This step removes a large amount of aggregates that are usually present in the preparation. The native complex elutes as two peaks, one corresponding to a size slightly above a dimer of the heterodimer and the other to a tetramer of the heterodimer (see Note 14). Both peaks show the same activity towards PCNA and can be pooled and stored at −80 °C.

3.3.2. Purification of ^{His}Rad5

Like Rad18, Rad5 is produced in *S. cerevisiae* (8), as expression in *E. coli* does not yield significant amounts of active, full-length protein.

1. Transform a protease-deficient *S. cerevisiae* strain, e.g., BJ5460, with plasmid pYES2-^{His}Rad5 (see Note 13) and cultivate on SC-Ura.
2. Follow the protocol described for Rad6-^{His}Rad18 in Subheading 3.3.1, steps 3–8. After elution from the Ni-NTA column, the preparation still contains a large number of contaminating proteins.

3. Apply the fractions containing ^{His}Rad5 to a heparin column equilibrated in buffer D + 200 mM NaCl. Wash with the same buffer and elute the protein with a linear gradient up to 500 mM NaCl. ^{His}Rad5 elutes at ca. 350 mM NaCl.
4. Optional: If the preparation still contains contaminants after this step, load onto a strong anion exchange column (Mono Q) equilibrated in buffer D + 100 mM NaCl, wash with the same buffer and elute with a gradient up to 500 mM NaCl. ^{His}Rad5 elutes at ca. 250 mM NaCl.
5. Pool and concentrate the relevant fractions and apply to a Superdex 200 gel filtration column equilibrated in buffer D + 150 mM NaCl (see Note 9). Rad5 elutes as a monomer at its expected size of ~134 kDa.

3.4. Preparation of Primed DNA

Monoubiquitylation of PCNA by Rad6-Rad18 requires the clamp to be loaded onto a DNA substrate by its cognate clamp loader, RFC (18, 19). RFC requires a 3′ primer-template junction or a nick in double-stranded DNA for loading PCNA. As the linkage between PCNA and DNA is only topological, PCNA can slide off linear DNA. Hence, the modification assay is most conveniently performed in the presence of RFC and nicked, circular plasmid DNA.

1. Treat 8 μg of an available plasmid (e.g., pBluescript) with a nicking enzyme that will act on the plasmid at least once, such as Nb.BsrDI, according to the manufacturer's instructions.
2. Remove the enzyme from the DNA by means of a DNA clean-up kit according to the manufacturer's instructions.
3. Measure the concentration of the DNA.

3.5. PCNA Monoubiquitylation Assays

PCNA monoubiquitylation assays are carried out in MAB in 10 μL or 20 μL reactions (see Note 15). A control reaction without E1 should be included. Other controls, such as the omission of Rad6-Rad18, are also useful. In order for the reaction to proceed, PCNA needs to be loaded onto a DNA substrate. Hence, RFC and a primed circular DNA are included in the reaction, but can be omitted for control purposes (see Note 16).

1. Add to a series of microcentrifuge tubes: 10× MAB, 2.5 nM (ca. 5 μg/mL) nicked plasmid DNA, 30 nM RFC, 1 mM ATP, and 50 nM PCNA.
2. Add 25 nM E1 (or buffer), 1 μM ubiquitin and 200 nM Rad6-Rad18 (see Note 17).
3. Incubate the reactions for 1 h at 30 °C and stop by addition of 0.2 volumes of 5× SDS loading buffer.
4. Incubate the tubes at 95 °C for 3 min to denature the proteins and analyze the reaction products on a 10 % SDS polyacrylamide gel, followed by Western blotting and development with an anti-PCNA antibody.

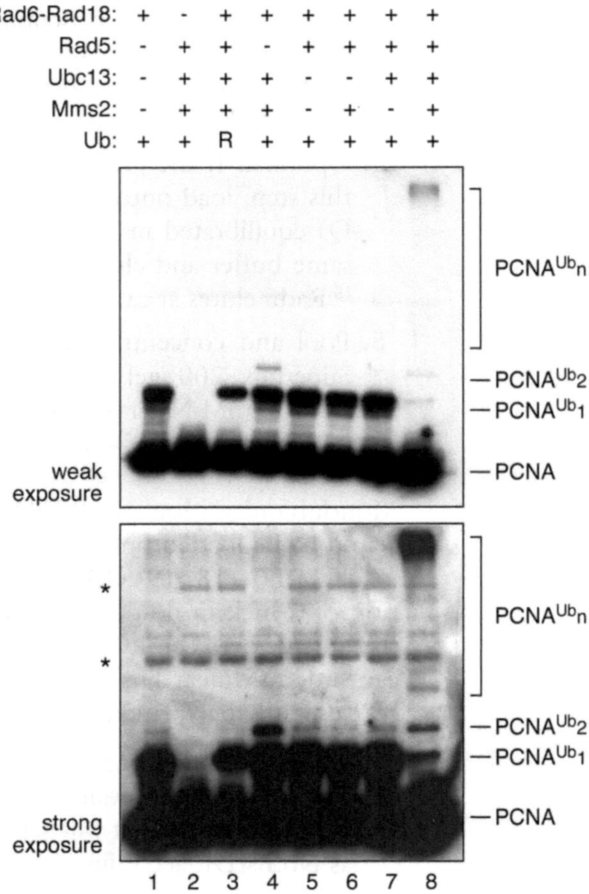

Fig. 1. Modification of PCNA by mono- and polyubiquitylation using purified enzymes. Reactions were carried out in the presence of a nicked plasmid and the clamp loader RFC in order to provide loaded PCNA as a substrate. All reactions contained E1 and ATP. Enzymes for PCNA monoubiquitylation (Rad6-Rad18) were added first where indicated, and the reaction was incubated for 40 min at 30 °C before addition of the polyubiquitylation factors (Ubc13-Mms2-Rad5) where indicated and further incubation for 40 min. Replacement of ubiquitin by a K63R mutant is indicated as "R". Products were detected by Western blotting with a PCNA-specific antibody. *Asterisks* indicate crossreactive bands visible upon prolonged exposure of the blots. This Figure was reproduced from ref. 8.

5. PCNA should be modified by Rad6-Rad18 with a single ubiquitin unit, visible as a distinct band running at a molecular weight of ca. 40 kDa (Fig. 1, lane 1). The reaction rarely goes to completion at this scale (see Note 18).

3.6. Purification of Monoubiquitylated PCNA

The monoubiquitylated form of PCNA can be purified, e.g., for DNA polymerase assays or for use in polyubiquitylation reactions. As shown in Fig. 1, the reaction is rarely complete. In order to enrich for the ubiquitylated form, His-tagged ubiquitin is therefore used in combination with untagged PCNA, which allows purification of the conjugate via Ni-NTA affinity chromatography.

1. Set up a PCNA monoubiquitylation reaction of 2 mL, scaling up the protocol described in Subheading 3.5 accordingly. Use ^{His}Ubiquitin and untagged PCNA.
2. Add 0.5 μL of benzonase stock solution and incubate at 30 °C for 15 min to degrade the DNA.
3. Dilute the reaction 1:1 with buffer E and pass over a 1 mL HiTrap™ heparin column equilibrated in buffer E + 100 mM NaCl. PCNA will be present in the unbound material. This step removes the Rad6-Rad18 complex.
4. Supplement the unbound material with 30 mM imidazole and perform Ni-NTA affinity chromatography as described in Subheading 3.1.4. Elute the ^{His}Ub-PCNA conjugate with buffer E containing 150 mM NaCl and 300 mM imidazole.
5. Apply the eluate to a HiTrap™ Q column equilibrated in buffer E + 200 mM NaCl and wash extensively. This step removes free ubiquitin from the preparation.
6. Develop the column with a linear gradient up to 500 mM NaCl. The ^{His}Ub-PCNA conjugate elutes at ca. 350 mM.
7. Pool the relevant fractions and dialyze into buffer E + 100 mM NaCl. The conjugate can be stored at −80 °C.

3.7. PCNA Polyubiquitylation Assays

Polyubiquitylation of PCNA proceeds with monoubiquitylated PCNA as an intermediate, but the action of Ubc13, Mms2, and Rad5 does not require PCNA to be loaded onto DNA (8). While mono- and polyubiquitylation assays can be performed in a single tube by either adding all components together in one reaction including RFC and DNA, or by preforming monoubiquitylated PCNA and adding the enzymes for the second step later, we describe here the use of purified monoubiquitylated PCNA (prepared as described in Subheading 3.6). Hence, no RFC or DNA are included in the reaction. As an alternative to monoubiquitylated PCNA, a linear fusion of ubiquitin to the N-terminus of PCNA can be used (see Note 19). The assays are carried out in 10 μL or 20 μL reaction volumes, and a control reaction without E1 should be included. Other controls, such as the omission of Ubc13, Mms2, or Rad5, or use of a ubiquitin (K63R) mutant, are also useful.

1. Add to a series of microcentrifuge tubes: 10× MAB, 1 mM ATP, and 50 nM ^{His}Ub-PCNA.
2. Add 25 nM E1 or buffer, 1 μM ubiquitin and 200 nM Ubc13, Mms2, and Rad5 each.
3. Incubate the reactions for 1 h at 30 °C and stop by addition of 0.2 volumes of 5× SDS loading buffer. Alternatively, perform a time course analysis by removing and denaturing samples at specified intervals.

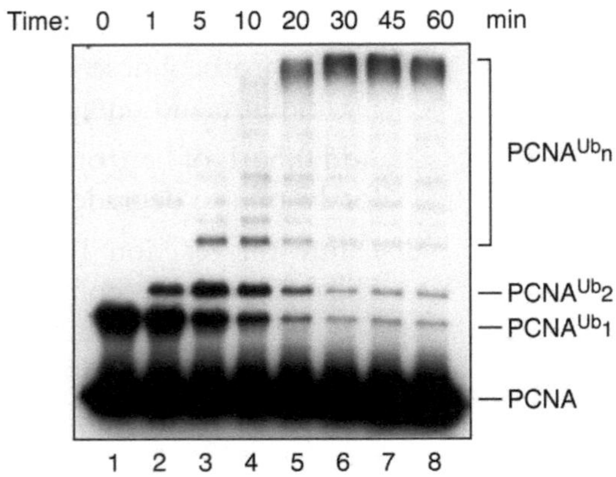

Fig. 2. Time course of PCNA polyubiquitylation. Reactions were set up with the monoubiquitylation enzymes as described above, and Ubc13, Mms2 and Rad5 were added after a 60 min incubation at 30 °C. Starting from this point, samples were taken at the indicated times and analyzed by Western blotting as described for Fig. 1. This Figure was reproduced from ref. 8.

4. Incubate the tubes at 95 °C for 3 min to denature the proteins and analyze the reaction products on a 10 % SDS polyacrylamide gel, followed by Western blotting and development with an anti-PCNA antibody.
5. Figure 1 shows that Ubc13, Mms2, and Rad5 are very efficient in chain extension, which results in complete conversion of monoubiquitylated to the polyubiquitylated PCNA within the time of the reaction (lane 8). Chain extension is very inefficient in the absence of Rad5 (lane 4). If a ubiquitin (K63R) mutant was used for monoubiquitylation, no extension is possible at all (lane 3). A time course of the reaction (Fig. 2) shows that chains are increasing in length with time until they are no longer resolved by the gel. This behavior is not observed in vivo, where no more than four ubiquitin moieties are detectable, possibly as a consequence of a balance between conjugation and deconjugation (13).

3.8. PCNA Sumoylation Assays

Although not strictly required, loading of PCNA onto DNA strongly enhances the efficiency of sumoylation by Ubc9, both in the presence and in the absence of Siz1 (12). Hence, RFC and nicked plasmid DNA are included in the reactions. A control reaction without E1 should always be included (see Note 16).

1. Add to a series of microcentrifuge tubes: 10× MAB, 2.5 nM nicked plasmid DNA, 30 nM RFC, 1 mM ATP, and 50 nM PCNA.
2. Add 100 nM Uba2-Aos1 or buffer, 50 nM Ubc9, 1 µM Smt3, and 25 nM Siz1. Prepare a master mix for as many components as possible in order to reduce variations due to inaccurate pipetting (see Note 17).

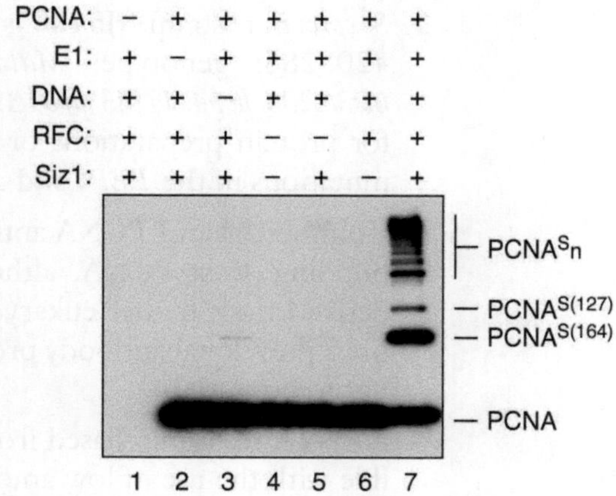

Fig. 3. Modification of PCNA by Smt3. In vitro sumoylation assays were performed with recombinant Ubc9 and SUMO in the presence or absence of PCNA, E1, RFC, Siz1 and circular, multiply primed ssDNA as indicated. Products were analyzed by Western blotting with PCNA-specific antibody. Note that full-length Siz1 was used in this particular assay. This figure was reproduced from ref. 12.

3. Incubate the reactions for 1 h at 30 °C and stop by addition of 0.2 volumes of 5× SDS loading buffer.
4. Incubate the tubes at 95 °C for 3 min to denature the proteins and analyze the reaction products on a 10 % SDS polyacrylamide gel, followed by Western blotting and development with an anti-PCNA antibody.
5. As shown in Fig. 3, sumoylation of PCNA proceeds efficiently only in the presence of DNA and RFC (compare lanes 3, 5, and 7). Use of lower enzyme concentration reduces, but does not abolish the formation of polymeric chains of Smt3 (12). Even the use of Smt3(K11,15,19R), where the consensus acceptor lysines on Smt3 have been mutated to arginine, does not completely prevent the formation of poly-Smt3 chains (13). At the concentrations used in this assay, Ubc9 alone exhibits minimal activity towards PCNA, but if the concentrations of both Ubc9 and PCNA are increased tenfold, modification at PCNA's consensus sumoylation site, K127, is observed (13). Time courses of the reaction (not shown) indicate that the majority of conjugates are formed within 5 min.

4. Notes

1. The plasmids for production of all proteins used in this chapter are described in Table 1, along with their sources and the approximate yield from the respective expression system. The plasmids prepared in our lab are available from the authors.

2. *S. cerevisiae* strain BJ5460 is a protease-deficient strain (ATCC #208285; genotype: *Mat***a** *ura3-52 trp1 lys2-801 leu2Δ1 his3Δ200 pep4::HIS3 prb1Δ1.6R can1 GAL*) commonly used for protein preparations or proteomic analysis, which carries mutations in the *PEP4* and *PRB1* genes.
3. Commercial anti-PCNA antibodies generally react poorly with budding yeast PCNA, although they work well for protein derived from higher eukaryotes or fission yeast. We therefore use a polyclonal antibody prepared in the lab, which is available upon request (14).
4. Ni-NTA resin purchased from GE Health Sciences is compatible with the use of low amounts (1 mM) of DTT and EDTA. Avoid their presence if using resin purchased from other sources, as they might strip the Ni^{2+} from the column.
5. All buffers listed in this section are given in their standard composition. For the purpose of ion exchange chromatography, they need to be supplemented with the appropriate salt (NaCl) concentration, which is given in the respective sections.
6. Ubiquitin from commercial sources is usually bovine and is compatible with all assays involving yeast conjugation factors and substrates, due to the high conservation of the protein. The preparations might contain a small amount of diubiquitin as a contaminant, which can be removed by gel filtration chromatography.
7. TB medium gives a much higher biomass than LB, but either medium is suitable for protein production.
8. The appropriate scale for each of the protein preparations naturally depends on the desired yield. Generally, purifications are easier to handle and result in a product of higher purity when performed on a larger scale, where protein loss is of little relevance and impure fractions can be discarded. Approximate yields are given in Table 1 and can serve as a guide for selecting appropriate culture volumes.
9. Generally, all lysis and purification steps are carried out at 4 °C. A gel filtration is usually performed as the last step of each purification. At this step, 10 % (w/v) glycerol can be used in the column buffer. When this is present, protein aliquots can be stored at −80 °C.
10. PCNA is produced with very high yield from *E. coli*. For gel filtration, avoid concentrations above 5 mg/mL in order to prevent the formation of higher aggregates. ^{His}PCNA does not show this aggregation behavior to the same extent as the untagged protein.
11. RFC is a 5-subunit complex that comprises Rfc1, Rfc2, Rfc3, Rfc4, and Rfc5. All are encoded on a single plasmid, pBL481-RFC (23). Rfc1 carries an N-terminal truncation, where amino

acids 3–273 have been replaced by a His_7-tag. This does not reduce in vitro activity towards PCNA, but improves solubility and may in some cases even enhance its activity, due to a lowered affinity for single-stranded DNA.

12. A protocol to purify the Rad6-Rad18 complex by co-expression in baculovirus-infected insect cells has been described as well (25), but the yeast-derived complex exhibits much improved solubility and activity in conjugation assays. Full-length budding yeast Rad18 cannot be purified in good yield in the absence of Rad6, presumably due to folding or solubility problems (26).

13. The use of the *leu2d* marker on the episomal vector used for overexpression can result in a higher yield compared to other selective markers. However, other markers are generally adequate to provide sufficient protein for biochemical assays. Using a freshly transformed strain generally improves the yield. When using galactose as a carbon source, the yeast cultures require very good aeration, which can be achieved by using baffled or Tunair flasks (IBI Scientific) or a fermentor.

14. The molecular mass of ^{His}Rad18 is 57 kDa. However, on SDS gels the protein runs at an apparent molecular mass of ca. 70 kDa. A small proportion of Rad18 is present as a stably monoubiquitylated form (25), which remains present throughout the purification. Rad6 runs at its expected molecular mass of 20 kDa.

15. For the purpose of the conjugation assays, all proteins, whether His-tagged or not, are referred to only by their proper names, unless otherwise noted.

16. A common control in ubiquitin or SUMO conjugation assays is the omission of ATP from the assay to prevent E1 activity. This control is not suitable for PCNA modification assays, because the RFC-catalyzed loading reaction requires ATP. Other useful controls are PCNA mutants in relevant acceptor lysines, K127R and/or K164R, or catalytically inactive versions of the E2 enzymes (8, 12, 13). A ubiquitin K63R mutant is instructive for polyubiquitylation assays, as this cannot be extended by Ubc13-Mms2 (27).

17. Whereas PCNA sumoylation proceeds with similar efficiency over a wide range of salt concentrations (30–250 mM NaCl) and pH ranges (pH 7.0–8.5), Rad6-Rad18-dependent monoubiquitylation reactions are much more sensitive to salt concentration. The final NaCl concentration should not be higher than 100 mM, taking into account the salt that is added through the addition of the protein preparations. In particular, since the Rad6-Rad18 concentration in the final preparation is usually rather low, a large volume is generally required (0.5–2 μL).

18. Monoubiquitylation of PCNA by budding yeast Rad6-Rad18 proceeds relatively slowly and inefficiently with increasing

amounts of product formed over ca. 60 min. Longer incubation times often do not increase the yield. This inefficiency has been observed previously, particularly in small-scale reactions (8, 18). One way of enhancing the efficiency of the reaction significantly is to use a linear Smt3 fusion to the N-terminus of PCNA (our unpublished results). Due to a SUMO-interaction motif present in Rad18, the sumoylated form of PCNA—or a linear fusion—is a much better substrate for the E3 than unmodified PCNA, and the reaction easily goes to completion even at a small scale. Once the reaction is complete, the SUMO tag can be removed by means of the SUMO-specific isopeptidase Ulp1 (28).

19. Whereas Ubc13-Mms2 alone are able to add a second ubiquitin onto monoubiquitylated PCNA, Rad5 is essential for efficient chain formation once monoubiquitylation has taken place (8). At this step, Rad18 is no longer required. Rad5 itself is insensitive to the placement of the first ubiquitin moiety on PCNA, so chains to be formed on linear fusions of ubiquitin to the N-terminus of PCNA. Note that fusions to the C-terminus are also modified, but with much lower efficiency, presumably due to the proximity of the fusion site to K63 of ubiquitin (8). Alternatively, a construct called "split PCNA" can be used as a mimic of monoubiquitylated PCNA that carries the ubiquitin moiety at the correct position (29).

Acknowledgments

We would like to thank Erica Johnson and Peter Burgers for supplying various expression plasmids. This work was supported by Cancer Research UK.

References

1. Bergink S, Jentsch S (2009) Principles of ubiquitin and SUMO modifications in DNA repair. Nature 458:461–467
2. Ulrich HD, Walden H (2010) Ubiquitin signalling in DNA replication and repair. Nat Rev Mol Cell Biol 11:479–489
3. Kerscher O, Felberbaum R, Hochstrasser M (2006) Modification of proteins by ubiquitin and ubiquitin-like proteins. Annu Rev Cell Dev Biol 22:159–180
4. Ardley HC, Robinson PA (2005) E3 ubiquitin ligases. Essays Biochem 41:15–30
5. Geiss-Friedlander R, Melchior F (2007) Concepts in sumoylation: a decade on. Nat Rev Mol Cell Biol 8:947–956
6. Ulrich HD (2009) Regulating post-translational modifications of the eukaryotic replication clamp PCNA. DNA Repair 8:461–469
7. Hoege C, Pfander B, Moldovan GL, Pyrowolakis G, Jentsch S (2002) *RAD6*-dependent DNA repair is linked to modification of PCNA by ubiquitin and SUMO. Nature 419:135–141
8. Parker JL, Ulrich HD (2009) Mechanistic analysis of PCNA poly-ubiquitylation by the ubiquitin protein ligases Rad18 and Rad5. EMBO J 28:3657–3666
9. Gangavarapu V, Haracska L, Unk I, Johnson RE, Prakash S, Prakash L (2006) Mms2-Ubc13-dependent and -independent roles of Rad5

ubiquitin ligase in postreplication repair and translesion DNA synthesis in *Saccharomyces cerevisiae*. Mol Cell Biol 26:7783–7790
10. Pages V, Bresson A, Acharya N, Prakash S, Fuchs RP, Prakash L (2008) Requirement of Rad5 for DNA polymerase zeta-dependent translesion synthesis in *Saccharomyces cerevisiae*. Genetics 180:73–82
11. Coulon S, Ramasubramanyan S, Alies C, Philippin G, Lehmann A, Fuchs RP (2010) Rad8Rad5/Mms2-Ubc13 ubiquitin ligase complex controls translesion synthesis in fission yeast. EMBO J 29:2048–2058
12. Parker JL, Bucceri A, Davies AA, Heidrich K, Windecker H, Ulrich HD (2008) SUMO modification of PCNA is controlled by DNA. EMBO J 27:2422–2431
13. Windecker H, Ulrich HD (2008) Architecture and assembly of poly-SUMO chains on PCNA in *Saccharomyces cerevisiae*. J Mol Biol 376:221–231
14. Papouli E, Chen S, Davies AA, Huttner D, Krejci L, Sung P, Ulrich HD (2005) Crosstalk between SUMO and ubiquitin on PCNA is mediated by recruitment of the helicase Srs2p. Mol Cell 19:123–133
15. Pfander B, Moldovan GL, Sacher M, Hoege C, Jentsch S (2005) SUMO-modified PCNA recruits Srs2 to prevent recombination during S phase. Nature 436:428–433
16. Parnas O, Zipin-Roitman A, Pfander B, Liefshitz B, Mazor Y, Ben-Aroya S, Jentsch S, Kupiec M (2010) Elg1, an alternative subunit of the RFC clamp loader, preferentially interacts with SUMOylated PCNA. EMBO J 29:s2611–2622
17. Moldovan GL, Pfander B, Jentsch S (2006) PCNA controls establishment of sister chromatid cohesion during S phase. Mol Cell 23:723–732
18. Garg P, Burgers PM (2005) Ubiquitinated proliferating cell nuclear antigen activates translesion DNA polymerases η and REV1. Proc Natl Acad Sci U S A 102:18361–18366
19. Haracska L, Unk I, Prakash L, Prakash S (2006) Ubiquitylation of yeast proliferating cell nuclear antigen and its implications for translesion DNA synthesis. Proc Natl Acad Sci U S A 103:6477–6482
20. Guthrie C, Fink GR (1991) Guide to yeast genetics and molecular biology, vol 194. Academic, San Diego
21. Sambrook J, Russell D (2001) Molecular cloning: a laboratory manual, 3rd edn. Cold Spring Harbor Laboratory Press, Cold Spring Harbor
22. Jentsch S, McGrath JP, Varshavsky A (1987) The yeast DNA repair gene *RAD6* encodes a ubiquitin-conjugating enzyme. Nature 329:131–134
23. Gomes XV, Gary SL, Burgers PM (2000) Overproduction in *Escherichia coli* and characterization of yeast replication factor C lacking the ligase homology domain. J Biol Chem 275:14541–14549
24. Ulrich HD (2003) Protein-protein interactions within an E2-RING finger complex. Implications for ubiquitin-dependent DNA damage repair. J Biol Chem 278:7051–7058
25. Davies AA, Huttner D, Daigaku Y, Chen S, Ulrich HD (2008) Activation of ubiquitin-dependent DNA damage bypass is mediated by Replication Protein A. Mol Cell 29:625–636
26. Bailly V, Lauder S, Prakash S, Prakash L (1997) Yeast DNA repair proteins Rad6 and Rad18 form a heterodimer that has ubiquitin conjugating, DNA binding, and ATP hydrolytic activities. J Biol Chem 272:23360–23365
27. Hofmann RM, Pickart CM (1999) Noncanonical *MMS2*-encoded ubiquitin-conjugating enzyme functions in assembly of novel polyubiquitin chains for DNA repair. Cell 96:645–653
28. Lee CD, Sun HC, Hu SM, Chiu CF, Homhuan A, Liang SM, Leng CH, Wang TF (2008) An improved SUMO fusion protein system for effective production of native proteins. Protein Sci 17:1241–1248
29. Freudenthal BD, Gakhar L, Ramaswamy S, Washington MT (2010) Structure of monoubiquitinated PCNA and implications for translesion synthesis and DNA polymerase exchange. Nat Struct Mol Biol 17:479–484
30. Bylebyl GR, Belichenko I, Johnson ES (2003) The SUMO isopeptidase Ulp2 prevents accumulation of SUMO chains in yeast. J Biol Chem 278:44113–44120
31. Johnson ES, Schwienhorst I, Dohmen RJ, Blobel G (1997) The ubiquitin-like protein Smt3p is activated for conjugation to other proteins by an Aos1p/Uba2p heterodimer. EMBO J 16:5509–5519
32. Johnson ES, Gupta AA (2001) An E3-like factor that promotes SUMO conjugation to the yeast septins. Cell 106:735–744
33. Johnson ES, Blobel G (1997) Ubc9p is the conjugating enzyme for the ubiquitin-like protein Smt3p. J Biol Chem 272:26799–26802

Chapter 38

Analysis of Inhibition of DNA Replication in Irradiated Cells Using the SV40 Based In Vitro Assay of DNA Replication

George Iliakis, Emil Mladenov, Ya Wang, and Hong Yan Wang

Abstract

The deleterious effects of DNA damage on DNA replication have been demonstrated in many model systems and the mechanisms of the resulting inhibition have been a research focus for at least 40 years. Moreover, recent studies have identified several major components of the S-phase checkpoint, providing a mechanistic background for understanding the basis of inhibition of the initiation and elongation steps of DNA replication after DNA damage. Yet several aspects of the underlying biochemical mechanisms remain unresolved including the characterization of the enzymatic activities involved in checkpoint activation and the coordination of this process with DNA repair. Helpful for the delineation of the mechanism of the S-phase checkpoint is the observation that factors inhibiting DNA replication in vivo can be found in active form in extracts prepared from irradiated cells, when these are tested using the simian virus 40 (SV40) assay for in vitro DNA replication. In this assay, replication of plasmid DNA carrying the minimal origin of SV40 DNA replication is achieved in vitro using cytoplasmic cell extracts and SV40 large tumor antigen (TAg) as the only noncellular protein. Here, we describe protocols developed to measure in vitro DNA replication with the purpose of analyzing its regulation after exposure to DNA damage. The procedures include the preparation of components of the in vitro DNA replication reaction including cytoplasmic extracts from cells that have sustained DNA damage. The assay provides a powerful tool for investigating the effect of distinct agents acting either by inducing lesions in the DNA, or by inhibiting the functions of checkpoint proteins. Nevertheless, the fact that several steps of DNA replication initiation are carried out in this in vitro assay by TAg and not the corresponding cellular factors, might be considered as a limitation of the approach.

Key words: DNA replication, S-phase Checkpoint, SV40, T-antigen, In vitro assay, Cell extract

1. Introduction

Inhibition of DNA replication in eukaryotic cells is one of the earliest effects of radiation to be reported and quantified. Elucidation of the mechanism causing this inhibition has been a research focus in several laboratories for at least four decades. A significant development for these efforts was the recognition that the mechanism of inhibition has a direct (*cis*-acting) and an indirect (*trans*-acting)

Lotte Bjergbæk (ed.), *DNA Repair Protocols*, Methods in Molecular Biology, vol. 920,
DOI 10.1007/978-1-61779-998-3_38, © Springer Science+Business Media New York 2012

component (for reviews, see ref. 1, 2). Whereas the direct component is thought to derive from radiation-induced DNA damage that alters chromatin structure and inhibits DNA replication in *cis*, the indirect component is attributed to the activation by DNA damage of regulatory processes that inhibit DNA replication in *trans*. The latter is equivalent to the activation of a checkpoint during the S-phase (2–7).

The realization that a checkpoint is activated in S-phase after induction of DNA damage has led to intensive studies aiming at its genetic and biochemical characterization. The best documented genetic alteration that affects the regulation of DNA replication in response to radiation exposure is found in individuals with the hereditary genetic disorder ataxia telagiectasia (AT). AT cells fail to inhibit DNA replication in response to radiation suggesting that the underlying gene, ATM, a key member of the PIKK family of protein kinases, is involved in the regulation of DNA replication (for reviews, see ref. 8, 9). Other experiments implicate ATR, another member of PIKK family in the regulation of DNA replication following the generation of other forms of DNA damage. Recent studies have identified several additional components of this checkpoint response and there is strong interest in its biochemical characterization.

Promising for the delineation of the mechanism of the S-phase checkpoint is the observation that factors that inhibit DNA replication in vivo can be found in active form in extracts prepared from irradiated cells, when these are tested for replication activity using the simian virus 40 (SV40) assay for in vitro DNA replication (10–12). In this assay, replication of plasmids carrying the minimal origin of SV40 DNA replication is achieved in vitro using cytoplasmic cell extracts and SV40 large tumor antigen (TAg) as the only noncellular protein (13–16). It is thought that cellular proteins function in this assay in the same manner as in vivo. A further advantage of the SV40 assay is its ability to initiate DNA replication at a defined origin, a function that cannot be achieved in other in vitro system, where only the elongation step of DNA replication can typically be studied.

The assay has been extremely successful in the field of DNA replication and has led to the characterization of a number of factors involved in eukaryotic DNA replication. In a similar way, the assay could also help in the biochemical characterization of important components of the regulatory pathway activated in response to DNA damage (e.g., (17, 18)). While the assay is limited by the fact that a single noncellular protein, TAg, carries out functions assigned to different families of proteins in eukaryotic cells (19), its potential for the biochemical characterization of checkpoint responses has not been exhausted and modifications mitigating some limitations may be possible.

Here, we describe protocols developed to measure in vitro DNA replication with the purpose of analyzing its regulation after

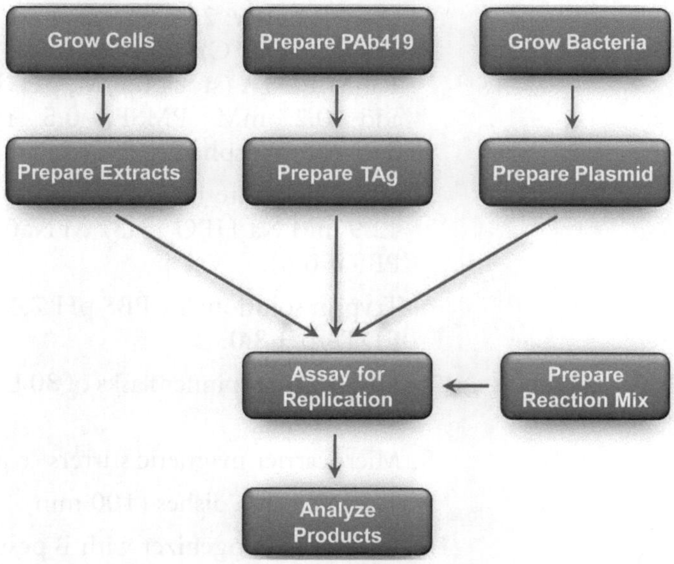

Fig. 1. Outline of the individual steps and preparations required for assembling DNA replication reactions using the SV40 system for in vitro DNA replication and for evaluating their outcome.

exposure to DNA damage. The required procedures include: (1) Preparation of cytoplasmic extract from cells that have sustained DNA damage; (2) Preparation of the SV40 large TAg; (3) Preparation of supercoiled plasmid DNA carrying SV40 origin of DNA replication; (4) Assembly of in vitro replication reactions; (5) Assay of DNA replication using incorporation of radioactive precursors, and of the DNA replication products using gel electrophoresis. Graphic outlines of the important steps of the procedures are summarized in Fig. 1.

2. Materials

2.1. Preparation of HeLa-Cell Extract

1. Minimum Essential Medium (MEM) modified for suspension cultures (S-MEM), supplemented with 5 % fetal bovine serum (FBS) or iron-supplemented bovine serum and antibiotics (penicillin 100 U/mL, streptomycin 100 μg/mL).
2. Hypotonic buffer solution: 10 mM HEPES-KOH, pH 7.5 (stock 0.6 M, pH 7.5 at room temperature (RT)), 1.5 mM $MgCl_2$ (stock 0.5 M), 5 mM KCl (stock 3 M). Immediately before use add 0.2 mM PMSF (stock 100 mM in ethanol), 0.5 mM dithiothreitol (DTT) (stock 1 M in H_2O; store at –20 °C), and 20 mM β-glycerophosphate.
3. High salt buffer: 100 mM HEPES-KOH, pH 7.5, 1.4 M KCl, and 15 mM $MgCl_2$.

4. Dialysis buffer: 25 mM Tris–HCl, pH 7.5 at 4 °C (stock 1 M, pH 7.5 at 4 °C), 10 % glycerol, 50 mM NaCl (stock 5 M), 1 mM EDTA (stock 0.5 M, pH 8.0). Immediately before use add 0.2 mM PMSF, 0.5 mM DTT, and 20 mM β-glycerophosphate.
5. 10× Phosphate-buffered saline (PBS): 14.7 mM KH_2PO_4, 42.9 mM Na_2HPO_4, 1.37 M NaCl, 26.8 mM KCl. (pH of 10× PBS is 6.8).
6. Trypsin solution: 1× PBS pH 7.4 (pH 1× PBS is 7.4), 10 mM EDTA pH 8.0.
7. Microcarrier spinner flasks of 30 L nominal volume (e.g., Bellco Glass Inc.).
8. Microcarrier magnetic stirrers (e.g., Bellco Glass Inc.).
9. Tissue culture dishes (100 mm).
10. Dounce homogenizer with B pestle, 50 mL.

2.2. Preparation of TAg

1. Hybridoma cell line PAb419; clone L19 generated by Harlow et al. should be used (20). It produces a monoclonal antibody that recognizes the amino terminal region of TAg and is used for the preparation of immunoaffinity columns employed in the purification of TAg. The cell line can be requested from Dr. Harlow, and is used to produce antibody that can be purified using standard procedures (21).
2. Insect cells *(Sf9)*. Available from ATCC.
3. Baculovirus 941T, *Autographa californica* expressing TAg. The 941T virus was constructed by Lanford et al. using a TAg cDNA (22) and can be requested from the author. The procedures used to prepare stocks of the virus and to measure its infectivity in pfu/mL are described in specialized protocols and the reader is referred to these publications for more information, e.g., (23).
4. TD buffer: 25 mM Tris–HCl pH 7.4, (stock 1 M, pH 7.4), 136 mM NaCl (stock: 5 M), 5.7 mM KCl (stock 3 M), 0.7 mM Na_2HPO_4 (stock 0.2 M).
5. Buffer B: 50 mM Tris–HCl, (stock 1 M, pH 8.0), 150 mM NaCl (stock 5 M), 1 mM EDTA (stock 0.5 M, pH 8.0), 10 % glycerol, 1 mM PMSF (stock 0.1 M), 1 mM DTT (stock 1 M).
6. Buffer C: 50 mM Tris–HCl pH 8.0, (stock 1 M, pH 8.0), 500 mM LiCl (stock 1 M), 1 mM EDTA (stock 0.5 M, pH 8.0), 10 % glycerol, 1 mM PMSF (stock 0.1 M), 1 mM DTT (stock 1 M).
7. Buffer D: 10 mM PIPES-NaOH, pH 7.4, (stock 1 M, pH 7.4, dissolved in 1 M NaOH), 5 mM NaCl (stock 5 M), 1 mM

EDTA (stock 0.5 M, pH 8.0), 10 % glycerol, 1 mM PMSF (stock 0.1 M), 1 mM DTT (stock 1 M).

8. Buffer E: 20 mM triethylamine, 10 % glycerol, pH 10.8. Prepare just before use.
9. Buffer F: 10 mM PIPES-NaOH, (stock 1 M, pH 7.0), 5 mM NaCl (stock 5 M), 0.1 mM EDTA (stock 0.5 M, pH 8.0), 10 % glycerol, 1 mM PMSF (stock 0.1 M), 1 mM DTT (stock 1 M).
10. Reagents: Sodium borate, dimethylpipelimidate, ethanolamine, triethylamine, merthiolate, NP-40.
11. Chromatography supplies: Protein A agarose, Sepharose 4B-Cl (Invitrogen, GE Healthcare), syringes (5 mL), or EconoColumns with ID of 0.75 cm (Bio-Rad).
12. Supplies and equipment for SDS-PAGE.
13. Two 1 L and one 250 mL microcarrier spinner flasks (Bellco).
14. Material for protein determination using the Bradford assay (a ready to use solution is available from Bio-Rad).

2.3. Preparation of Supercoiled Plasmid DNA Carrying the SV40 Origin of DNA Replication

Several plasmids that carry the minimum origin of SV40 DNA replication are available and can be used for this purpose, e.g., pSV01ΔEP (24); pSV010 (25); and pJLO (26). Large quantities of these plasmids can be prepared using cesium chloride ethidium bromide gradients. The description of these methods are beyond the scope of the present protocol and can be found in publications describing molecular biology protocols (23, 27).

2.4. Assembly of In Vitro Replication Reactions and Product Analysis

1. Replication reaction solution (5×): 200 mM HEPES-KOH, pH 7.5, 40 mM $MgCl_2$, 2.5 mM DTT, 200 mM Phosphocreatine, 15 mM ATP, 1 mM CTP, 1 mM GTP, 1 mM UTP, 0.5 mM dATP, 0.125 mM dCTP, 0.5 mM dGTP, 0.5 mM dTTP. Prepare 10 mL replication reaction solution and freeze in small aliquots.
2. Creatine phosphokinase stock: 2.5 mg/mL prepared in 50 % v/v glycerol. Phosphokinase together with phosphocreatine present in the replication reaction solution form an ATP regeneration system that is required for DNA replication.
3. Salmon sperm DNA.
4. Reagents: EDTA, trichloroacetic acid (TCA), sodium dodecyl sulfate (SDS).
5. Enzymes: Proteinase K, RNase A.
6. Scintillation counter.
7. Gel electrophoresis equipment.
8. Glass fiber filters (GF/C, Whatman).

3. Methods

3.1. Preparation of Cell Extract

The method described here is a modification of a method originally developed by Dignam et al. (28), and allows the preparation of cell extract from 10 L of cell suspension. Higher or lower amounts of extract can be prepared by appropriate scaling. Ionizing radiation or radiomimetic chemicals can be used for the generation of damage in the DNA (see Note 1). We obtained satisfactory results using a 3 h treatment with 0.5 μg/mL camptothecin, an effective inhibitor of type I topoisomerases. We also use routinely 10–40 Gy of 25 MV X-rays from a linear accelerator. The details for the latter treatment depend heavily upon the type of equipment used. It is therefore impractical to describe them here. Investigators with access to such equipment are advised to request the assistance of radiation physicists for dosimetry and other details on the irradiation protocol. Standard X-ray-producing equipment (50–250 kV) cannot be used to irradiate volumes of the magnitude described here due to their low penetration characteristics. Whatever the solution for the irradiation problem, it is important to keep in mind that disturbance in the cell culture has to be kept to a minimum before and after irradiation for reproducible results. Significant reductions in temperature, changes of the growth vessels, centrifugations, etc., should be avoided. Extracts can be processed for repair at a specific time after irradiation. The precise timing will depend upon the type of experiments and its specific goals. We prepare extracts 0–3 h after the treatment to induce DNA damage.

1. Grow HeLa cells at 37 °C for 3 days in twenty-five 100 mm tissue culture dishes prepared at an initial density of 3×10^6 cells/dish in 20 mL S-MEM supplemented with serum and antibiotics. The final density after 3 days of growth should be ~20×10^6 cells/dish, giving a total of 5×10^8 cells in 25 dishes (see Note 2).
2. During that step the cells are adherent and are attached to the growing vessel, therefore the trypsinization step is required. Detach the cells from all dishes using 0.05 % Trypsin in Trypsin solution and resuspend in 10 L of pre-warmed complete growth medium in a 30 L nominal volume microcarrier flask, thoroughly pre-gassed with 5 % CO_2 in air. The initial cell concentration should be ~5×10^4 cells/mL. Place in a warm room at 37 °C and provide adequate stirring (~40–60 rpm).
3. Allow cells to grow for 4 days, to a final concentration of 4–6×10^5 cells/mL. Do not exceed this concentration.
4. Collect cells by centrifugation (8 min at $2,500\times g$). Collection should be fast and is best done using a refrigerated centrifuge that can accept 1 L bottles (e.g., Beckman J6-MI). All further processing should be carried out at 0–4 °C.

5. Rinse twice in PBS and centrifuge (5 min at 500 × *g*). Determine the packed cell volume (pcv) (~12–15 mL total).
6. Resuspend cell pellet in 5 pcv volumes of hypotonic buffer solution and centrifuge quickly (5 min, 500 × *g*). Cells swell and the pcv approximately doubles.
7. Determine the new pcv. Resuspend cell pellet in 3 pcv of hypotonic buffer and disrupt in a Dounce homogenizer (20 strokes, pestle B). It is advisable to test cell disruption using a phase contrast microscope.
8. Add 0.11 volumes of high salt buffer and centrifuge 3,000 × *g* for 20 min.
9. Carefully remove supernatant and centrifuge at 100,000 × *g* for 1 h.
10. Place the resulting extract (S100) in dialysis tubing with an MW cutoff of 10–14 kDa and dialyze overnight against 50–100 volumes of dialysis buffer.
11. Collect extract. Centrifuge at 15,000 × *g* for 20 min to remove precipitated protein (see Note 1). Aliquot and snap freeze using liquid nitrogen. Immediately store at −70 °C and keep a small aliquot for determining protein concentration using the Bradford protein assay (Protein assay, Bio-Rad).

3.2. Preparation of TAg

Good quality TAg is essential for efficient replication in vitro of plasmids carrying the SV40 origin of DNA replication. Investigators can either obtain this protein from commercially available sources, or can prepare it in the laboratory using available reagents. The method of preparation described here is essentially the one described by Simanis and Lane (29), and can conveniently be separated in the preparation of the TAg immunoaffinity column and the purification of TAg from extracts of *sf9* cells infected with the baculovirus 941T which expresses TAg.

3.2.1. Preparation of TAg Immunoaffinity Column

1. Mix 5 mg PAb419 antibody with 2 mL wet protein A beads. Incubate at room temperature for 1 h with gentle rocking.
2. Wash the beads twice with 20 mL 0.2 M sodium borate (pH 9.0) by centrifugation at 1,000 × *g* for 5 min.
3. Resuspend the beads in 20 mL 0.2 M sodium borate (pH 9.0), mix and remove 100 μL bead-suspension for assaying coupling efficiency. Add solid dimethylpipelimidate to bring the final concentration to 20 mM.
4. Mix for 30 min at room temperature on a rocker and remove 100 μL suspension of the coupled beads.
5. Stop the coupling reaction by washing the beads with 20 mL 0.2 M ethanolamine (pH 8.0) and incubate for 2 h at room temperature in 0.2 M ethanolamine with gentle rocking.

6. Spin down and resuspend the beads in PBS. Add 0.01 % merthiolate if extensive storage is anticipated. At this point, beads are ready for use for the purification of TAg (if the quality control performed next is positive).
7. Check the efficiency of coupling by boiling in Laemmli sample buffer the samples of beads taken before and after coupling. Run the equivalent of 1 and 9 μL beads from both samples on a 10 % SDS-PAGE gel and stain with Coomassie Brilliant Blue G250. Due to the coupling procedure TAg is covalently attached to the Protein A agarose beads. Good coupling is indicated by bands indicating the mobility of IgG heavy chain (55 kDa) in the samples obtained before but not in the samples obtained after coupling procedure.
8. Prepare immunoaffinity column by pouring beads into a 5 mL syringe, or a 0.75 cm in diameter EconoColumn.

3.2.2. Purification of TAg from Extracts of sf9 Cells

1. Grow enough *sf9* cells to prepare 1 L of cell suspension at 2×10^5 cells/mL. Distribute the cell suspension in two 1 L microcarrier spinner flasks (500 mL cell suspension per spinner) and incubate at 27 °C under gentle spinning (1–1.5 revolutions per second). Allow cells to grow until they reach a concentration of 2×10^6 cells/mL (3–4 days) (see Note 3).
2. Spin cells for 5 min at $500 \times g$ and carefully return supernatant to spinner flasks.
3. Resuspend cells in enough volume of virus stock to reach a multiplicity of infection equal to 10 pfu per cell. Place cell suspension in a 250 mL spinner flask and allow attachment of virus to the cells by gentle stirring at 27 °C for 2 h.
4. Return cell suspension to the original spinner flasks and incubate for 48 h at 27 °C under gentle stirring to allow for protein expression. We have noted that good protein expression is also achieved by adding the viral stock directly into the cell culture and incubating for 48 h.
5. Harvest cells by centrifugation at $500 \times g$ for 5 min and resuspend in 25 mL TD buffer.
6. Centrifuge at $500 \times g$ for 5 min and resuspend in 30 mL buffer B. Add 10 % NP-40 to a final concentration of 0.5 %.
7. Place on ice for 30 min. Invert tube several times every 10 min.
8. Centrifuge for 10 min at $18{,}000 \times g$ in a corex tube. At this stage, cellular extract can be removed and quickly frozen at –70 °C for use at a later time for purification of TAg. When needed, thaw the extract quickly by immersing in 4 °C water bath. Do not allow extract to warm-up above 4 °C. All subsequent steps should be carried out in a cold room.

9. Save a 50 µL aliquot of the cell extract and load the remaining material onto a 2 mL Sepharose 4B-Cl column equilibrated with buffer B. Elution can be achieved either by gravity, or with the help of a peristaltic pump giving a flow rate of approximately 1 mL/min. This step retains proteins binding nonspecifically to sepharose media.
10. Allow the flow through of step 9 to pass through a 2 mL protein A agarose column equilibrated with buffer B. This can be achieved either by gravity, or with the help of a peristaltic pump giving a flow rate of approximately 1 mL/min. This step retains material binding nonspecifically to protein A.
11. Allow the flow through of step 10 to pass through a 2 mL immunoaffinity column of protein A agarose coupled with antibody PAb419. This can be achieved either by gravity, or with the help of a peristaltic pump at a flow rate of approximately 0.5 mL/min. This step retains TAg from the cell extract. Optionally, to increase the binding efficiency the loop may be designed to allow cellular extract containing TAg to pass through the column multiple times.
12. Repeat step 11. Save flow through.
13. Wash immunoaffinity column with 200 mL of buffer C. This is best achieved with a peristaltic pump giving a flow rate of approximately 2 mL/min.
14. Wash immunoaffinity column with 100 mL of buffer D. This is best achieved with a peristaltic pump giving a flow rate of approximately 2 mL/min.
15. Elute with buffer E. Collect 0.5 mL fractions in tubes containing 25 µL of 1.0 M PIPES-NaOH pH 7.0. Place fractions on ice (see Note 4).
16. Wash immunoaffinity column with 20 mL buffer E, and then with 40 mL buffer B. Column can be reused four to five times.
17. Measure protein concentration using the Bradford protein assay.
18. Combine fractions containing protein. Dialyze overnight against 2 L buffer F.
19. Test quality of purification by SDS-PAGE followed by silver staining. Test activity for in vitro SV40 DNA replication. The procedure yields 1–2 mg TAg per liter of *sf9* cell culture.

3.3. Preparation of Supercoiled Plasmid DNA Carrying the SV40 Origin of DNA Replication

Standard procedures can be used for the preparation of supercoiled plasmid DNA. It is preferable to purify the DNA using a two step purification procedure on $CsCl_2$/ethidium bromide gradient. Detailed protocols for this purpose are beyond the scope of the present protocol and can be found other sources of protocols (23, 27).

3.4. Assembly of In Vitro Replication Reactions and Evaluation of Replication Activity

1. Assemble 50 μL reactions by mixing in an Eppendorf tube kept on ice 5 μL reaction buffer, 2.5 μL creatine phosphate, 200–400 μg extract protein, 1 μg TAg, 0.3 μg superhelical plasmid DNA, 0.001 μCi/mL (α-^{32}P)dCTP; adjust volume to 50 μL with nuclease-free H_2O. Extracts from untreated and treated cells should be used in parallel so that the results obtained can be directly compared.
2. Incubate reactions at 37 °C for 1 h. Longer or shorter incubations can also be used if information on the kinetics of replication is desired.
3. Terminate reactions by adding EDTA to a final concentration of 20 mM.
4. Add 25 μg of denatured salmon sperm DNA and mix well.
5. Add 1 mL of cold 10 % TCA to precipitate nucleoprotein complexes. Mix well.
6. Collect precipitate onto Whatman® GF/C glass fiber filters. Wash three times with 10 mL cold 10 % TCA. Wash four times with 10 mL deionized water.
7. Add 5 mL scintillation fluid. Measure incorporated activity in a scintillation counter.

3.5. Analysis of the Replication Products

1. To analyze the DNA replication products by gel electrophoresis, add to the stopped replication reactions 0.1 % SDS.
2. Digest with RNase A (20 μg/mL) for 15 min at 37 °C.
3. Add proteinase K (200 μg/mL) and incubate at 37 °C for 30 min.
4. Purify DNA either by extraction in phenol/chloroform followed by precipitation in ethanol, or by using commercially available DNA purification systems.
5. Separate in 1 % agarose at 6.5 V/cm for 2 h. Electrophoretic conditions may need to be modified depending on the plasmid size used in the assay. For optimal resolution, reduce field intensity to 1 V/cm.

4. Notes

1. The effect of the DNA damage inducing agent on DNA replication in vitro can be variable. We found that this is usually due to the growth conditions (overgrown cultures), or to the absence of phosphatase inhibitors in the prepared extract. We routinely add β-glycerophosphate since we found it to significantly improve the reproducibility. Other phosphatase

inhibitors, as well as the use of protease inhibitors should be considered if reproducibility problems persist.

2. The preparation of a good extract depends strongly on the quality of the cells used. When cell growth is not optimal, or when cells overgrow, low replication activity may be obtained and the inhibition in extracts of treated cells may be suboptimal. To ensure optimal growth we routinely take the following measures: (a) Carefully test different batches of serum to find one with good growth characteristics. HeLa cells have a generation time of less than 20 h when grown as a monolayer, and less than 24 h when grown in suspension, under optimal growth conditions. (b) Use cells grown in dishes to start the suspension cultures for extract preparation. This helps to reduce clumping occurring after extensive growth in suspension. (c) We follow cell growth daily, and collect cells for extract preparation when they reach a concentration of $4–6 \times 10^5$ cells/mL. (d) We measure cell cycle distribution by flow cytometry. A high percentage of S-phase cells (~25 % for HeLa cells), suggests that the cell culture is still in an active state of growth.
3. Optimal cell growth is also a prerequisite of a successful preparation of TAg. We find that *sf9* cells grow more consistently if kept in suspension. Transfer from a monolayer state to a suspension state is usually associated with a shock that takes the cells some time to overcome.
4. We have observed that 20 mM triethylamine may not elute all bound TAg from the immunoaffinity column. If this proves to be the case, increase in triethylamine concentration (up to 100 mM), or alternative eluting methods (see ref. 21) should be considered. However, it should be kept in mind that such alternatives may reduce TAg activity.

References

1. Iliakis G (1997) Cell cycle regulation in irradiated and nonirradiated cells. Semin Oncol 24: 602–615
2. Iliakis G, Wang Y, Guan J, Wang H (2003) DNA damage checkpoint control in cells exposed to ionizing radiation. Oncogene 22:5834–5847
3. Zhou BB, Elledge SJ (2000) The DNA damage response: putting checkpoints in perspective. Nature 408:433–439
4. Nyberg KA, Michelson RJ, Putnam CW, Weinert TA (2002) Toward maintaining the genome: DNA damage and replication checkpoints. Annu Rev Genet 36:617–656
5. Branzei D, Foiani M (2009) The checkpoint response to replication stress. DNA Repair (Amst) 8:1038–1046
6. Muzi-Falconi M, Petrini J (2009) Checkpoint response to DNA damage. DNA Repair (Amst) 8:973–973
7. Zegerman P, Diffley JFX (2009) DNA replication as a target of the DNA damage checkpoint. DNA Repair (Amst) 8:1077–1088
8. Shiloh Y (2001) ATM and ATR: networking cellular responses to DNA damage. Curr Opin Genet Dev 11:71–77
9. Abraham RT (2001) Cell cycle checkpoint signaling through the ATM and ATR kinases. Genes Dev 15:2177–2196
10. Wang Y, Huq MS, Cheng X, Iliakis G (1995) Regulation of DNA replication in irradiated cells by *trans*-acting factors. Radiat Res 142: 169–175

11. Wang Y, Huq MS, Iliakis G (1996) Evidence for activities inhibiting *in trans* initiation of DNA replication in extract prepared from irradiated cells. Radiat Res 145:408–418
12. Wang Y, Perrault AR, Iliakis G (1997) Down-regulation of DNA replication in extracts of camptothecin-treated cells: activation of an S-phase checkpoint? Cancer Res 57:1654–1659
13. Challberg MD, Kelly TJ (1989) Animal virus DNA replication. Annu Rev Biochem 58: 671–717
14. Hurwitz J, Dean FB, Kwong AD, Lee SK (1990) The *in vitro* replication of DNA containing the SV40 origin. J Biol Chem 265: 18043–18046
15. Kelly TJ (1988) SV40 DNA replication. J Biol Chem 263:17889–17892
16. Stillman B (1989) Initiation of eukaryotic DNA replication in vitro. Annu Rev Cell Biol 5:197–245
17. Wang Y, Zhou XY, Wang H-Y, Iliakis G (1999) Roles of replication protein A and DNA-dependent protein kinase in the regulation of DNA replication following DNA damage. J Biol Chem 274:22060–22064
18. Wang Y, Guan J, Wang H, Wang Y, Leeper DB, Iliakis G (2001) Regulation of DNA replication after heat shock by RPA-nucleolin interactions. J Biol Chem 276:20579–20588
19. Mendez J, Stillman B (2003) Perpetuating the double helix: molecular machines at eukaryotic DNA replication origins. Bioessays 25: 1158–1167
20. Harlow E, Crawford LV, Pim DC, Williamson NM (1981) Monoclonal antibodies specific for simian virus 40 tumor antigens. J Virol 39: 861–869
21. Harlow E, Lane D (1999) Using antibodies. Cold Spring Harbor Laboratory Press, New York
22. Lanford RE (1988) Expression of simian virus 40 T antigen in insect cells using a baculovirus expression vector. Virology 167:72–81
23. Ausubel FM, Brent R, Kingston RE, Moore DD, Seidman JG, Smith JA, Struhl K (1994) Current protocols in molecular biology. Greene Publishing Associates and Wiley-Interscience, New York
24. Wobbe CR, Dean F, Weissbach L, Hurwitz J (1985) *In vitro* replication of duplex circular DNA containing the simian virus 40 DNA origin site. Proc Natl Acad Sci U S A 82:5710–5714
25. Prelich G, Stillman B (1988) Coordinated leading and lagging strand synthesis during SV40 DNA replication in vitro requires PCNA. Cell 53:117–126
26. Li JJ, Kelly TJ (1984) Simian virus 40 DNA replication in vitro. Proc Natl Acad Sci U S A 81:6973–6977
27. Sambrook J, Russell DW (2001) Molecular cloning, 3rd edn. Cold Spring Harbor Laboratory Press, New York
28. Dignam JD, Lebovitz RM, Roeder RG (1983) Accurate transcription initiation by RNA polymerase II in a soluble extract from isolated mammalian nuclei. Nucleic Acids Res 11: 1475–1489
29. Simanis V, Lane DP (1985) An immunoaffinity purification procedure for SV40 large T antigen. Virology 144:88–100

Chapter 39

CHK1 Kinase Activity Assay

Hong Yan Wang and Ya Wang

Abstract

Mammalian CHK1 is a Ser/Thr kinase that plays a critical role in the DNA damage-activated cell cycle checkpoint signaling pathway downstream of ATR (ATM and Rad3 related protein kinase). This chapter focuses on describing an assay to measure CHK1 activity in vitro. The basic mechanism of this assay is to observe the phosphorylated levels of a fragment of CDC25C containing the site that can be phosphorylated by CHK1 in vitro. This assay includes five major steps: (1) preparing extracts from the control or treated cells, (2) preparing substrate, (3) immunoprecipitating CHK1 protein from the cells, (4) assembling the kinase assay, (5) analyzing the phosphorylated level of the substrates by CHK1. Besides CHK1, CHK2 is another important checkpoint regulator that responds to DNA damage. Because CHK1 and CHK2 share some substrates such as CDC25C in vitro, this assay could also be used for a CHK2 activity assay, except that the CHK2 antibody will be replaced by the CHK1 antibody.

Key words: CHK1 kinase, CHK2 kinase, Checkpoint, DNA damage, Ionizing radiation, Camptothecin

1. Introduction

Checkpoint kinase 1 (CHK1) is an evolution conserved protein kinase that was first identified in fission yeast (1). The identification of a homologue in mammalian cells followed a few years later (2, 3). Although *chk1* is not an essential gene in fission yeast, it is essential in mice (1, 4, 5). CHK1 protein is an important component of the DNA damage checkpoint and of the DNA replication checkpoint (1, 4, 6–12). CHK1 responds to DNA damage inducers including ionizing radiation (IR), ultraviolet (UV) light (1, 8) and camptothecin (CPT) (13), as well as DNA replication

Lotte Bjergbæk (ed.), *DNA Repair Protocols*, Methods in Molecular Biology, vol. 920,
DOI 10.1007/978-1-61779-998-3_39, © Springer Science+Business Media New York 2012

inhibitors including hydroxyurea and aphidicolin (7, 9, 10). In *Xenopus* (*Xe*) and in human cells, Chk1 (CHK1) is phosphorylated and activated by ATR (ATM and Rad3-related protein kinase). The ATR/CHK1 pathway responds to agents that impair DNA replication, either directly (hydroxyurea, aphidicolin) or indirectly UV, IR, and CPT) (4, 12–16). Mammalian CHK1 is phosphorylated on serine 317 and 345 by ATR in vitro as well as in vivo (4, 14, 16). Mammalian CHK1 is involved in both S phase and G2 checkpoint responses after IR (4, 16–18). It has been proposed that CHK1 regulates the G2 checkpoint by phosphorylating the CDC25C protein phosphatase on residue(s) that facilitate the binding of 14-3-3 proteins, thus inactivating this phosphatase. As a result, CDC25C cannot further dephosphorylate the inhibitory phosphate group on CDK1 (original CDC2) and inactivate this kinase (2, 3, 9, 19, 20). In addition, it has been proposed that CHK1 regulates the S checkpoint by stabilizing CDC25A protein phosphatase (13, 16). CHK1-mediated checkpoint activation mainly facilitates homologous recombination repair (21, 22).

The recognition of CHK1 activation following DNA damage has led to increased studies of pathways regulating the checkpoint response. Although CHK1 is phosphorylated and activated by ATR, the phosphorylated CHK1 in the cells treated with some types of DNA damage reagents such as IR, is hard to detect by the regular gel (4). However, the changes of CHK1 activity following DNA damage are easily detected (13, 18, 23). We update the protocol to measure CHK1 activity following DNA damage induced by either IR or CPT (24). As CDC25C is a substrate of CHK1 (2), the basis of these protocols is to measure the phosphorylated levels of CDC25C by CHK1 in vitro. Besides CHK1, CHK2 is another important checkpoint regulator responding to DNA damage, which is phosphorylated and activated by ATM. Because CHK1 and CHK2 share some substrates such as CDC25C in vitro, this protocol could be used for CHK2 activity assay as well, except that the CHK2 antibody will be replaced by the CHK1 antibody.

The major procedures include the following:

1. Preparation of cellular extracts.
2. Preparation of substrate.
3. Immunoprecipitation (IP) of CHK1 from cellular extracts with the specific anti-CHK1 antibody.
4. Assembly of in vitro CHK1 kinase assay.
5. Analysis of the phosphorylated levels of substrate using polyacrylamide gel electrophoresis.

2. Materials

2.1. Preparation of Cellular Extracts

Investigators can either prepare cytoplasmic and nuclear extracts by using the kit NE-PER™ Nuclear and Cytoplasmic Extraction Reagents (Pierce) or prepare them in the laboratory using available reagents as follows:

1. Exponentially growing mammalian cells (human, rat or mouse).
2. Complete growth media.
3. Trypsin (Gibco BRL).
4. 5 mM KCl buffer: 10 mM HEPES, pH 7.5 (stock: 0.6 M, pH 7.5 at room temperature), 1.5 mM $MgCl_2$ (stock: 1 M), 5 mM KCl (stock: 3 M). Immediately before use, add 0.2 mM phenylmethylsulfonyl fluoride (PMSF) (stock: 100 mM in isopropanol), 0.5 mM dithiothreitol (DTT) (stock: 1 M in H_2O; store at –20 °C) and 20 mM glycerophosphate.
5. 0.8 M KCl buffer: 10 mM HEPES, pH 7.5, 1.5 mM $MgCl_2$, 0.8 M KCl.
6. 1.4 M KCl buffer: 10 mM HEPES, pH 7.5, 1.5 mM $MgCl_2$, 1.4 M KCl.
7. Spectrophotometer.

2.2. Preparation of Substrate: GST-Fusion Fragment of CDC25C Protein

1. BL21 bacterial cell strain (Novagen) (25).
2. Plasmid pGEX-2T (Amersham Biosciences) encoding human sequences of $CDC25C_{200-256}$ (2).
3. MicroSpin™ GST purification module (Amersham Biosciences).
4. Isopropysplthio-β-D-galactoside (IPTG) (Amersham Biosciences) (stock: 1 M in H_2O; store at –20 °C).
5. Bacteria incubator.
6. Alternatively, histone H1 (Sigma-Aldrich Inc.) could be used as a substrate of CHK1 (see Note 1).

2.3. Immunoprecipitation (IP) of CHK1 from Cellular Extracts

1. Cell extracts from Subheading 3.1.
2. Protein A-Sepharose slurry (RepliGen).
3. CHK1 antibody (Santa Cruz, *sc-7898*) (see Note 2).
4. IP buffer: 0.5 % NP-40, 1 mM Na_3VO_4, 5 mM NaF in phosphate-buffered saline (PBS) buffer (137 mM NaCl, 2.7 mM KCl, 10.6 mM Na_2HPO_4, 1.4 mM KH_2PO_4). Add 0.2 mM PMSF immediately before use.
5. Mini-shaker.

2.4. Assembly of In Vitro CHK1 Kinase Assay

1. IP products from Subheading 3.3.
2. Purified GST-CDC25C$_{200–256}$ or commercially obtained H1 from Subheading 2.2.
3. Kinase buffer: 10 mM HEPES, pH 7.5, 10 mM $MgCl_2$, 10 mM $MnCl_2$. Add 1 mM DTT immediately before use.
4. γ^{32}P– ATP.
5. Beta shield for radiation protection.
6. Heating blocks (30 and 100 °C).
7. 5× Protein loading buffer: 250 mM Tris–Cl, pH 6.8, 10 % SDS, 0.5 % bromophenol blue, 50 % glycerol, and 500 mM dithiothreitol. Keep the buffer at room temperature without dithiothreitol. Add dithiothreitol just before use.

2.5. Analysis of the Phosphorylating Levels of CDC25C Using Gel Electrophoresis

1. Acrylamide gel electrophoresis equipment with the Mini-gel accessories.
2. Prestained protein marker (Invitrogen, Inc., 10748).
3. Gel dryer (Bio-Rad).
4. PhosphorImager (Molecular Dynamics).

3. Methods

3.1. Preparation of Cellular Extracts

The method described here allows the preparation of cytoplasmic and nuclear extracts from mammalian cells of human, rat or mouse. The extracts can be prepared by using the kit NE-PER™ Nuclear and Cytoplasmic Extraction Reagents (Pierce) or in a laboratory by using available reagents as follows:

1. Grow mammalian cells in 100 mm dishes (at least five dishes for each point) at 70–80 % confluency (see Note 3).
2. Cell treatment: (1) IR: irradiate cells with ionizing radiation (10–20 Gy) and return the cells to 37 °C (optimal CHK1 activity is increased 3–6 h after IR). (2) CPT treatment: add CPT (1–2 μM) to the cell culture for 3–6 h.
3. Collect the treated and nontreated cells by trypsinization and centrifugation (4 °C, 5 min at 500×*g*). After the cells are trypsinized, the samples should be kept in cold (4 °C) conditions.
4. Rinse once in PBS and transfer the cells to Eppendorf tubes.
5. Wash once with the 5 mM KCl buffer in five packed cell volume (PCV) and spin the cells (15 s at 12,000×*g*), causing the PCV to approximately double. The following steps are based on the protocol we published previously (26).

6. Determine the new PCV. Resuspend the cell pellet in five PCV of hypotonic buffer.
7. Put the samples on ice for 10 min (the cells will swell). Disrupt the cell cytoplasmic membrane by three cycles of freeze–thaw: freeze the cells by putting the samples in methanol with dry ice and thaw the cells by putting the samples in a water bath (42 °C) (see Note 4). After this process, all the cytoplasmic membranes are 100 % disrupted while the nuclei retain their integrity (see Note 5).
8. Add 0.11 volumes of 1.4 M KCl buffer to the lysate and centrifuge the sample at 12,000 × *g*, 4 °C for 20 min. Additional high-salt buffer generates an isotonic condition (~140 mM KCl final concentration). Under these conditions, the nuclei shrink back to their original size. Transfer the supernatant (cytoplasmic extracts) to a clean Eppendorf tube on ice.
9. Add an equal volume of 0.8 M KCl buffer to the nuclei pellets. The final concentration of KCl in the sample is ~0.47 M. The high salt condition helps to extract proteins from the nuclei. Mix and shake the tube on the Mini-shaker at 4 °C for 30 min.
10. Centrifuge the sample at 12,000 × *g*, 4 °C for 20 min. Transfer the supernatant (nuclear extracts) to a clean Eppendorf tube on ice.
11. Measure the protein concentrations of the cytoplasmic and the nuclear extracts by a Bio-Rad assay (Bradford) according to the manufacturer's instructions.
12. Aliquot the sample (~1 mg/tube for cytoplasmic extracts and ~500 μg/tube for nuclear extracts) and freeze the sample at −80 °C (see Note 6).

3.2. Preparation of GST-Fusion Fragment of CDC25C Protein

Generate the GST fusion fragment of CDC25C protein first by PCR amplifying the cDNA fragment encoding amino acids 200–256 of CDC25C (from a human cDNA library, Novagen, Inc.) and then cloning into pGEX-2T vector according to the manufacturer's instructions.

1. Transform BL21 cells with the pGEX-2T vector containing the cDNA fragment encoding amino acids 200–256 of CDC25C according to the manufacturer's instructions.
2. Grow BL21 cells from one single colony picked from the selecting plate until OD_{600} (optical density) value is 0.6–0.8, and then add IPTG to the cell culture (final concentration 0.5 mM). After 2 h of expression induction, collect the cells according to the manufacturer's instructions.
3. Purify the GST-fusion CDC25C fragment protein by using the MicroSpin™, GST purification module (Amersham Biosciences) according to the manufacturer's instructions.

4. Measure the protein concentration, aliquot the sample (20 μg/tube), and freeze the sample at −80 °C.
5. Alternatively, histone H1 (Sigma-Aldrich Inc.) could be used as the substrate (see Note 1).

3.3. Immuno-precipitation (IP) of CHK1 from Cellular Extracts

1. Add 10 μL of 50 % (v/v) protein A-Sepharose slurry (RepliGen) to an Eppendorf tube. Wash the protein A-Sepharose slurry with IP buffer three times by centrifuging the sample at 12,000 × *g* for 10 s and aspirating the supernatant.
2. Add 5 μL of CHK1 antibody (about 1 μg) to the protein A-Sepharose slurry (see Note 7).
3. Add 250 μg of cytoplasmic or nuclear extracts (see Note 8) to the protein A-Sepharose slurry and bring the sample for a total final volume of 250–300 μL by adding lysis buffer to the same tube (see Note 9).
4. Rotate the tube at 4 °C for 4 h or overnight (see Note 10).

3.4. Assembly of In Vitro CHK1 Kinase Assay

1. Wash the sample once with cold lysis buffer and twice with cold kinase buffer without ^{32}P by centrifuging the sample (10 s at 12,000 × *g*) and aspirating the supernatants. Remove all of the liquid carefully after the final wash. The slurry contains the bound CHK1 antibody and CHK1 protein and any loss of this sample will affect the final results.
2. Add 20 μL of kinase buffer containing 2 μg of CDC25C protein or 1 μg of histone H1 (see Note 1) and 5 μCi $\gamma^{32}P$-ATP. In order to have an equal volume of substrate and isotope activity in each sample, mix kinase buffer, CDC25C or histone H1 protein and $\gamma^{32}P$-ATP in one tube, and then aliquot 20 μL of the mixture to each sample.
3. Quickly spin the sample (12,000 × *g*, for 5 s) (see Note 11) and incubate the reaction at 30 °C (heating block) for 30 min.
4. Stop the reaction by adding 6 μL of 5× protein loading buffer (see Note 12). Vortex the sample, spin the sample (12,000 × *g*, for 5 s) and boil the sample (in a heating block at 100 °C) for 5 min. After boiling, spin the sample again (12,000 × *g*, for 5 s). Now, the sample is ready for polyacrylamide gel electrophoresis.

3.5. Analysis of the Phosphorylated Levels of CDC25C or Histone H1 (See Note 1) Using Gel Electrophoresis

1. Prepare 12 % SDS-PAGE using 1 mm thick spacers (in order to load the whole sample).
2. Load the prestained protein marker (5 μL) in the first well and the sample in the next well, one piece of such Mini-gel will allow a load of nine samples in total (see Note 13).
3. Run the gel at 200 V for 40 min. To avoid the free isotope from running out of the gel and contaminating the buffer, keep the dye band in one-fourth from the bottom of the gel.

4. Maintain the gel between 68 and 28 kDa. The size of the CDC25C fragment or histone H1 is about 36 kDa (see Note 14). Cut off the gel below 28 kDa carefully. This piece of gel contains the dye bands and the free isotopes; put it in the isotope waste container.
5. Dry the gel by using a gel dryer (80 °C for 1 h).
6. Put the dried gel on a PhosphorImager plate for exposure.
7. After finishing the exposure (see Note 15), scan the plate in the PhosphorImager.
8. Analyze the levels of phosphorylation of CDC25C or histone H1 by using the quantitative software in the PhosphorImager (see Note 16).

4. Notes

1. Although histone H1 protein is usually used as a substrate of CDK1 or CDK2, it could also be phosphorylated by CHK1 in vitro. However, we observed that the phosphorylated level of histone H1 was opposite to that of CDC25C and CHK1 activity: higher CHK1 activity, higher phosphorylated level of CDC25C, lower phosphorylated level of histone H1 after DNA damage, suggesting that CHK1 might directly modify checkpoint activity without CDC25. More experiments are needed to elucidate the mechanism. It is suggested to use CDC25C fragment as a substrate control when to use histone H1 as the substrate to detect CHK1 activity in cells.
2. We have tested different CHK1 antibodies purchased from Santa Cruz Inc. (*sc-7898* and *sc-8408*). Although sc-8408 is good for western blot, it is not good for immunoprecipitation because the monoantibody (sc-8408) is against the N-terminal of CHK1, the kinase domain (27). Therefore, we suggest using the polyclonal CHK1 antibody (*sc-7898*) for CHK1 kinase assay.
3. CHK1 protein is expressed specifically during the S and G2 phase of the cell cycle (28). In order to obtain extracts with enough CHK1 protein, it is important to keep the cells in an exponentially growing condition. Too few cells (<50 % confluency of the culture dish) will result in less extract, and too many cells (about 100 % confluency of the culture dish) will result in extracts with less CHK1 protein.
4. Cells in 5 mM KCl hypotonic buffer are more easily broken than those in isotonic buffer. When doing the freeze–thaw cycle, do not leave the sample in the water bath (42 °C) too long. Shake the Eppendorf tube in water and when most of the sample has thawed, take the tube out and shake the tube at

room temperature until the sample thaws out completely. Immediately put the sample either on dry ice (continue the freeze–thaw cycle) or on ice (finish the freeze–thaw cycle).

5. At this step, it is suggested to confirm that 100 % cells are lysed by observing one drop of such cell lysis on a glass slide under the microscope.
6. Ideal concentration for the cytoplasmic extracts is >20 mg/mL and >10 mg/mL for the nuclear extracts. The nuclear extracts are prepared in the high salt buffer and high salt will affect IP; therefore, obtaining the high concentration of nuclear extracts (>10 mg/mL) could reduce the high salt effects (<25 μL volume of extracts containing 250 μg proteins required for the IP).
7. Cut off about 1 mm of the pipette tip with a scissors in order to easily pick up the protein A-Sepharose slurry. In order to add an equal volume of the slurry to each sample tube, take and wash the entire volume of the slurry in one Eppendorf tube. For example, if there are ten samples put 100 μL of the slurry into the Eppendorf tube and wash it with lysis buffer. Then add 1 mL of lysis buffer and 50 μL of CHK1 antibody to this tube. Mix well and aliquot (cut off about 1 mm of the tip with a scissors) the mixture to new prelabeled Eppendorf tubes.
8. The CHK1 activity of nuclear extracts but not the CHK1 activity of whole cell lyses correlates with the S and G2 checkpoint responses (13, 18). Thus, using nuclear extracts make the assay more sensitive. However, in order not to miss the changes in the CHK1 activity following DNA damage, it is better to examine the CHK1 activity separately from cytoplasmic and nuclear extracts but not from whole cell lyses.
9. The final volume of the sample depends upon the initial concentration of the extracts. The final volume of the sample should be more than ten times the volume of the extracts (<25 μL containing 250 μg) (see Note 5).
10. This IP step can be finished after 4 h of reaction. Depending on the time schedule, the reaction could be overnight without affecting the final result.
11. Do not shake the tube after spinning it, which prevents any slurry bead splashes to the wall above the level of the buffer in the tube. Keeping all the slurry beads in the buffer is important for a normal kinase reaction.
12. The 5× protein loading buffer is very sticky and it is easy to pick up the buffer by using a cut pipette tip. The purpose of using 5× protein loading buffer is to reduce the volume of the whole sample and allows loading the whole sample (~31 μL) to the well of the Mini-gel.
13. Load the whole sample ~31 μL (5 μL slurry, 20 μL reaction buffer, 6 μL 5× protein loading buffer) into the well.

14. The red band of the prestained marker is 68 kDa. Leave this band for recognizing the orientation of gel. Both the signals of CDC25C fragment and H1 protein are 36 kDa. Keep in mind that CHK1 could be autophosphorylated (29–31); therefore, the signals between 50 and 60 kDa might be the signals from autophosphorylated CHK1.
15. After 3 h exposure, the phosphorylation band should be visible on the PhosphorImager. It is suggested that the phosphorylation band be reexposured overnight to obtain better signals.
16. The time requirement for the whole procedure from preparing cell extracts to getting the PhosphorImager result is approximately 3 days. The work of day 1 includes preparing cell extracts and assembling the IP sample (IP over night). The work of day 2 includes assembling kinase assays, running gels, drying gels, and placing the dried gel in a PhosphorImager cassette. On day 3, the final results could be obtained by scanning the ^{32}P signals on the cassette plate with the PhosphorImager, analyzing the data with the software in the PhosphorImager and printing the data out.

Acknowledgments

This work was supported by NIH grant GM080771 (to Y.W.).

References

1. Walworth N, Davey S, Beach D (1993) Fission yeast chk1 protein kinase links the rad checkpoint pathway to cdc2. Nature 363:368–371
2. Sanchez Y, Wong C, Thoma RS, Richman R, Wu Z, Piwnica-Worms H, Elledge SJ (1997) Conservation of the Chk1 checkpoint pathway in mammals: linkage of DNA damage to Cdk regulation through Cdc25. Science 277:1497–1501
3. Peng C-Y, Graves PR, Thoma RS, Wu Z, Shaw AS, Piwnica-Worms H (1997) Mitotic and G2 checkpoint control: regulation of 14-3-3 protein binding by phosphorylation of Cdc25C on Serine-216. Science 277:1501–1505
4. Liu Q, Guntuku S, Cui XS, Matsuoka S, Cortez D, Tamai K, Luo G, Carattini-Rivera S, DeMayo F, Bradley A, Donehower LA, Elledge SJ (2000) Chk1 is an essential kinase that is regulated by Atr and required for the G_2/M DNA damage checkpoint. Genes Dev 14:1448–1459
5. Takata M, Sasaki MS, Tachiiri S, Fukushima T, Sonoda E, Schild D, Thompson LH, Takeda S (2001) Chromosome instability and defective recombinational repair in knockout mutants of the five Rad51 paralogs. Mol Cell Biol 21:2858–2866
6. Al-Khodairy F, Fotou E, Sheldrick KS, Griffiths DJF, Lehmann AR, Carr AM (1994) Identification and characterization of new elements involved in checkpoint and feedback controls in fission yeast. Mol Biol Cell 5:147–160
7. Lindsay HD, Griffiths DJF, Edwards RJ, Christensen PU, Murray JM, Osman F, Walworth N, Carr AM (1998) S-phase-specific activation of Cds1 kinase defines a subpathway of the checkpoint response in *Schizosaccharomyces pombe*. Genes Dev 12:382–395
8. Walworth NC, Bernards R (1996) *rad*-Dependent response of the chk1-encoded protein kinase at the DNA damage checkpoint. Science 271:353–356
9. Zeng Y, Forbes KC, Wu Z, Moreno S, Piwnica-Worms H, Enoch T (1998) Replication checkpoint requires phosphorylation of the

phosphatase Cdc25 by Cds1 or Chk1. Nature 395:507–510

10. Boddy MN, Furnari B, Mondesert O, Russell P (1998) Replication checkpoint enforced by kinases Cds1 and Chk1. Science 280: 909–912
11. Francesconi S, Grenon M, Bouvier D, Baldacci G (1997) p56chk1 protein kinase is required for the DNA replication checkpoint at 37 °C in fission yeast. EMBO J 16:1332–1341
12. Guo Z, Kumagai A, Wang SX, Dunphy WG (2000) Requirement for Atr in phosphorylation of Chk1 and cell cycle regulation in response to DNA replication blocks and UV-damaged DNA in Xenopus egg extracts. Genes Dev 14:2745–2756
13. Wang J-L, Wang X, Wang H, Iliakis G, Wang Y (2002) CHK1-regulated S-phase checkpoint response reduces camptothecin cytotoxicity. Cell Cycle 1:267–272
14. Zhao H, Piwnica-Worms H (2001) ATR-mediated checkpoint pathways regulate phosphorylation and activation of human Chk1. Mol Cell Biol 21:4129–4139
15. Cliby WA, Lewis KA, Lilly KK, Kaufmann SH (2002) S Phase and G2 arrests induced by topoisomerase I poisons are dependent on ATR kinase function. J Biol Chem 277:1599–1606
16. Zhao H, Watkins JL, Piwnica-Worms H (2002) Disruption of the checkpoint kinase 1/cell division cycle 25A pathway abrogates ionizing radiation-induced S and G2 checkpoints. Proc Natl Acad Sci U S A 99:14795–14800
17. Zhou XY, Wang X, Hu B, Guan J, Iliakis G, Wang Y (2002) An ATM-independent S phase checkpoint response involves CHK1 pathway. Cancer Res 62:1598–1603
18. Wang X, Li GC, Iliakis G, Wang Y (2002) Ku affects G2 checkpoint response following ionizing radiation. Cancer Res 62:6031–6034
19. Kumagai A, Guo Z, Emami KH, Wang SX, Dunphy WG (1998) The *Xenopus* chk1 protein kinase mediates a caffeine-sensitive pathway of checkpoint control in cell-free extracts. J Cell Biol 142:1559–1569
20. Zeng Y, Piwnica-Worms H (1999) DNA damage and replication checkpoints in fission yeast require nuclear exclusion of the Cdc25 phosphatase via 14-3-3 binding. Mol Cell Biol 19: 7410–7419
21. Hu B, Wang H, Wang X, Lu H-R, Huang C, Powell SN, Huebner K, Wang Y (2005) Fhit and CHK1 have opposing effects on homologous recombination repair. Cancer Res 65: 8613–8616
22. Sorensen CS, Hansen LT, Dziegielewski J, Syljuasen RG, Lundin C, Bartek J, Helleday T (2005) The cell-cycle checkpoint kinase Chk1 is required for mammalian homologous recombination repair. Nat Cell Biol 7:195–201
23. Wang H, Wang X, Zhou XY, Chen DJ, Li GC, Iliakis G, Wang Y (2002) Ku affects the ataxia and Rad 3-related/CHK1-dependent S phase checkpoint response after camptothecin treatment. Cancer Res 62:2483–2487
24. Wang Y, Wang H (2004) Methods and protocols, checkpoint control and cancer. Ed: Humana Press Inc., CHK1 kinase activity assay. Humana, Totowa, NJ p. 143–151
25. Yu Q, La Rose J, Zhang H, Takemura H, Kohn KW, Pommier Y (2002) UCN-01 inhibits p53 up-regulation and abrogates g-radiation-induced G_2-M checkpoint independently of p53 by targeting both of the checkpoint kinases, Chk2 and Chk1. Cancer Res 62:5743–5748
26. Wang Y, Wang H (1999) A rapid preparation of extracts for DNA replication *in vitro*. Radiat Res 151:59–64
27. Chen P, Luo C, Deng Y, Ryan K, Register J, Margosiak S, Tempczyk-Russell A, Nguyen B, Myers P, Lundgren K, Kan C-C, O'Connor PM (2000) Implications for Chk1 regulation: the 1.7 Å crystal structure of human cell cycle checkpoint kinase Chk1. Cell 100:681–692
28. Kaneko Y, Watanabe N, Morisaki H, Akita H, Fujimoto A, Tominaga K, Tachibana K, Ikeda K, Nakanishi M (1999) Cell cycle-dependent and ATM-independent expression of human Chk1 kinase. Oncogene 18:3673–3681
29. Wilsker D, Petermann E, Helleday T, Bunz F (2008) Essential function of Chk1 can be uncoupled from DNA damage checkpoint and replication control. Proc Natl Acad Sci U S A 105:20752–20757
30. Walton MI, Eve PD, Hayes A, Valenti M, De Haven Brandon A, Box G, Boxall KJ, Aherne GW, Eccles SA, Raynaud FI, Williams DH, Reader JC, Collins I, Garrett MD (2010) The preclinical pharmacology and therapeutic activity of the novel CHK1 inhibitor SAR-020106. Mol Cancer Ther 9:89–100
31. Morgan MA, Parsels LA, Zhao L, Parsels JD, Davis MA, Hassan MC, Arumugarajah S, Hylander-Gans L, Morosini D, Simeone DM, Canman CE, Normolle DP, Zabludoff SD, Maybaum J, Lawrence TS (2010) Mechanism of radiosensitization by the Chk1/2 inhibitor AZD7762 involves abrogation of the G2 checkpoint and inhibition of homologous recombinational DNA repair. Cancer Res 70:4972–4981

Chapter 40

Histone H2AX Phosphorylation: A Marker for DNA Damage

Arishya Sharma, Kamini Singh, and Alexandru Almasan

Abstract

The DNA damage response can be initiated in response to a variety of stress signals that are encountered during physiological processes or in response to exogenous cues, such as ionizing radiation or DNA-damaging therapeutic agents. A number of methods have been developed to examine the morphological, biochemical, and molecular changes that take place during the DNA damage response. When cells are exposed to ionizing radiation or DNA-damaging chemotherapeutic agents, double-stranded breaks (DSBs) are generated that rapidly result in the phosphorylation of histone H2A variant H2AX. Because phosphorylation of H2AX at Ser 139 (γ-H2AX) is abundant, fast, and correlates well with each DSB, it is the most sensitive marker that can be used to examine the DNA damage produced and the subsequent repair of the DNA lesion. γ-H2AX can be detected by immunoblotting and immunostaining using microscopic or flow cytometric detection. Since γ-H2AX can be also generated during DNA replication, as a consequence of apoptosis, or as it is found associated with residual DNA damage, it is important to determine the kinetics, number, size, and morphology of γ-H2AX-associated foci. This chapter describes a few standard protocols that we have successfully used in our laboratory for a number of experimental systems, primarily hematologic and epithelial cells grown in culture.

Key words: γ-H2AX, DNA damage, Flow cytometry, Immunofluorescence, Immunoblotting, Ionizing radiation, Lymphocytes, Epithelial cells

1. Introduction

The DNA damage response (DDR) represents an important component of the physiologic processes that occur during DNA replication and recombination, oxidative stress, as well as in response to the many cytotoxic therapeutics commonly used in cancer therapy (1). Among these, ionizing radiation is a prototypical DNA-damaging agent that has been used in many laboratories to define our mechanistic understanding of the DDR (2). The DDR represents a signal transduction pathway that senses DNA

Lotte Bjergbæk (ed.), *DNA Repair Protocols*, Methods in Molecular Biology, vol. 920,
DOI 10.1007/978-1-61779-998-3_40, © Springer Science+Business Media New York 2012

damage and replication stress and regulates physiological processes that involve critical cellular decisions whether to undergo DNA repair, enter terminal differentiation through senescence, or if the damage is more severe, undergo apoptosis. It is primarily mediated by proteins of the phosphatidylinositol 3-kinase-like protein kinase (PIKKs) family, ATM, ATR, and DNA-PK and by members of the poly (ADP-ribose) polymerase (PARP) family. Ataxia telangiectasia mutated (ATM) and DNA dependent protein kinase (DNA-PK) are activated by DNA-damaging agents, such as irradiation that create DSBs. Ataxia telangiectasia and Rad3-related (ATR), in complex with its partner protein ATRIP, is activated following recruitment to RPA-coated single-stranded (ss) DNA regions that are generated primarily in S-phase at stalled DNA replication forks (1). PARP1 and PARP2 are activated by SSBs and DSBs and catalyze the addition of poly (ADP-ribose) chains on proteins that recruit DDR factors to chromatin at DNA breaks. PARP1 is thought to mediate the initial accumulation of the MRE-Rad50-Nbs1 (MRN) complex at DSBs (3).

Recruitment of ATM by MRN and PARP1 could then contribute to the activation of the γ-H2AX complexes and stabilization of DDR factors at sites of DNA damage. The phosphoepitope of γ-H2AX directly binds MDC1, which further enhances MRN-ATM binding and contributes to a positive feedback loop that spreads γ-H2AX over mega-base domains around the DSBs. These initial signaling events activate one of the two DNA DSB repair pathways; nonhomologous end joining (NHEJ) is the major pathway, whereas homologous recombination is restricted to S- or G2-phases after DNA replication has been completed. γ-H2AX dephosphorylation by protein phosphatase 2A facilitates DSB repair (3). While ATM is the primary kinase that forms γ-H2AX at DSBs, ATR and DNA-PKcs, which are recruited to chromatin by ATRIP and Ku70/80, respectively, have been also implicated (4, 5).

Histone H2AX phosphorylation in response to DNA damage on serine 139 within its SQEY motif (6) at the site of DSBs allows the DNA damage signal to spread along the chromatin (7) covering large regions of the chromosome surrounding each DSB lesion, thus providing a platform for recruitment of other proteins that participate in DDR. Formation of foci of different nuclear proteins is a widely used method for assessing formation of DSBs and their repair; among these are 53BP1, Nbs1, Rad51, and BRCA1 (8), with γ-H2AX being the most frequently used.

γ-H2AX foci are formed within seconds after induction of DSBs, but since they are initially quite small and difficult to visualize, more reliably they are examined starting 15–30 min later (9). Since γ-H2AX spreads over a megabase chromatin region from the DSB site, discrete nuclear foci can be visualized using specific antibodies. The assay is highly sensitive, with virtually every γ-H2AX focus representing a single DSB (10). Once the repair of DSBs commences, dephosphorylation of γ-H2AX molecules leads to resolution of foci; therefore, the number of foci observed are informative about the extent of DNA repair (3). The major limitation associated with the use of this technique is the formation of non-DSBs associated γ-H2AX at ssDNA regions produced during DNA replication (11) (see Note 1).

Compared to other techniques used to detect DNA damage, such as the comet assay and pulsed-field gel electrophoresis (PFGE), γ-H2AX foci detection is much more sensitive. The same advantage applies compared to the nuclear foci formed by various DNA repair-enabling proteins, such as 53BP1, Nbs1, Rad50, Rad51, and Brca1 that are technically more challenging to detect because of their limiting sensitivity or more restricted occurrence (12). For example, while γ-H2AX can be detected in all phases of cell cycle, other foci, such as Rad51, are restricted to S- and G2-phases, being specific to homologous recombination. Impaired DSB repair, as determined by the kinetics of γ-H2AX foci provides also the best available correlation with radiosensitivity (9).

γ-H2AX can be measured using immunoblotting and immunostaining, which can be examined by microscopy or flow cytometry. Immunoblotting measures the total levels of γ-H2AX in whole cell or tissue extracts. It is the easiest to perform and relatively inexpensive. However, it is not informative regarding the type and nature of γ-H2AX foci and it is less sensitive than immunostaining. The advantage of using flow cytometry is that it allows for a rapid and accurate measurement of H2AX phosphorylation in individual cells and requires a relatively small cell number. In addition, the γ-H2AX expression, within the same cells, can be quantitated relative to DNA content, apoptosis induction, or any other cellular or molecular parameters of interest for which antibodies are available. Among γ-H2AX-based detection methods the most informative is immunostaining, which involves visualization of γ-H2AX foci in cultured cells or tissues. While immunostaining is more laborious and costly, its major advantage is its sensitivity and the quality of information generated that allows the study of DSB induction as well as the DNA repair kinetics. Detection of a single DSB is possible using microscopy. Additionally, it enables colocalization studies with other DNA repair proteins. In summary, γ-H2AX detection is a sensitive and undemanding technique for the measurement of DNA DSBs. Here we provide the protocols for the detection of γ-H2AX in mammalian cell lines, adherent or suspension as well as for peripheral blood lymphocytes (PBLs) using immunoblotting, microscopy, and flow cytometry. These protocols can also be extended to other cell types and treatments. Additional applications are described elsewhere (13, 14), with the therapeutic use of γ-H2AX as a clinical pharmacodynamic biomarker for human tissues having been recently established (15).

2. Materials

2.1. Isolation of PBLs

1. 50-mL conical centrifuge tubes.
2. Centrifuge (model GS-6R) with GH 3.8 swinging bucket rotor (Beckman Coulter) (see Note 2).
3. 2-mL pipettes.
4. Growth media for suspension cells: RPMI medium supplemented with 10 % fetal bovine serum (FBS), 100 U/mL penicillin, and 100 mg/mL streptomycin (Invitrogen).
5. Ficoll-Paque PLUS (Amersham).

2.2. Immunostaining in PBLs, Suspension and Adherent Cells

1. 15-mL conical centrifuge tubes.
2. 1× Phosphate-buffered saline (PBS), pH 7.4: 136.9 mM NaCl, 2.68 mM KCl, 8.1 mM Na_2HPO_4, 0.9 mM $CaCl_2$, 0.49 mM $MgCl_2$.
3. 2 % Paraformaldehyde solution (PFA): Mix 0.2 g PFA in 10 mL 200 mM HEPES, pH 7.5. Heat this mixture until the PFA has dissolved.
4. 0.1 % Triton-X 100 in PBS.
5. Microscopic superfrost slides (Fisher Scientific).
6. Coplin glass jars (Wheaton Industries, USA).
7. Glass coverslips (22 × 22 mm).
8. Liquid-repellent slide marker pen, PAP Pen (Cosmo Bio, USA).
9. Centrifuge with swinging bucket rotor.
10. Cytocentrifuge (model Shandon Cytospin 4, Fisher Scientific) and cytofunnels (Thermo Scientific EZ), required for only suspension cells.
11. 70 % Ethanol stored at −20 °C, must be chilled at the time of use.
12. Blocking solution (1× PBS containing 10 % FBS, stored at 4 °C, sterile filter) for diluting primary and secondary antibodies.
13. Primary antibodies: mouse monoclonal anti-γ-H2AX (Upstate, USA) (stored at −20 °C). Dilute in blocking solution before use.
14. Secondary antibodies: Goat anti-mouse Alexa Fluor 488-conjugated IgG (Molecular probes) (stored at 4 °C). Dilute in blocking solution before use.
15. 4′,6′-Diamidino-2-phenylindole hydrochloride (DAPI): 1 μg/mL final concentration.
16. Vectashield mounting medium for fluorescence (Vector Laboratories) and nail polish.
17. Leica TCS-SP-AOBS spectral laser scanning confocal microscope (Leica Microsystem) equipped with UV1364, two argon ion NA 1.4 (488 nm) lasers, and a krypton–argon ion laser (568 nm).

2.3. Western Blotting in PBLs, Suspension, and Adherent Cells

1. 1× PBS, pH 7.4, containing the protease inhibitor cocktail "Complete" (Roche Diagnostics, Mannheim, Germany) (only for PBLs).
2. 2× SDS (sodium dodecyl sulfate) protein gel loading solution: 0.5 M Tris–HCl, pH 6.8, 4.4 % (w/v) SDS, 20 % (v/v) glycerol, 2.5 % (v/v) beta-mercaptoethanol (β-ME), and bromophenol blue in deionized water (only for PBLs).
3. To lyse adherent and suspension cells following lysis buffer is required: 20 mM HEPES, pH 7.5, 1 mM EDTA, 150 mM NaCl, 1 % NP-40, 1 mM DTT containing the protease inhibitor cocktail "Complete" (Roche Diagnostics, Mannheim, Germany).
4. 6× SDS protein gel loading solution: 125 mM Tris–HCl, pH 6.8, 2 % SDS, 20 % glycerol, 0.2 % bromophenol blue in deionized water (for suspension and adherent cell lines) containing freshly added 5 % β-ME.
5. 1.5-mL microcentrifuge tubes.
6. 12 % Polyacrylamide gels.
7. Tris–glycine–SDS running buffer (10×): 30.3 g Tris base, 144 g glycine, and 10 g SDS in 1 L deionized water.
8. Prestained protein molecular weight marker.
9. Mini gel system (Bio-Rad Laboratories).
10. Tris–glycine transfer buffer: 100 mL (10×) running buffer, 200 mL methanol, 700 mL deionized water.
11. Nitrocellulose (NC) membrane (e.g., Schleicher and Schull).
12. PBS-T: 1× PBS containing 0.05 % Tween-20 (Invitrogen).
13. Nonfat milk powder.
14. Primary antibodies: Monoclonal anti-γ-H2AX (Upstate, USA) (stored at −20 °C) and rabbit polyclonal anti-H2AX (Cell Signaling) (stored at −20 °C). Dilute in blocking solution before use.
15. Secondary antibodies: Anti-rabbit horseradish peroxidase HRP-conjugated IgG and anti-rabbit HRP-conjugated (GE Healthcare).
16. Enhanced chemiluminescent (ECL) reagents (GE Healthcare).
17. High sensitivity chemiluminescent films (GE Healthcare).
18. Restoring western blot stripping buffer (Pierce Biotechnology).

2.4. Flow Cytometry in PBLs, Suspension and Adherent Cells

1. Primary antibodies: Mouse monoclonal anti-γ-H2AX (Upstate, USA) (stored at −20 °C). Dilute in blocking solution before use.
2. Secondary antibodies: Goat anti-mouse Alexa Fluor 488-conjugated IgG (Molecular Probes) (stored at 4 °C). Dilute in blocking solution before use.
3. 1× PBS.

4. 1 % Methanol-free formaldehyde dissolved in 1× PBS 1 % (w/v).
5. Solution of bovine serum albumin (BSA; Sigma) dissolved in 1× PBS.
6. 70 % Ethanol in deionized water.
7. Becton-Dickinson LSRII capable of 12-color and 14-parameter acquisition with five lasers (argon ion 488 nm, HeCd UV 355 nm, Violet 405 nm, HeNe 639 nm, and Green 532 nm).

3. Methods

3.1. Cell Culture and Treatments

We have primarily used tumor cell lines of hematologic origin, such as NALM-6, IM-9, Reh and amongst epithelial cells those from prostate cancer (2, 16). DNA damage is induced by irradiating cells using a Shepherd Mark II Cesium-137 irradiator. Alternatively, radiomimetic drugs (e.g., bleomycin, neocarzinostatin), topoisomerase inhibitors (e.g., etoposide, VP16), camptothecin and its derivatives (e.g., CPT-11 (16)), anthracyclines (e.g., doxorubicin), purine analogs (e.g., fludarabine) are added freshly to the culture medium (see Notes 3 and 4).

3.2. Isolation of PBLs

1. Dilute blood sample 2× with growth media in a 50-mL conical centrifuge tube.
2. Layer 1 volume of the blood-growth media mix on the top of 2 volume of Ficoll-Paque in a separate 50-mL centrifuge tube.
3. Centrifuge at 500 × *g* for 25–30 min at RT (see Note 5).
4. Discard the top layer. Carefully, collect the second layer from the top (white interphase) containing lymphocytes using a 2-mL pipette.
5. Wash the lymphocytes twice with 2 volumes of growth media by centrifuging at 450 × *g* for 7 min at RT. Discard the supernatant and resuspend the pellet in growth media for subsequent processing.
6. Dilute the lymphocytes to 4×10^6 cells/mL in growth media; they can be processed or treated immediately, or alternatively cultured at 37 °C for up to 3 days.

3.3. Immunocytochemical Detection of γ-H2AX in Mammalian Cells

The following protocol describes the immunostaining-based detection of γ-H2AX in mammalian adherent cell lines, suspension cell lines, and PBLs (modified from ref. 14). It can be used for in vivo studies monitoring patient response to irradiation, to DNA damage-inducing chemotherapeutics, and for in vitro experiments with mammalian cells to study DNA damage and repair.

3.3.1. Preparation of PBL Samples for Immunostaining

For immunostaining, at least 8–10 × 10^6 cells (4 × 10^6 cells/mL) are suspended in growth media in a 15-mL conical centrifuge tube and centrifuge at 450 × *g* for 5 min at RT. Decant the medium and add cold PBS. Mix very gently and spin it down (450 × *g* for 5 min). Remove the PBS carefully without disturbing the cell pellet.

3.3.2. Preparation of Suspension Cell Samples for Immunostaining

1. Seed 5 × 10^5 cells in six-well plates.
2. Add desired treatment and incubate in a tissue culture incubator at 37 °C.
3. Collect the treated and untreated cells by centrifugation at 125 × *g* for 5 min at RT.
4. Decant the medium and resuspend the cell pellet in cold 1× PBS very gently and spin it down at 125 × *g* for 5 min at RT. Remove the PBS carefully without disturbing the cell pellet.

3.3.3. Preparation of Adherent Cell Samples for Immunostaining

1. Place sterilized (by dipping in 70 % ethanol followed by passing through flame) glass coverslips into six-well plates or 60-mm dishes.
2. Plate 2 × 10^5 cells per well of six well plates with coverslips and allow to adhere and grow for 16–24 h in a tissue culture incubator at 37 °C before the treatment.
3. Treat the cells attached to coverslips with desired drug treatment by carefully replacing the growth media with treatment-containing media.

3.3.4. Fixation for Suspension Cells and PBLs

1. Fix with 500 μL of 2–4 % PFA solution.
2. Mix gently by vortexing or pipetting and incubate for 20 min at RT.
3. Add 5–10 mL of 1× PBS and centrifuge for 5 min at 500 × *g* (for PBLs) 125 × *g* (for cell lines) at RT. Discard the supernatant.
4. Repeat PBS washing two more times.
5. Resuspend the pellet in PBS containing 10 % FBS.
6. Pipette 200–300 μL of cell suspension into the EZ cytofunnel/slide assemblies and cytospin for 4 min at 80 × *g* at RT.
7. Use a PAP Pen to mark the boundary around the cells (see Note 6). After cytospin, air-dry the slides for 1 h at RT.
8. Cover the cells with 1× PBS for 15 min to rehydrate. Slides should not be dry at any time after this point (see Note 7).
9. Place in a glass Coplin jar and incubate with 70 % ethanol (prechilled at −20 °C) at RT for 20 min (see Note 8).

3.3.5. Fixation for Adherent Cells

1. Aspirate the medium and wash cells with 1× PBS once.
2. Add 1–2 mL of 2–4 % PFA solution into each well and fix for 20 min at RT.

3. Remove PFA and wash each well with 1× PBS for 5 min, three times.
4. Add 70 % ethanol (prechilled at −20 °C) to each well and incubate at RT for 20 min (see Note 9).

3.3.6. Staining

1. Wash the slide (or coverslip) bearing cells twice with PBS, for 5 min each time.
2. Incubate with 0.1 % Triton X-100 for 15 min at RT.
3. Quench any traces of PFA by incubating with 0.1 M glycine for 2 min twice.
4. Wash with PBS for 5 min, three times.
5. Block the slides (or coverslips) with 10 % goat serum for 60 min at RT.
6. Wash once with 1× PBS for 5 min.
7. Add 50–100 μL 10 % goat serum containing anti-γ-H2AX at 1:100 dilution on top of coverslip or each spot-bearing cells and incubate for 60–120 min at RT. Perform the incubation in a humidified chamber (see Note 10).
8. Wash with PBS for 5 min, three times.
9. Incubate with 50–100 μL 10 % FBS containing the secondary antibody at 1:500 dilution and incubate for 60 min at RT in a humidified chamber.
10. Wash with PBS for 5 min, three times.
11. Incubate with 100 μL 1 μg/mL DAPI for 10 min at RT.
12. Wash with PBS.
13. Add one drop of antifade mounting medium on each glass slide and invert carefully the glass coverslip onto it (for adherent cells). In case of suspension cells, add one drop of mounting (antifade) medium on each cytospin area, then cover it with the glass coverslip (see Notes 11 and 12).
14. Seal the edges of the coverslip with nail polish.

3.3.7. Imaging and Image Analysis

1. Count the γ-H2AX foci directly using a confocal microscope, typically counting foci in 50–100 nuclei.
2. A confocal microscope is used to capture whole nuclei. Each confocal image is taken with increments of 0.5 μm with *z*-sections condensed, so all detectable foci are visible in a single plane.
3. The number of γ-H2AX foci are counted manually by using the Image J counting analysis.
4. Alternatively, for a more quantitative analysis of the intensity of γ-H2AX foci, image-counting or intensity measuring software can be used, such as the Image Pro 6.2 Analyzer (Media Cybernetics, Bethesda, MD), IPLab (BD), or Image Quant (Molecular Dynamics).

3.4. Immunoblotting

The following protocol describes the detection of γ-H2AX levels relative to total H2AX in lysates from mammalian adherent cell lines, suspension cell lines, and PBLs (modified from ref. 14).

3.4.1. Protein Extraction from PBLs

1. Take 5×10^5 cells/mL of lymphocytes in a 1.5-mL microcentrifuge tube.
2. Spin the cells at 2,000 ×*g* for 5 min at 4 °C.
3. Wash the cells with 1 mL 1× PBS containing the complete protease inhibitor cocktail.
4. Centrifuge the cells at 2,000 ×*g* for 5 min at 4 °C and discard the supernatant.
5. Add 30 μL of 2× SDS protein gel loading solution containing 2.5 % β-ME. Add β-ME freshly to the loading solution.
6. Boil the specimens for 10 min.
7. Chill the specimens on ice for 5 min.
8. Vortex briefly and centrifuge the specimens at 16,000 ×*g* in a microfuge for 5 min at 4 °C.

3.4.2. Protein Extraction from Adherent or Suspension Cells

1. Harvest the treated/untreated cells (2×10^6) into a 15-mL conical centrifuge tube and centrifuge at 150 ×*g* for 5 min at RT.
2. Decant the medium and resuspend the cell pellet in cold 1× PBS and spin at 2,000 ×*g* for 5 min in a microfuge at 4 °C.
3. Aspirate the 1× PBS carefully without disturbing the cell pellet. Lyse the cells in a lysis buffer, with the cells incubated for 30 min on ice with occasional vortexing.
4. Centrifuge the cells at 16,000 ×*g* for 15 min at 4 °C and collect the supernatants.
5. For protein estimation, a spectrophotometric method using the Bio-Rad Protein Assay reagent (working solution, 1:10 dilution) may be used. 1 μL of the sample is mixed with 1 mL of diluted Bio-Rad protein assay reagent and the absorbance is measured at 595 nm. The concentration of unknown samples is determined from the BSA standard curve (drawn from the spectrophotometric readings of known concentrations of BSA).

3.4.3. SDS-Polyacrylamide Gel Electrophoresis

1. Dilute 10× Tris–glycine–SDS running buffer by mixing 100 mL of 10× Tris–glycine–SDS running buffer with 900 mL of deionized water to prepare 1× running buffer.
2. Place the 12 % gel in the running system and fill the system with 1× running buffer.
3. For PBLs, load 15 μL of supernatant samples mixed with SDS-sample buffer (finally 1×) containing β-ME. For cell lines, prepare 50–100 μg protein sample/well with SDS-sample buffer (to final 1×) containing β-ME. Boil for 5 min. Spin down the samples briefly and allow to cool for 5 min at RT. Load a

protein standard marker with the samples on a 12 % SDS PAGE to separate the proteins under denaturing conditions.

4. Perform electrophoresis at 100 V for 3 h, until the blue dye line reaches the end of the gel.

3.4.4. Western Blotting

1. Prepare 1× transfer buffer before the transfer step and keep at 4 °C so that it is chilled at the time of use.
2. After the electrophoresis run, take out the gel, cut and discard the stacking portion of the gel and place in prechilled transfer buffer for 10–15 min. Also equilibrate NC membrane in pre chilled transfer buffer.
3. Fill the mini-gel transfer tank and a glass tray with prechilled transfer buffer. Assemble the transfer cassette immersed in transfer buffer in a glass tray in the following order: sponges followed by wet filter paper, gel, NC membrane covered with wet filter paper, and sponges, ensuring that no bubbles are trapped in the resulting sandwich.
4. Place the cassette into the transfer apparatus, ensuring that the correct orientation [the cathode (black) is closer to the gel than the membrane].
5. Transfer the protein at 90 V for 1 h and use ice-pack during transfer to cool down the apparatus.
6. Prepare 5 % blocking buffer by adding 2.5 g of nonfat milk to 50 mL of PBS-T.
7. Once the transfer is complete, remove the cassette from the transfer apparatus and carefully disassemble and obtain the NC membrane immunoblotted with proteins.
8. Incubate the NC membrane in 10 mL of blocking buffer at RT for 1 h.
9. Prepare a 1:500 dilution of the anti-γ-H2AX primary mouse monoclonal antibody solution in blocking buffer in a 15-mL conical tube.
10. Incubate membrane with primary antibody at 4 °C overnight on shaker. Make sure that the protein side of the membrane is facing upwards.
11. Wash with PBS-T for 10 min, each for three times.
12. Prepare a 1:5,000 dilution of the secondary antibody solution in PBS-T. Incubate the membrane with the secondary antibody solution at RT for 1 h.
13. Wash with PBS-T for 10 min, each for three times.
14. Drain the excess PBS-T from the washed membrane. Prepare the detection reagent (according to the manufacturer's instructions). Cover the membrane with the detection reagent and incubate at RT for 2 min.

15. Drain the excess of detection reagent and wrap the membrane in a clear plastic membrane. Gently smooth away any air bubbles.
16. Place the wrapped membrane in an X-ray film cassette and expose to film.
17. After a satisfactory exposure of the γ-H2AX has been obtained, wash the blot with PBS-T for 5 min.
18. Strip the membrane and reprobe with the anti-H2AX antibody (1:2,000 dilution).
19. Repeat the probing and development steps as above.

3.5. Analysis of γ-H2AX in Mammalian Cells by Flow Cytometry

The following protocol can be used for detection of γ-H2AX by flow cytometry in mammalian adherent cell lines, suspension cell lines, and PBLs (modified from ref. 14).

3.5.1. Preparation of PBL Samples for Flow Cytometry

1. Use 10×10^6 lymphocytes per sample at 1×10^6 cells/mL.
2. Add desired treatment and incubate in a tissue culture incubator at 37 °C.
3. After treatment, transfer the cell suspension to a 15-mL conical tube.
4. Centrifuge at $500 \times g$ for 5 min at 4 °C and resuspend in 1 mL of ice-cold PBS and centrifuge again at $800 \times g$ for 5 min.

3.5.2. Preparation of Suspension and Adherent Cell Samples for Flow Cytometry

1. Seed 5×10^5 cells in six-well plates.
2. Add desired treatment and incubate in a tissue culture incubator at 37 °C.
3. Collect the treated and untreated cells by centrifugation at $125 \times g$ for 5 min. Decant the medium and resuspend the cell pellet in cold 1× PBS very gently and spin it down at $125 \times g$ for 5 min.
4. Remove the PBS carefully without disturbing the cell pellet.

3.5.3. Fixation

1. Resuspend the cell pellet in 0.5 mL of 1× PBS.
2. Add 5 mL of 1 % formaldehyde to this tube and fix for 15 min at 1–4 °C (on ice), then centrifuge at $450 \times g$ (for PBLs) or $125 \times g$ (for cell lines) for 5 min.
3. Suspend the pellet in 70 % ethanol for at least 2 h; the cells can be stored in 70 % ethanol for up to several days at −20 °C.

3.5.4. Staining

1. Centrifuge the cell suspension at $500 \times g$ for 5 min.
2. Wash twice with 1× PBS at $500 \times g$ for 5 min.
3. Add 2–3 mL 1 % BSA solution in 1× PBS to the cell pellet and centrifuge at $500 \times g$ for 5 min.

4. Resuspend the cell pellet in 100 μL of 1 % BSA and transfer to new tubes.
5. Add 1 μL (1:100) γ-H2AX primary antibody into the suspension, cover the tubes with parafilm to prevent drying, and incubate for 2 h at RT (see Note 7).
6. Add 2–3 mL 1 % BSA solution in 1× PBS to the cell pellet and centrifuge at 500 × *g* for 5 min.
7. Resuspend the cell pellet in 100 μL of 1 % BSA and transfer to new tubes.
8. Add 1 μL (1:100) Alexa Fluor 488 conjugated secondary antibody, cover the tubes to prevent drying, and incubate for 30 min at RT in the dark.
9. Analyze cells by flow cytometry on a Becton-Dickinson LSRII capable of 12 color and 14 parameter acquisition with five lasers (argon ion 488 nm, HeCd UV 355 nm, Violet 405 nm, HeNe 639 nm, and Green 532 nm) and data analysis may be done using Flow Jo software.

4. Notes

1. Factors considered for successful application and the associated limitations (9).
   (a) *Dose.* The dose of radiation necessary for γ-H2AX induction analysis depends on endogenous levels of foci. Foci induced by as low as mGy doses of radiation are detectable in primary fibroblasts and lymphocytes with low basal levels of foci. However, transformed cells often have high basal levels of foci due to genetic instability. Additionally, γ-H2AX foci at uncapped telomeres in senescent cells can produce γ-H2AX foci in the absence of DNA damage.
   (b) *Cell cycle progression during analysis.* An important consideration is the progression of cells through the cell cycle during analysis. Generation of ssDNA in S-phase can also potentially lead to non-DSB γ-H2AX formation. If left unrepaired, such ssDNA can be converted to toxic DSBs, for example in case of a block in homologous recombination when the ssDNA is trapped by Poly (ADP-ribose) polymerase (PARP) inhibitors (17). The use of quiescent (G0 phase) cells is, therefore, helpful albeit transformed cells can give high background due to their rapid cell cycle progression and their inability to enter the G0 phase. The use of BrdU or other cell cycle markers can be used to exclude replicating cells from the analysis.

2. The use of a swinging bucket rotor is recommended because centrifugation in a fixed angle rotor may result in cell loss as PFA fixation causes cells to stick to the side of the tube.
3. Ionizing radiation is administered in the clinic not only as external beam radiation (γ- and X-rays) but also as brachytherapy, which can be mimicked by use of low-dose rate irradiators (up to 10–50 cGy/h). As sources of ionizing radiation to induce DNA damage, most commonly used are: (1) a radioactive 137Cs γ-ray source, typically with a fixed dose rate of 2–3 Gy/min; we use a Shepherd Mark II Cesium-137 irradiator, (2) X-ray source; we use a Pantak HF320 Cabinet X-ray irradiator (320 kVP, 20 , half-value layer 2 mm Cu, East Haven, CT), (3) a low-dose-rate irradiator (LDRI), we use an iridium-192 irradiator (Ir-192 has a half-life of ~74 days). Loss from radioactive decay is appreciable, amounting to approximately 1 % per day. Radiomimetic drugs can also be used as a surrogate for radiation.
4. While traditionally considered as non-DNA damaging, other therapeutics may also induce DNA damage, even if only indirectly through apoptotic cell death, such as the cell death ligand-based therapeutic Apo2L/TRAIL (16) or Fas, through activation of the apoptotic nuclease caspase-activated DNAse (CAD)/DNA fragmentation factor 40 (DFF40) (18). Kinetic experiments can distinguish between early, direct DNA damage vs. an indirect effect, as a result of apoptotic nucleases. Blocking apoptosis, for example with a caspase inhibitor or overexpression of an antiapoptotic protein, can be used as an alternative approach.
5. The breaking function of the centrifuge must be deactivated for this step.
6. The PAP Pen is waterproof, therefore, it helps retain the solutions added onto cytospin within a boundary so that as little as 100 μL of antibody is sufficient to cover the cells.
7. Drying may result in damage of cells and increased immunofluorescence background.
8. The slides can be stored in 70 % ethanol for up to 1 week at 4 °C.
9. The coverslips can be stored in 70 % ethanol for up to several days at 4 °C.
10. It is recommended to avoid exposing the secondary antibody to light.
11. Remove excess PBS before adding mounting media, as dilution can reduce the antifade effect of the mounting media.
12. Avoid trapping any air bubble while mounting to prevent any light scattering caused during image acquisition.

Acknowledgments

This work was supported by a research grant from National Institutes of Health (CA127264-03 to A.A.).

References

1. Ciccia A, Elledge SJ (2011) The DNA damage response: making it safe to play with knives. Mol Cell 40:179–204
2. Plesca D, Mazumder S, Almasan A (2008) DNA damage response and apoptosis. Methods Enzymol 446:107–122
3. Chowdhury D, Keogh MC, Ishii H, Peterson CL, Buratowski S, Lieberman J (2005) gamma-H2AX dephosphorylation by protein phosphatase 2A facilitates DNA double-strand break repair. Mol Cell 20:801–809
4. Burma S, Chen BP, Murphy M, Kurimasa A, Chen DJ (2001) ATM phosphorylates histone H2AX in response to DNA double-strand breaks. J Biol Chem 276:42462–42467
5. Mukherjee B, Kessinger C, Kobayashi J, Chen BP, Chen DJ, Chatterjee A, Burma S (2006) DNA-PK phosphorylates histone H2AX during apoptotic DNA fragmentation in mammalian cells. DNA Repair (Amst) 5:575–590
6. Rogakou EP, Pilch DR, Orr AH, Ivanova VS, Bonner WM (1998) DNA double-stranded breaks induce histone H2AX phosphorylation on serine 139. J Biol Chem 273:5858–5868
7. Kinner A, Wu W, Staudt C, Iliakis G (2008) Gamma-H2AX in recognition and signaling of DNA double-strand breaks in the context of chromatin. Nucleic Acids Res 36:5678–5694
8. Paull TT, Rogakou EP, Yamazaki V, Kirchgessner CU, Gellert M, Bonner WM (2000) A critical role for histone H2AX in recruitment of repair factors to nuclear foci after DNA damage. Curr Biol 10:886–895
9. Lobrich M, Shibata A, Beucher A, Fisher A, Ensminger M, Goodarzi AA, Barton O, Jeggo PA (2010) gammaH2AX foci analysis for monitoring DNA double-strand break repair: strengths, limitations and optimization. Cell Cycle 9:662–669
10. Rothkamm K, Lobrich M (2003) Evidence for a lack of DNA double-strand break repair in human cells exposed to very low X-ray doses. Proc Natl Acad Sci U S A 100:5057–5062
11. Ewald B, Sampath D, Plunkett W (2007) H2AX phosphorylation marks gemcitabine-induced stalled replication forks and their collapse upon S-phase checkpoint abrogation. Mol Cancer Ther 6:1239–1248
12. Polo SE, Jackson SP (2011) Dynamics of DNA damage response proteins at DNA breaks: a focus on protein modifications. Genes Dev 25: 409–433
13. Nakamura A, Sedelnikova OA, Redon C, Pilch DR, Sinogeeva NI, Shroff R, Lichten M, Bonner WM (2006) Techniques for gamma-H2AX detection. Methods Enzymol 409:236–250
14. Redon CE, Nakamura AJ, Sordet O, Dickey JS, Gouliaeva K, Tabb B, Lawrence S, Kinders RJ, Bonner WM, Sedelnikova OA (2011) gamma-H2AX detection in peripheral blood lymphocytes, splenocytes, bone marrow, xenografts, and skin. Methods Mol Biol 682:249–270
15. Redon CE, Nakamura AJ, Zhang YW, Ji JJ, Bonner WM, Kinders RJ, Parchment RE, Doroshow JH, Pommier Y (2010) Histone gammaH2AX and poly(ADP-ribose) as clinical pharmacodynamic biomarkers. Clin Cancer Res 16:4532–4542
16. Ray S, Shyam S, Fraizer GC, Almasan A (2007) S-phase checkpoints regulate Apo2 ligand/TRAIL and CPT-11-induced apoptosis of prostate cancer cells. Mol Cancer Ther 6: 1368–1378
17. Strom CE, Johansson F, Uhlen M, Al-Khalili Szigyarto C, Erixon K, Helleday T (2011) Poly (ADP-ribose) polymerase (PARP) is not involved in base excision repair but PARP inhibition traps a single-strand intermediate. Nucleic Acids Res 9:3166–3175
18. Lovric MM, Hawkins CJ (2010) TRAIL treatment provokes mutations in surviving cells. Oncogene 29:5048–5060

INDEX

Lotte Bjergbæk (ed.), *DNA Repair Protocols*, Methods in Molecular Biology, vol. 920,
DOI 10.1007/978-1-61779-998-3, © Springer Science+Business Media New York 2012

MIX
Papier aus verantwortungsvollen Quellen
Paper from responsible sources
FSC® C105338

If you have any concerns about our products,
you can contact us on
ProductSafety@springernature.com

In case Publisher is established outside the EU,
the EU authorized representative is:
Springer Nature Customer Service Center GmbH
Europaplatz 3, 69115 Heidelberg, Germany

Printed by Libri Plureos GmbH
in Hamburg, Germany